Multiscale Modelling of Soft Matter

University of Groningen, The Netherlands
20–22 July 2009

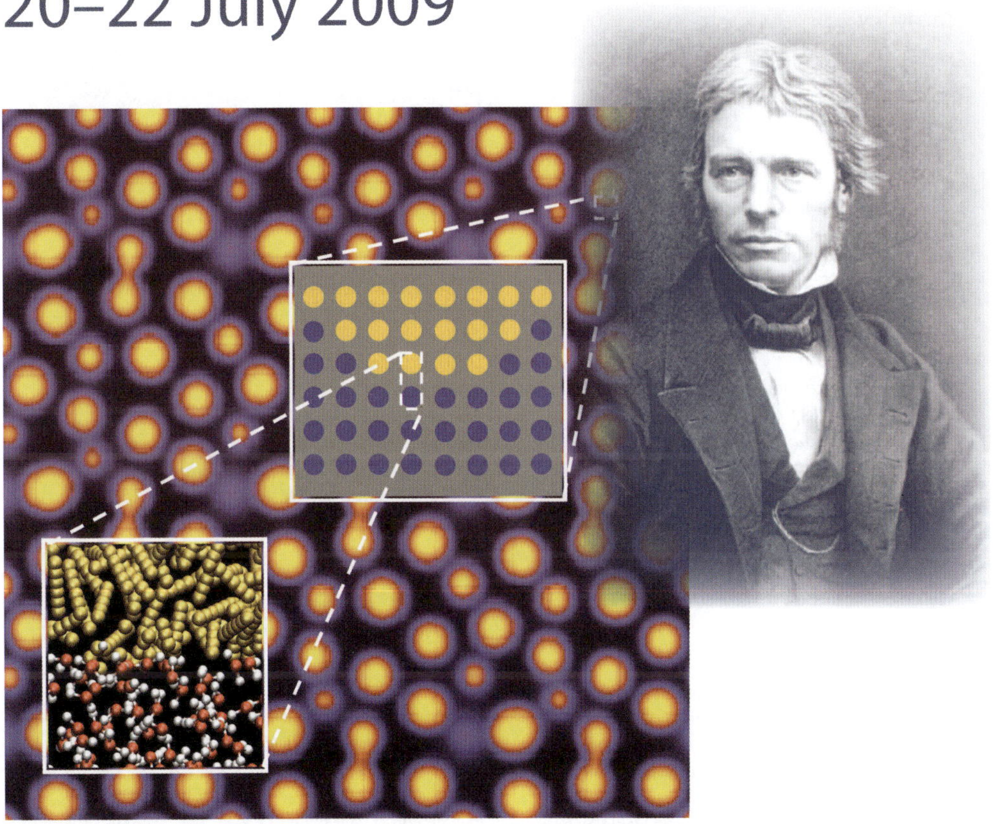

FARADAY DISCUSSIONS

Volume 144, 2010

RSC Publishing

Faraday Discussions (Print ISSN 1359-6640, Electronic ISSN 1364-5498) is published 4 times a year by the Royal Society of Chemistry, Thomas Graham House, Science Park, Milton Road, Cambridge, UK CB4 0WF.
Volume 144 ISBN-13: 978 1 84755 0392

2010 annual subscription price: print+electronic £622, US $1,160; electronic only £560, US $1,045. Customers in Canada will be subject to a surcharge to cover GST. Customers in the EU subscribing to the electronic version only will be charged VAT. All orders, with cheques made payable to the Royal Society of Chemistry, should be sent to RSC Distribution Services, c/o Portland Customer Services, Commerce Way, Colchester, Essex, UK CO2 8HP.
Tel +44 (0) 1206 226050;
E-mail sales@rscdistribution.org

If you take an institutional subscription to any RSC journal you are entitled to free, site-wide web access to that journal. You can arrange access *via* Internet Protocol (IP) address at www.rsc.org/ip. Customers should make payments by cheque in sterling payable on a UK clearing bank or in US dollars payable on a US clearing bank. Periodicals postage is paid at Rahway, NJ and at additional mailing offices. Airfreight and mailing in the USA by Mercury Airfreight International Ltd., 365 Blair Road, Avenel, NJ 07001, USA.

US Postmaster: send address changes to *Faraday Discussions*, c/o Mercury Airfreight International Ltd., 365 Blair Road, Avenel, NJ 07001. All despatches outside the UK by Consolidated Airfreight.

PRINTED IN THE UK

Faraday Discussions documents a long-established series of *Faraday Discussion* meetings which provide a unique international forum for the exchange of views and newly acquired results in developing areas of physical chemistry, biophysical chemistry and chemical physics.

Multiscale Modelling of Soft Matter

Faraday Discussions

www.rsc.org/faraday_d

A General Discussion on Multiscale Modelling of Soft Matter was held at the University of Groningen, Groningen, The Netherlands, on 20th, 21st and 22nd July 2009.

CONTENTS

ISSN 1359-6640; ISBN 978-1-84755-039-2

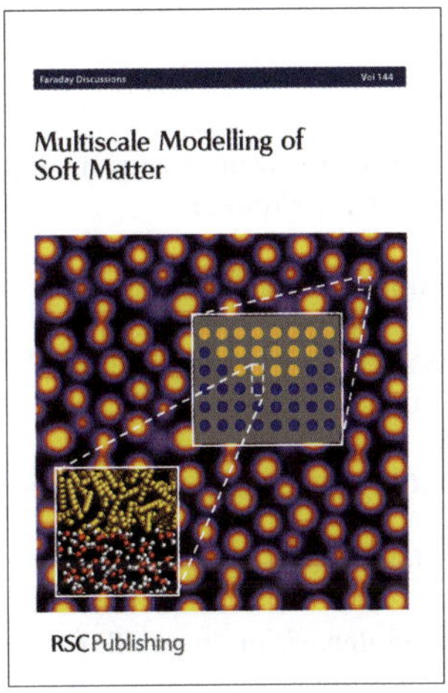

Cover
See Frenkel *et al.*, *Faraday Discuss.*, 2010, **144**, 223–243. Multiscale modelling of a complex system (oil/water mixture): The macroscopic structure and dynamics is predicted using a coarse-grained lattice model that has been constructed with input from atomistic simulations.

Image reproduced by permission of Professor Daan Frenkel, from *Faraday Discuss.*, 2010, **144**, 223.

INTRODUCTORY LECTURE

PAPERS AND DISCUSSIONS

CONCLUDING REMARKS

ADDITIONAL INFORMATION

Multiscale simulation of soft matter systems

Christine Peter and Kurt Kremer*

Received 22nd September 2009, Accepted 22nd September 2009
First published as an Advance Article on the web 30th September 2009
DOI: 10.1039/b919800h

This paper gives a short introduction to multiscale simulation approaches in soft matter science. This paper is based on and extended from a previous review.[1] (1. C. Peter and K. Kremer, *Soft Matter*, 2009, DOI:10.1039/b912027k.) It also includes a discussion of aspects of soft matter in general and a short account of one of the historically underlying concepts, namely renormalization group theory. Some different concepts and several typical problems are shortly addressed, including a (more personal) view on challenges and chances.

1 Introduction

Material properties of soft matter systems are determined by processes and interactions on a wide range of length and time scales. While these mutually influence each other, it is not straight forward to provide quantitative information and understanding without taking this properly into account. Although this holds for many physical systems, it is of special importance for soft matter, where the characteristic energy scale is the thermal energy k_BT. Unlike for electronic properties, where typically energies are measured in eV (1 eV $\approx$ 40 k_BT at $T = 300$ K), such low energies give rise to significant conformational and structural fluctuations. The materials are "soft" because of a characteristic low (non-bonded) energy density, which to a very first rough approximation resembles the elastic constants of the material. The locally relevant length scales of a few Å to a few nm, lead to the very low energy densities allowing for large thermally driven fluctuations. Thus simulating soft matter automatically means dealing with large spatial and/or conformational fluctuations, making equilibration in many cases particularly difficult. To put the energy scale in perspective and provide a guide for comparison of different experiments and simulation approaches, we present in Table 1 the thermal energy in different units, as they are typically used in different fields.

A typical covalent bond, *i.e.* carbon–carbon, has an energy of about 80 k_BT. If no chemical reaction comes into play, it can be considered as stable. In contrast, for

Table 1 The thermal energy k_BT at $T = 300$ K in a variety of units as they are frequently used in different fields of physics an chemistry

Research field	$k_BT \cong 4.1 \cdot 10^{-21}$ J
Electronic properties	$2.5 \cdot 10^{-2}$ eV
Quantum chemistry	$9.5 \cdot 10^{-4}$ E_H
Biophysics	4.1 pNnm
Spectroscopy	200 cm^{-1}
(Phys.) chemistry	0.6 kcal/mol
(Phys.) chemistry	2.5 kJ/mol

Max Planck Institute for Polymer Research, Mainz, Germany. E-mail: kremer@mpip-mainz.mpg.de

typical hydrogen bonds, energies vary between about 6 and 10 k_BT, respectively. Thus they can break and reform on a rather short time scale, yet fairly long for a molecular simulation. As a consequence one usually deals in soft matter simulations with systems of moderate size, *i.e.* usually less than about a million atoms, however for very long time periods. We here give a short introduction related to the many different approaches and concepts, which have been discussed during the recent Faraday Discussion. A complementary introduction focusing on biopolymers can be found in the Faraday Discussion 139.[2]

Molecular simulation approaches to soft matter problems that are determined by a wide range of scales demand for an equally wide range of simulation methods at various levels of resolution including a varying amount of degrees of freedom.[3,4] A variety of quantum mechanical methods are used to address electronic/energetic properties on a high-resolution microscopic level, however they are limited to short length and time scales. Classical atomistic force field methods as well as particle-based coarse grained approaches are capable of sampling microscopic to mesoscopic scales. Especially the latter which is also able to access large conformational fluctuations, yet these approaches still fail to cover many macroscopic phenomena. For this, one needs to go beyond (purely) particle based approaches and use for example the Lattice Boltzmann[5] or DPD[6,7] methods or other mesoscopic methods to include hydrodynamic effects. Fig. 1 illustrates the characteristic different regimes and the level of details the models include.

Many questions regarding soft matter systems can be studied with a single numerical approach on a single level of resolution. However, when it comes to a quantitative understanding of complex materials, approaches with a single level of resolution do not frequently suffice since the different levels of resolution are more intimately interwoven. "Multiscale simulation" refers to methods where different simulation hierarchies are combined and linked to obtain an approach that simultaneously addresses phenomena or properties of a given system at several levels of resolution and consequently on several time and length scales. Multiscale simulation approaches may operate in different ways in terms of combining the individual levels of resolution: (*i*) in sequential approaches the simulation models on different scales are treated separately by simply passing information (structures, parameters, energies *etc.*) from one level of resolution to the next, (*ii*) in hybrid simulations different levels of resolution are present simultaneously, thus requiring direct interaction between them, and (*iii*) adaptive methods allow for individual molecules to

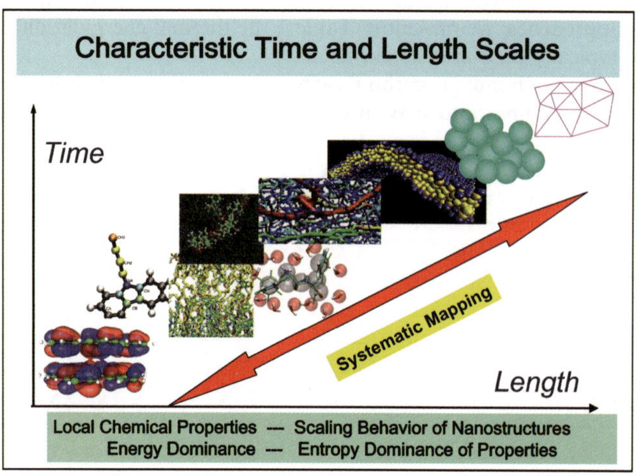

Fig. 1 Examples for characteristic numerical models for various length and time scales in soft matter systems.

adaptively switch between resolution levels on the fly—for example depending on their spatial coordinates. In either case, the exchange of information, interaction or particles requires a high level of consistency between the individual models.[8–13]

Simplified/generic coarse grained models which only account for a minimal set of properties of the (macro-)molecules of interest such as excluded volume, connectivity and a few basic types of interactions have since long been used and are perfectly well suited to study generic properties of soft matter systems. Since they reduce the computational complexity they allow for much longer effective time and length scales than more detailed models. Good examples are the investigation of scaling properties of polymeric systems,[14] both static and dynamic, as well as the investigation of biomembranes.[15,16] For example, for the problem of polymer melt dynamics such simulations, both molecular dynamics in continuum and Monte Carlo on lattices have been instrumental for a better understanding of the entanglement problem.[17–19] In order to link the results of such coarse grained simulations to real chemical systems one needs to appropriately devise the model parameters and interaction potentials. In multiscale simulations, where one wants to switch between resolution levels or use them next to each other, one has to go beyond scaled or fitted parameters because the levels of resolution need to be linked structurally and thermodynamically consistently. This requires a very careful development of CG models to avoid unphysical effects upon changes between scales. Here, we focus on methodologies to develop CG models based on an atomistic (force field) description.

Before we discuss a few ways to link different levels of resolution, let us shortly go back and mention the physical origin of coarse graining methods.

2 Coarse graining—general aspects

Dealing with the multiscale aspect of soft matter or more generally hierarchically structured materials can be done in many ways. One strategy, which has been followed by many disciplines from engineering to science is to devise independent models, which deal with typical aspects of a given scale. The link between the scales is then essentially given by the parameters characterizing each level. These parameters typically are closely linked to experiments. This is quite successful when it just comes to the description of material properties. A typical example which illustrates these rather complex structures can be studied by such an approach is the numerical investigation of the properties of lobster cuticles.[20] Here we want to go beyond that, however at the price that our systems, though already very complex, remain significantly simpler. In soft matter multiscale modelling the aim is not only to describe material properties but rather to understand the structural organisation and physical mechanisms which lead to morphologies, properties and eventually function. For this the different levels of description have to be much more intimately coupled.

Ideas linked to systematic coarse graining historically were linked to the fact that even for relatively simple systems it was simply not possible to perform all atom simulations. In addition, for many questions of concern, the value of this very detailed information coming from all atom simulations was questionable, since basic conceptional physical information could easily get lost. While physics is used all the time to sort out contributions in terms of small parameters a theoretical systematic link between a more local and a more global view was provided by renormalization group theory. The idea goes back to Kadanoff,[21] who introduced the block spin renormalization concept, which in variants is the basis for real space renormalization group treatments, which can directly be applied to polymers.[21–23] To illustrate this idea, let us look at a spin system as illustrated in Fig. 2.

Take a system of spins, which have only the two states, namely ± 1. When the system is divided into subcells of $(s \cdot a)^d$, a being the nearest neighbour distance and d the dimensionality, we can describe the free energy of the system in terms of cell variables, namely $F_{\text{cell}}(\varepsilon', H') = (as)^d F_{\text{site}}(\varepsilon, H)$, where the prefactor $(as)^d$ comes from the extensivity of the free energy. H is the Hamiltonian of the original system

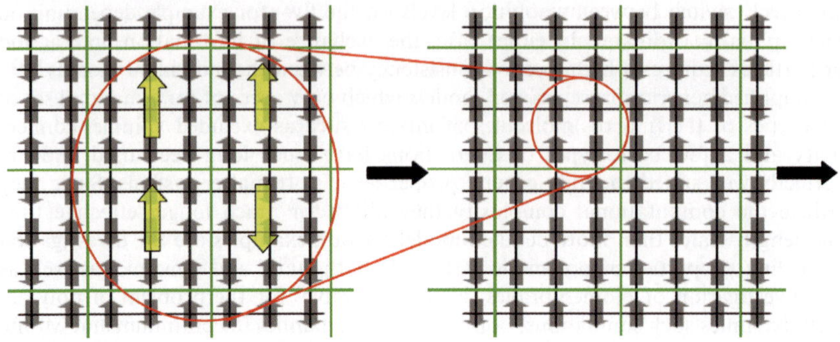

Fig. 2 Illustration of a simple majority rule real space renormalization step of a $d = 2$ Ising system, as it was introduced in the early 70s by L. P. Kadanoff based on the mathematical concepts of K. Wilson.[21,79] Blocks of 3×3 spins are mapped onto a single Ising spin based on a simple majority rule.

and H' the corresponding Hamiltonian on the basis of all cell interactions and $\varepsilon = (T - T_c)/T_c$, the normalized distance from the critical point. So far such a transformation is exact. Then however simple Ising spins (± 1) and nearest neighbour interactions transform into much more complicated longer ranged interactions and spins with many states. In practice usually one has to resort to an approximate treatment, as indicated in Fig. 2. Iterating such a procedure leads to the well known renormalization group flow diagrams. As de Gennes pointed out, polymers are a special case, which naturally suggest a renormalization along the back bone of the chains in a way that groups of monomers are lumped together into one monomer and this then is iterated. Fig. 3 shows a typical example of such a renormalization procedure based on a Monte Carlo simulation[23] of polymers made of hard spheres of diameter d and bonds of fixed length l.

In this context systematic coarse graining can be viewed as just one or two steps in such a renormalization group framework. Since this mapping step in almost all cases

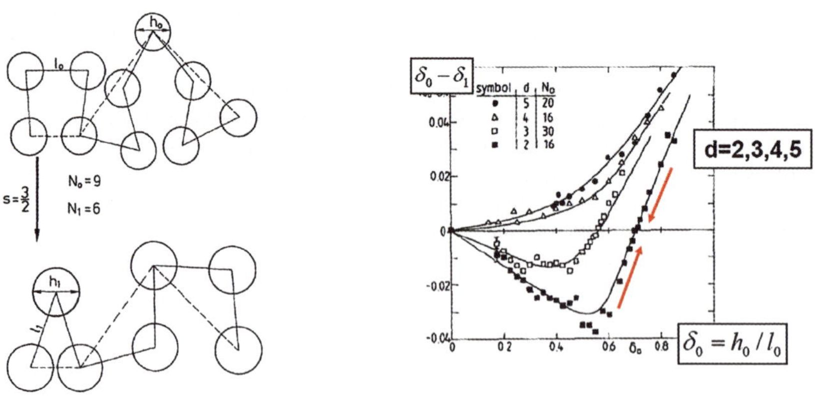

Fig. 3 Illustration of a Monte Carlo Renormalization Group Study of a hard sphere polymer model in spacial dimension $d = 2$ to 5. The relevant parameter is the ratio $\delta = h/l$ of sphere diameter and bond length. By matching the chain extensions (*i.e.* average squared end to end distance or radius of gyration) a renormalization flux diagram is generated, which leads to a stable fixed point $\delta^* = 0$ for $d = 4,5$ indicating random walk behavior (irrelevance of excluded volume) and non trivial stable fixed points of $\delta^* \approx 0.55$ and 0.7 respectively for $d = 3$ and 2, indicating the self avoiding walk structure. δ^* also indicates the optimal ratio of sphere diameter and bond length to minimize finite size corrections.[80]

requires significant approximations, it is also obvious that the free energy of a coarse grained system usually cannot be identical to that of the original system. For polymers, universality provides the criteria for which interactions are relevant and have to be properly transformed. This is because a polymer chain can be understood in very close relation to the problem of phase transitions. De Gennes showed that, within the so called n-vector model, a polymer chain of N steps can be seen as a path connecting lattice spins of the so called n-vector model. Since close to a critical point correlation lengths diverge, longer and longer paths, meaning walks of increasing length contribute. Thus there is the correspondence $(T - T_c)/T_c \propto 1/N$ and the conformations of these walks follow for $N \rightarrow \infty$ universal scaling laws leading, for example, to the nontrivial exponents $\nu = 3/(2 + d)$ for the end to end distance $\langle R^2(N) \rangle \propto N^{2\nu}$. For $d = 3$ the best field theoretic renormalization group studies and simulations give $\nu \approx 0.59$ instead of $\nu = 3/5$. While such general considerations give clear guidelines for coarse graining, they also illustrate limitations. Since only a few steps are performed and approximations are unavoidable in almost all cases, coarse grained systems also have to be studied carefully by themselves *i.e.* phase transitions pose special difficulties and there is *a priori* no reason that a coarse grained model, derived on the basis of a given scheme and some approximations, displays phase transitions at the very same temperature, pressure *etc.* as the underlying atomistic model. Second, the power of generic properties based on scaling laws relies on the proximity to asymptotics. While this often is reasonably fulfilled for long chain polymer melts or chains in solution, many of the current systems of interest certainly are not close to the asymptotic scaling regimes. Thus finite size corrections play a crucial role.

Alternatively one can view coarse graining procedures also as a special application of projector operator formalisms. Again the challenge is to define the optimal subspace of parameters, which on the one hand allow for a most efficient treatment of the systems and on the other hand do not exclude any aspect which is crucial for the question under study.

3 Linking levels of resolution: energies, forces and structures

Scale bridging requires systematic development of the individual models which are thermodynamically and/or structurally consistent. Many different approaches have been followed, both from the quantum mechanical to the classical level and from the classical all-atom level to a coarse grained description. For the latter we discuss here a few examples.

The derivation of interaction potentials between the coarse grained particles may be targeted at reproducing thermodynamic properties such as energies or free energies, for example partitioning data.[24,25] This approach has been particularly useful for simulating processes such as lipid membrane association where said properties play a decisive role.[26] On the other hand the energy based coarse graining approach does not *per se* guarantee reproduction of the structure of the system (for example of the underlying atomistic structure).[27] This may potentially cause problems and disruptions if one wishes to reinsert atomistic details into the CG structure. A recent alternative for (so far) rather simple model systems has been developed by Rutledge and Allen,[28–30] where the excess chemical potential of the degrees of freedom, which are averaged out by the coarse graining is properly accounted for.

In simple terms coarse graining methods are characterized by the physical quantities, which the models of different levels are supposed to reproduce as accurate as possible. Generally one can distinguish

- Structure based
- Force based
- Potential energy based

approaches, where the first ansatz most directly allows for a forward and backward mapping of the investigated systems. All three schemes face the problem of determining the coarse grained interactions based on the underlying more microscopic model. Again there are three methods, which are frequently employed. Based on detailed simulations of the high resolution system, interaction potentials are derived from

- (Iterative) Boltzmann inversion of distribution functions
- Inverse Monte Carlo sampling
- Force matching.

In principle all three methods of calculating interactions can be used for the different mapping schemes. A test comparing them for liquid SPC/E water, liquid methanol, liquid propane, and a single chain of hexane has been recently undertaken.[31]

Generally, structure-based methods provide CG interactions that reproduce a predefined target structure, often described by a set of radial distribution functions obtained from all-atom molecular simulations,[32–34] are well suited to reinsert atomistic coordinates. It is not clear whether they are equally well suited to reproduce thermodynamic properties of the system. Note that there is currently intensive research being carried out to investigate, whether—and if yes how—it is possible to derive coarse grained potentials that are both thermodynamically as well as structurally consistent with the underlying higher resolution description.[28,35]

An at first sight principally different methodology is the force matching method which has been applied to a multitude of soft matter, in particular biomolecular systems.[10,36] Here, the CG force field is determined such that the difference between the (instantaneous) CG forces and the forces in the underlying atomistic system are minimized. It can be shown that this method (in principle) determines a many-body multidimensional potential of mean force describing the CG representation of the system, thus being related to other structure-based CG methods, which usually rely on pair potentials of mean force.[37] The rather global (multibody) structural representation however bears the problem that the link to the underlying structure and the reproduction of local structural properties such as pair distributions may be rather weak. An exact reproduction of the underlying atomistic problem by force matching potentially requires the introduction of higher order interactions and forces.[31]

4 An example: structure-based coarse graining—from polymers to biomolecules

The general aim in structure-based coarse graining is to reproduce structural properties, either determined experimentally or from a higher resolution (atomistic) simulation. Below we will, in addition to references to the literature, give a few examples from work of the Mainz group, which range from classical amorphous polymers to relatively small biomolecules, as shown in Fig. 4. Though the latter are significantly smaller, the complexity compared to a isotropic homopolymer melt is significantly increased.[8,34,38–44]

Often the set of CG interaction functions is separated into bonded/covalent and nonbonded potentials. This approach relies on the assumption that the total potential energy U^{CG} can be separated into the respective contributions

$$U^{CG} = \sum U_B^{CG} + \sum U_{NB}^{CG} \qquad (1)$$

Following this separation ansatz we will first discuss the derivation of bonded (covalent, intramolecular) interaction potentials and possible implications of interdependet/correlated degrees of freedom. This separation ansatz is also important in the context of transferability of CG models. Assuming such a strict separation means that the intramolecular bonded/covalent CG interactions should be independent of the special scientific problem, *i.e.* of the surroundings of the molecules.

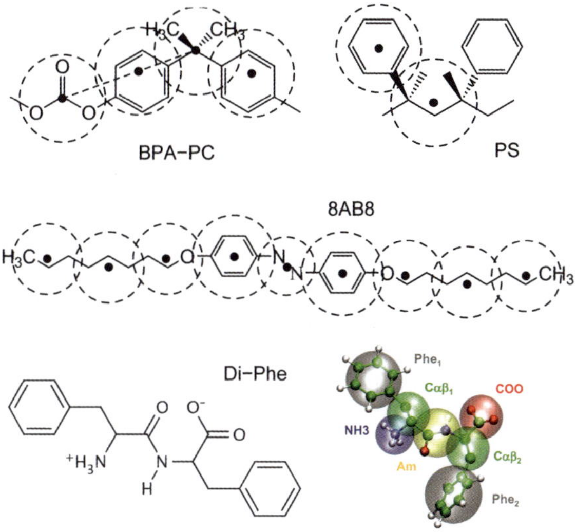

Fig. 4 Chemical structure and mapping schemes of some discussed CG examples: BPA-PC, Polystyrene (PS), the liquid crystalline compound 8AB8, and a small dipeptide (diphenylalanine).

4.1 Bonded/covalent interaction potentials

Bonded interactions are derived such that the local conformational statistics of the molecules is represented correctly in the CG model. These conformational distributions P^{CG} are usually characterized by specific bond lengths r, angles θ, and torsions ϕ between any pair, triple and quadruple of CG beads respectively, *i.e.* $P^{CG}(r, \theta, \phi, T)$. The assumption that the different CG internal degrees of freedom are uncorrelated, leads to the factorization of $P^{CG}(r, \theta, \phi, T)$ and reads $P^{CG}(r, \theta, \phi, T) = P^{CG}(r, T)P^{CG}(\theta, T)P^{CG}(\phi, T)$. This assumption has to be carefully checked (*i.e.* often certain combinations of CG bonds, angles and torsions are "forbidden" on the atomistic level). For a detailed discussion of how this can be achieved for the rather complex problem of different stereoregular subunits of polystyrene we refer to a recent study by Fritz *et al.*[42] The individual probability distributions $P^{CG}(r, T)$, $P^{CG}(\theta, T)$, and $P^{CG}(\phi, T)$ are then Boltzmann inverted to obtain the corresponding potentials:

$$U^{CG}(r, T) = -k_B T \ln (P^{CG}(r, T)/r^2) + C_r \tag{2}$$

$$U^{CG}(\theta, T) = -k_B T \ln (P^{CG}(\theta, T)/\sin(\theta)) + C_\theta \tag{3}$$

$$U^{CG}(\phi, T) = -k_B T \ln P^{CG}(\phi, T) + C_\phi \tag{4}$$

with C_r, C_θ, and C_ϕ being irrelevant constants used to set the minima of the respective potentials to zero. These potentials are in fact potentials of mean force, ergo free energies and consequently temperature dependent (not only due to the prefactor $k_B T$) and are either given in a tabulated form or determined by a fitting procedure.[34,39,40,45] Experience shows that a given parametrization is usually valid over typical temperature range of the order of ±10–20% (if no phase transition is within that range).

It is however also possible to determine the CG internal degrees of freedom based on distributions obtained from an atomistic simulation of the polymer melt.[33] In the latter case one obtains potentials for bonded and nonbonded interactions simultaneously based on the same melt (through iteration as described below).

Consequently all interaction functions are interdependent, *i.e.* there is no clear separation between covalent and nonbonded interaction potentials.

While the above clear separation of bonded and nonbonded interactions is desirable from a statistical mechanical point of view, the derivation of meaningful bonded potentials from an isolated single molecule requires that the conformational sampling of the isolated molecule and in the bulk (or solution) phase do not differ substantially. In biomolecular systems due to the peculiar nature of aqueous solutions (*i.e.* the presence of hydrogen bonds) this assumption can get problematic, as is illustrated by the dipeptide diphenylalanine.[43] For this dipeptide, bonded CG potentials were determined by Boltzmann inversion of the respective distributions obtained from conformational sampling of a single peptide in aqueous solution. Though bonded and nonbonded interactions are not as rigorously separated, covalent and nonbonded interactions were nevertheless separately and sequentially determined. The resulting CG model of the dipeptide (after adding also nonbonded potentials) turned out to very well reproduce the conformational equilibrium of the atomistic peptide.[43,44]

4.2 Nonbonded interaction potentials

Nonbonded interactions can be introduced in a variety of ways, depending of the system and the question one is studying. For amorphous polymers, where the density is known from experimental or atomistic simulations in many cases it is sufficient just to introduce an appropriate excluded volume for the CG beads.[8,38,39] This approach has been successful for studies of polycarbonate and polystyrene. In other cases nonbonded interaction potentials between coarse grained beads are derived based on the structure of isotropic liquids of small molecules (in the case of more complex molecules such as the liquid crystalline compound 8AB8, fragments of the target molecule are used). In the second scheme, the inverse Monte Carlo or the iterative Boltzmann inversion method[32,46] can be used to numerically generate a tabulated potential that precisely reproduces a given radial distribution function $g(r)$. It should be mentioned that the solution to the problem of finding a pair potential which exactly reproduces a given radial distribution function is unique.[47] However there usually exist different pair potentials which reproduce a given structure function within a hardly noticeable error. This can be used to impose additional constraints, *i.e.* to better reproduce thermodynamic quantities[46,48–50] without disrupting the local structure.

For molecules with many different CG beads (for example biological macromolecules) or in the case of liquid crystalline molecules with anisotropic structures the procedure to determine nonbonded interaction functions is slightly more involved. In these cases it is advantageous to split the target molecule into small fragments. Such fragment-based approaches have been successfully applied to the liquid crystalline compound 8AB8[34] (see also Fig. 4) or for the interaction of BPA-PC with metal surfaces.[38,51,52] However, there the problem of transferability of the fragment based potentials to the interactions of the large molecules has to be tested with great care.[52,53] Nevertheless, the procedure to derive CG potentials from chain fragments and low molecular weight liquids does open up the possibility to reuse certain CG potentials for reoccurring building blocks (such as alkyl or phenyl groups).

5 Backmapping

Various approaches have been employed to reinsert atomistic details into a CG structure or simulation trajectory. It should be noted that this "backmapping" or inverse mapping problem has in general no unique solution since every CG structure corresponds to an ensemble of atomistic microstates. For coarse grained polymeric melts it is possible to obtain backmapped atomistic structures by taking rigid all-atom chain fragments obtained from a correctly sampled distribution of all-atom

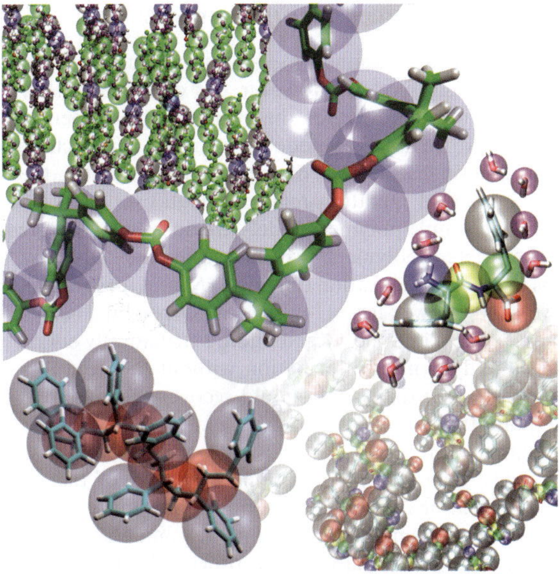

Fig. 5 Coarse grained superimposed with backmapped atomistic structures of a BPA-PC chain and a polystyrene oligomer, of liquid crystalline 8AB8 (upper left corner), of the aggregates formed by diphenylalanine (lower right corner) and a single diphenylalanine molecule with its water shell.

chain structures.[8,39,54–56] This works, if the structural relaxation and diffusion of molecules is slow compared to the local equilibration of the newly introduced atomistic coordinates. The case of more flexible low-molecular weight molecules requires the introduction of constraints in order to avoid the atomistic structure from drifting/diffusing too far from the CG.[34,43,44,57] Fig. 5 shows the result of the backmapping procedure for a few systems: a BPA-PC and a polystyrene chain, a snapshot of liquid crystalline 8AB8, a snapshot of the aggregates formed by diphenylalanine and a single diphenylalanine molecule with its water shell.

The combination of CG simulations with an efficient backmapping methodology is a powerful tool to efficiently simulate long time scale and large length scale soft matter processes and in the end to obtain well-equilibrated atomistic structures and trajectories. For example, the relevant time scale of many NMR experiments requires simulations beyond what is possible with atomistic models. Nevertheless, atomistic coordinates are often necessary to compare with experimental results, an important example for the use of backmapped CG trajectories.[54,58] In a slightly different manner one can also utilize inverse mapped structures in further computation. For example in order to obtain data for solubilities or permeabilities of small molecules in polymeric systems one can combine coarse grained simulations to obtain well equilibrated structures of the polymeric melt with atomistic free energy calculations based on the inverse mapped trajectories.[59–61] In a rather early study of phenol in BPA-PC phenol was introduced into remapped polycarbonate melts at a variety of temperatures. There it could be shown how the diffusion of the phenols coupled to the local fluctuations of the polymeric matrix.[62]

Another promising application of the combination of coarse grained simulations with a backmapping procedure is the possibility to validate the underlying atomistic force field—on time and length scales not accessible to atomistic simulations alone due to sampling problems.

6 Dynamics: coarse grained *versus* atomistic

The construction of a coarse grained model automatically determines the length scaling between the linked models. However, for dynamics this is not the case at all. We want to illustrate this here for the example of polymers. For both simple continuum as well as lattice polymer models it is known that such simulations reproduce the essential generic features of polymer dynamics; that is, the crossover from the Rouse to the entangled reptation regime, qualitatively and to a certain extent quantitatively.[19] Such simple polymer models are, in view of the present discussion, just another set of different polymers. Properly scaled they all follow the same rules. For short chains the longest relaxation time $\tau_R \propto N^2$ and for long chains in a polymer melt we observe a $N^{3.4}$ power law. However, a proper link between the atomistic representation of a system and the corresponding structurally coarse grained system can provide absolute dynamical information without the need to resort to generic scaling laws. Actually, eventually one should recover them as well. Thus one aims at a predictive quantitative modeling of diffusion, viscosity, rates, and correlation times, *etc.* of dynamic events.

This automatically generates the additional question of the minimal time and length scales CG simulations apply to. First it is important to realize that the coarse grained models are, from a simulation point of view, independent models with their own intrinsic dynamics. In the case of the previously discussed polystyrene and BPA-PC simulations one can deduce a typical simulation time scale, as it is traditionally done in MD simulations. Taking the strength of the interaction parameter in the nonbonded excluded volume interaction ε_{CG} (measured in units of the temperature), the average mass m_{CG} of the CG beads and the known length scales σ_{CG} one can determine the intrinsic time scale of the CG simulation† from $1\tau = 1(m_{CG}\sigma_{CG}{}^2/\varepsilon)^{1/2}$. This results for instance in $1\tau = 1.7$ ps for BPA-PC at 570 K and $1\tau = 1$ ps at $T = 463$ K for atactic polystyrene, respectively.[41,63] While these are the natural time scales of the CG model, this does not have to be the time scale of the underlying atomistic model at all. The CG interaction potentials are much smoother, barriers are lower *etc.*, resulting in significantly accelerated dynamics. Beyond the reduction of the number of degrees of freedom this is the main reason for the speed up due to coarse graining.

On the other hand, on length scales above the specific scale of the coarse graining we expect qualitatively the same behavior for the CG chains as for the atomistic chains, certainly on scales where generic properties dominate. This offers a direct way of deducing the time scaling between the CG model and the underlying atomistic model by matching the curves of the mean square displacements of the beads or the center of mass of the whole chain. Since the lengths are fixed by the mapping procedure itself the mean square displacements can be matched just by shifting the time scales. It is however important that the curves of the atomistic and CG mean square displacements not only meet in a point, but rather coincide from a characteristic point onwards, as will be discussed below. Fig. 6 shows a typical result for polystyrene.[64] This procedure leads for the two examples mentioned above to the time scaling of $1\tau = 400$ ps for PS at $T = 463$ K and 30 ps for BPA-PC at $T = 570$ K, respectively. Of course, as this example already illustrates different levels of coarse graining and different CG models lead to different time scales. Also it should be noted that the time scaling factor can be chain-length dependent, since the melt density varies as a function of chain length for polymers below the entanglement molecular weight (see Fig. 6).

The data, on which Fig. 6 is based on, however, reveals more important information. Looking at the displacements of the individual beads, rather than the center of

† Note that the resulting time scale τ depends on the choice of masses, if the CG model contains different beads. Thus there is some arbitrariness in the value of τ if one wants to use physical units.

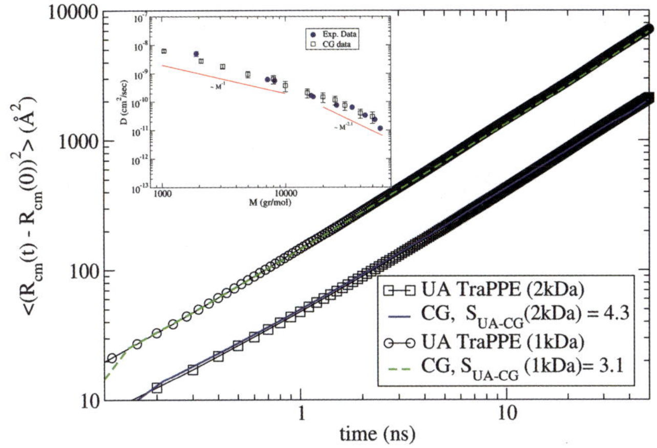

Fig. 6 Mean square displacements of centers of mass of coarse grained and atomistic (united-atom) polystyrene (PS) simulations as a function of time for two different chain lengths. By shifting the CG data along the time axis the time scaling can be obtained. To obtain the real time scaling in another step the united atom (UA; here hydrogens are lumped into carbon superatoms) and the atomistic (AA) simulations have to be compared in the same way. This is necessary since atomistic simulations for PS are extremely time consuming. Because of that two different atom based methods, AA and UA respectively, are needed to obtain a first scale factor from distances, which are too small for CG simulations. The inset shows the diffusion constants for PS obtained from CG simulations based on the resulting AA-CG time scaling in comparison to experiments for a molecular weight of up to M = 50 kDa (after ref. 41,64). Note that there is no adjustable parameter in the simulation data.

mass of the chains, one finds that the motion characteristics down to the characteristic scale of our coarse graining qualitatively and quantitatively, after the appropriate time scaling, agrees to the atomistic simulations. Thus the CG runs can be used to obtain realistic atomistic trajectories. In ref. 54 it was shown for BPA-PC that atomistic dynamic structure factors can be obtained from remapped CG runs in perfect match with short time atomistic data and by this, extend the dynamic information significantly. Recently, we have shown for polystyrene that this also quantitatively predicts correct bond orientation correlation times as obtained from NMR.[64]

Methods like this not only can be used to study the dynamics of homopolymer melts, but also the dynamics of additives in such melts[65,66] and allows quantitative predictions into experimentally extremely difficult to access regions. It should however be kept in mind that the time scaling factor for the additive might differ from that of the host polymer.

7 Variable/adaptive resolution methods

So far we have only mentioned methodologies, where the whole system is studied on a single level of resolution. In many cases, however, this is not desirable or even does not allow the investigation of important problems. To overcome this, methods have been developed over the last years which allow various levels of resolution within one simulation.

A classical problem in this context is crack propagation in solids. There the breaking of a chemical bond is treated by quantum mechanical methods. Beyond the immediate tip of the crack classical force field simulations suffice and beyond that the particles are coupled to continuum finite element like models, which properly can take care of more global stress and strain fields. The hierarchy of modelling is well defined and a given atom might at the beginning be treated quantum

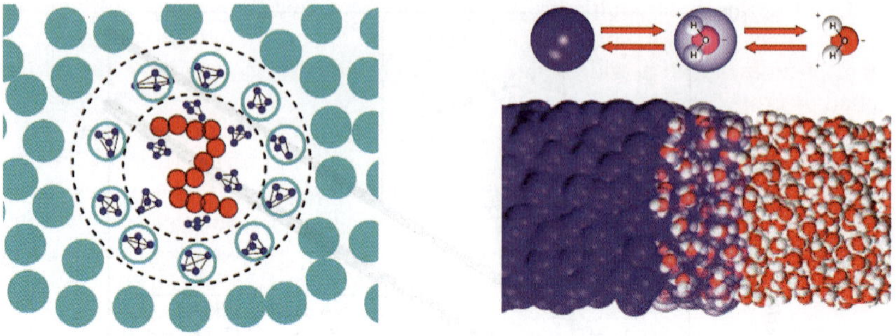

Fig. 7 Illustration of adaptive resolution simulations using the AdResS method. On the left is a small polymer chain in a good solvent of tetrahedral molecules.[81] They adjust their resolution depending on the region they are in, when they cross through the transition regime. The sphere with the "all atom" solvent, which also contains the polymer, diffuses with the chain. On the right the method is illustrated for the simulation of water.[82] The scheme on top indicates the change from an all atom (TIP3P) water to a single sphere model. The water molecules are free to move around and adjust their resolution to the region they are in.

mechanically, then classically and then eventually merges into the continuous description.[67,68]

For soft matter, especially biopolymers, a somewhat different approach has been very successful. Since there is no crystal structure and the systems are more strongly fluctuating it is decided from the very beginning which atoms are treated quantum chemically and which are treated by a classical force field. This so called QM/MM method turned out to very well describe local, quantum-effect dominated phenomena, while the surrounding can be treated classically.[69]

While such methods mark significant progress for a multiscale, or better dual or triple scale, description of materials, they all suffer from the problem that from the very beginning one has to determine, which atom/particle is going to be described on what level. There is no dynamic forward and backward exchange of resolution. Thus it is not possible to study liquids or strongly fluctuating systems by the above methods. One would like to be able to zoom in on demand and treat a dynamic equilibrium between different levels of resolution. Thus atoms/molecules should be allowed to change the level of resolution on the fly. By such a method one can focus on local phenomena, while keeping equilibrium with a greater surrounding, *i.e.* aggregation or adsorption/desorption phenomena. The idea is illustrated in Fig. 7. The condition of equilibrium and free exchange between regions of different levels of resolution poses special difficulties. Recently, an adaptive resolution simulation method (AdRess) has been developed, which allows for adaptive coupling of such different levels.[48,70,71] In a recent extension this method has also been coupled to an outer continuum, devising a way towards open system molecular dynamics.[72]

8 Conclusions and outlook

The general challenge that lies in coarse graining and generally in reducing the number of degrees of freedom of a computational model is to incorporate the (average) effect of the eliminated degrees of freedom into the lower resolution model. We have summarized different developments which are currently pursued with the aim to design coarse grained models that are both thermodynamically and structurally consistent with an underlying atomistic simulation model. This consistency is of particular importance for coarse grained models that are to be used in a multiscale simulation framework where different simulation hierarchies are combined and linked to obtain an approach that simultaneously addresses phenomena or properties of a given system at several levels of resolution and consequently on several time and length scales. Such multiscale approaches are of great importance in the investigation of complex soft matter systems such as biological and synthetic materials where phenomena on a wide range of scales "team up" to determine the overall (material) properties.

The development of these methods is one of the major methodological efforts in computational chemistry and physics. Scale-bridging approaches can operate on varying levels of "interaction" between the individual scales: the examples shown in the present article mainly combine models on different scales *via* treating them separately and sequentially, *i.e.* they pass information (in the present cases structures) between the levels of resolution. Other resolution exchange methods use the exchange (of the whole system) between simulations at different levels of resolution during the course of the simulation as a means to enhance sampling.[73–75] In the case of hybrid simulations, different levels of resolution are present simultaneously in one system. This is more complex than the sequential approach since interactions between entities at different levels of resolutions have to be devised. Hybrid approaches are widely used in the field of mixed quantum mechanical/classical simulations. The statistically mechanically consistent treatment of such problems has been addressed in methods that allow for individual molecules to adaptively switch between resolution levels on the fly.[48,70–72]

Even though we can now look back at more than a decade of intensive research along the lines discussed above, there are still a number of challenges which require a continued and even stronger effort. Specific problems/challenges are

- Non bonded interactions
- Time mapping
- Efficient and appropriate mapping schemes
- True multiscale simulations of disordered/fluctuating systems.

How delicate the problem of nonbonded interactions is can already be seen for the question of phase segregation in polymer mixtures. In order to hit the transition temperature an accuracy of the interactions of $O(k_BT/N)$ is needed (N being the chain length). This is beyond any predictive theoretical possibilities. Additional problems occur for solvent mediated or directional interactions, which are of special relevance for biomolecules and/or aggregation phenomena. One of the various reasons for these problems lies in the huge abundance of hydrogen atoms for which a classical description even at room temperature often is questionable. Actually, summing up typical uncertainties in the interactions along, *i.e.* a heteropolymer or a protein, one easily arrives at error bars of the size or even larger than the basin of attraction of the folded state.[76] Thus we expect in the future that appropriate coarse grained simulations might even be employed to improve underlying force fields.

The problem of time mapping is a key to understanding dynamical processes. Here however we encounter significant technical and principal problems. As shown for polymers, time scaling between different levels of resolution can be derived and leads to very good results. However such a single scaling factor does of course not work for very short times and is also chain length dependent. Thus the motion of a single additive molecule in a polymer matrix can follow a different time scaling than the matrix itself. This can for instance mean that, mapped back to an atomistic picture, the "different clocks" can lead to diverging time differences. To cure these problems significant further work is needed.

Both previous points are linked to the question of what is the most appropriate mapping scheme. So far groups of atoms are typically lumped together into one coarse grained particle and the appropriate bonded and nonbonded interactions are determined. The grouping is guided by the chemical structure and the question of investigation. For instance for many problems in polymer physics the crossing of chains cannot be allowed, while for some problems that is not crucial at all. So far this very much relies on the intuition and knowledge of the investigator and there are no systematic formal procedures. Here physics based formal concepts could be of tremendous help.

Taking all these aspects together we can expect very significant progress, when it comes to true multiscale simulations of strongly fluctuating and disordered systems. Here we do not only think of biological systems but also of other complex aggregates as they are studied, for example, in the context of organic electronics or of a molecularly based study of non equilibrium phenomena. With the coming new hardware, plain CPU time will probably, for the most part, not be the central issue in the future. However then we are, in a positive sense, back at the beginning, where ideas of coarse graining were more directly linked to the understanding of basic physical phenomena in soft matter physics.

Acknowledgements

We thank all present and past members of the multiscale modeling group at the Max Planck Institute for Polymer Research for fruitful collaborations and many stimulating discussions, in particular Luigi Delle Site, Vagelis Harmandaris, Matej Praprotnik, Nico van der Vegt and Alessandra Villa. We would like to acknowledge several teams developing multiscale modeling software packages, namely the Espresso and Espresso++ team,[77] the Gromacs developers,[78] and the VOTCA[31] team. CP acknowledges financial support by the German Science Foundation within the Emmy Noether Programme (grant PE 1625/1-1).

References

1 C. Peter and K. Kremer, *Soft Matter*, 2009, DOI: 10.1039/b912027k.
2 L. Mahadevan, *Faraday Discuss.*, 2008, **139**, 9–19.
3 H. J. C. Berendsen, *Simulating the Physical World*, Cambridge University Press, Cambridge, UK, 2007.
4 *Coarse-Graining of Condensed Phase and Biomolecular Systems*, ed. G. A. Voth, Chapman and Hall/CRC Press, Taylor and Francis Group, 2008.
5 B. Dunweg and A. J. C. Ladd, *Advanced Computer Simulation Approaches for Soft Matter Sciences III*, 2009, 221, pp. 89–166.
6 P. J. Hoogerbrugge and J. M. V. A. Koleman, *Europhys. Lett.*, 1992, **19**, 155–160.
7 R. D. Groot and P. B. Warren, *J. Chem. Phys.*, 1997, **107**, 4423–4435.
8 W. Tschöp, K. Kremer, J. Batoulis, T. Burger and O. Hahn, *Acta Polym.*, 1998, **49**, 61–74.
9 W. Tschöp, K. Kremer, O. Hahn, J. Batoulis and T. Burger, *Acta Polym.*, 1998, **49**, 75–79.
10 G. S. Ayton, W. G. Noid and G. A. Voth, *Curr. Opin. Struct. Biol.*, 2007, **17**, 192–198.
11 P. L. Freddolino, A. Arkhipov, A. Y. Shih, Y. Yin, Z. Chen and K. Schulten, *Coarse-Graining of Condensed Phase and Biomolecular Systems*, Chapman and Hall/CRC Press, Taylor and Francis Group, 2008.
12 S. J. Marrink, M. Fuhrmans, H. J. Risselada and X. Periole, *Coarse-Graining of Condensed Phase and Biomolecular Systems*, Chapman and Hall/CRC Press, Taylor and Francis Group, 2008.
13 N. F. A. van der Vegt, C. Peter and K. Kremer, *Coarse-Graining of Condensed Phase and Biomolecular Systems*, Chapman and Hall/CRC Press, Taylor and Francis Group, 2008.
14 K. Kremer and G. S. Grest, *J. Chem. Phys.*, 1990, **92**, 5057–5086.
15 M. Muller, K. Katsov and M. Schick, *Phys. Rep.*, 2006, **434**, 113–176.
16 B. J. Reynwar, G. Illya, V. A. Harmandaris, M. M. Müller, K. Kremer and M. Deserno, *Nature*, 2007, **447**, 461–464.
17 T. C. B. Mc Leish, *Adv. Phys.*, 2002, **51**, 1379.
18 R. Everaers, S. K. Sukumaran, G. S. Grest, C. Svaneborg, A. Sivasubramanian and K. Kremer, *Science*, 2004, **303**, 823–826.
19 K. Kremer, *Simulations in Condensed Matter: From Materials to Chemical Biology*, Springer, 2006, vol. 704.
20 H. Fabritius, A. Ziegler, S. Hild, S. Nikolov, M. Petrov, C. Sachs, L. Lymperakis, M. Friak, J. Neugebauer and D. Raabe, *Proceedings of Conference/Meeting: PLASTICITY*, St Thomas, Virgin Islands, USA, 2009.
21 L. P. Kadanoff, *Critical Phenomena*, ed. Green M. S., Academic Press, London, 1971.
22 P. G. De Gennes, *Scaling concepts in polymer physics*, Cornell University Press, Ithaca NY, 1978.
23 D. P. Landau and B.K., *A guide to Monte Carlo simulations in statistical physics*, Cambridge University Press, 2000.

24 S. J. Marrink, H. J. Risselada, S. Yefimov, D. P. Tieleman and A. H. de Vries, *J. Phys. Chem. B*, 2007, **111**, 7812–7824.
25 L. Monticelli, S. K. Kandasamy, X. Periole, R. G. Larson, D. P. Tieleman and S. J. Marrink, *J. Chem. Theory Comput.*, 2008, **4**, 819–834.
26 S. J. Marrink, A. H. de Vries and A. E. Mark, *J. Phys. Chem. B*, 2004, **108**, 750–760.
27 R. Baron, D. Trzesniak, A. H. de Vries, A. Elsener, S. J. Marrink and W. F. van Gunsteren, *ChemPhysChem*, 2007, **8**, 452–461.
28 E. C. Allen and G. C. Rutledge, *J. Chem. Phys.*, 2008, **128**, 154115.
29 E. C. Allen and G. C. Rutledge, *J. Chem. Phys.*, 2009, **130**, 034904.
30 E. C. Allen and G. C. Rutledge, *J. Chem. Phys.*, 2009, **130**, 204903.
31 V. Ruehle, C. Junghans, A. Lukyanov, K. Kremer and D. Andrienko, *J. Chem. Theory Comput.*, 2009, submitted.
32 A. P. Lyubartsev and A. Laaksonen, *Phys. Rev. E*, 1995, **52**, 3730–3737.
33 F. Müller-Plathe, *ChemPhysChem*, 2002, **3**, 754–769.
34 C. Peter, L. Delle Site and K. Kremer, *Soft Matter*, 2008, **4**, 859–869.
35 M. E. Johnson, T. Head-Gordon and A. A. Louis, *J. Chem. Phys.*, 2007, **126**, 144509.
36 S. Izvekov and G. A. Voth, *J. Phys. Chem. B*, 2005, **109**, 2469–2473.
37 W. Noid, J. Chu, G. Ayton and G. Voth, *J. Phys. Chem. B*, 2007, **111**, 4116–4127.
38 C. F. Abrams and K. Kremer, *Macromolecules*, 2003, **36**, 260–267.
39 V. A. Harmandaris, N. P. Adhikari, N. F. A. van der Vegt and K. Kremer, *Macromolecules*, 2006, **39**, 6708–6719.
40 V. A. Harmandaris, D. Reith, N. F. A. van der Vegt and K. Kremer, *Macromol. Chem. Phys.*, 2007, **208**, 2109–2120.
41 V. A. Harmandaris and K. Kremer, *Macromolecules*, 2009, **42**, 791–802.
42 D. Fritz, V. A. Harmandaris, K. Kremer and N. F. A. van der Vegt, *Macromolecules*, 2009, DOI: 10.1021/ma901242h.
43 A. Villa, C. Peter and N. F. A. van der Vegt, *Phys. Chem. Chem. Phys.*, 2009, **11**, 2077.
44 A. Villa, N. F. A. van der Vegt and C. Peter, *Phys. Chem. Chem. Phys.*, 2009, **11**, 2068.
45 G. Milano, S. Goudeau and F. Muller-Plathe, *J. Polym. Sci., Part B: Polym. Phys.*, 2005, **43**, 871–885.
46 D. Reith, M. Putz and F. Muller-Plathe, *J. Comput. Chem.*, 2003, **24**, 1624–1636.
47 R. L. Henderson, *Phys. Lett.*, 1974, **49**, 197–198.
48 M. Praprotnik, L. Delle Site and K. Kremer, *Phys. Rev. E: Stat., Nonlinear, Soft Matter Phys.*, 2006, **73**, 066701.
49 S. Jain, S. Garde and S. K. Kumar, *Ind. Eng. Chem. Res.*, 2006, **45**, 5614–5618.
50 H. Wang, C. Junghans and K. Kremer, *Eur. Phys. J. E*, 2009, **28**, 221–229.
51 L. Delle Site, C. F. Abrams, A. Alavi and K. Kremer, *Phys. Rev. Lett.*, 2002, **89**, 156103.
52 L. Delle Site, S. Leon and K. Kremer, *J. Am. Chem. Soc.*, 2004, **126**, 2944–2955.
53 J. D. McCoy and J. G. Curro, *Macromolecules*, 1998, **31**, 9362–9368.
54 B. Hess, S. Leon, N. van der Vegt and K. Kremer, *Soft Matter*, 2006, **2**, 409–414.
55 G. Santangelo, A. Di Matteo, F. Muller-Plathe and G. Milano, *J. Phys. Chem. B*, 2007, **111**, 2765–2773.
56 I. F. Thorpe, J. Zhou and G. A. Voth, *J. Phys. Chem. B*, 2008, **112**, 13079–13090.
57 X. Chen, P. Carbone, G. Santangelo, A. Di Matteo, G. Milano and F. Muller-Plathe, *Phys. Chem. Chem. Phys.*, 2009, **11**, 1977–88.
58 D. Stueber, T. Y. Yu, B. Hess, K. Kremer and J. Schaefer, *submitted to J. Chem Phys.*, 2009.
59 B. Hess, C. Peter, T. Ozal and N. F. A. van der Vegt, *Macromolecules*, 2008, **41**, 2283–2289.
60 T. Ozal, C. Peter, B. Hess and N. F. A. van der Vegt, *Macromolecules*, 2008, **41**, 5055–5061.
61 B. Hess and N. F. A. Van der Vegt, *Macromolecules*, 2008, **41**, 7281–7283.
62 O. Hahn, D. A. Mooney, F. Muller-Plathe and K. Kremer, *J. Chem. Phys.*, 1999, **111**, 6061–6068.
63 S. Leon, N. F. A. van der Vegt, L. Delle Site and K. Kremer, *Macromolecules*, 2005, **38**, 8078–8092.
64 V. A. Harmandaris and K. Kremer, *Soft Matter*, 2009, DOI: 10.1039/b905361a.
65 V. A. Harmandaris, N. P. Adhikari, N. F. A. van der Vegt, K. Kremer, B. A. Mann, R. Voelkel, H. Weiss and C. C. Liew, *Macromolecules*, 2007, **40**, 7026–7035.
66 D. Fritz, C. R. Herbers, K. Kremer and N. F. A. van der Vegt, *Soft Matter*, 2009, DOI: 10.1039/b911713j.
67 J. Q. Broughton, F. F. Abraham, N. Bernstein and E. Kaxiras, *Phys. Rev. B: Condens. Matter Mater. Phys.*, 1999, **60**, 2391–2403.
68 J. Rottler, S. Barsky and M. O. Robbins, *Phys. Rev. Lett.*, 2002, **89**, 148304.
69 H. M. Senn and W. Thiel, *Atomistic Approaches in Modern Biology. From Quantum Chemistry to Molecular Simulations*, Springer, Berlin, 2007, vol. 268, pp. 173–290.
70 M. Praprotnik, L. Delle Site and K. Kremer, *J. Chem. Phys.*, 2005, **123**, 224106.
71 M. Praprotnik, L. Delle Site and K. Kremer, *Annu. Rev. Phys. Chem.*, 2008, **59**, 545–571.

72 R. Delgado-Buscalioni, K. Kremer and M. Praprotnik, *J. Chem. Phys.*, 2008, **128**, 114110.
73 E. Lyman, F. M. Ytreberg and D. M. Zuckerman, *Phys. Rev. Lett.*, 2006, **96**, 028105.
74 D. M. Zuckerman, *Coarse-Graining of Condensed Phase and Biomolecular Systems*, Chapman and Hall/CRC Press, Taylor and Francis Group, 2008.
75 M. Christen and W. F. van Gunsteren, *J. Chem. Phys.*, 2006, **124**, 154106.
76 P. L. Freddolino, S. Park, B. Roux and S.K., *Biophys. J.*, 2009, **96**, 3772–3780.
77 H. J. Limbach, A. Arnold, B. A. Mann and C. Holm, *Comput. Phys. Commun.*, 2006, **174**, 704–727.
78 D. Van der Spoel, E. Lindahl, B. Hess, G. Groenhof, A. E. Mark and H. J. C. Berendsen, *J. Comput. Chem.*, 2005, **26**, 1701–1718.
79 K. G. Wilson and J. Kogut, *Phys. Rep.*, 1974, **12**, 75.
80 K.K., A. Baumgärtner and B.K., *Z. Phys. B: Condens. Matter*, 1981, **40**, 331.
81 M. Praprotnik, L. Delle Site and K. Kremer, *J. Chem. Phys.*, 2007, **126**, 134902.
82 M. Praprotnik, S. Matysiak, L. Delle Site, K. Kremer and C. Clementi, *J. Phys.: Condens. Matter*, 2007, **19**, 292201.

Fine-graining without coarse-graining: an easy and fast way to equilibrate dense polymer melts

Paola Carbone,*[a] Hossein Ali Karimi-Varzaneh*[b] and Florian Müller-Plathe*[b]

Received 4th February 2009, Accepted 5th March 2009
First published as an Advance Article on the web 4th August 2009
DOI: 10.1039/b902363a

A technique to prepare well-equilibrated polymer melts is presented. The method, named fine-graining, consists of two steps: the generation of continuum random walks characterized by different Kuhn lengths and the insertion of the atomistic units on the "parent" random walk chains. The procedure ensures a good equilibration at long as well as short length-scales and it is very easy to implement. Melts of polyethylene, atactic polystyrene and polyamide-66 are equilibrated with this technique and their long and short range structural properties can be successfully compared with previous simulation and experimental data.

Introduction

In materials science, atomistic simulations have become, in the last years, a powerful tool to predict properties, to verify theories or to explain experimental findings.[1–3] In particular, in polymer science the resort to molecular simulations has remarkably increased after the advent of high performance computers that allow the simulation of models composed of thousands of atoms in a reasonable time. However, despite the improvement in the hardware, modeling high molecular- weight polymer melts is still a computational challenge. Many striking features of polymers (including their viscoelastic behavior) depend indeed on the fact that the monomers are bonded together following a specific topology. The chain connectivity makes the conformational space to be sampled very broad and characterized by a complex energy landscape. Therefore, specific simulation techniques must be developed in order to efficiently explore the conformation space and gain information about the structure and dynamics of a polymer melt avoiding the system being trapped in the neighborhood of the initial configuration.

An effective way to simulate polymers must then match two conflicting requirements in the model description: (i) the structures and models need detailed atomistic resolution to capture subtleties in the polymer–polymer interactions and to allow comparison with experimental information. (ii) The prepared structures must cover a large spatial domain to capture the large-scale structural features and processes inherent in the polymeric materials. Due to computer-time limitations, one can either have the fine resolution or the large spatial dimension or an unsatisfactory compromise between the two.

[a]*School of Chemical Engineering and Analytical Science, University of Manchester, PO Box 88 Sackville St, Manchester, M60 1QD, United Kingdom. E-mail: paola.carbone@manchester.ac.uk*
[b]*Eduard-Zintl-Institut für Anorganische und Physikalische Chemie, Technische Universität Darmstadt, Petersenstrasse 20, D-64287 Darmstadt, Germany. E-mail: h.karimi@theo.chemie.tu-darmstadt.de; f.mueller-plathe@theo.chemie.tu-darmstadt.de*

Different approaches have been proposed in the past to efficiently relax high molecular weight polymer melts. These methods can be roughly divided into two classes based on the level of resolution of the model. The first consists of techniques where the atomistic models are subjected to specific algorithms (usually Monte Carlo (MC)) able to quickly sample the conformational energy surface and find a good local minimum.[4–7] An example of such methods is the advanced off lattice MC algorithm (end-bridging MC) proposed recently by Theodorou and coworkers,[8] where the MC moves involve large segments of the polymeric chain altering their connectivity. This connectivity-altering move (and its variant)[9] has proved to greatly enhance the sampling efficiency of the configurations. During such move, two chains are selected so that the end of one is within a certain bridging distance from a backbone segment of the other. A trimer centered at this latter backbone segment is excised from the second chain, thus defining two subchains. The end of one of these subchains is connected to the end of the first chain by constructing a bridging trimer, forming a new chain with prescribed molecular geometry (bond lengths and bond angles). The selection of the chains, the building of the bridging trimer, and the acceptance criteria of the method are designed so that the requirements of microscopic reversibility for proper sampling are satisfied. Using this technique various high molecular weight (M_w) polymers (polyethylene,[8,10] polyisoprene,[11] poly(ethylene-oxide)[12]) have been simulated and the results compared successfully with the experimental numbers.

Using again the MC technique, Neyertz and Brown[13] proposed a different set of moves (pivot MC) based on Flory's hypothesis, *i.e.*, that polymer configurations in a pure melt are very similar to those obtained by sampling an isolated molecule where only a certain number of specific near-neighbor atoms interact. Molecular configurations are generated using a pivot MC algorithm that efficiently samples the bond-angle and torsional phase space of isolated chains. Only interactions between backbone sites separated by a fixed number of backbone bonds (n_{bonds}) are considered at this stage. A collection of such chains are then randomly placed within the confines of a periodic simulation box and the excluded volume is introduced. The choice of the n_{bonds} value depends on the chemistry of the polymer chain. This method has been successfully applied to study melt of various unentangled polymers.[13–15]

Both MC methods have been proved to efficiently relax high M_W polymers but the first (the end bridging MC) is difficult to implement, and the second (the pivot MC) can be less efficient in equilibrating highly rigid polymers (where the value of n_{bonds} becomes very high) or polymers where long range interactions (like the electrostatic ones) do have a significant influence on the structure of the chain conformations in the bulk melt.

In a second class of methods, a coarser (than atomistic) model is subjected to a standard molecular dynamics (MD) simulation in order to relax the long range structure. The reduction of the degrees of freedom in the model is necessary to speed up the relaxation process and allows the use of a conventional MD algorithm to sample the conformational space. After relaxing the structure at the coarse-grained level, the omitted details (the atoms) are re-introduced (reverse-mapping procedure) into the model. This procedure has been used to successfully study entangled[16,17] and unentangled[15,18] polymer melts. Moreover, there is a further advantage in using the MD technique, since additional qualitative information about dynamic properties[19–21] can be gained from the resulting trajectories.

To achieve a quicker relaxation, an interesting combination of these two families of methods has also been proposed.[17,22] In particular, Spyriouni and coworkers[17] were able to equilibrate entangled atactic polystyrene melts applying the end-bridging MC algorithm to a coarse-grained model of polystyrene. The so relaxed coarse-grained structure was subsequently reverse-mapped to obtain the atomistic structures.

However, these latter methods present difficulties in the implementation (the development of the force field for the coarser model is not trivial[23]) and can lead to a wrong description of the local structure (after the atoms have been re-inserted) if the coarse-grained model has sampled conformational states forbidden to the atomistic one.

Here is presented a new and easy procedure to quickly equilibrate entangled linear polymers. The method proposes the use of a reverse-mapping (or fine-graining) technique to generate an atomistic polymer structure directly from a generic polymer model. This method (here called "fine-graining") requires only two steps: the generation of a random walk path (representing the averaged conformational state of a generic polymer chain in a melt) and the subsequent insertion of the atomistic details following the trace of the random walk. After inserting the missing atoms into the generic polymer structure, few simulation steps are further required to re-form backbone bonds among the monomers and to relax (locally) the newly formed atomic polymer chains. The method is applied to three types of amorphous polymer melts: polyethylene (PE), atactic polystyrene (a-PS) and polyamide-66 (PA-66). Results about global properties (gyration radius, end-to-end distance, radial distribution functions) and local structure (distribution of the dihedral angles, local reorientation of bonds) are compared with those obtained from other methods presented in literature and with experimental data.

Computational method

Continuum random walk

It is commonly accepted that in concentrated solution or melt polymers display random walk conformational properties on length-scales much larger than the monomer diameter. This means that there is no long range correlation between subsequent bonds when these two bonds are separated by a specific number of monomers (n_m) along the chain ("Flory ideality hypothesis").[24] The value of n_m (Kuhn length) defines the chain stiffness and its length (l_K), usually given in nm and not in monomer units, can be measured experimentally. Therefore, knowing the value of l_K for a specific polymer, it is possible to build a random walk chain whose monomer length is l_K.

Here, an off-lattice continuum random walk (CRW) algorithm is implemented to generate random walk chains characterized by different values of l_K. The values of l_K for the different polymers are taken from the literature; the algorithm allows the monomer length to oscillate around the specified average value $\langle l_K \rangle$ following a Gaussian distribution of standard deviation equal to 2% of $\langle l_K \rangle$. The width of the distribution is chosen considering a small fluctuation of the Kuhn length due to experimental error. The fluctuation is small enough to not affect the final results. The chains are randomly placed in a cubic simulation box whose dimensions correspond roughly to the experimental density of the polymer at a chosen temperature without paying attention to possible overlaps among the chains. The set of the correct density at this stage is not important as the correct value will be reached at the end of the reverse mapping procedure. A good guess (based on experimental data or previous simulation results) is necessary to ensure the correct equilibration procedure.

Reverse mapping

After generating the generic polymer chains by means of the CRW algorithm, atomistic chains are grown along their parent random walks (RW). The easiest way to do this is to populate each RW segment with a suitable atomistic fragment. Initially, the atomistic fragment inserted is treated as rigid and only one conformation for the backbone atoms (all-*trans*) is allowed. Thus, each atomistic moiety in its *trans* conformation is placed on each RW segment in a way that their geometric centers coincide (see Fig. 1a); subsequently, the atomic fragments are rotated around their center by a suitable angle to follow the RW tangent. The rotation angle is that

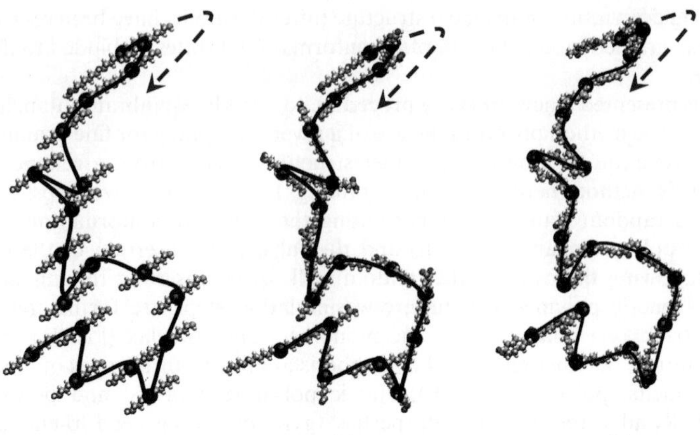

Fig. 1 Steps of the reverse-mapping procedure (the case of PA-66 is reported): (a) insertion of the atomistic fragments (colored beads) on the parent random walk (solid black line); (b) orientation of the atomistic fragments; (c) final configuration of the rebuilt atomistic model. The arrow indicates the direction of the growing chain.

between the vector formed by the centers of two subsequent RW segments and the one connecting the geometric center of the inserting fragment and its last atom (Fig. 1b). In order to keep the global structure of the chains generated by the CRW, the center of the RW segments and the inserted atomistic fragments are restrained to their positions during the relaxation steps with a restraining force constant of 1.0×10^9 kJ mol^{-1} nm^{-2}.

There is no established procedure to rebuild the backbone bonds among the inserted fragments and to relax (locally) the newly atomistic polymer chains; the recipe proposed here involves a set of several MD simulations, but other procedures, maybe more efficient, can be envisaged. During the MD simulations, temperature is controlled by the Berendsen thermostat with a coupling time equal to the value provided in the next section "Simulation parameters". Initially, the chains are treated and rebuilt separately. For each individual chain the following MD steps are performed:

(1) The non-bonded interactions are turned off (the parameters for the non-bonded interactions are set equal to zero) and a MD simulation with a very high bond constant (1.0×10^9 kJ mol^{-1} nm^{-2}) and small time step (0.00001 ps) is run for 100 000 time steps.

(2) The non-bonded interactions are then turned on and a simulation for 100 000 time steps with a soft-core potential is carried out (time step is set to 0.00001 ps). The soft-core potential is implemented in our simulation program YASP[25] in the following way: the short range part (0–d nm) of the non-bonded potential energy function is replaced by a cubic spline. The spline coefficients are chosen to satisfy four conditions: the spline matches the value (i) and the derivative (ii) of the original potential energy function at the crossover distance d; its derivative is zero at an interatomic distance r of zero (iii); most importantly, its value V_0 at $r = 0$ is finite. In all cases the parameters for the soft core potential are set as $d = 0.19$ nm and $V_0 = 15\,000$ kJ mol^{-1}.

(3) The value of V_0 is then increased to 50 000 kJ mol^{-1} and the simulation, with the time step 0.00001 ps, is run until the temperature reaches its target value (between 100 000 and 500 000 time steps are required depending on the polymer).

(4) The full non-bonded force field parameters are restored, the soft-core potential is turned off, and the system is initially equilibrated for 5 000 000 time steps.

The previous steps are necessary to rebuild a single atomistic chain from the different fragments inserted in the RW path. During the MD simulations, period boundary conditions are applied, but the geometric center of the atomistic fragments and the corresponding ones belonging to the RW segments are restrained in their positions. This means that only the local structure (bending and dihedral angles) is modified during the procedure. After that, all chains (still restrained to their parent RW) are placed in a simulation box of a size corresponding to the target density (say 0.75 g/cm^3 for PE) and, in order to remove possible overlaps between their atoms, the following steps are necessary:

(5) A soft-core (d = 0.19 nm and V_0 = 1000 kJ mol^{-1}) simulation is run for 100 000 time steps with a time step of 0.0001 ps; V_0 is then increased to 20 000 kJ·mol^{-1} and the simulation is carried on till the temperature reaches the target value. The bond constants are still set to a very high value (1.0×10^7 kJ mol^{-1} nm^{-2}) to maintain the topology of the chain.

(6) The full non-bonded potential is restored and a simulation is run for 500 000 time steps.

(7) Finally, all restraints applied on the center of the segments are relieved and another MD simulation is run for 100 000 time steps, initially with a time step of 0.00001 ps and then with a time step of 0.0001 ps. At this point, harmonic bonds are replaced by bond constants if required in the final force field.

All previous steps are run at constant volume. Now the system can be simulated in NPT and the normal simulation parameters (see next section "Simulation parameters") can be used.

As example the rebuilt structure of one chain of PA-66 can be seen in Fig. 1c.

Simulation parameters

Here the simulation details for the three polymers studied are briefly summarized. A more exhaustive description of the MD simulations can be found in the appropriate references reported below.

Polyethylene

The united-atom force field used both during the reverse-mapping steps and the MD runs is the one reported by Smit.[26] During the NPT simulations the cutoff radius is set to 1.0 nm; the temperature (450 K) and the isotropic pressure (1 atm) are kept constant by the Berendsen thermostat (with a coupling time of 0.2 ps) and barostat (with coupling time of 5 ps).[27] The bond constraints are maintained to a relative tolerance of 10^{-6} by the SHAKE procedure.[28] A timestep of 2 fs is used. The length of the production run ranges between 2 and 14 ns depending on the molecular weight (M_w).

Polystyrene

The all-atom force field used during the reverse-mapping steps and the MD runs is the one reported in ref. 29 and 30. During the NPT simulations the cutoff radius is set to 1.0 nm; temperature (500 K) and isotropic pressure (1 atm) are kept constant by the Berendsen thermostat (with a coupling time of 0.2 ps) and by the Berendsen barostat (with coupling time of 2.0 ps). The simulations are carried out with flexible bonds and a timestep of 2 fs. The production run is 8 ns for the highest M_w.

Polyamide-66

The all-atom force field parameters used can be found in ref. 14. Temperature (550 K) and pressure (1 atm) were controlled using the Berendsen algorithm with a coupling time of 0.2 ps and 5 ps, respectively. The equation of motion is integrated with a time step of 2 fs. The cutoff radius is to 0.9 nm, and a Verlet neighbor list with a pair cutoff of 1.0 nm is used, and updated by a link-cell scheme every 30 steps. Bond constraints are maintained to a relative tolerance of 10^{-6} by the SHAKE procedure. The length of the simulation is 2 ns.

Results

Polyethylene

Seven different molecular weights of linear monodisperse PE are simulated (see Table 1). For the creation for the CRW the Kuhn length is set equal to 1.6 nm, a value corresponding to the experimental one.[31] During the reverse-mapping procedure each segment of the random walk (RW) is replaced by eight PE monomers. The inserting fragment has all the backbone dihedrals (φ) in the *trans* conformation (φ = 180°). For each molecular weight studied the chains are placed in a simulation box that reproduced the same average density of 0.75 g/cm³. After the reverse-mapping, the atomistic models are subjected to MD simulations at constant pressure (1 bar).

The densities (reported as their reciprocal values) corresponding to the different molecular weights are shown in Fig. 2 and compared with the experimental data.

Table 1 Details of the polyethylene systems under investigation

Number of chains in the simulation box	Number of Kuhn segments per chain	Number of carbons per chain (N)	Total number of monomers in the box	$\langle R_e^2 \rangle$ (Å²)	$\langle R_g^2 \rangle$ (Å²)
15	40	640	4800	14678.5 ± 428.8	2849.3 ± 46.8
20	30	480	4800	8688.0 ± 267.3	1679.9 ± 36.9
30	20	320	4800	5828.0 ± 171.6	1324.1 ± 25.8
40	10	160	3200	3103.1 ± 72.9	570.4 ± 9.3
50	5	80	2000	1545.6 ± 33.7	277.2 ± 3.9
60	3	48	1440	871.0 ± 17.1	148.4 ± 1.8
70	2	32	1120	523.1 ± 8.4	84.1 ± 0.8

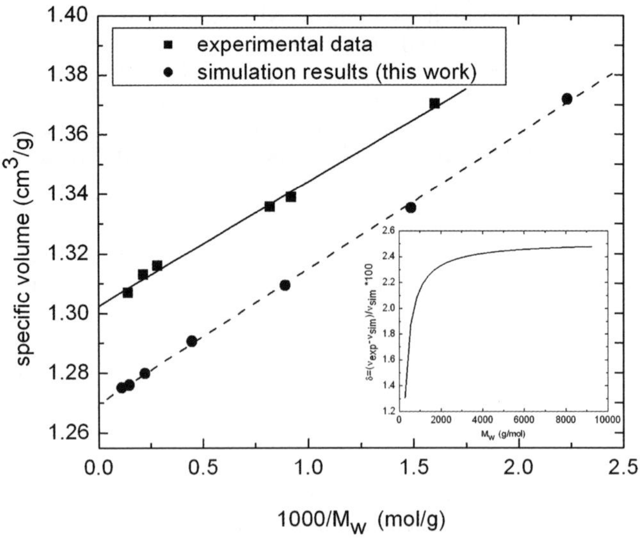

Fig. 2 Dependence of the specific volume (v) on inverse of the molecular weight (M_w) as predicted by this work and as measured experimentally.[8] The straight and the dashed lines show the fitted linear functions to the predicted and experimental points according to eqn (1) in the text. The inset shows in percentage the error between the experimental and the simulated values.

The overestimation of the density calculated from the simulations (for the percentage errors see inset of Fig. 2) is probably due to the atomistic force parameters that, as it has been verified,[8] are not very accurate in describing the volumetric properties and usually overestimate the density by *ca.* 3%. More importantly, the figure shows that, as expected for a polymer melt, the density values increases as the M_w increases until reaching a plateau for chain length exceeding 200 monomers. This trend in the density is caused by the reduced effect of the free volume of the chain ends when the molecular weight increases. The data can be fitted with the following equation

$$v = v_\infty + \frac{v_0}{M_w} \tag{1}$$

where v_∞ is the value of the specific volume (v) at infinite chain length and v_0 is a proportionality constant, describing the rate with which v changes with increasing M_w. Fitting the simulation data, v_∞ turns out to be equal to 1.270 cm³/g and $v_0 = 45.272$ cm³/mol; these numbers are in good agreement with the corresponding experimental results $v_\infty = 1.302$ cm³/g and $v_0 = 41.551$ cm³/mol.[32] The correct prediction of the trend of the density values with the polymer M_w shows that the fine-graining procedure is able to equilibrate the long-range thermodynamic properties of entangled and unentangled PE chains.

In order to analyse the local and global structural properties of the chains, the distribution of the backbone dihedral angles (averaged over the chains and the trajectories) and the X-ray structure factor (Fig. 3) are calculated. The probability of encountering a dihedral angle in *trans* conformation in the PE sample is $P_t = 0.647$ in perfect agreement with that found by Spyriouni *et al.*[33] calculated for PE (C_{16}) simulated using MC at the same temperature (T = 450 K) using the force field reported in ref. 34 ($P_t = 0.652$). This agreement shows that, even though the initial conformation of the inserted unit was set to be only *trans*, during the equilibration step the correct dihedral distribution is reached.

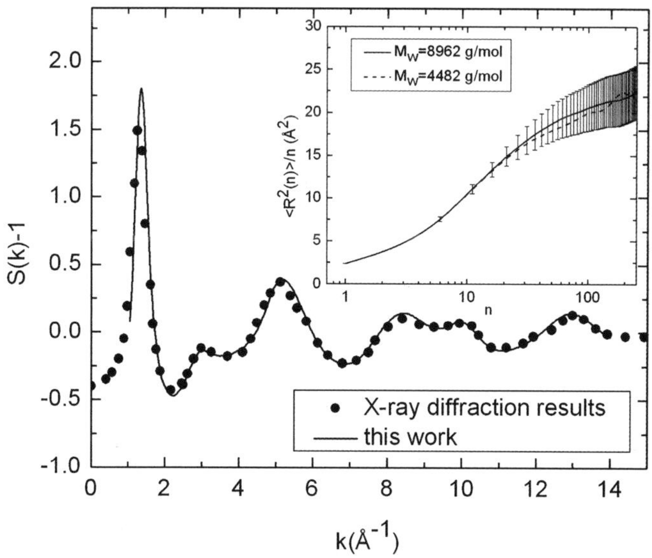

Fig. 3 X-Ray diffraction patterns of polyethylene from the simulation (T = 450 K, $M_w = 8960$ g/mol) and experiment[52] (T = 430 K, $M_w = 8806$ g/mol) at P = 1 atm. The inset shows the mean squared internal distances for PE of chain length $N = 640$ ($M_w = 8962$ g/mol) and $N = 320$ ($M_w = 4482$ g/mol). The error bars correspond to the highest M_w.

The partial structure factor can be calculated through the formula[35]

$$S_{\alpha\beta}(k) = \delta_{\alpha\beta} + 4\pi\sqrt{\rho_\alpha\rho_\beta}\int_0^b r^2\left[g_{\alpha\beta}(r) - 1\right]\frac{\sin(kr)}{kr}\,\mathrm{d}r \tag{2}$$

where $\rho_\alpha = \frac{N_\alpha}{V}$ is the number density of atoms of type α, $g_{\alpha\beta}(r)$ is the radial distribution function among the atoms α and β, $\delta_{\alpha\beta}$ is the Dirac delta function and b (the upper limit of the integral) is half of the simulation box length. The total X-ray structure factor is given by eqn (3):

$$S(k) = \sum_\alpha\sum_\beta C_\alpha C_\beta \frac{f_\alpha(k)f_\beta(k)}{\langle f(k)\rangle^2} S_{\alpha\beta}(k) \tag{3}$$

where C_α is the concentration of α atoms, f_α are the atomic form factors, and $\langle f(k)\rangle = \sum_\alpha C_\alpha f_\alpha(k)$. It must be noticed that for the PE, since the atomistic force field used is a united force field, there is only one type of particle and $\alpha \equiv \beta$. Fig. 3 shows an excellent agreement between the simulated and the experimental patterns and all the peaks both at low (corresponding to the intermolecular correlations) and at high (representing the intramolecular correlations) values of k are perfectly reproduced by the simulations. To assess the good equilibration of the polymer chain at intermediate length-scale, the mean square internal distances (MSID) are calculated for the PE samples.[36] The inset of Fig. 3 shows that the PE displays the Gaussian behavior as the MSID tends asymptotically to a constant value as N increases. Ahul *et al.*[36] pointed out that the MSID is a useful way to check whether deformations on the short and intermediate length-scales take place in the polymer chain. They identified the causes of such deformations in both the fast introduction of the excluded volume interactions and the inhomogeneous distribution of the density in the initial simulation box. In the present case the gradual introduction of the excluded volume interactions (spread over 1.5 ns) may have helped in overcoming the problem of the initial inhomogeneous distribution density and no evident deformations of the chain are visible.

Another important quantity to check in order to verify the ability of the fine-graining procedure to equilibrate polymer melts is the variation of the single chain conformational properties with the M_w. Fig. 4 shows the values of the end-to-end distance ($\langle R_e\rangle$) and gyration radius ($\langle R_g\rangle$) for different M_ws (values reported also in Table 1). The results are compared with those obtained for PE of similar M_ws equilibrated using the end-bridging MC (simulated at 450 K). The comparison is made even more interesting by the fact that for some samples the end-bridging MC and the fine-graining technique use the same force field reported by Smit.[26] The agreement between these data is impressive and it represents a major proof of the good equilibration achievable with the fine-graining method. The two sets of data follow the statistics of the ideal chains where $\langle R_g\rangle = \langle R_e\rangle/6$ and $\langle R_e\rangle \approx N^\upsilon$, and the Flory exponent υ is $\upsilon = 0.5$ (the straight line through the symbols in Fig. 4). From the value of $\langle R_e\rangle$ the characteristic ratio C_n can be calculated through the formula

$$C_n = \langle R_e^2\rangle/(n-1)l^2 \tag{4}$$

where n is the number of backbone bonds and l their averaged bond length (in this case 1.54 Å). For long chains C_n converges to a single value usually known as C_∞. The C_∞ can be then directly estimated from the slope of the line fitting $\langle R_e^2\rangle$ *versus* $(n-1)l^2$. The C_∞ obtained in such a way turns out to be 8.75, a value that is in good agreement with that obtained from the end-bridging MC method[8] using the force field developed by Mavrantzas and Theodorou[37] ($C_\infty = 9.13$) and with the value obtained more recently using the TRAPPE force field ($C_\infty = 8.26$).[38] The computed

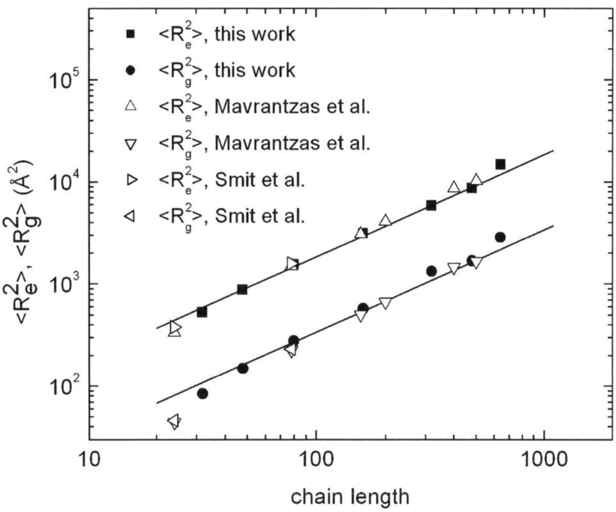

Fig. 4 Mean square end-to-end distance ($\langle R_e^2 \rangle$) and gyration radius ($\langle R_g^2 \rangle$) of polyethylene as function of the number of backbone atoms (N) in logarithmic scale. The results obtained from the fine-graining technique using the Smit force field[26] ("this work" in the legend) are compared with those obtained with end-bridging MC using the force fields reported in ref. 37 ("Mavrantzas *et al.*" in the legend) and in ref. 26 ("Smit *et al.*" in the legend). The straight lines represent the Flory predictions for $\langle R_e^2 \rangle$ and $\langle R_g^2 \rangle$ ($\langle R_e \rangle$ and $\langle R_g \rangle \approx N^{0.5}$). Error bars are at the size of the symbols.

value agrees quite well with recent experimental measurements which for molten PE yield $C_\infty = 7.8$.[39,40] The small overestimation made by the simulation is caused by the atomistic force employed that slightly overestimates the *trans* population of the dihedral angles along the PE chain.[9]

Atactic-polystyrene

The fine-graining procedure turns out to be a very effective and easy technique to equilibrate long PE chains. However, PE is topologically the simplest example of a polymer chain since, if an UA force field is used, the chain is made up only by spherical particles of the same van-der-Waals radius (except the first and the last monomers). On the contrary, PS shows a more complicated topology, first because of the steric pendant groups (the phenyl rings) attached to the backbone chain, second the methine carbons of the backbone are asymmetric and their relative chirality defines the tacticity of the polymer chains. Moreover, the PS chain is characterized by a long range correlation among the bonds (high Kuhn length value) and consequently a higher stiffness. Therefore it is worth to verify the effectiveness of the fine-graining procedure also in cases where the conformation of the inserting atomic moiety is not straightforward determined, more conformational states can be populated, and the stiffness of the chain is enhanced.

The Kuhn length for atactic PS is set to 2.0 nm[41] corresponding to roughly 8 monomers when the backbone dihedral angles are all in the *trans* conformation. Three different M_ws have been studied (see Table 2) all below the experimental entanglement molecular weight (16 600 g/mol)[31] that corresponds to roughly 320 backbone carbons; during the reverse-mapping steps atomistic segments, characterized each by a randomly different tacticity, are inserted in the RW monomers.

In literature different coarse-grained models of a-PS have been proposed with the aim of studying long range structural and dynamic properties.[16,42,43] The models differ in the number of degrees of freedom retained and in the procedure for

Table 2 Details of the polystyrene systems under investigation

Number of chains in the simulation box	Number of Kuhn segments per chain	Number of backbone's carbons per chain (N)	Number of monomers per chain	$\langle R_e^2 \rangle$ (Å^2)	$\langle R_g^2 \rangle$ (Å^2)
15	10	160	80	3811.1 ± 139.7	677.1 ± 17.9
11	13	208	104	3707.6 ± 93.1	783.3 ± 14.1
8	19	304	152	9563.9 ± 220.7	1737.9 ± 38.5

developing the effective interactions between the coarse-grained beads. Fig. 5 shows the values for $\langle R_e \rangle$ and $\langle R_g \rangle$ obtained from the fine-graining procedure compared with the numbers reported for coarse-grained models of PS of similar M_ws. The change in the structural parameters with increasing the M_w is similar in all models. It must be noticed anyway that the lower M_w models of this work ($N = 160$ and $N = 208$) show a moderate increase in $\langle R_g \rangle$ when the M_w increases and similar (considering the error) value of $\langle R_e \rangle$. A more conventional behavior is shown by the highest M_w ($N = 308$) where both $\langle R_g \rangle$ and $\langle R_e \rangle$ have higher values than the corresponding ones calculated for lower M_ws. Anyway, it is worth noticing that for low M_ws, a direct comparison between the absolute values of $\langle R_g \rangle$ and $\langle R_e \rangle$ obtained with the fine-graining method and the other models reported in literature can be done only in a qualitative way since a coarser model gives a rougher description of the molecular shape.

The good long range relaxation of the PS melt can be verified comparing the X-ray spectrum calculated using eqn (2) and (3) with the experimental pattern (Fig. 6). There are indeed substantial differences between the coherent scattering functions, $S(k)$, of the various polymers. For all polymers the "amorphous halo" is the

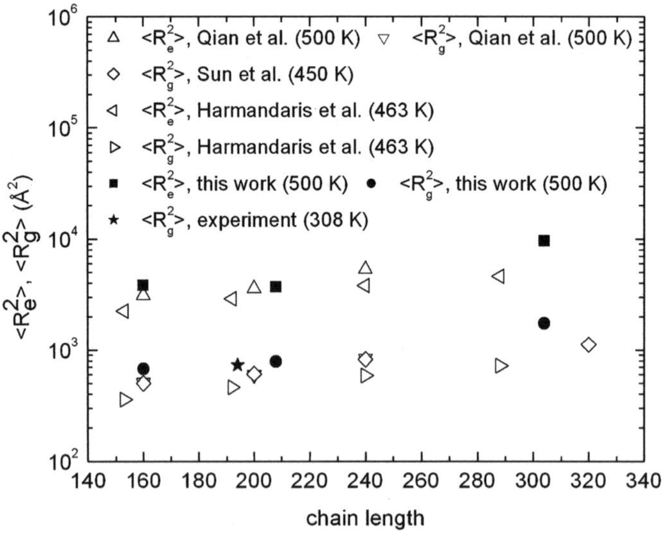

Fig. 5 Mean square end-to-end distance, $\langle R_e^2 \rangle$, and radius of gyration, $\langle R_g^2 \rangle$ of atactic polystyrene as a function of chain length N (T = 500 K). Error bars are at the size of the symbols. The results obtained from the fine-graining technique are compared with those obtained from the coarse-grained simulations of Qian *et al.*[30] (T = 500 K), Sun and Faller,[16] (T = 450 K) and Harmandaris and Kremer[21] (T = 463 K). The experimental value obtained from small-angle X-ray scattering in θ solution at T = 308 K is also reported.[53]

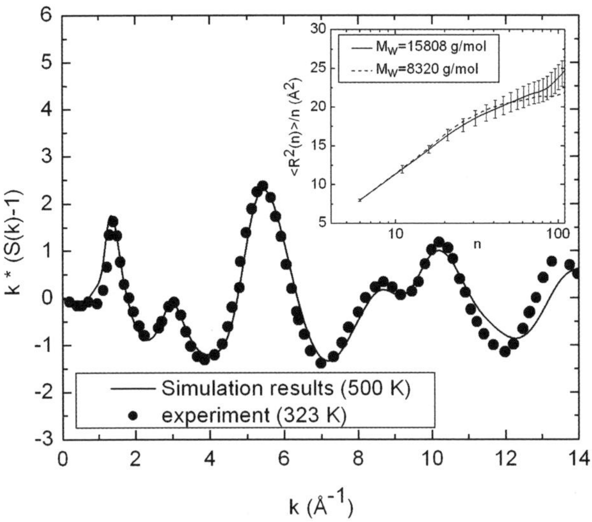

Fig. 6 The k-weighted total structure factor of atactic polystyrene from the simulation (N = 160) and experiment (N = 16).[54] The inset shows the mean squared internal distances for polystyrene chain of length N = 304 (M_w = 15 808 g/mol) and N = 160 (M_w = 8320 g/mol). The error bars correspond to the highest M_w.

prominent peak which occurs at $k < 2.0$ Å^{-1}, and this feature is observed to change in shape and position depending on the polymer. For example, this peak is narrow and symmetric for PE (Fig. 3) while for PS it shows two features: the main peak (1.5 Å^{-1}) is of a breadth similar to, and position slightly higher than, that of PE; moreover a small pre-peak, located at a lower value of k (0.75 Å^{-1}) (know as "polymerization peak") is also clearly visible. Fig. 3 and 6 show that even though the long length-scale equilibration has been performed simply by creating RWs characterized by different Kuhn lengths, after the reverse-mapping scheme the simulated X-ray shows the correct low-k peaks characteristic of different polymers. The MSID analysis for two M_ws of the PS samples is reported in the inset of Fig. 6. Here, due to the small value of M_w, the asymptotic behavior showed by PE cannot be seen. In order to check possible deformations of the polymer chains, the analysis is made on two M_ws (N = 160 and N = 304). The two curves show a similar behavior at short and intermediate length-scales independent of the total chain length (N) and there is no evidence of stretching of the chain for $N < 30$. Other simulations will be carried on in order to investigate the behavior of the chain as the M_w further increases. Another important analysis that can be done on polystyrene is to study the relative backbone-backbone or phenyl-phenyl positions in the simulation box. A convenient way is to compare the partial X-ray structure factors with the ones calculated from previous simulations reported by Ayyagari *et al.*[44] that, by means of MD simulations, calculated the intra- and intermolecular contributions to the X-ray structure factor in order to clarify the interactions responsible for the presence of two peaks in the "amorphous halo" region and their temperature-dependence position. Fig. 7 shows the Kratky plot of the individual contributions as well as the total $S(k)$ for a-PS at 500 K as obtained after the fine-graining procedure. The agreement with the results found by Ayyagari (not shown in the figure) is remarkable: the phenyl-phenyl and phenyl-backbone correlations turn out to be responsible for the presence of the high-k peak, while the low-k peak is due mainly to the backbone-backbone contributions. This perfect agreement among the total and partial X-ray spectra leads to the important conclusion that the a-PS melt is well equilibrated at different length-scales and that the relative positions of the atoms

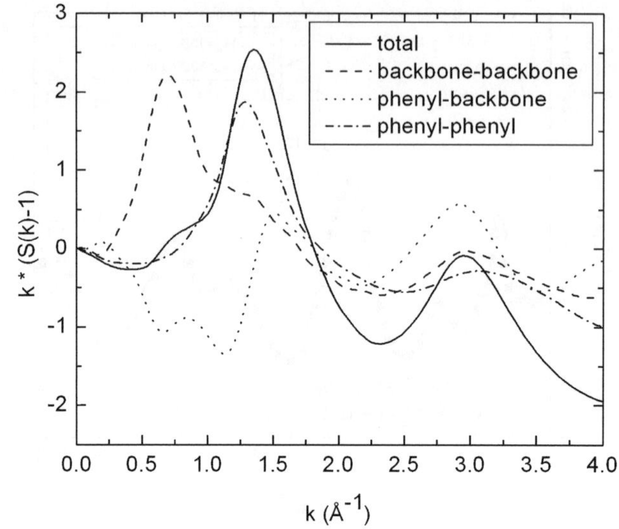

Fig. 7 Kratky plot of the contribution of phenyl-phenyl, phenyl-backbone and backbone-backbone correlations to the X-ray structure factor of polystyrene ($N = 160$) at 500 K.

belonging to the backbone atoms and to the pendant group is correct. To analyze the equilibration of the chain at a shorter length-scale, the orientation distribution functions describing the mutual orientation of aromatic rings is calculated. A convenient measure is the cosine of the angle between the plane normals **u** of two rings. In order not to distinguish between two symmetrically equivalent orientations (one ring turned by 180°), the absolute value of the scalar product $|\mathbf{u}\cdot\mathbf{u}|$ is used. This is 1 for two coplanar rings, 0 for a T-shaped arrangement, and 1/2 for a random distribution of orientations. In Fig. 8, this orientational distribution function (ODF) as a function of the average distance between the pair of rings is reported. At a short distance (less than 0.5 nm) the phenyl rings are predominantly co-planar, while for larger distance

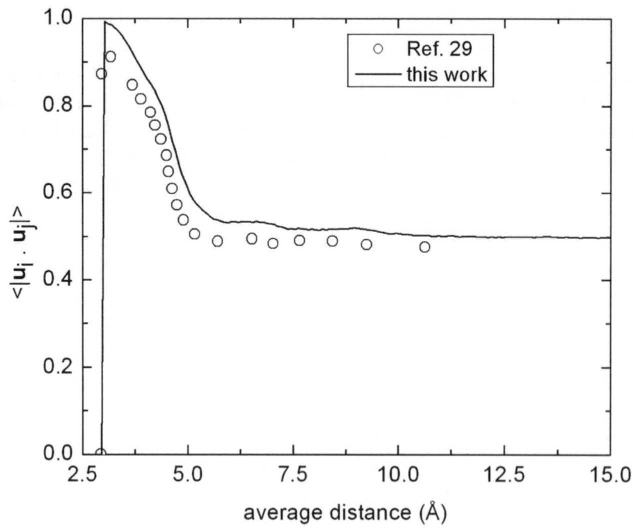

Fig. 8 Comparison between the orientation distribution function describing the mutual orientation of the plane normals of phenyl rings calculated for a-PS with $N = 160$ ("this work") and from ref. 29.

the co-planarity is quickly lost and no further specific orientation can be seen. This result agrees with that found in previous atomistic simulations of a-PS bulk simulated with the same force field.[29]

To study the local relaxation the dihedral distribution of the backbone angles is calculated from the simulations (see Table 3). From NMR measurements on a-PS, Dunbar[45] estimated an amount of dihedral angles in the *trans* (t) conformation of around 68 ± 10% at room temperature. The result obtained for the PS with $N = 160$ for the overall *trans* content is in agreement with the experimental result, but the nearly symmetric distribution obtained for g (most favourable conformation) and g' (highly unfavoured conformation since responsible for the so called "pentane effect") is probably caused by the steric hindrance of the pendant rings that locks the model in a unfavourable conformations. To avoid this problem, longer MD simulations after the reverse-mapping steps are probably necessary to relax the local structure completely. It must be noticed anyway that a very similar distribution has been obtained from standard MD atomistic simulations carried out on ten monomers of a-PS using the same atomistic force field employed here.[30] The simulations predicted around 60% for the *trans* conformation, 21% for the g and 19% for the g' state. This can suggest that the reason for the fact that the g and g' conformations are similarly populated, stems on the force field used more than on the equilibration procedure. Another important analysis that can be done in order to check the correct local conformation is the distribution of two successive backbone dihedrals. Considerable deviations may occur between the distributions gained interpreting NMR data[46] and the predictions obtained from the atomistic simulations. Robyr *et al.*[47] were able to reproduce the realistic a-PS chain conformations using a modified RIS model but the correct reproduction of the dyad conformations from simulations is still far off from being achieved even when using sophisticated procedures to equilibrate the chain.[17,48] The dyad distribution obtained from the fine-graining procedure is also not in perfect agreement with the experimental findings, but the results are anyway encouraging. The NMR experiments[46] show that at least 50% of the racemo dyads are close to the tt state, and more than 8% is in the $tg'/g't$ conformations, while in the case of meso dyads a large amount of dihedrals are near the tg/gt conformations (>80%) and a small amount (<10%) near the tt state. For the racemic dyad the simulations predict 71% of tt and 7.6% in the $tg'/g't$ while 19% of tt and 40% of tg/gt conformations are found for the meso dyad. The discrepancy between the simulations and the experimental

Table 3 Overall dihedral distribution and dyad conformations calculated at 500 K for atactic polystyrene. All molecular weights studied show similar probabilities

	States	Probability %
Overall distribution	t	67
	g	17
	g'	16
Meso dyad	tt	19.0
	tg/gt	40.4
	gg	1.0
	$g'g/gg'$	1.7
	$tg'/g't$	38.1
	$g'g'$	0
Racemic dyad	tt	71.2
	tg/gt	7.6
	gg	6.5
	$g'g/gg'$	2.3
	$tg'/g't$	7.6
	$g'g'$	4.5

finding for the meso dyad (underestimation of the amount of the *tg*/*gt* conformations and the overestimation of the *tt* conformers) is remarkably similar to that reported by Mulder *et al.*[48] (19.8% of *tt* and 64.9% of *tg*/*gt*) and Spyriouni *et al.*[17] (60% *gt*/*tg*) which used an atomistic force field different than the one used in this paper. However, further investigations will be carried out to verify the atomistic force field used here and to monitor changes in the torsion distribution when longer MD simulations are performed.

Polyamide-66

The last polymer on which the fine-graining technique is tested belongs to the family of polyamides where a widespread hydrogen bond network connecting the amide groups dominates their properties even at high temperature. Moreover, the experimental value of the Kuhn length is not available. Therefore the l_K value (l_K = 1.8 nm) obtained from previous atomistic simulations[14] is used to prepare the RWs. In this case each RW segment is repopulated with 22 backbone atoms (including four amide groups), corresponding to one and half monomers (NHCO–$(CH_2)_4$–CONH–$(CH_2)_6$–NHCO–$(CH_2)_4$–CONH) with all the backbone dihedral angles in the *trans* conformation. Due to the lack of experimental data, the simulation results are compared with previous MD simulations[14] run with the same force field employed here; moreover, at the moment the simulations will be restricted only to one M_w (M_w = 4540 g/mol) barely behind the critical mass reported in literature as 4700 g/mol[49] above which entanglements start to be formed. In ref. 14 the PA66 melt has been equilibrated using the pivot MC method (PMC) of Neyertz and Brown[13] using $n_{bonds} = 4$ (backbone bonds) implying that two neighboring amide groups do not interact with each other. Once the single chains have been generated using the PMC algorithm, the procedure followed to equilibrate the melt is very similar to that followed in this paper: the chains are submitted to periodic boundary conditions and the full intermolecular interactions have been introduced gradually. Table 4 compares the values of several properties as obtained from the fine-graining and from the PMC technique. The density and, most importantly, the averaged (on the trajectory and the polymer chains) number of hydrogen bonds are in perfect agreement; the single chain property, $\langle R_e \rangle$, also matches ($\langle R_e \rangle = 68.5 \pm 12.8$ Å from the PMC and $\langle R_e \rangle = 66.5 \pm 13.3$ Å from the fine-graining). Both results are in reasonable agreement with the experimentally available value, obtained for a θ solution, estimated as 61 ± 6 Å for 4500 g/mol chains.[50]

The overall good relaxation of the melt can be seen in Fig. 9 that compares the $S(k)$ calculated from eqn (2) and (3) obtained from the fine-graining technique and from the PMC after 2 ns of MD simulations. Both the "amorphous halo", at

Table 4 Comparison between the results obtained from the fine-graining procedure ("this work") after 2 ns of MD of 12 chains and the ones obtained using the pivot MC method (PMC) after 20 ns of MD of 24 chains of polyamide-66 at 550 K

	This work	PMC
Density (kg/m³)	986.3 ± 3.2	981.0 ± 1.5
$\langle R_e^2 \rangle$ (Å²)	4422.4 ± 176.3	4699.4 ± 164.2
$\langle R_g^2 \rangle$ (Å²)	872.8 ± 18.3	935.0 ± 19.0
Fraction of hydrogen bonds [a]	0.49 ± 0.01	0.49 ± 0.02

[a] The hydrogen bonds are defined according to a geometrical criterion stating that the distance between the hydrogen of the donor group and the acceptor O has to be <0.297 nm and the donor-hydrogen-acceptor angle >130°.[20] The "fraction of hydrogen bond" corresponds to the fraction of amide groups hydrogen bonded.

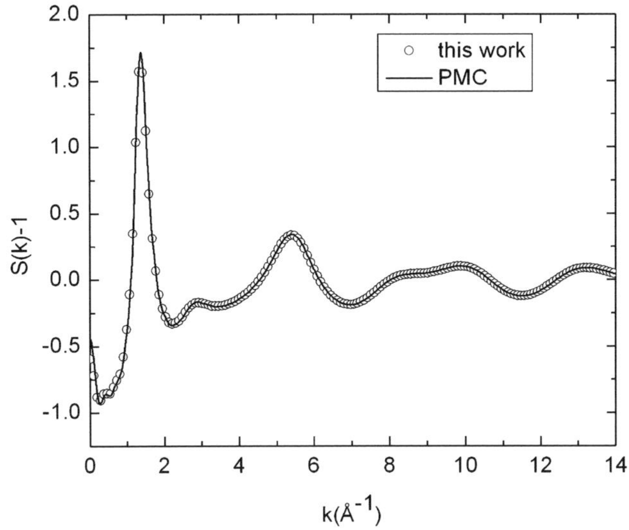

Fig. 9 Total structure factor of polyamide-66 chains with 20 monomers per chain. Comparison between the results obtained using the fine-graining technique ("this work") for 12 chains and the pivot MC ("PMC") method for 24 chains followed by MD relaxation. Both simulations use the same force field and temperature (T = 550 K).

$k < 0.2$ nm^{-1}, and the high-k peaks show the same spacing. Since the presence of hydrogen bonds between amide groups determines most of the conformational properties of PA-66 even at high temperatures, it is interesting to analyze the distribution of these functional groups within the simulation box in more detail. Fig. 10 shows the inter-molecular radial distribution functions (RDF) of the carbonyl carbon atoms and compares the results obtained using the PMC and the fine-graining methods. The first peak (at ~0.5 nm) of the RDF corresponds to the averaged inter-chain spacing between two carbonyl groups and it is related to the presence

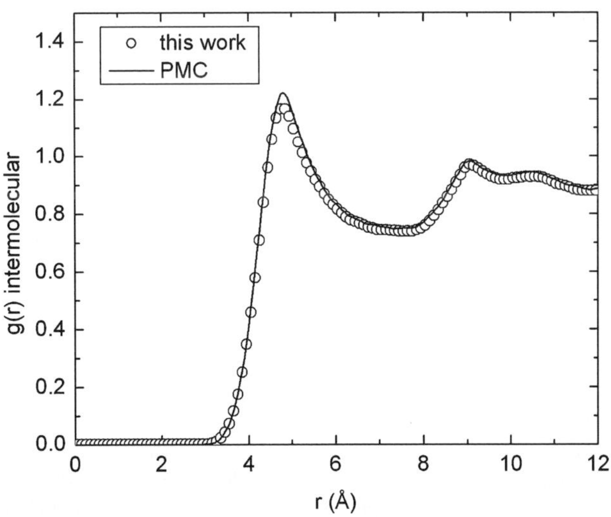

Fig. 10 Intermolecular radial distribution functions calculated among the carbonyl carbons of the polyamide-66 chains with 20 monomers per chain. The comparison between the results obtained using the fine-graining technique ("this work") and the pivot MC ("PMC") method is shown. (T = 550 K).

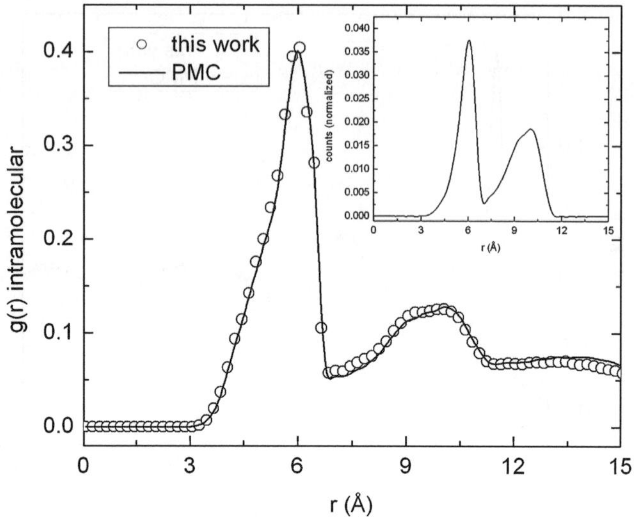

Fig. 11 Intramolecular radial distribution function of the carbonyl carbons of the polyamide-66 chains with 20 monomers per chain. The comparison between the results obtained using the fine-graining technique ("this work") and the pivot MC ("PMC") method followed by MD relaxation is shown (T = 550 K). The inset is the distribution of distance between subsequent carbonyl carbon.

of the intermolecular hydrogen bonds connecting the amide groups. The PMC and the fine-graining methods reproduce the same distribution and the perfect overlap among the peaks confirms the agreement between the values of the fraction of hydrogen bonds reported in Table 4. The last analysis concerns the conformation of the single chain investigated looking at the intramolecular RDF between the carbonyl carbons (Fig. 11). The RDF presents two peaks centered at 0.6 and 0.9 nm. In order to understand their molecular origin, the distribution of the distance among the carbonyl carbons within the chain is calculated and reported in the inset of Fig. 11. This distribution considers only the distance among subsequent carbonyl carbons. The distribution is double peaked and the peaks lay at the same spacing as the RDF ones. Thus, it is evident that the first peak of the RDF at ~0.6 nm corresponds to the carbonyl carbon pairs separated by four methylenic carbons, while the second one (at ~0.9 nm) is mainly due to the carbons separated by six methylenic groups. Also in this case, the PMC and the fine-graining techniques show that even at a short length-scale, the PA-66 chains equilibrated with the MC approach and the fine-graining ones have adopted the same local conformations.

Conclusions

A technique to relax high M_w polymer melts has been proposed. The method, named fine-graining, acts on two different length-scales to equilibrate the melt: at the long length-scale the chains are described as a random walk whose Kuhn length is chosen depending on the polymer under study. Then, following the parent random walk, the atomistic details are introduced by replacing the RW segments with the corresponding atomistic fragments of the appropriate number of monomers. At the short length-scale the equilibration consists of a simple procedure involving a series of MD simulations. A similar approach has been used by Kotelyanskii *et al.*[51] where lattice self-avoiding random walks are generated to completely occupy a cubic lattice. The random walks are then decorated assigning to each lattice site a specific building block (of size smaller than a monomer). The rebuilt atomistic structure is then annealed and equilibrated. The fundamental difference among the fine-graining and the procedure proposed by Kotelyanskii is that the fine-graining exploits the "Flory ideality hypothesis" generating a non-self avoiding random walk and re-introducing the atoms at a length-scale equal to the Kuhn length. It is worth noting that the reverse mapping procedure presented here is not unique and others may be tried.

The technique has been successfully applied to equilibrate three different polymers (polyethylene, atactic polystyrene and polyamide-66) characterized by different molecular weights, chain flexibility (Kuhn length value), chemistry and topology. The method has proved to be very competitive due mainly to its easy implementation. In particular, the perfect agreement between structural properties (computed at different length-scales) obtained with the fine-graining procedure and other methods (*i.e.* end-bridging MC or PMC) but simulated with the same simulation details (force field, M_w of the polymers), is very impressive, and it gives clear proof of the reliability of the fine-graining method.

In the future, some important aspects of the technique will be further investigated. First the choice of an accurate atomistic force field is probably a crucial point for the success of the procedure since the formation of the backbone bonds connecting the inserting atomistic fragments and the relaxation of the local conformations, rely on atomistic MD simulations. Moreover, from the results it turned out that the length of the final NPT MD trajectory is not a major concern for a flexible chain, such as for PE, where the correct dihedral distributions is obtained after a short MD run, but it may be a problem for a more rigid polymer such as PS. In the latter case, indeed the presence of bulky pendant groups may require long MD simulations with soft-core potential or an "*ad hoc*" choice of the internal conformation of the atomistic fragment to relax the local structure properly. In the particular case of PS, a possible improvement could also be the use of a fragment library containing polymer segments in different configurations or the reduction of the torsional barriers during the reverse mapping steps, together with the soft core potential, leading to a faster equilibration of the dihedral angles. Another point that may be of interest is the effect that the correct choice of the Kuhn length value has on the final results. This can be important especially in those cases where the experimental numbers are not available to check whether the Kuhn length available for polymers with similar chemistry could be used or if an accurate estimation of the value (from experiments or molecular simulations) is necessary.

Acknowledgements

The authors thank Dr Hu-Jun Qian for the help in setting up the simulations of polystyrene and for the useful discussions.

References

1 B. J. Reynwar, G. Illya, V. A. Harmandaris, M. M. Müller, K. Kremer and M. Deserno, *Nature*, 2007, **447**, 461–464.
2 G. Srinivas, D. E. Discher and M. L. Klein, *Nat. Mater.*, 2004, **3**, 638–644.
3 R. Everaers, S. K. Sukumaran, G. S. Grest, C. Svaneborg, A. Sivasubramanian and K. Kremer, *Science*, 2004, **303**, 823–826.
4 J. J. Depablo, M. Laso and U. W. Suter, *J. Chem. Phys.*, 1992, **96**, 2395–2403.
5 J. I. Siepmann and D. Frenkel, *Mol. Phys.*, 1992, **75**, 59–70.
6 L. R. Dodd, T. D. Boone and D. N. Theodorou, *Mol. Phys.*, 1993, **78**, 961–996.
7 A. Uhlherr, *Macromolecules*, 2000, **33**, 1351–1360.
8 V. G. Mavrantzas, T. D. Boone, E. Zervopoulou and D. N. Theodorou, *Macromolecules*, 1999, **32**, 5072–5096.
9 N. C. Karayiannis, A. E. Giannousaki, V. G. Mavrantzas and D. N. Theodorou, *J. Chem. Phys.*, 2002, **117**, 5465–5479.
10 O. Alexiadis, V. G. Mavrantzas, R. Khare, J. Beckers and A. R. C. Baljon, *Macromolecules*, 2008, **41**, 987–996.

11 M. Doxastakis, V. G. Mavrantzas and D. N. Theodorou, *J. Chem. Phys.*, 2001, **115**, 11339–11351.
12 C. D. Wick and D. N. Theodorou, *Macromolecules*, 2004, **37**, 7026–7033.
13 S. Neyertz and D. Brown, *J. Chem. Phys.*, 2001, **115**, 708–717.
14 S. Goudeau, M. Charlot, C. Vergelati and F. Müller-Plathe, *Macromolecules*, 2004, **37**, 8072–8081.
15 S. Queyroy, S. Neyertz, D. Brown and F. Müller-Plathe, *Macromolecules*, 2004, **37**, 7338–7350.
16 Q. Sun and R. Faller, *Macromolecules*, 2006, **39**, 812–820.
17 T. Spyriouni, C. Tzoumanekas, D. Theodorou, F. Müller-Plathe and G. Milano, *Macromolecules*, 2007, **40**, 3876–3885.
18 P. Carbone, H. A. Karimi-Varzaneh, X. Y. Chen and F. Müller-Plathe, *J. Chem. Phys.*, 2008, **128**.
19 X. Y. Chen, P. Carbone, W. L. Cavalcanti, G. Milano and F. Müller-Plathe, *Macromolecules*, 2007, **40**, 8087–8095.
20 H. A. Karimi-Varzaneh, P. Carbone and F. Müller-Plathe, *Macromolecules*, 2008, **41**.
21 V. A. Harmandaris and K. Kremer, *Macromolecules*, 2009, **42**, 791–802.
22 K. Kamio, K. Moorthi and D. N. Theodorou, *Macromolecules*, 2007, **40**, 710–722.
23 F. Müller-Plathe, *ChemPhysChem*, 2002, **3**, 754–769.
24 M. Rubinstein and R. H. Colby, *Polymer Physics*, Oxford University Press, 2007.
25 F. Müller-Plathe, *ChemPhysChem*, 1993, **78**, 77–94.
26 B. Smit, S. Karaborni and I. J. Siepmann, *J. Chem. Phys.*, 1995, **102**, 2126.
27 H. J. C. Berendsen, J. P. M. Postma, W. F. van Gusteren, A. Di Nola and J. R. Haak, *J. Chem. Phys.*, 1984, **81**, 3684.
28 J.-P. Ryckaert, G. Ciccotti and H. J. C. Berendsen, *J. Comput. Phys.*, 1977, **23**, 327.
29 F. MüllerPlathe, *Macromolecules*, 1996, **29**, 4782–4791.
30 H. J. Qian, P. Carbone, X. Y. Chen, H. A. Karimi-Varzaneh, C. C. Liew and F. Müller-Plathet, *Macromolecules*, 2008, **41**, 9919–9929.
31 *Physical Properties of Polymers Handbook*, ed. J. E. Mark, Springer, 2007.
32 D. J. Kinning, E. L. Thomas and J. M. Ottino, *Macromolecules*, 1987, **20**, 1129–1133.
33 T. Spyriouni, I. G. Economou and D. N. Theodorou, *Macromolecules*, 1997, **30**, 4744–4755.
34 L. R. Dodd and D. N. Theodorou, *Adv. Polym. Sci.*, 1994, **116**, 249.
35 B. L. Bhargava, R. Devane, M. L. Klein and S. Balasubramanian, *Soft Matter*, 2007, **3**, 1395–1400.
36 R. Auhl, R. Everaers, G. S. Grest, K. Kremer and S. J. Plimpton, *J. Chem. Phys.*, 2003, **119**, 12718–12728.
37 V. G. Mavrantzas and D. N. Theodorou, *Macromolecules*, 1998, **31**, 6310–6332.
38 K. Foteinopoulou, N. C. Karayiannis, M. Laso and M. Kröger, *J. Phys. Chem. B*, 2009, **113**, 442–455.
39 J. C. Horton, G. L. Squires, A. T. Boothroyd, L. J. Fetters, A. R. Rennie, C. J. Glinka and R. A. Robinson, *Macromolecules*, 1989, **22**, 681–686.
40 L. J. Fetters, W. W. Graessley, R. Krishnamoorti and D. J. Lohse, *Macromolecules*, 1997, **30**, 4973–4977.
41 J. P. Cotton, D. Decker, H. Benoit, B. Farnoux, J. Higgins, G. Jannink, R. Ober, C. Picot and Jd. Cloizeau, *Macromolecules*, 1974, **7**, 863–872.
42 V. A. Harmandaris, N. P. Adhikari, N. F. A. van der Vegt and K. Kremer, *Macromolecules*, 2006, **39**, 6708–6719.
43 G. Milano and F. Müller-Plathe, *J. Phys. Chem. B*, 2005, **109**, 18609–18619.
44 C. Ayyagari, D. Bedrov and G. D. Smith, *Macromolecules*, 2000, **33**, 6194–6199.
45 M. G. Dunbar, B. M. Novak and K. Schmidt-Rohr, *Solid State Nucl. Magn. Reson.*, 1998, **12**, 119–137.
46 P. Robyr, Z. Gan and U. W. Suter, *Macromolecules*, 1998, **31**, 8918–8923.
47 P. Robyr, M. Müller and U. W. Suter, *Macromolecules*, 1999, **32**, 8681–8684.
48 T. Mulder, V. A. Harmandaris, A. V. Lyulin, N. F. A. van der Vegt, K. Kremer and M. A. J. Michels, *Macromolecules*, 2009, **42**, 384–391.
49 Y.-H. Zang and P. J. Carreau, *J. Appl. Polym. Sci.*, 1991, **42**, 1965.
50 B. D. Viers, in *Polymer Data Handbook*, Oxford University Press, New York, 1999.
51 M. Kotelyanskii, N. J. Wagner and M. E. Paulaitis, *Macromolecules*, 1996, **29**, 8497–8506.
52 G. H. Kevin, D. M. John, G. C. John, S. S. Kenneth, H. N. Alfred and H. Anton, *J. Chem. Phys.*, 1991, **94**, 4659–4662.
53 T. Konishi, T. Yoshizaki, T. Saito, Y. Einaga and H. Yamakawa, *Macromolecules*, 1990, **23**, 290–297.
54 J. D. Londono, A. Habenschuss, J. G. Curro and J. J. Rajasekaran, *J. Polym. Sci., Part B: Polym. Phys.*, 1996, **34**, 3055.

Systematic coarse-graining of molecular models by the Newton inversion method

Alexander Lyubartsev,* Alexander Mirzoev, LiJun Chen and Aatto Laaksonen*

Received 23rd January 2009, Accepted 29th April 2009
First published as an Advance Article on the web 7th August 2009
DOI: 10.1039/b901511f

Systematic construction of coarse-grained molecular models from detailed atomistic simulations, and even from *ab initio* simulations is discussed. Atomistic simulations are first performed to extract structural information about the system, which is then used to determine effective potentials for a coarse-grained model of the same system. The statistical-mechanical equations expressing the canonical properties in terms of potential parameters can be inverted and solved numerically according to the iterative Newton scheme. In our previous applications, known as the Inverse Monte Carlo, radial distribution functions were inverted to reconstruct pair potential, while in a more general approach the targets can be other canonical averages. We have considered several examples of coarse-graining; for the united atom water model we suggest an easy way to overcome the known problem of high pressure. Further, we have developed coarse-grained models for L- and D-prolines, dissolved here in an organic solvent (dimethylsulfoxide), keeping their enantiomeric properties from the corresponding all-atom proline model. Finally, we have revisited the previously developed coarse-grained lipid model based on an updated all-atomic force field. We use this model in large-scale meso-scale simulations demonstrating spontaneous formation of different structures, such as vesicles, micelles, and multi-lamellar structures, depending on thermodynamical conditions.

1 Simulating the real world

Three rather practical issues are always crucial in planning to perform computer experiments. All are equally important and therefore assigned collectively based on the available computing resources at the moment. We could consider them spanning an operational space with three "orthogonal" axes as bases:

1. System *size*
2. Motional *time* scale†
3. The *accuracy* of the model

Within this space we wish to perform the simulations in an optimal region to keep the computations feasible in one hand, while expecting to obtain reliable results in the other hand. In other words a reasonable compromise is always sought.

Possibly because of the incredibly fast development in hardware technology, very little effort was put in to improve the methods and the models during the first decades of computer simulations. Only during the last two decades have new simulation methods been introduced to stretch the *time* and *length* scales much further.

Division of Physical Chemistry, Arrhenius Laboratory, Stockholm University, SE-106 91 Stockholm, Sweden. E-mail: sasha@physc.su.se; aatto@physc.su.se

† The number of moves in Monte Carlo (MC) simulations.

New schemes are also proposed to increase the *accuracy* by bringing MD and MC to the domains of quantum mechanics (QM) or to the first principles of physics. This allows simulations free of any empirical parameters (such as potential functions or molecular mechanical (MM) force fields) as input. The most widespread technique of this kind of method is the Car–Parrinello molecular dynamics.[1] Hybrid methods mixing QM and MM based schemes[2,3] are common tools today. The most interesting development may still be the schemes beyond atomistic resolution to model meso- and nano-scale systems and soft matters.[4,5] There are now reliable simulation methods available to treat a system at three levels of physical description (QM, MM and mesoscopic soft matter), where the accuracy is successively decreasing while allowing the system length and time scales to be increased. Examples of these are Car–Parrinello molecular dynamics, classical atomistic MD based on MM force fields, and dissipative particle dynamics (DPD). In common terminology, models beyond the atomistic resolution are a result of coarse-graining (CG). There is no unique way to do coarse-graining within an off-lattice framework. For heterogeneous systems like biological molecules normally fairly *ad hoc* approaches have been used to parameterize CG potentials,[6–8] while for homogeneous systems, like in materials design, finite element and grid-based models are commonly employed. In the case of biological systems a coarse-grained description of water molecules surrounding biomolecules represents a great challenge. Such simplifications may include implicit description of the solvent with the help of solvent-mediated potentials[9] or coarse-grained representation of solvent molecules.[8] The problem is, however, how we can accurately enough specify the interaction potentials for such coarse-grained models.

We will discuss here a straight-forward hierarchical multiscale modeling approach enabling us to link together different levels (physical models) of simulations by consecutively removing non-interesting degrees of freedom. The approach is based on a previously developed Inverse Monte Carlo scheme for reconstruction of pair potentials from known RDFs which can be computed from radial distribution functions. Previously this approach was illustrated for parameterization of effective potentials for water molecules and water-ion interactions starting from RDF obtained in *ab initio* CPMD simulations,[10,11] effective solvent mediated ion-ion[12] and ion-DNA[13] potentials as well as parameterization of a coarse-grained lipid model.[14] This paper reports further development in the field. It the next section, we describe a general approach enabling us to reconstruct parameters of the interaction potential (or the potential as a whole) from canonical averages computed from detailed simulations (or known from the experiment). As a simple example, we consider removing hydrogen atoms from a water molecule and derive effective potential for a single-site model, which provides realistic phase behavior of such a model. In the next section, we derive effective solvent-mediated potentials for united atom proline molecules in implicit DMSO solvent. Finally, we describe new updates on effective potentials for implicit solvent lipid models, and demonstrate applications of this in a description of behavior of lipid assemblies.

2 Hierarchical multiscale modelling approach

We begin from a quantum-mechanical first-principles simulation with the system size which we can afford. Here a Car–Parrinello molecular dynamics can be employed or any other *ab initio* molecular dynamics with quantum-mechanical computation of forces. Results of such *ab initio* simulations are used to parameterize an atomistic molecular mechanical force field which allows an increase of the size of the system and the simulation time by several orders of magnitude. On the next level, molecular dynamics simulation with *ab initio* derived force field will provide data for parameterization of the next level coarse-grained model for mesoscopic simulations. This idea of hierarchical, from the first principle modeling of matter can in principle be continued to further levels, but evidently the main problem to be solved in order to make this idea feasible, is to find a way on how to use the results obtained from a more detailed, more fundamental model, to derive interaction parameters for simulation on the next, coarse-grained level. A general formalism on how to do this can be depicted as follows.

Let us consider transition from atomistic to coarse-grained level of molecular modeling. We divide all the degrees of freedom of the detailed system into important ones $\{R_I, I = 1, \ldots, N\}$, which we want to keep on the coarse-grained level, and unimportant degrees of freedom $\{r_i, i = 1, \ldots, n\}$ (which can be solvent coordinates, local details on macromolecular structure *etc.*). The "important" degrees of freedom may be real atom coordinates of the detailed system, or some combinations of them, for example centers of masses of some atomic groups. We can then define the effective N-body coarse-grained Hamiltonian (potential energy) which is also the N-body mean force potential and which is equal to the free energy of the removed degrees of freedom:[9,15]

$$\beta H^{\mathrm{PMF}}(R_1, \ldots, R_N) = -\int \mathrm{d}r_1, \ldots \mathrm{d}r_n \exp(-\beta H(R_1, \ldots, R_N, r_1, \ldots, r_n)) \quad (1)$$

where $\beta = 1/kT$ and $H(R_1, \ldots, R_N, r_1, \ldots, r_n)$ is Hamiltonian of the atomistic system. Simulation with Hamiltonian eqn (1) provides the same structural and thermodynamic properties of the coarse-grained system as those observed in atomistic simulation.‡[15] Practical simulations employing N-body potential are however unfeasible, that is why we need to approximate it by a more convenient expression, for example as a sum of effective pair potentials:

$$H^{\mathrm{PMF}}(R_1, \ldots, R_N) \approx \sum_{I<J} V_{IJ}^{\mathrm{eff}}(R_{IJ}) \quad (2)$$

with $R_{IJ} = |R_I - R_J|$. It is however not necessary to be limited by pair potentials, any practically usable expression can be given in eqn (2), for example angle or torsion potential terms for macromolecular CG models. Even orientational and three- and higher-order terms can be considered. However, from the computational point of view, we are interested just in pair-wise solutions: the very aim of coarse-graining is the computational speed-up, and use of many-body potentials would greatly hamper this goal.

The task of building a coarse-grained force field can be thus reformulated to find "as best as possible" approximation according to eqn (2). What is "the best approximation" can be determined however in different ways. One can determine a set of target properties which one may wish the CG model to keep. These properties can be either of microscopic character, as forces or instantaneous energies, or canonical averages as radial distribution functions, average energies, pressure. For example, minimizing the force difference coming from both sides of eqn (2) (weighted with the Boltzmann factor) is equivalent to the force matching method.[16–18] Similarly, comparing the energy, one can formulate a potential matching method.[19] If the target is a radial distribution function, the pair potential in eqn (2) can be reconstructed by the Inverse Monte Carlo method.[9] In principle, other properties of interest or any combination of them can be used for parameterization of effective potentials.

Assume our effective potentials are determined by a set of parameters $\{\lambda_i\}$, and the set of target properties (which we know from the atomistic simulations) is $\{A_j\}$. If we know the set of $\{\lambda_i\}$, we can always, simulating the system directly, compute average properties $\{\langle A_j \rangle\}$. The inverse problem, finding parameters $\{\lambda_i\}$ from averages $\{\langle A_j \rangle\}$, is less trivial. We can consider the relationship between $\{\lambda_i\}$ and $\{\langle A_j \rangle\}$ as a nonlinear multidimensional equation, and use the Newton inversion method (known also as Newton–Raphson method) to solve it iteratively.

‡ For thermodynamic properties, implicit dependence of H^{PMF} on temperature and density should be taken into account.

At each iteration of the Newton inversion, we need to compute the matrix of derivatives (Jacobian), showing how different potential parameters affect different averages. It is not difficult, using expression for the averages in the canonical ensemble, to arrive at the following formula:

$$\frac{\partial\langle A_j\rangle}{\partial\lambda_i} = -\beta\left(\left\langle\frac{\partial H}{\partial\lambda_i}A_j\right\rangle - \left\langle\frac{\partial H}{\partial\lambda_i}\right\rangle\langle A_j\rangle\right) \tag{3}$$

The Jacobian eqn (3) can be used to find parameters $\{\lambda_i\}$ corresponding to target values of $\{\langle A_j\rangle\}$ solving the system of linear equations:

$$\Delta\langle A_j\rangle = \sum_i \frac{\partial\langle A_j\rangle}{\partial\lambda_i}\Delta\lambda_i + O(\Delta\lambda^2) \tag{4}$$

where the second order corrections are neglected. One starts from some initial potential determined by a trial set of parameters $\{\lambda_i^{(0)}\}$, runs a simulation, and computes deviation of computed values $\{\langle A_j^{(0)}\rangle\}$ from the target values A^*:

$$\Delta\langle A_j\rangle = A^* - \{\langle A_j^{(0)}\rangle\} \tag{5}$$

as well as the Jacobian eqn (3). Then, from the system of linear eqn (4), one finds corrections to the potential parameters $\Delta\lambda_i$. The procedure is repeated with the new parameter set $\lambda_i^{(1)} = \lambda_i^{(0)} + \Delta\lambda_i$ until convergence is reached. If initial approximation $\{\lambda_i^{(0)}\}$ is poor, some regularization of the iterative procedure might be necessary, in which the difference in eqn (5) is multiplied by some factor between 0 and 1.

The procedure described above can be readily implemented if the number of potential parameters is equal to the number of target properties. If the number of properties exceeds the number of potential parameters, the problem can be solved in the variational sense, by finding the set of $\{\lambda_i\}$ which provides the least possible deviation of the computed properties from the target values. Note, that to get the solution of the variational problem one also needs computation of derivatives according to eqn (3).

An important example of the described above approach is the case when parameters $\{\lambda_i\}$ are the values of the pair potential in a number of points covering the whole range of distances, and the target properties are the values of RDF in the same set of points. Then the inverse problem is reformulated as finding the pair interaction potential which reconstructs the given radial distribution functions. It is known that the solution of such an inverse problem is unique,[20] a statement which holds even in the multi-component case for the relationship between a set of RDFs and the corresponding set of pair potentials. The equations given above become equivalent in this case to the inverse Monte Carlo approach described by us in previous publications.[9,21]

Another example where the Newton inversion for finding potential parameters can be applied is a problem of simultaneous fitting Lennard–Jones parameters σ and ε to two known thermodynamic observables, for example average energy and pressure. Such a problem was considered in ref. 22 using the weak coupling approach.

While the Newton inversion procedure was depicted above for the transition from the atomistic to CG level, it can work in the same way for the connection between *ab initio* and atomistic levels, with the only difference that the quantum-mechanical energy surface is used instead on *N*-body potential of mean force in eqn (2).

In the following, we shall concentrate on reconstruction of pair-wise potentials from RDFs which are computed in simulations on a more detailed level. In case of having several different CG sites in the system, indexes i, j in eqn (3–5) run both over all pairs of sites and distance points, and corresponding $\langle A_i\rangle$ represent a complete set of pair distribution functions. We include into the described scheme some other potential terms which are important for the description of macromolecular structures: covalent bonds, angles, and torsion potentials. Then indexes i, j run even over possible (discretized) values of the bond lengths, angles and torsions, with λ_i values representing the bond, angle or torsion potentials and $\langle A_i \rangle$ representing the corresponding distributions over bond lengths, angles or torsions. In all cases, average values of $\langle A_i \rangle$ can be acquired from a detailed simulation of a small system.

There exist a few other approaches which can be used to invert RDFs as well as bond or angle distributions. Soper[23] introduced an empirical potential structure refinement method (also known as the "Iterative Boltzmann Inversion") in which the pair potential is corrected after each iteration according to the mean field approximation:

$$V^{(i+1)} = V^{(i)} + kT \ln \frac{g^{(i)}(r)}{g_{\text{ref}}(r)} \tag{6}$$

Correction of potential according to eqn (6) is straightforward to implement, and such an approach was used in a number of studies.[24–27] In cases when several different types of coarse-grained sites, and correspondingly several different potentials are involved, cross-correlations between RDFs according to eqn (3–5) need to be taken into account in order to provide convergence.

In some cases it is possible to solve the inverse problem using numerical solutions of the liquid theory equations, for example the Hypernetted-Chain (HNC) approximation.[28] In the case of solvent-mediated potentials between ions, HNC solution was found to provide very accurate solutions of the inverse problem, coinciding with the results obtained by the inverse MC simulations.[29] We should also mention a few other works devoted to the inverse problem.[30–34]

3 Applications

3.1 A united atom water model

Water is the most common solvent. It is also the main substance in all biological organisms. Still, water is a difficult liquid in computer modeling and simulations. On the atomistic level, there exist a large number of models, most of them parameterized empirically, which more or less reasonably describe properties of water. Often explicit water can also be replaced by continuum models, from the generalized Born theory to effective solvent-mediated potentials. There is also a current interest to develop intermediate models in between atomistic and continuum models, presenting water as a liquid of particles, but without atomistic details. Example of such "coarse-grained" models may be the one-site model used in DPD simulations of water,[35] or the one used in the Martini force field.[8] These models are purely empirical. Here we present an alternative coarse-grained water model where the parameters are derived from atomistic simulations.

We consider here the simplest possible way to coarse-grain liquid water, presenting each molecule as a spherically symmetrical particle interacting with others by a distance-dependent pair potential. Such a potential can readily be obtained by inversion of the oxygen-oxygen RDF obtained in atomistic simulations of water (see Fig. 1). Corresponding RDF between molecular center-of-mass points can also be used. Similar water potentials have been presented in a number of publications recently.[36–38] Our water potential is displayed in Fig. 2. This potential reproduces perfectly the oxygen-oxygen water RDF from atomistic simulations (Fig. 1) if simulations are run exactly at the same density. The problem is, however, that the pressure in the CG system at normal density (1 g/cm^3) is typically at the level 8000–9000 atm.[38] One can clearly see from the potential shape in Fig. 2, that such a potential is essentially repulsive, with only a very weak attractive minimum of about −0.5 kJ/mol at distances corresponding to second neighbors. One would

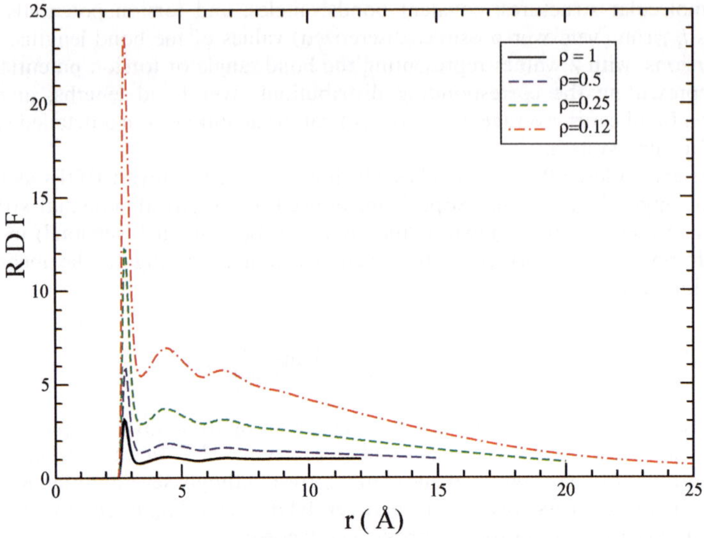

Fig. 1 Oxygen-oxygen RDF of water simulated at constant volume corresponding to densities 1, 0.5, 0.25 and 0.12 g/cm³.

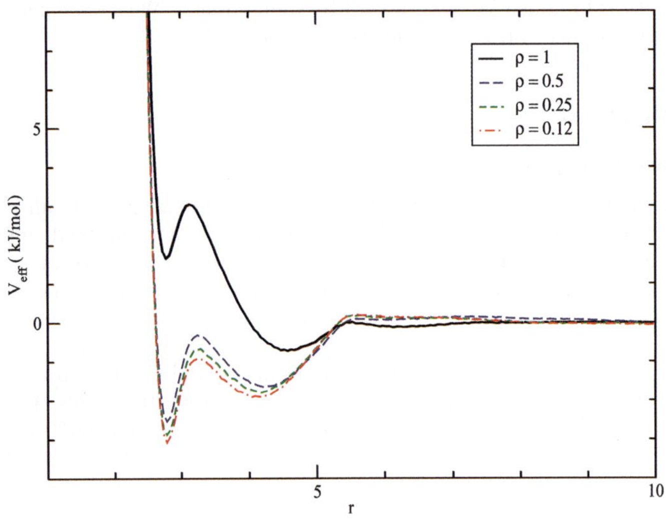

Fig. 2 Effective potentials for coarse-grained water obtained by inversion of RDFs shown in Fig. 1.

need to apply a high pressure to force the particles to approach each other to obtain the distances corresponding to those at normal liquid density. An undesirable consequence of this is that the system will expand immediately if there is space available, making it behave as a strongly compressed gas rather than a liquid. Simulations of various other properties within this model are questionable because of the very high internal pressure.

In a recent paper,[38] a method was suggested to build a "pressure consistent" one-site water model where the potential obtained by inversion of RDF from atomistic simulation was complemented by a slowly decreasing function, varying which one can get correct pressure with only a minor effect on RDF. Here we describe an alternative approach to obtain a pressure-consistent effective potential avoiding the use of a fitting function. This example also illustrates problems related to concentration dependence of structure-derived effective potentials.

Assume that we run atomistic simulations of water using an SPC-like potential, in a volume several times larger than that corresponding to normal density. One can then expect two phases of water where molecules will condense in one place forming a liquid of nearly correct density, while the rest of the available volume will be mainly empty, representing the vapor phase. If we then use the RDF obtained from the phase-separated system to build the effective potential for the CG model, we can expect that the CG model would behave similarly (in order to maintain atomistic RDF), it will condense forming the same structure as the atomistic model. The important point is that if the system is phase separated into liquid and gas phases, the total pressure must be low. Thus we have reached two goals: we get a single-site model which correctly reproduces liquid structure and also provides a realistic pressure.

The idea described above was tested for the flexible SPC water model.[39] Simulations were run for 500 molecules at densities 1, 0.5, 0.25 and 0.12 g/cm^3, and the corresponding oxygen-oxygen RDFs are displayed in Fig. 1. In all cases except in the first, a phase separation was observed. This can be seen from the snapshots, and also concluded from the fact that first and second RDF maxima increase nearly inversely proportional to the density. These RDFs were used as input to the inverse procedure, with the same number of molecules and the volume as the corresponding atomistic simulations. The results are shown in Fig. 2. One can clearly see that, for all phase separated systems, the effective potentials are very similar to each other, and they drop down substantially lower than the potential generated at bulk water density. The shape of these potentials is also similar to that obtained in paper[38] after addition of the term correcting the pressure.

The potentials generated from the RDFs simulated in low density systems, were also tested for the normal density 1 g/cm^3. In these simulations, the potentials were cut after 10 Å (note that deviations of these potentials from zero are less than 0.15 kJ/mol already after 6 Å). The result is shown in Fig. 3. One can see

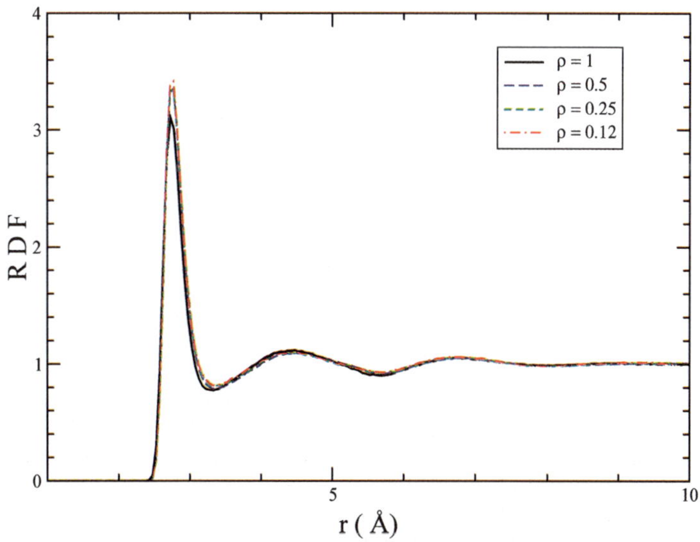

Fig. 3 RDFs of coarse-grained water presented by potentials given in Fig. 2 obtained at density 1 g/cm^3.

Table 1 Computed pressure at density 1 g/cm^3, and density corresponding to pressure 1 atm, for coarse-grained water potentials obtained from atomistic simulations of water at density ρ_0

ρ_0 (Atomistic simulations)	Pressure (bar)	Density (g/cm^3) at pressure 1 atm
1	9050	<0.01
0.5	2370	0.71
0.25	60	0.99
0.12	−2070	1.17

that all these potentials give RDF practically overlapping with the result of atomistic simulation. One can note only a slight increase of the first RDF maximum. This means that changes in the effective potential, because of the initial lower densities, do not noticeably affect results after the correct density is restored.

A more important feature of the potentials obtained from lower density simulations is that they now provide more realistic pressures at the normal density. The values for pressure at density 1 g/cm^3, for each of the four considered potentials, are given in Table 1. Although it is still different from the normal pressure 1 atm, one can see a very substantial improvement from the potential obtained by the direct inversion of bulk RDF. The system generated from atomistic simulation at 0.25 g/cm^3 density provides practically correct pressure. The too low negative pressure in the case of potential generated in 0.12 g/cm^3 simulation can be a result of the effect of the surface tension in the drop of water, which causes a higher pressure and thus higher density inside the drop simulated at the atomistic level, and which manifests in lowering pressure when the surface is no longer present at density 1 g/cm^3. Observe that all the three potentials generated in lower density simulations, describe a liquid state. By a change of the density, the pressure can be brought to 1 atm. This feature makes such potentials suitable for constant-pressure simulations, as well as in the simulation of macromolecular assemblies in contact with water.

3.2 Coarse-grained simulations of an equimolar mixture of L- and D-proline in DMSO

A molecular point of view, based on simulations, can help to better understand the crystallization behavior and the relative stability of the racemic and enantiopure compounds. In this study we investigated L- and D-prolines (see Fig. 4) dissolved in dimethyl sulfoxide (DMSO) to study nucleation and crystallization behavior of the racemates in supersaturated solution. In this investigation we have focused on 0% enantiomeric excess solution of the two stereo isomers. This study is linked with asymmetric proline-mediated catalysis taking place in DMSO. In crystal and in solution, proline exists in a zwitterionic form becoming engaged in a network of strong hydrogen bonds due to the charged amine and carboxyl groups. Because hydrogen bonds in computer simulations are produced entirely by Coulombic interactions the simulation results are very sensitive to the used partial charges of the atoms. The electrostatic potentials of proline were calculated using the RESP method. The new RESP fitted charges, combined with the AMBER force field parameters, were imported into M.DynaMix[40] package to do the molecular dynamics simulations. Molecular dynamics simulations of proline and DMSO mixtures (a box containing 1000 DMSO molecules and 30 proline molecules, giving proline concentration about 0.4 M) were performed. Nose-Hoover thermostat and barostat were employed. A double time-step algorithm (0.2 and 2.0 fs for the short and long step, respectively) was implemented and the temperature is set at $T = 278$K.

We have removed all the hydrogen atoms and substituted the COO– group by a single site in our coarse-grained proline model. Totally, the coarse-grained proline was described by 6 sites, of which three carbon united atoms of the ring were

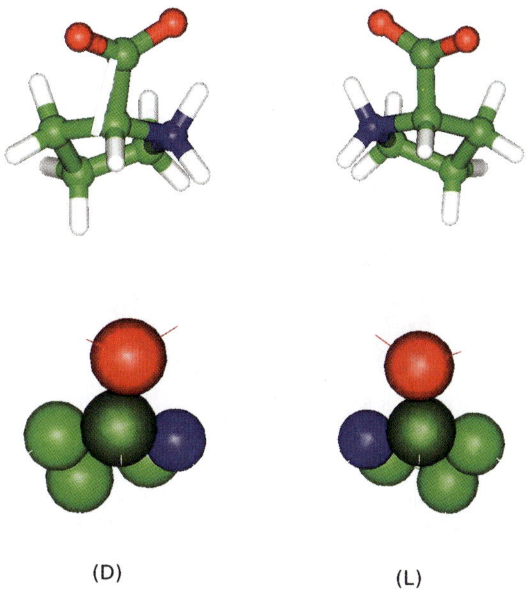

Fig. 4 Atomistic and coarse-grained representations of D- and L-prolines.

considered as equivalent (see Fig 4). Thus, we have four different coarse-grained sites on the proline molecule.

The RDFs between the coarse-grained sites of the prolines dissolved in DMSO were first determined in atomistic simulations. They were used to compute the effective solvent-mediated potentials between all the pairs of the coarse-grained proline sites. The enantiometric character of the proline molecules was kept by using intramolecular bond potentials, which were also reproduced from the distribution of the corresponding bond lengths. Also, the charges at the nitrogen and COO sites were kept in order to maintain the zwitterionic character of the coarse-grained model. In the coarse-grained model, the charges interact *via* Coulombic potential with the dielectric permittivity of DMSO $\varepsilon = 46$.

In simulations without explicit DMSO solvent the derived effective potentials provide the very same structural properties of proline clusters as those observed in atomistic simulations. An example of such clustering is shown in Fig. 5, where a similar aggregate is quickly formed of the D- and L-prolines both in all-atom

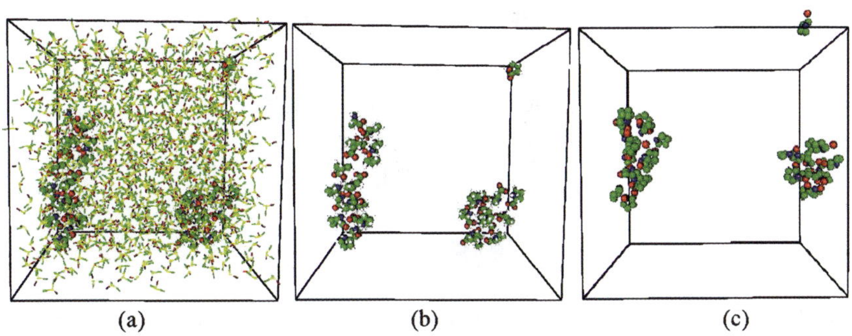

Fig. 5 Cluster of proline molecules obtained in molecular dynamics with explicit DMSO solvent: (a) DMSO are shown, (b) DMSO are not shown, and cluster formed in coarse-grained simulation with implicit solvent: (c).

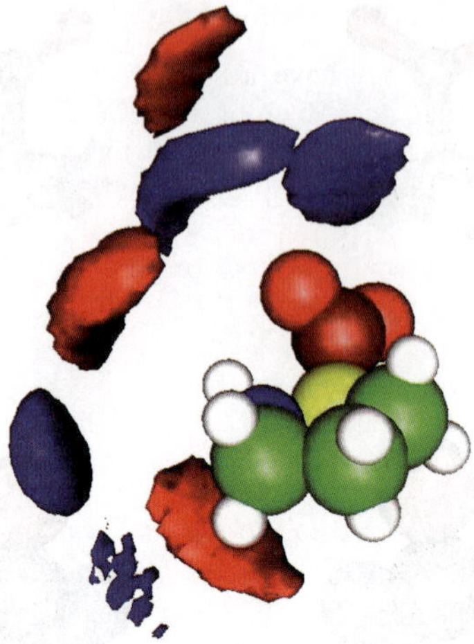

Fig. 6 Spatial distribution functions of CO (red) and NH (blue) CG-atoms around D-proline (red spheres - CO group, green - CH_2 groups, blue - NH group, sphere - CH group). Results are based on full atomistic MD calculation. CO- and NH- SDF shown in the figure has intensity value of 250 with maximum close to 2000.

simulations with explicit DMSO solvent (a and b) and in corresponding simulations using the six-site coarse-grained model with implicit DMSO solvent (c). Similar to the example of the one-site water model, we can suggest that effective potentials, derived from atomistic simulations of the "phase separated" system (that is, when the proline molecules were gathered to a cluster) would provide not only local structure of the proline cluster, but also a reasonable description of their osmotic pressure and affinity. To illustrate that the CG model contains the atomistic features of proline in implicit solvent we display the spatial distribution functions for the carbonyl and amine groups. In Fig. 6 these are calculated in all-atom simulations while in Fig. 7 they are calculated in CG simulations, correspondingly.

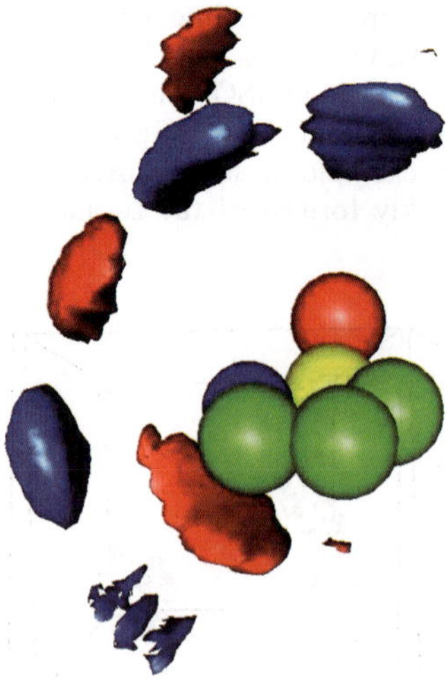

Fig. 7 Spatial distribution functions of CO (red) and NH (blue) CG-atoms around D-proline (red sphere CO group, green - CH_2 groups, blue - NH group, yellow - CH group). Results of MC calculation based on CG-potentials obtained using IMC procedure. CO-related SDF shown on the figure has intensity value of 220 with maximum of 1406 and NH-SDF has intensity value of 200 with maximum of 890.

3.3 Coarse-grained phospholipid model for bilayers and vesicles

Simulations of lipid membranes have attracted much attention during the last decade due to the fact that such membranes form outer shells of living cells. However, atomistic simulation of even a small piece of membrane consisting of about 100 lipids and surrounding water is a computational challenge, while many actual biophysical problems, such as studies of membrane mechanical properties, fusion, morphology, rafts formation, *etc.*, require consideration of substantially larger membrane fragments. For investigation of all these phenomena in molecular simulations, coarse-grained level of modeling provides practically the only possible choice. In recent years, coarse-grained lipid model based on Martini force field was used to model self-assembly of lipids into bilayers, micelles and vesicles.[8,41]

As another example of our multiscale modeling scheme, the effective solvent-mediated potentials for 10-sites CG model of DMPC lipid molecule (see Fig. 8) have been constructed.[14] The CG site-site RDFs were computed from atomistic molecular dynamics simulations of 16 lipids dissolved in water and described by the CHARMM-27 force field. In the subsequent development, the effective potentials derived in work[14] were re-parameterized after re-computation of CG site-site RDFs according to the recent modification of the CHARMM27 force field described in ref. 42. This modification of the CHARMM force field has been done with the primary aim to improve agreement with experiments for atomistic simulations of lipid bilayers, and to reproduce correctly the average area per lipid in particular. The principle point of this modification was recalculation of the atomic charges on the lipid head group by the *ab initio* Hartree–Fock method. The reparametrized DMPC model was simulated in 100 ns atomistic simulations, at a 1 : 100 DMPC/water molar ratio.

Similar to the described above cases of water simulated at low density, or proline molecules in DMSO, the simulated lipids were gathered in an unordered cluster of irregular structure, from which the site-site RDFs were determined. They were used in the inverse procedure to generate the effective potentials. The described above modification of the atomistic lipid potentials had a rather limited effect on the computed effective potential and they appear rather similar to those given in a previous paper.[14]

Large-scale Langevin molecular dynamics simulations of the CG lipid model, with the number of lipids in the range of 400–5000 and the system size of 200–500 Å, have been carried out using an ESPResSo package.[43] First, it was demonstrated that the coarse-grained model provides the same structure of a plane bilayer as the atomistic model. Then a number of other simulations were performed. It was shown that depending on conditions, lipids being initially thrown randomly in the simulation cell, organize themselves in different structures. If the number of lipids is small (less that 1000) the prevailing resulting structure is a bicell (a piece of bilayer of

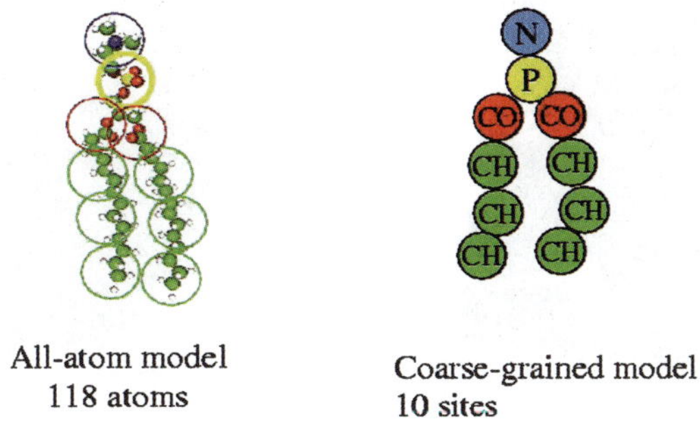

Fig. 8 Atomistic and coarse-grained representations of DMPC. Lipid-equivalent sites are shown in the same color.

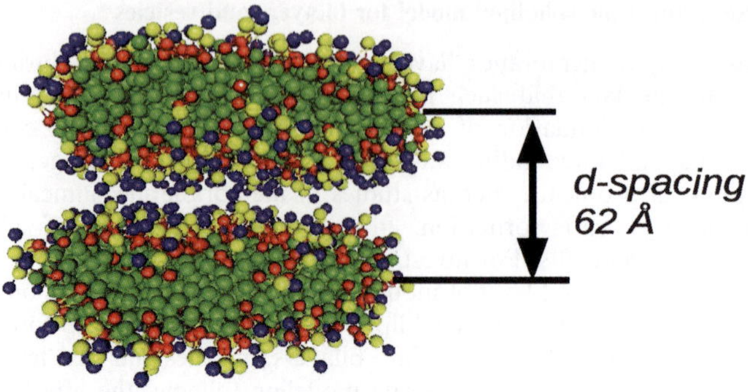

Fig. 9 Two discoid bicells, one over another, formed from an initially random state of 400 CG lipids in a 200 Å box. The choline groups of lipids are shown in blue, phosphates yellow, ester groups red and hydrocarbon tails green.

discoid shape). As an intermediate structure, several bicells in a stack one above the other (Fig. 9) were observed. It is quite remarkable that the average distance between the middle planes of two layers was about 62 Å, in perfect agreement with the experimental *d*-spacing of a multi-lamellar DMPC bilayer stack.[44]

A larger number of lipids (with 1500 lipids in a 400 Å box and 3500 lipids in a 500 Å box) was found to spontaneously form a spherical vesicle. The observed vesicle formation took place on a 100 ns time scale. Note however that in these simulations, the Langevin friction parameter γ was set to the very low value 0.001. With such a value of the parameter, the equations of motion are close to the Newtonian ones, while the role of the friction parameter is only to control the temperature. The dynamics in this case is artificially accelerated, which allows reaching equilibrium faster. It might be possible to adjust the Langevin friction parameter to reproduce lipid diffusion observed in atomistic simulation and thus also reproduce real-time description of lipid dynamics.

Another simulation has been performed at higher lipid concentration corresponding to 5000 CG lipids in a smaller periodic box, 300 Å. The size of the system does not allow formation of a single bicell or vesicle. The starting structure again was random. Initially, during the first 100 ns, a tendency to form multi-lamellar bilayer structures was observed, see Fig. 10a. This structure has been evolving further, and finally, after about 300 ns, a multi-lamellar vesicle consisting of almost ideally

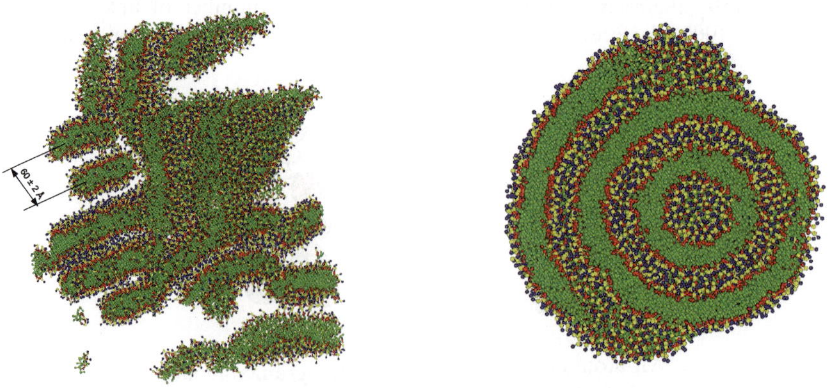

Fig. 10 Formation of multi-lamellar structures in a Langevin dynamics simulation of 5000 coarse-grained DMPC lipids in a periodic cubic box of 300 Å. Snapshots after 50 ns (left) and 300 ns (right; cut-plane through the middle of the vesicle is shown).

spherical layers was formed, see Fig. 10b. The third outer layer is not completely filled because there were not enough lipids to fill it. This structure was stable during the remaining 200 ns of simulation.

Again, it should be noticed that both in the case of bicells (Fig. 9) and multi-lamellar structures (Fig. 10), the distance between the mid-planes of neighboring layers is 60–62 Å, which means that there is space of about 20 Å between the surfaces of neighboring layers, filled by water. Also, the multi-lamellar vesicle in Fig. 10b occupies about one third of the total cell volume, which means in fact phase separation on a dense multi-lamellar phase and empty space representing bulk water. It is known that in real life solutions, lipids undergo such phase separation, on pure water phase and concentrated, lamellar phase with about 30 water molecules per lipid, which corresponds to about 20 Å water layers between lipid bilayers. Although our CG model does not include explicit water, it perfectly reproduces experimental phase behavior of phospholipid solutions, including a proper molar ratio of lipids in the lamellar phase. Thus the coarse-grained model, developed exclusively from all-atomic simulation data, reproduces well all the basic features of lipids in water solution.

4 Conclusions

A hierarchical true multiscale modelling approach presented here links together three levels of molecular modelling: *ab initio* molecular dynamics, classical molecular dynamics and meso-scale simulations. The method providing the link between these levels is the Newton inversion which in fact is equivalent to the inverse Monte Carlo approach previously developed by us. On the *ab initio* level, Car–Parrinello simulations, or alternatively, a highly efficient and accurate *ab initio* tight-binding (AITB) scheme, recently developed by us,[45] can be used in the future. The described methodology provides a consistent scheme to build molecular models for different scales without the need of empirical fitting of parameters. Some elements of this scheme were also demonstrated. Of course, there is still a very long way to go to define properties of molecules or materials exclusively *in silico*, and some tuning of the models against available experimental data is always an option. Also, transferability of the coarse-grained potentials needs to be checked in every case.

An interesting feature of the inverse procedure, which we demonstrated in the examples, is that when the detailed system was forced into the condition of phase separation, the derived "coarse-grained" system tried to reproduce the same phase separated structure, which implied the maintaining of some basic thermodynamic properties.

We hope that the suggested approach would increase the fraction of "*ab initio*" derived features in molecular models in expense of "*ad hoc*" or "empirically fitted" ones, which would enhance reliability of molecular simulations, increase their predictive power and open possibilities to address new, truly "large-scale" problems which are not yet considered in the molecular simulation domain.

Acknowledgements

This work has been supported by grants from the Swedish Science Council (VR) and computing grants from the Swedish National Infrastructure for Computing (SNIC).

References

1 R. Car and M. Parrinello, *Phys. Rev. Lett.*, 1985, **55**, 2471.
2 A. Warshel and M. Levitt, *J. Mol. Biol.*, 1976, **103**, 227.
3 M. J. Field, P. A. Bash and M. Karplus, *J. Comput. Chem.*, 1990, **11**, 700.
4 P. Espanol, *Phys. Rev. E*, 1995, **52**, 1734.
5 R. Benzi, S. Succi and M. Vergassola, *Phys. Rep.*, 1992, **222**, 145.

6 R. D. Groot, *Langmuir*, 2000, **16**, 7493.
7 H. Noguchi and M. P. Takasu, *J. Chem. Phys.*, 2001, **115**, 9547.
8 S. J. Marrink, A. H. de Vries and A. E. Mark, *J. Phys. Chem. B*, 2004, **108**, 750.
9 A. P. Lyubartsev and Laaksonen, *Phys. Rev. E*, 1995, **52**, 3730.
10 A. P. Lyubartsev and A. Laaksonen, *Chem. Phys. Lett.*, 2000, **325**, 15.
11 A. P. Lyubartsev, K. Laasonen and A. Laaksonen, *J. Chem. Phys.*, 2001, **114**, 3120.
12 A. P. Lyubartsev and Laaksonen, *Phys. Rev. E*, 1997, **55**, 5689.
13 A. P. Lyubartsev and A. Laaksonen, *J. Chem. Phys.*, 1999, **111**, 11207.
14 A. P. Lyubartsev, *Eur. Biophys. J.*, 2005, **35**, 53.
15 W. G. Noid, J.-W. Chu, G. S. Ayton, V. Krishna, S. Izvekov, G. A. Voth, A. Das and H. C. Andersen, *J. Chem. Phys.*, 2008, **128**, 244114.
16 F. Ercolessi and J. Adams, *Europhys. Lett.*, 1994, **26**, 583.
17 S. Izvekov, M. Parrinello, C. J. Burnham and G. A. Voth, *J. Chem. Phys.*, 2004, **120**, 10896.
18 S. Izvekov and G. A. Voth, *J. Phys. Chem. B*, 2005, **109**, 2469.
19 G. Toth, *J. Phys.: Condens. Matter*, 2007, **19**, 335222.
20 R. L. Henderson, *Phys. Lett. A*, 1974, **49**, 197.
21 A. P. Lyubartsev and A. Laaksonen, On the reduction of molecular degrees of freedom in computer simulations in *Novel Methods in Soft Matter Simulations*, ed. M. Karttunen, I. Vattulainen, A. Lukkarinen, Springer, Berlin, 2004, pp. 219–244.
22 S. L. Njo, W. F. van Gunsteren and F. Müller-Plathe, *J. Chem. Phys.*, 1995, **102**, 6199.
23 A. K. Soper, *Chem. Phys.*, 1996, **202**, 295.
24 J. C. Shelley, M. Y. Shelley, R. C. Reeder, S. Bandyopadhyay and M. L. Klein, *J. Phys. Chem. B*, 2001, **105**, 4464.
25 D. Reith, M. Putz and F. Muller-Plathe, *J. Comput. Chem.*, 2003, **24**, 1624.
26 D. Reith, M. Putz and F. Muller-Plathe, *J. Phys. Chem. B*, 2005, **109**, 2469.
27 V. A. Harmandaris, N. P. Adhikari, N. F. A. van der Vegt and K. Kremer, *Macromolecules*, 2006, **39**, 6708.
28 L. Reatto, D. Levesque and J. J. Weis, *Phys. Rev. A: At., Mol., Opt. Phys.*, 1986, **33**, 3451.
29 A. P. Lyubartsev and S. Marcelia, *Phys. Rev. E: Stat., Nonlinear, Soft Matter Phys.*, 2002, **65**, 041202.
30 M. Ostheimer and H. Bertagnolli, *Mol. Simul.*, 1989, **3**, 227.
31 Y. Rosenfeld and G. Kahl, *J. Phys.: Condens. Matter*, 1997, **9**, L89.
32 G. Toth and A. Baranyai, *J. Mol. Liq.*, 2000, **85**, 3.
33 R. L. C. Akkermans and W. L. Briels, *J. Chem. Phys.*, 2001, **114**, 1020.
34 N. G. Almarza and E. Lomba, *Phys. Rev. E: Stat., Nonlinear, Soft Matter Phys.*, 2003, **68**, 011202.
35 R. D. Groot, Applications of Dissipative particle Dynamics in *Novel Methods in Soft Matter Simulations*, ed. M. Karttunen, I. Vattulainen, A. Lukkarinen, Springer, Berlin, 2004, pp. 5–37.
36 A. Eriksson, M. N. Jacobi, J. Nyström and K. Tunstrøm, *J. Chem. Phys.*, 2008, **129**, 024106.
37 M. Praprotnik, L. D. Site and K. Kremer, *Annu. Rev. Phys. Chem.*, 2008, **59**, 545.
38 H. Wang, C. Junghans and K. Kremer, *Eur. Phys. J. E*, 2009, **28**, 221.
39 K. Toukan and A. Rahman, *Phys. Rev. B: Condens. Matter Mater. Phys.*, 1985, **31**, 2643.
40 A. P. Lyubartsev and A. Laaksonen, *Comput. Phys. Commun.*, 2000, **128**, 565.
41 S. J. Marrink and A. E. Mark, *J. Am. Chem. Soc.*, 2003, **125**, 15233–15242.
42 C.-J. Högberg, A. M. Nikitin and A. P. Lyubartsev, *J. Comput. Chem.*, 2008, **29**, 2359.
43 H. Limbach, A. Arnold, B. A. Mann and C. Holm, *Comput. Phys. Commun.*, 2006, **174**, 704.
44 N. Kucerka, Y. Liu, N. Chu, H. I. Petrache, S. Tristram-Nagle and J. F. Nagle, *Biophys. J.*, 2005, **88**, 2626.
45 Y. Tu, S. P. Jacobsson and A. Laaksonen, *Phys. Rev. B: Condens. Matter Mater. Phys.*, 2006, **74**, 205104.

Mesoscale modelling of polyelectrolyte electrophoresis

Kai Grass*[a] and Christian Holm*[abc]

Received 30th January 2009, Accepted 20th February 2009
First published as an Advance Article on the web 10th August 2009
DOI: 10.1039/b902011j

The electrophoretic behaviour of flexible polyelectrolyte chains ranging from single monomers up to long fragments of a hundred repeat units is studied by a mesoscopic simulation approach. Abstracting from the atomistic details of the polyelectrolyte and the fluid, a coarse-grained molecular dynamics model connected to a mesoscopic fluid described by the Lattice-Boltzmann approach is used to investigate free-solution electrophoresis. Our study demonstrates the importance of hydrodynamic interactions for the electrophoretic motion of polyelectrolytes and quantifies the influence of surrounding ions. The length-dependence of the electrophoretic mobility can be understood by evaluating the scaling behavior of the effective charge and the effective friction. The perfect agreement of our results with experimental measurements shows that all chemical details and fluid structure can be safely neglected, and a suitable coarse-grained approach can yield an accurate description of the physics of the problem, provided that electrostatic and hydrodynamic interactions between all entities in the system, *i.e.*, the polyelectrolyte, dissociated counterions, additional salt and the solvent, are properly accounted for. Our model is able to bridge the single molecule regime of a few nm up to macromolecules with contour lengths of more than 100 nm, a length scale that is currently not accessible to atomistic simulations.

1 Introduction

Nowadays, electrophoresis methods are widely used to separate biomolecules[1,2] such as peptides, proteins, DNA, as well as synthetic polymers.[3,4] In order to be able to improve the processes involved in current electrophoretic separation methods it is a prerequisite to gain a thorough understanding of the behaviour of polyelectrolytes (PEs) in an externally applied electric field. Several theories[5–8] have been developed to describe PE electrophoresis and successfully describe qualitatively the experimentally observed behaviour of various PEs under bulk conditions. However, the mobility of small oligomeric PEs shows under low salt conditions a non-monotonic behaviour that current theories have not been able to explain. Here, computer simulations can provide detailed analysis and lead to a better understanding of this behaviour.[9]

In a recent publication,[10] we employed a mesoscopic coarse-grained model using molecular dynamics simulations in connection with a Lattice-Boltzmann (LB) algorithm to extend the theoretical understanding on a more detailed level, and in

[a]*Frankfurt Institute for Advanced Studies, Goethe University, Ruth-Moufang-Str. 1, 60438 Frankfurt/Main, Germany. E-mail: grass@fias.uni-frankfurt.de*
[b]*Max-Planck-Institut für Polymerforschung, Ackermannweg 10, 55128 Mainz, Germany*
[c]*Institute for Computational Physics, University of Stuttgart, Pfaffenwaldring 27, 70569 Stuttgart, Germany. E-mail: holm@icp.uni-stuttgart.de*

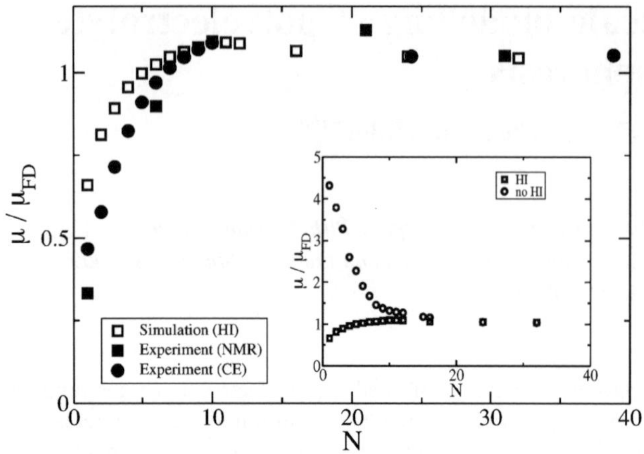

Fig. 1 The normalized electrophoretic mobility μ/μ_{FD} as a function of the number of repeat units N for simulation data including hydrodynamic interactions (HI), and experimental data coming from capillary electrophoresis (CE) and from electrophoretic NMR. The inset compares to simulation data obtained with a model neglecting hydrodynamic interactions.

particular, we investigated the role of hydrodynamic interactions in these systems. Our results were able to match the free-solution electrophoretic mobility μ of short polyelectrolyte chains, here polystyrene sulfonate (PSS), as a function of the number of repeat units N with quantitative agreement to experiments as shown in Fig. 1. Since the three data sets have different solvent viscosities the mobility is normalized by the corresponding constant mobility for long chains, the so-called free-draining mobility, μ_{FD}. The electrophoretic mobility increases for short oligomers, reaches a maximum for intermediate degrees of polymerization, and slowly decreases towards a plateau value for long chains. To understand this observation, the hydrodynamic interactions were investigated in detail and we found that they are actually the major driving force for the length dependent mobility for short and intermediate chain lengths. The constant mobility for long chains can be attributed to an effective screening of hydrodynamic interactions, which leads to the so-called free-draining behavior. The inset in Fig. 1 shows a comparison to a coarse-grained simulation that neglects hydrodynamic interactions. This leads to a qualitatively completely different behavior, showing a monotonically decreasing mobility. Agreement to the experimentally observed behaviour is only achieved as long as hydrodynamic interactions are included correctly as has been shown in detail in our previous investigations.[11,12]

In this article, we will extend our work by studying the electrophoresis of generic flexible polyelectrolyte chains ranging from single monomers to long fragments of hundred repeat units. Abstracting from the atomistic details of the polyelectrolyte and the fluid, a coarse-grained molecular dynamics model connected to a mesoscopic fluid described by the Lattice-Boltzmann approach is used to investigate the free-solution behavior under varying salt concentration.

In the next section we will introduce the employed simulation model. In Section 3, the main results of this study are presented and discussed. We conclude with final remarks in Section 4.

2 Model

We employ molecular dynamics (MD) simulations using the ESPResSo package[13] to study the behaviour of linear polyelectrolytes (PE) of different lengths. The PEs are modelled by a totally flexible bead-spring model. The monomers are connected to each other by finitely extensible nonlinear elastic (FENE) bonds[14]

$$U_{\mathrm{FENE}}(r) = \frac{1}{2}kR^2 \ln\left(1 - \left(\frac{r}{R}\right)^2\right)$$

with stiffness $k = 30\varepsilon_0$, and maximum extension $R = 1.5\sigma_0$, where r is the distance between the interacting monomers. Additionally, a truncated Lennard-Jones or WCA potential[15]

$$U_{\mathrm{LJ}}(r < r_c) = \varepsilon_0\left(\left(\frac{\sigma_0}{r}\right)^{12} - \left(\frac{\sigma_0}{r}\right)^6 + \frac{1}{4}\right)$$

is used for excluded volume interactions between all monomers. A cutoff value of $r_c = \sqrt[6]{2}\sigma_0$ ensures a purely repulsive potential. All dissociated counterions and additional salt ions are modelled by appropriately charged spheres using the same WCA potential.

Here, ε_0 and σ_0 define the energy and length scale of the simulations. We use $\varepsilon_0 = k_BT$, *i.e.* the energy of the system is expressed in terms of the thermal energy. The length scale σ_0 defines the size of the monomers and the dimension of the system. For this study, σ_0 is chosen to be 4 Å. Different polyelectrolytes can be mapped by changing σ_0. Unless mentioned otherwise, all observables are expressed in reduced simulation units, and we will not use σ_0 and ε_0 explicitly from now on.

The chain length is varied from $N = 1$ to $N = 128$ and all chain monomers carry a negative electric charge $q = -1e_0$, where e_0 is the elementary charge. For charge neutrality, N monovalent counterions of charge $+1e_0$ are added. Additional monovalent salt is added to the simulation, corresponding to concentrations between $c_s = 0$ mM and $c_s = 160$ mM. The later concentration being equivalent to a particle density of the salt ions of $\rho_s = 0.01$.

A homogeneous electric field with reduced field strength $E = 0.1$ is applied in x-direction creating a force $F_E = qE$ on all charged particles, and thus inducing an electrophoretic mobility. It has been carefully checked that the field strength is within the linear response regime, *i.e.*, it does not influence the chain conformation or the distribution of the surrounding ions.[12]

Full electrostatic interactions are calculated with the P3M algorithm using the implementation of Deserno *et al.*[16] The Bjerrum length

$$l_B = e_0^2/(4\pi\varepsilon_0\varepsilon_r k_BT) = 1.8$$

in simulation units corresponds to 7.1 Å, the Bjerrum length in water at room temperature. This means that the effect of the surrounding water is modelled implicitly by simply using the dielectric properties of water, having a relative dielectric constant of $\varepsilon_r \approx 80$.

The simulations are carried out under periodic boundary conditions in a cubic simulation box. The size L of the box is varied to realize a constant monomer concentration of $c_{PE} = 16$ mM independent of chain length. This is equivalent to a monomer density $\rho_{PE} = 0.001$.

We include hydrodynamic interactions by using a Lattice-Boltzmann algorithm[17] that is interacting with the beads in the MD simulations *via* a frictional coupling introduced by Ahlrichs *et al.*[18] The mesoscopic LB fluid is described by a velocity field generated by discrete momentum distributions on a spatial grid rather than explicit fluid particles. We use an implementation of the D3Q19 model with a kinematic viscosity $\nu = 1.0$, and a fluid density $\rho = 1.0$. The resulting fluid has a dynamic viscosity $\eta = \rho\nu = 1.0$. The simulation box is discretised by a grid with spacing $a = 1.0$. As usual in a standard Langevin approach, the particle-fluid interaction is realised by a dissipative force. This force depends on the difference between the particle velocity $\mathbf{v}$ and the fluid velocity at the particle position $\mathbf{u}$:

$$\mathbf{F}_R = -\Gamma_{bare}(\mathbf{v} - \mathbf{u}).$$

Here, the coupling constant takes on the value $\Gamma_{bare} = 20.0$. Additional random fluctuations introduced to the particles and fluid act as a thermostat. The interaction between particles and fluid conserve total momentum, and this algorithm has been shown to yield correct long-range hydrodynamic interaction between individual particles.[18]

Additionally, a second type of MD simulation is used which is based on the Langevin equations of motions with a velocity dependent dissipative and a random term in addition to the interparticle forces. Together, both additional terms implicitly model the effects of a solvent surrounding the particles: the dissipative force,

$$\mathbf{F}_D = -\Gamma_0\mathbf{v},$$

with $\Gamma_0 = 1.0$ provides local friction and the non-correlated zero-mean Gaussian random forces,

$$\mathbf{F}_R = \xi(t),$$

mimic thermal kicks (Brownian motion). In order to fulfil the fluctuation–dissipation theorem, dissipative and random force have to be coupled together: $\langle\xi_i(t)\cdot\xi_j(t')\rangle = 6\Gamma_0 k_B T\delta_{ij}\delta(t\text{–}t')$. This approach only offers local particle–fluid interactions, and therefore destroys long-range hydrodynamic interactions. Nevertheless one can use it to compute the effective charge as has been presented in ref. 11,12. This effective charge is used to obtain the effect friction in the presence of hydrodynamic interactions and illustrates their importance for the electrophoretic mobility.

All simulations are carried out with a MD time step $\tau_{MD} = 0.01$ and LB time step $\tau_{LB} = 0.05$. After an equilibration time of 10^6 steps, 10^7 steps are used for generating the data. The time-series of four independent simulations are analyzed using autocorrelation functions to estimate the statistical errors as detailed by Wolff.[19] Error bars of the order of the symbol size or smaller are omitted in the figures.

3 Results and discussion

3.1 Electrophoretic mobility

We determine the electrophoretic mobility μ as the ratio between the measured center of mass velocity v_{PE} and the magnitude of the electric field E:

$$\mu = \frac{v}{E}$$

For comparison, the results are normalized by the monomer mobility μ_1.

Fig. 2 displays the characteristic behaviour of flexible polyelectrolytes for vanishing salt concentration $c_s = 0$ mM: initially, the electrophoretic mobility increases with N to reach a maximum at intermediate chain lengths and then slowly decays towards a constant value for long chains. This constant value, often called the free-draining limit μ_{FD}, can be explained by the length independence of the ratio between effective charge and effective friction for long chains as we will show in this article.

In the presence of added salt, the long chain mobility is reduced, which is consistent with the experimentally observed behavior.[20] Furthermore, the shape of the curve is influenced, and the maximum at intermediate chains is suppressed for increased salt concentration. At $c_s = 160$ mM the maximum disappears and the measured mobility becomes length independent within the resolution of the simulation. A further increase of the added salt concentration leads to a further

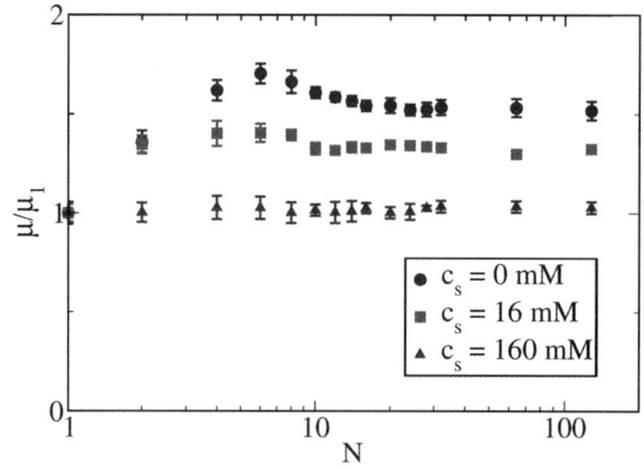

Fig. 2 The normalized electrophoretic mobility μ/μ_1 of polyelectrolyte chains of length N for three different salt concentrations using the LB algorithm. The added salt not only influences the absolute mobility, but likewise changes the characteristic shape of the mobility with respect to chain length N.

reduction of the limiting mobility μ_{FD}, not shown here, while the monomer mobility μ_1 remains almost unchanged. This leads eventually to an inverted length-dependence with a monotonic decrease of the mobility towards the limiting value.

In the simple local force picture, the constant center of mass velocity v_{PE} that determines the electrophoretic mobility is a direct result of the cancellation of two acting forces: the electric driving force $F_E = Q_{eff}E$ is cancelled by the solvent friction or drag force $F_D = \Gamma_{eff}v_{PE}$. Here, Q_{eff} is the effective charge of the polyelectrolyte, which can be thought of as the bare charge of the polyelectrolyte reduced by oppositely charged ions in solution that associate to the polyelectrolyte chain. The association of counterions to a PE chain is known as counterion condensation.[21,22] The compound formed by the polyelectrolyte and the associated ions is moved through the solvent under the influence of the external field and experiences a Stokesian drag force with an effective friction coefficient Γ_{eff} that is *a priori* unknown. In the steady state both forces balance and the mobility is given by

$$\mu = \frac{v}{E} = \frac{Q_{eff}}{\Gamma_{eff}}$$

Next, let us compare the results of Fig. 2 to the case when long-range hydrodynamic interactions between the particles are neglected in simulations, *i.e.*, by using a standard Langevin thermostat. The results are shown in Fig. 3. One immediately notices that the observed electrophoretic mobility differs significantly from the behaviour observed in Fig. 2. Independent of the salt concentration, the mobility decreases monotonically with chain length and slowly approaches a constant value for long chains which is independent of the salt concentration. This difference to the experimental observations and to the LB simulation including hydrodynamics will be analyzed in detail in the following sections.

3.2 Effective charge

To analyze the observed influence of the added salt on the polyelectrolyte mobility, we will determine the effective charge, and can then calculate $\Gamma_{eff} = Q_{eff}/\mu$ to obtain an estimate for the effective friction of the polyelectrolyte-ion compound. A word of care has to be taken here, since the value of the effective charge depends on definition. Qualitatively one can differentiate between a static definition and a dynamic definition.[23] In our case it is obviously a dynamic definition. In ref. 11,12, we

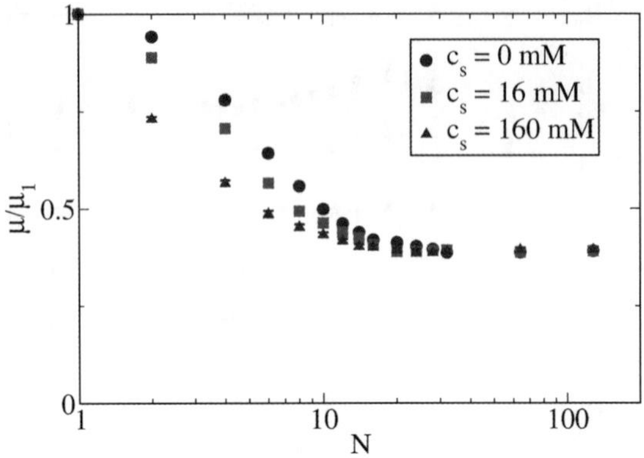

Fig. 3 The normalized electrophoretic mobility μ/μ_1 for different chain length N at varying salt concentrations c_s without hydrodynamic interactions differs significantly from the behaviour observed in Fig. 2. The mobility shows a salt-dependent monotonic decrease for short chains and a salt-independent constant value for long chains.

introduced several static and dynamic estimators for the effective charge and showed their equivalence at vanishing salt concentration. Here, three of them will be reviewed and applied to the case of added salt.

The local force picture described above can be used to estimate the effective charge of the polyelectrolyte based on the measurement of the electrophoretic mobility in the absence of hydrodynamic interactions. Let N_{CI} be the number of associated counterions reducing the bare charge of the polyelectrolyte which is equal to N. The effective charge is then given by

$$Q_{\text{eff}} = N - N_{CI}$$

Without long-range hydrodynamic interactions, the interaction of each particle with the solvent is purely local and directly given by the friction constant Γ_0 of the Langevin algorithm. The total effective friction of the polyelectrolyte and the ions is then:

$$\Gamma_{\text{eff}} = \Gamma_0(N + N_{CI})$$

This results in an expression for the electrophoretic mobility

$$\mu = \frac{N - N_{CI}}{\Gamma_0(N + N_{CI})}$$

from which an expression for N_{CI} is obtained. Therefore we can express the effective charge purely as a function of the mobility measurements shown in Fig. 3 and on our input value for Γ_0, independent of the knowledge of the value of N_{CI} by the following expression:

$$Q_{\text{eff}}^{(1)} = N\left(1 - \frac{1 - \mu\Gamma_0}{1 + \mu\Gamma_0}\right)$$

An alternative way of characterizing the associated ions is presented by Lobaskin *et al.*,[24] who suggested to determine the ion velocity with respect to the distance to the center of mass of the polyelectrolyte. For this method we use the LB algorithm to include hydrodynamical interactions, and the result for a chain of $N = 64$ at

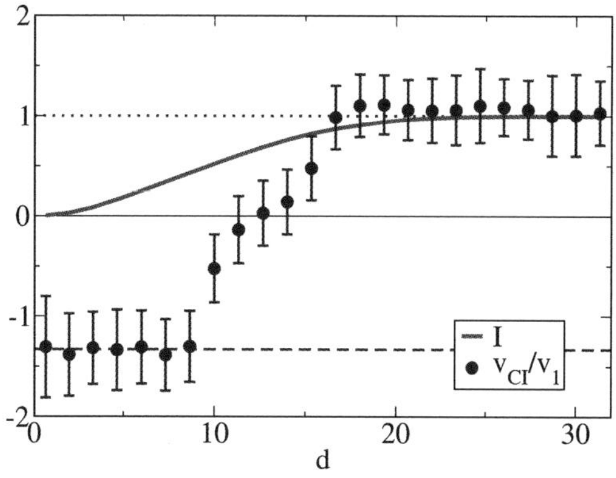

Fig. 4 The average ion velocity in the direction of the electric field v_{CI} (here for a chain with $N = 64$ monomers, salt concentration $c_s = 16$ mM, and hydrodynamics included) depends on the distance d to the center of mass of the polyelectrolyte. Ions close to the center co-move with the chain's velocity (dashed line), whereas ions far away from the center move with the single particle velocity $v_1 = \mu_1 E$ into the opposite direction. The distance d_0 at which $v_{CI}(d_0) = 0$ is used to separate co-moving, associated ions from non-associated ones. The solid line shows the integrated fraction of charges I that is found up to the distance d of the center of mass.

$c_s = 16$ mM can be inspected in Fig. 4. The average ion velocity in the direction of the electric field v_{CI} is a function of the distance d to the center of mass of the polyelectrolyte chain and in general depends on the chain length N and the salt concentration c_s. As shown, ions close to the center move with the chain at negative speed, whereas ions far away from the center move with the single particle velocity $v_1 = \mu_1 E$ in the opposite direction. The association of ions to the chain is strong enough to move them against the electric field. For every chain length and every salt concentration, the distance d_0 at which $v_{CI}(d_0) = 0$ is used to separate co-moving, associated ions from non-associated ones.

We use this distance d_0 to define the effective charge by summing up the total charge in the system found within this distance to the center of mass of the polyelectrolyte:

$$Q_{\text{eff}}^{(2)} = N(1 - I(d_0)),$$

where $I(d_0)$ is the integrated fraction of neutralizing charges found by adding the number of counterions and positively charged salt ions reduced by the number of negatively or like-charged salt ions. Far away from the center of mass of the chain, the total bare charge of the polyelectrolyte is neutralized and $I = 1$.

The effective charge Q_{eff} as obtained from both estimators is presented in Fig. 5a. Initially, Q_{eff} is close to the bare charge N, but as ion condensation sets in, the effective charge is reduced. Longer chains show a linear increase of their charge close to the Manning prediction for counterion condensation in the salt free case $Q_{\text{eff}} = (1/\xi)N$, where Manning parameter $\xi = l_B/b$ is the ratio between the Bjerrum length and the charge spacing along the polyelectrolyte backbone. For the model used here $b = 0.9$ and therefore $\xi = 2.0$. We note that there is no apparent dependence of the effective charge for long polyelectrolyte chains on the salt concentrations when measured by the dynamic effective charge estimators presented here.

Fig. 5b plots the effective charge per monomer, Q_{eff}/N. Here, the influence of the salt concentration for short and intermediate chain length can be seen. The higher the concentration of the added salt, the faster the electric charge of the polyelectrolyte is reduced by condensed counterions. For long chains, the charge per monomer is again independent of the salt concentration and comparable to the Manning

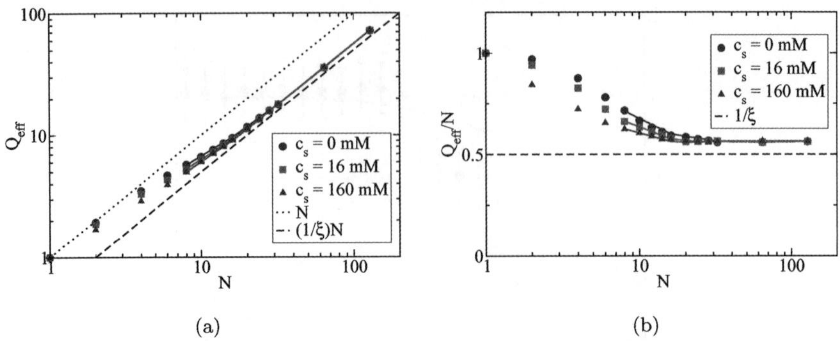

Fig. 5 (a) The effective charge Q_{eff} as a function of chain length N (symbols for $Q_{\mathrm{eff}}^{(1)}$, lines for $Q_{\mathrm{eff}}^{(2)}$). Both charge estimators show good agreement. Initially, Q_{eff} is close to the bare charge N (dotted line), but as ion condensation sets in, the effective charge is reduced. Longer chains show a linear increase of their charge close to the Manning prediction $(1/\xi)N$ (dashed line). (b) The effective charge per monomer $Q_{\mathrm{eff}/N}$ is influenced by the salt concentration for short and intermediate chains. The higher the concentration of the added salt, the faster the electric charge of the polyelectrolyte is reduced by condensed ions. For long chains, the charge per monomer is again independent of the salt concentration and comparable to the Manning prediction $1/\xi$ (dashed line).

prediction $1/\xi$. The difference for short and intermediate chains at different salt concentrations can be attributed to stronger association of counterions with increasing salt concentrations. For short chains, smaller than the Debye length, effects due to the finite size play a leading role in the ability to condense counterions.[25–27]

Additionally, Fig. 5 shows the equivalence of the two dynamic estimators $Q_{\mathrm{eff}}^{(1)}$ and $Q_{\mathrm{eff}}^{(2)}$ independently of the presence or absence of hydrodynamic interactions also in the presence of additional salt. This new observation supports the applicability and importance of these charge estimators for the study of polyelectrolytes during electrophoresis.

The charge estimators $Q_{\mathrm{eff}}^{(1)}$ and $Q_{\mathrm{eff}}^{(2)}$ measure the effective charge of the moving polyelectrolyte and its surrounding counterions. Therefore, they measure the effective dynamic charge of the polyelectrolyte. Similarly, it is possible to define a static estimate of the effective charge using the following simple method

$$Q_{\mathrm{eff}}^{(3)} = N_{\mathrm{PE}} - N_{\mathrm{CI}}(d < d_0),$$

where $N_{\mathrm{CI}}(d < d_0)$ is the average number of counterions that can be found within a distance d to the closest monomer. Here, the threshold d_0 chosen to be $d_0 = 2\sigma_0$.

The second static charge estimator used in ref. 12 based on the inflection criterion to estimate the threshold of counterion condensation[28–30] breaks down in the presence of high salt concentrations and therefore can not be applied here.

In Fig. 6, we compare the static charge estimate $Q_{\mathrm{eff}}^{(3)}$ for varying salt concentrations to the dynamic charge estimate obtained by $Q_{\mathrm{eff}}^{(1)}$. Unlike the dynamic estimate the static charge estimate shows a strong dependence on the salt concentration. While both estimators agree for vanishing salt concentration as previously shown in ref. 12, the static charge estimate shows a decrease with higher salt concentration, hence an increase in counterion condensation, as could be expected from a mean-field comparison,[30] and eventually it falls below the Manning prediction.

The higher salt concentrations increase the number of counterions in the close vicinity of the chain as measured by $Q_{\mathrm{eff}}^{(3)}$. At the same time, the electrostatic interactions in the system are reduced due to electrostatic screening, which also reduces the strength of the coupling between the polyelectrolyte and the counterions. The independence of the dynamic effective charge on the salt concentration for long chains as

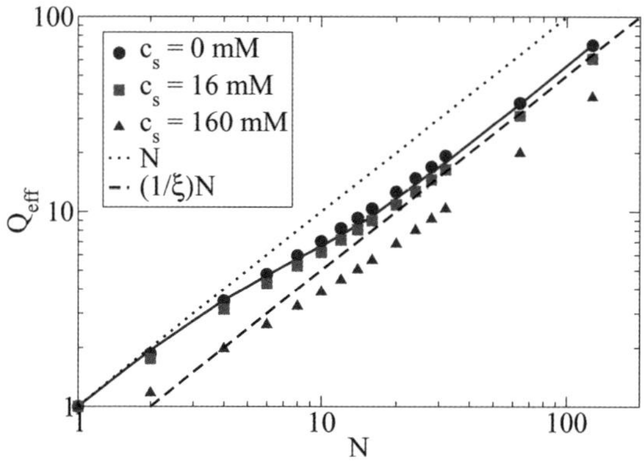

Fig. 6 The effective charge as measured by the static estimator $Q_{\text{eff}}^{(3)}$ shows a strong dependence on the salt concentration. At $c_S = 0$ mM the static estimate agrees with the dynamic estimates (solid line). The higher the salt concentration, the lower the static charge estimate. For comparison, the bare charge N (dotted line) and the Manning prediction (dashed-line) are plotted.

shown in Fig. 5 has to be understood as the cancellation of both effects: with increasing salt concentration more counterions in close vicinity to the polyelectrolyte are influenced by the chain but the strength of the interactions is reduced in such a way that the combined action remains unchanged and yields a concentration independent of dynamic effective charge.

In the following, we will use the dynamic effective charge to calculate the effective friction of the polyelectrolyte-ion compound.

3.3 Effective friction

When long-range hydrodynamic interactions are present, the effective friction of the polyelectrolyte and the associated counterions cannot be given in a simple analytic form. We therefore obtain it from the measurements of the mobility and the effective charge presented above:

$$\Gamma_{\text{eff}} = \frac{Q_{\text{eff}}}{\mu}$$

In Fig. 7a, the effective charge Γ_{eff} is displayed as a function of chain length N for different salt concentrations c_s. The friction increases monotonically with chain length.

Neglecting the contribution of the counterions the effective friction can be obtained from the hydrodynamic radius R_h of the polyelectrolyte defined by:

$$\left\langle \frac{1}{R_h} \right\rangle = \frac{1}{N} \sum_{i \neq j} \left\langle \frac{1}{\|\vec{r}_i - \vec{r}_j\|} \right\rangle$$

Here,

$$\vec{r}_i$$

is the position of the i-th chain monomer, and

$$\vec{r}_{\text{cm}}$$

the center of mass of the polyelectrolyte chain. The angular brackets $\langle \ldots \rangle$ indicate an ensemble average. The hydrodynamic radius is expected to exhibit a power law scaling $R_h \sim (N - 1)^{\nu}$, where the scaling exponent ν depends on the system. For an uncharged polymer with ideal chain behaviour one should get $\nu \approx 0.588$ (Flory exponent),[31] whereas for a fully charged polyelectrolyte without electrostatic screening one expects $\nu = 1$. Depending on the salt concentration, we obtain values between $\nu \approx 0.66$ for $c_s = 0$ mM and $\nu \approx 0.59$ for $c_s = 160$ mM (not shown here).

Initially, the friction increases with N as given by the hydrodynamic size of the polyelectrolyte

$$\Gamma = 6\pi\eta R_h \propto N^{0.61}$$

With the onset of counterion condensation the friction exceeds the value of the bare polyelectrolyte and for long chains becomes linear in N. The higher the concentration of the additional salt, the earlier the transition between the two regimes is observed. We furthermore note that the absolute friction value is increased with the addition of external salt.

The role of the additional salt can be best understood when looking at the effective friction per monomer as presented in Fig. 7b. Γ_{eff}/N shows an initial decrease with chain length which can be understood by hydrodynamic shielding: the monomers of short polyelectrolyte chains are hydrodynamically coupled and shield each other from the effect of the solvent. This reduces the friction per monomer below the value of a single particle. The decrease in friction due to the hydrodynamic shielding is stronger, the lower the salt concentration is. The presence of ions in the vicinity of the chain monomers reduces the hydrodynamic coupling. For longer length scale, *i.e.*, for longer chains, the ions effectively decouple different parts of the polyelectrolyte chain such that the friction per monomer becomes a constant value. The chain length N_{FD} for which this transition occurs is dependent on the salt concentration. The higher the salt concentration, *i.e.*, the shorter the Debye length in the system, the more confined is the hydrodynamic shielding effect and the earlier the effective friction becomes constant.

The role of hydrodynamic interactions for the effective friction of the polyelectrolyte can be seen by comparing Fig. 7 to Fig. 8 and shows that the effective friction obtained with the Langevin algorithm neglects long-range hydrodynamic interactions. In Fig. 8a, the initial increase of the effective friction is super linear, but linear scaling is reached for longer chains. The absolute friction value for long chains is

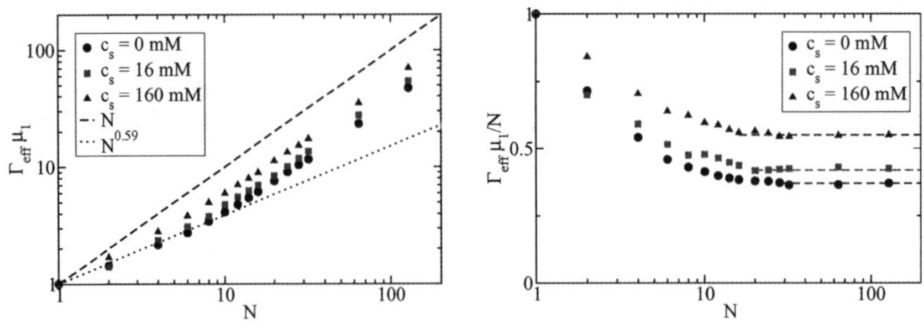

(a) (b)

Fig. 7 (a) The normalized effective friction $\Gamma_{eff}\mu_1$ as a function of chain length N for different salt concentrations c_s using the LB algorithm. Initially, the friction increases as given by the hydrodynamic size of the polyelectrolyte $\Gamma = \sim N^{0.59}$ (dotted line). With the onset of counterion condensation the friction exceeds the value of the bare polyelectrolyte and for long chains becomes linear in N (dashed line). The absolute friction value is increased with the addition of external salt. (b) The normalized effective friction per monomer $\Gamma_{eff}\mu_1/N$ shows an initial decrease with chain length that is stronger the lower the salt concentration is. From a concentration dependent value of N_{FD} onwards, the friction per monomer becomes a constant value that increases with increasing salt concentration (indicated by dashed lines).

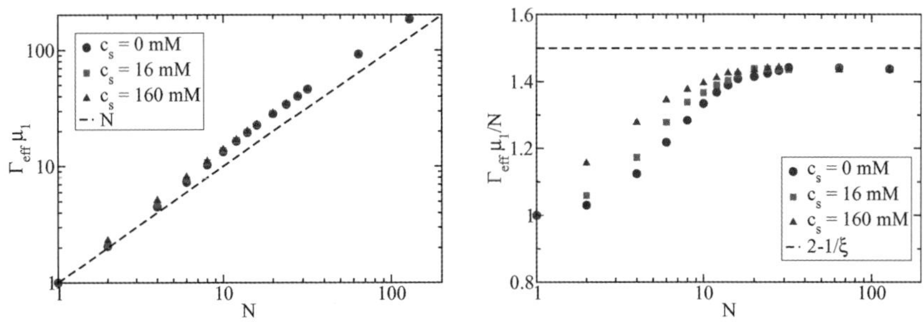

(a) (b)

Fig. 8 (a) The normalized effective friction $\Gamma_{\rm eff}\mu_1$ as a function of chain length N for different salt concentrations c_s without long-range hydrodynamic interactions. Initially, the increase of the effective friction is super linear, but linear scaling (dotted line) is reached for longer chains. For these chains, the absolute friction value is independent of the addition of external salt. (b) The normalized effective friction per monomer $\Gamma_{\rm eff}\mu_1/N$ shows an initial increase with chain length that is stronger, the higher the salt concentration is. For longer chains, a plateau value is reached which is independent of the salt concentration and can be compared to the predicted value based on the counterion condensation theory (dashed line).

independent of the salt concentration. This can be understood by realizing that the total effective friction of a polyelectrolyte in the Langevin algorithm is only based on the local friction parameter Γ_0 and the number of co-moving particles, *i.e.*, the sum of N monomers and $N_{\rm CI}$ condensed counterions: $\Gamma_{\rm eff} = \Gamma_0(N + N_{\rm CI})$. As shown in Fig. 5 the effective charge, and therefore also $N_{\rm CI}$ and $\Gamma_{\rm eff}$, is only influenced by the salt concentration for short and intermediate chains, but not for long chains. Fig. 8b shows the increase of the effective friction per monomer from $\Gamma_{\rm eff}(1) = 1/\mu_1 \approx \Gamma_0$ to a constant value for long chains, which is comparable to the plateau value predicted using counterion condensation theory: $\Gamma_{\rm eff}/N = \Gamma_0(2 - 1/\xi)$.

Fig. 9 schematically illustrates how counterions and salt in the vicinity of polyelectrolyte chains influence the hydrodynamic interactions during electrophoresis. Fig. 9a indicates the regime, where all parts of the chain can interact *via* hydrodynamic interactions. The individual chain segments provide hydrodynamic shielding to each other. During electrophoresis, see Fig. 9, the counterions within the polyelectrolyte limit the range of the hydrodynamic interaction. The hydrodynamic screening length depends on the ion concentration in the vicinity of the chain.

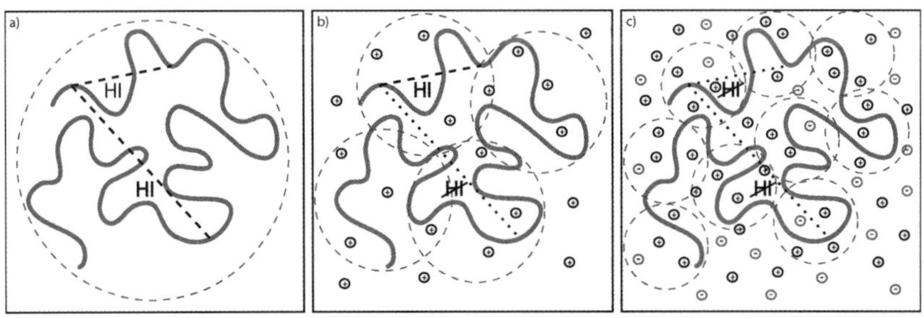

Fig. 9 Illustration of the influence of surrounding ions on to the long-range hydrodynamic interactions between different parts of a polyelectrolyte chain. (a) For an uncharged polymer, the hydrodynamic interactions are unscreened and all chain monomers can interact with each other. (b) The presence of counterions during electrophoresis of polyelectrolytes limits the hydrodynamic interaction. (c) The more salt is added to the system, the higher is the ion density in the vicinity of the chain, which reduces the hydrodynamic interaction range even further, so that most parts of the chain appear to be hydrodynamically decoupled.

This relationship between the ion density and the hydrodynamic screening length was previously suggested by different authors.[32,33]

The connection between electrostatic screening and hydrodynamic screening can be easily motivated by the following reasoning: the Debye length is the length-scale on which the charge of the polyelectrolyte is screened by the surrounding ions. When looking at this object from the outside, the total force exerted by the applied electric field is zero, *i.e.*, no momentum is transferred to the polyelectrolyte–ion complex. Due to momentum conservation, the interaction with the fluid has to result in a vanishing total force.

The counterions that associate with the polyelectrolyte influence the solvent flow around it, effectively cancelling the beneficial shielding effects. When additional salt is added to the system, the like charged salt ions likewise contribute to this effect as shown in Fig. 9c. The higher the ion concentration is, *i.e.*, the shorter the Debye length is, the shorter is the length scale on which different polyelectrolyte monomers can interact hydrodynamically. On a length scale comparable to the Debye length in the system, different parts of the polyelectrolyte become decoupled. For longer chains, the effective friction per segment does not depend on the length of the polyelectrolyte any more. Consequently, the effective friction per monomer becomes independent of the length of the polyelectrolyte chain, as seen in Fig. 7b.

4 Conclusion

We presented a detailed study of the electrophoretic behaviour of flexible polyelectrolyte chains by means of a mesoscopic coarse-grained molecular dynamics model including full hydrodynamic and electrostatic interactions.

The electrophoretic mobility exhibits a characteristic length dependence for short polyelectrolyte chains and a constant length independent value for long chains. We showed that both the shape and the constant long chain value, depend on the salt concentration of the solution if hydrodynamic interactions were properly accounted for. The long chain mobility was then found to be decreasing with increasing salt concentration, in agreement with experimental observations.

Direct measurements of the effective charge by two independent estimators showed that the dynamic effective charge for long chains is independent of the salt concentration. We therefore conclude that the dependence of the long chain mobility on the salt concentration is not due to a reduced effective charge but has to be attributed to a change in the effective friction. On the other hand, the effective charge for short and intermediate chains is influenced by the salt concentration which explains the different behaviour of the electrophoretic mobility in this regime.

We note that a static estimate of the effective charge shows a dependence on the salt concentration leading to different charge estimates for finite salt concentrations: a static and a dynamic effective charge.

We showed that the effective friction of the polyelectrolyte is strongly influenced by the presence of ions in the solution. For short chains and low salt concentrations no counterions are associated with the polyelectrolyte chain and the effective friction is given by the hydrodynamic radius. The presence of ions in the vicinity of the chain reduces the hydrodynamic shielding between the chain monomers and leads to an increased friction. The longer the chains and the higher the salt concentration, the more the shielding is reduced, until for chains longer than a specific length N_{FD} the friction becomes linear with chain length. In this regime, different parts of the chain are effectively decoupled.

From this, the specific behaviour of the electrophoretic mobility as observed in experiments can be understood: the hydrodynamic shielding between the monomers allows for an initial increase in the mobility. The onset of counterion condensation counteracts this increase as it reduces the effective charge and at the same time increases the effective friction. For long chains, charge and friction both become linearly dependent on chain length which therefore results in the well-known constant mobility, or free-draining limit.

The presence of salt reduces the length scale on which the chain monomers can interact hydrodynamically. This reduces the initial hydrodynamic shielding and therefore suppresses the mobility maximum. At the same time, the total friction is increased leading to a reduced long-chain mobility. Salt concentrations exceeding the ones in this simulation can cause a total decoupling of the individual chain monomers, which can then be simulated without hydrodynamic interactions. We expect a length-dependence of the mobility as shown in Fig. 3 and an effective friction per monomer, *cf.* Fig. 7b, that does not depend on N.

The study shows that chemical details and fluid structure can be neglected, and a higher level of abstraction yields an accurate description of the physics of the problem, as long as electrostatic and hydrodynamic interactions between all entities in the system, *i.e.*, the polyelectrolyte, dissociated counterions, additional salt and the solvent, are properly accounted for. In this way we were able to model a process bridging the single molecule regime of a few nm up to macromolecules with contour lengths of more than 100 nm, a regime currently not accessible to atomistic simulations.

Acknowledgements

Funds from the Volkswagen foundation, the DAAD, and DFG under the TR6 are gratefully acknowledged. All simulations were carried out on the compute cluster of the Center for Scientific Computing (CSC) at Goethe University Frankfurt/Main.

References

1 *Capillary Electrophoresis in Analytical Biotechnology*, ed. P. G. Righetti, CRC Press, Boca Raton, 1996.
2 V. Dolnik, *Electrophoresis*, 2006, **27**, 126–141.
3 H. Cottet, C. Simo, W. Vayaboury and A. Cifuentes, *J. Chromatogr., A*, 2005, **1068**, 59–73.
4 H. Cottet and P. Gareil, in *CE from small ions to Macromolecules, in the Series Methods in Molecular Biology, Molecular Medicine and Biotechnology*, ed. P. Schmitt-Kopplin and J. M. Walker, Humana Press, NJ, USA, 2007, ch. Separation of synthetic (co)polymers by capillary electrophoresis techniques.
5 J.-L. Barrat and J.-F. Joanny, *Adv. Chem. Phys.*, 1996, **94**, 1–66.
6 M. Muthukumar, *Electrophoresis*, 1996, **17**, 1167–1172.
7 A. R. Volkel and J. Noolandi, *J. Chem. Phys.*, 1995, **102**, 5506–5511.
8 U. Mohanty and N. C. Stellwagen, *Biopolymers*, 1999, **49**, 209–214.
9 G. W. Slater, C. Holm, M. V. Chubynsky, H. W. de Haan, A. Dubé, K. Grass, O. A. Hickey, C. Kingsburry, D. Sean, T. N. Shendruk and L. Zhan, *Electrophoresis*, 2009, **30**, 792–818.
10 K. Grass, U. Böhme, U. Scheler, H. Cottet and C. Holm, *Phys. Rev. Lett.*, 2008, **100**, 096104.
11 K. Grass and C. Holm, *J. Phys.: Condens. Matter*, 2008, **20**, 494217.
12 K. Grass and C. Holm, *Soft Matter*, 2009, **5**, 2079–2092.
13 H. J. Limbach, A. Arnold, B. A. Mann and C. Holm, *Comput. Phys. Commun.*, 2006, **174**, 704–727.
14 T. Soddemann, B. Dünweg and K. Kremer, *Eur. Phys. J. E*, 2001, **6**, 409.
15 J. D. Weeks, D. Chandler and H. C. Andersen, *J. Chem. Phys.*, 1971, **54**, 5237.
16 M. Deserno and C. Holm, *J. Chem. Phys.*, 1998, **109**, 7678.
17 G. R. McNamara and G. Zanetti, *Phys. Rev. Lett.*, 1988, **61**, 2332–2335.
18 P. Ahlrichs and B. Dünweg, *J. Chem. Phys.*, 1999, **111**, 8225–8239.
19 U. Wolff, *Comput. Phys. Commun.*, 2004, **156**, 143–153.
20 D. A. Hoagland, E. Arvanitidou and C. Welch, *Macromolecules*, 1999, **32**, 6180–6190.
21 G. Manning, *J. Chem. Phys.*, 1969, **51**, 924–933.
22 F. Oosawa, *Polyelectrolytes*, Marcel Dekker, New York, 1971.
23 P. Wette, H. Schöpe and T. Palberg, *J. Chem. Phys.*, 2002, **116**, 10981.
24 V. Lobaskin, B. Dünweg and C. Holm, *J. Phys.: Condens. Matter*, 2004, **16**, S4063–S4073.
25 H. J. Limbach and C. Holm, *J. Chem. Phys.*, 2001, **114**, 9674–9682.

26 D. Antypov and C. Holm, *Phys. Rev. Lett.*, 2006, **96**, 088302.
27 D. Antypov and C. Holm, *Macromolecules*, 2007, **40**, 731–738.
28 L. Belloni, M. Drifford and P. Turq, *Chem. Phys.*, 1984, **83**, 147.
29 L. Belloni, *Colloids Surf.*, 1998, **A 140**, 227.
30 M. Deserno, C. Holm and S. May, *Macromolecules*, 2000, **33**, 199–206.
31 M. A. Moore and A. J. Bray, *J. Phys. A: Math. Gen.*, 1978, **11**, 1353–1359.
32 D. Long and A. Ajdari, *Eur. Phys. J. E*, 2001, **4**, 29–32.
33 M. Tanaka and A. Y. Grosberg, *J. Chem. Phys.*, 2001, **115**, 567–574.

Kinetic Monte Carlo simulations of flow-induced nucleation in polymer melts

Richard S. Graham[*a] **and Peter D. Olmsted**[b]

Received 26th January 2009, Accepted 27th February 2009
First published as an Advance Article on the web 15th August 2009
DOI: 10.1039/b901606f

We derive a kinetic Monte Carlo algorithm to simulate flow-induced nucleation in polymer melts. The crystallisation kinetics are modified by both stretching and orientation of the amorphous chains under flow, which is modelled by a recent non-linear tube theory. Rotation of the crystallites under flow is modelled by a simultaneous Brownian dynamics simulation. Our kinetic Monte Carlo approach is highly efficient at simulating nucleation and is tractable even at low under-cooling. The simulations predict enhanced nucleation under both transient and steady state shear. Furthermore the model predicts the growth of shish-like elongated nuclei for sufficiently fast flows, which grow by a purely kinetic mechanism.

1 Introduction

Semi-crystalline polymers make up a very significant fraction of the world's production of synthetic polymers. Unlike simple molecules, the connectivity of polymer molecules causes them to crystallise into a composite structure of crystalline and amorphous regions. The proportion of amorphous and crystalline material, along with the arrangement and orientation of the crystals, is collectively known as the morphology. The crystal morphology strongly influences strength, toughness, permeability, surface texture, transparency and almost any other property of practical interest. Furthermore, polymer crystallisation is radically influenced by the types of flow that are ubiquitous in polymer processing. Such flows drastically enhance the rate at which polymers crystallise and have a profound effect on their morphology. Flow distorts the configuration of polymer chains and, it is believed, this distortion breaks down the kinetic barriers to crystallisation and directs the resulting morphology. However, the molecular mechanisms underlying these processes have yet to be established. As a result, there is no predictive molecular model of flow-induced crystallisation (FIC). The impact of such a model on the polymer industry would be considerable since it would allow control over the crystalline properties of polymer products by simply tailoring their processing conditions.

It has long been known that flow can radically enhance the rate of polymer nucleation and can cause the formation of highly aligned, elongated crystals, known as shish-kebabs.[1] There have been numerous recent experimental studies, focusing mainly on entangled polymers. These have quantified the effect of flow on nucleation,[2] have illuminated details of shish-kebab formation[3–5] and have highlighted the role of blend concentration,[6] molecular architecture[7] and molecular relaxation

[a]*School of Mathematical Sciences, University of Nottingham, Nottingham, NG7 2RD, UK. E-mail: richard.graham@nottingham.ac.uk*
[b]*School of Physics and Astronomy, University of Leeds, Leeds, LS2 9JT, UK. E-mail: p.d.olmsted@leeds.ac.uk*

time.[8] Almost all of these experiments have been performed at low undercooling, where crystallisation in the absence of flow is too slow to be measured, as often the most pronounced flow-induced effects are seen in this temperature regime.

Existing theoretical approaches for FIC fall broadly into two extremes of coarse-graining, highly simplified semi-empirical models[9–13] and detailed simulations at the level of molecular dynamics[14,15] or simulations with slightly higher coarse-graining.[16–19] Simplified coarse-grained models follow a semi-empirical approach,[9–13] containing some molecular elements but also requiring *ad hoc* arguments to describe experimental phenomena. Such approaches also require numerous pre-averaging or closure approximations, which neglect the delicate coupling between the underlying stochastic processes. Finally, these models invariably use oversimplified flow models that cannot predict the full range of molecular deformation in a melt. The accuracy of the constitutive model is tacitly assumed to be ensured by simple parameter fitting to rheological measurements. However, this approach cannot accurately model the melt's high molecular weight tail, which is known to dominate the FIC behaviour.[6–8] At the opposite extreme of coarse-graining, molecular dynamics (MD) simulations[14,15] provide a more rigorous approach to crystallisation. While these MD simulations have provided much useful information on the growth phase, they can only access a limited range of timescales and so are not well-suited to model nucleation. Faster alternatives, with higher coarse-graining, include kinetic Monte Carlo[16–18] and Langevin dynamics simulations.[19] Nevertheless, these simulations still have difficulty modelling primary nucleation, forcing them to focus on high degrees of undercooling. In contrast, recent experiments show that insight into low undercooling is essential to both a fundamental and practical understanding of FIC,[3,5,7,8] as the most pronounced flow-induced effects occur at these temperatures. Finally, the algorithms for predicting the effect of flow on the non-crystalline chain configurations used in these simulations has not been verified against relevant measurements such as mechanical stresses[20] and neutron scattering.[21] It appears that, to progress, an intermediate level of coarse-graining, between semi-empirical approaches and detailed simulations, is required.

In this paper we develop a coarse-grained simulation algorithm for flow-induced nucleation in polymers. The algorithm is intended to capture the dominant physical processes while remaining tractable at all relevant temperatures, including low undercooling. Simulating nucleation is problematic, especially at low undercooling, because it is a rare event. Much simulation time can be spent resolving the evolution of small nuclei, far from the critical size. In contrast to previous Monte Carlo approaches[16–18] we address this issue by using a kinetic Monte Carlo algorithm with variable step size,[22] known in some fields as the Gillespie algorithm.[23] Although, this method has successfully been applied to quiescent crystal growth in dilute polymers,[24] its intrinsically adaptive nature is especially suited to nucleation; large time-steps are automatically taken when the nucleus is small whereas larger nuclei receive more time resolution. In addition we use a recently derived molecular flow model,[25] which has been extensively validated against data from non-crystalline polymer fluids under strong flow for both mechanical stress[20,25–27] and small angle neutron scattering.[21,28–30]

2 Simulation algorithm

Our coarse-grained simulation for flow induced crystal nucleation in polymer melts aims to describe how bulk flow affects nucleation density, nucleus aspect ratio and nucleus orientation. The algorithm tracks the time evolution of three processes. The first is stretching and orientation of the amorphous chains due to flow. This is calculated using the Graham-Likhtman and Milner-McLeish (GLaMM) model.[25] The second part is the attachment and detachment of chain segments to the nucleus, which were modelled by a kinetic Monte Carlo simulation. The final part is rotation of the crystal nucleus by the bulk flow, and this is tracked by Brownian dynamics simulation.

The amorphous chain stretching is computed first for the whole simulation time, through a single deterministic run of the GLaMM model, the results of which are stored to provide modified kinetics in the subsequent nucleation simulations. The Monte Carlo simulations and the Brownian dynamics rotation simulations are then run concurrently, with multiple runs being used to resolve the statistics. The effect of crystallisation on the amorphous chain dynamics is neglected and, thus, information from the nucleation process is not fed-back into the model for the amorphous chain dynamics. While this may seem to be a crude approximation, experiments on entangled star polymers are beginning to indicate that star arm relaxation is quite similar to linears under non-linear flow,[31,32] justifying somewhat this assumption for chains participating in a single crystallite. Also essential to the model derivation is the separation of timescales between the local crystallisation dynamics τ_0 and chain dynamics at the entanglement lengthscale τ_e. We denote this ratio of the flow and crystallisation timescales $\mathcal{S} = \tau_e/\tau_0$. Local crystallisation is taken to be fast compared to the Rouse time of an entanglement segment ($\tau_0 << \tau_e$). Also, although the flow rate $\dot{\gamma}$ is non-linear, it will be small compared to the entanglement relaxation time ($\dot{\gamma}\tau_e < 1$).

2.1 The GLaMM model

In the GLaMM model a linear chain is divided into Z entanglement segments, each containing N_e Kuhn steps ("monomers") of length b. The GLaMM model computes the following tube tangent correlation function,

$$\mathbf{f}(s, s') = \left\langle \frac{\partial \mathbf{R}(s)}{\partial s} \frac{\partial \mathbf{R}(s')}{\partial s'} \right\rangle, \tag{1}$$

where $\mathbf{R}(s)$ is the space-curve describing the tube shape, $s = 0\ldots Z$ is a continuous variable, labelling tube segments and all lengths are measured in terms of $\sqrt{N_e}b$, which is the tube diameter. The numerical solution of the GLaMM model provides $\mathbf{f}(s,s)$ for each entanglement segment. This tangent vector correlation $\mathbf{f}(s,s)$ is effectively the local end-to-end vector correlation of each tube segment $\mathbf{f} \approx \left\langle \frac{\Delta \mathbf{R}}{\Delta s} \frac{\Delta \mathbf{R}}{\Delta s} \right\rangle = \langle \mathbf{rr} \rangle$ since $\mathbf{r} = \Delta \mathbf{R}$ and $\Delta s = 1$. By this method the GLaMM model provides a set of microscopic strains for each subchain through $\mathbf{f}_i$, where i labels the tube segment number, running from 0 to Z (see Fig. 1). Finite chain extensibility is included by replacing the Gaussian spring force with Cohen's approximation[33] for the non-linear spring force. We use $N_e = 100$ throughout this paper.

2.1.1 Segment free energy and orientation. The deformation of the amorphous chains, contained in $\mathbf{f}$, modifies the nucleation kinetics in two ways: the increase in monomer free energy upon stretching reduces the entropic penalty for crystallisation; and monomer alignment modifies the probability of compatible alignment

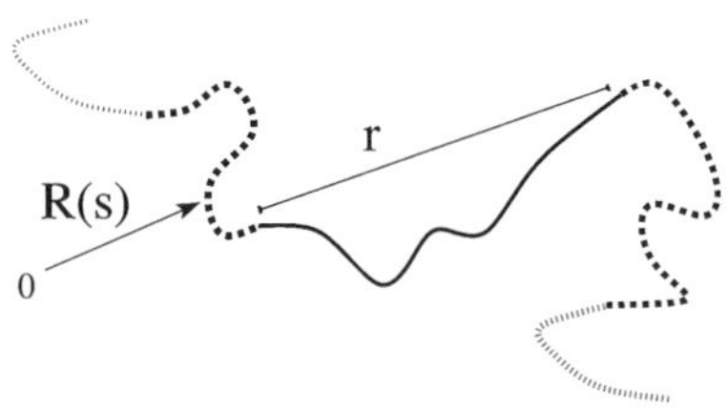

Fig. 1 The end-to-end vector of a tube segment, which defines the microscopic strain $\langle \mathbf{rr} \rangle$.

with the nucleus. In appendix A we show that the change in elastic free energy, $\Delta F^{\rm el}$, and the distribution of monomer orientations, $w(\theta)$, for a chain subjected to the ensemble average constraint $\mathbf{f} = \langle \mathbf{rr} \rangle$ can be accurately approximated by essentially the expressions for a fixed end-to-end vector with $|\mathbf{r}|$ replaced by $\sqrt{\mathrm{Tr}\,\mathbf{f}}$,

$$\Delta F^{\rm el}(\mathbf{f}) = \frac{1}{2}\mathrm{Tr}\,\mathbf{f} - \frac{1}{2}\mathrm{Tr}\,\ln\mathbf{f} - N_{\rm e}\ln\left(1 - \frac{\mathrm{Tr}\,\mathbf{f}}{N_{\rm e}}\right) - \Gamma, \tag{2}$$

$$w(\theta) = \frac{\mathscr{L}^{-1}(\sqrt{\mathrm{Tr}\,\mathbf{f}}/N_{\rm e})}{4\pi\sinh(\mathscr{L}^{-1}\sqrt{\mathrm{Tr}\,\mathbf{f}}/N_{\rm e}))}\cosh\left(\mathscr{L}^{-1}\left(\frac{\sqrt{\mathrm{Tr}\,\mathbf{f}}}{N_{\rm e}}\right)\cos\theta\right), \tag{3}$$

where $\mathscr{L}^{-1}$ is the inverse Langevin function, Γ is a constant chosen such that $\Delta F^{\rm el} = 0$ for an equilibrium coil and θ is the angle between the monomer and the principle axis of $\mathbf{f}$. We use Cohen's approximation for the inverse Langevin function,[33]

$$\mathscr{L}^{-1}(x) \approx x\frac{3 - x^2}{1 - x^2}. \tag{4}$$

2.2 Description of the nucleus

In our coarse-grained simulations we take the minimal nucleus description required to model anisotropic nucleation. The nucleus comprises of a collection of crystallised "monomers" (Kuhn steps) arranged in stems. Each stem is formed from a single chain and the simulation tracks the monomer number of the top and bottom crystallised monomer in each stem ($n_{\rm top}$, $n_{\rm bot}$). This defines the total number of stems $N_{\rm s}$ and the total number of monomers $N_{\rm T}$. We assume the nucleus to be spheroidal and use $N_{\rm T}$ and $N_{\rm s}$ to provide the two radii. We take the crystallised Kuhn segments dimensions to be $b_{\rm l} \times b_{\rm w} \times b_{\rm w}$, with the $b_{\rm l}$ length always parallel to the spheroid polar radius L (see Fig. 2). Thus equatorial radius W is given by

$$W = b_{\rm w}\frac{1}{\sqrt{\pi}}\sqrt{N_{\rm s}}, \tag{5}$$

since the cross sectional area about W is πW^2 and this area contains $N_{\rm s}$ stems. The volume is determined by the total number of monomers

$$V = {b_{\rm w}}^2 b_{\rm l} N_{\rm T}. \tag{6}$$

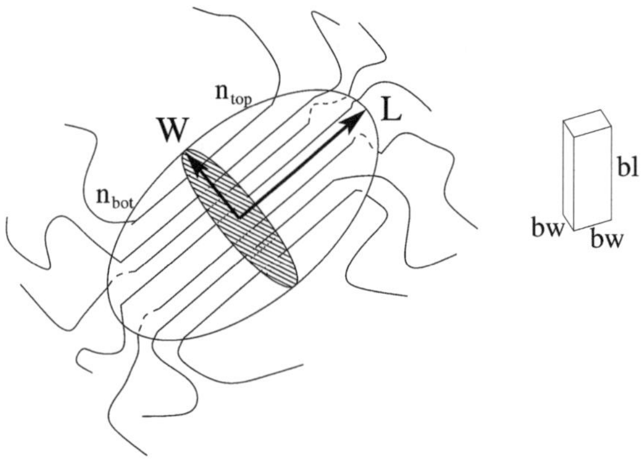

Fig. 2 Definition of the spheroid nucleus, along with the dimensions of a single Kuhn segment.

The polar radius L is obtained from the total volume ($V = 4/3\pi L W^2$), leading to

$$L = \frac{3}{4} b_l \frac{N_T}{N_s}. \tag{7}$$

Later on we will require the crystal aspect ratio $\rho = L/W$,

$$\rho = a_r \frac{N_T}{N_s^{3/2}}, \tag{8}$$

where $a_r = \frac{3\sqrt{\pi}}{4}\frac{b_l}{b_w}$ is a dimensionless prefactor that controls the aspect ratio for a given N_s and N_T. We also simulate the nucleus orientation through $\hat{\mathbf{v}}$, a unit vector parallel to the nucleus polar radius.

2.2.1 Surface area. The surface area for a spheroid is computed as follows. The ellipticity ε for prolate and oblate spheroids in terms of a_r is

$$\varepsilon_{\text{prolate}} = \sqrt{1 - \frac{1}{a_r^2}\frac{N_s^3}{N_T^2}}, \quad \varepsilon_{\text{oblate}} = \sqrt{1 - a_r^2 \frac{N_T^2}{N_s^3}}. \tag{9}$$

For a prolate spheroid the surface area, S is then given by

$$\begin{aligned} S &= \pi\left(2W^2 + 2\tfrac{LW}{\varepsilon}\arcsin\varepsilon\right) \\ &= b_w^2\left(2N_s + 2a_r\frac{N_T}{\sqrt{N_s}}\frac{1}{\varepsilon}\arcsin\varepsilon\right), \end{aligned} \tag{10}$$

and for an oblate spheroid

$$\begin{aligned} S &= \pi\left(2W^2 + \tfrac{L^2}{\varepsilon}\ln\left(\frac{1+\varepsilon}{1-\varepsilon}\right)\right) \\ &= b_w^2\left(2N_s + a_r^2\frac{N_T^2}{N_s^2}\frac{1}{\varepsilon}\ln\left(\frac{1+\varepsilon}{1-\varepsilon}\right)\right). \end{aligned} \tag{11}$$

2.2.2 Nucleus free energy. As in classical nucleation theory the nucleus free energy is a balance of the free energy of crystallisation proportional to the nucleus volume with a free energy penalty proportional to the spheroid surface area, $\mathcal{F}^*_{\text{nuc}}(N_T, N_s) = -\varepsilon^*_B V + \mu^*_S S$, where ε^*_B and μ^*_S are the dimensional bulk free energy per unit volume and surface free energies per unit area, respectively. This can be rewritten as

$$\mathcal{F}_{\text{nuc}}(N_T, N_s) = -\varepsilon_B N_T + \mu_S \tilde{S}, \tag{12}$$

where $\mathcal{F}_{\text{nuc}} = \frac{\mathcal{F}^*_{\text{nuc}}}{k_B T}$, $\varepsilon_B = \frac{b_l b_w^2}{k_B T}\varepsilon^*_B$, $\mu_S = \frac{b_w^2}{k_B T}\mu^*_S$ and $S = b_w^2 \tilde{S}$. These dimensionless parameters control the free energy landscape of the nucleation process. For the remainder of this paper all free energies will be expressed in units of $k_B T$. Note that, from the definitions above, the free energy is determined by just the number of stems, total number of monomers and the aspect ratio parameter a_r.

2.3 Kinetic Monte Carlo moves

In the kinetic Monte Carlo simulations two types of basic moves are possible, stem addition and stem lengthening, both of which add a single monomer (see Fig. 3) and have a corresponding reverse move.

2.3.1 Stem lengthening. Stem lengthening involves attaching a new monomer to the top or bottom of the crystal and so increasing the length of an existing stem. The

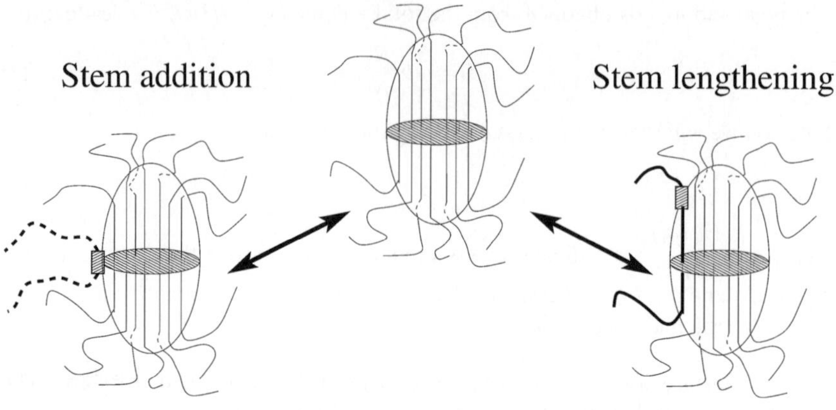

Fig. 3 The basic Monte Carlo moves of stem addition and stem lengthening.

number of monomers increases, but the number of stems remains fixed. Thus, the change in free energy upon adding (+) and removing (−) a monomer from an existing stem is

$$\Delta\mathcal{F}^{+}_{\text{length}} = \mathcal{F}(N_{\text{T}} + 1,N_{\text{s}}) - \mathcal{F}(N_{\text{T}},N_{\text{s}}), \tag{13}$$

$$\Delta\mathcal{F}^{-}_{\text{length}} = \mathcal{F}(N_{\text{T}},N_{\text{s}}) - \mathcal{F}(N_{\text{T}} - 1,N_{\text{s}}). \tag{14}$$

The corresponding move rates k are

$$k^{+}_{\text{length}} = \frac{1}{\tau_0}\min\left(1, \exp(-\Delta\mathcal{F}^{+}_{\text{length}})\right), \tag{15}$$

$$k^{-}_{\text{length}} = \frac{1}{\tau_0}\min\left(1, \exp(-\Delta\mathcal{F}^{-}_{\text{length}})\right), \tag{16}$$

where τ_0 is the timescale of a basic crystallisation step, which can be thought of as an activation energy barrier for hopping between crystal states. This timescale is assumed to be constant, regardless of the shape or size of the nucleus. Each stem contributes four moves to the total rate sum, addition and removal at both the top and bottom of the stem. Thus the contribution from stem lengthening, per stem, to the total sum over rates is

$$K_{\text{length}} = \sum_{i=1}^{N_{\text{s}}}\left(\left(k^{+\text{top}}_{\text{length}}\right)_i + \left(k^{-\text{top}}_{\text{length}}\right)_i + \left(k^{+\text{bot}}_{\text{length}}\right)_i + \left(k^{-\text{bot}}_{\text{length}}\right)_i\right). \tag{17}$$

If a stem contains only one monomer then the removal move is counted as stem removal and so ${k_{\text{length}}}^{-\text{top}}$ and ${k_{\text{length}}}^{-\text{bot}}$ are both set to zero. On a successful lengthening move the added monomer is always the next monomer along the chain that forms the stem as the stem "zips-up" the chain. Thus after a lengthening move the appropriate monomer number at the top or bottom of the stem (n_{top} or n_{bot}) is incremented up or down.

2.3.2 Stem addition. Stem addition involves attaching a new stem to the side of the nucleus. Both the number of stems and the number of monomers increase by one, so the change in free energy is

$$\Delta\mathscr{F}^{+}_{\text{stem}} = \mathscr{F}(N_T + 1, N_s + 1) - \mathscr{F}(N_T, N_s), \tag{18}$$

$$\Delta\mathscr{F}^{-}_{\text{stem}} = \mathscr{F}(N_T, N_s) - \mathscr{F}(N_T - 1, N_s - 1). \tag{19}$$

Similarly to stem lengthening the basic move rates are

$$k^{+}_{\text{stem}} = \frac{1}{\tau_0}\min\left(1, \exp\left(-\Delta\mathscr{F}^{+}_{\text{stem}}\right)\right), \tag{20}$$

$$k^{-}_{\text{stem}} = \frac{1}{\tau_0}\min\left(1, \exp\left(-\Delta\mathscr{F}^{-}_{\text{stem}}\right)\right), \tag{21}$$

For quiescent simulations the monomer number of the newly attached monomer can be chosen randomly from any point along the attached chain.

The rate of stem addition will be proportional to the nucleus surface area available to attach new stems. As shown in Fig. 4 there will be substantial excluded volume from the dangling amorphous chains around the nucleus, which prevents stem addition across much of the nucleus. We assume that stem addition is restricted to a band around the spheroid equator, thus the area available for stem addition f_{add} depends on the equatorial circumference, and is given by

$$f_{\text{add}}(N_s) = 2a\sqrt{\pi}\sqrt{N_s}, \tag{22}$$

where a is a constant of proportionality. A value of $a = 0.4$ gives spherical nuclei in the quiescent limit. The stem removal rate is set to obey detailed balance. Therefore only stems containing a single monomer are candidates for a stem removal move and these have a basic removal rate, k^{-}_{stem}. This rate is multiplied by a factor of $f_{\text{add}}(N_s - 1)/N_s$, which is the probability that the given stem is at the surface of the nucleus.

2.3.3 Effect of flow. As discussed above, flow modifies the crystallisation kinetics because chain stretching reduces the entropic penalty for crystallisation and chain orientation changes the probability of compatible alignment between the nucleus and any attaching monomers. To account for chain stretching, the change in chain free energy per monomer on stretching is subtracted from the free energy change of a Monte Carlo move. We deal with orientation with a similar approach to ref. 12. We assume that, in order to attach to the nucleus, the candidate monomer must be oriented within a solid tolerance angle Ω of the nucleus orientation. All

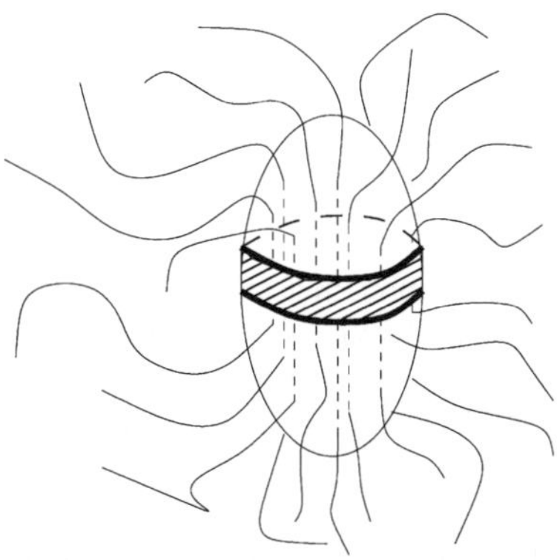

Fig. 4 Available surface area for stem addition.

monomers outside this angle are unable to attach. If Ω is small then the fraction of monomers within this angle is $w(\theta)\Omega$, where θ is the angle between the nucleus polar radius and the sub-chain principle strain axis.

Flow causes a linear chain to stretch and align by different amounts at different points along its contour. Thus, even for a monodisperse melt we must deal with a distribution of segment types, with varying degrees of stretch. We take a melt containing S different species, with species i having volume fraction ϕ_i. For a monodisperse polymer of Z entanglements there are Z species, each with concentration $\phi_i = 1/Z$. The GLaMM model provides $\mathbf{f} = \langle \mathbf{rr} \rangle$ for each species, from which we compute the change in elastic free energy $\Delta F^{\rm el}{}_i$ and the fraction of monomers in species i that are compatibly aligned with the nucleus Θ_i, using eqn (2) and (3) for each species i.

With these modifications the rate of stem attachment for species i becomes

$$\left(k^{+}_{\rm stem}\right)_i = \frac{1}{\tau_0}\phi_i\Theta_i \min\left(1, \exp\left[-\left(\Delta\mathscr{F}^{+}_{\rm stem} - \Delta\mathscr{F}^{\rm el}_i/N_{\rm e}\right)\right]\right), \qquad (23)$$

where $\Delta F^{\rm el}{}_i$ is the elastic energy of species i, $\Theta_i = 4\pi w(\theta_i)$ to give agreement with eqn (20) in the quiescent limit, and a constant of $\ln(\Omega/4\pi)$ has been added to $\varepsilon_{\rm B}$.

Similarly, the rate of removing stem j becomes

$$\left(k^{-}_{\rm stem}\right)_j = \begin{cases} \dfrac{1}{\tau_0}\min\left(1, \exp\left[-\left(\Delta\mathscr{F}^{-}_{\rm stem} + \Delta F^{\rm el}_j/N_{\rm e}\right)\right]\right), & \text{If stem contains 1 monomer} \\ 0 & \text{Otherwise} \end{cases} \qquad (24)$$

where $\Delta F^{\rm el}{}_j$ is the elastic energy of the monomer species in stem j. Note that there are no concentration terms for the removal rate.

The rate of lengthening moves at the top of stem j is given by

$$\left(k^{+{\rm top}}_{\rm length}\right)_j = \frac{1}{\tau_0}\Theta_j \min\left(1, \exp\left[-\left(\Delta\mathscr{F}^{+}_{\rm length} - \Delta F^{\rm el}_j/N_{\rm e}\right)\right]\right), \qquad (25)$$

where $\Delta F^{\rm el}{}_j$ and Θ_j are calculated for the species of the next monomer to be added along the chain at the top of stem k, which is $n_{\rm top} + 1$. Note that for stem lengthening only the next monomer along the chain forming the stem can crystallise. The concentration of this monomer at the nucleus surface where the lengthening event occurs is taken to be unity, hence the species concentration ϕ_j does not appear in eqn (25). The effective concentration due to relative alignment between the crystal and the chain is included since correct alignment with of the monomer still required. The lengthening rate from the bottom of the stem is calculated similarly. The removal rate from the top of stem j is given by

$$\left(k^{-{\rm top}}_{\rm length}\right)_j = \begin{cases} 0 & \text{If stem contains 1 monomer} \\ \dfrac{1}{\tau_0}\min\left(1, \exp\left[-\left(\Delta\mathscr{F}^{-}_{\rm length} + \Delta F^{\rm el}_j/N_{\rm e}\right)\right]\right) & \text{Otherwise} \end{cases} \qquad (26)$$

where $\Delta F^{\rm el}$ is calculated for the monomer at the top of stem j. Note that, as before, the concentrations, ϕ_j and Θ_j play no part in the detachment moves. Also, if the stem has only one remaining monomer then the move becomes a stem removal move and is accounted for in eqn (24). The removal rate from the bottom of the stem is calculated similarly.

2.3.4 Sum over all move rates. The variable time step kinetic Monte Carlo algorithm requires a sum over all possible moves.[22,23] This sum can be written as

$$K_{\mathrm{Total}} = f_{\mathrm{add}}(N_s) \sum_{i=1}^{S} (k^{+}_{\mathrm{stem}})_i + \frac{f_{\mathrm{add}}(N_s - 1)}{N_s} \sum_{j=1}^{N_s} (k^{-}_{\mathrm{stem}})_j$$

$$+ \sum_{j=1}^{N_s} \left(\left(k^{+\mathrm{top}}_{\mathrm{length}}\right)_j + \left(k^{-\mathrm{top}}_{\mathrm{length}}\right)_j + \left(k^{+\mathrm{bot}}_{\mathrm{length}}\right)_j + \left(k^{-}_{\mathrm{length}}\right)_j \right). \tag{27}$$

The first term is a sum over all species and corresponds to stem addition; the second is a sum over all stems and accounts for stem removal; the final sum is over all stems and accounts for lengthening and shortening of each stem.

2.4 Kinetic Monte Carlo time-stepping

With the sum over all possible moves K_{Total} computed from eqn (27), a kinetic Monte Carlo timestep can be taken. A single move is performed at random, with the probability of each move being picked being weighted by its rate. That is, the probability of move i is,

$$P_i = \frac{k_i}{K_{\mathrm{Total}}}. \tag{28}$$

If the move is a stem attachment move, the species and entanglement segment number are known, but the exact monomer to be attached is not and the attached monomer is chosen randomly and uniformly from all monomers in the given entanglement segment. Finally, time is incremented by a stochastically determined interval given by

$$\Delta t = -\frac{\ln\zeta}{K_{\mathrm{Total}}}, \tag{29}$$

where ζ is chosen uniformly on 0, 1.[23] As time proceeds the amorphous chain configuration changes, as pre-calculated by the GLaMM model. For a timestep of Δt, the GLaMM configuration is incremented forward an interval of $\Delta t/S$. Then the values of ΔF_i^{el} and Θ_i are updated, and the change in free energy of the moves recalculated.

2.4.1 Dummy Monte Carlo move. For particularly small nuclei with high energy barriers the sum over possible moves can be very small. In this case the chain deformation may be evolving faster than the Monte Carlo time step and so will need to be updated more frequently. Although kinetic Monte Carlo algorithms are available for time-dependent barriers[22] we use a simple solution at this stage. We allow a dummy Monte Carlo move with rate k_0. If this dummy move is selected the crystal configuration is not changed, but time is still incremented and then the tube configuration can be updated. We choose k_0 to be similar to the rate at which the tube configuration evolves, namely $1/\tau_e$, to ensure that the tube configuration is updated sufficiently often. For large enough k_0 the simulation results are independent of k_0 and all results presented herein are converged with respect to increases in k_0.

2.5 Brownian dynamics rotation algorithm

After the kinetic Monte Carlo timestep, the rotational dynamics of the nucleus under flow are iterated over Δt, using a Brownian dynamics algorithm.

2.5.1 Convection term. The rotation rate of a dilute rigid spheroid in a Newtonian liquid is given by Leal and Hinch[34] using the Jeffery algorithm.[35] A volume conserving deformation, with a velocity gradient tensor κ, can be split into a symmetric and antisymmetric part

$$\boldsymbol{\kappa} = \mathbf{E} + \boldsymbol{\Omega},\ \mathbf{E}^{\mathrm{T}} = \mathbf{E},\ \boldsymbol{\Omega}^{\mathrm{T}} = -\boldsymbol{\Omega}. \tag{30}$$

The bulk flow causes $\hat{\mathbf{v}}$, a unit vector parallel to the nucleus polar radius, to evolve according to

$$\frac{\mathrm{d}\hat{\mathbf{v}}}{\mathrm{d}t} = \boldsymbol{\Omega}\cdot\hat{\mathbf{v}} + G\left[\mathbf{E}\cdot\hat{\mathbf{v}} - \hat{\mathbf{v}}\left(\hat{\mathbf{v}}\cdot\mathbf{E}\cdot\hat{\mathbf{v}}\right)\right], \tag{31}$$

where G is the shape factor. For a spheroid of aspect ratio ρ this is given by

$$G = \frac{\rho^2 - 1}{\rho^2 + 1}. \tag{32}$$

Thus for a timestep Δt the convection term is computed first from $\hat{\mathbf{v}}(t + \Delta t) = \hat{\mathbf{v}} + \Delta t \frac{\mathrm{d}\hat{\mathbf{v}}}{\mathrm{d}t}(t)$.

2.5.2 Brownian rotation step. For the Brownian diffusion step, an axis of rotation is chosen by generating a random unit vector, $\hat{\mathbf{u}}$. Then a random angle ϕ is generated from a Gaussian distribution with the following moments,

$$\langle\phi\rangle = 0 \quad \langle\phi^2\rangle = \frac{6\Delta t}{\tau_{\mathrm{rot}}}, \tag{33}$$

where τ_{rot} is the crystal rotation relaxation time, related to the rotational diffusion constant by $D_{\mathrm{rot}} = 1/\tau_{\mathrm{rot}}$. The crystal vector is then rotated through an angle ϕ around the axis $\hat{\mathbf{u}}$, using the rotation formula

$$\hat{\mathbf{v}}(t + \Delta t) = \hat{\mathbf{v}}\cos\phi + \hat{\mathbf{u}}(\hat{\mathbf{u}}\cdot\hat{\mathbf{v}})(1 - \cos\phi) + (\hat{\mathbf{v}} \times \hat{\mathbf{u}})\sin\phi. \tag{34}$$

The rotation relaxation time will increase with both the size and the aspect ratio of the spheroid. Leal and Hinch[36] use a result from the Jeffery algorithm,[35]

$$D_{\mathrm{rot}} = \frac{k_{\mathrm{B}}T}{4\eta_{\mathrm{s}}VH(\rho)}, \tag{35}$$

where η_{s} is the solvent viscosity, V is the spheroid volume, ρ is the spheroid aspect ratio and

$$H(\rho) = \frac{\rho^2 + 1}{\rho^3 \int\limits_0^\infty \frac{1}{(\rho^2 + \lambda)^{3/2}(1 + \lambda)}\mathrm{d}\lambda + \rho \int\limits_0^\infty \frac{1}{(\rho^2 + \lambda)^{1/2}(1 + \lambda)^2}\mathrm{d}\lambda}. \tag{36}$$

The integrals in eqn (36) can both be performed analytically for the two cases $\rho > 1$ and $0 < \rho < 1$. Defining $\tau_{\mathrm{rot}} = 1/D_{\mathrm{rot}}$ and using our expression for the nucleus volume (eqn (6)) we obtain

$$\tau_{\mathrm{rot}} = \frac{4\eta_{\mathrm{s}}b_{\mathrm{w}}^2 b_{\mathrm{l}}}{k_{\mathrm{B}}T} N_{\mathrm{T}}H(\rho). \tag{37}$$

There is some uncertainty in the pre-factor in this expression when used in our modelling since the solvent is non-Newtonian, the nucleus is not a perfect spheroid and the dangling amorphous chains attached to the nucleus will contribute some drag. Therefore we replace this pre-factor with an unknown dimensionless parameter α, such that

$$\tau_{\mathrm{rot}} = \alpha\tau_0 N_{\mathrm{T}}H(\rho), \tag{38}$$

which maintains Jeffery's scaling of the rotational diffusion with volume and aspect ratio and ensures $\tau_{rot} \sim \tau_0$ when $N_T = 1$.

3 Results

In our kinetic MC simulations we follow a single nucleus beginning with a single monomer. The algorithm is especially effective at simulating nucleation at low undercooling since for small nuclei the sum of rates is small, meaning that large time steps are automatically taken.

3.1 Quiescent results

The simulated quiescent nucleation time can be accurately described by an analytical calculation of the free energy landscape. Without flow the amorphous chains of all species have the same equilibrium configuration and therefore the same move rates. In effect a single species of concentration $\phi = 1$ is available to attach to the nucleus. This simplification allows the nucleus free energy to be calculated by the following method. The crystal nucleus is defined by the number of stems, N_s, and the total number of monomers, N_T, in the crystal. For each N_T and N_s, there will be several ways of distributing the N_T monomers amongst the N_s strands, subject to the constraint that each stem must contain at least one monomer. Thus after one monomer has been placed in each stem, the number of ways of distributing the remaining $N_T - N_s$ monomers is,

$$\omega(N_T, N_s) = \frac{(N_T - 1)!}{(N_T - N_s)!(N_s - 1)!}. \tag{39}$$

Thus the free energy of all possible nuclei with N_T monomers and N_s stems is given by $f(N_T,N_s) = \mathscr{F}(N_T,N_s) - \ln(\omega(N_T,N_s))$. In principle, this can be used in a two-dimensional diffusion calculation of the first passage of time for a particle over this two-dimensional barrier. However, below we will show that the simulated nucleation times can be accurately approximated by projecting this landscape onto one degree of freedom, the total number of monomers N_T. The partition function Z_N for a nucleus of N_T monomers is obtained by summing over all possible strand numbers,

$$Z_N = \sum_{n=1}^{N_s} \omega(N_T, n)\exp(-\mathscr{F}_{nuc}(N_T, n)), \tag{40}$$

where $\mathscr{F}_{nuc}(N_T,N_s)$ is given by eqn (12). The total free energy Δf for a nucleus of N_T monomers is

$$\Delta f(N_T) = -\ln Z_N + \ln Z_1, \tag{41}$$

where the free energy is set to zero for a crystal of one monomer. This free energy can also be extracted from a single long simulation by logging the fractional amount of time spent with each number of total monomers. The fraction of time spent with N_T monomers t_{NT}/t_{total} is proportional to the Boltzmann factor of Δf,

$$\frac{t_{N_T}}{t_{total}} = A\exp(-\Delta f(N_T)), \tag{42}$$

from which $\Delta f(N_T)$ can be obtained. The constant A is set so that $\Delta f(1) = 0$. Fig. 5 shows the agreement between the calculated and simulated free energies and illustrates how the critical nucleus size n^* and the nucleation barrier Δf^* can be extracted from these calculations. In these simulations, moves that grow the nucleus beyond some maximum number of monomers $N_{max} > n^*$ are prevented to allow good resolution of the landscape around n^*.

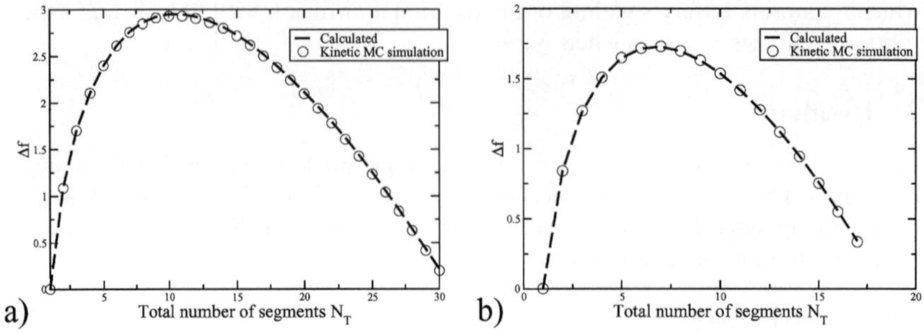

Fig. 5 Comparison of calculated and simulated quiescent free energy landscape for two sets of parameters. $\varepsilon_B = 0.3$ in both cases and $\mu_S = 0.25$, 0.22 for a and b, respectively.

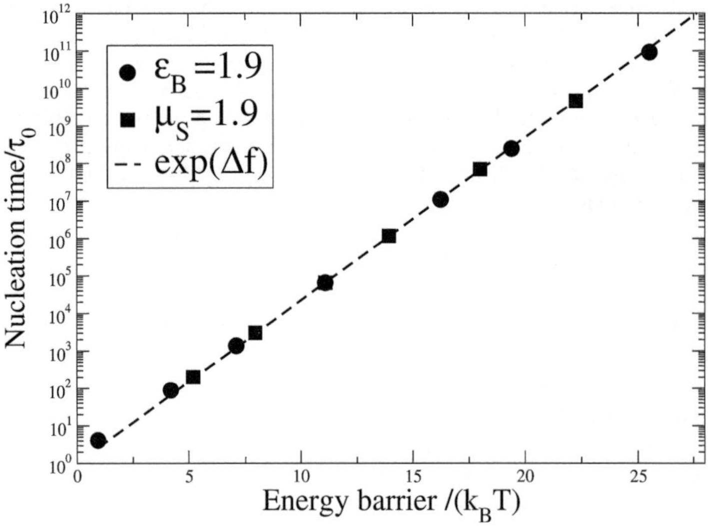

Fig. 6 Quiescent nucleation rate against nucleation barrier.

A series of simulations provides the nucleation time τ_N, which is taken as the first time the polar and equatorial radii simultaneously exceed the critical radius $r^* = \sqrt[3]{3n^*/4\pi}$. The average simulated nucleation time is well described by the Boltzmann factor of the nucleation barrier, $\langle\tau_N\rangle = \tau_0\exp(\Delta f^*)$, over a wide range of free energy parameters, as shown in Fig. 6. Fig. 6 also illustrates the efficiency of the algorithm at simulating nucleation even for very large nucleation barriers. We obtained good statistics for nucleation over a barrier of $\sim 25k_BT$ in $\sim$50 h on a single processor, giving a nucleation time of $\sim 10^{11}\tau_0$. Finally, the distribution of nucleation times is Poissonian, meaning that a well-defined quiescent nucleation rate is given by $\dot{N}_0 = 1/\langle\tau_N\rangle$.

3.2 Flow results

3.2.1 Enhanced nucleation. We examine first shear that is slow compared to the nucleus angular relaxation time. By setting α sufficiently small that $\dot{\gamma}\tau_{\rm rot}^{n^*} \ll 1$, where $\tau_{\rm rot}^{n^*} = \alpha n^* H(1)\tau_0$ is the angular relaxation time of a critical nucleus, no significant nucleus alignment occurs before the nucleus reaches the critical size and so alignment effects make no contribution to the nucleation time. We simulate the effect of steady shear by holding the amorphous chains fixed at the steady state GLaMM model predictions for a given shear rate throughout the whole simulation and

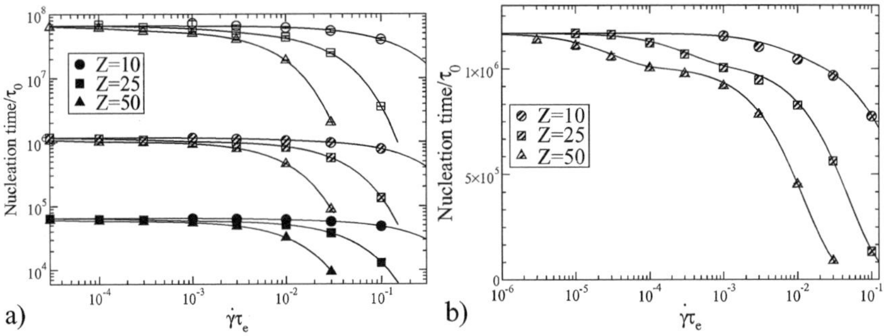

Fig. 7 (a) The steady shear nucleation time against shear rate for a range of chain lengths (Z) and free energy parameters: $\varepsilon_B = 1.9$ (solid symbols) $\varepsilon_B = 1.7$ (striped symbols) $\varepsilon_B = 1.5$ (open symbols). (b) A closer view of the $\varepsilon_B = 1.7$ data, showing the double exponential behaviour. In both cases the lines are from fitting eqn (43).

simulating the average nucleation time. Fig. 7(a) shows the average nucleation time against shear rate for a range of free energy parameters, with the surface energy fixed at $\mu_S = 1.9$ and the bulk free energy varied to simulate the effect of varying undercooling. Values of $\varepsilon_B = 1.9$, 1.7 and 1.5 were used, leading to nucleation barriers of $\Delta f^* = 11.1$, 13.9 and 18.0, respectively. The results as plotted are independent of the separation of shear and crystallisation timescale S. Fig. 7(b) shows the $\Delta f^* = 13.9$ data on a shorter y-axis, which shows the double exponential behaviour of the nucleation time. All of the data in Fig. 7 have this double exponential shape and can be fitted by the following semi-empirical expression

$$\frac{\tau_N}{\tau_0} = \Phi_z \exp\left(-\frac{\dot{\gamma}\tau_d}{W_d^*}\right) + (1 - \Phi_z)\exp\left(-\frac{\dot{\gamma}\tau_R}{W_R^*}\right). \quad (43)$$

The fitting parameters W^*_d and W^*_R are characteristic Weissenberg numbers with respect to the chain reptation and Rouse times, respectively, which define the Weissenberg numbers for the onset of enhanced nucleation due to tube orientation and chain stretch, respectively. The two exponentials correspond to the effect of tube orientation and chain stretching on the nucleation time. The former is controlled by the chain reptation time τ_d, which was calculated using the Likhtman and McLeish model,[37] and the latter is the chain Rouse time $\tau_R = Z^2\tau_e$. The balance of these two contributions is controlled by Φ_Z, which is the fractional reduction in the nucleation time due to orientation. Φ_Z is also fitted to the simulation data. Tables 1 and 2 contain the parameters obtained by fitting the data in Fig. 7. The Weissenberg numbers depend only on Δf^* and not Z, indicating that the contributions from tube orientation and chain stretching scale with Z in the same way as the appropriate relaxation time. Table 1 shows that, while the reptation Weissenberg number has a weak dependence of Δf^*, the Rouse Weissenberg number reduces significantly as Δf^* rises, indicating that the nucleation becomes more sensitive to chain stretching as the degree of undercooling is reduced. Table 2 shows that Φ_Z depends weakly on both Δf^*, increasing slightly with both increasing Z and Δf^*.

Table 1 Weissenberg numbers used to fit the steady state nucleation results

Δf	W^*_d	W^*_R
11.1	6.7	41.5
13.9	6.9	31.1
18.0	8.0	23.3

Table 2 Values of Φ_Z obtained by fitting to the steady state nucleation data. Φ_Z is the fractional reduction of the nucleation time due to tube alignment for a range of degrees of undercooling and molecular weights

Z	$\Phi_Z[\Delta f^* = 11.1]$	$\Phi_Z[\Delta f^* = 13.9]$	$\Phi_Z[\Delta f^* = 18.0]$
10	0.055	0.08	0.08
25	0.09	0.12	0.15
50	0.105	0.13	0.15

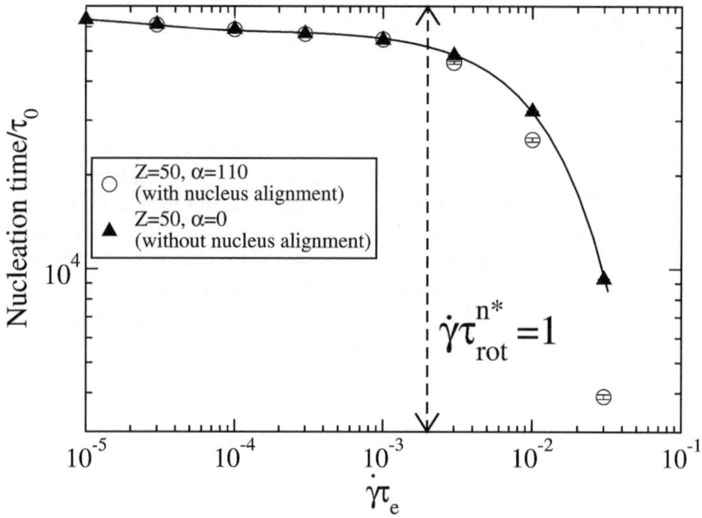

Fig. 8 Steady state nucleation time with and without the effect of nucleus rotation, $\mathcal{S} = 10$ in both cases.

Fig. 8 shows the effect of nucleus rotation on the nucleation time. The data for $\Delta f = 11.1$ with $\alpha = 0$ is compared with data where $\alpha = 110$. For the larger value of α the critical nucleus Weissenberg number exceeds 1 for shear rates $\dot{\gamma}\tau_e > 0.03$, as shown in Fig. 8. Beyond this shear rate the $\alpha = 110$ data show a progressive departure from the $\alpha = 0$ data as nucleus alignment now occurs below the critical nucleus size and so contributes to accelerating the nucleation. When nucleus rotation is included the decay of the nucleation time with shear rate is faster than exponential.

The simulations can also compute the nucleation time during a transient flow, by updating the chain configuration during the simulations as detailed in section 2.4. A nucleation rate $\dot{N}$ can be obtained from these data *via*

$$\dot{N}(t) \approx \frac{1}{1 - n(t)} \frac{n(t + \Delta t) - n(t - \Delta t)}{2\Delta t}, \tag{44}$$

where $n(t)$ is the cumulative fraction of successful nucleation events at time t. Fig. 9 shows this transient nucleation rate for a 25 entanglement monodisperse linear melt under start-up shear at $\dot{\gamma}\tau_e = 0.1$, both with and without nucleus rotation. The nucleation rate rises from the quiescent value up to a maximum occurring at around the

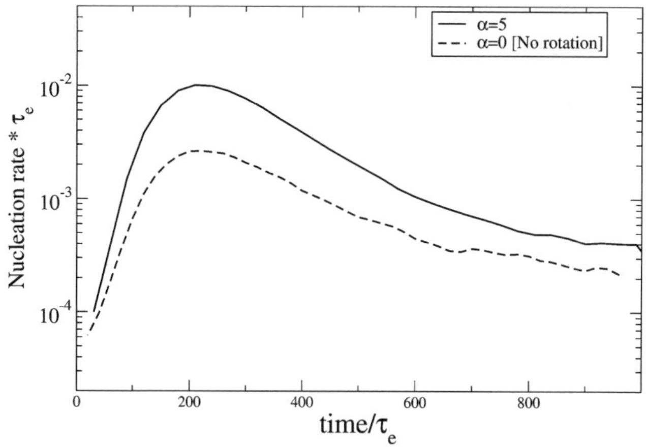

Fig. 9 Transient nucleation rate for a 25 entanglement monodisperse linear melt with $\varepsilon_B = 1.9$ and $\mu_S = 1.9$ under start-up shear at $\dot{\gamma}\tau_e = 0.1$; $\mathcal{S} = 3$ for both curves.

same time as the overshoot in the shear stress, indicating the strong correlation between nucleation rate and chain configuration. When nucleus alignment is included ($\alpha = 5.0$), the overshoot in the nucleation rate is increased as monomer alignment also makes a substantial contribution at this high shear rate. In both cases the transient maximum nucleation rate is a factor of ~10 higher than the steady state value.

3.2.2 Shish nuclei. Our model predicts strongly anisotropic growth under certain conditions. In our simulations shish-like nuclei are especially prevalent in melts of short chains blended with a small amount of very long chains, a system widely used in experiments to enhance shish formation.[4–8] We simulated a melt of

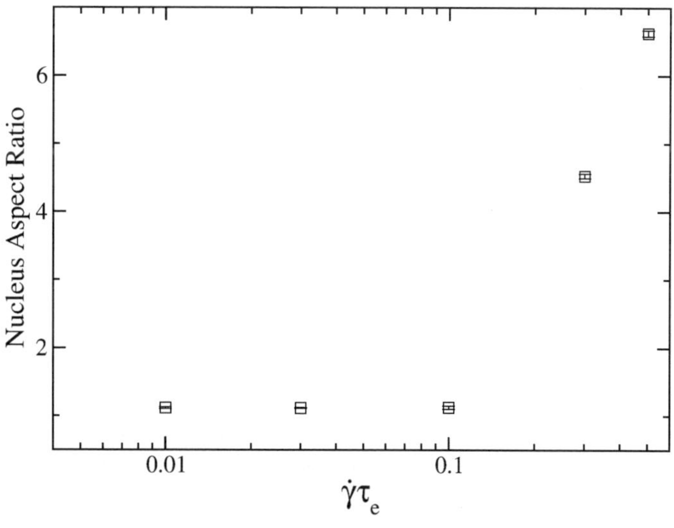

Fig. 10 Nucleus aspect ratio at the point of nucleation against shear rate for a 2% high molecular weight blend, with $\alpha = 5.0$ and $\varepsilon_B = 3.0$ $\mu_S = 2.5$ and a fixed shear time of $120\tau_e$.

15 entanglement chains blended with 2wt% of 52 entanglement chains, with the flow predictions provided by a recent generalisation of the GLaMM model to bimodal blends.[30] A surface energy of $\mu_S = 2.5$ was used as we found that rare nucleation enhances the anisotropy. Fig. 10 shows the average nucleus aspect ratio (L/W) at the point of nucleation, against shear rate. High shear rates produce very elongated nuclei, due to a purely kinetic mechanism where the shish length grows faster than the width. The shish widen by adding new stems for which any monomer from the melt can attach. In contrast, the shish length increases by adding monomers along an existing stem. Therefore the concentration of monomers from stretched segments at the growth surface is greater for lengthening than for widening, provided the nucleus contains a disproportionate number of stretched segments. Fast flow conditions are required for this disparity to overcome the significant surface area cost of elongated nuclei.

4 Conclusions

We present an efficient kinetic Monte Carlo algorithm suitable for modelling polymer nucleation even at low undercooling. The configuration of the non-crystalline chains under flow is computed using the deterministic GLaMM model, the nucleation is modelled by a kinetic Monte Carlo algorithm and the nucleus rotation is followed by a Brownian dynamics simulation. Flow modifies the basic Monte Carlo move rates in two ways: the entropic penalty for crystallisation is reduced by the flow induced change in chain free energy and the probability of compatible alignment between the nucleus and an attaching monomer is modified by flow induced molecular alignment.

The model confirms that the changes in chain free energy produced by non-linear flow are sufficiently large to produce a drastic enhancement of the nucleation rate. Our simulations of steady shear show that the reduction in nucleation time can be separated into two contributions; from tube orientation and chain stretching, with chain stretching being the dominant effect. The characteristic shear rate for the onset of these two processes have the same scaling with molecular weight as the appropriate relaxation times from the tube model. Furthermore, the free energy changes are also sufficient to induce a purely kinetic mechanisms of shish growth, in which the lengthening of existing stems is sufficiently faster than the rate of attachment of new stems to produce highly elongated nuclei, despite the high surface energy cost of this morphology.

While the model has many of the qualitative features of experiments on FIC, a direct quantitative comparison is not possible since virtually all literature measurements are on polydisperse materials and many of the features seen in the simulations will be smoothed out by this polydispersity. Generalising the crystallisation algorithm to polydisperse systems is straightforward as a range of effective species, with separate degrees of stretching, is already required for monodisperse melts. The limiting factor is in flow modelling of polydisperse melts. There is currently no sufficiently detailed model for non-linear flows of polydisperse melts at the level of the GLaMM model, although this is an area of intensive work and such models are nascent. This step would make possible an extensive comparison with literature data, allow the effect of molecular weight distribution on FIC to be modelled and would lead to a model of FIC suitable for industrial polymer resins.

A Constrained average

The flow-induced deformation of the non-crystalline chains modified the kinetic Monte Carlo moves through both the increase in monomer free energy upon stretching and changes in the monomer orientation. To calculate the change in elastic free energy, ΔF^{el}, for a chain subjected to the ensemble average constraint $\mathbf{f} = \langle \mathbf{rr} \rangle$, Olmsted and Milner introduced a field conjugate to $\mathbf{f}$ to the Gaussian chain partition function and made a Legendre transform.[38] Although this calculation is not possible analytically for finitely extensible chains, we expect $\Delta F^{\mathrm{el}}(\langle \mathbf{rr} \rangle)$ to converge towards $\Delta F^{\mathrm{el}}(\mathbf{r}^2)$ at large stretching since fluctuations are suppressed in highly stretched chains by the steep gradients in the free energy. Thus we expect an expression that crosses over from the Olmsted and Milner result for small Tr **f** to Cohen's[33] approximation with $\mathbf{r}^2 = \mathrm{Tr}\,\mathbf{f}$ at high stretching, to describe the free energy's dependence on **f**,

$$\Delta F^{\mathrm{el}} = \frac{1}{2}\mathrm{Tr}\,\mathbf{f} - \frac{1}{2}\mathrm{Tr}\,\ln\mathbf{f} - N_{\mathrm{e}}\ln\left(1 - \frac{\mathrm{Tr}\,\mathbf{f}}{N_{\mathrm{e}}}\right) + \Gamma, \quad (45)$$

where Γ is a constant chosen such that $\Delta F^{\mathrm{el}} = 0$ for an equilibrium coil $\mathbf{f} = 1/3\mathbf{I}$. We refer to this expression as the modified Cohen formula. We expect a similar result to hold for the monomer orientation.

A.1 Numerical free energy change. To test this idea we perform the required Legendre transform numerically and compare the results to eqn (45). We show that the result holds for a uniaxial deformation and then assume that it will also be true for biaxial deformations such as shear. Introducing the tensorial field $\mathbf{\Pi}$ conjugate to the end-to-end vector **r** into the finitely extensible chain partition function gives

$$\mathcal{Z}[\mathbf{\Pi}] = \int P(r, N_{\mathrm{e}})\exp(\mathbf{r}.\Pi.\mathbf{r})\mathrm{d}\mathbf{r}, \quad (46)$$

where $P(r, N_{\mathrm{e}})$ is the probability of end-to-end vector r for a finitely extensible chain of N_{e} steps. From here the chain free energy can be defined

$$\mathscr{F}[\mathbf{\Pi}] = -\ln\mathscr{Z}[\mathbf{\Pi}]. \quad (47)$$

Differentiating eqn (47) and using eqn (46) leads to

$$\frac{\partial\mathscr{F}}{\partial\Pi_{ij}} = W_{ij}, \quad (48)$$

where $W_{ij} = -\langle r_i r_j \rangle$. Thus $\mathbf{\Pi}$ and **W** are a conjugate pair and the free energy can be expressed as a function of **W** by a Legendre transform. The transformed free energy, F^{el}, can be written as

$$F^{\mathrm{el}}[\mathbf{W}] = \mathscr{F}[\mathbf{\Pi}(\mathbf{W})] - \mathbf{\Pi} : \mathbf{W}. \quad (49)$$

We will perform this transform numerically for a uniaxial deformation, which **W** is diagonal and $W_{yy} = W_{zz}$. The final part implies that $\Pi_{yy} = \Pi_{zz}$ so we must numerically find the solution of the following coupled system of non-linear equations

$$W_{xx} = \frac{\partial\mathscr{F}}{\partial\Pi_{xx}}\left[\Pi_{xx}, \Pi_{yy}, \Pi_{yy}\right], \quad (50)$$

$$W_{yy} = \frac{\partial\mathscr{F}}{\partial\Pi_{yy}}\left[\Pi_{xx}, \Pi_{yy}, \Pi_{yy}\right], \quad (51)$$

to obtain Π_{xx} and Π_{yy} from a given W_{xx} and W_{yy}. We first seek an expression for $\mathcal{Z}$. Switching the integral in eqn (47) into spherical polar co-ordinates, with the axis in the x-direction gives

$$\begin{aligned}\mathcal{Z}[\mathbf{\Pi}] = &\int_0^{\sqrt{N_{\mathrm{e}}}} r^2 P(r, N_{\mathrm{e}}) \int_0^{\pi}\int_0^{2\pi} \\ &\times\exp\left[r^2\left(\Pi_{xx}\cos^2\theta + \Pi_{yy}\sin^2\theta\cos^2\phi + \Pi_{zz}\sin^2\theta\cos^2\phi\right)\right]\sin\theta\mathrm{d}\phi\mathrm{d}\theta\mathrm{d}r,\end{aligned} \quad (52)$$

Using $\Pi_{yy} = \Pi_{zz}$ significantly simplifies the exponential and allows the integral over ϕ to be performed

$$\mathcal{Z}[\mathbf{\Pi}] = 2\pi \int_0^{\sqrt{N_e}} r^2 P(r, N_e) \int_{-1}^{1} \exp\left[r^2\left(\Pi_{xx}u^2 + \Pi_{yy}\left(1-u^2\right)\right)\right] \mathrm{d}u\mathrm{d}r, \tag{53}$$

where the substitution $u = \cos\theta$ has been made. The inner integral can now be performed.

$$\mathcal{Z}[\mathbf{\Pi}] = \frac{2\pi^{3/2}}{\sqrt{\Pi_{xx} - \Pi_{yy}}} \int_0^{\sqrt{N_e}} rP(r, N_e)e^{\Pi_{yy}r^2} \mathrm{Erfi}\left(\sqrt{\Pi_{xx} - \Pi_{yy}}r\right)\mathrm{d}r, \tag{54}$$

which uses the imaginary error function $\mathrm{Erfi}(x) = 2\exp(x^2)D(x)/\sqrt{\pi}$, where $D(x)$ is Dawson's integral.[39] This leaves a one dimensional integral over r which can be evaluated numerically. Differentiation of eqn (54) leads to

$$\frac{\partial \mathcal{Z}}{\partial \Pi_{xx}} = -\frac{Z}{2\left(\Pi_{xx} - \Pi_{yy}\right)} + \frac{2\pi}{\Pi_{xx} - \Pi_{yy}} \int_0^{\sqrt{N_e}} r^2 P(r, N_e)e^{\Pi_{xx}r^2} \mathrm{d}r, \tag{55}$$

which can be evaluated numerically, to give W_{xx} since

$$W_{xx} = \frac{1}{Z}\frac{\partial \mathcal{Z}}{\partial \Pi_{xx}}. \tag{56}$$

The differentiation required to produce W_{yy} (eqn (51)) must be performed before the $\Pi_{xx} = \Pi_{yy}$ result is used. Therefore differentiating eqn (52) gives

$$\begin{aligned} \frac{\partial \mathcal{Z}}{\partial \Pi_{yy}} &= \int_0^{\sqrt{N_e}} r^4 P(r, N_e) \int_0^{\pi}\int_0^{2\pi} \sin^2\theta \cos^2\phi \\ &\times \exp\left[r^2\left(\Pi_{xx}\cos^2\theta + \Pi_{yy}\sin^2\theta\cos^2\phi + \Pi_{zz}\sin^2\theta\cos^2\phi\right)\right]\sin\theta\,\mathrm{d}\phi\mathrm{d}\theta\mathrm{d}r. \end{aligned} \tag{57}$$

Similarly to above, using $\Pi_{yy} = \Pi_{zz}$, performing the integral over ϕ and substituting $u = \cos\theta$ gives

$$\frac{\partial \mathcal{Z}}{\partial \Pi_{yy}} = \pi \int_0^{\sqrt{N_e}} r^4 P(r, N_e) \int_{-1}^{1} (1-u^2)\exp\left[r^2\left(\Pi_{xx}u^2 + \Pi_{yy}\left(1-u^2\right)\right)\right]\mathrm{d}u\mathrm{d}r. \tag{58}$$

Again, despite appearances, the integral over u can be performed to give

$$\begin{aligned} \frac{\partial \mathcal{Z}}{\partial \Pi_{yy}} &= \frac{\pi}{2\left(\Pi_{xx} - \Pi_{yy}\right)^2} \int_0^{\sqrt{N_e}} P(r, N_e)r\Big[-2r\left(\Pi_{xx} - \Pi_{yy}\right)e^{\Pi_{xx}r} \\ &+ \sqrt{\Pi_{xx} - \Pi_{yy}}e^{\Pi_{yy}r}\sqrt{\pi}\left(1 + 2\left(\Pi_{xx} - \Pi_{yy}\right)r^2\right)\mathrm{Erfi}\left(\sqrt{\Pi_{xx} - \Pi_{yy}}\,r\right)\Big]\mathrm{d}r. \end{aligned} \tag{59}$$

In order to evaluate the integrals for Z and $\partial Z/\partial \Pi_{xx}$ and $\partial Z/\partial \Pi_{yy}$ we require the probability distribution for a finitely extensible chain $P(r, N_e)$. Ref. 40 gives this as

$$P(r, N_e) = A \exp\left[-\sqrt{N_e}\, r \mathcal{L}^{-1}\left(\frac{r}{\sqrt{N_e}}\right)\right]\left(\frac{\sinh \mathcal{L}^{-1}\left(\frac{r}{\sqrt{N_e}}\right)}{\mathcal{L}^{-1}\left(\frac{r}{\sqrt{N_e}}\right)}\right)^{N_e}, \quad (60)$$

where we use Cohen's approximation to the inverse Langevin function $\mathcal{L}^{-1}$ (eqn (4)) and A is a normalisation constant whose value is not needed since it merely adds a constant to the free energy. To perform the Legendre transform, eqn (50) and (51) are solved numerically using Broyden's method[41] with numerical evaluations of eqn (54), (55) and (59). This provides Π_{xx} and Π_{yy} corresponding to a prescribed $\langle r_x^2 \rangle$ and $\langle r_y^2 \rangle$ pair, which can then be used to calculate F^{el} with eqn (49). The numerical results shown in Fig. 11 were generated by incrementally increasing $\langle r_x^2 \rangle$, using the previous value of Π_{xx} and Π_{yy} to provide a good initial guess for the subsequent calculation. Our results depend only on $\langle \mathbf{r}^2 \rangle$ not the individual values of $\langle r_x^2 \rangle$ and $\langle r_y^2 \rangle$.

A.2 Monomer orientation. Similarly, we expect that the distribution of monomer orientations, $w(\theta)$, can be approximated by modifying the expression for $w(\theta)$[12] with a direct constraint on $\mathbf{r}$. That is, we expect to accurately approximate the monomer orientation distribution for a chain with a constrained average end-to-end vector, $\mathbf{f}$ by

$$w(\theta) = \frac{\mathcal{L}^{-1}(\sqrt{\mathrm{Tr}\,\mathbf{f}}/N_e)}{4\pi \sinh\left(\mathcal{L}^{-1}\sqrt{\mathrm{Tr}\,\mathbf{f}}/N_e\right)\right)} \cosh\left(\mathcal{L}^{-1}\left(\frac{\sqrt{\mathrm{Tr}\,\mathbf{f}}}{N_e}\right)\cos\theta\right), \quad (61)$$

where $\mathcal{L}^{-1}$ is the inverse Langevin function and θ is the angle with the principle axis of $\mathbf{f}$. To check this we numerically calculate the true values of $w(\theta)$ by extending the methods above and compare this to eqn (61). We begin by constraining the partition function so that one of the monomers $\mathbf{u}_j$ has vector $\mathbf{w}$ (with $|\mathbf{w}| = 1|$)

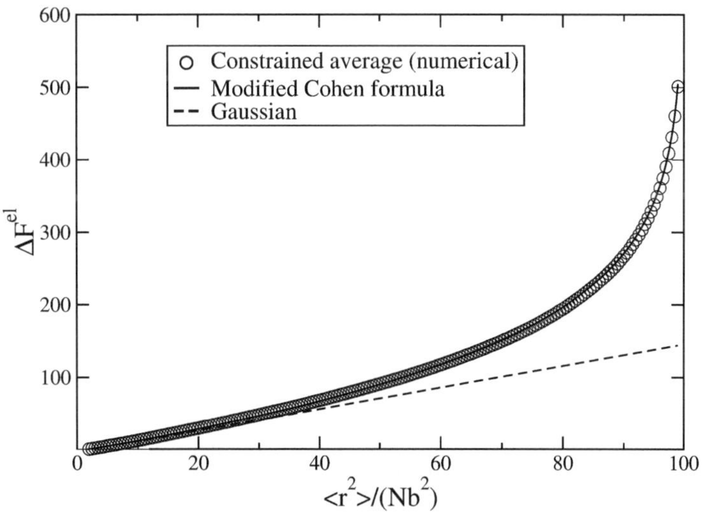

Fig. 11 Comparison between numerical computations of the free energy of a uniaxially deformed finitely extensible chain of constrained average end-to-end vector, with $N_e = 100$, and the modified Cohen formula (eqn (45)). Also shown is the exact result for Gaussian chains.[38]

$$\mathcal{Z}[\mathbf{\Pi},\mathbf{w}] = \int\int \delta(\mathbf{u}_j - \mathbf{w})\, W_0(\mathbf{u}_j) P(\mathbf{r}-\mathbf{w}, N_e - 1)\exp(\mathbf{r}.\Pi.\mathbf{r})\mathrm{d}\mathbf{r}\mathrm{d}\mathbf{u}_j, \tag{62}$$

where $W_0(\mathbf{u}) = \delta(|\mathbf{u}| - 1)/4\pi$ is the probability distribution of a freely rotating monomer.[38] Carrying out the integral over $\mathbf{u}_j$ and substituting $\mathbf{r}' = \mathbf{r} - \mathbf{w}$ gives

$$\mathcal{Z}[\mathbf{\Pi},\mathbf{w}] = \frac{1}{4\pi}\int P(|\mathbf{r}'|, N_e - 1)\exp((\mathbf{r}'+\mathbf{w}).\Pi.(\mathbf{r}'+\mathbf{w}))\mathrm{d}\mathbf{r}'. \tag{63}$$

We switch to spherical polar co-ordinates along the x-axis. However, because of the uniaxial symmetry of $\mathbf{\Pi}$ we can choose the azimuthal angle ϕ so that $\phi_{\mathbf{w}} = 0$. In this co-ordinate system $w_y = \sqrt{1-w_x}$ and

$$\begin{aligned}(\mathbf{r}'+\mathbf{w}).\mathbf{\Pi}.(\mathbf{r}'+\mathbf{w}) &= r'^2\left(\Pi_{xx}\cos^2\theta + \Pi_{yy}\sin^2\theta\right) + w_x^2\Pi_{xx} + \left(1-w_x^2\right)\Pi_{yy} \\ &\quad + 2r'\left(\Pi_{xx}w_x\cos\theta + \Pi_{yy}\sqrt{1-w_x}\sin\theta\cos\phi\right).\end{aligned} \tag{64}$$

Substituting in eqn (64) gives

$$\begin{aligned}\mathcal{Z}[\mathbf{\Pi},\mathbf{w}] = &\frac{1}{4\pi}\exp\left[w_x^2\Pi_{xx} + \left(1-w_x^2\right)\Pi_{yy}\right] \\ &\int_0^{\frac{N_e-1}{\sqrt{N_e}}} r'^2 P(r', N_e - 1)\int_0^{\pi}\exp\left[r'^2\left(\Pi_{xx}\cos^2\theta + \Pi_{yy}\sin^2\theta\right) + 2r'\Pi_{xx}w_x\cos\theta\right] \\ &\int_0^{2\pi}\exp\left[2r'\Pi_{yy}\sqrt{1-w_x}\sin\theta\cos\phi\right]\mathrm{d}\phi\,\sin\theta\mathrm{d}\theta\mathrm{d}r'.\end{aligned} \tag{65}$$

Carrying out the integral over ϕ and substituting $u = \cos\theta$ gives

$$\begin{aligned}\mathcal{Z}[\mathbf{\Pi},\mathbf{w}] = &\frac{1}{2}\exp\left[w_x^2\,\Pi_{xx} + \left(1-w_x^2\right)\Pi_{yy}\right]\int_0^{\frac{N_e-1}{\sqrt{N_e}}} r'^2 P(r', N_e - 1) \\ &\int_{-1}^{1}\exp\left[r'^2\left(\Pi_{xx}u^2 + \Pi_{yy}\left(1-u^2\right)\right) + 2r'\Pi_{xx}w_x u\right] \\ &\times I_0\left(2r'\Pi_{yy}\sqrt{1-w_x^2}\,\sqrt{1-u^2}\right)\mathrm{d}u\mathrm{d}r',\end{aligned} \tag{66}$$

where I_0 is the modified Bessel function of the first kind.[39] The probability of a monomer making an angle θ with the x-axis is given by

$$w(\theta) = \frac{2\pi}{\mathcal{Z}[\mathbf{\Pi}]}\mathcal{Z}[\mathbf{\Pi}, w_x = \cos(\theta)], \tag{67}$$

where the factor of 2π arises by integrating eqn (66) over all possible azimuthal angles of w_x. For a given $\langle r_x^2\rangle$ and $\langle r_y^2\rangle$ pair, the corresponding Π_{xx} and Π_{yy} are already known from the Legendre transform and this allows eqn (66) to be evaluated by numerical computation of the remaining double integral, leading to $w(\theta)$. Eqn (61) agrees closely with numerical evaluations, as expected.

Acknowledgements

We acknowledge the EPSRC for funding through the Micro Polymer Processing 2 grant (GR/T11807/01). We thank the members of the Micro Polymer Processing 2 project for useful discussions, especially Luigi Balzano, Rudi Steenbakkers, Gerrit Peters, Oleksandr Mykhaylyk, Tony Ryan and Tom McLeish. RSG acknowledges a travel award from the Royal Society.

References

1 A. Keller and H. Kolnaar, in *Processing of Polymers*, ed. H. Meijer, Wiley-VCH, Weinheim, 1997, vol. 18, p. 189.
2 I. Coccorullo, R. Pantani and G. Titomanlio, *Macromolecules*, 2008, **41**(23), 9214–9223.
3 S. Kimata, T. Sakurai, Y. Nozue, T. Kasahara, N. Yamaguchi, T. Karino, M. Shibayama and J. A. Kornfield, *Science*, 2007, **316**, 1014–1017.
4 B. S. Hsiao, L. Yang, R. H. Somani, C. A. Avila-Orta and L. Zhu, *Phys. Rev. Lett.*, 2005, **94**, 117802.
5 L. Balzano, N. Kukalyekar, S. Rastogi, G. W. M. Peters and J. C. Chadwick, *Phys. Rev. Lett.*, 2008, **100**, 048302.
6 M. Seki, D. W. Thurman, J. P. Oberhauser and J. A. Kornfield, *Macromolecules*, 2002, **35**, 2583–2594.
7 E. L. Heeley, C. M. Fernyhough, R. S. Graham, P. D. Olmsted, N. J. Inkson, J. Embery, D. J. Groves, T. C. B. McLeish, A. C. Morgovan, F. Meneau, W. Bras and A. J. Ryan, *Macromolecules*, 2006, **39**(15), 5058–5071.
8 O. O. Mykhaylyk, P. Chambon, R. S. Graham, J. P. A. Fairclough, P. D. Olmsted and A. J. Ryan, *Macromolecules*, 2008, **41**(6), 1901–1904.
9 H. Zuidema, G. W. M. Peters and H. E. H. Meijer, *Macromol. Theory Simul.*, 2001, **10**, 447–460.
10 S. Coppola, N. Grizzuti and P. L. Maffettone, *Macromolecules*, 2001, **34**, 5030–5036.
11 R. Zheng and P. Kennedy, *J. Rheol.*, 2004, **48**(4), 823–842.
12 L. Jarecki, in *Progress in Understanding of Polymer Crystallization*, ed. G. Reiter and G. R. Strobl, Springer, Berlin, 2007, vol. 714 of *Lecture Notes in Physics*, pp. 65–86.
13 J. van Meerveld, M. Hütter and G. Peters, *J. Non-Newtonian Fluid Mech.*, 2007, **150**, 177–196.
14 N. Waheed, M. J. Ko and G. C. Rutledge, *Polymer*, 2005, **46**, 8689–8702.
15 M. J. Ko, N. Waheed, M. S. Lavine and G. C. Rutledge, *J. Chem. Phys.*, 2004, **121**, 2823–2832.
16 W. Hu, D. Frenkel and V. Mathot, *Macromolecules*, 2002, **35**, 7172–7174.
17 W. Hu and D. Frenkel, *Adv. Polym. Sci.*, 2005, **191**, 1–35.
18 J. Zhang and M. Muthukumar, *J. Chem. Phys.*, 2007, **126**, 234904.
19 I. Dukovski and M. Muthukumar, *J. Chem. Phys.*, 2003, **118**, 6648–6655.
20 D. Auhl, J. Ramirez, A. E. Likhtman, P. Chambon and C. Fernyhough, *J. Rheol.*, 2008, **52**(3), 801.
21 J. Bent, L. R. Hutchings, R. W. Richards, T. Gough, R. Spares, P. D. Coates, I. Grillo, O. G. Harlen, D. J. Read, R. S. Graham, A. E. Likhtman, D. J. Groves, T. M. Nicholson and T. C. B. McLeish, *Science*, 2003, **301**, 1691–1695.
22 J. Lukkien, J. Segers, P. Hilbers, R. Gelten and A. Jansen, *Phys. Rev. E*, 1998, **58**, 2598–2610.
23 D. T. Gillespie, *J. Phys. Chem.*, 1977, **81**(25), 2340–2361.
24 J. P. K. Doye and D. Frenkel, *Phys. Rev. Lett.*, 1998, **81**, 2160–2163.
25 R. S. Graham, A. E. Likhtman, T. C. B. McLeish and S. T. Milner, *J. Rheol.*, 2003, **47**(5), 1171–1200.
26 M. W. Collis, A. K. Lele, M. R. Mackley, R. S. Graham, D. J. Groves, A. E. Likhtman, T. M. Nicholson, O. G. Harlen, T. C. B. McLeish, L. R. Hutchings, C. M. Fernyhough and R. N. Young, *J. Rheol.*, 2005, **49**, 501–522.
27 R. S. Graham and T. C. B. McLeish, *J. Non-Newtonian Fluid Mech.*, 2008, **150**(1), 11–18.
28 A. Blanchard, R. S. Graham, M. Heinrich, W. Pyckhout-Hintzen, D. Richter, A. E. Likhtman, T. C. B. McLeish, D. J. Read, E. Straube and J. Kohlbrecher, *Phys. Rev. Lett.*, 2005, **95**, 166001.
29 R. S. Graham, J. Bent, L. R. Hutchings, R. W. Richards, D. J. Groves, J. Embery, T. M. Nicholson, T. C. B. McLeish, A. E. Likhtman, O. G. Harlen, D. J. Read, T. Gough, R. Spares, P. D. Coates and I. Grillo, *Macromolecules*, 2006, **39**(7), 2700–2709.
30 R. S. Graham, J. Bent, N. Clarke, L. R. Hutchings, R. W. Richards, T. Gough, D. M. Hoyle, O. G. Harlen, I. Grillo, D. Auhl and T. C. B. McLeish, *Soft Matter*, 2009, **5**, 2383–2389.
31 A. Tezel, L. G. Leal and T. C. B. McLeish, *Macromolecules*, 2005, **38**, 1451–1455.
32 A. K. Tezel, J. P. Oberhauser, R. S. Graham, K. Jagannathan, T. C. B. McLeish and L. G. Leal, *J. Rheol.*, 2009, in press.

33 A. Cohen, *Rheol. Acta*, 1991, **30**, 270.
34 L. G. Leal and E. J. Hinch, *J. Fluid Mech.*, 1972, **55**, 745.
35 G. B. Jeffery, *Proc. R. Soc. London, Ser. A*, 1922, **102**, 161–179.
36 L. G. Leal and E. J. Hinch, *J. Fluid Mech.*, 1971, **46**(27), 685–703.
37 A. E. Likhtman and T. C. B. McLeish, *Macromolecules*, 2002, **35**(16), 6332–6343.
38 P. D. Olmsted and S. T. Milner, *Macromolecules*, 1994, **27**, 6648–6660.
39 M. Abramowitz and I. A. Stegun, *Handbook of Mathematical Functions*, Dover Publications, 1965.
40 L. R. G. Treloar, *The Physics of Rubber Elasticity*, Oxford University Press, Oxford, 1975.
41 W. H. Press, B. P. Flannery, S. A. Teukolsky, and W. T. Vetterling, *Numerical Recipes in C: The Art of Scientific Computing*, Cambridge University Press, Cambridge, 1992.

General discussion

Professor De Pablo opened the discussion of the paper by Professor Kremer: In the correlation between plateau modules and l_k/p that was shown, do you have to predict both the entanglement length and the density of the polymeric material?

Professor Kremer replied: The plot of the plateau modulus I showed during the talk[1] gives the normalized plateau modulus $G^\circ\ l_k^3/k_BT$ as a unique (within the error bars) function of the ratio l_k/p. l_k is the Kuhn length of the chain and p the so called packing length. In amorphous polymer melts p is the typical average strand–strand distance. Thus to obtain $G^\circ\ l_k^3/k_BT$ one only needs l_k/p. To determine G° from that plot l_k is needed as well. The definition of the packing length $p = N/(\rho\langle R^2\rangle)$, N being the chain length, includes the density ρ. Thus, if ρ and l_k (or $\langle R^2\rangle$ are known, the entanglement length and the plateau modulus is known to a rather good accuracy.

1 R. Everaers, S. K. Sukumaran, G. S. Grest, C. Svaneborg, A. Sivasubramanian and K. Kremer, *Science*, 2004, **303**, 823

Professor Español commented: You have mentioned that each coarse-grained model has its own thermodynamics. What do you mean exactly?

Professor Kremer responded: Of course thermodynamics is the same for all systems. What I meant by this comment to your question, was that a coarse grained model typically does NOT have the same free energy as the underlying fine grained model. Taking the analogy to the renormalization group procedure one sees that with each iteration "irrelevant" contributions are neglected. A similar argument would hold, if one follows your projector operator scheme. This should be kept in mind, when coarse and fine grained models for the same system are (quantitatively) compared.

Professor Allen said: In coarse graining the intramolecular degrees of freedom (e.g. bond lengths and angles), you indicated that if they are not decoupled then you must be very careful. Can you give an example of where this is important, and how you handle it carefully?

Professor Kremer replied: Using Boltzmann inversion of distribution functions is a convenient way to generate coarse grained (cg) potentials from underlying atomistic probability distribution functions. The basis of this is the assumption, that distribution functions for different cg degrees of freedom factorize, *i.e.* are independent of each other. If this is the case, *i.e.* bonded and nonbonded interactions can be treated separately and the bonded cg interactions usually can also be transferred to other physical situations as well. An example where this works very well is an amorphous melt of polycarbonate (PC). The repeat unit is a rigid fairly large object (carbonate–carbonate distance about 11 Å) with a very flexible joint at the carbonate group. In that case the angular distribution functions at the carbonate group, as determined from the simulation of a single joint in vacuum, can be used to determine the cg angular potentials of such a chain in a melt with very good accuracy. The situation for polystyrene (PS), which has a more simple and smaller repeat unit, is however already much more complicated due to the longer range interaction of the benzene side group along the backbone of the chain. Another level of complication occurs, if specific nonbonded interactions like hydrogen bonds strongly influence the local macromolecular conformations. Then the assumption of the factorizations of the distributions functions of different cg variables can easily lead to cg models, which are not connected to the underlying chemistry in a reasonable way. Then additional interactions along the backbone of the chains have to be considered.[1] This is of special importance for many biomolecular systems.

1. D. Fritz, V. A. Harmandaris, K. Kremer and N. van der Vegt, *Macromolecules*, 2009, DOI: 10.1021/ma901242h

Dr Ensing asked: You explain in your lecture that the course grained dynamics is inherently too fast so that you have to introduce a (single) scaling factor to obtain estimates for dynamical properties. My question is whether you would not actually need a scaling factor for each specific dynamical property that is of interest, as each dynamical process has its own specific dependence on the underlying missing atomistic variables?

Professor Kremer responded: Of course the remark is correct and there is not simply one time scaling factor for different dynamical processes, when one compares the dynamics of the microscopic system to that of the coarse grained (CG) model. This however also was clearly stated during the lecture. Different processes can lead to different time scalings within one cg model. This actually is one of the big challenges for multi-scale modelling. We have discussed this in the literature for melts of polycarbonate, polystyrene and melts with low molecular weight additives (*c.f.* references in the introductory paper). What however was found for amorphous polymer melts, was that on length scales above the typical coarse graining unit (about 10 Å) properties like the chain diffusion, bead mean square displacements and structural relaxation along the backbone of the chains can well be described by a single scaling factor. Whether this can be extended at all to completely different systems, *i.e.* partially ordered polymer systems, is part of the above mentioned challenge.

Professor Müller-Plathe remarked: Which are the challenges to the coarse-graining community for the next 10–20 years?

Professor Kremer answered: I have some difficulty making predictions for such a long period of time. What one can say is that multi-scale simulations in biosciences but especially also in material science will become even more important than they are now. This is simply because analytic theory can only treat rather idealized limiting cases (which nevertheless will be important anchor points for the simulations) and experiments will provide many more systems, however, with a rather limited characterization when it comes to atomistic datails. Taking the typical soft matter systems one challenge will be to follow the dynamics of mid size systems for a long time. First studies along that line are underway (see also the introductory paper). This is a precondition for many further developments. Otherwise plain CPU power will not be a central issue for most problems in the future, considering the next generation computers. Thus I think multi-scale modelling will focus more on identifying structures and mechanisms rather than replacing giant all atom simulations. Scientifically, the link to quantum mechanics and fields like organic electronics will see a significant rise of activities as well as non equilibrium phenomena. They all pose major challenges and we are just at the beginning.

Professor Allen opened the discussion of the paper by Professor Müller-Plathe: You mentioned some problems of conformational equilibration for atactic polystyrene. Can this be improved, for example by including more chemical detail or conformational information at an earlier stage?

Professor Müller-Plathe responded: Yes, equilibration can be improved. There is no requirement to insert the Kuhn fragments in an all-*trans* conformation, which we did in this pilot study for simplicity. One can use more realistic conformations. These can be obtained, for example, from a library of fragment conformations. Such an ensemble of structures may be extracted from previous simulations of short-chain oligomers. Examples of this approach can be found elsewhere.[1,2]

1 H. A. Karimi-Varzaneh, P. Carbone, and F. Müller-Plathe, *J. Chem. Phys.*, 2008, **129**, 154904
2 X. Chen, P. Carbone, G. Santangelo, A. Di Matteo, G. Milano, and F. Müller-Plathe, *Phys. Chem. Chem. Phys.*, 2009, **11**, 1977–1988

Professor Bolhuis asked: When the random walk simulations are fine grained and relaxed they should, in principle, obey the Boltzmann distribution. This is however, not guaranteed by your approach. Did you test whether the final distribution is really the equilibrium one, and if not, do you expect this to be a problem?

Professor Müller-Plathe replied: We did not test whether the final structures obey Boltzmann statistics. Moreover, I would be very pessimistic for the chance to come up with a good enough statistic of structures prepared by our procedure to allow a meaningful determination of a distribution. For the moment, we are very happy that we reproduce the atomistic structure as probed by X-rays and neutrons.

Professor van der Vegt said: The inverse-mapped conformations all have a ratio of the mean square end-to-end distance to the mean square radius of gyration smaller than 6.0 (Tables 1, 2, and 4 in the paper). This remains the case for polyethylene chains up to 40 Kuhn segments (Table 1 in the paper) where I expect that the chains are long enough to obey random walk statistics. Considering that you start off from continuum random walks, does a ratio <6 indicate that your procedures used to equilibrate distance scales of several chemical repeat units lead to a perturbation of the global scale dimensions?

Professor Müller-Plathe responded: The explanation lies already at the level of the random walks, *i.e.* before any fine-graining. We have been using a small number of random walks and the discrepancy between the mean-square end-to-end distance R_e and the mean-square radius of gyration R_g is due to insufficient statistics. In the case of polyethylene, we used 15 random walks. Their average ratio (R_e^2/R_g^2) is 4.4 with a standard deviation of 2.3.

Professor M. Müller said: You mentioned in your presentation the possibility of using an equilibrated melt of a coarse-grained model instead of randomly placing chains in a period box (as described in the conference paper). Both approaches significantly differ because the latter does not capture intermolecular correlations, *i.e.* there is no correlation hole in the intermolecular pair correlation function, *etc.* The intermolecular correlations of randomly placed molecules will be similar to those of an ideal gas. Since the intermolecular correlations extend to the size of the polymer coil, a time on the order of the Rouse time will be required to establish them.

In particular, I would expect the structure factor of such a randomly assembled system to resemble the form factor of an individual chain rather than the structure factor of a dense, nearly-incompressible polymer melt. This would also have consequences for the compressibility (as determined by the small wave vector limit of the structure factor). Are these effects negligible because the system size is not large compared to the extension of a molecule and a single chain is wrapped back *via* the periodic boundary conditions?

Professor Müller-Plathe replied: You raise different points here. Let me take them in turn.

1. If by intermolecular correlations you mean those between the centres of mass of the chains, they may be those of an ideal gas. The correlations between chain segments will certainly not be that of an ideal gas. The excluded volume and a short atomistic relaxation will see to that.

2. For many "atomistic" quantities (*i.e.* those calculated from atom positions) but not for all, the centre-of-mass correlations will be irrelevant. Even in its present form, the method may find practical applications. It goes without saying that backmapping from an equilibrated melt of random walks rather than from randomly assembled random walks, as mentioned in the talk, will take care of the centre-of-mass correlations as well. It is the better method and it will be tried in the future.

3. The question of compressibility raises the point, how much a compressibility calculated from the structure factor is worth in practice. If you fill a simulation box with excluded volume objects of the right size to the right density, they will produce the more or less correct compressibility as (dV/dp), no matter what the structure factor says. It is not even important whether the hard objects are connected into a polymer chain.

4. With regard to system size and periodic boundary conditions, I have no answer, as we did not try larger (very much larger!) systems.

Dr Ensing said: In your paper you show the rotational dynamics of frozen atomistic functions centered with their centers of mass at the coarse grained particle positions. I referred to our reverse mapping procedure, in which we used a simplified energy function of local atomistic bonds and angles to also rotate atomistic functions around their centers of mass.[1] My question is whether your method to introduce a vector and rotate the atoms to have this vector align with the bead–bead vector is more efficient than to minimize atomistic interactions and I was also wondering how your method would work for non-linear (branched) chains, as in that case there are more than one bead–bead vectors to choose from.

1 S. O. Nielsen, B. Ensing, P. B. Moore, and M. L. Klein, Coarse-grain to atomistic mapping algorithm: a tool for multi-scale simulations, in *Advances in Hierarchical and Multi-Scale Simulations of Materials*, ed. S. Mohanty and R. B. Ross, Taylor and Francis, 2006.

Professor Müller-Plathe responded: It might not have become clear that we are not doing any rotational dynamics in the sense of molecular dynamics simulations. We are only placing our atomistic (Kuhn-length) fragments on the sites of the coarse-grained monomers and then rotating them roughly into a sensible orientation. The rest is taken care of by ordinary Cartesian molecular dynamics. As our fragments are flexible (*e.g.* 8 carbons in the case of polyethylene), a purely rotational dynamics would not relax them sufficiently.

It should be pointed out, however, that in cases where smaller, rigid atomistic fragments have been inserted into a coarse-grained model, these fragments have been moved by rigid-body rotations implemented by quaternions. This work has been pioneered by Giuseppe Milano.[1]

I am quite confident that our method would also work for branch points in the polymer, even though we have not tried it yet. The reason is that the rotation is only used as a pre-orientation of the fragment and the main relaxation is done by Cartesian MD anyway. This would also work for branch points. Here, one would of course use all 3 rotational degrees of freedom to pre-orient the fragment into the best position with its 3 neighbours. In the present application of a non-branching fragment, only two rotational degrees are really needed, the third one is undetermined.

1 G. Santangelo, A. Di Matteo, F. Müller-Plathe, and G. Milano, *J. Phys. Chem. B*, 2007, **111**, 2765–2773.

Dr Milano responded: Regarding a possible reverse mapping procedure able to rotate atomistic units to have a more reasonable guess of the atomistic structure, I would like to recall a recent paper[1] in which is described a method based on a fully geometrical approach (rotations based on quaternion algebra) and not involving potential energy and force evaluations that allows a very fast and efficient reconstruction of the atomistic detail.

1 G. Santangelo, A. Di Matteo, F. Müller-Plathe and G. Milano, *J. Phys. Chem. B*, 2007, **111**, 2765–2773.

Dr Carbone remarked: During the reverse mapping procedure we need not only to rebuild the bonds among the atomistic fragments but also to minimize locally their conformations.

In this situation the minimization of the atomistic interactions is not enough since this would only create the backbone bonds but would not solve the problem of the local minimization of the fragments.

The two steps procedure adopted in the manuscript (rotation of the frozen atomistic moieties followed by a soft core molecular dynamic simulation) solves the problem for the presented cases.

As the method is based on the concept of persistence length, its use for branched polymers, may not be possible.

Dr Ensing opened the discussion of the paper by Dr Lyubartsev: One solution to the slow convergence problem, of fitting a pair potential against a target radial distribution function, that has been proposed is to add other target properties, such as the pressure, to fit against simultaneously. In the case of using pressure as an additional target property, no extra reference calculations have to be done, as the pressure can be evaluated from the same atomistic reference calculation as is used for the pressure. Is the method that you propose not more demanding as it requires several reference calculations at different densities?

Professor Lyubartsev answered: The possibility to fit simultaneously RDF and some other properties, for example pressure, exists, though in this case, the solution can be found in the variational sense, to minimize the difference between simulated and target RDF, pressure and perhaps other properties. This was briefly mentioned in the text but not yet implemented.

Professor Voth remarked: Your Newton inversion coarse-graining methodology seems very interesting. However, will it be convergent for complex systems such as proteins or even for small peptides in water? Can you comment on this? Have you ever demonstrated this approach for anything more complicated than a lipid bilayer?

Professor Lyubartsev responded: We have not tried yet to apply this approach for proteins and small peptides, and CG model of lipids is probably the most "complex" system for which it was applied (other cases were atomistic water–water and water ion potentials from CPMD simulations, solvent-mediated ion–ion and ion–DNA potentials) . What is concerned convergence of the method for peptide-like systems (in the sense, to find the potential which reproduces RDFs obtained in atomistic simulations), I do not see principal problems. But whether obtained in this way effective potentials will be useful for modelling of, *e.g.*, protein folding is an open question.

Dr Wilson asked: Could you comment on the approximate functional form of the numerical water potential obtained in Fig. 2 of the paper? Specifically, (i) what is the steepness of the repulsive part of the potential (*c.f.* the traditional $1/r^{12}$ used in some coarse-grained water potentials such as MARTINI), (ii) the second potential well is presumably down to H-bonding but do you feel we should be looking to add orientational dependence to this to make an effective single site coarse-grained water?

Professor Lyubartsev answered: (i) I did not checked this question explicitly, but it seems quite plausible that the repulsive part of the potential at shorter than 2.8 Å distances follows $1/r^{12}$ behaviour which is inherited from the repulsive part of the OO potential in the atomistic SPC water. (2) It is the first minimum at about 2.8 Å which corresponds to H-bonded waters, while the second could be probably interpreted as effective dipole–dipole attraction. Naturally, one can add the orientational dependence to the potential, but then it computationally will not be much "cheaper" than the original three-site SPC water.

Dr van der Sman ask: How can the water potential be applied to mesoscopic continuum models (like Lattice Boltzmann) or DPD? I am particularly interested in how to model hydrophobic walls. When water flows through micron sized holes, it matters whether the pore is hydrophilic or hydrophobic (which show deviations from Navier–Stokes). In Lattice Boltzmann or DPD one can define free energy functionals (from which chemical potentials are derived). How does the coarse-grained effective potential for water fit in this framework?

Professor Lyubartsev replied: About the application of the coarse-grained water model (derived from atomistic simulations) in DPD, I refer to a recent paper.[1] I am a bit doubtful that such potential can be useful in Lattice Boltzmann, it implies a somewhat higher level of coarse-graining. The derived potential can be used for DPD simulations of bulk water. If you are interested in behaviour of water near hydrophilic or hydrophobic surface, this potential should be complemented by a potential describing water-surface interactions (which in principle could be derived in a similar way from atomistic simulations).

1 A. Eriksson, M. N. Jacobi, J. Nyström and K. Tunstrøm, *J. Chem. Phys.*, 2009, **130**, 164509.

Dr Milano addressed Professor Lyubartsev and Professor Kremer: Can you critically compare iterative Boltzmann inversion and reverse Monte Carlo pointing out limits and advantages of these two approaches?

Professor Lyubartsev answered: There are advantages and limitations in both of the approaches. Iterative Boltzmann inversion is simpler to implement (only RDF is needed, without cross-correlation matrix), it can begin converging starting virtually from any initial potential. However, it may have difficulty converging accurately (that is, reproducing the target RDF exactly) for complex multicomponent systems. The inverse Monte Carlo/inverse Newton may provide better convergence and better reproduction of the target RDF, but it works well if the initial potential is already a reasonable good guess of the exact solution. Also, the inverse MC requires generally more MC steps per iteration. The practical advise may be to start the fitting process using the iterative Boltzmann, and after it stops to approach to the target RDF, switch to the inverse MC to provide a more accurate fitting.

Professor Kremer answered: In principle both methods apply well to the same problems and give almost identical and eventually undistinguishable results. We have tested and compared them in detail for some test cases.[1] In short, iterative Boltzmann inversion displays a significant slower convergence than inverse Monte Carlo methods. On the other hand, inverse Boltzmann is much less affected by finite size effects compared to inverse Monte Carlo, which is due to nonlocal updates in the latter approach. For a practical test we found that for the coarse graining of methanol a simulation box of size three times the cutoff in the considered radial distribution functions is not sufficient for inverse Monte Carlo while it is for iterative Boltzmann. Thus, if one has large systems at hand, inverse Monte Carlo is very fast and efficient, while for smaller systems iterative Boltzmann methods apply better.

1 V. Rühle, C. Junghans, A. Lukyanov, K. Kremer and D. Andrienko, *J. Chem. Theory Comput.*, 2009, submitted.

Professor Holm addressed Professor Lyubartsev and Professor Kremer: To follow up the previous discussion. Does anyone know a case where the results of Boltzmann inversion and Newton inversion can be different. I understood that this could happen to the neglect of cross-correlations in the Boltzmann inversion, but wonder how the system has to be composed of, so that this matters.

Professor Lyubartsev answered: If both methods converged (that is, they reproduce the same target RDF), the result for the potential should be in principle the same within the statistical error. Practically, the result for potential may be different since sometimes small changes of RDF correspond to large changes in the potentials even for a one-component system (*e.g.* Fig. 2 and 3 in the paper). Another question, is whether Boltzmann inversion always converge—but I do not have much experience on that.

Professor Müller-Plathe commented: The unique correspondence between pair potential and radial distribution function works robustly in one way (potential → RDF) but not in the other (RDF → potential) in practice. Visibly different potentials give RDFs which are identical to within line thickness. Therefore, the RDFs may not have enough precision to allow the construction of a unique potential.

Professor Lyubartsev answered: Right, there are many examples of such behaviour that some changes in the potential do not (almost) affect the RDF. Such changes of the potential are described by eigen values of matrix (eqn (4) and see also ref. 21 of this paper).[1] Noteworthy, the possibility to change potential without noticeable effect on RDF can be used to fit other properties of interest, such as pressure in the CG water model in this paper (though I am not aware about more regular studies of this possibility).

1 A. P. Lyubartsev and A. Laaksonen, *On the reduction of molecular degrees of freedom in computer simulationsin Novel Methods in Soft Matter Simulations*, ed. M. Karttunen, I. Vattulainen, A. Lukkarinen, Springer, Berlin, 2004, pp. 219–244.

Professor Theodorou asked: If I understand correctly, in your water example you have used an averaged $g(r)$ between liquid and gas phases to define your effective potential. Is this $g(r)$ a well-defined quantity? Also, have you tried to estimate a vapour pressure or boiling temperature for water in the context of your coarse-grained model?

Professor Lyubartsev answered: Yes, at fixed N and V, $g(r)$ is well defined by the standard expression in the canonical ensemble (and can be computed in simulations as usual). In some sense it is a mixture of liquid, gas phase and the interface. In these simulations we did not try to evaluate vapour density/pressure since it is supposed to be low and thus its reliable estimation would require very long MD simulations.

Professor Marrink said: According to classical theory, the forces acting between membranes are of four types: electrostatic, van der Waals, undulation, and hydration. The first two are explained by Derjaguin–Landau–Verwey–Overbeek (DLVO) theory, the third one by Helfrich. The origin of the fourth one is less well explained, but is relatively short-ranged. The van der Waals force is always attractive and displays a power law dependence. The three other forces are repulsive between most membranes. The undulation force also displays a power law dependence, whereas both the electrostatic and hydration force decay exponentially with distance.

For neutral membranes, the experimentally observed swelling limit is explained by the balance between the two long range forces, *i.e.* the van der Waals attraction and the repulsion arising form suppressed undulations.

In the simulations presented in the paper, you observe that the experimental swelling limit is reproduced for a DMPC multi-layered vesicle. However, the nature of the interactions in the model is built-in to be short ranged (about 1.5 nm, compared to the swelling limit of more than 6 nm), excluding a correct description of the long range van der Waals forces crucial for the extent to which multilamellar samples would swell. It therefore seems that (i) either the agreement is fortuitous, *i.e.* the right behavior is obtained for the wrong reasons, or (ii) that the classical theory needs revision.

Professor Lyubartsev responded: This CG lipid model includes both short-range interactions (which are indeed limited by about 1.5 nm) and long-range electrostatic interactions coming from charges on N and P sites of the lipid headgroup, which interact by the true Coulombic potential scaled by the dielectric constant (70 in this specific case). The dipoles from surfaces of two bilayers interact with each other, and produce the effective attractive force, exactly according to the mechanism of appearance of attractive van der Waals forces due to dipole–dipole interactions. This attractive force counterweighted by an average repulsive short-range force (due to hydration, effectively accounted for by the effective solvent-mediated potentials) and undulations, which are also present in the model. So there is no need to revise the classical theory.

The presence of charges in the implicit solvent coarse-grained models and their treatment in the inverse procedure was described in previous papers (see ref. 14, 21 in the paper),[1,2] unfortunately mentioning of this was missed in the present paper.

1 A. P. Lyubartsev, *Eur. Biophys. J.*, 2005, **35**, 53.
2 A. P. Lyubartsev and A. Laaksonen, *On the reduction of molecular degrees of freedom in computer simulationsin Novel Methods in Soft Matter Simulations*, ed. M. Karttunen, I. Vattulainen, A. Lukkarinen, Springer, Berlin, 2004, pp. 219–244.

Professor Allen said: You commented on the dynamics of vesicle assembly using Langevin dynamics with a low friction coefficient. It may be important to use a momentum-conserving set of dissipative and random forces (as in DPD) and indeed the authors of your paper's ref. 36 have published some recent papers discussing the tuning of (pair) friction terms to reproduce dynamics, in the context of a potential coarse-graining scheme.[1] To do this, while maintaining temperature control, it is helpful to combine deterministic (low friction) and stochastic (high friction) methods;[2] papers on possible momentum-conserving thermostats of both kinds have also appeared.[3,4]

1 A. Eriksson, M. N. Jacobi, J. Nystrom, K. Tunstrom, *J. Chem. Phys.*, 2008, **129**, 024106.
2 S. D. Stoyanov and R. D. Groot, *J. Chem. Phys.*, 2005, **122**, 114112.
3 C. P. Lowe, *Europhys. Lett.*, 1999, **47**, 145.
4 M. P. Allen and F. Schmid, *Mol. Simul.*, 2007, **33**, 21.

Professor Lyubartsev responded: I agree that this is a good way to reproduce the dynamics of coarse-grained models.

Professor Español commented: The Newton inversion method from the paper by Professor Lyubartsev allows one to construct coarse-grained potentials that best reproduce targeted microscopic averaged information. Let me show that the method can be reformulated as a *tuning on the fly* of the parameters of the coarse-grained potential according to certain targeted microscopic information, while running the coarse-grained dynamics.

Let z be the microstate of the system, consisting on the positions and momenta of the atoms of the system. Consider a number of coarse-variables $A(z)$ which describe the system at a coarse-level and which take numerical values α. These variables can be, for example, some of the microscopic degrees of freedom or combinations of them like the positions and velocities of centers of mass of portions of complex molecules. The equilibrium distribution function of the coarse variables is given by

$$P^{eq}(\alpha) = \int dz \frac{1}{Z} \exp\left\{-\beta H(z)\right\} \delta(A(z) - \alpha) \equiv \frac{1}{Z'} \exp\left\{- \beta H^{eff}(\alpha)\right\} \quad (1)$$

where $H(z)$ is the microscopic Hamiltonian of the system and the partition functions Z,Z' are the usual normalization factors of the microscopic and coarse-grained probabilities. The effective Hamiltonian $H^{eff}(\alpha)$ is introduced as a way to re-write the equilibrium probability in an appealing way. Depending on the context $\beta H^{eff}(\alpha)$ may be understood as a free-energy or as an entropy, like in Einstein fluctuation theory.

From eqn (1), any equilibrium average of functions $F(\alpha)$ of the coarse-grained variables can be computed in the following two equivalent ways

$$\begin{aligned} \langle F \rangle^{eq} &= \int dz \frac{1}{z} \exp\{-\beta H(z)\} F(A(z)) \\ &= \int d\alpha \frac{1}{Z'} \exp\{ - \beta H^{eff}(\alpha)\} F(\alpha) \end{aligned} \quad (2)$$

The effective Hamiltonian is, in general, a function in the multidimensional space spanned by the coarse-variables α. As such, it is a very difficult object to compute because sampling a high dimensional space requires unaffordable computer power. Therefore, it is necessary to summarize the functional form into simplified functions $H^{eff}(\alpha;\lambda)$, typically of the pair-wise form, that may depend on a set of parameters λ. These parameters should be chosen as to produce "the best results". The Newton inversion method is one method to achieve this. The idea is that given the parameters λ of the effective Hamiltonian it is always possible to compute some target observables like, for example, a radial distribution function of CoM. The target observables will be averages $o^{eq} = \langle O \rangle^{\lambda}$ of some functions $O(A(z))$ of the coarse variables. The notation $\langle ... \rangle^{\lambda}$ denotes the equilibrium average computed with the parametrized coarse-grained Hamiltonian, *i.e.* the second eqn (2) with the λ dependent effective Hamiltonian $H^{eff}(\alpha;\lambda)$. These observables o^{eq} can be, in turn, computed from the exact underlying molecular dynamics (*i.e.* from the first eqn (2)). This gives in general a connection between o^{eq} and λ in the form $o(\lambda)$. However, one is interested in the inverse of this function, $\lambda(o)$, that provides the best set of parameters λ that produces the correct averages of the observables o^{eq}. The method proposed in Lyubartsev's paper uses the iterative Newton-Raphson method to obtain the best set for λ. The purpose of this note is to reinterpret the method in such a way that it may be easier to implement, while at the same time it offers an appealing simplicity. The whole discussion is simplified when we assume that the number of parameters is equal to the number of observables. This assumption is taken in the following.

The essential idea of the method of Lyubartsev's paper is to run coarse-grained simulations with tentative effective potentials, and improve upon the effective potential by driving the parameters to better estimates. In a very succinct way let us write the dynamics of the coarse-grained model as

$$\partial_t \alpha(t) = F(\alpha;\lambda) \quad (3)$$

where the partial time derivative is denoted by ∂_t. For the case that the coarse variables are positions and momenta of selected atoms or of CoM of groups of atoms $\alpha = \{\mathbf{R}_\mu;\mathbf{P}_\mu\}$, the above equations are Hamilton's equations with the effective Hamiltonian $H^{\text{eff}}(z;\lambda)$. This is

$$\begin{aligned}\partial_t \boldsymbol{R}_\mu &= \frac{\partial H^{\text{eff}}}{\partial \boldsymbol{P}_\mu}(\alpha;\lambda)\\ \partial_t \boldsymbol{P}_\mu &= \frac{\partial H^{\text{eff}}}{\partial \boldsymbol{R}_\mu}(\alpha;\lambda)\end{aligned} \tag{4}$$

This dynamic allows one to compute the equilibrium averages $\langle O\rangle^\lambda$ of the observables and also of the Jacobian matrix, which is explicitly given by

$$\begin{aligned}J(\lambda) &\equiv \frac{\partial\langle O\rangle^\lambda}{\partial\lambda}\\ &= -\beta\left[\left\langle\frac{\partial H^{\text{eff}}}{\partial\lambda}O\right\rangle^\lambda - \left\langle\frac{\partial H^{\text{eff}}}{\partial\lambda}\right\rangle\langle O\rangle^\lambda\right]\end{aligned} \tag{5}$$

This Jacobian matrix, being the negative of a covariance, has a negative definite character and is, therefore, invertible. According to the Newton-Raphson method, one improves the value of λ to $\lambda' = \lambda + \Delta\lambda$ where $\Delta\lambda$ is obtained from the solution of the linear system of equations

$$\langle O\rangle^{\text{eq}} - \langle O\rangle^\lambda = J(\lambda)\Delta\lambda \tag{6}$$

Here $\langle O\rangle^{\text{eq}}$ is the targeted equilibrium average, which is computed from the microscopic dynamics of the system and it is taken as input of the method. Eqn (6) can be formally solved as

$$\Delta\lambda = J^{-1}(\lambda)(\langle O\rangle^{\text{eq}} - \langle O\rangle^\lambda) \tag{7}$$

The whole iterative process suggests an "evolution" in the parameter space. In fact, we may write this dynamic in continuum time as

$$\partial_t\lambda(t) = \frac{1}{T}J^{-1}(\lambda)\left(\langle O\rangle^{eq} - \langle O\rangle^\lambda\right) \tag{8}$$

where T is a convenient time scale. Of course, in order to compute the time evolution of λ through eqn (7) we need to compute, for every λ encountered in the evolution, the averages $\langle O\rangle^\lambda$ and the Jacobian matrix $J(\lambda)$ which require the running of the coarse grained dynamics.[2]

Our proposal is to perform the averages $\langle\ldots\rangle^\lambda$ required in eqn (8) at the same time as the evolution of λ takes place. To this end, we enlarge the state space of the coarse variables with the set of parameters and propose the following coupled dynamics in the enlarged space

$$\begin{aligned}\partial_t\alpha &= F(\alpha;\lambda)\\ \partial_t\lambda &= \frac{1}{T}\left(\langle O\rangle^{\text{eq}} - O(\alpha)\right)\end{aligned} \tag{9}$$

The first equation is just the coarse-grained evolution at a given set of parameters in eqn (3), while the second equation is suggested by the form of eqn (8). Note, though, that there are no averages involved in eqn (9) (apart from the input target equilibrium average $\langle O\rangle^{\text{eq}}$ which should be computed previously). We could include a definite positive matrix multiplying the term $\langle O\rangle^{\text{eq}} - O(\alpha)$ in order to represent the factor $J(\lambda)^{-1}$. A calculation of the inverse $J(\lambda)^{-1}$ is, however, cumbersome. In addition, for the rest of the argument the actual form of this matrix is irrelevant, although it could affect the performance of the method. Therefore, at this stage we choose the simplest positive definite matrix, the identity.

The image we have in mind is that we evolve the coarse-grained simulation with a set of slowly varying parameters λ in the effective Hamiltonian that evolve in such a way to reduce as much as possible the differences between the targeted equilibrium average from the actual average. Therefore, the time scale T that governs the scale of evolution of λ is a time scale which is "large". We may think of T as the time span over which the required averages in eqn (7) are computed through time averages.

In order to be more precise and see that eqn (9) will do the job, we rescale time with $T = \varepsilon^{-1}$ and rewrite eqn (9) in a standard way in the analysis of stiff ODEs

$$\begin{aligned} \partial_t \lambda &= g(\alpha) \\ \partial_t \alpha &= \frac{1}{\varepsilon} F(\alpha; \lambda) \end{aligned} \tag{10}$$

where $g(\alpha) = \langle O \rangle^{\text{eq}} - O(\alpha)$. In the limit $\varepsilon \to 0$, the variables α are very fast as compared with the variables λ. We will assume that the dynamics of the fast variables is ergodic, implying that for every fixed λ the second equation in (10) has an equilibrium measure. This is a rather natural assumption and it is, of course, implicitly assumed when computing equilibrium averages from time averages of the parametrized coarse-grained dynamics. We call $P^{\text{eq}}(\alpha;\lambda)$ the equilibrium probability of α conditional to the parameters λ. Now, according to a general theorem of averaging due to Anosov,[1] when the dynamics of α is ergodic, in the limit $\varepsilon \to 0$ the slow variables λ evolve according to the following closed equation $\partial_t \lambda = G(\lambda)$ (11)
where the function $G(\lambda)$ is defined as

$$G(\lambda) \equiv \int \mathrm{d}\alpha P^{\text{eq}}(\alpha; \lambda) g(\alpha) \tag{12}$$

In order to close the argument, note that if the dynamic eqn (10) has a stationary state λ^* for which $\partial_t \lambda = 0$, then, the value λ^* satisfies $G(\lambda^*) = 0$ or, in an equivalent way,

$$\int \mathrm{d}\alpha P(\alpha; \lambda*) O(\alpha) = \langle O \rangle^{\text{eq}} \tag{13}$$

this is $\langle O \rangle^{\lambda^*} = \langle O \rangle^{\text{eq}}$ (14)

In summary, by running a sufficiently long coarse-grained simulation in which the parameters λ of the effective Hamiltonian depend on time and evolve according to eqn (9), in the limit of large T and for large times, the parameters λ should converge towards the value λ^* that gives the correct matching between the coarse-grained averages and the targeted averages. The beauty of the method is that by using the enlarged set of eqn (9) and running a single simulation, one recovers the parametrization of the coarse-grained model that reproduces the targeted averages.

Note that the reversible dynamics at the coarse level in eqn (3) is not the correct dynamics of the coarse variables. In general, the dynamics should include also friction and noise.[2] This does not compromise the present method nor the Newton inversion method, because these methods are directed exclusively to find approximate and manageable expressions for the effective Hamiltonian. Because the method relies on slow evolution and invariant measures, it allows to find these purely static quantities by using just the reversible part of the coarse dynamics. It is a very interesting open question whether a similar idea of running the full coarse-dynamics (*i.e.* reversible and irreversible parts included) can be used to tune on the fly the parameters in the coarse dynamics in order to target both static and dynamic quantities.

1 D. Givon, R. Kupferman, and A. Stuart, *Nonlinearity*, 2004, **17**, R55.
2 C. Hijon, P. Español, E. van den Eijnden, and R. Delgado-Buscalioni, *Faraday Discuss.*, 2010, **144**, DOI: 10.1039/b902479b.

Professor Theodorou continued the discussion of the paper by Professor Müller-Plathe: Inputs to your equilibration method are the Kuhn length and the density of the polymer under investigation. Do you have a recommendation for estimating these quantities, in cases where they are not known *a priori*?

Professor Müller-Plathe replied: Alas, I have no recommendations, which are cleverer than the common practice: use atomistic simulations with a good force field on oligomers of different size and extrapolate to long chain lengths. Or use the atomistic end-bridging or double-bridging MC of Theodorou and coworkers to simulate long chains directly. But then you have solved the problem of generating equilibrated atomistic structures, too.

Professor Theodorou asked: Does the quality of structures you generate depend on the rate at which you introduce excluded volume interactions in your MD relaxation process?

Professor Müller-Plathe responded: It probably does, although we have not checked this systematically. We have used a rather slow, and possibly unnecessarily slow, route for turning on the potentials, in order to be on the safe side.

Professor Frenkel continued the discussion of the paper by Dr Lyubartsev: Your Fig. 1 shows an example of $g(r)$ in the two-phase region. Surprisingly, you find that you can derive consistent effective potentials (Fig. 2). However, it is not obvious to me that $g(r)$ in a two-phase region is a well-defined quantity.

Professor Lyubartsev answered: If N and V are specified, $g(r)$ is fully defined (standard expression in the canonical ensemble). Naturally, it depends on N and V (and even on V at fixed density in the case of two-phase separation). The effective potentials are computed at the same N and V as the corresponding $g(r)$, so the size effects are cancelled out. That is why the effective potentials appear more consistent (less size or density dependent) than $g(r)$.

Professor Español opened the discussion of the paper by Dr Graham: The model presented is an "algorithmic" model that can be understood as a kinetic model. Is it possible to formulate a well-defined continuum model whose numerical implementation would correspond, in certain limits, to the kinetic model shown? This would be useful in order to have independent analytical predictions (at least in simple situations) from the continuum model and get, possibly, additional insight.

Dr Graham answered: Producing continuum models is likely to be a useful application of this model. Our recent simulations show that the nucleation rate is a simple universal function of chain stretching, which it may be possible to derive analytically. Such a result would be very useful in deriving simple continuum models of flow-induced crystallisation that would be suitable for computational modelling of polymer processing. Our algorithm's potential to link detailed simulations to simple continuum models is an interesting avenue.

Professor Kremer asked: The results suggest the option for a truly quantitative comparison of simulations to experiments and/or simulations. For this it would be good to compare the results to already existing experiments. How well defined and accessible are the different paramaters, which enter the theory? Also I would like to learn more about details of the underlying simulation model for the polymer chain.

Dr Graham responded: Quantitative comparison with data is a truly pressing issue and such a comparison with experimental data is, in principle, possible. Directly observed nucleation rates under flow, such as recent measurements by Coccorullo,

Pantani and Titomanlio[1] are good candidates for this comparison. There are some difficulties as most experiments are made on polydisperse samples and, unfortunately, no suitable model for non-linear flow of polydisperse materials currently exists. This could be solved by improvements in either polymer synthesis or molecular flow modelling, both of which are nascent. Most of the model parameters can be determined by independent quiescent measurements, with rheology providing the molecular relaxation times and quiescent crystallisation giving the crystallisation parameters. Some issues arise because typical characterisation techniques such as size exclusion chromatography and linear rheological measurements are generally substantially harder for semi-crystalline polymers than for amorphous polymers. A combination of comprehensive characterisation and crystallisation measurements would provide a challenging quantitative test for the model.

The model for the underlying polymer chains was originally derived to model mechanical stresses and neutron scattering from flowing amorphous polymers. It does not simulate individual chains; instead it models the deterministic evolution of the ensemble averaged chain configurations. It does, however, resolve the variation of molecular deformation with position along the polymer chains. Currently, we do not feedback information about the evolving crystallisation into the model for the amorphous chain dynamics, although it would be possible to relax this assumption in the future.

1 I. Coccorullo, R. Pantani, G. Titomanlio, *Macromolecules*, 2008, **41**(23), 9214–9223.

Professor De Pablo opened the discussion of the paper by Professor Holm: Could you please elaborate on your choice of a Lattice–Boltzmann technique for description of hydrodynamic interactions over a method based on Green-tensor based hydrodynamics?

Professor Holm replied: Its the method we know best. I see no real reason for this study to prefer one mesoscopic hydrodynamics method over the other. The ones I am familiar with, namely LB, DPD, SRD will all perform more or less similar in speed and accuracy, at least to my knowledge. We have performed a comparison of the LB method with the DPD method for a simple Poiseulle flow, showing no differences,[1] and I would expect similar results for ANY mesoscopic hydrodynamic method. I think, also Green-tensor based hydrodynamics would do equally well, and there are FFT accelerated methods available which makes it probably equally suited for this problem. The advantage of the Lattice–Boltzmann method is, that if we increase the particle concentration and keep the simulation box fixed, the cost of the LB algorithm stays roughly constant.

1 J. Smiatek, M. Sega, C. Holm, U. D. Schiller and F. Schmid, *J. Chem. Phys.*, 2009, **130**, 244702.

Dr van der Sman said: You take excluded volume into account for the polyelectrolytes and counter-ions in their interactions *via* the Lennard–Jones potential. Why don't you take excluded volume (excluded by polymers and solutes) into account in the model for the solvent phase. In concentrated systems their volume fraction can be significant. In two-fluid models of soft matter, as by Fielding and Olmsted (PRE,2003) excluded volume of dispersed phase is taken into account.

Professor Holm answered: This is not exactly in the spirit of the Lattice–Boltzmann method we are using here. Its a one-component, incompressible lattice fluid, and we are investigating basically singe-chain mobility properties, hence we are working under dilute conditions. The effects you have probably in mind might play a role in very concentrated solutions, which are not considered here, where the excluded volume effects are much larger (probably micrometre spheres). The excluded volume interactions have definitely an effect for the short range interactions, but then the granularity of a real solvent, like hydrogen bonds for water, would also play a prominent role on short distances. The Lattice–Boltzmann method is designed to reproduce the long range hydrodynamic interactions correctly, and for this purpose it works marvellously well.

Professor Frenkel continued the discussion of the paper by Dr Graham: Can you comment on the formation of the "Kebab" part of the Shish-kebab structures that you mention? Your article focuses on the hydrodynamic enhancement of shish formation only.

Dr Graham responded: Our model focuses on shish formation as this process appears to be most strongly affected by flow. In particular, we aimed to produce a framework to predict the flow conditions required for shish formation, to which the details of kebab growth are not essential. A lowering of our coarse-graining scheme would be needed to model kebab formation. We note that simulations by Hu, Frenkel and Mathot[1] have shown that unstretched chains will form kebabs around a single aligned chain. Thus we would expect kebabs to grow around our elongated nuclei if we switched to a similar simulation scheme.

1 W. B. Hu, D. Frenkel, V. B. F. Mathot, *Macromolecules*, 2002, **35**, 7172–7174.

Dr Boek said: You have observed that, in your model, nucleation increases with increasing shear rate. What is the effect of shear on the size of the critical nuclei? In a recent experimental study on colloidal asphaltene aggregation and fragmentation in a shear field,[1] the size of colloidal aggregates is measured as a function of increasing shear rate. In these experiments, the aggregate size first increases due to enhanced collision rates, then decreases due to shear forces eroding the aggregate. Are you observing similar effects in your model?

1 N. H. G Rahmani, J. H. Masliyah and T. Dabros, *AIChE J.*, 2003, **49**, 1645.

Dr Graham answered: In our approach the critical nucleus size is the size required for the nucleus to reach the maximum of the free energy barrier to nucleation. That is, the size beyond which growth is spontaneous. In our model, flow increases the thermodynamic driving force without affecting the surface energy cost. This changes the balance of the bulk free energy gain of crystallisation with the surface area cost, meaning that smaller crystals will be stable. Thus flow decreases the critical nucleus size.

Professor Bolhuis asked: The model you have developed assumes that classical nucleation theory holds also for polymer crystal nuclei. This means that the surface energy (or surface tension) is assumed to be isotropic, *i.e.* identical everywhere on the surface of the nucleus. How correct is this assumption for polymer crystallisation, with or without flow? For instance, would making this assumption make a qualitative difference for the nucleation rate?

Dr Graham replied: Yes, an isotropic surface energy is assumed. In effect, this neglects the excluded volume of dangling amorphous chains emerging from the crystallised stems and the potential for surface folds in these chains. This is clearly important for extended regions of crystal lamellae but may be less so for crystal nuclei around the critical size as the excluded volume restriction is lower for smaller crystallites. This effect could be accounted for within our approach by adding an additional free energy penalty that depends on the number of stems in the nucleus. Even with this effect included, nucleation would still be dominated by the overall barrier height so no qualitative changes in the nucleation rate would be expected. This will, however, make shish nuclei more favourable and so could make an important contribution to the transition to oriented crystals.

Professor Deserno continued the discussion of the paper by Professor Holm: In your model both the polyelectrolyte monomers as well as the ions are represented by beads which have the same friction coefficient in the solvent. Now this does not necessarily have to be true in reality. What would happen to the force balance inside a polyelectrolyte-counterions coil when these friction coefficients are different?

Professor Holm replied: I think if the friction coefficient for both ion species did not have the same value, we would observe almost the same effects. The observed mobilities would change quantitatively, but not qualitatively. However, we have not investigated this case, and it might be a worthwhile option.

Professor Jackson asked: The dielectric of the solution is very sensitive to the addition of salt. How important is this feature and have you taken it into account in your studies?

Professor Holm answered: We did not take the change of the dielectric constant with salt concentration into account. I think this is actually an excellent question which needs to be investigated in more detail in the future. In recent work[1] we found that for simple electrolyte solutions (NaCl) the dielectric constant can decrease by a factor of 2–3 for salt concentrations of up to four molar. The problem to take these effects into account in simulations with spatially varying dielectric constant is that there is no fast solver available that can calculate the interactions for periodic boundary conditions. This is actually a problem we are currently working on. However, for the considered salt concentrations, I do not think that salt effects will change the dielectric very much (no big quantitative difference is expected), nor will they change the results qualitatively. The PE chains are still relatively spherically symmetric objects, and dielectric mismatches will average out.

1 B. Hess, C. Holm and N. van der Vegt, *Phys. Rev. Lett.*, 2006, **96**, 147801.

Professor Marrink remarked: In all simulations, an electric field of reduced strength $E = 0.1$ was used. You write that it has been checked that this field strength is small enough as to not affect the chain conformation or distribution of surrounding ions.

Two related questions:

(1) How does this field strength compare to the experimentally used fields?

(2) Can you also exclude the possibility that the field is affecting the hydrodynamics around the polyelectrolyte? One could imagine that, in an extreme case, turbulent patterns may arise that would strongly change the electrophoretic mobility although they may not affect the chain conformation so much.

Professor Holm replied: (1) Our applied field values are still higher than the experimentally used ones. This has, however, no influence on the measured mobility, since *via* simultaneous measurements under zero field, and using the Green–Kubo formalism, we made sure that this field value lies well within the linear response regime, and since the mobility is the quotient of observed velocity over applied field strength, the mobility in this regime is independent of the applied field strength.

(2) The simulations are performed in the low Reynolds number regime, and hence one does not really expect to find a turbulent regime here. Already the fact that we are in the linear regime for the mobility should guarantee this. Some time back, when we investigated electrophoresis of small nano-colloids in the linear regime,[1] we looked at the flowline pattern of the fluid, which gave no sign of turbulence. I am sure that we see even smoother flowlines in the case of the PE electrophoresis, although we have not looked specifically at the flowlines. Only in extremely high fields, well beyond the linear regime, one could possibly see this. However, as stated before, we were not interested in these extreme cases, but opted rather to make contact with experiments.

1 V. Lobaskin, B. Dünweg, C. Holm, *J. Phys.: Condens. Matter*, 2004, **16**, S4063–S4073.

Professor Frenkel asked: Have you looked at the non-linear response of poly-electrolyte solution to an applied electrical field?

Professor Holm responded: No, that problem was never in the focus of our intention. On the contrary, we always made sure that we were strictly confined to the linear regime, since we were mainly interested in the electrokinetic behavior which is experimentally accessible. There the applied voltages are so small, that you are orders of magnitude away from the non-linear regime. The non-linear regime, however, has been investigated by R. Winkler an S. Frank in a recent EPL by using the stochastic rotation dynamics (SRD) method.[1]

1 S. Frank and R. G. Winkler, *EPL*, 2008, **83**, 38004.

Professor Kremer said: Both simulations of polyelectrolytes as well as simulations including hydrodynamic effects can be subject to severe finite size effects. Since the present simulations deal, in some cases, with low or no salt content, both the electrostatic screening as well as the hydrodynamic screening lengths relative to the periodic box size might be critical for the comparison to experiment. To exclude this please give the precise box/system sizes and estimates of the electrostatic and hydrodynamic screening lengths.

Professor Holm answered: We actually checked explicitly for finite size effects, by varying the number of chains at fixed monomer concentrations, and found no effects for our investigated system sizes. The box sizes we used were ranging roughly from L approx 50 σ for the longest chain to L approx = 10 σ for a single monomer. The associated Debye lengths are ranging from approx 7 σ in the salt free-case to approx 2 σ in the salt case, and we have seen that the hydrodynamic screening lengths are of the same order of magnitude. Let us keep in mind that hydrodynamic interactions are screened in the case of electrophoresis (and this is notably different from the case of sedimentation!), and therefore, one would already expect that finite size effects for these small screening lengths should not be visible for our investigated situations.

Dr van der Sman queried: You talk about effective friction coefficients for the polyelectrolytes, and ions being modified by the electrostatic interactions. Why is this enhanced friction not reflected in the Brownian motion, and does this mean a violation of the fluctuation–dissipation theorem?

Professor Holm replied: One has to carefully distinguish here between the friction that the polymer and ion beads have with the solvent, and which is determined *via* the viscous coupling of the particles to the Lattice Boltzmann fluid within the algorithm (hence a fixed number), and the effective friction that is defined *via* the local force balance equation on page 5 in the paper. The algorithm we are using obeys the fluctuation–dissipation theorem, hence there cannot be any violation of it by construction. What I meant by modified friction, is the fact that the quantity which we call effective friction, depends on electrostatic interactions in the way that ions that comove with the chain, also contribute to the observed enhanced effective friction by slowing down the mobility. *Via* independant measurements of the effective charge we could quantify this, since from mobility measurement and effective charge measurement one obtains a value for the effective friction. Also in the chain diffusion this effect is visible, and can also be interpreted as being the effect of the electrostatic coupling of the counter ions to the chain charges.

Professor Theodorou continued the discussion of the paper by Dr Graham: This was a very interesting kinetic Monte Carlo approach for nucleation in polymer melts. I guess, in a quiescent polymer melt, your model would give a spherical nucleus consisting of parallel chains. How is this reconciled with the spherulitic morphology one sees in practice. Can you envision extensions of your modelling approach that might lead to spherulitic morphologies? These morphologies perhaps suggest that the nucleus looks more like a lamella than a sphere.

Dr Graham answered: A simple modification of the surface energy term to account for the different faces of a lamella would change the simulated nuclei from a spherical to a lamella morphology. This change is unlikely to result in qualitative changes in the nucleation rate predictions as the nucleation rate is dominated by the height of the nucleation barrier, with the overall shape of the free energy landscape having a lesser effect. This feature may, however, be important to correctly predict the details of shish nucleation. The spherulitic morphology, which results from the later stages of crystal growth, is beyond what can be captured by our current level of coarse-graining. Our coarse-graining scheme focuses on nucleation as flow has the most dramatic effect on this process.

Professor Allen continued the discussion of the paper by Dr Graham: You concentrate in your paper on shear flow. What are the differences when the type of flow is changed in the experiments (*e.g.* extensional flow) and does your scheme handle this equally well?

Dr Graham responded: Shear is the mostly widely studied flow geometry in flow induced crystallisation (FIC) experiments as it is more straightforward to realise than extensional flow. On the other hand, extensional flow is ubiquitous in industrial polymer processing and so a useful model of FIC must cover this flow. The primary difference between extension and shear in this context is that, when comparing identical deformation rates, extension generally stretches the amorphous chains more than shear. Thus extension results in faster nucleation and a greater tendency to produce shish nuclei, an effect that is seen in our model. Our model also suggests some subtle differences between extension and shear. We predict that the distribution of stretch along a chain in extensional flow is more even than in shear flow. This means the "zipping-up" mechanism that we propose for shish formation is more effective in extensional flow, although it is unclear whether this effect is strong enough to be detected in experiments.

Professor Lowen continued the discussion of the paper by Professor Holm: How does the maximum in the mobility as a function of chain length change if the flexibility of the chains is reduced? Does it occur for semiflexible or stiff polyelectrolytes (and for colloidal spheres) at low salt concentrations?

Professor Holm replied: The maximum in the mobility will decrease in the case of stiffer chains. It should also occur for semiflexible PEs in low salt concentrations. For colloidal spheres the answer depends on the precise prescription on how you scale up the charges, which can be done either at fixed colloidal radius or at fixed charge density. For fixed colloidal radius we did some work with V. Lobaskin and B. Dünweg[1] some years ago where we also observed a maximum in the mobility at some certain value of the bar charge. At this point charge condensation does not increase the effective charge any more, which then would result in a constant mobility if one assumes that the effective friction would stay constant. However, due to the increase in double layer thickness, the effective friction also increases slightly, yielding a slightly decreasing mobility for increasing N, hence it also gives a maximum in the mobility. However, I think the comparison is not quite fair, since the colloid keeps its form, whereas the polyelectrolyte first stays always at the same bare linear charge density (no counterion condensation occurs yet), and also its conformation can change.

1 V. Lobaskin, B. Dünweg and C. Holm, *J. Phys.: Condens. Matter*, 2004, **16**, S4063–S4073.

Dr van der Sman continued the discussion of the paper by Professor Holm: Why should you not take into account explicit lubrication forces between ions and polyelectrolytes? If polyelectrolytes are partially charged like proteins, they can still coil up to globular coils and have sizes of about 10 nm.

Professor Holm answered: Polyelectrolytes only have like charges on their backbone. Therefore they do not coil up, but are rather extended objects with a fractal dimension close to one. Only in the case of poor solvent polyelectrolytes, where you have some sort of microphase separation on a single chain into partially collapsed domains, yielding so-called pearl-necklace structures. These kind of polyelectrolytes were not investigated here. The lattice spacing of the Lattice Boltzmann algorithm is the same as the bead diameters of all beads in our system, polyelectrolyte beads as well as counterion beads, and have the value 1σ (which is the Lennard–Jones unit in our system). This corresponds to about 4 Å, meaning we work here very much with microscopic length scales, where solvent granularity already plays a big role in real systems. Therefore I believe that for these system sizes, lubrication forces play no role, our objects are simply too small.

Professor Frenkel commented: As poly-electrolytes are not made of smooth spheres, the lubrication picture is not the one to use when Lattice–Boltzmann reaches the resolution limit.

Professor Holm answered: Thank you for the comment, Dan. This answers basically Ruuds question.

Simulations of theoretically informed coarse grain models of polymeric systems

François A. Detcheverry,[a] Darin Q. Pike,[a] Paul F. Nealey,[a] Marcus Müller[b] and Juan J. de Pablo[a]

Received 5th February 2009, Accepted 22nd April 2009
First published as an Advance Article on the web 4th September 2009
DOI: 10.1039/b902283j

Simulations of theoretically informed coarse grain models, where the interaction energy is given by a functional of the local density, are discussed in the context of polymeric melts. Two different implementations are presented by addressing two examples. The first relies on a grid-based representation of non-bonded interactions and focuses on the concept of density multiplication in block copolymer lithography. Monte Carlo simulations are used in a high-throughput manner to explore the parameter space, and to identify morphologies amenable to lithographic fabrication. In the second example, which focuses on the order–disorder transition of block copolymers, the constraints imposed by a grid are removed, thereby enabling simulations in arbitrary ensembles and direct calculation of local stresses and free energies.

1 Introduction

The theoretical and computational description of self-assembled block copolymers[1] is of considerable interest, and issues of central importance, such as the temperature of the order–disorder transition (ODT), or the morphologies that may arise in multiblock materials are still under investigation.[2] Efficient simulation methods to predict block copolymer morphology and properties would also be useful for development of emerging applications of copolymers in nanoscale fabrication. In the context of electronic devices, current efforts aim to exploit the spontaneous self-assembly of block copolymers to overcome the limitations of traditional lithographic techniques, and obtain controlled nanostructures with characteristic dimensions in the range of a few tens of nanometres.[3–9] In addition to the periodic structures that are useful for memory storage devices, non-regular geometries, such as T-junctions and bends, are needed for the fabrication of integrated circuits and are actively pursued.[10] "Block copolymer lithography" has been shown to offer considerable potential for creation of features smaller than those accessible to traditional lithographic techniques. Smaller features, however, require that shorter polymers be used. As characteristic length scales approach a few nanometres, the role of fluctuations—and the need for theoretical descriptions that account for them—is expected to become increasingly important.

Given that ordered morphologies have a natural periodicity ranging from 5 to 500 nanometres, an atomistic description of block copolymers is impractical and coarse-graining strategies must be adopted. A wide variety of coarse grain models available in the literature can be broadly classified into two categories. In particle-based approaches, the polymeric chains are represented by particles which interact through a potential energy function. Thermodynamic properties are computed with Monte

[a]*Department of Chemical and Biological Engineering, University of Wisconsin, Madison, WI, 53706, USA*
[b]*Institut für Theoretische Physik, Georg-August Universität, 37077 Göttingen, Germany*

Carlo (MC) simulations, and the dynamics of individual chains can be studied using molecular dynamics (MD) or dissipative particle dynamics (DPD) approaches,[11,12] to name a few. These approaches include fluctuation effects. As an additional advantage, particle-based techniques can handle complex molecular architectures in a relatively straightforward manner.

In field-based models, the degrees of freedom are the fields of local density and the interaction energy of the system is written as a functional of those fields. The properties of this "standard model" of block copolymers[13,14] have been studied by a variety of analytical and numerical methods, the most common being the self-consistent field theory (SCFT).[15,16] Compared to most particle-based representations, field-based descriptions involve a higher level of coarse-graining, since the chains are described only implicitly and are assumed to be in equilibrium with the fields at all times. As a result, the description of dynamics requires the introduction of Onsager coefficients and often involves more drastic approximations than particle-based models. With only a few exceptions—notably the field-theoretic simulations (FTS),[17] which are more computationally demanding—field-based models are usually studied with methods where fluctuations are neglected.

Particle-based and field-based approaches are complementary since each of them can efficiently address a class of materials or properties. However, because the underlying models are different, a comparison between those approaches is seldom straightforward. Ideally, one would prefer to describe all properties using the same model; furthermore, that model could or should be treated either with particle-based methods (MC and MD) or with field-based treatments (SCFT or FTS). With this idea in mind, a number of studies[8,18–22] have sought to introduce particle-based descriptions where the interactions between chains are taken into account with a functional of the densities, as is done in traditional field-theoretic approaches.[14,15] For conciseness, we refer to such approaches as "Theoretically Informed Coarse Grain" (TICG) simulations.

In what follows we describe the ideas and some applications of TICG simulations. For completeness, we do so in the context of two examples. The first examines the ability of block copolymer materials to multiply the density of patterns created lithographically on a surface. "Density multiplication" or "pattern interpolation" is of interest in copolymer lithography,[9] and TICG simulations are used to explore the morphologies that might arise in experiments as a function of a number of key parameters. Predicted morphologies can subsequently be used for interpretation of experimental micrographs and for design of actual nanoscale patterning strategies. In this application, an efficient grid implementation is adopted that facilitates consideration of hundreds of different scenarios, thereby providing a broad overview of the behavior of the system. While the grid implementation provides a computationally efficient calculation of interactions in dense systems, it suffers from a number of shortcomings. In particular, simulations are restricted to constant volume ensembles, and forces between particles are not well defined. Our second example presents a method that circumvents those difficulties, and that shows how TICG simulations can be carried out in a variety of ensembles to directly compute thermodynamic and structural properties. Since forces between particles are uniquely defined, mechanical properties such as the local stress become accessible, thereby offering interesting prospects for design of materials with specific mechanical characteristics. To illustrate the capabilities of the method, we consider a symmetric block copolymer with experimentally relevant invariant degree of polymerization $\overline{\mathcal{N}} \approx 10^4$. We calculate the distribution of local stresses and locate the ODT by computing the Gibbs free energy of the material as a function of temperature.

2 Method

The model considered in this work consists of n chains of an AB diblock copolymer. The volume is denoted by V and the temperature by T. Each molecule is represented by a collection of N beads connected by harmonic springs. Denoting by $\mathbf{r}_l(s)$ the position of the sth bead in the lth chain, the bonded energy is:

$$\frac{\mathcal{H}_{\text{b}}}{k_{\text{B}}T} = \frac{3}{2}\sum_{l=1}^{n}\sum_{s=1}^{N-1}\frac{[\mathbf{r}_l(s+1) - \mathbf{r}_l(s)]^2}{b^2} \tag{1}$$

where k_{B} is Boltzmann's constant and b^2 is the mean squared bond length in an ideal chain. The non-bonded energy is given by:

$$\frac{\mathcal{H}_{\text{nb}}}{k_{\text{B}}T} = \sqrt{\bar{\mathcal{N}}}\int_V \frac{\mathrm{d}\mathbf{r}}{R_e^3}\left[\chi_{\text{o}}N\phi_{\text{A}}\phi_{\text{B}} + \frac{\kappa_{\text{o}}N}{2}(\phi_{\text{A}} + \phi_{\text{B}})^2\right] \tag{2}$$

where $R_e^2 = (N-1)b^2$ is the mean squared end-to-end distance of an ideal chain, $\phi_{\text{K}}(\mathbf{r})$ denotes the local, dimensionless density of beads with type $K = A$ or B and $\bar{\mathcal{N}} = (\rho_{\text{o}}R_e^3/N)^2$, where ρ_{o} is the bead number density, is the invariant degree of polymerization (which in a melt is proportional to the molecular weight). The first term of eqn (2) describes the incompatibility between unlike beads. The second term is inspired by Helfand's quadratic approximation[14] and gives the melt a finite compressibility. The assumption of strict incompressibility that is often adopted in traditional field-theoretic models is not made here.

When considering confined block copolymers or thin films, a neutral surface can be represented as a hard wall that beads are not allowed to penetrate. The effect of a non neutral substrate can be modeled by the surface energy $\mathcal{H}_{\text{S}} = \sum_{i=1}^{nN} U_{\text{S}}(\mathbf{r}_{\text{i}},K_{\text{i}})$, where $\mathbf{r}_{\text{i}}$ is the position of bead i, K_{i} is its type, and U_{S} is a potential acting on each bead:

$$\frac{U_{\text{S}}(\mathbf{r},K)}{k_{\text{B}}T} = -\frac{\Lambda_{\text{S}}^{\text{K}}(x,y)}{d_{\text{S}}/R_e}\exp\left[-\frac{z^2}{2\,d_{\text{S}}^2}\right] \tag{3}$$

Here, z is the distance from the substrate and $\Lambda_{\text{S}}^{\text{K}}N$ and d_{S} determines the strength and range of interaction. For a homogenous substrate $\Lambda_{\text{S}}^{\text{K}}$ is constant; for a patterned substrate $\Lambda_{\text{S}}^{\text{K}}$ depends on position.

To completely specify the Hamiltonian $\mathcal{H} = \mathcal{H}_{\text{b}} + \mathcal{H}_{\text{nb}} + \mathcal{H}_{\text{S}}$, one needs a suitable means of defining the local densities from the positions of the beads. In the simplest grid implementation, the simulation box is subdivided into cubic cells of width ΔL. The local densities in each cell are computed by simply counting the number of beads. More details about the model and the grid method as implemented here can be found in ref. 22.

As alluded to earlier, resorting to a grid is computationally efficient but introduces a number of limitations. The interaction between beads depends on the grid position, and one cannot implement calculations at constant pressure or stress in a straightforward manner. To avoid the use of a grid, a density cloud $w(\mathbf{r})$ is assigned to each bead,[18] and a continuous density field is then defined as:

$$\phi_{\text{A}}(\mathbf{r}) = \sum_i \delta_{\text{AK}_\text{i}}\, w(\mathbf{r} - \mathbf{r}_{\text{i}}) \tag{4}$$

where the sum is over all beads i, and δ is the Kronecker delta. Considering for simplicity a single term $\mathcal{H}$ of the functional in eqn (2), one can write:

$$\begin{aligned}\mathcal{H} &= \int_V \mathrm{d}\mathbf{r}\, \phi_{\text{A}}(\mathbf{r})\, \phi_{\text{B}}(\mathbf{r}) \\ \mathcal{H} &= \sum_{i,j}\delta_{\text{AK}_\text{i}}\delta_{\text{BK}_\text{j}}\int_V \mathrm{d}\mathbf{r}\, w(\mathbf{r} - \mathbf{r}_{\text{i}})\, w(\mathbf{r} - \mathbf{r}_{\text{j}})\end{aligned} \tag{5}$$

where the integral on the right-hand side is now denoted by $I(\mathbf{r}_{ij})$, and $\mathbf{r}_{ij} = \mathbf{r}_i - \mathbf{r}_j$. Up to a constant, the non-bonded contribution defined in eqn (2) is thus rewritten as $\mathscr{H}_{nb} = \sum_{j>i} U_{ij,nb}$, with the non-bonded inter-bead potential given by:

$$\frac{U_{ij,nb}}{k_B T} = \frac{\sqrt{\mathcal{N}}}{R_e^3}\left[\chi_o N\left(1 - \delta_{K_i K_j}\right) + \kappa_o N\right] \; I\left(\mathbf{r}_{ij}\right) \tag{6}$$

Any non-bonded energy $\mathscr{H}_{nb}$ that is a polynomial functional of the local densities can be converted into an inter-bead interaction potential, with terms of order p yielding p-body potentials. Fortunately, most block copolymer models involve only quadratic terms, which translate into pairwise additive interactions. The simulation now requires that all pairs of beads that are sufficiently close to interact be identified. The gridless implementation is therefore more computationally demanding than the grid approach described above but, as shown later, it offers many benefits, particularly the fact that it enables simulations in a variety of ensembles.

Since the forces between beads are now well-defined ($\mathbf{f}_{ij} = -\nabla U_{ij}$, where $U_{ij} = U_{ij,nb} + U_{ij,b}$), the local stress tensor $\sigma(\mathbf{r})$ can be computed from:

$$\frac{\sigma_{ab}(\mathbf{r})}{k_B T} = \sum_i \delta(\mathbf{r} - \mathbf{r}_i)\delta_{ab} + \sum_{i,j>i} r_{ij,a} \, f_{ij,b} \int_0^1 \mathrm{d}s \, \delta\left(\mathbf{r} - \mathbf{r}_i - s\,\mathbf{r}_{ij}\right) \tag{7}$$

where $a, b \in \{x, y, z\}$. The global stress is $\sigma_{ab} = \int_V d\mathbf{r} \, \sigma_{ab}(\mathbf{r})/V$ and the pressure is $P = \mathrm{Tr}(\sigma)/3$.

The behavior of the copolymer model considered here is determined by four physical invariants: R_e, $\chi_o N$, $\kappa_o N$, and $\overline{\mathcal{N}}$. The parameter $\overline{\mathcal{N}}$ depends on the molecular weight. In previous studies and in this work, $\kappa_o N$ is chosen so that the local density remains close to one, with the largest deviation smaller than 10%. Matching the isothermal compressibility of the simulated system to the experimental value of typical copolymers is possible, but would require values of $\kappa_o N$ that are two or three orders of magnitude larger, leading to lower acceptance ratios and longer simulations (see ref. 23 for further discussion). Furthermore, the field-level of description that we envisage here is not meant to capture local packing effects that ultimately determine compressibility effects. To keep both the spirit and advantages of a coarse grain description, we therefore prefer to work at moderate $\kappa_o N$. The parameter $\chi_o N$ that arises in eqn (6) is referred to as the "bare" $\chi_o N$. Because of short-range correlation effects,[20] it is different from the Flory–Huggins parameter χN. There are several possibilities to estimate χN, one of which is presented later in this work. The contour discretization N (*i.e.*, the number of beads used to represent a chain) and the cloud functions $w(\mathbf{r})$ are discretization parameters. In what follows it is shown that their influence on the results can be controlled through judicious choices.

The morphology and thermodynamic properties of the model outlined above can be determined from MC simulations, which sample configurations according to their Boltzmann weight through suitable acceptance criteria for the ensemble of interest. In the canonical ensemble, for instance, Metropolis sampling leads to acceptance criteria of the form: $\mathscr{P}_{acc} = \min[1, \exp(-\Delta U/(k_B T))]$, where the change in energy between the initial and trial configurations, ΔU, includes the bonded, non-bonded, and surface contributions. For the implementations considered in this work, trial moves include displacement of individual beads, reptation moves, whole-chain translation moves, and inverting the block sequence while keeping the chain backbone fixed. With the gridless implementation, the choice of simulation ensemble is unrestricted. Working in the *nPT* ensemble is particularly convenient because the size of the simulation box adjusts spontaneously to the natural periodicity of the relevant block copolymer morphology. An additional advantage is that the relevant thermodynamic potential (the Gibbs free energy per chain) is simply the chemical potential, which, in the absence of harsh, short-ranged interactions, can be computed in a straightforward manner by the Widom insertion method and configurational bias techniques.[23,24]

3 Results

3.1 Density multiplication in copolymer lithography—grid implementation

Our first application is concerned with density multiplication. The aim is to create a sparse chemical pattern on a substrate through the use of time-intensive patterning techniques, such as e-beam writing, and to spin-coat a copolymer film on top of that pattern that will then self-assemble to fill the gaps between the surface features in an orderly manner. Density multiplication has been demonstrated before in the context of surface "dots" and cylindrical morphologies,[9] but emerging applications are also likely to require density multiplication in the context of surface "stripes" and lamellar morphologies. The experimental system considered here is shown schematically in Fig. 1 (see ref. 9 for details of the experimental process). A PS-*b*-PMMA diblock with molecular weight 74 kg mol^{-1} and lamellar spacing $L_0 = 45$ nm, is spin-coated on a patterned substrate. The film thickness is $L_z \simeq L_0$. The pattern consists of stripes of width W, forming an array of period $2L_0$. The background is a silicon wafer covered by a PS brush layer, that interacts preferentially with the PS block. On the stripes, the PS brush has been oxidized and wets preferentially the PMMA block. It is experimentally challenging to obtain accurate estimates of the surface energy, and it is therefore of interest to use theory to map out the behavior of the system as a function of the interaction between the blocks and the patterned substrate.

For the TICG simulations, we chose a symmetric diblock copolymer, with $\chi_o N = 25$, $\kappa_o N = 35$, and $\overline{\mathcal{N}} = 110^2$. The discretization parameters are $N = 32$ and $\Delta L = 0.166\ R_e$. Matching the lamellar spacings in experiment and in simulation gives the length scale $R_e \simeq 27$ nm. The block/substrate interactions are specified by the four parameters $\Lambda_b^A N$, $\Lambda_s^A N$, $\Lambda_b^B N$ and $\Lambda_s^B N$, where A and B refer to PMMA and PS respectively, 'b' refers to the "background" of the pattern and 's' refers to the stripes. For simplicity, we assume $\Lambda_b^A N = -\Lambda_b^B N$ and $\Lambda_s^B N = -\Lambda_s^A N$ and, from now on, the two interaction parameters are denoted by Λ_b and Λ_s.

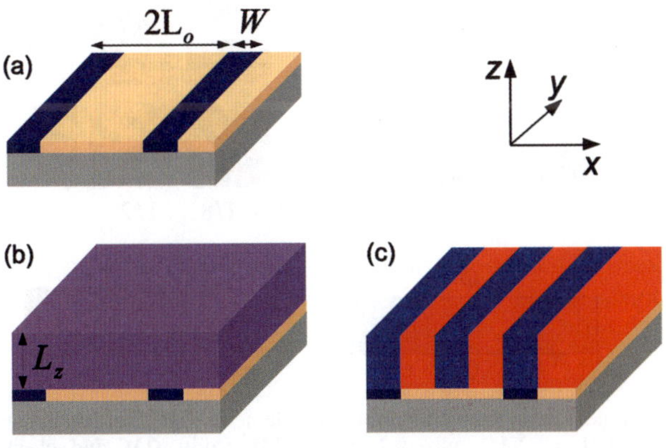

Fig. 1 Schematics of the experimental system. The copolymer is a PS-*b*-PMMA, with lamellar spacing L_0. (a) The chemical pattern. The background is a PS brush and interacts preferentially with the PS block. In each stripe, the brush has been oxidized, and interacts preferentially with the PMMA block. (b) A thin film of thickness L_z is spin-coated and annealed. (c) The self-assembled morphology, showing interpolation of lamellae.

Using TICG simulations, our goal is to identify conditions that permit successful pattern interpolation; to that end, the three pattern parameters are varied systematically, with W ranging from 0 to 0.875 L_0, and Λ_b and Λ_s ranging from 0 to 2. This maximum value is sufficient to enforce that the pattern be completely covered by its preferred block. Stronger interactions induce some unrealistic effects, such as a significant increase in density near the substrate. Three film thicknesses ($L_z = 0.9$, 0.95, and $1L_0$) are considered, leading to more than 400 cases, each representing a different combination of parameters. Simulations of such wide-ranging scope would be difficult within the context of more costly models involving pairwise interactions or atomistic representations. For each combination, several simulations are run in a box of dimensions $L_x \times L_y \times L_z$, with $L_x = 4L_0$ and $L_y = 4$ or $8L_0$, depending on the case. Most simulated systems are equilibrated for 10^5 MC cycles. The final morphology is generally reached in the early stages of the simulation.

Fig. 2 presents the five morphologies observed in our simulations. To make contact with experimental micrographs, the average composition in the x–y plane is plotted for a layer of thickness d starting at the top surface of the film. Since the depth probed by transmission electron microscopy is not exactly known, three depths are depicted in Fig. 2, namely $d/L_z = 1/8$, 1/2, and 1. The average composition in the x–z plane is also shown. The first three morphologies—horizontal, vertical and mixed lamellae—share the lamellar nature of the bulk morphology of symmetric copolymers. The latter consists of lamellae that are horizontal above the wetting

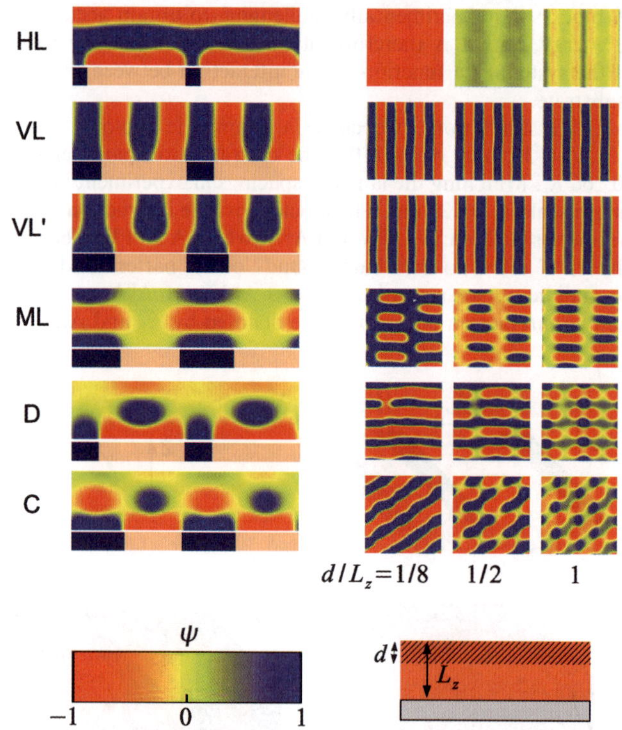

Fig. 2 Morphologies seen in simulation: horizontal lamellae (HL), vertical lamellae without or with asymmetry (VL or VL′), mixed lamellae (ML), dots (D), and checkerboard (C). Figures show the local composition $\psi = (\phi_A - \phi_B)/(\phi_A + \phi_B)$. In the left column, each colormap plots the average composition in the x–z plane (*i.e.*, resulting from an average taken over the y direction). The (x, y, z) axis are defined in Fig. 1. Colormaps in the right column show the average composition in the x–y plane, for a layer of thickness d at the top of the film, with $d/L_z = 1/8$, 1/2, and 1. Note that the scale is different in the two columns.

stripes and vertical elsewhere. Some vertical lamellae exhibit a significant difference in the width of the two A domains, and are called asymmetric. The "dots" and "checkerboard" morphologies differ from previous cases in that background and stripes are always entirely covered by their preferred block. In the case of dots, the B domain forms above each A line some "bridges", whose upper parts merge together to form lamellar domains. For the checkerboard, the copolymer domains alternatively replicate the pattern or its opposite, depending on the distance from the substrate.

While the interplay between the three pattern parameters W, Λ_s and Λ_b is complex, some qualitative trends can be identified. For a neutral background ($\Lambda_b = 0$), the main observation is the transition from vertical lamellae to mixed lamellae, as W or Λ_s is increased. On the other hand, a strongly selective background ($0.75 \lesssim \Lambda_b$) mainly leads to dots, except for the narrowest stripe ($W = 0.25L_0$) which can yield vertical lamellae. Finally, for $\Lambda_b = 0.25$, vertical lamellae are found for all Λ_s as long as $W \leqslant 0.75L_0$. From a practical point of view, this is the optimal background since it leads to successful interpolation, while allowing the maximal tolerance in the patterning of the stripe. Overall, the dot morphology is most common, followed by the vertical lamellae; mixed lamellae are restricted to backgrounds close to neutral and checkerboards arise only for large W, Λ_b and Λ_s.

The "high-throughput" or exploratory simulations outlined above serve to provide a general overview of the behavior of the system. In order to arrive at quantitative phase boundaries between any two morphologies—say α and β—it is necessary to compute the free energy difference $F_\beta - F_\alpha$. This can be achieved by thermodynamic integration, provided a reversible path can be identified between morphologies α and β. To construct such a path, we use an external field that artificially constrains the system in the desired state and enforces a gradual transformation of one morphology into the other. Specifically, the external field $U_{ext}(\mathbf{r})$ acts in a way similar to the substrate field U_S, but with the difference that it is now three-dimensional. From the average local densities of a morphology α, one can define an α field that, if strong enough, drives the system into the α morphology and maintains it there. The integration path considered here comprises three branches. First, starting from an α morphology, the system is subjected to an α field of increasing strength. Then, using a linear interpolation between fields α and β, the α morphology is gradually converted into the β morphology. Finally, the β field is slowly turned off. Thermodynamic integration along this external field path provides $F_\beta - F_\alpha$ for a given set of materials characteristics; simple thermodynamic integration can then be used to map the free energy difference in the entire region where both morphologies are stable. The approach outlined above is an extension of that discussed in ref. 21 and 25, where it was used to compute the free energy difference between the disordered and ordered states of a symmetric diblock copolymer. Here the two states represent ordered morphologies, one of them being metastable.

For concreteness, we focus on the competition between the lamellar and dot morphologies, since the former is the desired structure and the latter is the most commonly found throughout the parameter space explored in our simulations. From now on, $\Delta F = F_{dot} - F_{lam}$ denotes the free energy difference between dots and lamellae, expressed in k_BT units per chain. In Fig. 3, we plot ΔF as a function of the stripe and background selectivity, for a fixed stripe width $W = 0.75L_0$. The coexistence curve, along which $\Delta F = 0$, is shown in Fig. 4. The predominant equilibrium morphology consists of lamellae, suggesting that interpolation is extremely robust. The dots are the equilibrium morphology only when the stripes are weakly selective and the background is strongly selective.

Fig. 4 also shows the morphology that arises most frequently in simulations: somewhat unexpectedly, dots are found in most of the parameter space considered. Only for weakly selective backgrounds are lamellae obtained. For $\Lambda_b \simeq 0.375$, different realizations lead to lamellae or dots, or to the two morphologies within the same simulation box. These results were obtained with simulations where the

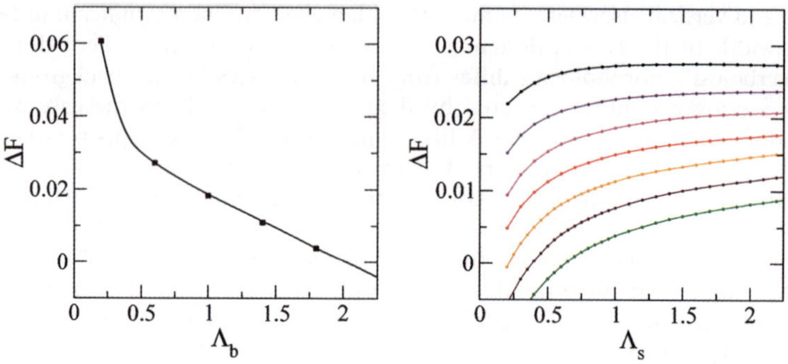

Fig. 3 Free energy difference between dot and lamellae, in k_BT units per chain, as a function of the stripe and background selectivities. (Left) $\Lambda_s = 1$. The squares (■) show ΔF obtained by thermodynamic integration along the external field path. The line is ΔF computed by thermodynamic integration along Λ_b, taking $\Lambda_b = 0.6$ as the reference point. (Right) ΔF computed by thermodynamic integration along Λ_s for $\Lambda_b = 0.6, 0.8, 1, 1.2, 1.4, 1.6, 1.8$ (from top to bottom).

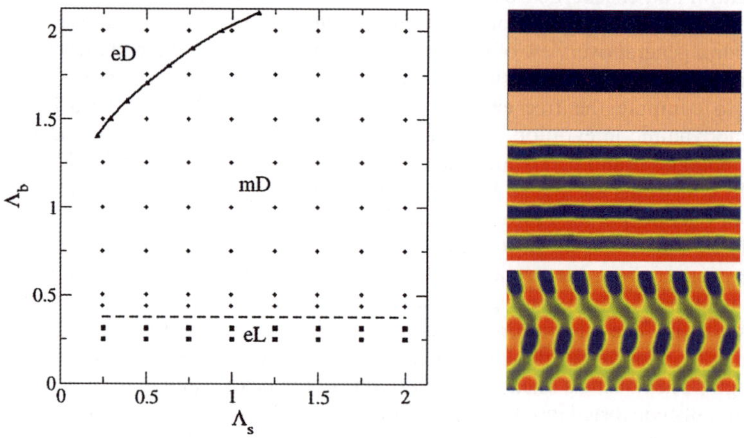

Fig. 4 (Left) Influence of pattern selectivities on the formation of dots and lamellae. The stripe width is $W = 0.75\ L_0$. The solid line is the equilibrium coexistence line between the two morphologies ($\Delta F = 0$). Points indicate which structure predominantly forms in simulation: lamellae (■) or dots (○). There are thus three areas: equilibrium dots (eD), metatable dots (mD) and equilibrium lamellae (eL). The approximate boundary between the last two is indicated by the dashed line. For neutral stripes or neutral background ($\Lambda_s = 0$ or $\Lambda_b = 0$), either the lamellae or the dots are unstable. (Right) Top view of the pattern and average composition in the *x–y* plane for the lamellar and dot morphologies (the legend is the same as that of Fig. 2).

position and configuration of a chain are drastically altered by the global MC moves mentioned above. Simulations with local moves only (displacement of a single bead and reptation) appear to yield similar results; the formation of lamellae is simply slower.

The morphologies observed in our simulations are therefore dominated by metastable dots (quantitatively ΔF is small; for instance, $\Delta F \simeq 0.02$ for $\Lambda_s = \Lambda_b = 1$.) We have verified that the metastability is not an artifact of the dimensions of the box. To understand why the dot morphology is kinetically favored, one can look at the early stages of self-assembly. Shortly after the simulation starts, each pattern area is quickly covered by a layer of its preferred block. The lamellae morphology requires that an A domain forms above the background, which implies a partial removal of this layer. This process is not required for dots—or at least to a much lesser extent—thereby facilitating their formation.

3.2 Order–disorder transition in block copolymers—gridless implementation

In the previous example we resorted to a grid-based implementation that significantly reduced the computational demands of our calculations, thereby permitting consideration of hundreds of different conditions in a reasonable amount of computer time. In the following example, a gridless implementation is adopted. While more demanding, it permits direct evaluation of free energies and phase boundaries. Our aim is to determine the order–disorder temperature for a symmetric diblock copolymer in the presence of fluctuations.

Unless otherwise noted, the invariant degree of polymerization is $\overline{\mathcal{N}} = 128^2$, which is representative of a typical experimental system† and $\kappa_o N = 50$. As regards the density cloud, a computationally simple choice is to use a square function $w(\mathbf{r}) = C\, w(x)w(y)w(z)$, where $w(u) = 1$ if $|u| < \Delta L/2$ and 0 otherwise. The normalization constant C is fixed by the condition that the average density be unity if the chains are ideal, *i.e.* $N\sqrt{\overline{\mathcal{N}}}\int \frac{\mathbf{dr}}{R_e^3} w(\mathbf{r}) = 1$. Note that C is constant throughout the simulation and independent of the ensemble considered. With the square function used here, the non-bonded interaction energy between two beads is proportional to the volume overlap between the two clouds. When changing only one coordinate, the force decreases linearly from its maximum (complete overlap) to zero (no overlap). The interaction range ΔL determines the average number of beads a given bead interacts with, $n_{\text{int}} = \rho_o(2\Delta L)^3 = N\sqrt{\overline{\mathcal{N}}}(2\Delta L/R_e)^3$. When changing the chain discretization N, n_{int} is kept constant and as a result, ΔL depends on N and $\overline{\mathcal{N}}$. In this work, we set $n_{\text{int}} \approx 14$ so that a given density cloud overlaps with a large number of neighboring clouds. Under these conditions, the particular form of the density cloud is not expected to play a significant role. Indeed, we have verified that within the statistical uncertainty of our calculations, the results obtained with a spherical density cloud are indistinguishable from those obtained with a square function.

Fig. 5 shows the density profiles corresponding to $\chi_o N = 35$ from TICG simulations and SCFT calculations in the nVT ensemble. Here it is assumed that the ratio $\chi N/\chi_o N$ is the same for the block copolymer considered here and for the corresponding blend of homopolymers. The ratio $\chi N/\chi_o N$ can then be calculated by matching the exchange chemical potential of the simulated blend to the semi-grand canonical equation of state (see ref. 26 for details). This procedure leads to $\chi N/\chi_o N \approx 0.82$ and $\chi N = 28.7$ in Fig. 5. Whether a single χ parameter can simultaneously describe the thermodynamics of homopolymer blends and block copolymers is an open question[27] that goes beyond the scope of this manuscript. The density profiles obtained with discretization $N = 32$ ($\Delta L = 0.075R_e$) and $N = 64$ ($\Delta L = 0.06R_e$) are indistinguishable. The equilibrium lamellar spacing L_0, which corresponds to an isotropic global stress tensor, is $1.78R_e$ and $1.79R_e$ for $N = 32$ and 64, respectively. Both numbers are slightly above the SCFT result $L_0 = 1.75R_e$.

Information pertaining to mechanical properties of an ordered morphology is of considerable interest, particularly in lithographic applications where the fabrication process relies on the stability of distinct nanoscopic domains. Predictions of such properties by means of SCFT calculations have been scarce.[28,29] For a lamellar morphology, a recent study[30] presented flat profiles for the tangential stresses and a normal stress with a minimum at the interface and two maxima in its vicinity‡. Using the gridless implementation outlined above, one obtains the local stress

† For instance, a PS-PMMA block copolymer with molecular weight around 10^3 kg mol^{-1}.
‡ In ref. 30, the melt was assumed to be incompressible, whereas our system has a finite compressibility.

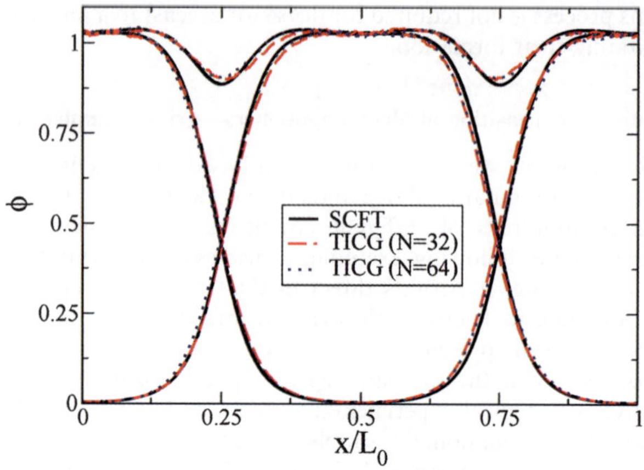

Fig. 5 Density profiles in a lamellar phase of a symmetric diblock, computed with TICG simulations and SCFT, in the nVT ensemble, for $\chi N = 28.7$ ($\chi_o N = 35$). The x coordinates are normalized by the lamellar spacing L_0. The system size for the simulation is $2L_0 \times L_0 \times L_0$.

profiles shown in Fig. 6. The normal stress is constant, as required by mechanical stability.[31] The tangential stresses are not constant but exhibit a minimum at the interface, surrounded by one maximum on each side.

The order–disorder transition temperature can be directly determined from TICG simulations in the nPT ensemble, by computing the Gibbs free energy of each phase. The chemical potential of a chain (or Gibbs free energy per chain) is computed with the Widom insertion method, augmented by a configurational bias scheme.[32,33] The chemical potential can be broken into an ideal and an excess contribution: $\mu = \mu^{id} + \mu^{ex}$. In the nVT ensemble, μ^{id} depends only on the density and μ^{ex} is computed from:

$$\exp\left(-\frac{\mu^{ex}}{k_B T}\right) = \left\langle \exp\left(-\frac{\Delta U_g}{k_B T}\right)\right\rangle \tag{8}$$

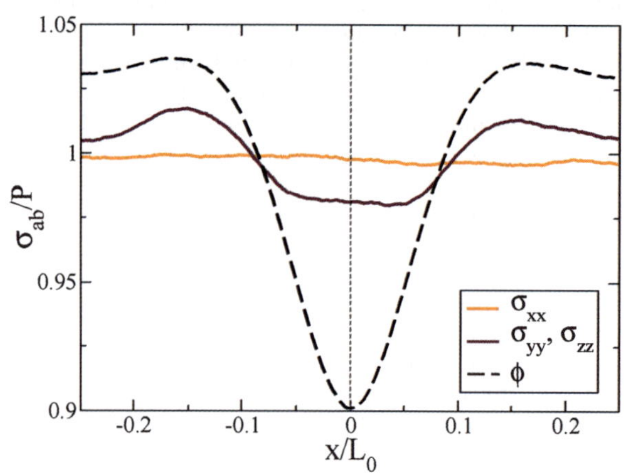

Fig. 6 Local stress in the lamellar phase of a symmetric diblock. Each component is divided by the total pressure $P = 19.4\ k_B T/b^3$. The A/B interface is located at $x = 0$. The total density ϕ of the melt is also shown (long-dashed line).

where ΔU_g represents the energy of interaction of a "ghost" molecule inserted into the system. When shifting to the nPT ensemble, the ideal term becomes a function of pressure, and the excess term is given by:

$$\exp\left(-\frac{\mu^{ex}}{k_B T}\right) = \left\langle \frac{PV}{(n+1)k_B T} \exp\left(-\frac{\Delta U_g}{k_B T}\right) \right\rangle \tag{9}$$

where, as before, n is the number of chains in the simulation. In contrast to models where harsh repulsive interactions prevent significant overlap of the particles, the softer nature of the potential employed here facilitates insertion of relatively long chains into the system, thereby enabling accurate estimation of μ^{ex}.

The chemical potential was computed as a function of $\chi_o N$ for the three cases described in Table 1. Case I corresponds to the default parameters ($\bar{\mathcal{N}} = 128^2$, $\kappa_o N = 50$, and $N = 32$). Case II is meant to gauge the effect of finite chain discretization: all physical invariants have the same value as in case I, but the chain discretization is $N = 64$ instead of $N = 32$. Case III examines the influence of molecular weight: the degree of polymerization $\bar{\mathcal{N}}$ is approximately twice as large as in case I; the other parameters are chosen so that the bead interaction potentials are the same as in case I. For all systems, the pressure is set to $P = 18\ k_B T/b^3$, the value extracted from nVT simulations for case I at $\chi_o N = 0$.

Fig. 7 shows μ^{ex} as a function of $\chi_o N$. In all three cases, branches for the disordered and ordered phases are approximately linear in the vicinity of the ODT, with distinctly different slopes. By employing local moves only, a metastable lamellar structure could be observed, which has a higher chemical potential than the disordered phase. Both observations are consistent with the first-order character of the ODT.[21,34,35] The nominal $(\chi_o N)_{ODT}$ is estimated as 16.8, 16.3 and 16.5 for cases I to III, respectively. In order to make a meaningful comparison with previous results from the literature, it is necessary to take into account two effects. The first effect stems from short-ranged correlations and, as explained above, is accounted for with the ratio $\chi N/\chi_o N$ (= 0.82, 0.86, 0.82 for cases I to III). The second effect comes from the finite discretization of the chains. To evaluate the corresponding correction, we considered chains in a homopolymeric melt ($\chi_o N = 0$), computed the partial structure factors S_{AA} and S_{AB} of blocks and estimated the location of the ODT by using the random phase approximation:[36]

$$\chi_{ODT}^{RPA} N = \min_{\mathbf{q}} \left[\frac{N}{S_{AA}(\mathbf{q}) - S_{AB}(\mathbf{q})} \right] \tag{10}$$

This yields $\chi_{ODT}^{RPA} N = 10.1$ and 10.3 for chain discretizations $N = 32$ and 64, respectively. The resulting ratio between the observed first-order transition and the mean-field prediction is $\chi_{ODT} N/\chi_{ODT}^{RPA} N = 1.36$, 1.36, and 1.31 for cases I to III (Table 1). We note that the results for two different discretizations are similar (cases I and II) and that, as expected, a higher molecular weight leads to a smaller deviation from the mean-field prediction (case III).

Table 1 Parameters and ODT values for the three systems considered (see text)

Case	$\bar{\mathcal{N}}$	$\kappa_o N$	N	$(\chi_o N)_{ODT}$	$\chi N/\chi_o N$	$\chi_{ODT}^{RPA} N$	$\chi_{ODT} N/\chi_{ODT}^{RPA} N$
I	128^2	50	32	16.8	0.82	10.1	1.36
II	128^2	50	64	16.3	0.86	10.3	1.36
III	185^2	100	64	16.5	0.82	10.3	1.31

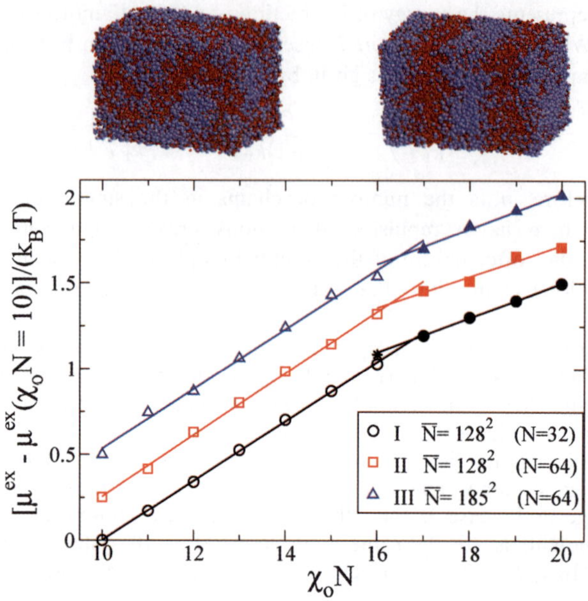

Fig. 7 Excess chemical potential as a function of $\chi_o N$ for the three cases described in Table 1. The curves have been shifted for clarity. Empty and filled symbols denote the disordered and lamellar morphologies, respectively. The star corresponds to a metastable lamellar state. From linear fits to the data, $(\chi_o N)_{ODT}$ is estimated as 16.8, 16.3, and 16.5 for cases I to III. The errors are comparable to the symbol size. The snapshots show the disordered phase at $\chi_o N = 16$ and the ordered phase at $\chi_o N = 17$.

Predictions for the ODT from previous studies and from this work are shown in Fig. 8. The simulations studied involve different molecular weights (invariant degrees of polymerization), but a comparison can be made by resorting to the prediction of a Hartree analysis of fluctuations:[37]

$$\frac{\chi_{ODT} N}{\chi_{ODT}^{RPA} N} - 1 = 3.9\ \overline{\mathcal{N}}^{-1/3} \tag{11}$$

TICG simulations yield a deviation from the mean-field value that is larger than that predicted by eqn (11); such deviations are also observed at lower $\overline{\mathcal{N}}$ in MC simulations of short self-avoiding chains on a lattice,[35] and at higher $\overline{\mathcal{N}}$ in recent field-theoretic simulations.[38] Altogether, results from simulations are roughly consistent with the power law of eqn (11), but suggest a larger coefficient (see Fig. 8).

Since the compressibility is finite, the density of the copolymer melt is not constant and varies with temperature, as shown in Fig. 9. When increasing $\chi_o N$, the average density steadily decreases up to the vicinity of the ODT, where it reaches a minimum that is approximately 5% below the homopolymer value ($\chi_o N = 0$). Then, as lamellae form and the A and B beads begin to separate from each other, the density steadily rises. Leibler's mean-field theory assumes that chain configurations are ideal, even at the ODT. TICG simulations reveal deviations from ideal behavior; to quantify those deviations, we have computed the radius of gyration R_g and the distance R_{AB} between the centers of mass of block A and block B. Both quantities, normalized by their value for an ideal chain ($\chi_o N = 0$, $\kappa_o N = 0$) are shown in Fig. 9. As the ODT is approached and a local structure begins to emerge, there is a departure from ideal behavior and the values of R_g and R_{AB} show significant increases. These observations are consistent with experiments,[39] MD simulations of

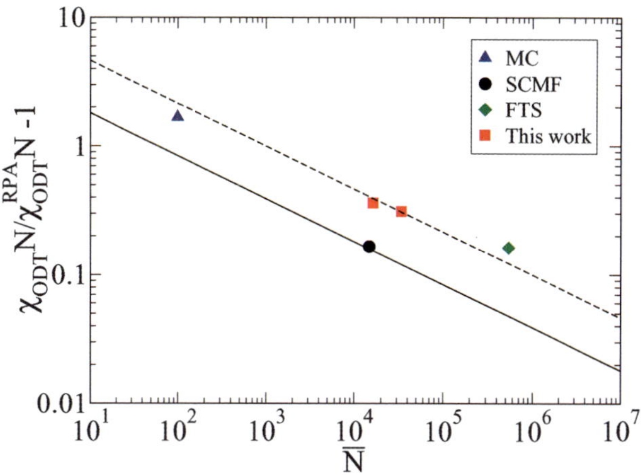

Fig. 8 Order–disorder transition and its deviation from the RPA value from various studies. Note that the different studies employ different definitions for the Flory–Huggins parameter. MC, SCMF, and FTS refer to ref. 35, 21, and 38, respectively. The lines are given by the equation $\chi_{ODT}N/\chi_{ODT}^{RPA}N - 1 = c\bar{\mathcal{N}}^{-1/3}$. The continuous line is the prediction from a Hartree analysis of fluctuations, where $c = 3.9$; the dashed line corresponds to $c = 10$.

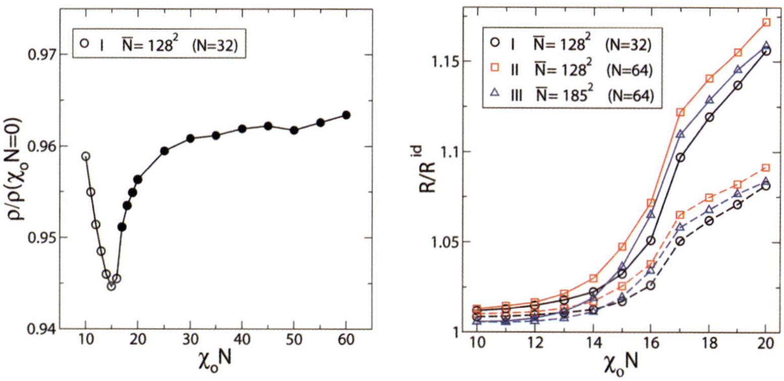

Fig. 9 (Left) Average density as a function of $\chi_o N$. (Right) Properties of single chain conformation: R_g (dashed lines) and R_{AB} (solid lines) are defined in the text. Both quantities are normalized by their value for an ideal chain. Lines are guides for the eye.

continuous models,[40] and MC simulations of lattice models,[41] that have also reported deviations from Gaussian behavior.

4 Conclusion

New particle-based simulations of theoretically informed coarse grain (TICG) models have enabled studies of relatively complex polymeric systems over large domains with molecular-level resolution, including the effects of fluctuations. The essence of such methods is that the interactions between the chains are derived from functionals of the local density. Such methods can be used in computationally efficient grid implementations, or in more demanding but more flexible gridless versions. In this work the usefulness and attributes of both implementations have been discussed in the context of two examples. The first example involved block copolymer films on nanopatterned surfaces for which experimental data are limited and, when available, are difficult to interpret. It was shown how TICG simulations can be employed in a high-throughput manner to explore wide ranges of parameter space. The second example illustrated how gridless implementations can be used for direct simulations of phase behavior in arbitrary ensembles. It was shown that local stresses, chemical potentials and free energies can all be calculated in models for which such quantities have traditionally been difficult to extract. The methods outlined in this work offer promise in the study of homopolymers, multiblock copolymers and nanocomposites, as well as in other complex fluids that exhibit internal structure over relatively long length scales, including bilayer membranes and liquid crystals.

Acknowledgements

The authors are grateful for the financial support from the National Science Foundation through the Nanoscale Science and Engineering Center (NSEC) at the University of Wisconsin (DMR-0425880). Support from the Semiconductor Research Corporation and the Volkswagen foundation is also gratefully acknowledged. The calculations presented in this work were performed in the Grid Laboratory of Wisconsin (GLOW).

References

1 I. W. Hamley, *The Physics of Block Copolymers*, Oxford University Press, Oxford, 1998.
2 D. C. Morse, *Ann. Phys.*, 2006, **321**(10), 2318–2389.
3 S. O. Kim, H. H. Solak, M. P. Stoykovich, N. J. Ferrier, J. J. de Pablo and P. F. Nealey, *Nature*, 2003, **424**(6947), 411–414.
4 M. P. Stoykovich, M. Müller, S. O. Kim, H. H. Solak, E. W. Edwards, J. J. de Pablo and P. F. Nealey, *Science*, 2005, **308**(5727), 1442–1446.
5 K. C. Daoulas, M. Müller, M. P. Stoykovich, S. M. Park, Y. J. Papakonstantopoulos, J. J. de Pablo, P. F. Nealey and H. H. Solak, *Phys. Rev. Lett.*, 2006, **96**(3), 036104.
6 M. P. Stoykovich, E. W. Edwards, H. H. Solak and P. F. Nealey, *Phys. Rev. Lett.*, 2006, **97**(14), 147802.
7 C. T. Black, *ACS Nano*, 2007, **1**(3), 147–150.
8 H. Kang, F. A. Detcheverry, A. N. Mangham, M. P. Stoykovich, K. C. Daoulas, R. J. Hamers, M. Müller, J. J. de Pablo and P. F. Nealey, *Phys. Rev. Lett.*, 2008, **100**, 148303.
9 R. Ruiz, H. Kang, F. A. Detcheverry, E. Dobisz, D. S. Kercher, T. R. Albrecht, J. J. de Pablo and P. F. Nealey, *Science*, 2008, **321**(5891), 936–939.
10 M. P. Stoykovich, H. Kang, K. C. Daoulas, G. Liu, C.-C. Liu, J. J. de Pablo, M. Müller and P. F. Nealey, *ACS Nano*, 2007, **1**(3), 168–175.
11 R. D. Groot, *Lect. Notes Phys.*, 2004, **640**, 5–58.
12 I. Pagonabarraga and D. Frenkel, *J. Chem. Phys.*, 2001, **115**(11), 5015–5026.
13 M. Matsen, *J. Phys.: Condens. Matter*, 2002, **14**(2), R21–R47.
14 E. Helfand, *J. Chem. Phys.*, 1975, **62**(3), 999–1005.
15 G. H. Fredrickson, *The Equilibrium Theory of Inhomogeneous Polymers*, Clarendon Press, Oxford, 2006.
16 M. Müller and F. Schmid, *Advanced Computer Simulation Approaches for Soft Matter Sciences II, Vol. 185 of Advances in Polymer Science*, Spring-Verlag, Berlin Heidelberg, 2005, p. 1.
17 A. Alexander-Katz and G. H. Fredrickson, *Macromolecules*, 2007, **40**(11), 4075–4087.
18 M. Laradji, H. Guo and M. J. Zuckermann, *Phys. Rev. E*, 1994, **49**(4), 3199–3206.
19 M. Müller and G. D. Smith, *J. Polym. Sci., Part B: Polym. Phys.*, 2005, **43**(8), 934–958.
20 K. C. Daoulas and M. Müller, *J. Chem. Phys.*, 2006, **125**(18), 184904.
21 M. Müller and K. C. Daoulas, *J. Chem. Phys.*, 2008, **128**(2), 024903.
22 F. A. Detcheverry, H. Kang, K. C. Daoulas, M. Müller, P. F. Nealey and J. J. de Pablo, *Macromolecules*, 2008, **41**(13), 4989–5001.
23 D. Q. Pike, F. A. Detcheverry, P. F. Nealey, M. Müller and J. J. de Pablo, *J. Chem. Phys.*, 2009, **131**(8), 084903.
24 F. A. Detcheverry, D. Q. Pike, P. F. Nealey, M. Müller and J. J. de Pablo, *Phys. Rev. Lett.*, 2009, **102**(19), 197801.
25 M. Müller, K. C. Daoulas and Y. Norizoe, *Phys. Chem. Chem. Phys.*, 2009, **11**, 2087–2097.

26 M. Müller, *Macromol. Theory Simul.*, 1999, **8**(4), 343–374.
27 W. W. Maurer, F. S. Bates, T. P. Lodge, K. Almdal, K. Mortensen and G. H. Fredrickson, *J. Chem. Phys.*, 1998, **108**(7), 2989–3000.
28 C. A. Tyler and D. C. Morse, *Macromolecules*, 2003, **36**(21), 8184–8188.
29 J. L. Barrat, G. H. Fredrickson and S. W. Sides, *J. Phys. Chem. B*, 2005, **109**(14), 6694–6700.
30 P. Maniadis, T. Lookman, E. M. Kober and K. O. Rasmussen, *Phys. Rev. Lett.*, 2007, **99**(4), 048302.
31 F. Varnik, J. Baschnagel and K. Binder, *J. Chem. Phys.*, 2000, **113**(10), 4444–4453.
32 J. Siepmann and D. Frenkel, *Mol. Phys.*, 1992, **75**(1), 59–70.
33 J. J. de Pablo, M. Laso and U. W. Suter, *J. Chem. Phys.*, 1992, **96**(8), 6157–6162.
34 H. X. Guo and K. Kremer, *J. Chem. Phys.*, 2003, **118**(16), 7714–7723.
35 O. N. Vassiliev and M. W. Matsen, *J. Chem. Phys.*, 2003, **118**(16), 7700–7713.
36 L. Leibler, *Macromolecules*, 1980, **13**(6), 1602–1617.
37 G. H. Fredrickson and E. Helfand, *J. Chem. Phys.*, 1987, **87**(1), 697–705; A. M. Mayes and M. Olvera de la Cruz, *J. Chem. Phys.*, 1991, **95**(6), 4670–4677.
38 E. M. Lennon, K. Katsov and G. H. Fredrickson, *Phys. Rev. Lett.*, 2008, **101**(13), 138302.
39 K. Almdal, J. H. Rosedale, F. S. Bates, G. D. Wignall and G. H. Fredrickson, *Phys. Rev. Lett.*, 1990, **65**(9), 1112.
40 M. Murat, G. S. Grest and K. Kremer, *Macromolecules*, 1999, **32**(3), 595–609.
41 H. Fried and K. Binder, *J. Chem. Phys.*, 1991, **94**(12), 8349–8366.

31 [illegible]
32 [illegible]
33 [illegible] Chem. [illegible] 1992, [illegible]
34 [illegible] 2005, [illegible]
35 [illegible]
36 [illegible]
37 [illegible]
41 [illegible]
42 [illegible]

A simple coarse-grained model for self-assembling silk-like protein fibers

Marieke Schor, Bernd Ensing and Peter G. Bolhuis*

Received 26th January 2009, Accepted 3rd April 2009
First published as an Advance Article on the web 8th September 2009
DOI: 10.1039/b901608b

Collagen-silk-collagen triblock polypeptides can self-assemble at low pH into nanometer thin fibers with a length in the order of micrometers. Previously we predicted, *via* all-atom simulations, the structure of the folded silk domain to be a β-roll. In this work we develop a simple coarse-grained model of the silk domain to enable a numerical study of the fiber's properties and formation on a larger length and time scale. As an initial coarse-grained model for the fiber forming protein we chose the model of Brown *et al.*, *Proc. Natl. Acad. Sci. U. S. A.*, 2003, **100**, 10712–10717. We adapted this model, and optimized its parameters to reproduce the all-atom molecular dynamics simulation structural data. The unknown strength of the attraction between the beads representing the residues is optimized by computing the Potential of Mean Force for unfolding a strand of the β-roll, using non-equilibrium steered MD simulations in combination with the Jarzynski relation. Using these optimized parameters we observed spontaneous folding of a short peptide. The coarse-grained β-roll, as well as a much larger stack (a fiber) of β-rolls, were found to be stable. Moreover, the predicted fiber persistence length is in agreement with experiment. The efficacy of the mapping of a coarse-grained system onto an all-atom simulation is discussed. The approach opens the way for large-scale simulations of fibers, based on molecular structure, and allows investigation of their nucleation, growth, cross-linking mechanism, network dynamics, and rheology.

1 Introduction

Self-assembling, protein-like polymers have a high potential as supra-molecular materials.[3,4] Prediction of structure and kinetics of the self-assembly process is crucial to control and improve the design of such novel soft materials. Because specificity of self-assembly is important, protein elements with a high β-sheet content, such as those of the *Bombyx mori* silk fibroin,[4] are promising building blocks for these nanomaterials. One of the silk-inspired self-assembling elements is the glycine-alanine rich (GAGAGAGX) repeat, in which the eighth amino acid (X) is chosen to disrupt the tight, hydrogen-bonded fibroin-like packing promoted by the Gly-Ala repeats.[5–8] Smeenk *et al.*, showed that triblock co-polymers of such silk-inspired polypeptides with hydrophilic polyethylene oxide (PEO) outer blocks, form highly defined fibrils when crystallized from a methanol/formic acid mixture.[6] Recently, Werten *et al.* synthesized a triblock copolymer with a middle block containing repeats of the silk-like octapeptide, with glutamic acid (E) on the X position,[9] flanked by two hydrophilic, non-aggregating, collagen-like blocks. When the glutamic acid is neutralized by lowering the pH in aqueous solution, this block copolymer

Van't Hoff Institute for Molecular Sciences, Universiteit van Amsterdam, Nieuwe Achtergracht 166, 1018 WV Amsterdam, The Netherlands. E-mail: p.g.bolhuis@uva.nl

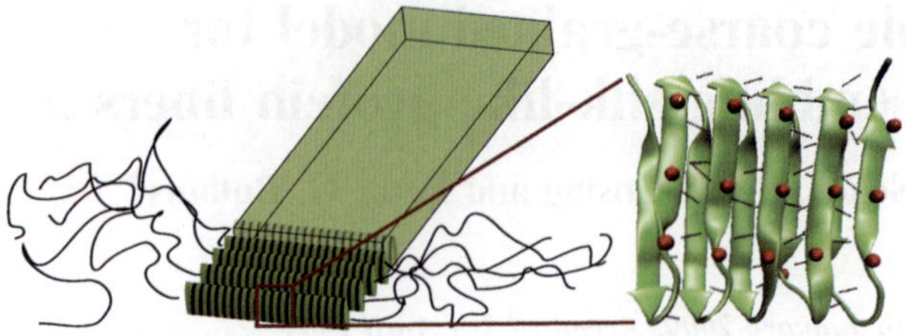

Fig. 1 Left: Schematic picture of the fiber. The central silk part of the triblock copolymer forms β-rolls (five are shown explicitly) which stack on top of each other to form long fibers (shown as a block). The collagen (black lines) forms a corona (not shown for clarity) and protects the hydrophilic residues. Right: The β-roll structure obtained through REMD simulations (1) consists of parallel β-strands. Alanine side chains (rendered as beads) are located on both outsides of the β-roll resulting in two hydrophobic surfaces responsible for assembly into fibers.

self-assembles into fibers.[10] The middle, silk-based, block forms the core of the fiber, whereas the outer collagen-like blocks form a corona around this core (see Fig. 1). While the formation of the fiber is thought to occur *via* folding into β-roll units followed by a nucleation and growth process, the coupling of folding and assembly that exists in reality renders the mechanism in fact poorly understood. In addition, the presence of the collagen is crucial in order to avoid aggregation into random globular structures, and possible further complicates the mechanism of the fiber formation.

In this paper we develop a coarse-grained protein model with the aim to investigate the structure and formation of a silk-like fiber. Although the coarse-graining scheme is probably more generally applicable, we specifically focus on the above mentioned collagen-silk-collagen triblock copolymer with a $(GAGAGAE)_{48}$ repeat as the silk part. In a previous publication we predicted, using all-atom molecular dynamics simulations, that the silk-block folds into plate-like objects consisting of the β-roll structure (see Fig. 1).[1] Not only were these simulated β-rolls stable in solution, but they were even more stable when assembled into a stack. Moreover, we showed that in water the common β-sheet structure was considerably less stable than the β-roll. These results were corroborated by experimental data in the form of SAXS and CD spectra.[10] Unfortunately, the all-atom simulations can only address rather limited system sizes (typically in the order of 10^5 atoms) and the number of atoms grows quickly with the fiber size. In addition, because in an all-atom MD simulation the time step is limited by the fast bond vibrations to 1–2 fs, the MD simulations cannot assess time scales above 100 ns. Yet, such large length and long time scales are required if we want to investigate fiber properties and eventually understand the mechanism of their self-assembly.

Coarse-graining is a viable way to access the required length and time scales. In a coarse-grained model several atoms are lumped together in a single particle, interacting *via* a simplified potential. This simplification emerges from integrating out of degrees of freedom. The Go-model, originally introduced in the 1970s[11] is among the most popular coarse-grained models. In this model the protein is represented by a chain of beads, interacting *via* effective potentials, designed such that the model protein folds in the *a priori* chosen native state. Simple coarse-grained protein models that are not based on a *a priori* chosen native state include the Honeycutt–Thirumalai model,[12,13] the Head-Gordon[2] and Hall[14] models. Other recently developed coarse-grained force fields include the non-native Go-like model of Clementi *et al.*,[15] OPEP,[16] Martini,[17] and UNRES.[18] Application of Molecular or Langevin dynamics allows for prediction of protein behavior without relying on knowledge of the native state. Most of these force fields have been devised to reproduce several properties, such as structure, or thermodynamics. Accurate coarse-graining methods that can reproduce thermodynamic observables include the inversion of the radial distribution function structure,[19] force matching,[20] and thermodynamic partitioning[21] but turn out not to be straightforwardly applicable to proteins undergoing a conformational change.[17]

Here we aim to employ a simple model that is able to predict the structures as well as the folding behavior correctly. Our first attempt is based on the Head-Gordon (HG) model,[2] which in turn is derived from the Honeycutt–Thirumalai model.[12] In this model, the protein is represented as a chain of beads positioned at the α-carbon atoms. The HG model can describe secondary structure formation, and has been applied to amyloid aggregation amongst others.[22] As the model has only been applied to common β-sheet forming systems, it needs to be optimized for the β-roll forming silk-like protein we study here. The parameterization of the bond, bending and dihedral potentials can be extracted from the all-atom simulation of the folded structures, and of a short peptide in solution, respectively. Furthermore, the strength of the non-bonded interaction between residues is a key-determining factor for the folding and unfolding behavior. This effective attraction is caused by hydrogen bond formation, electrostatics, as well as by the hydrophobic effect. The deeper the attractive potential well, the more the equilibrium shifts toward the folded state. We would like the coarse-grained model to mimic the equilibrium behavior of the all atom system. This can be achieved by explicitly computing the equilibrium between folded and unfolded states and tuning the interaction strength until they match. However, this is a computationally expensive procedure, as it requires full equilibration of both systems. The replica exchange molecular dynamics method[23] can accelerate these long folding and unfolding processes, but, while certainly a step in the right direction, converging the free energy landscape is not trivial and still computationally intensive.[24]

An alternative, more effective approach would be to map the potential of mean force (PMF) as a function of a relevant order parameter, such as the distance between strands. While extracting a PMF from a straightforward MD run is inefficient, non-Boltzmann sampling methods such as umbrella sampling,[25] metadynamics,[26] hyperdynamics,[27] local elevation[28] and conformational flooding,[29] can in principle efficiently assess the PMF or free energy along the reaction coordinate of interest. Here, we will compute the PMF from non-equilibrium MD simulation of single molecule pulling 'experiments', also known as Steered MD (SMD). These pulling experiments yield a series of force-extension curves that can be converted into a PMF using the Jarzynski equality.[30,31] In the late 1990s Jarzynski[32] proved a remarkable relation that enables the computation of an equilibrium free energy from the work done in non-equilibrium trajectories. While the individual non-equilibrium coarse-grained trajectories are not likely to describe the dynamical behavior of the atomistic system correctly, the PMF is a true equilibrium quantity that can be used to map the coarse-grained models onto the atomistic simulation. The advantage of this procedure is that it is simple to implement, as most popular MD packages have an SMD module that allows such non-equilibrium pulling experiments. The idea is to optimize the parameterization of the coarse-grained model in such a way that it reproduces the atomistic PMF as good as possible for this coarse-grained model. Another advantage of SMD is that it allows for an easy assessment of the statistical errors from the averaging over a series of pulling experiments as shown in the Results section. This also allows for a quick estimate from a few preliminary pulling runs of the PMF, the computation cost to reach convergence, the quality of the reaction coordinate and the optimal pulling velocity for both the coarse-grained and all atoms model.

Using the coarse-grained model we test the stability of the β-roll, as well as that of a stack of two rolls and a larger fiber. From the flexibility of this large fiber we can then estimate the persistence length and compare it to experimental values.

The remainder of the paper is organized as follows: In section 2, we describe the development of the model. The mapping of the structural parameters is discussed in section 2.2, while the mapping of the unfolding PMF is described in section 2.3. In section 3 we apply the coarse-grained model to several large systems. We end with a concluding discussion in section 4.

2 Model development

2.1 Adapting the Head-Gordon model

The Head-Gordon (HG) model[2] is one of the few coarse-grained models that is able to accurately reproduce protein folding and self-assembly.[2,22] In this model, the amino acid alphabet is reduced to only three letters: neutral (N), hydrophilic (L) and hydrophobic (B). Each amino acid of a protein is represented by one bead, whose center is located at the position of the alpha carbon. The solvent is treated implicitly and is left out of the force field description. The potential energy of the system is given by a sum of bond angle, dihedral, and non-bonded interaction terms, respectively:

$$\begin{aligned} V &= \sum_{\theta}\frac{1}{2}k_{\theta}(\theta-\theta_0)^2 \\ &+ \sum_{\phi}\left[A(1+\cos\phi)+B(1-\cos\phi)+C(1+\cos3\phi)+D\left(1+\cos\left(\phi+\frac{\pi}{4}\right)\right)\right] \\ &+ \sum_{i,j\geq i+3} 4\varepsilon_h S_1\left[\left(\frac{\sigma}{r}\right)^{12}-S_2\left(\frac{\sigma}{r}\right)^{6}\right] \end{aligned} \tag{1}$$

The three sums run over all bond angles, dihedrals and pair interactions respectively. In this expression θ denotes the bond-angle, θ_0 the equilibrium value of the bond angle. ϕ is the dihedral angle, σ is the diameter of the beads, and r is the distance between bead centers. A, B, C, D, S_1, S_2, ε_h, k_θ, and θ_0 are parameters determining the force field. Note that some of these parameters can vary, depending on the dihedral index, or combination of bead types. The HG model uses reduced units with all bond lengths fixed to 1. Also the bead size σ is chosen to be unity. The angle bending constant k_θ is set to 20 ε_h and the equilibrium angle to $\theta_0 = 105°$. For β-sheets the dihedral constants are set to $A = 0.9\ \varepsilon_h$, $C = 1.2\ \varepsilon_h$ and $B = D = 0$; whereas for turns they are $A = B = D = 0$ and $C = 0.2\ \varepsilon_h$. Non-bonded interactions are determined by the constants S_1, and S_2. For *B–B* interactions $S_1 = 1$ and $S_2 = 1$; for *N–L*, *N–N* and *N–B* interactions $S_1 = 1$ and $S_2 = 0$; and for *L–L* and *L–B* interactions $S_1 = 1/3$ and $S_2 = -1$. In the simulations the bond length is constrained using the RATTLE algorithm.[33] The equation of motion is integrated using MD in the canonical ensemble as detailed in section 5.

In this paper we adapted the above model to achieve a more accurate description of our fiber-forming proteins. From our previous all-atom replica exchange molecular dynamics (REMD) simulations, we concluded that the central silk-based part of the triblock co-polymer forms a β-roll in solution.[1] This β-roll consists of two interconnected sheets, with strands within one sheet parallel to each other. The two sheets are connected with each other through reverse type II β-turns. Compared to the more common β-sheet, a β-roll is more compact and has increased possibilities for hydrogen bonding. This rather unusual structure agrees well with the experimental dimensions measured for the fibers formed by the block co-polymers, and also agrees with the CD spectra.[1] When coarse-graining this protein according to the HG model, the original sequence *E*(GAGAGAGE)$_n$ is substituted by *L*(NBNBNBNL)$_n$, where n denotes the number of times the sequence is repeated. Our first attempt to simulate the β-roll using the HG model, starting from a fully formed β-roll ($n = 10$), resulted in immediate unfolding even at such low temperatures as $T = 10$ K. This was found to be caused by the fact that neutral beads repel each other quite strongly.

We therefore modeled Gly as a hydrophobic residue instead of a neutral one, and replaced all N beads by B, thus changing the sequence into $L(B_7L)_n$. This adaptation only turned out a partial solution, as during the simulation the turn region rearranged from a reverse type II β turn to a standard type II β turn. This was to be expected as the turn dihedrals in the HG model are parameterized to describe a standard β-turn.[2]

After these preliminary results we decided to re-parameterize the model based on atomistic simulations of an 81 residue ($n = 10$) β-roll in explicit water. We refrained from doing the full 384-residue silk-domain as equilibrating such a large system using atom MD in explicit water would be computationally too expensive. From a 50 ns all-atom MD simulation of the β-roll several distributions were extracted and used to optimize the coarse-grained model, as will be discussed in the following section. The subsequent tuning of the non-local interaction strength by comparing several PMFs for the coarse-grained model with the atomistic simulations will be the subject of section 2.3.

2.2 Fitting distributions from atomistic simulation

A 50 ns atomistic simulation of a 81 residue β-roll in water at 300 K yielded several key distributions (See Methods for a description of the simulation details). Bond lengths, measured as the distance between C-alpha atoms of connected amino acids, are on average 3.84 ± 0.12 Å. This rather narrow distribution justifies the use of constrained bonds in the original HG model. The dihedral angles between four subsequent C_α atoms in the strand regions are on average 180° ± 30°, whereas in the turn region the dihedral angles are around −30° ± 20°.

While the dihedral angles of the strand regions are well described by the original HG model, the turn regions are not. Therefore we reparameterized the dihedral angle potential for the turn regions. The MD simulation of the fully formed β-roll does not sample all possible dihedral angles. A better sampling of the dihedrals is obtained from a separate 10 ns all-atom simulation of a short GAGEGAG peptide, representing the turn region (see Fig. 2 for a schematic representation of a single hairpin in the β-roll). The three resulting distributions $P(\phi)$ of the dihedral angle ϕ belonging to the GAGE, AGEG and EGAG turn sequences are shown in Fig. 3 and are quite similar to each other. We can extract an effective potential from this data by taking the negative logarithm $\beta V(\phi) = -\ln P(\phi)$ of the average

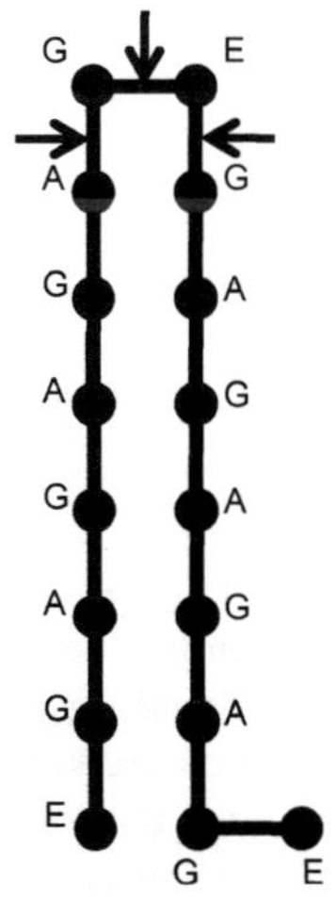

Fig. 2 Schematic picture of one single hairpin in the β-roll to clarify which three dihedrals are involved in the reverse β-turns. The arrows point at the GAGE, AGEG, GEGA dihedrals.

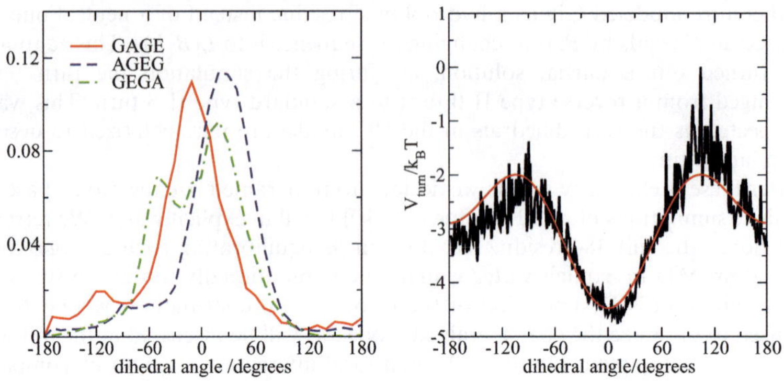

Fig. 3 Left: the distribution of the turn dihedrals in an all-atom simulation of the short peptide GAGEGAG. Right: The dihedral potential is fitted (solid red line) to the negative logarithm of the average distribution.

dihedral distribution. To obtain one expression for the effective dihedral potential for the turn region, we fitted the negative logarithm of the average dihedral distribution to a single quadratic polynomial in cosϕ, resulting in the following function for the dihedral potential energy (see Fig. 3)

$$V_{\text{turn}}(\phi) = \varepsilon_{\text{h}}(2.4 - 0.8\cos\phi - 1.6\cos^2\phi) \tag{2}$$

This coarse-grained potential will reproduce the measure distribution of the short peptide, and at the same time is not inconsistent with the average dihedral angles in the β-roll.

Furthermore, distance distributions between strands i and $i + 2$ were calculated as the distance between C_α atoms of the central alanine residues. Also, the distance between neighboring glutamate residues was measured. Both distances are on average 4.8 ± 0.5 Å. This distance can be used to set the range of the nonbonded Lennard-Jones interaction.

Combining these β-roll specific measurements with the original HG model, the adapted total potential energy of the system now reads:

$$\begin{aligned} V &= \sum_b \frac{1}{2}k_b(b-b_0)^2 + \sum_\theta \frac{1}{2}k_\theta(\theta-\theta_0)^2 \\ &+ \sum_\phi \left[A + B\cos\phi + C\cos^2\phi + D\cos^3\phi\right] + V_{\text{LJ}} \end{aligned} \tag{3}$$

$$V_{\text{LJ}} = \sum_{i,j\geq i+3} 4k_{\text{LJ}}S_1\left[\left(\frac{\sigma}{r}\right)^{12} - S_2\left(\frac{\sigma}{r}\right)^6\right] \tag{4}$$

Analogous to eqn (1), the sums run over, respectively, the bonds, angles, dihedrals and the pair interaction. Note that the parameters A, B, C, D are different from the original model as expressed by eqn (1). Lengths are in Å, angles and dihedral angles in degrees and the energy scale is set by $\varepsilon_{\text{h}} = 2495$ J mol^{-1}. Note also that, as the software package used for the coarse-grained molecular dynamics (CM3D (34)) employs a multi-timestep algorithm, the bond length potentials are now explicitly included in the energy functions, whereas in the original HG model they were constrained with the RATTLE algorithm. The bond potential force constant is set to $k_{\text{b}} = 33\,\varepsilon_{\text{h}}$, and $b_0 = 3.84$ Å, consistent with the measured all-atom C_α–C_α distance distributions, and the bending angle parameters $k_\theta = 20\ \varepsilon_h$ and $\theta_0 = 105°$ are taken from the HG model (2). For the β-strands the dihedral potential parameters are $A = 2.1\ \varepsilon_h$, $B = -2.7\ \varepsilon_h$, $C = 0$ and $D = 4.8\ \varepsilon_h$, which is equivalent to the values in the HG model. For the turn regions the dihedral parameters follow from eqn (3) and are $A = 2.4\ \varepsilon_h$, $B = -0.8\ \varepsilon_h$, $C = -1.6\ \varepsilon_h$ and $D = 0\ \varepsilon_h$. The non-bonded interaction parameter σ sets the range of the interaction and was initially set to $\sigma = 4.2$ Å such that the minimum of the LJ-potential corresponds to the average distance of 4.8 Å between neighboring strands. The strength of the nonbonded interaction k_{LJ} is for the time being set to $k_{LJ} = 4\ \varepsilon_h$ but will be further optimized in the next section.

In this work, we treat both alanine and glycine as hydrophobic (attractive) beads ($S_1 = 1$, $S_2 = 1$). As the protonation of the glutamate residues is supposed to trigger folding and aggregation by reducing repulsive interaction and introducing the possibility of hydrogen bond formation, we also treat the Glu-Glu interaction as attractive (hydrophobic). However, to mimic their hydrophilic nature the glutamates behave as neutral (repulsive) beads ($S_1 = 1$, $S_2 = 0$) when interacting with glycine or alanine residues.

Short coarse-grained simulations using the CM3D package with the above settings resulted in a twisted β-roll (see Fig. 4). This twist can be quantified by the angle between residue position vectors $\boldsymbol{r}_1$ (Glu17-Glu65) and $\boldsymbol{r}_2$ (Glu25-Glu73). For the coarse-grained model, the twist angle of the β-roll is approximately −78°, much larger than was observed in the 50 ns atomistic simulation where the twist angle was around −40°. This large twist angle can be explained by the long-ranged nature of the Lennard-Jones interactions employed in the HG model. This long range induces an additional (unphysical) attraction between non-neighboring strands, which forces the β-roll to twist. To prevent such spurious attraction, we can shorten the range of the non-bonded potential by scaling and shifting the original Lennard-Jones as follows:

$$V_{LJ} = \sum_{i,j \geq i+3} 4k_{LJ}S_1\left[\left(\frac{\sigma}{r-\sigma_0}\right)^{12} - S_2\left(\frac{\sigma}{r-\sigma_0}\right)^{6}\right] \tag{5}$$

Here, decreasing σ decreases the absolute range of the potential, and σ_0 is used to shift the potential to match the size of the bead as given by the original potential. Fig. 5 shows both the original and the shifted-scaled LJ potentials.

Using a shorter ranged shifted potential in the coarse-grained simulations resulted in a smaller twist in the β-roll structure. We found that a combination of $\sigma_0 = 2$ Å and $\sigma = 2.45$ Å results in a twist angle similar to that in the atomistic simulation

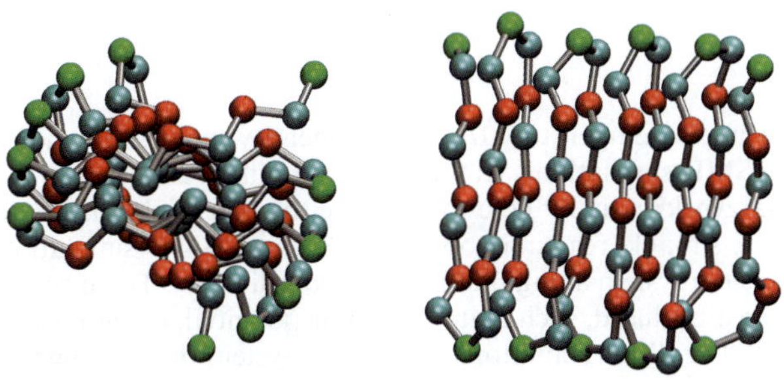

Fig. 4 Effect of shifting the long-ranged interactions on the β-roll structure. Left: When the long ranged interactions are computed without shifting ($\sigma_0 = 0$; $\sigma = 4.2$ Å) the structure shows a twist angle of approximately 78°. Right: When shifted with $\sigma_0 = 2$ Å and $\sigma = 2.45$ Å the twist angle in the β-roll is comparable to the twist in the atomistic structure. In the coarse-grained representations, alanine is red, glycine blue and glutamate green.

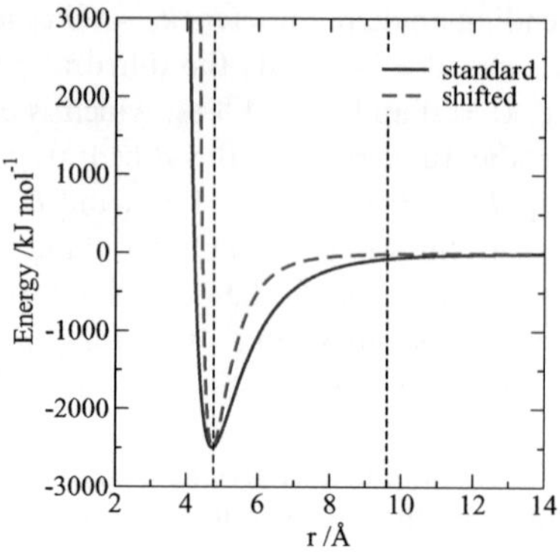

Fig. 5 Shifting the Lennard-Jones potential has the effect of shortening the range of the potential. The position of the nearest neighbor strand and the second nearest neighbor strand are given by the dashed vertical lines. Note that in the standard LJ potential the second nearest neighbor still experiences a non-negligible attraction.

(see Fig. 4c). The coarse-grained twist angle is now only $-20°$ while the twist in the all-atom simulations is $-40°$. In our previous atomistic simulations we observed that the twist in the β-roll is reduced when two or more rolls form a stack (1). The same trend is seen in the coarse-grained simulations.

To summarize, based on all-atom simulations of a β-roll and of a short peptide in solution we have optimized the coarse-grained model to reproduce the structural properties of the β-roll. Our model is given by eqn (3) in combination with (5). The parameter that determines the binding strength between the strands, and thus the folding–unfolding equilibrium k_{LJ} is not yet optimized. This will be done in the next section.

2.3 Matching coarse-grained and atomistic PMFs

For our initial simulations we use a $k_{LJ} = 4\,\varepsilon_h$, which is equivalent to $4k_BT$, where k_B is Boltzmann's constant and T the temperature in Kelvin. This is, however, a rather rough estimate. A better estimate for k_{LJ}, follows from comparing the potentials of mean force (PMF) obtained from steered MD (SMD) simulations of the atomistic and the coarse-grained model. To compute these PMFs from SMD, we follow the method of Park and Schulten.[30,31]

Jarzynskis equality[32] relates the equilibrium free energy difference between two states, ΔF, to the average of the exponential of the work, W, done by a non-equilibrium process to drive the system from one state to the other

$$e^{-\beta\Delta F} = \langle e^{-\beta W} \rangle \quad (6)$$

where β is the inverse temperature $1/k_BT$. This identity enables the calculation of free energies from SMD simulations. Rather than computing ΔF between two states, we can use the Jarzynski equality here to compute the potential of mean force $\Phi(\xi)$ as a function of a suitable reaction coordinate ξ. In an SMD simulation, a guiding potential steers the system along that reaction coordinate ξ. To do so, an extra variable λ has to be introduced, such that the guiding potential, or spring, constrains ξ to be close to λ.[31] The SMD Hamiltonian $\tilde{H}$ of the system then becomes

$$\tilde{H}(\boldsymbol{r},\boldsymbol{p},\lambda) = H(\boldsymbol{r},\boldsymbol{p}) + \frac{k}{2}\left(\xi(\boldsymbol{r}) - \lambda\right)^2 \quad (7)$$

where $H(\boldsymbol{r}, \boldsymbol{p})$ is the original unbiased Hamiltonian—with $\boldsymbol{r}$ and $\boldsymbol{p}$ the positions and momenta of all atoms, respectively—and the quadratic term on the right side is the harmonic guiding potential that keeps the reaction coordinate close to the position of the driving potential. The free energy follows from integrating the Hamiltonian $\tilde{H}$ and projecting along ξ. For sufficiently stiff springs (large k), however, the reaction coordinate ξ follows λ closely, and therefore [31]

$$F(\lambda) \approx \Phi(\lambda) \tag{8}$$

In SMD the guiding potential moves at a certain fixed velocity v

$$\lambda_t = \lambda_0 + vt \tag{9}$$

In this case, the work W done on the system over the entire trajectory can be calculated from integrating over the position of the reaction coordinate as a function of time.

$$W_{0\to t} = \int_0^t \mathrm{d}t' \frac{\mathrm{d}\lambda_{t'}}{\mathrm{d}t'} \left(\frac{\partial \tilde{H}}{\partial \lambda}\right) \tag{10}$$

$$= kv \int_0^t \mathrm{d}t' \left[\xi(r_{t'}) - \lambda_0 - vt'\right] \tag{11}$$

Now combining eqn (8) and Jarzynski's equality eqn (6) the PMF can be calculated as:

$$\Phi(\lambda_t) = \Phi_{\lambda_0} - \frac{1}{\beta}\ln\langle\exp(-\beta W_{0\to t})\rangle \tag{12}$$

where the brackets denote an average over an ensemble of trajectories starting from the initial state at λ_0 to the final state at $\lambda_t = \lambda_0 + vt$. The exponential average is dominated by those trajectories that have the lowest values for work W. These are however rarely sampled, especially at typical pulling velocities. Moreover, the equality strictly only holds in the case when the sampling is done over an infinite number of pulling simulations, that is, when

$$\lim_{M\to\infty} -\frac{1}{\beta}\ln\frac{1}{M}\sum_{i=1}^{M}\exp(-\beta W_{0\to t}) = -\frac{1}{\beta}\ln\langle\exp(-\beta W_{0\to t})\rangle \tag{13}$$

For a finite number of work estimates, the non-linear average is biased due to the convexity of the logarithmic function. Park and Schulten[30,31] derive approximate expressions, obtained by expanding the last term in eqn (12) in cumulants

$$\ln\langle e^{-\beta W}\rangle = -\beta\langle W\rangle + \frac{\beta^2}{2}\left(\langle W^2\rangle - \langle W\rangle^2\right) - \frac{\beta^3}{6}\left(\langle W^3\rangle - 3\langle W^2\rangle\langle W\rangle + 2\langle W\rangle^3\right) + \ldots \tag{14}$$

and they note that for slow enough pulling velocities, the work distribution is Gaussian, in which case the terms higher than the second order cumulant are zero. Interestingly, they also found that the statistical error for the finite-sampling average using the second order cumulant is smaller than that for the full exponential average. We will therefore evaluate both averages to assess whether the used pulling velocity is slow enough to remain in the regime of a Gaussian distribution of the work, and use the second cumulant average to obtain the PMF with lowest statistical noise from a limited number of pulling simulation. Other methods that make use of

Jarzynski's equality include the method proposed by Hummer and Szabo,[35] which was developed for weak springs, and combination with path sampling methods.[36] A more rigorous approach would be to use both forward and backward paths in combination with the Crooks fluctuation theorem.[37]

We pull one terminal strand away from the rest of the protein by increasing the distance $\xi = |r_{E1} - r_{E49}|$ between residues Glu1 and Glu49 from 13.6 Å to 29 Å as indicated in Fig. 6. As the β-roll is highly symmetric, *i.e.* all strands feel the same forces from their neighboring strands, we assume that pulling away more than one strand will not yield additional information. Both the atomistic and the coarse-grained SMD simulations were performed with a spring constant of $k = 100$ kJ mol^{-1} Å^{-2} and a pulling velocity of $v = 5$ Å ns^{-1}. To remain close to equilibrium, the system was allowed to equilibrate after each ns for about 0.5 ns. The total trajectory was 3 ns. Twenty two atomistic trajectories were generated and converted as described above, using, respectively, the unweighted $\langle W \rangle$, the 2nd order cumulant expression, and the exponentially weighted expression $\langle \exp(-\beta W) \rangle$. The resulting PMFs are plotted in Fig. 7. The three differently weighted PMFs are reasonably similar, indicating a reasonable convergence. The atomistic PMF starts at a distance of 13.6 Å with a value of zero, and then almost monotonically increases with the distance. This indicates that the force on the strand that is being pulled away exerted by the rest of the protein is almost constant. This suggests a zipper-like unfolding, where as one residue is being pulled away from the main block, the next residue is already starting to feel that force.

For the coarse-grained model, different values for the strength of the non-local interactions $k_{LJ} = 2\ \varepsilon_h$, $3\ \varepsilon_h$, $4\ \varepsilon_h$ and $5\ \varepsilon_h$ are tested. For each value 20 trajectories were generated and analyzed in the same way as was done for the atomistic simulations. The statistical error in the PMF is estimated by taking 4 block averages. The resulting PMFs for the three weighting expressions are plotted in Fig. (7).

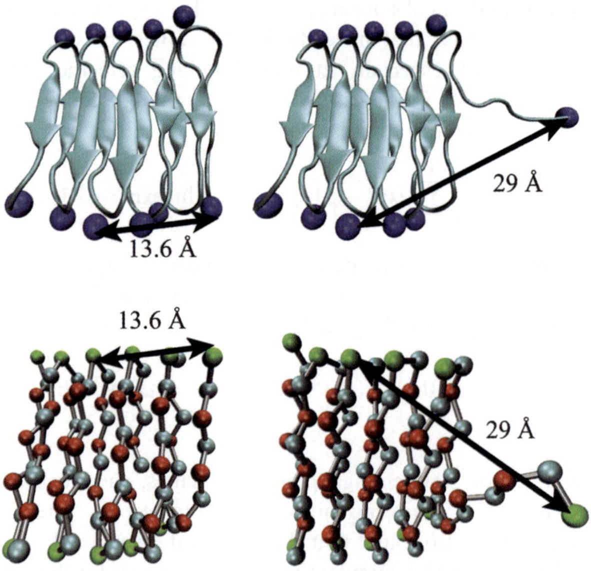

Fig. 6 Structures obtained in the atomistic and coarse-grained SMD simulations. Top row: two ribbon structures showing the initial and final configuration of the atomistic protein, Bottom row: two structures in bead representation showing the coarse-grained initial and final structure ($k_{LJ} = 3\ \varepsilon_h$). The arrows indicate the distance parameter ξ used for the pulling simulation.

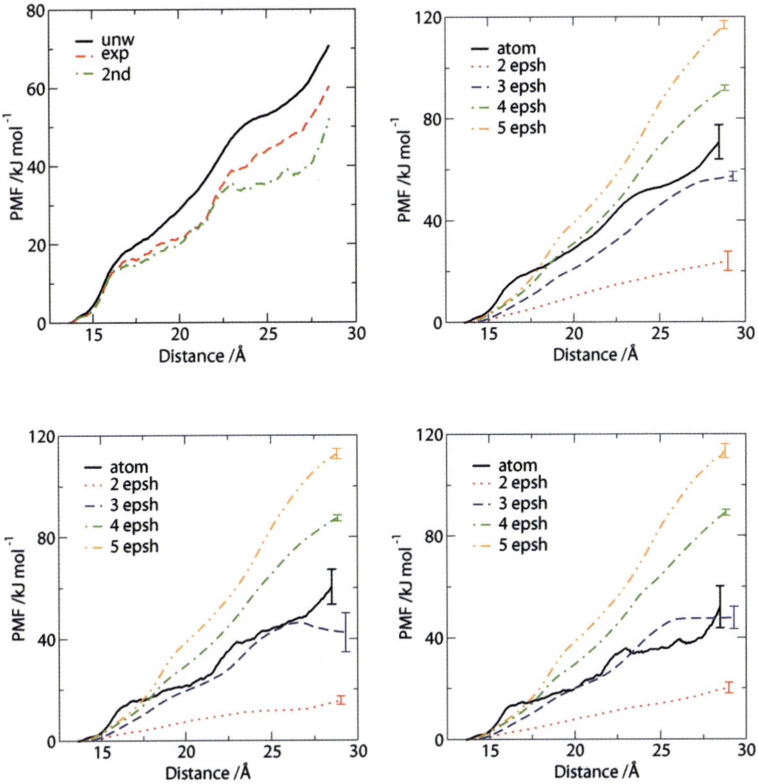

Fig. 7 Comparison of the Potentials of Mean Force as a function of the distance ξ of the atomistic and coarse-grained SMD simulations. Top left: the PMFs obtained from the atomistic runs, for the unweighted, exponentially averaged and second order cumulant expressions. The other three panels compare the coarse-grained PMFs with the atomistic one for these different averaging expressions for several values of k_{LJ}. The top right panel shows the linearly averaged work (unweighted). Bottom left: the exponentially weighted work. Bottom right: the second order cumulant expression. The error bars indicate the error in the PMFs from the simulations, and were obtained from block averaging the trajectories. Note that each subtrajectory is averaged individually.

From comparing the PMFs of the atomistic and coarse-grained simulations it follows that the $k_{LJ} = 2\ \varepsilon_h$ curve has a systematically too small slope, while the slope of the $k_{LJ} = 5\ \varepsilon_h$ is systematically too high. For $k_{LJ} = 3\ \varepsilon_h$ and $k_{LJ} = 4\ \varepsilon_h$ the curves are more similar to the atomistic data. Whereas for the unweighted work expression the $k_{LJ} = 4\ \varepsilon_h$ seems to be best, the atomistic data in Fig. 7 suggest that the exponential and 2nd order cumulants are more trustworthy. We conclude that $k_{LJ} = 3\ \varepsilon_h$ is closest to the atomistic PMF for the exponential and 2nd order cumulant, and will therefore use the value $k_{LJ} = 3\ \varepsilon_h$ in the simulations of stacks of silk-based blocks.

3 Coarse-grained simulation of the β roll and large stacks

Using the optimized coarse-grained model, we ran a 50 ns coarse-grained simulation of one 81 residue β-roll and a stack of two such rolls. For the single β-roll a twist angle of −19.4° (−39.6° for atomistic resolution) and a RMSD of 1.84 Å (*versus* 1.40 Å atomistic) were observed. For the stack the twist angle is less: −9.4° (*versus* −13.1° atomistic). The RMSD is also smaller: 1.52 Å (1.28 Å atomistic). The distance between the two β-roll centers of mass in the stack is on average 0.95 ± 0.5 nm, which agrees well with the roll-height of 0.92 ± 0.7 nm in the atomistic simulation. These results show that the coarse-grained model is accurate enough to reproduce the mechanical stability, dimensions and twist of the β-roll. An open question is whether the coarse-grained model can also predict the folded state from an unfolded conformation. A simulation of the entire 81-residue polypeptide started from an unfolded conformation resulted in an unstructured aggregate. To test the folding behavior we therefore simulated a shorter peptide: $E(\text{GAGAGAGE})_3$. Starting from a short, unfolded conformation, the peptide quickly folds into a double hairpin. Such downhill folding seems to be consistent with the measured PMF. However, as glycine and alanine are parameterized with the same bead type, two (very similar) structures are formed: hairpins with all glycines on the inside as in the atomistic simulations and β-rolls with all glycines on the outside. This behavior follows from the symmetric sequence in the coarse-grained model, *i.e.* there is no difference between the C and N terminus. In a future version of the model one could use a distinct bead type for the alanine residues in order to restore the asymmetry of the real polypeptide.

The aim of the coarse-grained model is to allow investigations of the fibers at the nm–μm length scale. From the above mentioned simulations it follows that a stack of two rolls is stable and behaves comparable to its atomistic counterpart for at least 50 ns. To study the behavior of longer fibers, we stacked 24 coarse-grained β-rolls (Fig. (8)), equilibrated the system, and simulated the fiber for 10 ns at room temperature. This simulation shows that the coarse-grained β-rolls can indeed form a long stable fiber. The linear dimension of the coarse-grained stack is about 21.6 nm, corresponding to a height of 0.94 nm per β-roll, slightly more compressed than the stack of two rolls. Furthermore, the fiber bends slightly in an oscillatory fashion, indicating that it is not infinitely stiff, but has finite bending rigidity. We can estimate the bending rigidity, or persistence length by looking at the deviation of the fibers position along the long axis. The fiber is aligned along

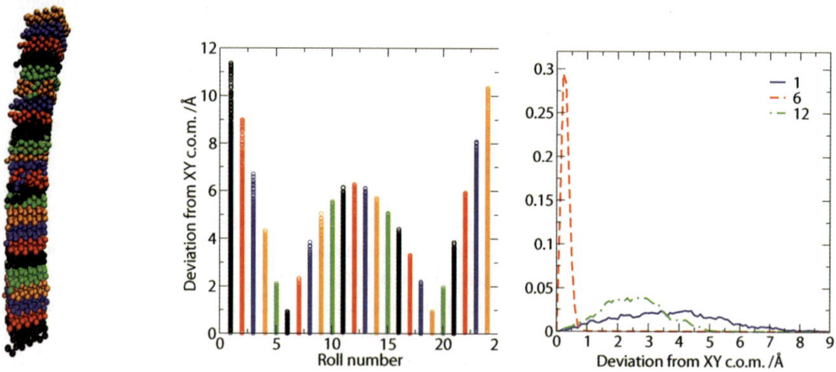

Fig. 8 An indication of the mesoscopic bending behavior of a long fiber, as obtained from a 10 ns coarse-grained simulation of a stack of 24 β-rolls. Left: a snapshot of the stack at maximum bending. Middle: the average deviation from a straight fiber obtained by plotting the deviation of the center of mass of each β-roll with respect to the *z*-axis as a function of the *z* position for every time step. From this plot, and from looking at the trajectory, it is clear that rolls on the outside of the fiber are most mobile. The fiber oscillates in its lowest mode, as is clear from the presence of two nodes (roll 6 and roll 19, counting from the bottom roll). Right: Histogram of the distributions of the deviation from the *z*-axis for rolls 1, 6 and 11. Roll 1 shows indeed a wide distribution of distances, whereas roll 6 (the node) has a narrow distribution. Roll 11 shows an intermediate range. The average deviation taken from these distributions allows an estimate of the persistence length.

the z-axis. Plotting the distribution of the absolute deviation of the center of mass of each roll in the stack from the z-axis as a function of z (Fig. 8) yields information about the bending behavior and flexibility of the fiber. From this distribution it follows that the fiber oscillates in its lowest mode, with two nodes around roll 6 and roll 19. As was to be expected, rolls on the outside show a slightly higher deviation than rolls on the inside. From this oscillation it is possible to derive the persistence length l_P. Assuming that the long fiber behaves like a worm like chain, the mean square end-to-end distance $\langle R^2 \rangle$ is given by[38]

$$\langle R^2 \rangle = 2Ll_p \left[1 - \frac{l_p}{L} \left(1 - e^{-L/l_p} \right) \right] \tag{15}$$

where l_p is the persistence length, and L is the contour length of the fiber. Because $L \ll l_p$ in this case, a simple expansion gives.

$$\langle R^2 \rangle \approx L^2 - \frac{1}{3} \frac{L^3}{l_p} \tag{16}$$

Realizing that $\langle R^2 \rangle$ can also be related to the average deviation from a straight fiber d by $L^2 = \langle R^2 \rangle + d^2$, one can solve for the persistence length,

$$l_p = \frac{1}{3} \frac{L^3}{d^2} \tag{17}$$

The fiber length L in the simulation is about $L = 21.6$ nm, and the average deviation between the end points about $d = 1.2$ nm. The latter value follows from the histograms in Fig. 8, and realizing that while the endpoints deviate about 3.5 Å from the z-axis, the middle part of the fiber moves 2.5 Å in the opposite direction. This 6 Å difference between the middle and one endpoint translates into a 12 Å average endpoint deviation from a straight fiber. These numbers result in an estimate for the persistence length of $l_p = 2.33$ μm. This is very reasonable when comparing the experimental transmission electron and atomic force microscopy results, which show that the fiber has a persistence length of the order of a few microns.[10] We note that these simulation results cannot be compared directly to the experiments because the silk domain is much smaller than the 384 residue block that is used in the experiments and, moreover, the collagen blocks are missing. Nevertheless, these results show that even a simple coarse-grained model can already assess mesoscopic properties such as the persistence length.

4 Discussion and conclusions

We have developed a coarse-grained protein model, which enables description of fiber assembly and fiber–fiber interaction on the nm–μm length scale. This model is based on fitting to structural data, as well as to unfolding PMFs. This coarse-grained procedure, while still very rough, fits entirely in the philosophy of coarse-graining that a simplified model can and should reproduce several important (thermodynamic or structural) properties, but might, and, in fact, will not necessarily be able to reproduce *all* equilibrium quantities. The procedure we advocate here reproduces the structure of the folded state, and approximates the equilibrium between folded and unfolded states. We conjecture that this method is applicable to the folding behavior of other proteins. We will however first need to test this on the folding–unfolding equilibrium behavior of our coarse-grained model, *e.g.* by replica exchange MD. Nevertheless, besides all these caveats we have shown here that even a simple model can reproduce some of the interesting mesoscopic properties of a long protein fiber.

As we have about two orders of magnitude fewer particles in the coarse-grained simulation with respect to the all-atom simulation, the gain in efficiency is at least also of that order. Further gains in efficiency can be achieved by using a larger time-step, something that we have not yet investigated.

In a future version of the coarse-grained model we can expand the model to include the Ala side-chain. In addition, we will test the influence of the solvent, fiber length, and the effect of including collagen. Using such a comprehensive description of the system we will try to reproduce the overall equilibrium properties of the fiber, as well as fiber–fiber interactions, and eventually the rheology of fibers. In this way we can span the large gaps in time and length scales that lie between the molecular structure of proteins and the mesoscopic–macroscopic level of protein fibers. Indeed, models like these open the way for large-scale simulations of fibers, based on molecular structure, and allows investigation of their nucleation, growth, cross-linking mechanism, network dynamics, and rheology.

5 Methods

The GROMOS96 force field[39] in combination with the SPC water model was used for all atomistic simulations. The GROMACS package version 3.3.2[40] was used for these simulations, including the steered MD simulations. The pressure was kept at 1 bar using Parinello–Rahman coupling and the temperature was kept at 298 K using a Nosé–Hoover thermostat. Bonds were constrained with LINCS, allowing for a 2 fs timestep. Electrostatics were treated with PME.

The initial configuration for the β-roll was taken from our previous simulation studies (ref. 1). After energy minimization, a short, protein restrained run was performed to allow for equilibration of the solvent, followed by a 1 ns MD simulation to equilibrate the whole system. For the 50 ns simulations of one β-roll or a stack of 2 β-roll dodecahedral boxes with volumes of 90 nm^3 and 150 nm^3, respectively were used. The SMD simulations were performed in a cubic box with a volume of 115 nm^3.

All coarse-grained simulations were run with the CM3D program.[34] As the program uses a multi-timestep, bonds are not constrained. All simulations were run in the NVT ensemble, at 300 K in a cubic box of 1000 nm^3, only the stack of 24 β-rolls was run in a cubic box of 27 000 nm^3. A 2 fs timestep was used for all simulations. The temperature was kept constant with a Nosé–Hoover thermostat. Starting structures were generated by extracting the C_α coordinates from an atomistic simulation and simulating for 1 ns at low temperature (T = 10 K). Structures were visualized with VMD.[41]

Acknowledgements

The authors acknowledge Christopher Lowe for helpful discussions on the persistence length of fibers. The authors would like to thank Berk Hess for pointing out an error in the PMF calculations, which was corrected in the proofs of this paper.

References

1 M. Schor, A. A. Martens, F. A. de Wolf, M. A. Cohen Stuart and P. G. Bolhuis, *Soft Matter*, 2009, **5**, 2658–2665.
2 S. Brown, N. J. Fawzi and T. Head-Gordon, *Proc. Natl. Acad. Sci. U. S. A.*, 2003, **100**, 10712–10717.
3 S. Zhang, *Nat. Biotechnol.*, 2003, **21**, 1171–1178.
4 H. M. König and A. F. M. Kilbinger, *Angew. Chem., Int. Ed.*, 2007, **46**, 8334–8340.
5 M. T. Krejchi, S. J. Cooper, Y. Deguchi, E. D. Atkins, M. J. Fournier, T. L. Mason and D. A. Tirrell, *Macromolecules*, 1997, **30**, 5012–5024.
6 J. M. Smeenk, M. B. J. Otten, J. Thies, D. A. Tirrell, H. G. Stunnenberg and J. C. M. van Hest, *Angew. Chem., Int. Ed.*, 2005, **44**, 1968–1971.

7 N. I. Topilina, S. Higashiya, N. Rana, V. V. Ermolenkov, C. Kossow, A. Carlsen, S. C. Ngo, C. C. Wells, E. T. Eisenbraun and K. A. Dunn, *Biomacromolecules*, 2006, **7**, 1104–1111.
8 M. T. Krejchi, E. D. Atkins, A. J. Waddon, M. J. Fournier, T. L. Mason and D. A. Tirrell, *Science*, 1994, **265**, 1427–1432.
9 M. W. T. Werten, A. P. H. A. Moers, T. H. Vong, H. Zuilhof, J. C. M. van Hest and F. A. de Wolf, *Biomacromolecules*, 2008, **9**, 1705–1771.
10 A. A. Martens, G. Portale, M. W. T. Werten, R. J. de Vries, G. Eggingk, M. A. Cohen Stuart and F. A. de Wolf, *Macromolecules*, 2009, **42**, 1002–1009.
11 H. Taketomi, Y. Ueda and N. Go, *Int. J. Pept. Protein Res.*, 1975, **7**, 445–459.
12 J. D. Honeycutt and D. Thirumalai, *Biopolymers*, 1992, **32**, 695–709.
13 A. I. Jewett, A. Baumketner and J. E. Shea, *Proc. Natl. Acad. Sci. U. S. A.*, 2004, **101**, 13192–13197.
14 A. V. Smith and C. K. Hall, *J. Mol. Biol.*, 2001, **312**, 187–202.
15 C. Clementi, A. E. Garcia and J. N. Onuchic, *J. Mol. Biol.*, 2003, **326**, 933–954.
16 P. Derreumaux and N. Mousseau, *J. Chem. Phys.*, 2007, **126**, 025101.
17 L. Monticelli, S. K. Kandasamy, X. Periole, R. G. Larson, D. P. Tieleman and S. J. Marrink, *J. Chem. Theory Comput.*, 2008, **4**, 819–834.
18 A. Liwo, M. Khalili and H. A. Scheraga, *Proc. Natl. Acad. Sci. U. S. A.*, 2005, **102**, 2362–2367.
19 P. G. Bolhuis and A. A. Louis, *Macromolecules*, 2002, **35**, 1860–1869.
20 S. Izvekov and G. A. Voth, *J. Phys. Chem. B*, 2005, **109**, 2469–2473.
21 S. J. Marrink and et al., *J. Phys. Chem. B*, 2007, **111**, 7812–7824.
22 N. J. Fawzi, E.-H. Yap, Y. Okabe, K. L. Kohlstedt, S. P. Brown and T. Head-Gordon, *Acc. Chem. Res.*, 2008, **41**, 1037–1047.
23 Y. Sugita and Y. Okamoto, *Chem. Phys. Lett.*, 1999, **314**, 141–151.
24 J. Juraszek and P. G. Bolhuis, *Proc. Natl. Acad. Sci. U. S. A.*, 2006, **103**, 15859–15864.
25 G. M. Torrie and J. P. Valleau, *Chem. Phys. Lett.*, 1974, **28**, 578–581.
26 A. Laio and M. Parrinello, *Proc. Natl. Acad. Sci. U. S. A.*, 2002, **99**, 12562–12566.
27 A. F. Voter, *Phys. Rev. Lett.*, 1997, **78**, 3908–3911.
28 T. Huber, A. Torda and W. van Gunsteren, *J. Comput.-Aided Mol. Des.*, 1994, **8**, 695–708.
29 H. Grubmuller, *Phys. Rev. E*, 1995, **52**, 2893–2906.
30 S. Park, F. Khalili-Araghi, E. Tajkhorsid and K. Schulten, *J. Chem. Phys.*, 2003, **119**, 3559–3566.
31 S. Park and K. Schulten, *J. Chem. Phys.*, 2004, **120**, 5946–5961.
32 C. Jarzynski, *Phys. Rev. Lett.*, 1997, **78**, 2690–2693.
33 H. C. Andersen, *J. Comput. Phys.*, 1983, **52**, 24–34.
34 http://www.cmm.upenn.edu/resources/indexsoft.html.
35 G. Hummer and A. Szabo, *Proc. Natl. Acad. Sci. U. S. A.*, 2001, **98**, 3658–3661.
36 P. L. Geissler and C. Dellago, *J. Phys. Chem. B*, 2004, **108**, 6667–6672.
37 G. E. Crooks, *J. Stat. Phys.*, 1998, **90**, 1481–1487.
38 M. Doi and S. F. Edwards, *The Theory of Polymer Dynamics*, Oxford University Press, Oxford, 1986.
39 C. Oostenbrink, A. Villa, A. E. Mark and W. F. van Gunsteren, *J. Comput. Chem.*, 2004, **25**, 1656–1676.
40 D. Van der Spoel, E. Lindahl, B. Hess, G. Groenhof, A. E. Mark and J. H. C. Berendsen, *J. Comput. Chem.*, 2005, **26**, 1701–1718.
41 W. Humphrey, A. Dalke and K. Schulten, *J. Mol. Graphics*, 1996, **14**, 33–38.

Phase behavior of low-functionality, telechelic star block copolymers

Federica Lo Verso,*[a] Athanassios Z. Panagiotopoulos*[b] and Christos N. Likos*[c]

Received 12th March 2009, Accepted 8th April 2009
First published as an Advance Article on the web 26th August 2009
DOI: 10.1039/b905073f

We apply state-of-the-art, Grand Canonical Monte Carlo simulations to determine the self-organization and phase behavior of solutions of block copolymer stars. The latter consist of f AB-block copolymers with N monomers each, which contain a solvophilic block A and solvophobic block B, and which are tethered on a common center on their A-side. We vary the degree of polymerization N and the relative composition of the block copolymer arms and investigate the interplay between macrophase and microphase separation in the system. Preliminary results of the effect of increasing the number of arms, f of the stars are also presented.

1 Introduction

Star polymers consisting of f homopolymer chains attached to a common center have been thoroughly studied in the last two decades by a combination of experimental[1,2] and theoretical[3] efforts. For the case of athermal solvent conditions, their conformations[4] and interactions[5] are by now well-understood and have led to a number of recent discoveries pertaining to their vitrification and rheological properties.[6–9] Less is known about star polymers that are *chemically asymmetric*, *i.e.*, their arms are either chemically distinct from each other (inter-arm asymmetry) or they contain blocks of different chemistry (intra-arm asymmetry).[10] These are known as *miktoarm* or μ-arm stars and several synthesis techniques have appeared in the literature.[11] In this work, we focus on a particular case of miktoarm stars with intra-arm asymmetry, for which each chain is an AB-block copolymer with the solvophilic part A anchored at their inner part and the solvophobic part B at their periphery. Such macromolecular aggregates are also termed *telechelic* star polymers.[12]

Telechelics are defined as polymeric molecules with reactive terminal groups able to create intra- as well as inter-molecular bonds in the appropriate solvent. In the last twenty years, the methods of analysis and synthesis of this class of systems have considerably increased.[12,13] This increased interest is motivated by the large number of applications, ranging from medical devices to daily consumer products,[12] and which include detergency, catalysis, food and cosmetics, and oil recovery. A salient feature of telechelics is the macromolecular size and structure comparable with those of biological systems. In addition, due to the presence of the reactive groups, significant changes in their conformation can be induced by varying pH, solvent quality,

[a]*Institut für Physik, Johannes-Gutenberg-Universität Mainz, Staudinger Weg 7, D-55099 Mainz, Germany. E-mail: loverso@uni-mainz.de*
[b]*Department of Chemical Engineering and PRISM, Princeton University, Princeton, NJ, 08540, USA. E-mail: azp@princeton.edu*
[c]*Institute for Theoretical Physics II: Soft Matter, Heinrich-Heine-Universität Düsseldorf, D-40225 Düsseldorf, Germany. E-mail: likos@thphy.uni-duesseldorf.de*

temperature, pressure and other solution conditions. Several molecular geometries, more or less complex, originate from irreversible synthesis or reversible self-assembly of block copolymers.[14,15] In this context, the simplest example of telechelics is the single tri-block copolymer, important as precursor of chain extenders, to build grafted surfaces or for synthesis of different size and block-sequences of multi-copolymers. A second example are tree-like structures such as dendritic micelles, with the hydrophilic external shell and the hydrophobic inner core. The core can create a pocket capable of capturing hydrophobic molecules, relevant in the context of drug delivery systems. As a last example, without exhausting the range of possibilities, we mention star-shaped poly(L-lactide)s with pyrene lipophilic groups employed as slow-release drug delivery systems, and for surgical implants, due to the good mechanical properties, biocompatibility and biological degradability of the lactides. The high sensitivity to environmental factors and to chemical synthesis conditions suggests that many parameters can be tuned in order to change and to study their inter- and intra-molecular aggregation properties.

From a theoretical/simulation point of view, it is highly desirable to simplify the problem focusing only on a few parameters at a time in such a way as to have a systematic comprehension of the different contributions to the molecular configuration and to be able to apply the suitable theoretical method of investigation. In the analysis of telechelic star polymers, the first simplification which has been done is to consider separately two different density regimes, namely dilute solutions, and in particular the limit of a single molecule,[16–19] and concentrated solutions of stars.[20] At low densities the intra-molecular association processes depend on the length and number of chains per molecule, as well as on the number of attractive tails per arm. The important physical parameters are thus the number of arms (functionality) f, the degree of polymerization N of each arm as well as a characteristic reduced temperature $T^* = k_B T/\varepsilon$, where k_B is Boltzmann's constant, T is the absolute (room) temperature and ε the typical depth of the attractive potential that acts between the monomers of the solvophobic, B-endgroups. In the range of parameters $f \leq 10$, $10 \leq N \leq 200$ and $0.01 \leq T^* \leq 1.5$, it has been found that the probability of having collapsed, watermelon structures is higher the lower the N and T^*, and the higher f are.[16–19] For higher arm numbers, $f > 10$, and/or higher molecular weight, the molecules form partial watermelon structures at low temperature, while they prefer the open star-like configuration at high temperature. In the specific case of a single molecule in solution, the studies were performed employing Gaussian self-consistent theory and simulations[16,17] as well as monomer resolved molecular dynamic simulations and scaling theory.[18,19] The latter allowed analysis of the effects of the elastic, excluded volume and attractive contribution to the free energy of the star.

Departing from these investigations, attention was subsequently turned to the inter-molecular aggregation processes, increasing the density of the system and in particular changing the intensity of the attraction between chains, *i.e.*, considering a variable number of stickers per arm.[20] As a first step, the arm number and the total-number of monomers per chain have been fixed: the aggregation behavior of trifunctional molecules with $N = 10$ has been analyzed and a competition between micellization and fluid–fluid phase separation was established. The relevant parameters are here the ratio between attractive and repulsive monomers per arm. If the number of repulsive units exceeds the number of stickers the coexistence phase region disappears and micellization phenomena are dominant. In concentrated solutions, a number of supramolecular structures show up, such as bicontinuous phases, in the high-density side of the coexistence curve, or aggregates of interconnected micelles at moderate concentrations.

In this work, we extend previous studies in a number of ways. We first consider the change in the conformation of the single star increasing the concentration, and in particular the dependence of the radius of gyration on the same. Second, we increase the number of monomers per arm and we analyze the effect on the interplay between micellization and fluid–fluid phase separation; the goal is to understand how the entropic and excluded volume penalty influenced the aggregation processes. Finally, we increase the functionality f of the stars to qualitatively examine the effects of the latter on phase separation and micellization.

The paper is organized as follows: in section 2 we describe the model, the method of analysis and the simulation technique employed. Our results are presented in section 3 and in particular: results for $f = 3$ and $N = 10$ in section 3.1; the effects of changing the degree of polymerization per arm, $f = 3$, $N = 20$, are presented in section 3.2, together with the effects of changing the functionality to $f = 6$. In section 4 we present our conclusions and an overview of open questions and future problems.

2 Interaction model and method of analysis

As anticipated above, the different mechanisms of inter- and intra-molecular aggregation and the subsequent competition between micellization and fluid–fluid separation as well as the appearance of lamellar structures in telechelics are driven by several factors such as the length of the chains, the internal rigidity, the temperature, the strength of the attractive interactions, the size ratio and the sequences of block-copolymers, and the density regime.[12,21] The high number of molecules and the related computational efforts connected to high concentrations, force us to consider a different numerical study with respect to ref. 18, based on lattice simulations. Following the scheme described in ref. 20 we employ Grand Canonical Monte Carlo simulations (GCMC) on a three-dimensional cubic lattice with a coordination number of 26, to enhance the flexibility of the chains, and periodic boundary conditions in all three directions. The relative position vectors for the interactions are (1, 0, 0), (1, 1, 0), (1, 1, 1) and their reflections along the principal axes. The simulation cell has a fixed volume $V = L \times L \times L$; here we consider different box sizes, $L = 20, 30, 40, 50$, depending on the length of the star chains and/or in order to verify the presence of a fluid–fluid phase transition (see below for details). As in the previous investigations,[20] we applied the Larson model for amphiphile–oil–water systems.[22] In our specific case we considered N molecules labeled as $H(H_nT_m)_f$, *i.e.*, consisting of f chains chemically linked to a common core-H site. Every chain contains n solvophilic head monomers (H) and m solvophobic tail monomers (T) and every lattice site is occupied either by a (H or T) monomer or by a solvent (S) molecule. In terms of the interaction parameters, the head sites and the solvent/vacuum particles are the same, H = S. The whole system can be described by a single energy parameter, $\varepsilon = 2\varepsilon_{HT} - \varepsilon_{HH} - \varepsilon_{TT}$, where ε_{HH} and ε_{HT} are set to zero, while $\varepsilon_{TT} = -1$. As a consequence, we define the reduced temperature as $T^* = k_BT/\varepsilon$. We studied systems with a total number of monomers per arm $N = 10$ and 20 for $f = 3$, and $N = 10$ for $f = 6$.

As in ref. 20, we performed a mixture of displacement, single regrowth and cluster moves and we used the *athermal* version of the configurational-bias sampling method,[23,24] to enable insertions and removals of the monomers. In a simulation step, the centers of the stars are inserted randomly following the usual rules for self-avoiding walks.[25] Subsequent star molecules are placed on unoccupied lattice positions computing the *Rosenbluth weight* for each growth/removal step,[23,25] for details see ref. 20. The number of MC steps per run was roughly between 10^6 and 10^9, depending on the region of the phase diagram we had to investigate and the number of monomers/size of the lattice we considered. Chemical potential and temperature are parameters of the simulation. For a high number of stickers per molecule, we investigated the fluid–fluid phase coexistence curve, which has been determined *via* histogram-reweighing techniques.[26,27] In order to obtain accurate critical parameters, we employed finite size scaling methods.[28–30]

Without entering into the technical details of the simulation, we would like here only to emphasize a few important and practical aspects connected to our analysis. First of all, subcritical isotherms show hysteresis effects, *i.e.*, the states sampled at a given temperature and chemical potential depend on the initial conditions. The approach used to avoid this problem was to link states by providing connections

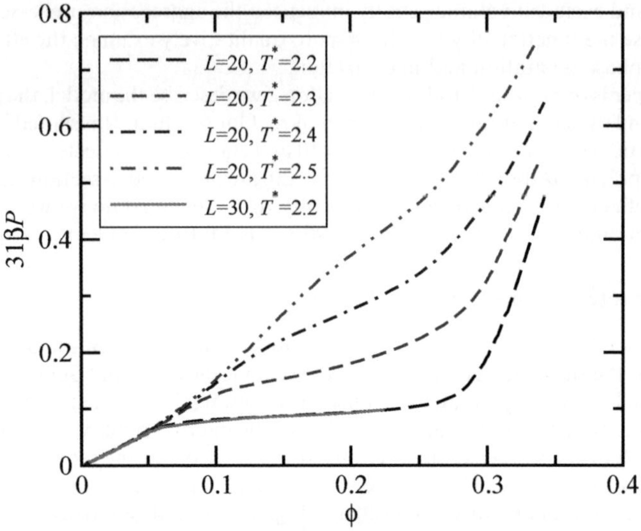

Fig. 1 Pressure–concentration curves for a $H(H_8T_2)_3$-system at various temperatures and for different system sizes, as obtained by simulation of our model. Plotted is the pressure multiplied by the number of monomers per star, so that at low values of ϕ a straight line with slope unity (ideal gas law) results.

through a supercritical path. It should be pointed out that hysteresis effects also have consequences for systems in which phase separation is suppressed and micellization takes place instead. Moreover the presence of micellar aggregates results in significant equilibration problems which become more pronounced as the system size increases. As a consequence small systems have computational advantages for the calculations and they permit the determination of the relevant quantities with low statistical uncertainties over a broad range of temperature.[31] Further, we employed the size dependence of the phase diagrams and/or of the osmotic pressure to determine whether a system undergoes macrophase or microphase separation.[32] In Fig. 1 and 2 we show some examples obtained during our study. In a finite-size system, the

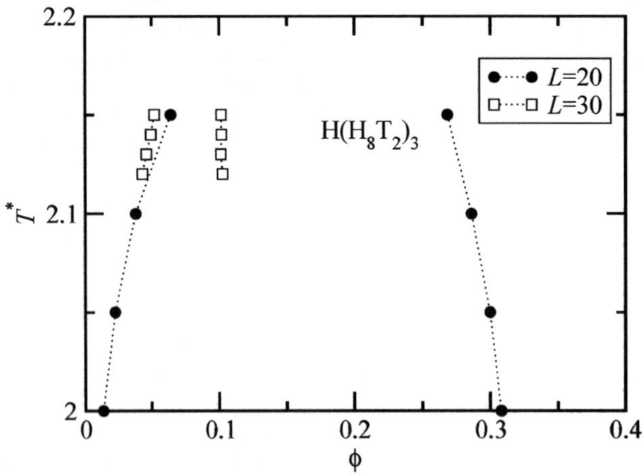

Fig. 2 Pseudo-coexistence curves for a $H(H_8T_2)_3$-system that does not undergo phase separation, as obtained by simulation of systems with two different sizes. The shrinking of the apparent coexistence curves for $L = 30$ demonstrates the fact that no true coexistence takes place in the thermodynamic limit.

finite slope of the osmotic pressure approaches a horizontal line as the system-size increases, indicating a first-order transition between two phases at different concentrations. In contrast, if the system undergoes micellization, we do not observe any change in the slope of the osmotic pressure *vs.* concentration line. This can be amply seen in Fig. 1, which pertains to trifunctional telechelic stars with $N = 10$ monomers per arm, with 8 solvophilic inner monomers and 2 solvophobic end-monomers per arm, written as $H(H_8T_2)_3$. This system undergoes micellization rather than phase separation,[20] and hence the pressure curve for $T^* = 2.2$ does not differ between a simulated system in a $L = 20$ and a $L = 30$ box. The flat part of the pressure curve at that temperature might lead to the erroneous conclusion that the system phase-separates. Fig. 2 demonstrates that the opposite is true: performing the determination of (pseudo) phase boundaries for the two different system sizes leads to a dramatic shrinking of the former as L grows and establishes the absence of macroscopic phase separation.

3 Results

3.1 Stars with $f = 3$ and $N = 10$

We considered trifunctional telechelic block copolymers of the type $H(H_nT_m)_3$ with $m = 2, 4, 5, 6, 8, 10$ and $N = n + m = 10$. Before extending the study described in ref. 20, we close the gap with the single-molecule analysis of ref. 18. We performed lattice simulation on the extremely diluted regime, so that the simulation box contained on average one molecule. We focus on the molecule $H(H_8T_2)_3$, which is quite similar to the model of ref. 18: our purpose here is to test once again the possibility of watermelon structures as well as the ability of the lattice approach employed here to reproduce the same physics as the continuum model of ref. 18. To this end, we monitor during the lattice simulation the fraction of end-monomers that appears aggregated with end-monomers from other chains of the star, where *aggregated* means that other end-monomers are within the range of the attractive monomer–monomer interaction. We classify a configuration as *triplet* if end-monomers form three different chains aggregate, *doublet* if two chains only participate in the aggregation and *singlet* if no interchain aggregation takes place. Accordingly a triplet is the same as the watermelon configuration and a singlet corresponds to the open star, in the terminology of ref. 18 and 19. The results are shown in Fig. 3. In full agreement with the off-lattice, Molecular Dynamics and scaling theory analyses,[18,19] we see at low temperatures that the fraction of triplets is unity, *i.e.*, the molecules are with certainty in the watermelon configuration. As T grows, doublets and singlets appear but the former are never the conformation with the highest probability and, at $T^* \gtrsim 0.8$, the open star configuration takes over. We remind here that trifunctional telechelic star polymers have also been considered from the experimental point of view.[33–36] In particular for di- and tri-ω-functionalized zwitterionic star polybutadienes rheological and scattering data showed a variety of self-organizing processes including system-spanning networks and collapsed, soft-sphere conformations. The network characteristic and/or the preference for intra-molecular association depends on the molecular weight of the arms. The trends nicely agree with the theoretical results.[18]

As found in ref. 20, $H(H_nT_m)_3$-systems with $N = 10$, show a qualitative change in their macroscopic behavior when $m < n$: whereas for $m \geq n$ macrophase separation takes place, for $m < n$ the system forms well-defined, micelles which develop mutual interconnections as the density grows. In Fig. 4 we observe a typical tails aggregation process for phase separating systems, increasing the density: clockwise we have a monomer concentration $\phi = 0.03, 0.104, 0.22$, and 0.51 for $T^* = 6.9$ and a $H(H_4T_6)_3$ system. From the snapshots we can observe the appearance of bicontinuous phases. Decreasing the temperature (here $T^* = 5.8$), on the right side of the coexistence curve, the telechelic stars with n/m close to one form layers (see

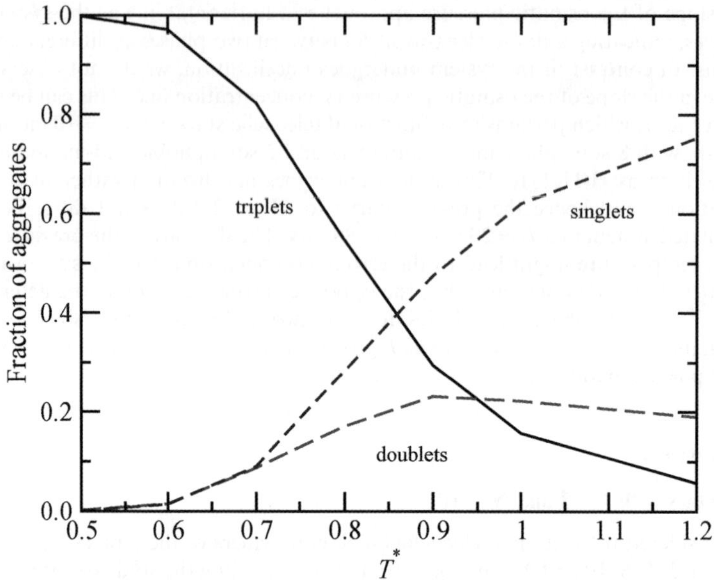

Fig. 3 The fraction of interchain aggregates as a function of temperature for isolated $H(H_8T_2)_3$-molecules, as obtained by the lattice simulation.

Fig. 5), whereby the specific geometry of the tails inside such an aggregate have to be analyzed in detail.

To investigate the influence of the above mentioned scenario on the conformation of the molecules, we measured the gyration radius R_g of the same as a function of the

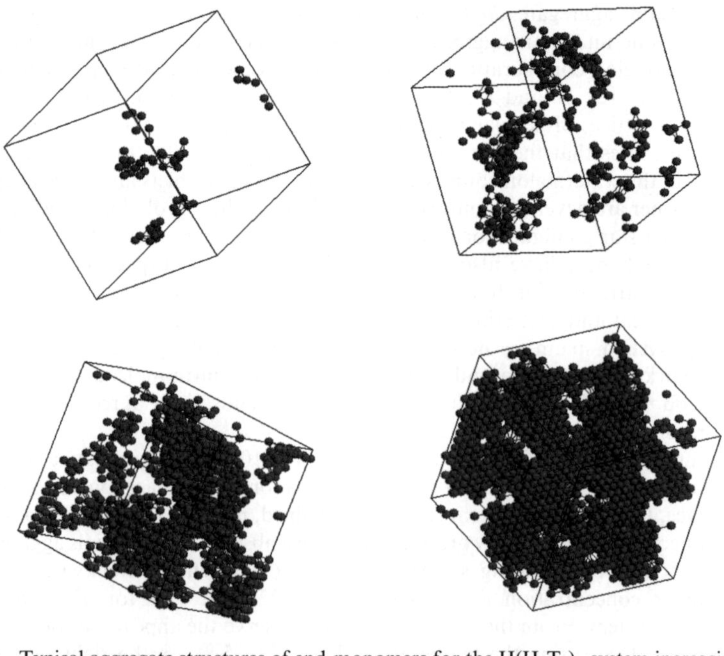

Fig. 4 Typical aggregate structures of end-monomers for the $H(H_4T_6)_3$-system increasing the density, at $T^* = 6.9$: clockwise we have $\phi = 0.03$, 0.104, 0.22, 0.51.

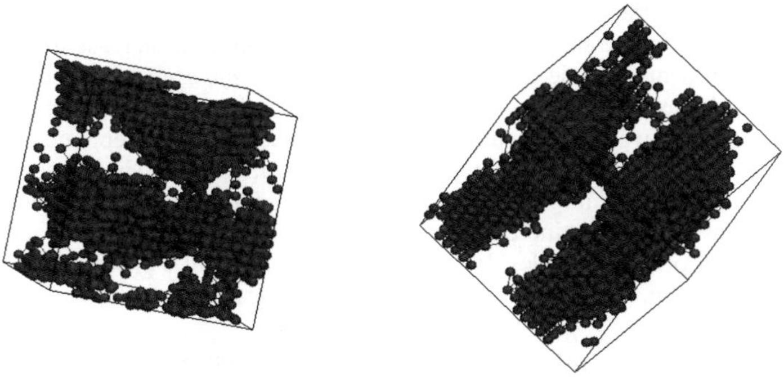

Fig. 5 End-monomers aggregate for the $H(H_4T_6)_3$-system. $T^* = 5.8$, from the left $\phi = 0.44$ and $\phi = 0.56$.

monomer concentration ϕ. The results are shown in Fig. 6. Here, the temperature is fixed to the same value considered for the morphological analysis in ref. 20, *i.e.*, slightly supercritical for the phase-separating systems. In particular we considered $T^* = 15.25$, $T^* = 10.25$, $T^* = 6.9$, $T^* = 5.8$, $T^* = 4.4$, and $T^* = 2.3$ for $n = 0, 2, 4, 5, 6$, and 8, respectively.

The conformation of the star depends on the interplay between attraction and excluded-volume interactions, as well as on the overall phase behavior of the solution. For fixed density, the radius of gyration increases with increasing n. This is a trivial intra-molecular effect due to the increase of the number of solvophilic heads per chain, which leads to a dominance of the excluded-volume interactions against the attractive ones.

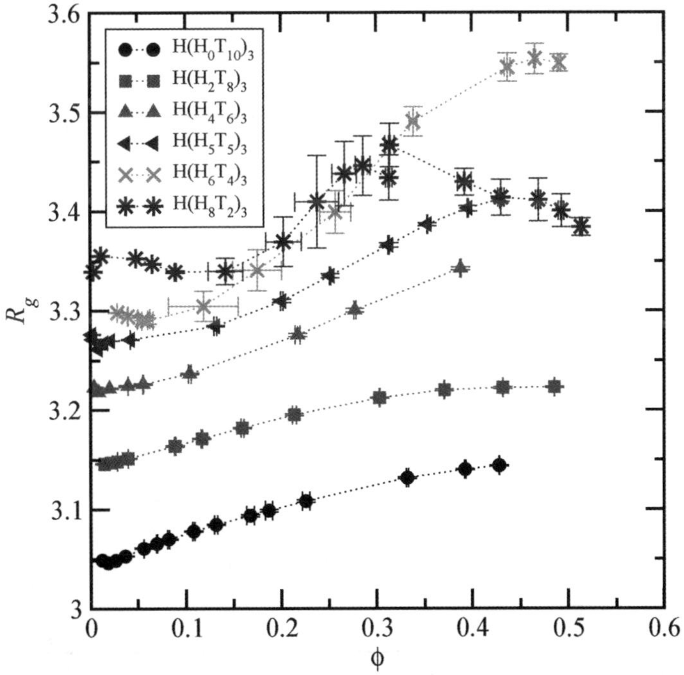

Fig. 6 The radius of gyration R_g of the stars as a function of concentration for different systems. $T^* = 15.25$, $T^* = 10.25$, $T^* = 6.9$, $T^* = 5.8$, $T^* = 4.4$, $T^* = 2.3$ for $n = 0, 2, 4, 5, 6, 8$ respectively.

The effects of the concentration are more intriguing. For phase-separating systems, $m \geq n$, an increase in ϕ brings about a concomitant increase in the radius of gyration in a monotonic fashion. This effect is *the opposite* of the screening of excluded volume interactions for self-avoiding polymers, in which case R_g decreases as ϕ grows. As the number of molecules in the system increases, the possibility of intermolecular aggregation becomes favorable, since it entails the possibility of maintaining the energy gain by contacts between endgroups while offering an entropy increase due to the possibilities of the chains to wander towards neighboring molecules. In this fashion, the molecules grow in average size with concentration, albeit in moderate amounts, typically by 10% for a range $0 \leq \phi \leq 0.5$.

The behavior of R_g with f for the two micelle-forming systems, $H(H_6T_4)_3$ and $H(H_8T_2)_3$ is more complicated. First, at low ϕ there is a slight non-monotonicity, with R_g initially growing and subsequently decreasing again at about the critical micelle concentration (c.m.c.), which has been determined as $\phi_{cmc} = 0.070 \pm 0.002$ for $H(H_6T_4)_3$ and $\phi_{cmc} = 0.092 \pm 0.002$ for $H(H_8T_2)_3$.[20] This is caused by the shrinking of the stars that self-organize into micelles. Thereafter, the size grows again due to the formation of well-shaped, spherical micelles, a process that causes the bonds in the inner of the micelles to stretch in order to accommodate more and more arms. For the more attractive, $H(H_6T_4)_3$-system the growth of the micelle size seems to persist at least up to $\phi = 0.5$, although some signs of saturation are clear. For the less attractive, $H(H_8T_2)_3$-system, the full stretching of the bonds in the inner of the micelles reaches a limit at $\phi \gtrsim 0.3$. Thereafter, the overlap of the solvophilic coronae of different micelles brings about a shrinkage of micelle size, akin to the one observed for polymers in good solvents due to screening of the excluded-volume interactions. Subsequently, R_g decreases markedly again.

We have also determined the c.m.c. of micelle-forming systems by looking at the pressure curves, as in Fig. 1, and identifying it with the point at which there is a change in its slope. Indeed, for isolated molecules in a dilute solution, one expects that the ideal-gas law $(3N + 1)\beta P = \phi$ holds, a feature fully confirmed by Fig. 1. On the contrary, when micelles form, the slope of the curve will become much smaller.

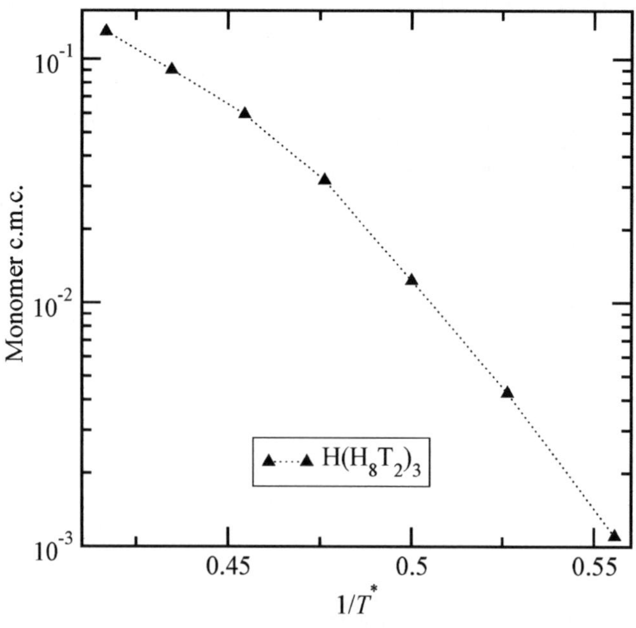

Fig. 7 The critical micelle concentration of the $H(H_8T_2)_3$-system as a function of temperature.

Fig. 7 shows the dependence of the c.m.c. on temperature for the system $H(H_8T_2)_3$. The c.m.c. is indeed a very small number, which increases by two orders of magnitude for a moderate change of T^*, from 1.8 to 2.4. The model reproduces, therefore, salient features of micelle-forming systems.

3.2 Increasing the degree of polymerization and the functionality

We now turn our attention to the effects of changing the degree of polymerization per arm, N, while keeping the functionality constant at $f = 3$. We analyzed the same n/m ratios considered in ref. 20, namely the pairs $n = 0$ and $m = 20$, $n = 4$ and $m = 16$, and $n = 8$ and $m = 12$ ($N = 20$), corresponding to $n = 0$ and $m = 10$, $n = 2$ and $m = 8$, $n = 4$ and $m = 6$ ($N = 10$) respectively. In Fig. 8 we present the coexistence curves we obtained for this case and $L = 40$ (full symbols). Again, we find phase separation when the majority of (terminal) groups are attractive. Moreover, exactly as for $N = 10$ the critical temperature and density move to lower values by decreasing the number of tails per arm. In the same figure we also plot the coexistence curves obtained for $N = 10$: there the same symbols (empty and full) mark the same n/m ratio. Let us now focus on each couple of curves represented by the same symbol. For $N = 20$, the critical temperature is always higher than for $N = 10$, as expected from the increased number of attractive stickers. The gap between each couple ($N = 10$, $N = 20$) of critical temperatures is bigger the higher n/m is. However the analysis of the system-size dependence of the coexistence regions points out that for $n = 8$ the system seems to be really close to a micellization/phase separation boundary. This case deserves further investigation.

The spatial patterns formed by the aggregating end-monomers can be analyzed morphologically by employing the Minkowski functionals,[37–39] as for the case $N = 10$.[20] In Fig. 9 we show the exposed area per volume, V_1/V, the integral mean curvature density V_2/V as well as the Euler characteristic per site, V_3/V, for

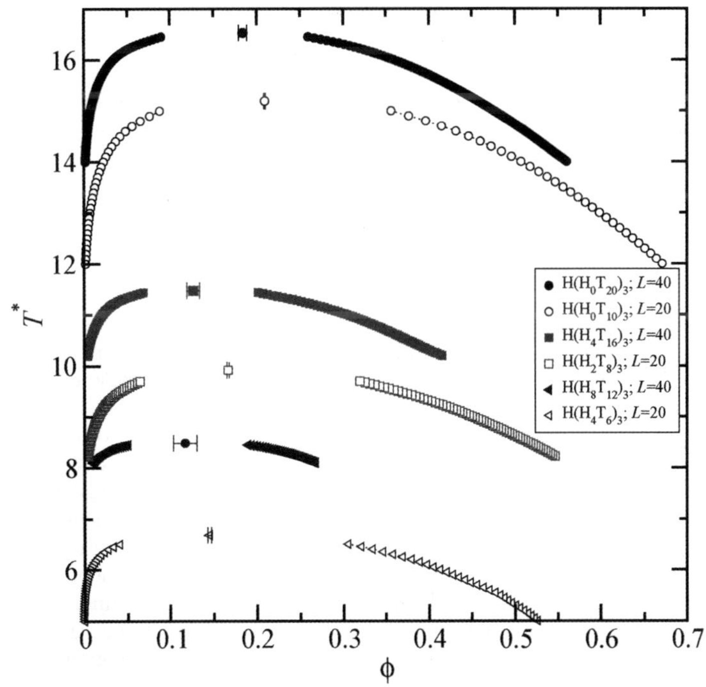

Fig. 8 Phase coexistence boundaries for trifunctional, telechelic star block copolymers for different block compositions and number of monomers per arm.

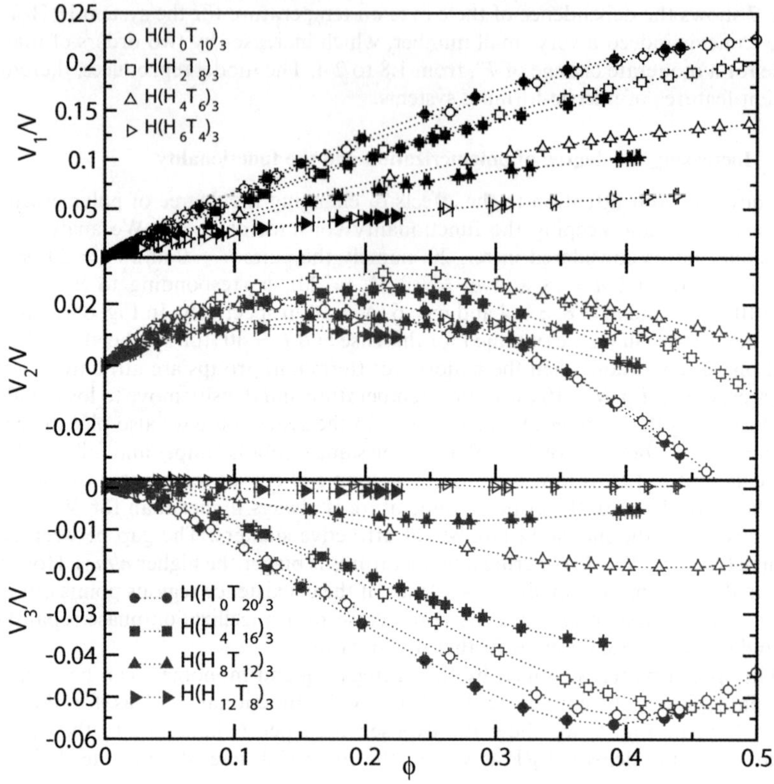

Fig. 9 The Minkowski functionals of trifunctional, telechelic star block copolymer solutions with $N = 10$, 20 monomers per arm.

$N = 10$ and $N = 20$. The case $n = 0$ does not seem to be affected by the increased number of monomers. Also for $n > 0$ the stars with the highest molecular weight show a general behavior very similar to the one observed for $N = 10$. In particular, the negative values of the Euler characteristic point to percolating, bicontinuous structures formed in the mixture, whose 'spongy' nature ($V_3 < 0$) diminishes as the number of stickers becomes lower. Indeed a negative Euler characteristic occurs only for systems which phase separate, the negative value starts at lower and lower densities increasing the number of tails per chain. We can observe that for $N = 20$ and $n/m = 2/3$ the curve is really close to the $x = 0$ axis.

The trends observed so far from the morphological analyses and phase behaviour of the solution suggest that for $n/m > 1$, micelle formation arises, and macrophase separation disappears. As a qualitative evidence we present in Fig. 10 a few snapshots for $H(H_{12}T_8)_3$, increasing ϕ ($T^* = 6.5$): the end-monomers of different stars seem to organize in micellar clusters. The increased molecular weight brings about a more irregular shape with respect to the almost spherical aggregates we found for the same n/m value at $N = 10$, *i.e.*, the $H(H_6T_4)_3$-system. The morphological curves agree with the previous qualitative analysis. In particular the Euler characteristic is positive and close to zero for low/intermediate densities. The negative value for the concentrated system is related to the progressive interconnection between micelles and the consequent formation of bicontinuous structures. Moreover the low value of V_2/V is in line with the non-spherical shape of the aggregates.

The dependence of the radius of gyration R_g on ϕ is shown for this case in Fig. 11 and it is very similar to that for $N = 10$ shown in Fig. 6. Also in this case at fixed density, the radius of gyration increases with increasing number of heads per chain,

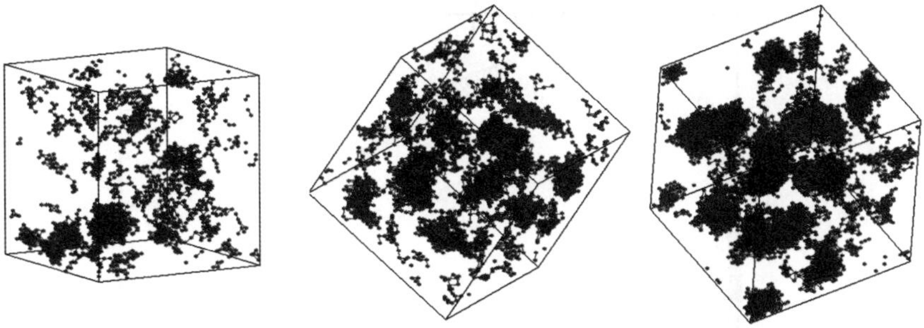

Fig. 10 Clusters of end-monomers for $H(H_{12}T_8)_3$ ($T^* = 6.5$). From the left $\phi = 0.089$, 0.154, 0.22.

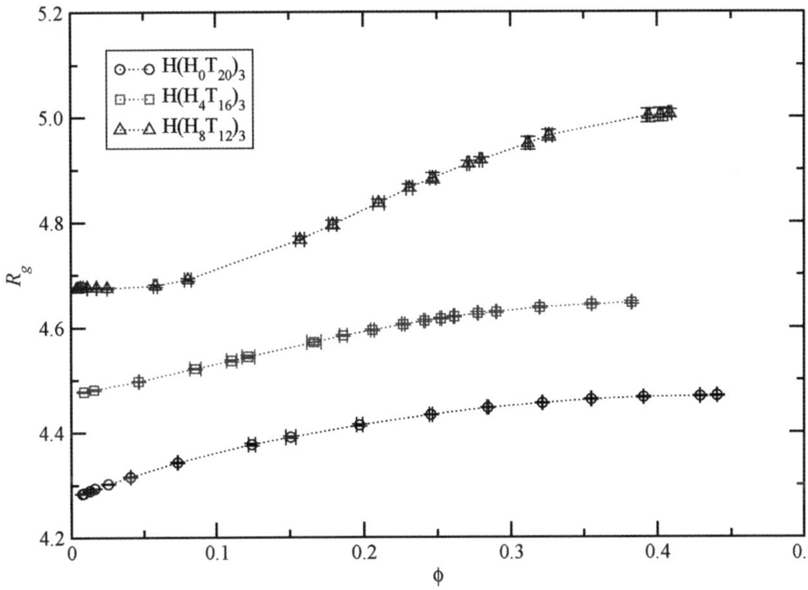

Fig. 11 Dependence of the radius of gyration R_g on the monomer concentration for the same systems as in Fig. 9 ($N = 20$).

which results into a strengthening of the entropic contributions to the star free energy against the enthalpic ones. Again, for phase-separating systems $m \geq n$, the radius of gyration increases monotonically with density. Around $\phi = 0.4$ a sign of saturation is present.

Finally, we consider the effects of a change of functionality, $f = 6$ but with the original degree of polymerization per arm, $N = 10$. In this fashion, a given star has the same total number of monomers as a trifunctional one with $N = 10$. Results for the phase coexistence are shown in Fig. 12 for $n = 0$, where it can be seen that indeed a $f = 6$, $N = 10$ star is akin to a $f = 3$, $N = 20$ star as far as its phase behavior is concerned. For these three cases the morphological coefficients show the same behaviour (see Fig. 13).

In Fig. 14 we present the comparison of the radius of gyration. The monotonic behaviour is very similar to the $N = 10$-case. Furthermore, for $f = 6$, R_g is bigger than for $f = 3$. This is related to the additional stretch of the chains consequent to the accommodation of a bigger arm number in the center of the star. These are preliminary results: a systematic investigation of the different n/m-cases is in progress.

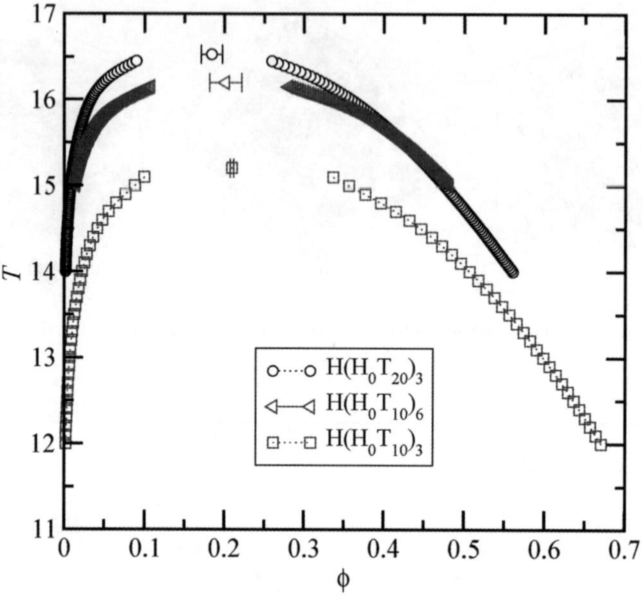

Fig. 12 Comparison of the coexistence curves for three $n = 0$ systems, differing in the number of monomers per arm ($N = 10, 20$) and in the functionality ($f = 3, 6$).

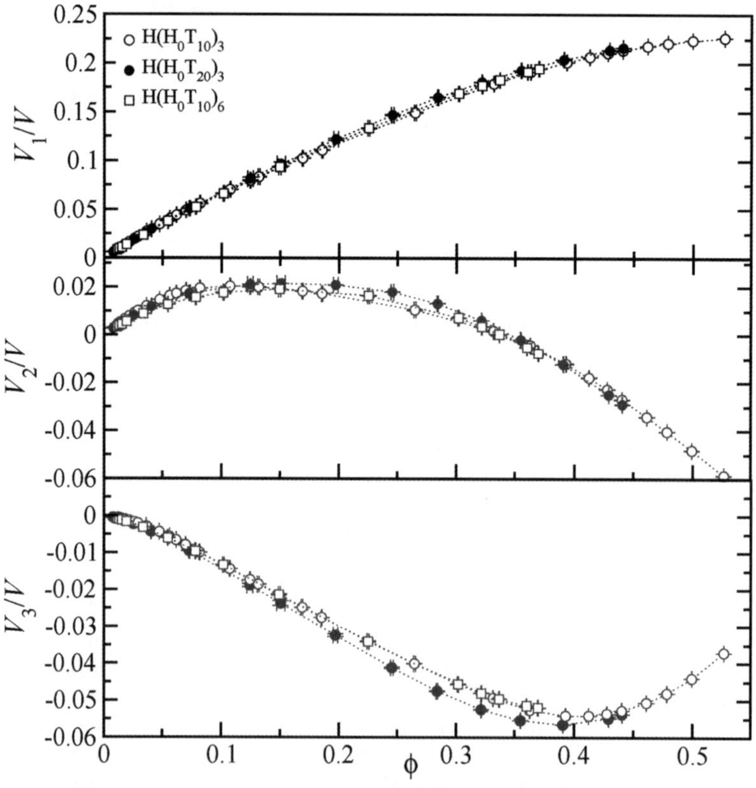

Fig. 13 Comparison of the Minkowski functionals of the $n = 0$ telechelic star block copolymer solutions ($N = 10, 20$ and $f = 3$, $N = 10$ and $f = 6$).

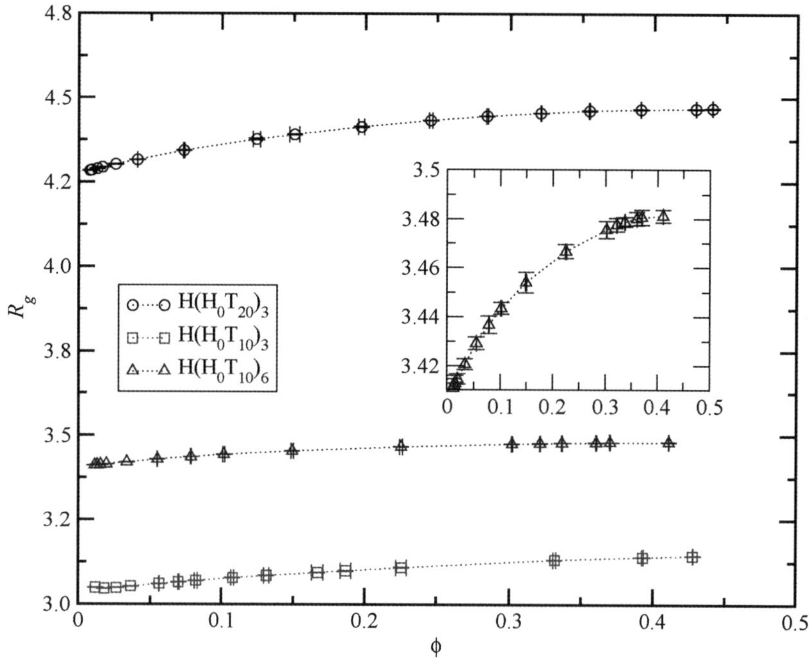

Fig. 14 Comparison of the radius of gyration for three systems, differing in the number of monomers per arm ($N = 10, 20$) and in the functionality ($f = 3, 6$).

4 Summary and outlook

Macromolecules with end-functionalized groups, akin to surfactants and to block-copolymers, can be synthesized with different architectures, sequences between the various groups and molecular weight of the respective block units. These parameters and the thermodynamical condition of the solution influence the intra- and inter-molecular aggregation processes. We focused here on telechelic block copolymer stars, using theory and lattice simulations. The simulation results we found are consistent with the previous off-lattice study.[18] In particular we varied the degree of polymerization ($N = 10, 20$) and the relative composition of the block copolymers and investigated the interplay between macrophase- and microphase separation in the system. For the same n/m ratios the general trends are quite insensitive to the change of the number of monomers. Indeed for $n/m < 1$ the different samples phase separate; further, the critical temperature and densities shift to lower values increasing the number of heads. For $N = 20$, the critical temperature is always higher than for $N = 10$, as expected from the increased number of functional groups. Moreover the coexistence curves appear flatter with respect to $N = 10$. When n exceeds m the system undergoes micellization phenomena and the increased chain length forces a more irregular shape of the aggregates.

Moreover, we considered the dependence of the radius of gyration on concentration. The conformation of the single star is indeed strongly dependent on the aggregation processes which characterize the specific n/m ratio. Competition among the entropic and excluded volume penalty and the end-monomers attraction determine the conformations. To better understand how the different contributions influence the aggregation mechanisms, we studied the radius of gyration of the single star. For phase separating systems, *i.e.*, $n < m$, R_g increases monotonically with concentration, due to the inter-molecular attraction and, at high enough monomer volume fractions ϕ, a sign of saturation appears. For $n > m$, at high enough concentration, the overlap of the solvophilic coronae of different micelles brings about a *shrinkage* in micelle size. Finally, we increased the functionality f of the stars to qualitatively examine the effects of the latter on phase separation and micellization. For $n = 0$, $f = 6$ and $N = 10$ the coexistence curve and critical parameters are akin to the $f = 3$ and $N = 20$ case. Here further studies with several n/m values are necessary to understand the interplay between macrophase and microphase separation.

The results presented in this work can be extended to considering other chain molecular weights and different arm numbers as well as by deducing a micelle–micelle effective interaction for micelle-forming systems.[40,41] We are also studying the case of partial functionalization of the star, *i.e.*, the situation where only some of the chains are end-capped with reactive groups. This analysis is going to encompass experimental results.[33–36] Future challenges include the modification of the bending ability of the block-copolymers backbone[42] and the analysis of telechelic stars at interfaces. By tackling these aspects, the focus of theoretical studies will approach addressing practical demands, such as medical and biological applications[12] associated with telechelic polymers.

Acknowledgements

FLV has been supported by the Foundation Blanceflor Boncompagni-Ludovisi, née Bildt, and thanks M. Della Morte for helpful discussions, A. L. Ferguson for technical support and the Erwin Schrödinger Institute (ESI, Vienna) for its hospitality. AZP acknowledges support from the Princeton Center for Complex Materials (NSF grant DMR-0819860), and CNL acknowledges Princeton University and the ESI for their hospitality.

References

1 G. S. Grest, L. J. Fetters, J. S. Huang and D. Richter, *Adv. Chem. Phys.*, 1996, **XCIV**, 67.
2 D. Vlassopoulos, *J. Polym. Sci., Part B: Polym. Phys.*, 2004, **42**, 2931.
3 C. N. Likos, *Soft Matter*, 2006, **2**, 478.
4 M. Daoud and J. P. Cotton, *J. Phys. (Paris)*, 1982, **43**, 531.
5 C. N. Likos, H. Löwen, M. Watzlawek, B. Abbas, O. Jucknischke, J. Allgaier and D. Richter, *Phys. Rev. Lett.*, 1998, **80**, 4450.
6 D. Vlassopoulos, G. Fytas, T. Pakula and J. Roovers, *J. Phys.: Condens. Matter*, 2001, **13**, R855.
7 M. Kapnistos, D. Vlassopoulos, G. Fytas, K. Mortensen, G. Fleischer and J. Roovers, *Phys. Rev. Lett.*, 2000, **85**, 4072.
8 C. Mayer, E. Zaccarelli, E. Stiakakis, C. N. Likos, F. Sciortino, J. Roovers, A. Munam, M. Gauthier, N. Hadjichristidis, H. Iatrou, P. Tartaglia, H. Löwen and D. Vlassopoulos, *Nat. Mater.*, 2008, **7**, 780.
9 C. Mayer, F. Sciortino, C. N. Likos, P. Tartaglia, H. Löwen and E. Zaccarelli, *Macromolecules*, 2009, **42**, 423.
10 A. Ramzi, M. Prager, D. Richter, V. Efstathiadis, N. Hadjichristidis, R. N. Young and J. B. Allgaier, *Macromolecules*, 1997, **30**, 7171.
11 N. Hadjichristidis, *J. Polym. Sci., Part A: Polym. Chem.*, 1999, **37**, 857.
12 F. Lo Verso and C. N. Likos, *Polymer*, 2008, **49**, 1425.
13 C. Glotzer and M. J. Solomon, *Nat. Mater.*, 2007, **6**, 557.
14 M. Muthukumar, C. K. Ober and E. L. Thomas, *Science*, 1997, **277**, 1225.
15 S. A. Jenekhe and X. L. Chen, *Science*, 1998, **279**, 1903.
16 F. Ganazzoli, Y. A. Kuznetsov and E. G. Timoshenko, *Macromol. Theory Simul.*, 2001, **10**, 325.
17 R. Connolly, E. G. Timoshenko and Y. A. Kuznetsov, *J. Chem. Phys.*, 2003, **119**, 8736.
18 F. Lo Verso, C. N. Likos, C. Mayer and H. Löwen, *Phys. Rev. Lett.*, 2006, **96**, 187802.
19 F. Lo Verso, C. N. Likos and H. Löwen, *J. Phys. Chem. C*, 2007, **111**, 15803.
20 F. Lo Verso, A. Z. Panagiotopoulos and C. N. Likos, *Phys. Rev. E Rapid Commun.*, 2009, **79**, 010401.
21 S.-Y. Kim, A. Z. Panagiotopoulos and M. A. Floriano, *Mol. Phys.*, 2002, **100**, 2213.
22 R. G. Larson, L. E. Scriven and H. T. Davis, *J. Chem. Phys.*, 1985, **83**, 2411; R. G. Larson, *J. Chem. Phys.*, 1988, **89**, 1642; R. G. Larson, *J. Chem. Phys.*, 1989, **91**, 2479; R. G. Larson, *J. Chem. Phys.*, 1992, **96**, 7904.

23 D. Frenkel, G. C. A. M. Mooij and B. Smit, *J. Phys.: Condens. Matter*, 1992, **4**, 3053.
24 J. J. de Pablo, M. Laso, J. I. Siepmann and U. W. Suter, *Mol. Phys.*, 1993, **80**, 55.
25 D. Frenkel and B. Smit, *Understanding Molecular Simulation*, Academic Press, London, 1996.
26 A. M. Ferrenberg and R. H. Swendsen, *Phys. Rev. Lett.*, 1988, **61**, 2635; A. M. Ferrenberg and R. H. Swendsen, *Phys. Rev. Lett.*, 1989, **63**, 1195.
27 A. Z. Panagiotopoulos, *J. Phys.: Condens. Matter*, 2000, **12**, R25.
28 For a review, see: *Finite-size Scaling and Numerical Simulation of Statistical Systems*, ed. V. PrivmanWorld Scientific, Singapore, 1990.
29 A. D. Bruce and N. B. Wilding, *Phys. Rev. Lett.*, 1992, **68**, 193.
30 N. B. Wilding, *Phys. Rev. E*, 1995, **51**, 2079.
31 M. A. Floriano, E. Caponetti and A. Z. Panagiotopoulos, *Langmuir*, 1999, **15**, 3143.
32 A. Z. Panagiotopoulos, M. A. Floriano and S. K. Kumar, *Langmuir*, 2002, **18**, 2940.
33 M. Pitsikalis, N. Hadjichristidis and J. W. Mays, *Macromolecules*, 1996, **29**, 179.
34 D. Vlassopoulos, T. Pakula, G. Fytas, M. Pitsikalis and N. Hadjichristidis, *J. Chem. Phys.*, 1999, **111**, 1760.
35 D. Vlassopoulos, M. Pitsikalis and N. Hadjichristidis, *Macromolecules*, 2000, **33**, 9740.
36 M. Pitsikalis and N. Hadjichristidis, *Macromolecules*, 1995, **28**, 3904.
37 C. N. Likos, K. R. Mecke and H. Wagner, *J. Chem. Phys.*, 1995, **102**, 2079.
38 N. Hoffmann, F. Ebert, C. N. Likos, H. Löwen and G. Maret, *Phys. Rev. Lett.*, 2006, **97**, 078301.
39 N. Hoffmann, C. N. Likos and H. Löwen, *J. Phys.: Condens. Matter*, 2006, **18**, 10193.
40 C. Pierleoni, C. Addison, J.-P. Hansen and V. Krakoviack, *Phys. Rev. Lett.*, 2006, **96**, 128302.
41 B. Capone, C. Pierleoni, J.-P. Hansen and V. Krakoviack, *J. Phys. Chem. B*, 2009, **113**, 3629.
42 V. Firetto, M. A. Floriano and A. Z. Panagiotopoulos, *Langmuir*, 2006, **22**, 6514.

Mesoscopic modelling of colloids in chiral nematics

Miha Ravnik,*[a] Gareth P. Alexander,[b] Julia M. Yeomans[c] and Slobodan Zumer[ad]

Received 1st May 2009, Accepted 18th May 2009
First published as an Advance Article on the web 24th August 2009
DOI: 10.1039/b908676e

We present numerical modelling of colloidal particles in chiral nematics with cubic symmetry (blue phases) within the framework of the Landau-de Gennes free energy. The interaction potential of a single, nano-sized colloidal particle with a $-1/2$ disclination line is calculated as a generic trapping mechanism for particles within the cholesteric blue phases. The interaction potential is shown to be highly anisotropic and have threefold rotational symmetry. We discuss the equilibration of the colloidal texture with respect to particle positions and the unit cell size of the blue phase. We also describe how preservation of the liquid crystal volume and the number of particles allows blue phase colloidal structures with different unit cell sizes and configurations to be compared numerically.

1 Introduction

Liquid crystal colloids have attracted considerable attention for the past decade or more. The coupling of nano-, submicron- and micron-sized particles with the orientational order of the liquid crystal gives rise to qualitatively new phenomenology and material properties, motivating the development of new optical elements, photonic structures,[1] and metamaterials.[2] Assisted and true self-assembly of a variety of colloidal structures, such as linear chains,[3,4] clusters,[5,6] two-dimensional colloidal crystals,[7,8] and two-dimensional hexagonal interface ordering[9,10] have all been reported in the nematic liquid crystal. In chiral cholesteric liquid crystalline materials, it has been reported in particle stabilised gels,[11] cholesteric displays based on colloidal self-assembly[12] and cholesteric colloids with ferroelectric nanoparticles.[13]

The assembly of colloidal structures arises from the inherent anisotropy of liquid crystals and their elastic response to perturbations. Immersing colloidal particles in a liquid crystal perturbs its orientational order and generates structural forces acting on the colloids to reduce the elastic deformation. An important feature, governing much of the response of the liquid crystal to the invading particles, are the topological defects that the liquid crystal exhibits. These are points, or lines, at which the orientation changes discontinuously and the order is ill-defined.[14,15] Colloidal particles treated to promote homeotropic surface anchoring effectively act like topological point defects of charge $+1$,[16] known as radial hedgehogs, and are ordinarily accompanied by the formation of a topological defect in the nematic to compensate

[a]*Faculty of Mathematics and Physics, University of Ljubljana, Jadranska 19, 1000 Ljubljana, Slovenia. E-mail: miha.ravnik@fmf.uni-lj.si*
[b]*Department of Physics and Astronomy, University of Pennsylvania, 209 South 33rd Street, Philadelphia, PA, 19104-6396, USA*
[c]*Rudolf Peierls Centre for Theoretical Physics, 1 Keble Road, OxfordOX1 3NP, UK*
[d]*J. Stefan Institute, Jamova 39, 1000 Ljubljana, Slovenia*

for this topological charge. In a uniform nematic confined to a planar or homeotropic cell the net topological charge is zero and therefore particles are compensated by hyperbolic point defects of charge −1 in a dipole configuration. Topologically, a −1 point defect is equivalent to a defect loop with winding number −1/2,[17] allowing for the possibility of another configuration known as a 'saturn ring' defect. Entangled colloidal structures form when a single defect loop is shared by several colloidal particles.[18] These defects, through the long range distortions they create in the host liquid crystal, mediate effective interactions between inclusions, thereby controlling their assembly into chains or 2D lattices.[7]

However, the liquid crystal may also contain defects for other reasons, arising for example from boundary conditions, the application of external fields, or, as in the cholesteric blue phases, through the addition of chiral dopants. In the case of strong surface anchoring and micron-sized particles, the interaction of colloidal particles with the elastic distortion field of a disclination is complicated, with a non-radial force that is attractive or repulsive depending on the sign of the disclination and on the parallel or antiparallel orientation of the colloid-hedgehog dipole.[19,20] By contrast, in the limit of very weak anchoring, the potential energy of the colloid may be approximated by the saving of elastic distortion energy from the liquid crystal region excluded by the presence of the particle. This leads to a purely radial force that is always attractive, regardless of the sign of the disclination, and at large distances has magnitude $4\pi K s^2 a^3/3R^3$, where K is an elastic constant, s is the winding number of the disclination (strength), a is the particle radius and R is the separation. The disclination thus acts as an attractant for colloids with weak surface anchoring, and in equilibrium the two are bound together by an energy $\sim Ka$ that can be several thousand times $k_B T$ for particles of ~100 nm radius. This has two natural consequences: first, it lowers the energy cost of disclination lines and secondly it provides a means for controlling the positioning of colloidal particles. Indeed, these general principles have been demonstrated in cholesteric liquid crystals where colloidal particles were shown to stabilise an 'oily streak' texture of defect lines by localising at the nodes of the network.[11]

A more intricate setting in which to further explore the interactions between colloids and defects is provided by the crystalline blue phases of highly chiral liquid crystals.[21] The blue phases are remarkable mesophases that occur in a narrow temperature range between the familiar isotropic and cholesteric phases of highly chiral compounds. There can be as many as three thermodynamically distinct blue phases, depending on the strength of the chirality and occurring in the order blue phase III, blue phase II, blue phase I upon lowering the temperature. Blue phase III has an amorphous structure with the same symmetry as the isotropic fluid, whereas blue phase I and blue phase II both exhibit cubic orientational order, corresponding to the space groups $O^{8-}(I4_132)$ and $O^2(P4_232)$, respectively, with lattice constants of several hundred nanometres. This unique combination of full fluidity and three-dimensional crystalline orientational order makes the cubic blue phases highly appealing for use as fast light modulators, photonic crystals or tunable lasers.[22–24] The texture of the blue phases may be understood in terms of a delicate balance between energetically favourable regions of double twist order, in which the molecules display helical ordering along two orthogonal directions instead of the more usual single twist of the ordinary cholesteric helix, and the energetic cost of disclination lines that naturally occur when this local order is extended globally. The result is a regular, periodic array of double twist cylinders threaded by a lattice of defect lines. The large energetic cost of the reduced order within the cores of the disclination lines restricts the thermodynamic stability of the blue phases to typically less than 1 K just below the isotropic phase, providing severe limitations to any potential device applications. However, recently blue phases with substantially extended stability ranges, of more than 40 K including room temperature, have been reported in systems using the photo-crosslinking of polymers[25] or pure mixtures of bimesogenic molecules.[26]

Here, we describe phenomenological numerical modelling of colloidal dispersions in the cholesteric blue phases based on the Landau-de Gennes free energy. We consider equilibration with respect to the particle positions and the blue phase unit cell size. The interaction of particles with the $-1/2$ disclination lines of the blue phase is identified as the basic mechanism for the equilibration of particle positions. We study how the size of the blue phase unit cell varies as a function of the radius of the colloidal particles, taking care to correctly preserve the volume of liquid crystal and number of particles. Finally, true trapping of colloidal particles in the defect lines of both blue phase I and blue phase II is demonstrated.

2 Theory and modelling

In order to simulate the interactions of colloidal particles with a liquid crystal we employ a continuum, mean field Landau-de Gennes approach based upon the tensor order parameter, Q_{ij}, description of liquid crystals.[14] The order parameter is a real, traceless, symmetric, 3×3 matrix, given at the microscopic scale by the second moment of the molecular orientational distribution function and at the macroscopic scale by the anisotropic part of the dielectric tensor. The description of liquid crystals using the full **Q**-tensor order parameter naturally accounts for both the variations in the magnitude of the order and the biaxiality inherent in disclinations and the cholesteric blue phases. The liquid crystalline director, n_i, is then given by the eigenvector of Q_{ij} corresponding to its maximal eigenvalue. Using the full order parameter tensor proves to be very advantageous in strongly confined geometries with numerous complex defect structures because it intrinsically incorporates liquid crystalline defects and no ansätze are needed as with the director field approach.

The Landau-de Gennes free energy for a colloidal liquid crystal system is constructed out of the invariants of Q_{ij} and its derivatives, and takes the form

$$F = \int_{LC} \left\{ \frac{A_0(1-\gamma/3)}{2} Q_{ij}Q_{ij} - \frac{A_0\gamma}{3} Q_{ij}Q_{jk}Q_{ki} + \frac{A_0\gamma}{4}(Q_{ij}Q_{ij})^2 \right\} dV + \int_{LC} \left\{ \frac{L}{2} \frac{\partial Q_{ij}}{\partial x_k} \frac{\partial Q_{ij}}{\partial x_k} + 2q_0 L \varepsilon_{ikl} Q_{ij} \frac{\partial Q_{lj}}{\partial x_k} \right\} dV + \int_{CS} \left\{ \frac{W}{2}(Q_{ij} - Q^0_{ij})(Q_{ij} - Q^0_{ij}) \right\} dS \tag{1}$$

where LC denotes an integral over the volume occupied by the liquid crystal only and CS over the surfaces of the colloidal particles. The first term is the bulk free energy and describes a first order transition (at $\gamma = 2.7$) between an isotropic and a nematic phase, with A_0 a constant with the dimensions of an energy density and γ an effective dimensionless temperature. The second term describes the energy cost of distortions in the liquid crystalline order; L is an elastic constant (for simplicity we restrict ourselves to the one elastic constant approximation), q_0 is a chiral parameter related to the pitch of the cholesteric helix by $p = 2\pi/q_0$ and ε_{ikl} is the fully antisymmetric alternating tensor equal to +1 (−1) if *i*, *k*, *l* is an even (odd) permutation of 1, 2, 3 and zero otherwise. The final term describes the interaction of the liquid crystal with the surfaces of the colloids *via* a Rapini–Papoular like type term, where W represents the strength of the uniform surface anchoring and $Q_{ij}{}^0$ is the order parameter tensor preferred by the surface, which we take here to be homeotropic with a magnitude given by the value for a uniaxial bulk nematic. When modelling particles with degenerate planar surface anchoring, the surface free energy (last term in eqn (1)) has to be rewritten in terms of the surface projections of the order parameter tensor.[27]

In an equilibrium configuration of minimum free energy the order parameter satisfies the Euler–Lagrange equations

$$\begin{aligned} L\nabla^2 Q_{ij} - 2q_0 L\left(\varepsilon_{ikl}\frac{\partial Q_{lj}}{\partial x_k} + \varepsilon_{jkl}\frac{\partial Q_{li}}{\partial x_k}\right) - A_0(1-\gamma/3)Q_{ij} \\ -A_0\gamma\left(Q_{ik}Q_{kj} - \frac{1}{3}Q_{kl}Q_{kl}\delta_{ij}\right) - A_0\gamma Q_{kl}Q_{kl}Q_{ij} = 0 \end{aligned} \tag{2}$$

throughout the bulk of the liquid crystal, together with the boundary conditions on the particle surfaces

$$\nu_k\left[-L\frac{\partial Q_{ij}}{\partial x_k} + q_0 L\left(\varepsilon_{ikl}Q_{lj} + \varepsilon_{jkl}Q_{li}\right)\right] = 0 \tag{3}$$

where ν_k is the unit outward normal to the colloid's surface. We find solutions to these equations by allowing the order parameter to evolve towards equilibrium according to a Ginzburg–Landau relaxational equation

$$\frac{\partial Q_{ij}}{\partial t} = -\Gamma\left(\frac{\delta F}{\delta Q_{ij}} - \frac{1}{3}\,\mathrm{tr}\left[\frac{\delta F}{\delta Q_{kl}}\right]\delta_{ij}\right) \tag{4}$$

where Γ is a rotational diffusion constant, tr denotes the tensorial trace and $\delta F/\delta Q_{ij}$ is the functional derivative of the free energy. The term in brackets on the right hand side is known as the molecular field and yields precisely the Euler–Lagrange eqn (2) in the bulk and the appropriate boundary conditions (eqn (3)) on the particle surfaces. Eqn (4) is solved using an explicit Euler finite difference algorithm on a cubic mesh with periodic boundary conditions in all three directions.[28] Here, we are neglecting the effects of fluid flow, which may effect the kinematics of the formation of the colloidal blue phase, but are not expected to influence the equilibrium configurations. The mesh resolution is chosen in accordance to the material correlation length and is typically taken to be 5 nm. Unless stated otherwise, the following parameter values characteristic for a typical chiral nematic are used:[21,29] $L = 2.5 \times 10^{-11}$ N, $A_0 = 1.02 \times 10^5$ J/m^3, ($\gamma = 3.375$, $p_0 = 2\pi/q_0 = 0.566$ μm) in blue phase I, and ($\gamma = 2.755$, $p_0 = 2\pi/q_0 = 0.616$ μm) in blue phase II.

In order to produce the desired blue phase texture, initial conditions were chosen for the tensor Q_{ij} corresponding to the analytic expressions in the high chirality limit ($q_0 \to \infty$).[30] These expressions capture the essential symmetry of the desired texture and in subsequent numerical evolution the order parameter relaxes to the local free energy minimum with this symmetry, even when the particular texture is only metastable. Structural transitions are only observed when the texture is fully unstable. For blue phase I the initial condition comprises the set of fundamental wave vectors corresponding to a texture with body centred cubic symmetry and is given by

$$\begin{aligned} Q_{xx} = \mathscr{A}\Big\{ & -\sin(ky/\sqrt{2})\cos(kx/\sqrt{2}) - \sin(kx/\sqrt{2})\cos(kz/\sqrt{2}) \\ & + 2\sin(kz/\sqrt{2})\cos(ky/\sqrt{2})\Big\} \end{aligned} \tag{5}$$

$$\begin{aligned} Q_{xy} = \mathscr{A}\Big\{ & -\sqrt{2}\sin(kx/\sqrt{2})\sin(kz/\sqrt{2}) - \sqrt{2}\cos(ky/\sqrt{2})\cos(kz/\sqrt{2}) \\ & + \sin(kx/\sqrt{2})\cos(ky/\sqrt{2})\Big\} \end{aligned} \tag{6}$$

with the other components obtained by cyclic permutation. The magnitude of the amplitude $\mathscr{A}$ for the initialisation is not crucial to the subsequent evolution, however its sign is. When $\mathscr{A} > 0$ the texture known as O^{8-} is obtained, whilst for $\mathscr{A} < 0$ we obtain the texture O^{8+}, differing from O^{8-} by an interchange of the locations of the double twist cylinders and disclination lines.[21,30] The texture relevant to blue phase I is O^{8-} and so we choose $\mathscr{A} > 0$ and typically use a value of $\mathscr{A} = 0.2$. For blue phase II the initial condition is again given by restricting the fundamental set of wave vectors, this time with simple cubic symmetry, and is given by

$$Q_{xx} = \mathscr{A}\{\cos(kz) - \cos(ky)\} \tag{7}$$

$$Q_{xy} = \mathscr{A}\{\sin(kz)\} \tag{8}$$

with the other components obtained by cyclic permutation as before. Again we use a typical value of $\mathscr{A} = 0.2$ for the amplitude of the initialisation.

In both blue phases the size of the unit cell is set by the strength of the chirality, with the magnitude of the fundamental wave vector of the blue phase given by $k = 2q_0\, r$, where r is a pure number known as the redshift.[29] Typical values for the redshift are $r = 0.68$ for blue phase I and $r = 0.86$ for blue phase II.[31] In a system without particles the optimum value of the redshift leading to a full minimisation of the free energy can be calculated during a single simulation through a simple rescaling procedure,[31] however, when the colloidal particles are added this option is no longer available to us. Thus the value of the redshift in this case was obtained by performing a series of simulations, identical except for the total number of mesh points to vary the size of the unit cell, and finding the minimum value of the free energy per unit volume over this series of runs.

The colloidal particles are initialised numerically as shells with a thickness equal to the mesh resolution. Within the thickness of the shell, mesh points obey the surface eqn (3) with the surface normal ν_k and homeotropic direction imposed by constructing the surface order parameter tensor $Q_{ij}{}^0$ to point to the true geometrical centre of the particle. This allocation method avoids the effective surface roughness of the discretised particle, as may be seen from the smooth profile of the scalar degree of order close to the particle's surface. For particles with radius $a = 40$ nm, a mesh resolution of 5 nm allows for a numerical error of less than $\pm 5\%$ in the allocation of the particle surface and hence also in the calculation of the equilibrium surface free energy. In the regime of weak anchoring this numerical error is reduced even further, since the free energy is dominated by the elastic contributions and the variation in the magnitude of the order. Outside the shell region the liquid crystal is governed by the bulk eqn (2), while in the interior no calculations are performed.

3 Results

We show in Fig. 1 the equilibrium textures of both blue phase I (Fig. 1a) and blue phase II (Fig. 1b) without particles. The network of disclination lines in each case is shown in red, while the director field and arrangement of double twist cylinders is indicated in grey. We show both a large image revealing the periodic nature of the texture and a single conventional cubic unit cell. In blue phase I the defect network consists of an array of isolated disclination lines of topological strength $-1/2$. Similarly, in blue phase II the network is also composed of an array of strength $-1/2$ disclination lines, but in this case the lines meet at nodal points, of local topological strength -1, corresponding to the corners and centre of the conventional unit cell. The overall defect texture is then a bi-continuous, interpenetrating double diamond network.

The $-1/2$ disclination lines in a blue phase act as effective trapping sites for colloidal particles. To investigate the form of the interaction, we performed a generic calculation of the interaction potential of a single colloidal particle in the hyperbolic director field of a $-1/2$ disclination line in a regular achiral nematic ($q_0 = 0$). For true

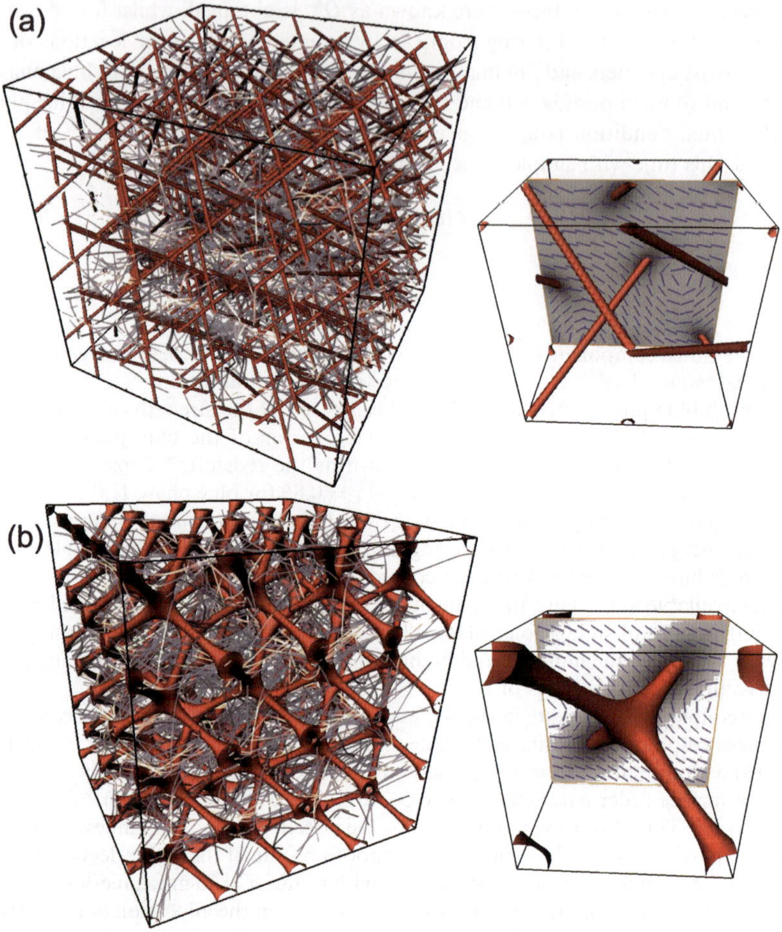

Fig. 1 Defect networks (in red) and director lines for (a) blue phase I and (b) blue phase II. Note the twisted cylinders. Insets show the corresponding conventional unit cells. In blue, the projection of the director on the given plane is visualised. Note the hyperbolic director pattern around the defects characteristic of $-1/2$ disclination lines. Defect lines are visualized as isosurfaces of the scalar degree of order $S = 0.23$ (BP I) and $S = 0.1$ (BP II).

nano-sized particles, the pitch of the blue phases is typically much larger than the particle–disclination line separation so that the chirality of the liquid crystal can be reasonably omitted. To pin the disclination line, the hyperbolic director profile $\boldsymbol{n} = (\cos(\phi/2), -\sin(\phi/2), 0)$ was imposed at the boundary of the simulation box (ϕ is the polar angle). Fig. 2a shows a homeotropic particle with radius $a = 15$ nm approaching the disclination line. The corresponding potential is shown in Fig. 2c. Three attractive and three repulsive directions are found, as observed by Pires and Galerne.[19] The profile of the interaction potential nicely reflects the three-fold rotational symmetry of the underlying director profile, characteristic for the $-1/2$ disclination line. To generalise the behaviour we also investigated the interaction potential for particles with degenerate planar anchoring (Fig. 2b). Modelling details for particles with a planar surface can be found in ref. 32. The interaction potential of planar particles with the $-1/2$ disclination line shows qualitatively similar behaviour as for homeotropic particles, except that it is effectively rotated by 60°. Trapping by the defect may be understood simply on the grounds that the

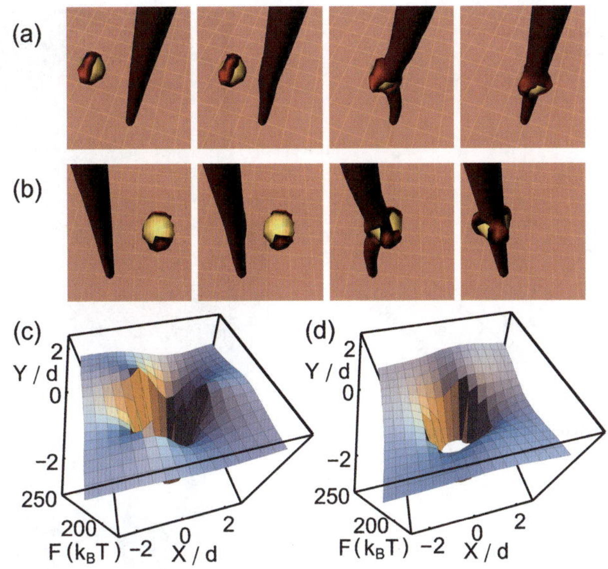

Fig. 2 Trapping of nano-sized particles in a $-1/2$ disclination line. (a) Particle with homeotropic anchoring at various positions ($X \leq 0$, $Y = 0$). The last panel corresponds to ($X = 0$, $Y = 0$). (b) Particle with planar anchoring at various positions ($X \geq 0$, $Y = 0$). The last panel corresponds to ($X = 0$, $Y = 0$). (c,d) The interaction potential of the (a) homeotropic particle and (b) planar particle with the $-1/2$ disclination line. For both types of anchoring the particles are attracted towards the disclination and are strongly bound once they intersect it. Parameter values: surface anchoring $W = 10^{-3}$J/m^2, diameter $d = 30$ nm, $q_0 = 0$, and material parameters as in ref. 18. Defects are drawn as isosurfaces of the nematic degree of order $S = 0.5$ ($S_{\text{bulk}} = 0.533$).

colloid's own volume replaces part of the disclination line and thus removes part of the elastic energy cost associated with it. The free energy saving scales as $\sim Ka$ and, for nano-sized particles with $a \sim 15$ nm, is at least an order of magnitude greater than k_BT. Thus the disclination acts as a trapping potential for particles, which once bound will not be able to break free from purely thermal effects.[11]

Even in nematics the details of the interactions between a particle and a distant disclination can be complicated as they are sensitive to the nature of the anchoring conditions on the particle's surface and its orientation relative to the location of the defect. However, the general tendency of disclination lines to trap particles still holds for blue phases. But even restricting our attention to those configurations in which all the particles are trapped within the blue phase defect network still leaves many free variables. For example, in addition to the type and strength of the surface anchoring we can also vary the placement of the particles within the defect network, the size of the particle relative to the blue phase unit cell and the number of particles per unit cell.

For the particle positions, we show in Fig. 3 the effective trapping potential experienced by an individual colloid. To reduce the effects of periodic boundary conditions we take a simulation box containing $3 \times 3 \times 3$ blue phase unit cells and consider only displacements of a single particle in the central unit cell. Here we have to be careful in performing our simulations for in the same way that we are free to move the initial position of the colloidal particle, so too is the blue phase network able to shift its position relative to the particle in order to reduce the free energy. Thus, if only one colloid is used the equilibrium texture obtained is always the same, because the periodic boundary conditions allow the blue phase to freely adjust its position. We therefore used an entire array of particles, placed at positions

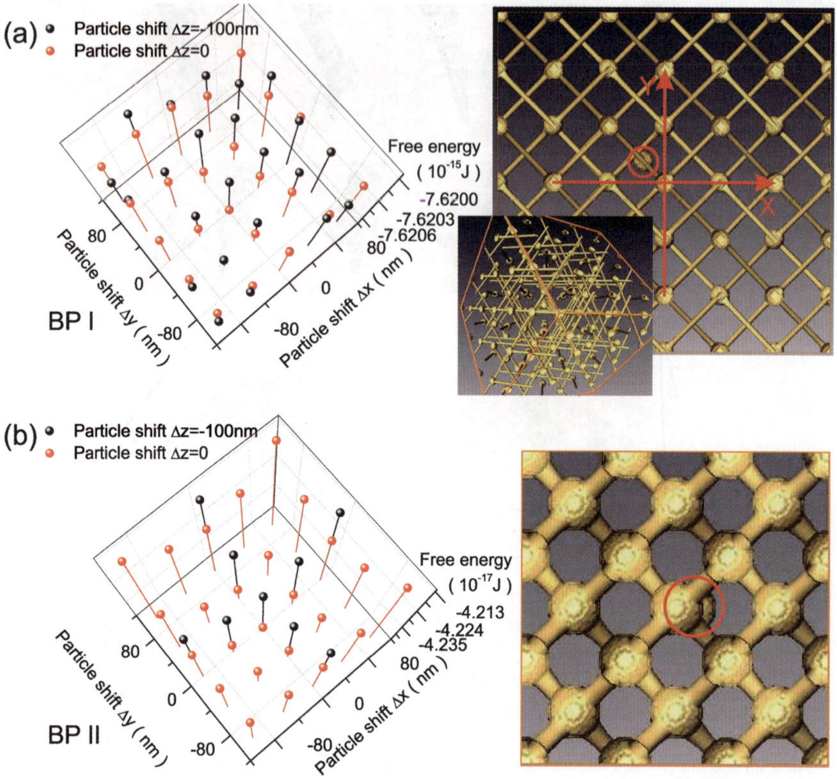

Fig. 3 Effective trapping potential experienced by a single colloidal particle within an array of particles in (a) blue phase I and (b) blue phase II. Shifts from the symmetric positions within the array in all directions (Δx, Δy, Δz) result in an increase of the total free energy. Panels on the right show shifts of a single particle from the symmetric arrays of particles in blue phase I and blue phase II.

of high symmetry of the blue phase defect network to effectively pin it in place. In blue phase I we used a configuration with four particles per unit cell (1 in the corners and 3 in the centres of the faces), and in blue phase II a configuration with two particles within the unit cell (1 in the corners and 1 in centre of the unit cell). In both cases the particles were all at symmetry-equivalent positions within the blue phase network. A single particle in the central unit cell was then selected and freely displaced relative to this network to study the trapping potential of the blue phase. In all cases we found that displacements away from the core of the defect, or from positions of high symmetry of the blue phase, led to an increase in the free energy.

To obtain a full minimisation of the energy it is important to not only relax the order parameter Q_{ij} towards its equilibrium value, but also to allow the size of the blue phase unit cell to vary as well. We do this by performing a series of simulations increasing and decreasing the size of the cubic unit cell in the simulation whilst keeping all other simulation parameters and the size of the colloid fixed. When resizing the unit cell, it is important to retain the relative position of the particle centre with respect to the mesh in order to avoid 'ghost' contributions to the total free energy due to the different allocations of the particle's surface. Thus, we resize the unit cell size of the blue phases by increasing the number of mesh points, yet keeping the resolution of the numerical mesh unchanged. Fig. 4 shows the free energy profile for various unit cell sizes of blue phase I and blue phase II. Particles of size $2a = 100$

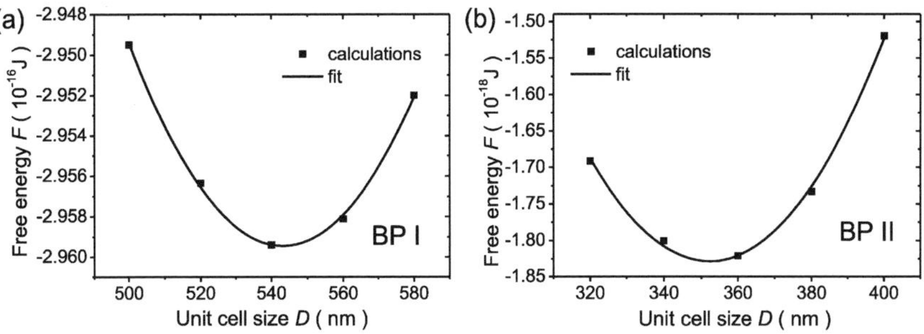

Fig. 4 Equilibration of the cubic unit cell size: (a) blue phase I with four immersed particles per unit cell and (b) blue phase II with two particles per unit cell. Results of the calculations are fitted with $F = F_0 + A(D - D_0)^2$ to obtain a more precise value of the equilibrium unit cell size D_0 and equilibrium free energy F_0.

nm were positioned as before with four particles per unit cell in blue phase I and two per unit cell in blue phase II. The results of numerical calculations were fitted by a parabolic function in order to obtain the equilibrium cell size at a precision below mesh resolution. In blue phase I the minimum occurs at 543 nm and represents a decrease of 0.2% as compared to the size of the unit cell without particles (544 nm). In blue phase II we find a similar shift of 0.9% from 355 nm without particles to 352 nm when the particles are added.

An important aspect in the equilibration of the blue phase unit cell size with immersed particles is that the total volume of the liquid crystalline material and number of particles have to be preserved in order to allow a fair comparison of free energies. When resizing the unit cell size of the blue phases, the volume of the liquid crystal changes and to avoid this it would be natural to use the free energy density, *i.e.* free energy of a given unit cell size divided by its volume. Yet by doing so, one assumes that there is a reservoir of particles (the simulations assume that the system is periodic in 3D) which steadily fills the voids or compensates for the escape of particles from the material when the unit cell shrinks or expands, respectively. However, this is not necessarily the case in experiments. To preserve both the volume and number of the particles, we therefore first equilibrate the unit cell size of a given blue phase with no particles by observing the free energy density. The volume of such an equilibrated blue phase V_{BP} (see Fig. 5) is next set as the reference volume at a given temperature and chirality. Now, when the unit cell size of the blue phase *with particles* V_{col} changes, a compensating volume V' is calculated to maintain the constant value $V_{BP} = V_{col} + V'$. Finally, to compare the free energies of blue phase unit cells doped with particles, that have various unit cell sizes, the free energy contribution corresponding to a volume portion V' of an equilibrated blue phase

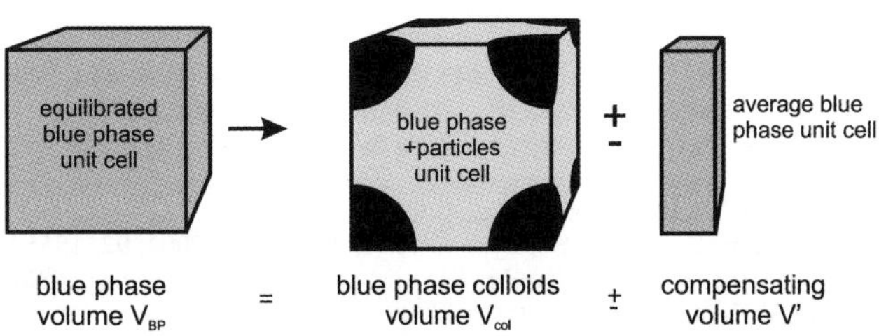

Fig. 5 Comparison of free energies in terms of preserving the total volume and the number of particles. A volume V_{BP} of an equilibrated blue phase is set as the reference volume. The average free energy density of an equilibrated blue phase with *no particles* multiplied by V' is added to the free energy of the blue phase *with particles*.

unit cell is added/subtracted from the original free energies. Such a comparison of the free energies assumes that in the blue phases doped with particles, regions exist where there are no particles. Experimentally, this would correspond to volume concentrations with fewer than one particle per unit cell on average.

4 Conclusion

We have demonstrated phenomenological modelling of colloidal particles in the cholesteric blue phases based on the Landau-de Gennes free energy. The key equilibration mechanisms and degrees of freedom governing the interaction of the colloids with the blue phase are addressed. First, we calculate the interaction of nano-sized particles with a $-1/2$ disclination under the approximation of a locally achiral medium, treating both the cases of homeotropic and planar anchoring. The interaction potential can be as large as several $100k_BT$ for particles with a diameter of 30 nm and surface anchoring $W = 10^{-3}$J/m^2 and promotes a trapping of the particles within the disclination line. Second, we considered the positioning of an individual particle within an array of colloids dispersed in both blue phase I and blue phase II. Symmetric configurations were used, with four particles per unit cell in blue phase I and two particles per unit cell in blue phase II, to preserve the original symmetry of the texture. Any shift of a single particle from these initial symmetric positions resulted in an increase of the total free energy. Third, we studied the change of the blue phase unit cell size in response to the added particles. To provide a fair comparison of the free energies of colloidal blue phases with different unit cell sizes, appropriate compensating volumes of an equilibrated blue phase without particles must be added.

Doping of blue phases with particles opens a new route to colloidal interactions mediated by the intrinsic defects of the liquid crystal. The large number of parameters for both the particles and the liquid crystal, and the interplay between these parameters, represent a substantial challenge for numerical modelling and simulations. However, insight into the material response at sub-micron and nanometre scales is crucial for the understanding and development of these materials and numerical modelling provides a powerful and efficient tool for such studies. Our future work will focus on the material aspects of blue phases doped with particles and possible arrangements of particles which can arise as a result of complex particle-defect interactions.

Acknowledgements

MR and SZ acknowledge financial support from ARRS P1-0099 program and 215851-2 EC ITN Marie Curie research network Hierarchy. GPA would like to thank Randall Kamien for discussions and acknowledges partial support from NSF Grant DMR05-47320.

References

1 Y. A. Vlasov, X.-Z. Bo, J. C. Sturm and D. J. Norris, *Nature*, 2001, **414**, 289.
2 D. R. Smith, J. B. Pendry and M. C. K. Wiltshire, *Science*, 2004, **305**, 788.
3 P. Poulin, H. Stark, T. C. Lubensky and D. A. Weitz, *Science*, 1997, **275**, 1770.
4 J. C. Loudet, P. Barois and P. Poulin, *Nature*, 2000, **407**, 611.
5 P. Poulin and D. A. Weitz, *Phys. Rev. E*, 1998, **57**, 626.
6 M. Yada, J. Yamamoto and H. Yokoyama, *Phys. Rev. Lett.*, 2004, **92**, 185501.
7 I. Musevic, M. Skarabot, U. Tkalec, M. Ravnik and S. Zumer, *Science*, 2006, **313**, 954.
8 I. Musevic and M. Skarabot, *Soft Matter*, 2008, **4**, 195.
9 V. G. Nazarenko, A. B. Nych and B. I. Lev, *Phys. Rev. Lett.*, 2001, **87**, 075504.
10 I. I. Smalyukh, S. Chernyshuk, B. I. Lev, A. B. Nych, U. Ognysta, V. G. Nazarenko and O. D. Lavrentovich, *Phys. Rev. Lett.*, 2004, **93**, 117801.
11 M. Zapotocky, L. Ramos, P. Poulin, T. C. Lubensky and D. A. Weitz, *Science*, 1999, **283**, 209.

12 K. Chari, C. M. Rankin, D. M. Johnson, T. N. Blanton and R. G. Capurso, *Appl. Phys. Lett.*, 2006, **88**, 043502.
13 O. Kurochkin, O. Buchnev, A. Iljin, S. K. Park, S. B. Kwon, O. Grabar and Y. Reznikov, *J. Opt. A: Pure Appl. Opt.*, 2009, **11**, 024003.
14 P. G. de Gennes and J. Prost, *The Physics of Liquid Crystals*, Oxford University Press, Oxford, 2nd edn, 1993.
15 M. Kléman, *Points, Lines, and Walls: in Liquid Crystals, Magnetic Systems, and Various Ordered Media*, Wiley, New York, 1983.
16 M. Kléman and O. D. Lavrentovich, *Philos. Mag.*, 2006, **86**, 4117.
17 N. D. Mermin, *Rev. Mod. Phys.*, 1979, **51**, 591.
18 M. Ravnik, M. Skarabot, S. Zumer, U. Tkalec, I. Poberaj, D. Babic, N. Osterman and I. Musevic, *Phys. Rev. Lett.*, 2007, **99**, 247801.
19 D. Pines, J.-B. Fleury and Y. Galerne, *Phys. Rev. Lett.*, 2007, **98**, 247801.
20 M. Skarabot, M. Ravnik, S. Zumer, U. Tkalec, I. Poberaj, D. Babic and I. Musevic, *Phys. Rev. E*, 2008, **77**, 061706.
21 D. C. Wright and N. D. Mermin, *Rev. Mod. Phys.*, 1989, **61**, 385.
22 W. Cao, A. Muńoz, P. Palffy-Muhoray and B. Taheri, *Nat. Mater.*, 2002, **1**, 111–113.
23 Y. Hisakado, H. Kikuchi, T. Nagamura and T. Kajiyama, *Adv. Mater.*, 2005, **17**, 96.
24 S. Yokoyama, S. Mashiko, H. Kikuchi, K. Uchida and T. Nagamura, *Adv. Mater.*, 2006, **18**, 48.
25 H. Kikuchi, M. Yokota, Y. Hisakado, H. Yang and T. Kajiyama, *Nat. Mater.*, 2002, **1**, 64.
26 H. J. Coles and M. N. Pivnenko, *Nature*, 2005, **436**, 997.
27 J. B. Fournier and P. Galatola, *Europhys. Lett.*, 2005, **72**, 403.
28 W. H. Press, B. P. Flannery, S. A. Teukolsky, and W. T. Vetterling, *Numerical Recipes*, Cambridge University Press, Cambridge, 1986.
29 H. Grebel, R. M. Hornreich and S. Shtrikman, *Phys. Rev. A*, 1983, **28**, 1114.
30 A. Dupuis, D. Marenduzzo and J. M. Yeomans, *Phys. Rev. E*, 2005, **71**, 011703.
31 G. P. Alexander and J. M. Yeomans, *Phys. Rev. E*, 2006, **74**, 061706.
32 M. Vilfan, N. Osterman, M. Copic, M. Ravnik, S. Zumer, J. Kotar, D. Babic and I. Poberaj, *Phys. Rev. Lett.*, 2008, **101**, 237801.

A molecular level simulation of a twisted nematic cell†

Matteo Ricci,[a] Marco Mazzeo,‡[a] Roberto Berardi,[a] Paolo Pasini[b] and Claudio Zannoni*[a]

Received 27th January 2009, Accepted 25th February 2009
First published as an Advance Article on the web 18th August 2009
DOI: 10.1039/b901784d

We have performed a Monte Carlo simulation of a sub-micrometric twisted nematic cell with nearly 10^6 particles using an off-lattice molecular model of a liquid crystal. This computer experiment is a proof of principle that molecular models can be pushed to the limit of the system sizes addressable with finite element models thus bridging the mesoscopic gap for multiscale modelling while providing a direct molecular level view of the working of the display. This approach, that allows a direct prediction of molecular organisations, properties, and responses of device systems without the requirement of prior estimate or knowledge of material properties (*e.g.* elastic constants), is particularly important in view of simulating materials and devices for which these quantities are not known. Results for the molecular organisation are discussed, with particular regard to its helical nature in the field-*off* state.

1 Introduction

One of the most successful stories in advanced materials must be that of liquid crystals (LC) displays.[1] The basic concept behind the most classical device, the twisted nematic (TN) display,[2] is that a pixel is activated by a change of molecular organisation in the few micrometres thick cell from a surface dominated one (*off* state) to a field-driven one (*on* state). Thus, an initial configuration of the local preferred direction (the director) is established between two perpendicular aligning surfaces (*e.g.* rubbed glass or polyimide) that confine the LC. An experimental fact is that linearly polarised light is going through the pixel in this *off* state and this is compatible with a microscopic helical configuration of the local director orientation which rotates the plane of polarisation. If the chosen LC has a positive dielectric anisotropy and a suitable voltage is applied across the cell in correspondence of a pixel, then polarised light does not go through, compatibly with a monodomain organisation. When the field is switched off the original organisation is re-established thanks to elastic restoring forces and surface interactions.

To the best of our knowledge there is little evidence that the molecular level organisation in absence of a field is actually a uniform helix. For instance, the classic

[a]*Dipartimento di Chimica Fisica e Inorganica, and INSTM-CRIMSON, Università di Bologna, viale Risorgimento 4, 40136 Bologna, Italy. E-mail: claudio.zannoni@unibo.it*
[b]*Istituto Nazionale di Fisica Nucleare (INFN), Sezione di Bologna, via Irnerio 46, 40126 Bologna, Italy*

† Electronic supplementary information (ESI) available: Supplementary movie 1. See DOI: 10.1039/b901784d

‡ Current address: Centre for Computational Science, University College of London, 20 Gordon St., London WC1H OAJ, UK; Email: E-mail: m.mazzeo@ucl.ac.uk.

textbook picture is that of uniformly twisted layers that do not actually exist as such as the organisation remains that of a nematic without positional order or layering.

Moreover, the way the twisted organisation is established is not obvious, for instance after an *off–on–off* cycle does the reorganisation re-start from the centre of the cell or from the surface? Is a uniform helix really formed? Or how helical is the structure?

Computer simulations of similar TN device setup have been performed by using lattice models,[3–7] with fixed spin positions. However, simulation studies based on models where particles have both positional and orientational degrees of freedom (off-lattice) have not been reported so far.

In this work we tackle some of these questions by setting up a molecular resolution off-lattice model of a TN cell containing $\mathcal{O}(10^6)$ particles and simulating its behaviour using a Monte Carlo (MC) method. This corresponds to samples two orders of magnitude larger than the typical ones currently studied with off-lattice molecular models, and thus presents some significant computational challenges with present day computers.

In this paper, after introducing our model $\pi/2$ TN cell, we shall discuss the technical details of the specific parallel MC code we have developed for this purpose. In the second part of the manuscript we shall show our simulation results both for the spontaneous formation of a supermolecular helical structure in the *off* state, and the commutation of the central pixel to the *on* state by an electric field applied across the TN cell.

2 Model

Standard modelling of LC electro-optical devices uses continuum descriptions[8,9] relying on input parameters for elastic constants, viscosities, and other material constants. These data are not available for novel materials (*e.g.* biaxial nematics[10]). Atomistic simulations,[11] where molecules are described at full atomic resolution, would be the natural alternative approach to use, but the sample sizes required to realistically model the distance and time scales of a TN cell are currently well beyond available and foreseeable computational resources. At the other extreme we find lattice models,[3–5] where every site represents a uniformly aligned monodomain formed by hundreds of molecules. In this case the simplicity of the model allows the simulation of mesoscopic size systems, even though at the price of disregarding possibly relevant positional degrees of freedom. Instead, we are not aware of any computer simulation study based on off-lattice models where molecules are explicitly described with simple geometrical shapes and suitable interaction energies. Here we have modelled the LC molecules inside the cell using the Gay–Berne (GB) potential[11–13] which is *de facto* a standard and well established model for liquid crystalline systems, giving the correct temperature dependence of order parameters, and the sequence of phase transitions for many model mesogens.[14] The GB pair potential is an anisotropic Lennard–Jones which can be written as

$$U_{\mathrm{GB}}(\mathbf{r},\hat{\mathbf{u}}_1,\hat{\mathbf{u}}_2) = 4\varepsilon_0\varepsilon(\mathbf{r},\hat{\mathbf{u}}_1,\hat{\mathbf{u}}_2)[u^{12}(\mathbf{r},\hat{\mathbf{u}}_1,\hat{\mathbf{u}}_2) - u^6(\mathbf{r},\hat{\mathbf{u}}_1,\hat{\mathbf{u}}_2)] \quad (1)$$

where the terms $u(\mathbf{r},\hat{\mathbf{u}}_1,\hat{\mathbf{u}}_2) \equiv \sigma_c/(r - \sigma(\mathbf{r},\hat{\mathbf{u}}_1,\hat{\mathbf{u}}_2) + \sigma_c)$, contain the anisotropic contact function $\sigma(\mathbf{r},\hat{\mathbf{u}}_1,\hat{\mathbf{u}}_2)$ which estimates the contact distance between two ellipsoids.[15,16] The contact function depends both on the axes σ_x, σ_y, and σ_z of the ellipsoid, and on the orientations $\hat{\mathbf{u}}_1$, $\hat{\mathbf{u}}_2$ for the long molecular axes (as unit vectors), and the intermolecular vector $\mathbf{r} = \mathbf{r}_2 - \mathbf{r}_1$, with $\mathbf{r}_i = (x_i, y_i, z_i)$.

The interaction function $\varepsilon(\mathbf{r},\hat{\mathbf{u}}_1,\hat{\mathbf{u}}_2)$ defines the attractive well depth, and depends on the axes σ_i, on the three interaction coefficients ε_x, ε_y, and ε_z defining the relative energy for the *side-by-side*, *face-to-face*, and *end-to-end* configurations of a pair of particles,[17,18] and on the orientations. The empirical parameters σ_c, μ, and ν modify the width and depth of the attractive wells. The constants σ_0, and ε_0 define the length and energy scales. We represent the nematic molecules as uniaxial GB elongated ellipsoids, and use a parameterisation already employed in past works,[19] namely $\sigma_x = \sigma_y = \sigma_c = 1\sigma_0$, $\sigma_z = 3\sigma_0$, and $\varepsilon_x = \varepsilon_y = 1\varepsilon_0$, $\varepsilon_z = 0.2\varepsilon_0$, and $\mu = 1$, $\nu = 3$, giving isotropic (I), a wide nematic (N) phase, and smectic (Sm) at low temperatures. Here we are interested in the nematic mesophase, so we have worked at constant dimensionless temperature $T^* = k_B T/\varepsilon_0 = 2.8$, and density $\rho^* = \sigma_0{}^3 N/V = 0.3$, with $\sigma_0 \approx 6 \times 10^{-10}$ m, and $\varepsilon_0 \approx k_B 100$ K, which correspond to a room temperature nematic organisation with average bulk order parameter $\langle P_2 \rangle = 0.806 \pm 0.012$.[19]

A geometrical scheme of the sample setup for the TN cell simulation is given in Fig. 1. The internal dimensions of the TN cell were set to $L_x = L_y = 167.7906\sigma_0$, and $L_z = 93.8604\sigma_0$, which roughly correspond to a $0.100 \times 0.100 \times 0.067$ $(\mu m)^3$ volume, available to the $N_{lc} = 787320$ GB particles modelling the nematic fluid.

To model a π/2 TN cell with planar aligning surfaces we have used periodic boundaries along the laboratory *X*, and *Y* directions, and confining surfaces along *Z* consisting of two slabs cut from a lattice of parallel GB particles with thickness $w_z = 9\sigma_0$ corresponding to nine crystalline layers. The positions and orientations of the confining GB particles were kept fixed during the MC simulation. The two surfaces were prepared with orthogonal directions of molecular alignment, and an additional small pretilt angle of $\theta = 5°$, was imposed on both surfaces to remove energetical degeneracy.[20] The surface particles had the same parameterisation of those modelling the nematic fluid, *i.e.* we assumed the surface–fluid interactions to be of the same entity of the fluid–fluid ones. The overall MC sample size was $N = 944\,784$ GB particles (with $N_{lc} = 787\,320$ for the nematic, and $N_w = 157\,464$ for the surfaces).

3 Computational methodology

The MC or molecular dynamics (MD) simulation of a realistic TN display is a very challenging task because the several micrometres size, and the milliseconds response times of the device are well beyond the conventional spatial and temporal ranges accessible with standard computer simulations. We have chosen to employ the MC method under canonical conditions which, even for molecules much more

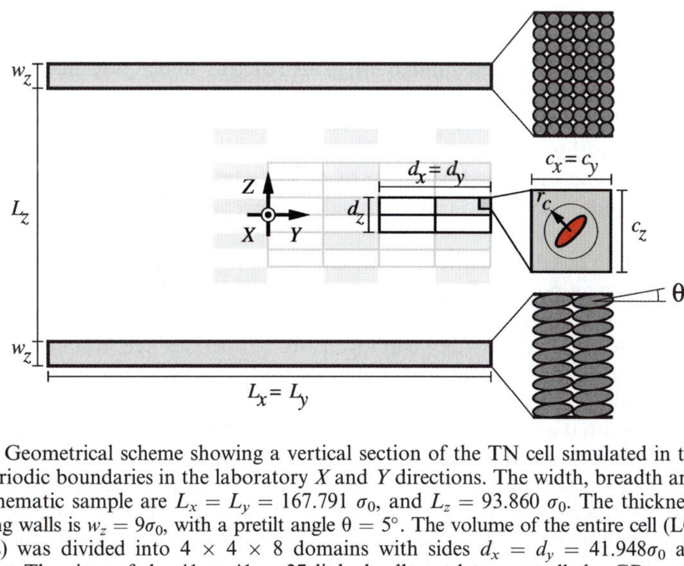

Fig. 1 Geometrical scheme showing a vertical section of the TN cell simulated in this work with periodic boundaries in the laboratory *X* and *Y* directions. The width, breadth and height of the nematic sample are $L_x = L_y = 167.791\ \sigma_0$, and $L_z = 93.860\ \sigma_0$. The thickness of the confining walls is $w_z = 9\sigma_0$, with a pretilt angle $\theta = 5°$. The volume of the entire cell (LC fluid + surfaces) was divided into $4 \times 4 \times 8$ domains with sides $d_x = d_y = 41.948\sigma_0$ and $d_z = 13.983\sigma_0$. The sizes of the $41 \times 41 \times 27$ linked-cells used to map all the GB particles are $c_x = c_y = 4.092\sigma_0$, and $c_z = 4.143\sigma_0$. The GB potential cutoff radius $r_c = 4\sigma_0$ is also shown.

complex than the simple particles used here, is expected to have some advantages in exploring phase space.[21,22] However, performance problems arise when using MC methods with very large systems because the standard Metropolis algorithm for updating molecular configurations proceeds by generating random sequences of single-particle moves in an intrinsically sequential way. General multi-particle moves are possible but unless performed with a biasing scheme[23] they may suffer from poor sampling statistics. Recently, effective parallel MC techniques with modified Markov chains for large off-lattice samples have been introduced[24,25] relying on uncorrelated particles which can be moved simultaneously by different processors working in parallel. Here we have developed a special purpose parallel code based on an algorithm similar to that described in ref. 24,25. The basis of the method is that if the sample is sufficiently large and the pair interaction potential is short-ranged (*i.e.* with a dependence on distance r^{-n} with $n > 3$) it is possible to introduce a potential cutoff distance r_c. This cutoff is possible also if the effective long-range interactions are shielded,[26] and $g(r)$ levels off to the asymptotic limit of 1 for small distances r, and the system is far from first order phase transitions. In this case two (or more) not directly interacting particles (*i.e.* at distance $r > r_c$) can be moved independently without affecting the physics of the system. This is thus a special case of a multi-particle move which does not bias the standard Metropolis sampling scheme and we have taken advantage of this.

The case of GB interactions is fairly favourable since they fall off as r^{-6}, and this allows the partitionin of the sample into non-interacting sub-domains. In detail, for a MC simulation concurrently using P processors, the whole cell (including the surfaces) is subdivided into $P = N_x \times N_y \times N_z$ virtual domains with sides d_x, d_y, and d_z equal or larger than $r_c + r_{max}$, where r_{max} is the width of the uniform distribution used to sample random translational MC moves. Every processor independently evolves particles only in a given domain (and also uses its own sequence of random numbers). Each domain is further subdivided into eight local sub-domains (three-dimensional octants), labelled according to their position. At each MC sweep a sub-domain label is collectively selected, and each processor attempts the MC move of a randomly selected particle in its local sub-domain. Due to this spatial decomposition the $r > r_c + r_{max}$ constraint is always fulfilled by all $P(P - 1)/2$ pairs, even for the largest random displacements, and clashes are prevented by design. After a MC move the inter-processor exchange of updated coordinates is made only to the seven CPUs managing the domains adjacent to the sub-domains where the particle moves have been accepted.[24,25] This concurrent P-particle Markov chain becomes computationally advantageous if an efficient underlying mapping algorithm does the necessary bookkeeping of the various domains and sub-domains populations, allowing a fast computation of pair interactions. In practice, a standard linked-cells algorithm with particle binning based on a $41 \times 41 \times 27$ partition of the whole TN cell, and optimised handling of periodic boundary conditions, has been used, corresponding to linked-cells with sides $c_x = c_y = 4.092\sigma_0$, and $c_z = 4.143\sigma_0$ (see Fig. 1).

This algorithm has demonstrated sufficient scalability and we have run all our MC simulations using a pool of $P = 128$ processors with $N_x = N_y = 4$ and $N_z = 8$, and using the MPI message passing libraries. It might be worth noting that, even though we have focused and successfully employed MC, in perspective MD could also be used, as several efficient parallel MD engines for off-lattice molecular models are becoming publicly available. For instance, the biaxial[17,18] and RE-squared[15] variants of the GB potential are now included in the LAMMPS[27,28] open-source code.

To compute the optical image of transmitted polarised light across the simulated cell (including crossed polarisers), we have used the Stokes–Mueller 4×4 matrix formalism as described in ref. 3–5,29,30. Even though the rigorous solution of the Maxwell equations[31] would be necessary to reproduce finer effects, the simple geometrical optics based on Stokes–Mueller matrices[29,30] has proved to be quite adequate and sufficient to provide an overall description of the optical transmission of anisotropic planar media like a TN cell. In our model calculation we have considered the refractive indexes for the 5-cyano-biphenyl (5CB), namely the ordinary $n_o = 1.5$, and extraordinary $n_e = 1.7$, and a transmitted wave length of $\lambda = 500$ nm.

4 Simulation results

The analysis of the MC simulation results is divided into two stages. First we show the results of the relaxation experiment leading to the spontaneous formation of a twisted organisation. Then we move to the response to an external switching field and the computation of optical transmission properties.

4.1 Spontaneous helix formation

We have investigated the spontaneous formation of a helical organisation inside our TN cell. The canonical, NVT MC simulation has been started from a configuration of rod-like GB particles uniformly aligned along the laboratory Z axis (*i.e.* perpendicular to the confining surfaces, see Fig. 2b). This unfavourable starting configuration resembles the one obtained ideally after switching off a field that has aligned the nematic (dark pixel). The sample was allowed to relax by performing 1.76×10^6 sweeps, each consisting of N attempted rotational and translational MC moves. The response under an external switching field described in the next section was simulated running additional 1.12×10^6 sweeps starting from the final configuration of the relaxation process (see Fig. 2c). The sampling ranges for translational $[-t_{\max}/2, t_{\max}/2]$ and orientational $[-\theta_{\max}/2, \theta_{\max}/2]$ moves were automatically adjusted during the MC run to give an average acceptance ratio of 0.4 for the roto-translational moves, resulting in typical $t_{\max} = 0.08\sigma_0$, and $\theta_{\max} = 13°$ values. Orientational moves were randomly sampled using the standard Barker–Watts[32] algorithm.

The relaxation process was followed by monitoring the evolution of the average potential energy per particle $\langle U_{GB} \rangle$ of the nematic sample, and the second rank order parameter $\langle P_2 \rangle = \langle (3(\hat{\mathbf{u}}_i \cdot \hat{\mathbf{n}})^2 - 1)/2 \rangle$, computed with respect to the overall sample director $\hat{\mathbf{n}}$ (given as a unit vector) using the standard algorithm of diagonalisation of the order matrix.[33] The simulation was continued until no systematic drift was observed in these observables (see Fig. 3a and 3b). In Fig. 3a we see that the potential energy grows in few MC ksweeps from $-11.24\varepsilon_0$ to $\approx -7.6\varepsilon_0$, then relaxes to a practically constant value. Also for the $\langle P_2 \rangle$ plot of Fig. 3b the starting uniform director distribution is destroyed in a few MC ksweeps, and the overall $\langle P_2 \rangle$ drops down from 1 to ≈ 0.5.

More insight on the molecular organisation across the cell and its evolution was obtained dividing the TN cell into N_s virtual parallel slabs of thickness $\Delta z = L_z/N_s$

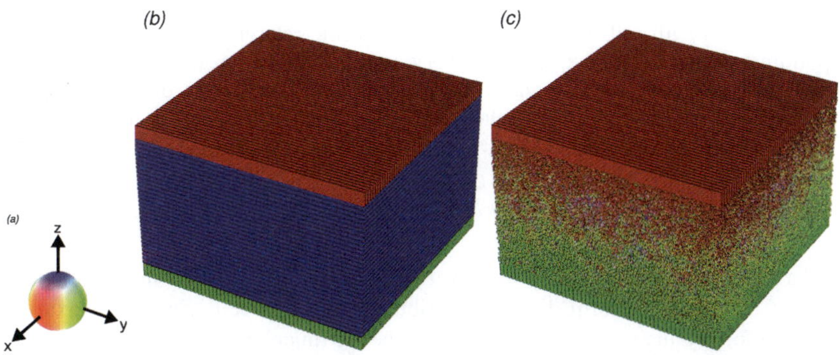

Fig. 2 Snapshots of the starting (b) and final (c) MC configurations of the TN cell after a relaxation of 1.76×10^6 sweeps. GB particles are colour coded according to the orientation of molecular $\mathbf{u}_i$ axis with respect to the laboratory frame using the 3D palette of plate (a).

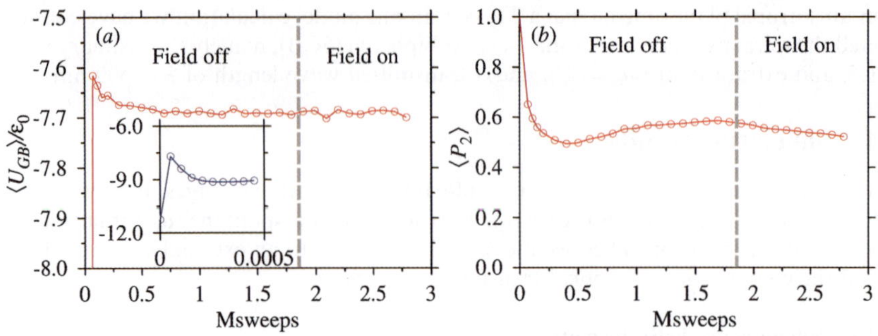

Fig. 3 Average GB energy per particle $\langle U_{\mathrm{GB}}\rangle/\varepsilon_0$ (a), and nematic order parameter $\langle P_2\rangle$ computed with respect to the overall cell director $\hat{\mathbf{n}}$ (b) as a function of the number of millions of MC sweeps. The vertical dashed line at 1.76×10^6 sweeps marks the end of the relaxation and the beginning of the virtual switching experiment of the central pixel.

(here $N_s = 30$) to compute specific local order parameters $\langle P_2\rangle_s$ and orientational correlation functions between pairs of particles belonging to different slabs (*i.e.* with different z_i elevation). In our cell each of the $N_s = 30$ slabs has thickness $\Delta z = 3.729\sigma_0$, *i.e.* approximately 25% thicker than the length of the mesogenic particles, and contains on average $\approx$ 26 100 GB particles. After the relaxation of the uniformly aligned initial configuration the order parameters $\langle P_2\rangle_s$ relative to the local director $\hat{\mathbf{n}}_i$ of each planar slab remain essentially constant across the sample and over the simulation run (not shown here). In this part of the MC evolution the GB particles which are not in proximity of the surfaces equilibrate to a plateau $\langle P_2\rangle_s \approx 0.79$, which is within the error bar of the $\langle P_2\rangle = 0.806 \pm 0.012$ average value found for the bulk system[19] at the same temperature and density. Thus, the presence of the aligning surfaces and the formation of the twisted helical structure does not affect, for samples of this size, the magnitude of the orientational ordering in the centre of the cell,[34] but only the direction of the local director. This result is consistent with a physical picture of gradual re-alignment of small ordered regions/clusters whose local properties are not largely perturbed by the overall helical organisation.[35]

The same is also true for the density distributions of centres of mass of the nematic particles $g(z)$ (not shown here) which show that during the entire relaxation stage the sample was nematic and devoid of positional ordering, apart from approximately 4 to 5 surface-induced smectic layers in the proximity of the planar surfaces. This is fairly commonly observed in simulations of nematics close to aligning surfaces.[36,37]

The formation of a helical structure can be monitored by computing order parameters giving the orientation of the local directors with respect to that of a perfectly helical organisation (with a pitch equal to four times the $\pi/2$ cell thickness L_z). The centre of every ith slab has assigned to it a specific reference orientation $\hat{\mathbf{t}}_i$ corresponding to the ideal helix, and for every configuration a local director $\hat{\mathbf{n}}_i$ can be obtained from the standard algorithm of diagonalisation of an order matrix for the molecular $\hat{\mathbf{u}}_i$ axes. The local helical order parameters $\langle P_2\rangle_h$ (ref. 3) are defined as

$$\langle P_2\rangle_h = \frac{1}{N_s}\sum_{i=1}^{N_s} P_2(\hat{\mathbf{n}}_i \cdot \hat{\mathbf{t}}_i) \qquad (2)$$

and a fully formed helix corresponds to $\langle P_2\rangle_h \approx 1$. In Fig. 4 we show how the disruption of the initial uniform director distribution takes place during the relaxation from uniformly aligned to twisted nematic. The starting phase of the process appears to be the alignment of the nematic slabs closest to the surfaces, leaving the central portion of the $\langle P_2\rangle_h$ profiles essentially unaffected. This transformation takes place in a few thousands MC sweeps, and gradually propagates from both directions

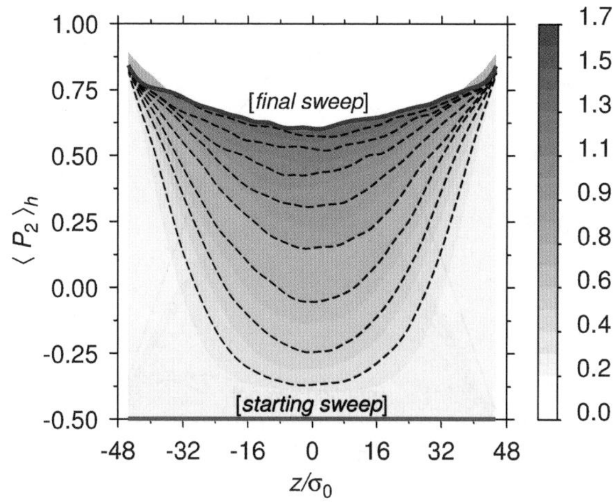

Fig. 4 Evolution profiles for the local helical order parameter $\langle P_2 \rangle_h$ (see eqn (2)) from the MC relaxation of the uniformly aligned cell. The bottom thick (red) line is for the starting ordered configuration, while the top (blue) one is for twisted organisation after 1.76×10^6 sweeps. Intermediate profiles are shown both as dashed lines every 2×10^5 sweeps and as a greyscale background (see lateral palette). For a perfect helix $\langle P_2 \rangle_h = 1$.

towards the centre of the cell. The ensuing reorganisation process of the central region is much longer (still at constant $\langle P_2 \rangle_s$), and the number of MC sweeps required to reach $\langle P_2 \rangle_h \geq 0.75$ is now of the order of 10^6 sweeps (approximately two orders of magnitude longer). These order parameter values are significantly smaller than 1, hinting that the simulated cell does not attain a perfectly uniform helical organisation across the sample, and the local directors $\hat{\mathbf{n}}_i$ have some sort of deviation from the geometrical pattern described by the $\hat{\mathbf{t}}_i$ orientations. This can be quantified by computing the chiral correlations[37] between nematic particles at a certain elevation z with those near the aligning surfaces

$$S^{221}(z) = -\frac{\sqrt{3}}{\sqrt{10}\left(L_x L_y \rho\right)^2} \left\langle \delta(z - z_j + Z_1)(\hat{\mathbf{u}}_i \cdot \hat{\mathbf{u}}_j \times \hat{\mathbf{z}}_{ij})(\hat{\mathbf{u}}_i \cdot \hat{\mathbf{u}}_j) \right\rangle_{i,j} \tag{3}$$

where the subscript i refers to the particles in the first slab (centred at laboratory Z_1), while j spans the remaining GB particles in the sample. The unit vector $\hat{\mathbf{z}}_{ij}$ is parallel to the laboratory Z axis when $z_j > z_i$ (as in our case), otherwise it is antiparallel. We see in Fig. 5 how the chiral correlation gradually and steadily develops during the relaxation starting from the surfaces towards the cell centre, and as the simulation progresses, it slowly approaches (some sort of) asymptotic limiting profile which is however still far from the ideal helical one. We also see that the small initial structural asymmetry around $Z = 32\sigma_0$ due to a local fluctuation is eventually lost, as expected, at the end of the relaxation process.

Yet another way of characterising the director distribution in the various planar slabs is that of monitoring the azimuthal angles ϕ_i formed by the projection of the local directors $\hat{\mathbf{n}}_i$ on the XY and YZ laboratory planes. The azimuthal angles ϕ_{xy}, and ϕ_{yz} change during the relaxation process, and Fig. 6a shows how the initial $\phi_{xy} = 0°$ constant value evolves and eventually approaches fairly closely the ideal linear profile for a geometrically perfect helical organisation. The deviations from the ideal helical structure are due to the tilt out of plane of the local directors $\hat{\mathbf{n}}_i$, shown in Fig. 6b, where we see that even after 1.76×10^6 sweeps the central slabs have a residual director tilt $\phi_{yz} \approx 25°$. As will be shown later this reflects into the

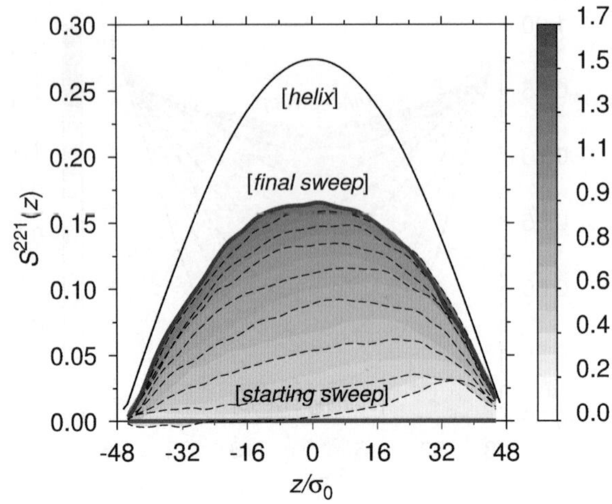

Fig. 5 Evolution profiles for the chiral correlation function $S^{221}(z)$ (see eqn (3)) from the MC relaxation of the uniformly aligned cell. The thin continuous line is the reference profile for a perfect helix. See the caption of Fig. 4 for additional details.

response to an external field mainly arising from the central slabs. For both azimuthal angles profiles the changes taking place during the MC evolution start in the slabs at the surface boundaries and then propagate towards the central ones.

A direct, molecular level, representation of these arguments is given by the sample snapshots of Fig. 7 which clearly show how the structural transformation takes place under the effect of the two opposite driving forces given by the tendency to align parallel to the closest surface, and that of minimising the elastic strain imposed by the $\pi/2$ cell twist. We see from the colour coding of the orientations that the director distribution is not perfectly uniform and small local fluctuations are present. To make the relationship to the actual optical performance more clear, we also show images (Fig. 7) of the linearly polarised light transmitted across the TN cell, including the crossed polarisers, computed as described in ref. 3–5. These show how a transparent state is obtained only in the final part of the MC relaxation, and that the local transmittance can fluctuate several times from opaque to transparent and back before the final bright state is achieved. These snapshots also

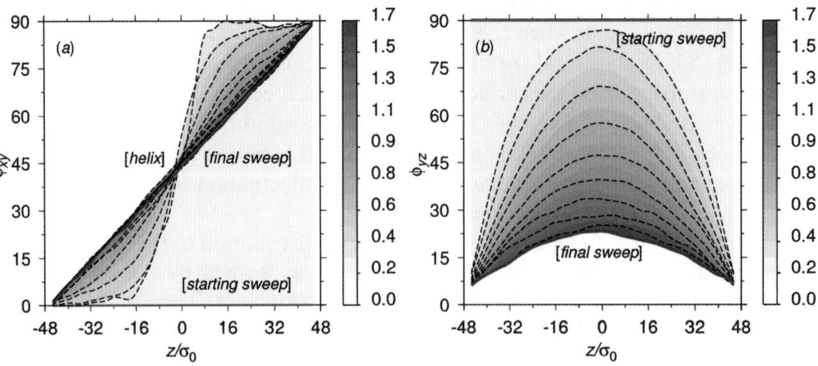

Fig. 6 Evolution profiles for the azimuthal angles ϕ_{xy} (a), and ϕ_{yz} (b) defined in the text from the MC relaxation of the uniformly aligned cell. The reference lines for a perfect helix are a straight one bisecting the first quadrant $\phi_{xy} = z$ (a), and a horizontal one $\phi_{yz} = 0$ (b). See the caption of Fig. 4 for additional details.

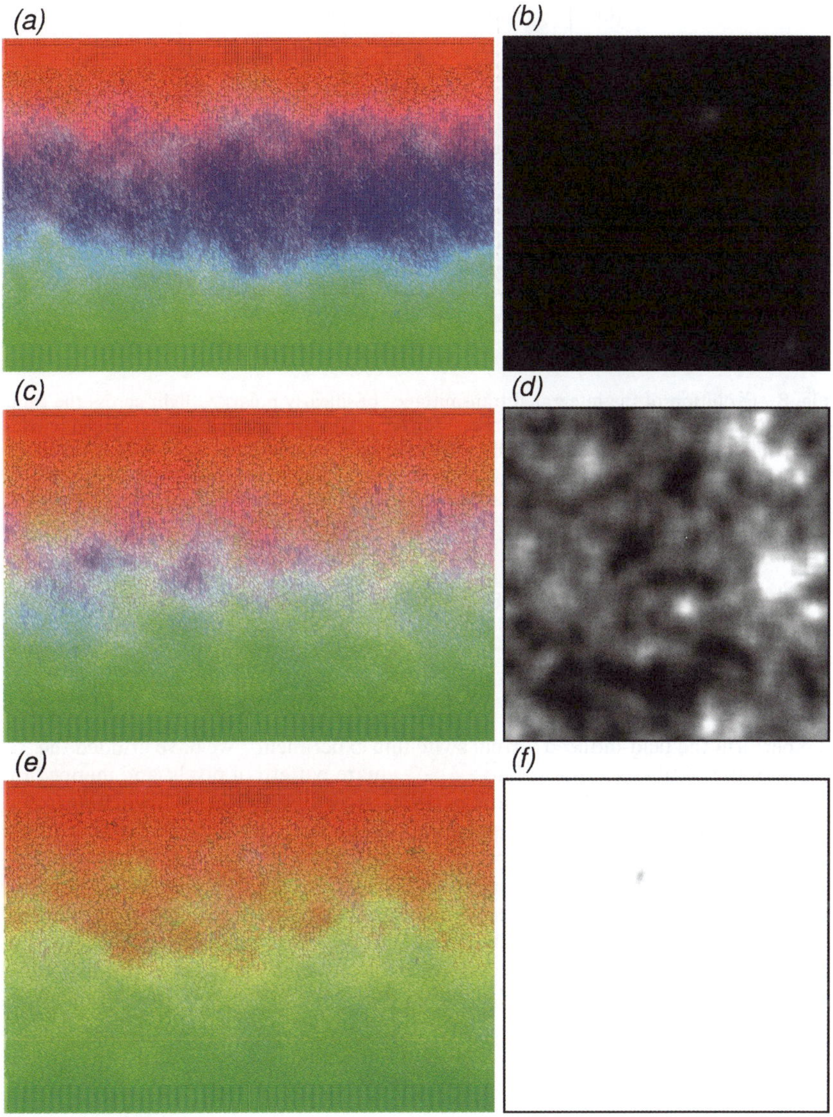

Fig. 7 Transverse views of the TN cell from the MC relaxation after 0.46×10^6 sweeps (a), 0.86×10^6 sweeps (c), and 1.63×10^6 sweeps (e) as seen from the laboratory X axis. GB particles are colour coded according to their orientations using the 3D palette of Fig. 2, and smoothly rendered to reduce the visual effect of the edges. The corresponding transmittance maps of linearly polarised light across the cell and crossed polarisers (plates b, d, and f), as seen from the laboratory Z axis, are represented by using a linear greyscale palette ranging from 0 (opaque) to 1 (transparent). The overall cell transmittances are 3.5% (b), 30% (d), and 89% (f).

show that whatever the structure within the cell, the particles close to the confining surfaces remain parallel to the alignment direction. These arguments are made more quantitative in Fig. 8 where we plot the total cell transmittance both for the field-*off* MC relaxation from uniformly aligned, and the virtual switching experiment described in the next section. In particular, the leftmost portion of Fig. 8 shows a two-regimes relaxation, characterised by an initial low-transmittance state (lower than 5%), which is maintained for approximately 5×10^5 sweeps, and is followed by a second stage with quite a larger positive slope of the transmittance curve lasting for

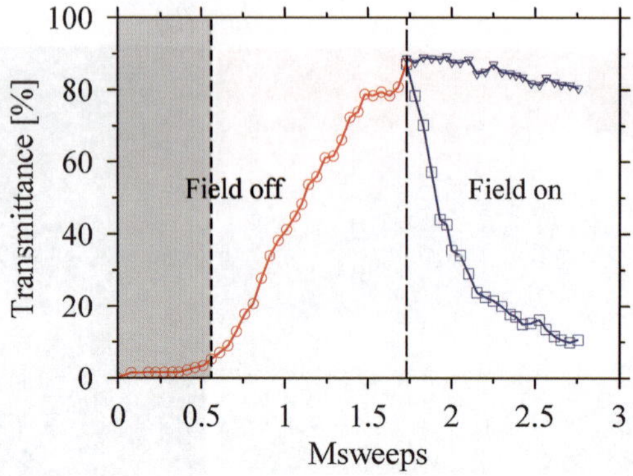

Fig. 8 Evolution of the integrated transmittance of linearly polarised light across the simulated TN cell (including crossed polarisers) for the relaxation from uniformly aligned (circles, red), and for the central pixel switching (squares, blue). The transmittance of the external pixels during the field-*on* experiment is also given (triangles, blue). The vertical short-dashed line corresponds to a 5% transmittance, while the long-dashed one at 1.76×10^6 sweeps marks the end of the relaxation and the beginning of the switching experiment.

roughly 1×10^6 sweeps until it begins levelling off in correspondence of a 90% transmittance of the entire cell.

4.2 Central pixel switching

To perform the field-induced virtual switching experiment[35] we have gridded the XY plane of the confining surfaces into a 3×3 square pattern of pixels, and mapped the x_i, and y_i coordinates of the nematic particles accordingly. The MC switching experiment is characterised by the application, in the cell region corresponding to the central pixel, of a (electric) field parallel to the laboratory Z axis, and has been performed by running 1.12×10^6 sweeps starting from the last configuration of the previous relaxation process. The beginning of the virtual experiment is indicated in Fig. 3 and 8 by a vertical black line at 1.76×10^6 sweeps. In practice, particles in the central pixel were subjected to an extra potential energy term U_{F} modelling the dielectric (second rank) coupling of the molecular $\hat{\mathbf{u}}_i$ axis with the external field, according to the equation

$$U_{\mathrm{F}} = -\xi_{\mathrm{F}}\left[\frac{3}{2}(\hat{\mathbf{u}}_i \cdot \hat{\mathbf{Z}})^2 - \frac{1}{2}\right] \tag{4}$$

where the field-coupling constant is defined as $\xi_{\mathrm{F}} = \varepsilon_0 \Delta\varepsilon V_0 E^2/3$, with $\Delta\varepsilon$ the susceptibility anisotropy, V_0 the molecular volume, $\hat{\mathbf{Z}}$ the direction of the electric field, and ε_0 the vacuum permettivity. The coupling coefficient we used was $\xi_{\mathrm{F}} = 0.05\varepsilon_0$, roughly corresponding (with $\Delta\varepsilon = 12$) to an applied field intensity of $E \simeq 7.6 \times 10^7$ Vm^{-1}, approximately one order of magnitude bigger than those used in real TN devices, and nonetheless giving a U_{F} energy much lower (typically < 4.5%) than the U_{GB} term. In Fig. 3a we see that the GB energy does not drift during the switching experiment, and this is consistent with a cartoon of a weak coupling between external field and nematic molecules. We also see that the overall order parameter of Fig. 3b exhibits a rather small decrease after the central pixel re-alignment has taken place.

Again, we have used local averages and correlations to characterise the switching process. In Fig. 9 we report the profiles for the helical $\langle P_2 \rangle_h$, and the field direction $\langle P_2 \rangle_e$ order parameters. We see that the central slabs of the switched pixel do not

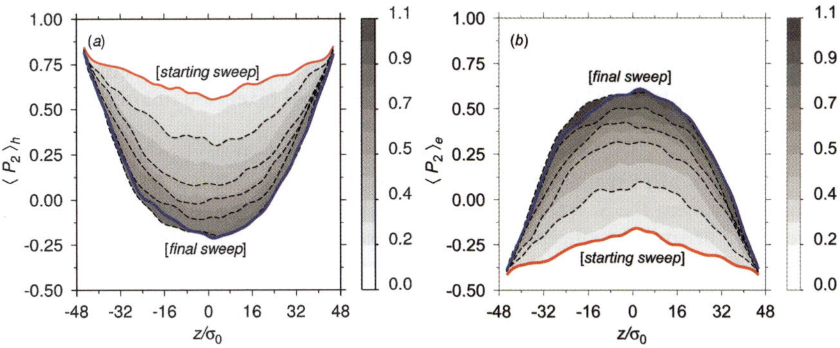

Fig. 9 Evolution profiles for the local helical $\langle P_2 \rangle_h$ (a), and for the field direction $\langle P_2 \rangle_e$ (b) order parameters relative to the central field-*on* pixel during the virtual MC switching experiment. The thick lines represent the starting twisted configuration (red), and the final aligned one (blue) after 1.12×10^6 sweeps. Intermediate values are shown both as dashed lines every 2×10^5 sweeps and as a greyscale background (see lateral palettes). For the reference profile of a geometrically perfect helix $\langle P_2 \rangle_h = 1$, and $\langle P_2 \rangle_e = 0$.

completely align along the laboratory Z direction (*i.e.* they do not attain the limiting $-1/2$ value for $\langle P_2 \rangle_h$, and 1 for $\langle P_2 \rangle_e$) and a certain amount of orientational disorder remains. In particular, the central slab has the largest degree vertical alignment ($\langle P_2 \rangle_e \approx 0.6$) but is still lower than the extent ($\langle P_2 \rangle \approx 0.8$) found in a bulk nematic system at the same temperature and density.[19] On the other hand, the slabs closer to the confining surfaces maintained during the entire switching process a positive $\langle P_2 \rangle_h$ (Fig. 9a), and a negative $\langle P_2 \rangle_e$ (Fig. 9b). The 8 external pixels are not globally influenced by the field, and the profiles of the local order parameters do not change during the virtual switching experiment (not shown here), *i.e.* boundary effects are negligible for this 3×3 switching pattern.

Fig. 10a shows an interesting residual helical structuring in ϕ_{xy} (*i.e.* in the n_x, and n_y components of the local directors) which corresponds to the initial helical distribution prior to the switching on of the external field. However, the noisy profiles of Fig. 10a suggest that the plotted ϕ_{xy} values are an artificial amplification of small and fluctuating events, and at the time being we are not able to discriminate between a true *"memory"* effect and boundary effects originating from the *on* pixel being completely surrounded by *off* ones (other switching patterns may produce different

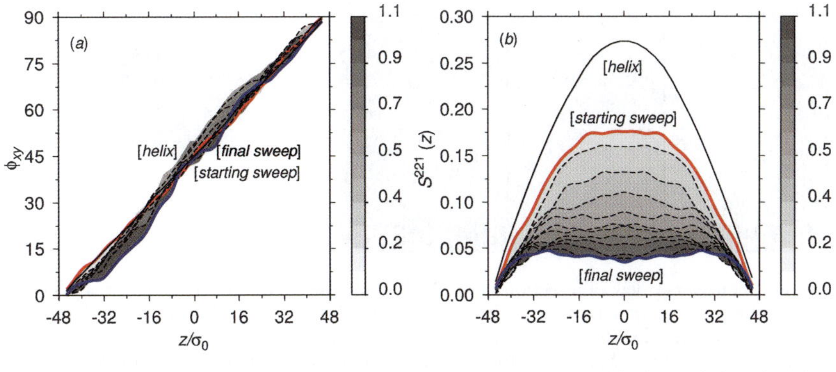

Fig. 10 Evolution profiles for the azimuthal angles ϕ_{xy} (a), and chiral correlation function $S^{221}(z)$ (b) for the central field-*on* pixel during the virtual MC switching experiment. See Fig. 6 and 9 for details.

behaviour). This phenomenon may be important in the restoring of the helical configuration after the aligning field is switched off, since this process should be governed by the twist elastic constant, and only indirectly by the aligning affect due to the confining surfaces. The ϕ_{xy} profiles for the external pixels are unaffected by the switching experiment (not shown here).

The chiral pair correlations in the central pixel are destroyed in approximately 5×10^5 sweeps after switching on the field (see Fig. 10b). In the second half of the virtual experiment the $S^{221}(z)$ levels to a broad M-like profile, and the residual chiral correlations are mainly concentrated into the external slabs where the slope of $S^{221}(z)$ is larger. The chiral pair-correlation profiles for the 8 external pixels are unaffected by the switching experiment (not shown here).

The organisation at the end of the virtual experiment is graphically rendered by the snapshot of Fig. 11a, where the TN cell has been vertically sliced at $Y = 0$ to expose the orientational structure of the central pixel. The colour coding of the particles shows that the internal core is fairly aligned with the laboratory Z axis, while the intermediate nematic and boundary slabs have not responded to the same extent to the external perturbation. In our simulation there is no gap between pixels[20] and we see that the aligned molecules of the field-*on* pixel tend to align those of the neighbouring ones, and the snapshot of the cross-section in Fig. 11a also shows that the central oriented region has a somewhat globular shape, which slightly spreads to the neighbouring pixels where the external field is absent. This can also be seen from Fig. 11b giving the transmittance along Z of linearly polarised light across the whole TN cell. We see that the dark opaque area extends irregularly over the boundaries of the central pixel, which also has a non uniform degree of light extinction, with an average value of 91%.

The commutation from transparent to dark is shown in Fig. 12 where we see how the transmittance of the central pixel decreases during the virtual MC experiment. The dark area develops in the core of the pixel, where the effects due to the external pixels are smaller, and gradually intensifies and enlarges towards the boundaries to cover the entire field-*on* cross-section and beyond. The total cell transmittances for both central and external pixels are given by the right portion of Fig. 8. While light passing through the external pixels decreases less than 10% from the starting value, for the central pixel it smoothly diminishes from 90% to 10% in 1×10^6 sweeps.

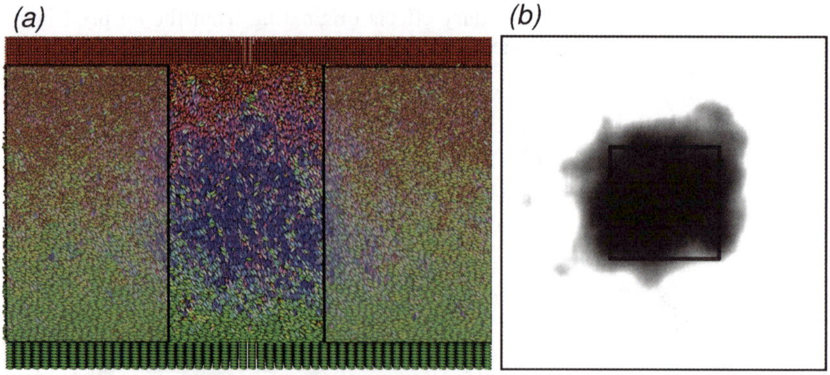

Fig. 11 Transverse view of a section of the MC configuration (after 1.12×10^6 sweeps) from the switching *on* experiment of the central pixel (a). To show the organisation of the *on* pixel its particles are rendered without a smoothing filter. Particles are colour coded according to their orientations using the 3D palette of Fig. 2. The transmittance map of linearly polarised light is shown in plate (b) (see Fig. 7 for details). Vertical black bars (a) and blue lines (b) delimit the region affected by the external field. The central pixel transmittance is 9%, while the average for the external ones is 79% (b).

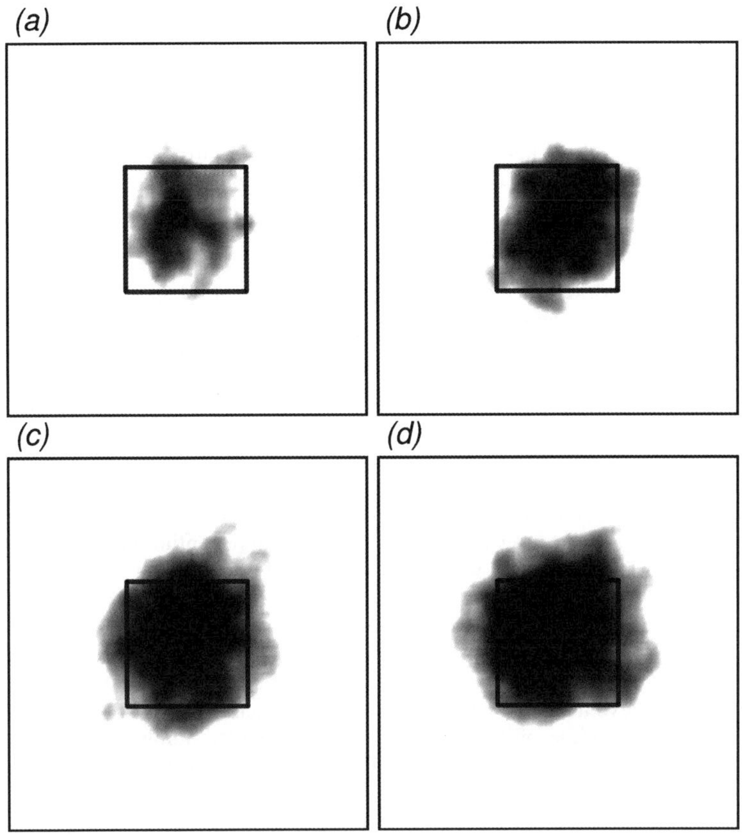

Fig. 12 Transmittance maps of linearly polarised light across the cell after 0.2×10^6 sweeps (a), 0.4×10^6 sweeps (b), 0.6×10^6 sweeps (c), and 0.8×10^6 sweeps (d). See Fig. 11 for details. The central pixel transmittances are 43% (a), 29% (b), 19% (c), and 15% (d), while the average external ones are 88% (a), 88% (b), 84% (c), and 81% (d).

5 Discussion and conclusions

This paper has two main objectives. The first is to show that fairly massive parallel computing can now be applied to push the application of coarse-grained molecular resolution simulations to the modelling of fairly realistic devices. We have successfully employed to this end a specially written MC code to perform an extensive simulation of a TN cell containing $\mathcal{O}(10^6)$ GB mesogenic particles.

The second objective is to gain insight on the actual microscopic working of an important electro-optical device that is commonly assumed to correspond to a uniformly helical structure letting plane polarised light through a pixel, and turning, under the effect of a suitable field, to a vertical monodomain that has the effect of shutting the transmission through crossed polarisers.

We have found that this picture is somewhat incomplete and perhaps even misleading. We have seen from MC simulations that a uniform helix is not formed at once when an initial vertical monodomain is relaxed under the effect of the two aligning surfaces and the nematic intermolecular interactions. Rather, two uniformly aligned domains start to grow under the orienting effect of the two perpendicular surfaces and only when the two domains meet is the twisted structure formed with the consequent passage of light.

Since we perform Monte Carlo simulations the natural evolution variable is not real time but MC sweeps. Nonetheless, we believe these MC simulations have some predictive power not only for the initial and final equilibrium states of a virtual experiment, but also for the evolution between the two states. Since our MC moves correspond to single-particle physical movements such as translations and rotations, we can essentially map the MC evolution onto a stochastic jump process for the molecules. Thus we may expect that the ratio between the number of MC sweeps required for the *on–off* and *off–on* processes can be approximately compared to the experimental one for a device of similar geometry, and measured in real time units. In particular, we also expect that the initial low-transmittance stage we have observed (lasting approximately for 1/3 of the entire relaxation run), may be experimentally investigated. Indeed similar experiments have been performed by Toriumi and coworkers[38–40] using a variant of time-resolved spectroscopic ellipsometry (TRSE) where the polarisation of light modulated by the interfacial layer of a nematic (5CB) in an antiparallel homogeneously aligned cell has been measured. In that case, two regimes were also found for the relaxation of the director distribution of the nematic, previously homeotropically aligned with a field, to the homogeneous state determined by the boundaries. The two regimes, successfully modelled in ref. 40 with a Frank-type continuum theory consist, similarly to what we have found here (albeit for a different cell geometry), of a fast relaxation of the nematic close to the surface and a much slower one for the one in the cell centre. The fact that such an agreement is found starting from both molecular and continuum descriptions is a practical confirmation that we have essentially closed the gap between the two treatments.

As for the twisted structure obtained, we have examined its character in detail, finding that it approximates but never fully assumes a truly helical organisation and we have quantified this with a helical order parameter and a chiral correlation function. We have also studied the field-*on* realignment process and found that it starts from the middle of the TN cell taking a smaller number of MC sweeps ($\approx 1.1 \times 10^6$ sweeps) to complete than the previous field-*off* relaxation. We find that even when applying the field to a square area of the display, the resulting black pixel is somewhat more fuzzy as a consequence, at least for our cell configuration, of the aligning effect that the region of field-aligned molecules exerts on the neighbouring molecules not subject to the field and *vice versa*. However, the process does not lead to significant changes of the overall order and structure, or average transmittance, of the domain surrounding the field-*on* one.

As a final statement, even if the effort and resource needed to perform these types of simulations is at the moment huge, we would like to point out some features that could make the technique quite important in the near future. The first is the possibility of studying non-uniform, rough, or nanostructured confining surfaces, or surfaces containing a given concentration of defects and seeing their effect on the device. The second is the possibility of studying more exotic displays, *e.g.* those currently planned with biaxial nematics,[41,10] where the chemical structure, or the molecular organisation and its changes under the effect of an applied field are less known and obvious, while at the same time the material constants are very numerous and unknown. In these cases, the molecular technique presented here should allow the prediction of the workings and the performance of a device starting only from the knowledge of the molecular parameters, avoiding the use of continuum level theories.

6 Acknowledgements

We thank the EU-STREP *"Biaxial Nematic Devices"* (BIND) FP7-216025 for financial support, and *"Distributed European infrastructure for supercomputing applications"* (DEISA) FP6-508830, and particularly CINECA (Italy) and CSC (Finland), for computational resources within the DEISA Extreme Computing Initiative.

References

1 P. Semenza, *Nat. Photonics*, 2007, **1**, 267–268.
2 M. Schadt and W. Helfrich, *Appl. Phys. Lett.*, 1971, **18**, 127–128.
3 E. Berggren, C. Zannoni, C. Chiccoli, P. Pasini and F. Semeria, *Int. J. Mod. Phys. C*, 1995, **6**, 135–141.
4 C. Chiccoli, P. Pasini, S. Guzzetti and C. Zannoni, *Int. J. Mod. Phys. C*, 1998, **9**, 409–419.
5 C. Chiccoli, S. Guzzetti, P. Pasini and C. Zannoni, *Mol. Cryst. Liq. Cryst.*, 2001, **360**, 119–129.
6 R. Memmer and O. Fliegans, *Phys. Chem. Chem. Phys.*, 2003, **5**, 558–566.
7 L. V. Mirantsev and E. G. Virga, *Phys. Rev. E: Stat., Nonlinear, Soft Matter Phys.*, 2007, **76**, 021703.1–7.
8 D. W. Berreman, *Philos. Trans. R. Soc. London, Ser. A*, 1983, **309**, 203–216.
9 F. H. Yu and H. S. Kwok, *J. Opt. Soc. Am. A*, 1999, **16**, 2772–2780.
10 R. Berardi, L. Muccioli, S. Orlandi, M. Ricci and C. Zannoni, *J. Phys.: Condens. Matter*, 2008, **20**, 463101.1–16.
11 M. R. Wilson, *Int. Rev. Phys. Chem.*, 2005, **24**, 421–455.
12 J. G. Gay and B. J. Berne, *J. Chem. Phys.*, 1981, **74**, 3316–3319.
13 C. M. Care and D. J. Cleaver, *Rep. Prog. Phys.*, 2005, **68**, 2665–2700.
14 C. Zannoni, *J. Mater. Chem.*, 2001, **11**, 2637–2646.
15 R. Everaers and M. R. Ejtehadi, *Phys. Rev. E: Stat., Nonlinear, Soft Matter Phys.*, 2003, **67**, 041710.1–8.
16 X. Zheng and P. Palffy-Muhoray, *Phys. Rev. E: Stat., Nonlinear, Soft Matter Phys.*, 2007, **75**, 061709.1–6.
17 R. Berardi, C. Fava and C. Zannoni, *Chem. Phys. Lett.*, 1995, **236**, 462–468.
18 R. Berardi, C. Fava and C. Zannoni, *Chem. Phys. Lett.*, 1998, **297**, 8–14.
19 R. Berardi, A. P. J. Emerson and C. Zannoni, *J. Chem. Soc., Faraday Trans.*, 1993, **89**, 4069–4078.
20 D.-K. Yang and S.-T. Wu, *Fundamentals of Liquid Crystal Devices*, John Wiley and Sons, New York, 2006.
21 W. L. Jorgensen and J. Tirado-Rives, *J. Phys. Chem.*, 1996, **100**, 14508–14513.
22 J. P. Ulmschneider, M. B. Ulmschneider and A. Di Nola, *J. Phys. Chem. B*, 2006, **110**, 16733–16742.
23 D. Frenkel and B. Smit, *Understanding Molecular Simulations: From Algorithms to Applications*, Academic Press, 2nd edn, 2001.
24 G. S. Heffelfinger and M. E. Lewitt, *J. Comput. Chem.*, 1996, **17**, 250–265.
25 G. S. Heffelfinger, *Comput. Phys. Commun.*, 2000, **128**, 219–237.
26 D. Wolf, P. Keblinski, S. R. Phillpot and J. Eggebrecht, *J. Chem. Phys.*, 1999, **110**, 8254–8282.
27 S. J. Plimpton, *J. Comput. Phys.*, 1995, **117**, 1–19.
28 S. J. Plimpton, 2008, URL, http://lammps.sandia.gov/.
29 F. Xu, H.-S. Kitzerow and P. P. Crooker, *Phys. Rev. A*, 1992, **46**, 6535–6540.
30 W. A. Shurcliff, *Polarized Light: Production and Use*, Harvard University Press, Cambridge, MA, 1962.
31 B. Witzigmann, P. Regli and W. Fichtner, *J. Opt. Soc. Am. A*, 1998, **15**, 753–757.
32 J. A. Barker and R. O. Watts, *Chem. Phys. Lett.*, 1969, **3**, 144–145.
33 C. Zannoni in *The Molecular Physics of Liquid Crystals*, Academic Press, 1979, pp. 51–83.
34 T. Mima, T. Narumi, S. Kameoka and K. Yasuoka, *Mol. Simul.*, 2008, **34**, 761–773.
35 R. Berardi, L. Muccioli and C. Zannoni, *J. Chem. Phys.*, 2008, **128**, 024905.
36 D. Micheletti, L. Muccioli, R. Berardi, M. Ricci and C. Zannoni, *J. Chem. Phys.*, 2005, **123**, 224705.1–10.
37 R. Berardi, H.-G. Kuball, R. Memmer and C. Zannoni, *J. Chem. Soc., Faraday Trans.*, 1998, **94**, 1229–1234.
38 T. Tadokoro, T. Fukazawa and H. Toriumi, *Jpn. J. Appl. Phys.*, 1997, **36**, L1207–L1210.
39 T. Fukazawa, T. Tadokoro, H. Toriumi, T. Akahane and M. Kimura, *Thin Solid Films*, 1998, **313–314**, 799–802.
40 T. Tadokoro, H. Toriumi, S. Okutani, M. Kimura and T. Akahane, *Jpn. J. Appl. Phys.*, 2003, **42**, 4552–4563.
41 R. Berardi and C. Zannoni, *J. Chem. Phys.*, 2000, **113**, 5971–5979.

Lyotropic self-assembly mechanism of T-shaped polyphilic molecules

Andrew J. Crane and Erich A. Müller*

Received 26th January 2009, Accepted 20th February 2009
First published as an Advance Article on the web 10th August 2009
DOI: 10.1039/b901601e

We present coarse-grained molecular dynamics simulations of mixtures of a model of T-shaped polyphilic bolaamphiphile liquid crystal molecules with a solvent. Based on the premise that the most important features of the liquid structure stem from the balance between the close range repulsions and the strong directional forces typical of hydrogen bonding and association, we have employed a coarse-graining approach that simplifies and minimises the attractions present in the system. The model consists of six fused rigid spheres, where the two end spheres have a significant attraction amongst themselves while the rest are repulsive in nature. A weakly self-attracting lateral chain consisting of fully flexible tangently bonded spheres is attached to one of the central spheres. Thus, the T-shaped molecule is composed of three mutually repulsive segments which allow the pure system to self-assemble into a liquid crystalline honeycomb columnar phase. The stability of the columnar phase is probed by the sequential addition of a solvent that has affinity with only one of the segments of the molecule. Our coarse-graining technique allows us to observe dynamically not only the 1st level nanoscale segregation but also the 2nd level reorganization which leads to the formation of replicated periodic structures. It is seen how this latter structuring takes place at times which are an order of magnitude longer than the former, and by itself explains the practical limitations of studying self-assembly with more detailed atomistic models. Mobility coefficients (related to diffusion constants), order parameters and direct visualization of the configurations are used to present a phase diagram for the solvated system in the whole concentration range. At low solvent density, the solvent solvates the honeycomb structure, but does not alter the order significantly. At modest volume fractions of solvent, the solvent mostly segregates into a distinct phase, while the T-shaped molecules retain a phase with columnar structure. At very large solvent concentrations, the T-shaped molecules form structureless aggregates, while at the infinite dilution limit present themselves as dimers and monomers.

1 Introduction

1.1 Coarse-grained modelling

In the process of describing fluids on a microscopic level, different length and time scales provide complementary information that ultimately allow a full picture of the physical phenomena to be made. Both in the quantum and full atomistic levels of matter description, reasonably well accepted and employed methodologies are

Department of Chemical Engineering, Imperial College London, UK. E-mail: e.muller@imperial.ac.uk

now in place. As an example of the latter, molecular dynamics (MD) modelling is an integral part of the research in many areas of nanomaterials, biotechnology and chemical engineering. Although appealing, an all-atom MD approach to studying supramolecular assembly of macromolecules is hampered by both long time spans associated with certain self-assembly processes and/or the increased computational demand brought by the complex nature of the molecules involved. Both aspects confabulate to consume computational resources well beyond those available now and in the foreseeable future. On the other end of the spectrum, the description of fluids and soft matter from the continuous macroscopic level is also well established with methodologies like computational fluid dynamics (CFD). However, as these methods are based on macroscopically-derived phenomenological expressions, the description of the structure of the fluid is lost and the driving forces for molecular self-assembly are not explicitly present. Between these two scales, there is a myriad of competing techniques that attempt to describe soft matter by considering the discontinuous nature of matter whilst placing aside enough fine details in order to make the modelling computationally tractable. These approaches all fall under the umbrella of coarse-graining (CG) techniques. A full review of CG techniques for fluids and soft matter is far beyond the scope of this paper, and the reader is referred to recent reviews on the topic[1,2] for an introduction. It suffices to say that most common techniques start with an atomistic description of a molecule, grouping elementary atoms and charges into "superatoms". Different methodologies exist to map the properties of these superatoms with the expected properties of the parent atomistic model. Non-exhaustive examples are to angle-average the pairwise potentials,[3] to perform matching to structural properties,[4] or to match the forces between groups of atoms.[5] Uniquely relevant to this discussion is a recent paper by Bates and Walker[6] where dissipative particle dynamics (DPD) is both described and employed to explore the self-assembly of model pure fluids closely related to the ones of interest here.

Our approach to describing complex soft matter follows a route that has historically been successful in describing simple atomistic and molecular fluids. In these approaches the structure of the fluid is described by spherical hard cores at a density that may be matched to the experimental density.[7,8] Attractions between the fluid particles, resulting from dispersion interactions, electrostatics and possible hydrogen bonding are treated as perturbations, which are added sequentially. This overall approach has been the underlying motivation behind the development of modern theories of liquid state, such as SAFT,[9] which are able to link intermolecular potentials with macroscopic properties.

In common with low-level coarse-graining techniques, we represent molecules as made of beads, each accounting for a group of atoms. Consequently, if an accurate description of a particular model were desired, the properties of these beads could be mapped to experimental physical properties (*e.g.* radial distribution functions, densities, *etc.*). This has not been attempted in this work, as we are interested in presenting a proof-of-concept rather than a particular application. We wish to make the statement that the use of a simplistic, coarse-grained model helps to highlight the "crucial" characteristics that allow for the experimentally observed complex mesophase behaviour. By decreasing proportionally the strength of the interactions of all segments of the molecule, we can increase substantially the efficiency of the calculation, without changing the basic structure of the fluid phase, which will depend directly on the interplay between the repulsive cores of the molecules and secondly on the specific nature of the hydrogen bonds and/or long-range electrostatics.

The case model employed here is inspired by a recent series of papers by Tschierske *et al.* where the synthesis and characterisation of polyphilic molecules is detailed (for a recent review see ref. 10). Examples are the T-shaped bolaamphiphiles described therein. These unique T-shaped molecules all share the same broad aspects: a liquid crystal core where the end groups have been substituted, and a side chain of variable length grafted to it. The relative incompatibility between each

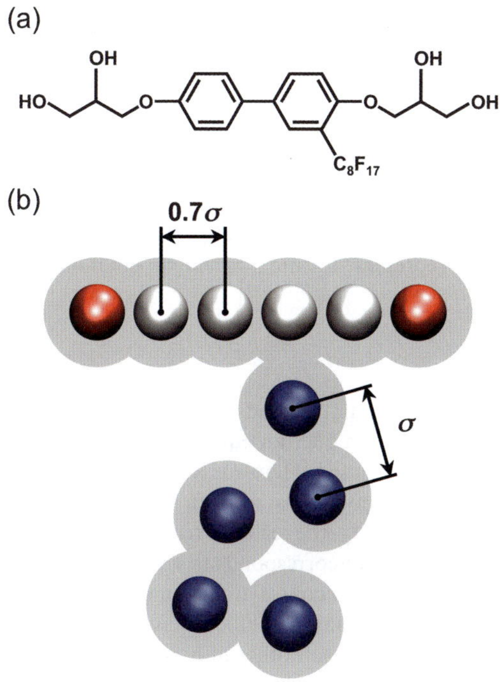

Fig. 1 (a) A typical T-shaped polyphilic molecule that exhibits the Col_h phase is based on a 4,4′-bis(2,3-dihydroxypropoxy-biphenyl) molecule with a grafted perfluorooctane chain (compound 4/8 of ref. 12). (b) Schematic of model used to describe T-shaped bolaamphiphiles.

molecular component cause these compounds to have rich phase behaviour, even as pure substances. Experimental analysis, through polarised light optical microscopy, differential scanning calorimetry and X-ray scattering, have suggested over a dozen unique self-assembled liquid structures that form depending on the length and chemical properties of the grafted chains. Amongst the phases observed there exist columnar phases where the molecules segregate into honey-combed hexagonal cylinders characterized by "walls" composed of rigid liquid crystal cores, with hydrogen bonded vertices and grafted chains filling the voids of the columns. A unique feature of structure is that it is a fluid, self-organized system. An example of such a molecule is the grafted 4,4′-bis(2,3-dihydroxypropoxy-biphenyl) molecule, shown in Fig. 1(a). With an appropriate side chain, this molecule self-assembles into a 2D columnar hexagonal structure (Col_h).[11,12] It is reported from experiments that the formation of these structures is not sensitive to the exact nature of the side chain, and is apparently governed largely by the volume ratio f of the side chain to the rigid liquid crystalline (LC) core. Similarly, the position of the graft and the actual detail of the polar end groups seems also relatively unimportant.[13] It is possible that the geometry of these phases is driven by both micro-segregation of incompatible units and minimisation of free volume and consequently may be non-specific. For that reason, no attempt was made here to fit the model to a unique molecule, but rather to describe the generic phase behaviour.

2 Molecular model

We use here a simple CG model that exemplifies the leitmotiv of our proposal. The model has been used before to map out the phase diagram of T-shaped bolaamphiphiles grafted with side chains of various length,[14] and will be briefly described here for completeness. In this model, six beads are kept in a rigid linear configuration with an inter-bead distance of 0.7σ, where σ is a characteristic length of the model, defined to be roughly the diameter of a bead. The resulting LC core has an aspect ratio of 4.5. A flexible chain of five beads is attached to the third bead of the rigid unit. This allows the stretched-out lateral chain to have roughly the same length as the rigid core, and a ratio of lateral chain volume to rigid core volume f of 0.44, in communion with the overall dimensions of the original target molecules. The basic topology of this model molecule is depicted in Fig. 1(b). Bonded interactions between beads in the flexible chain and in the link between the flexible chain and the rigid unit are represented using a harmonic potential of the form

$$U^{\mathrm{Har}} = \frac{1}{2}k_{\mathrm{sp}}(r - r_{\mathrm{o}})^2 \quad (1)$$

where r is the distance separating the beads, r_{o} is the equilibrium separation and k_{sp} is the spring constant. In the model an equilibrium distance, $r_{\mathrm{o}} = \sigma$, and spring constant, $k_{\mathrm{sp}} = 50(\varepsilon/\sigma^2)$, were used. Here ε is a characteristic energy of the model, defined in terms of the pair potentials between beads, described later. Whilst the side chain has complete flexibility, the reader is reminded that this is a coarse-grained representation, where each bead corresponds to a group of atoms. From this point of view, it is entirely analogous to coarse-grained models typically used for polymer systems.[15] In order to mimic bolaamphiphilic behaviour, the two end spheres of the rigid unit are defined to attract each other in a preferential way. The lateral chain beads are also defined to have self-attraction, although the strength of this is less than the end bead counterparts. Hence the beads in the model may be categorised into three distinct types. Type 1 beads, located at each extreme of the rigid unit, seek to represent strongly interacting hydrogen bond-like sites. Type 2 beads, constituting the remainder of the rigid unit, represent weakly interacting sites typical of the polyphenyl core. Finally, type 3 beads, located in the lateral chain, represent medium-strength interaction sites, typical of perfluoroalkane chains. The actual detail of the form of the non-bonded inter-bead interaction potentials seem at this level irrelevant as different soft-core potentials are expected to lead to the same qualitative mesogenic features in CG model fluids.[16] To this end, for simplicity, all six bead pair potential energies were defined through the Lennard-Jones cut and shifted potential (LJCS)

$$U_{\mathrm{AB}}^{\mathrm{LJCS}}\left(r; C_{\mathrm{ij}}\right) = \begin{cases} U_{\mathrm{AB}}^{\mathrm{LJ}}(r) - U_{\mathrm{AB}}^{\mathrm{LJ}}(C_{\mathrm{AB}}) & \text{for} \quad r < C_{\mathrm{AB}} \\ 0 & \text{for} \quad r \geq C_{\mathrm{AB}} \end{cases} \quad (2)$$

with

$$U_{\mathrm{AB}}^{\mathrm{LJ}}(r) = 4\varepsilon_{\mathrm{AB}}\left[\left(\frac{\sigma_{\mathrm{AB}}}{r}\right)^{12} - \left(\frac{\sigma_{\mathrm{AB}}}{r}\right)^{6}\right] \quad (3)$$

where, A and B define the bead type, r is the bead separation, $\varepsilon_{\mathrm{AB}}$ is an energy parameter defining the potential well depth, σ_{AB} is the length parameter defining the range of the potential, and C_{AB} is the cut and shifted distance. The parameters used for each pair potential are summarised in Table 1. In this parameterisation all beads have the same value σ. The cut-off parameter of $C_{\mathrm{AB}} = \sigma_{\mathrm{AB}}{}^{1/6}$ used for all cross interactions and those between type 2 beads corresponds to a cut and shift at the minimum of the Lennard-Jones potentials, consequently these correspond to purely repulsive potentials (typically known as the Weeks-Chandler-Andersen potential). To model lyotropic systems, novel to this study, solvents are for simplicity modelled

Table 1 Table defining LJCS parameters used for each non-bonded pair potential, parameters listed as (ε_{AB},σ_{AB},C_{AB})

	Type 1	Type 2	Type 3
Type 1	ε, σ, 2σ	ε, σ, $2^{1/6}\sigma$	ε, σ, $2^{1/6}\sigma$
Type 2	ε, σ, $2^{1/6}\sigma$	ε, σ, $2^{1/6}\sigma$	ε, σ, $2^{1/6}\sigma$
Type 3	ε, σ, $2^{1/6}\sigma$	ε, σ, $2^{1/6}\sigma$	$\varepsilon/2$, σ, 2σ

as non-bonded beads of type 2 that would roughly correspond to aromatic solvents (*e.g.* benzene).

3 Metrics

The structured phases exhibited during the simulations were monitored through the computation of order parameters and mobility coefficients from simulation configuration data. These allowed a more quantitative understanding of the different phases to be obtained and resulted in an improved location of phase transition points.

3.1 Columnar director type metrics

A number of metrics to aid understanding system behaviour may be found by as signing a director axis to the phase structure. A natural director associated with the hexagonal columnar phase may be viewed as the vector orthogonal to their layers, as indicated in Fig. 2. For our model, the set of unit vectors, $\{\mathbf{u}_1, \mathbf{u}_2, \ldots, \mathbf{u}_N\}$, associated with the bolaamphiphilic rigid unit axes were used to define the orientation of each molecule in the system. With this arrangement, the director described is specified as the unit vector, $\mathbf{n}$, that is maximally orthogonal to the set of molecular orientation vectors. Concisely this may be written as

$$\min(\|\mathbf{U}\mathbf{n}\|) \text{ subject to } \|\mathbf{n}\| = 1 \tag{4}$$

where $\mathbf{U}$ represents the N by 3 matrix containing the molecule orientation unit vectors, $\mathbf{u}_i$. The solution to this total least square problem is obtained through determining the singular value decomposition factorisation of matrix $\mathbf{U}$, and assigning the right singular vector corresponding to the smallest singular value as the director, $\mathbf{n}$. Using this director it is possible to define a planar and planar orientational order parameter. For columnar systems the director was also used to study variations of bead densities in planes orthogonal to the director. This type of metric was useful in rationalising the behaviour of low solvent fraction mixtures described later, with respect to the pure system. The order parameters and methodology to calculate director orthogonal density profiles are described in the following sections.

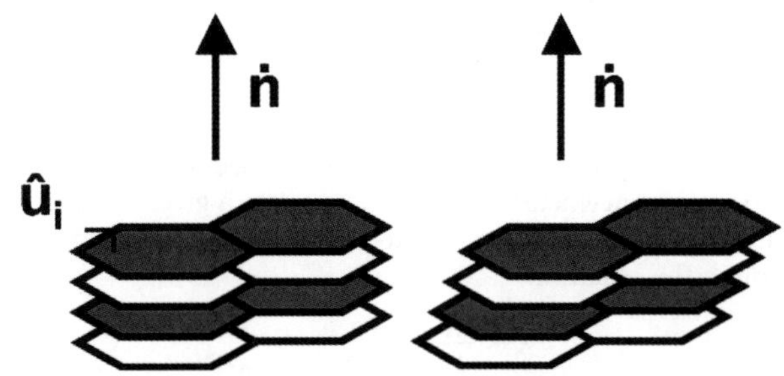

Fig. 2 Definition of the **n** director, orthogonal to the planes of molecules conforming the columns.

3.1.1 The planar order parameter. The planar order parameter, S_2, measures how orthogonal the molecular orientation vectors are to the director vector, and is defined by the equation

$$S_2 = \frac{\sum_{j=1}^{N} 3\sin^2\phi_j - 2}{N} \tag{5}$$

where ϕ_j is the angle between the director, $\mathbf{n}$, and the orientation vector, $\mathbf{u}_j$, of the jth molecular rigid unit. This order parameter is similar to the P_2 order parameter used to study orientation order of rod-like liquid crystalline systems.[17] For an isotropic system, the director and orientation vectors are uncorrelated, and S_2 by definition takes a value of 0. Conversely for an ordered system where all the molecular orientation vectors are orthogonal to the director, S_2 takes a value of 1. The order parameter monotonically increases with increasing system order between these extremes.

3.1.2 The planar orientational order parameter. To monitor the geometry of the columnar phases as seen from an observer looking into the director vector (*i.e.* square/hexagonal columns), the planar orientational order parameter, ψ_k, is defined through the equation

$$\psi_k = \left| \frac{1}{N} \sum_{j=1}^{N} \exp\left(ik\theta_j\right) \right| \tag{6}$$

where θ_j is the angle between the vector given by the projection of the orientation vector, $\mathbf{u}_j$, onto the director orthogonal plane, and a fixed arbitrary axis orthogonal to the director. The coefficient k takes the value 4 and 6 for the square and hexagonal planar orientational order parameters, respectively. As with the planar order parameter, it is defined such that it monotonically increases between 0 and 1 on going from the unordered to fully ordered state.

3.1.3 Director orthogonal density profiles. In the liquid crystalline hexagonal columnar phases seen during simulations of our model, the director axes is observed to be coincident with the axes defining the column direction. Consequently it is possible to use the director in order to calculate average bead density profiles across layers orthogonal to the columns. This is achieved by first projecting the positions of each bead in the simulation box onto an arbitrary plane orthogonal to the director. This plane is then discretised in small bins and the number of particles in each grid square counted. As the number of particles in each gridded square on the plane is dependent on the thickness of the simulation box which is projected through onto the orthogonal plane, these particle counts must be corrected to obtain a full density profile. The densities, ρ^*, reported here are in units beads per σ^3. By looking at the density profiles of each particle type this method allows insight into the structure that is difficult to obtain by simple inspection of simulation movies.

3.2 The mobility coefficient

To quantify the mobility of the bolaamphiphile molecules in the various phases, mean squared bead displacements were calculated. For a given number of time steps, s, each of time $\Delta\tau$, the mean squared bead displacement over time $s\Delta\tau$, $R_{s\Delta\tau}$, was calculated through the equation

$$R_{s\Delta\tau} = \frac{\sum_{j=1}^{N} \sum_{k=1}^{11} \sum_{i=0}^{t_{\text{tot}}-s} \left| \Delta \boldsymbol{r}_{i\Delta\tau,(i+s)\Delta\tau}{}^{j,k} \right|^2}{11N(t_{\text{tot}} - s + 1)} \tag{7}$$

where $\Delta\mathbf{r}_{i\Delta\tau,(i+s)\Delta\tau}^{j,k}$ is the displacement of the kth bead in the jth molecule, between time $i\Delta\tau$ and $(i + s)\Delta\tau$, and t_{tot} is the total number of time steps in the simulation. By determining the rate of change of this displacement with time, a quantity defined as the mobility coefficient was obtained. As the simulations were performed on coarse-grained models the reduced time does not reflect real time, therefore this quantity is related in a non-trivial way to the real values of the diffusion coefficient of the system. In spite of that, the mobility coefficient varied by several orders of magnitude between the isotropic fluid phases and the solid phases, and is thus a definitive measure of the fluidity of the system. Details of the numerical procedure to obtain the rate of change of the mean squared displacement with time are given elsewhere.[14]

4 Results for the pure fluid

Simulations for the pure T-shaped molecule were performed at a range of temperatures to determine regions of crystalline, liquid crystalline and isotropic phase behaviour. Full details of these simulations may be found elsewhere,[14] however some relevant results are summarised here for completeness. Simulations were performed in continuum space *via* molecular dynamics using the simulation suite DL_POLY.[18]

The isotension-isothermal ensemble, with the number of molecules, N, the temperature, T, and the stress tensor, σ, held fixed was used for most simulations.[19] This is a deviation from the use of the more common isobaric-isothermal (NPT) ensemble. Some of the simulations reported herein encounter non-homogeneities in the fluid phases, *i.e.* microphase separation, interfaces and or anisotropies associated with the liquid crystal-like ordering. During NPT simulations of simple molecular species the invariant average of the stress tensor diagonal (mean stress) is maintained close to the desired bulk pressure by affine transformation simulation box volume changes. Resulting from molecular mobility the principle stresses may relax to being roughly equivalent during the simulation *i.e.* pressure is equal in all directions. For NPT simulations of systems that are anisotropic in equilibrium, such as our own, it may not be possible for the system to organise in the simulation box such that the principle stresses are equivalent; *e.g.* an interface would contribute to an additional stress in the plane of the interface. Consequently whilst the mean stress is maintained at the desired bulk pressure, the stress deviator tensor will contain non-zero entries. This results in a direction dependent pressure, causing a simulation box deformation force to exist, and maintaining a non-equilibrium state. By performing a $N\sigma T$ simulation each of the diagonal stress tensor components are maintained close to the desired pressure, and the non-diagonals are maintained close to zero by both size (affine transformation) and shape (shear transformation) of the simulation box. In practice, this also allows for the simulation box to adapt to the periodicity of some of the self-assembled structures. We point out that our simulation boxes are rather large compared to the observed periodicity, so it is unlikely that the box artificially imposes the formation of these structures.

The temperature, $T^* = Tk_b/\varepsilon$, where k_b is Boltzmann's constant, and pressure, $P^* = P\sigma^3/\varepsilon$, are expressed throughout this work in the standard reduced form that reflect the energy and length scales of the model. Time is reported here in actual time, which is related to reduced time by the formula, $t^* = t(m\sigma^2/\varepsilon)^{-1/2}$, where m is defined as the mass associated with a bead. In our simulations $(m\sigma^2/\varepsilon)^{-1/2}$ took a value of around 800 fs^{-1}.

4.1 Order parameter and mobility coefficient analysis

The results for the pure T-shaped molecule are summarised in Fig. 3 and show how the order parameters and mobility coefficient change with temperature. On reviewing this diagram passing from high to low temperature, the transition from isotropic to a hexagonal columnar phase is clearly delineated by a jump from near zero values of S_2 and ψ_6 to values of around 0.72 and 0.36, respectively, at $T^* = 0.87$. On

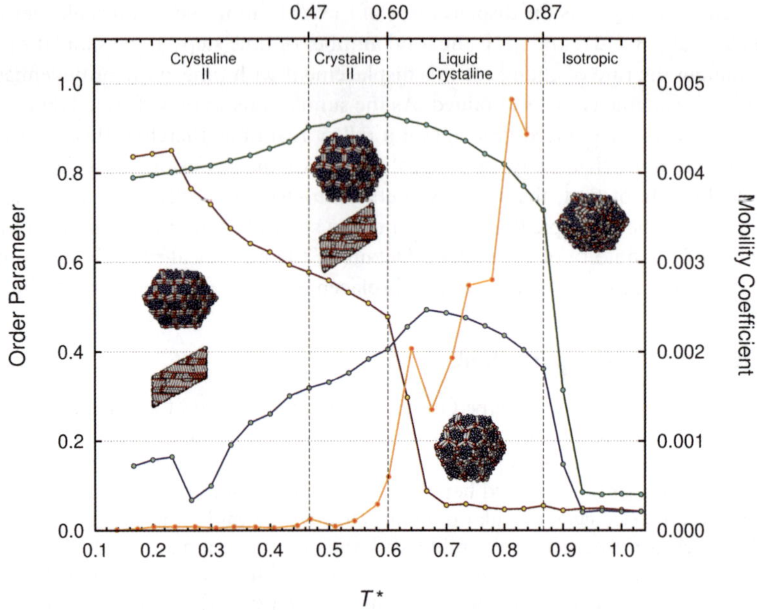

Fig. 3 Order parameters (left ordinate) and mobility coefficient (right ordinate) for the pure T-shaped bolaamphiphile as a function of temperature. Here the green line is the planar order parameter, S_2, the maroon line is the square columnar order parameter, ψ_4, the blue line is the hexagonal columnar order parameter, ψ_6. The yellow line is the mobility coefficient.

decreasing the temperature further, both S_2 and ψ_6, slowly increase until $T^* = 0.60$, where there is a steep increase in the ψ_4. Below this temperature the ψ_4 parameter slowly increases, whilst S_2 and ψ_6 slowly decrease. The non-zero values of mobility coefficient in the range $T^* = 0.60$–0.87, in addition to the previously reportedself-assembly from isotropic to hexagonally structured phase in an MD simulation performed at $T^* = 0.73$, are clearly suggestive that the hexagonally columnar phase in this temperature range is liquid crystalline in nature. These latter conditions were used as starting points for considering the effects of solvent (next section).

4.2 Director orthogonal density profiles

The director orthogonal density profiles for all bead types combined (global local densities) and densities for the individual bead types are displayed in Fig. 4. Here, the hexagonal geometry of the pure fluid phase is clearly visible. Reviewing the all bead density profile one may observe that the highest density regions (red) are associated with the position of the rigid rod type 2 beads at the columnar walls. Similarly, the lowest density regions (green) correspond to the interior part of the columns near the vertices. There is a steric hindrance for the chain molecules in this region to access the corners of the hexagonal columns. Further information may be obtained by reviewing the individual bead density profiles. The segregation of the different parts of the molecule is clearly seen. The density of type 3 (lateral side chain) beads is lower in the central regions of the columns. It suggests a capacity for the structure to hold solvent with similar physical properties in this region or longer chains.

5 Results for mixture simulations

To investigate bolaamphiphile and solvent mixtures, MD simulations were performed at two system sizes; a series of large simulations to investigate low solvent

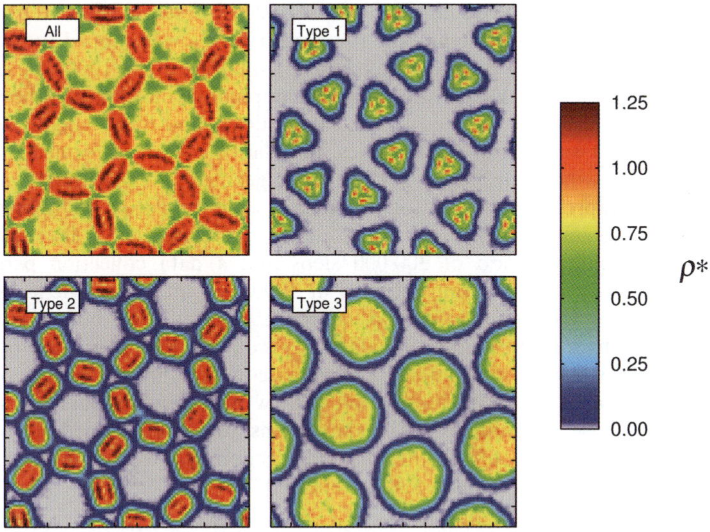

Fig. 4 Director orthogonal density, ρ^*, profiles for all particles types (global density), and for each of the particle types, taken from an MD simulation of a pure bolaamphiphile system that forms a hexagonal columnar phase.

fractions and a series of smaller simulations to scan the full solvent fraction range. The purpose of the low solvent fraction simulations was to find out the effect of the solvent on the stability of the hexagonal columnar geometry, determine localisation of solvents in the structure, and determine if and at which composition the Col_h phase breaks down. The director orthogonal density profiles were key to this study. The second series of simulations were performed to understand the global phase behaviour across the full range of compositions.

All simulations were performed with molecular dynamics in the $N\sigma T$ ensemble at a temperature $T^* = 0.67$ and pressure $P^* = 0.16$, corresponding to a state point in the liquid crystalline range for the pure bolaamphiphile system. For both simulations series, an identity exchange procedure was used. Taking output configuration data of a previous MD run we converted the identity of a random fraction of bolaamphiphiles molecules into solvent particles (bead by bead) and used this as a non-equilibrated input configuration. This procedure allowed a systematic route to change solvent fraction whilst retaining total particle numbers, effectively modelling a sequential addition process. With this method, a two stage equilibration was required. In the first stage the maximum force exerted on all particles was capped. This limitation on particle accelerations, allowed slow reorganisation of solvent and relaxation from the energetically unfavourable initial configurations created by the conversion of the bolaamphiphilic rod into discrete solvent molecules. The second equilibration stage, without force capping, allowed the system to relax into a stable phase. As the number of bolaamphiphiles converted between simulation steps was relatively small, relaxation into a stable phase was rapid, thus allowing comparatively small equilibration times. As this approach conserves the total number of beads in the system, compositions are reported as fraction υ of the total number of beads forming part of bolaamphiphile molecules with respect to the total number of beads. Volume fraction was not used as a measure of composition due to the ill-defined nature of the volume of softly interacting particles, however υ would be proportional to the volume fraction of bolaamphiphiles.

For the large scale simulations, an initial system of 4096 bolaamphiphile molecules (45 056 beads) equilibrated in the hexagonal columnar liquid phase was obtained by MD simulation. The mixture simulations were performed with a timestep of 10 fs, over an equilibration period of 200 000 steps, followed by a data collection period of 100 000 steps. The force on particles was capped for the first 1000 steps. Configuration data was sampled for analysis every 1000 steps. After each mixture simulation, the identity of 50 bolaamphiphile molecules in the output configuration data were converted to solvent particles, and this configuration file used for the next simulation in the series. Simulations were performed up until solvent fractions where the hexagonal columnar geometry of the phase was lost. An analogous approach was used for the smaller scale simulations, however these simulations were initiated from an equilibrated pure system of 2000 molecules, had an equilibration period of 300 000 steps and a data collection period of 500 000 steps, with configuration sampled at 5000 step intervals. The same time step was used for the simulations, and again the identity of 50 bolaamphiphiles was switched between simulations. Simulations were performed from pure bolaamphiphile to pure solvent compositions. The increase in equilibration time steps were to allow sufficient time for the formation of new phases when passing through composition phase boundaries. All simulations were performed using the DL_POLY simulation suite.[18] In the following section the results for the mixture simulations of the three solvent types are reported.

5.1 Sequential addition of solvent (identity change)

Simulation snapshots taken from the type 2 solvent/bolaamphiphile mixture composition scan series are presented in Fig. 5. At the lowest solvent concentrations simulated, $\upsilon = 0.975$, the hexagonal columnar structure remained intact, with solvent localising in the corners of the hexagonal columns. The director orthogonal density profile of the pure hexagonal columnar bolaamphiphile phase, show lower bead density in this region, consequently this solvent positioning is effective in space filling, without disrupting bonding between the flexible side chain beads or end rod beads. Inspection of the director orthogonal solvent density profile demonstrates solvent localisation more clearly. Profiles are shown for a few low concentration solvent mixtures in Fig. 6. With increasing solvent fraction, it eventually becomes impossible for the hexagonal columnar structure to accommodate all the solvent. At a bolaamphiphile concentration of approximately $\upsilon = 0.950$, two phases form,

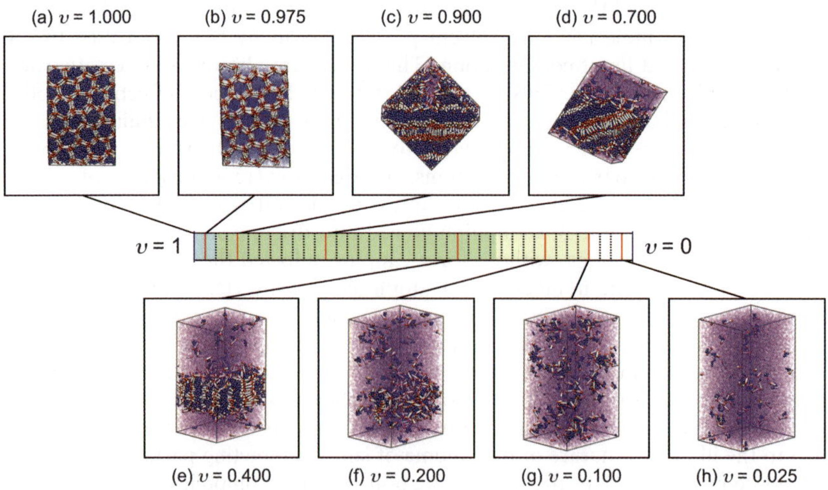

Fig. 5 Snapshots taken from MD simulations at various concentrations of bolaamphiphile, υ, in a solvent (purple). Different colours in the concentration graphic delineate the observed stages of mesophase segregation described in the main text.

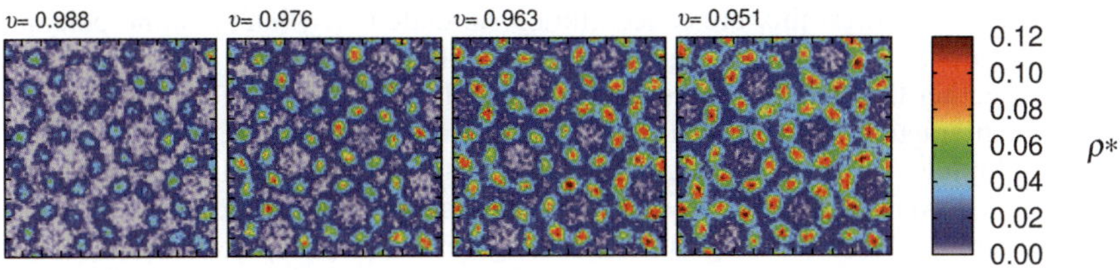

Fig. 6 Director orthogonal density, ρ^*, profiles of solvent beads in low solvent concentration mixtures.

one a solvent rich liquid phase, and the other a bolaamphiphile rich phase which retains the hexagonal order. Fig. 5(c) shows an example of this phase. In this snapshot it may be seen that solvent rich phase is positioned such that the interface between phases is at the ends of the hexagonal columns. With this arrangement there is minimal disruption to the hexagonal columns, and far less breaking of type 1 bead non-bonding interactions, than would occur if the solvent were positioned between hexagonal columns. This arrangement minimizes the free energy penalty of the interface. On decreasing bolaamphiphile concentration further a slab of the hexagonal columnar bolaamphiphile structure eventually forms across the simulation cell. Fig. 5(d) and (e) show snapshots of simulations at bolaamphiphile concentrations $\upsilon = 0.700$ and $\upsilon = 0.400$. Similar as in higher concentrations, the column ends of the bolaamphiphile rich phase are exposed to the interface. In the snapshots of the slab arrangement it may be seen that there are bolaamphiphiles in the solvent phase, *i.e.* the phase segregation is not absolute. Furthermore, from inspection of the simulation movies (not included) it may be seen that bolaamphiphiles are dynamically interchanging between phases *e.g.* it is seen that bolaamphiphiles positioned in columnar phase centrally between the interfaces at one time are later found in the solvent. This is a consequence of the fluidity of hexagonal columnar phase and its liquid crystalline nature. At a bolaamphiphile composition fraction of between $\upsilon = 0.325$ and $\upsilon = 0.300$ the hexagonal columnar structure in the bolaamphiphile rich phase was found to break down. Fig. 5(f) shows a snapshot from the $\upsilon = 0.200$ simulation. At this state point there is no discernable structure in bolaamphiphile rich region, although there is general arrangement of the individual molecules such that their lateral flexible side chains type 3 beads and rigid rod end type 1 beads are close in space. It is hard to classify the bolaamphiphile rich region as a separate phase, and labelling it as an amorphous bolaamphiphile micelle-like phase may be more appropriate. With decreasing bolaamphiphile concentration the bolaamphiphile micelle-like structure eventually disappears at around $\upsilon = 0.100$, and below this concentration the phase is essentially homogeneous. It must be noted, however,

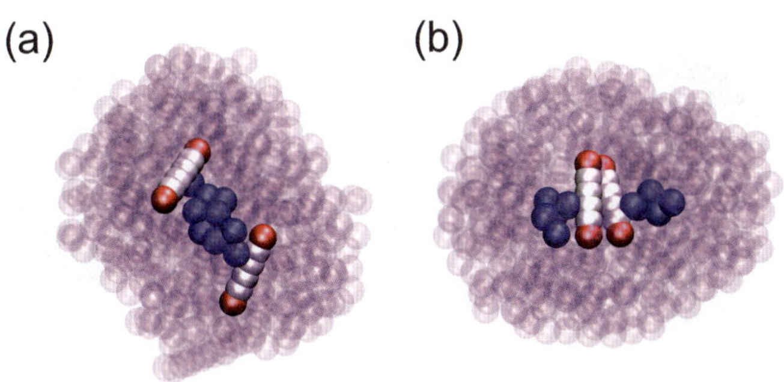

Fig. 7 Examples of bolaamphiphiles dimers found at low bolaamphiphile concentrations (a) a frequently encountered variety of dimer bonded through the lateral side chain and (b) a less common dimer with rods aligned such that type 1 groups bond.

that even at the lowest mixture concentration simulated, $\upsilon = 0.025$, bound bolaamphiphile pairs are found. From inspection these pairs appear more commonly to be bound together through a lateral side chain. The flexibility of the side chain and the multiple interaction sites, mean that there exists many arrangements of two close bolaamphiphiles to bond through their side chains, consequently their presence apparently lowers the total free energy of the system. This is in contrast to pairs of bolaamphiphiles interacting through the rigid rod end groups, which are not so common. Simulation snapshots of these configurations are shown in Fig. 7.

5.2 Phase separation of bolaamphiphile and type 2 solvent mixture

To establish whether the phases described by the sequential addition methodology are real equilibrium phases or metastable states influenced by the persistence of phases used as starting points, we describe here a simulation initiated from an isotropic solvent-bolaamphiphile mixture. The system was chosen to contain 2746 bolaamphiphiles and 14 850 solvents beads ($\upsilon = 0.67$), at $T^* = 0.67$ and $P^* = 0.16$. From the results of the previous section this corresponds to a microphase segregated system consisting of a solvent rich liquid region, and a hexagonal columnar bolaamphiphile rich phase. For this system, approximately 4.5 M simulation steps of 10 fs were performed. No force capping was used in these simulations, but the first 10% of the simulation was run with a time step of 1 fs. Data was sampled at 5000 step intervals.

To monitor phase separation of solvent bolaamphiphile mixture simulations the distribution of solvent densities found in different regions of the box was studied as a function of time. To determine the distribution at a given time the simulation box was divided into congruent volumes of similar shape to the simulation box, the number of solvent particles found in each counted, and a histogram of densities produced.

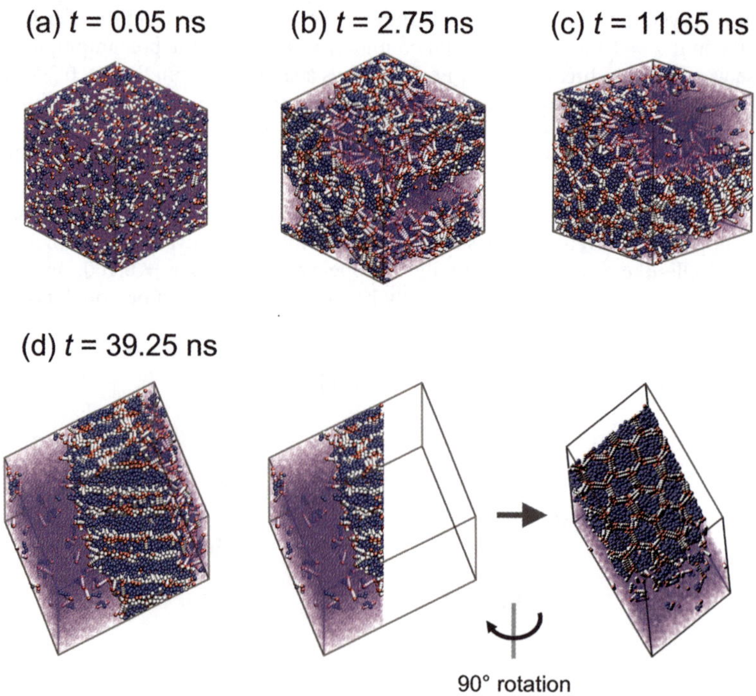

Fig. 8 Snapshots taken from the MD simulation initiated from a homogeneous isotropic solvent–bolaamphiphile mixture with $\upsilon = 0.67$.

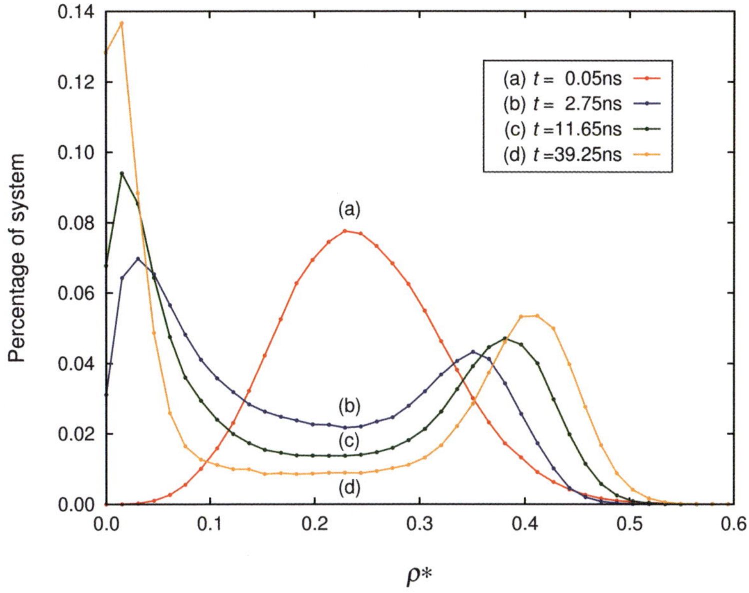

Fig. 9 Solvent density distributions taken from the MD simulation initiated from an isotropic-homogeneous solvent–bolaamphiphile mixture with $\upsilon = 0.67$. Points (a–d) correspond to the states depicted in Fig. 8.

The results of the simulations are summarised in Fig. 8 and Fig. 9 with snapshots of the simulation and the solvent density distribution at different stages given, respectively. Also, Fig. 10 shows the variation of S_2 and ψ_6 with time. At a time $t = 0.05$ ns (point (a) in Fig. 8–10) the snapshot of the simulations, the near normal

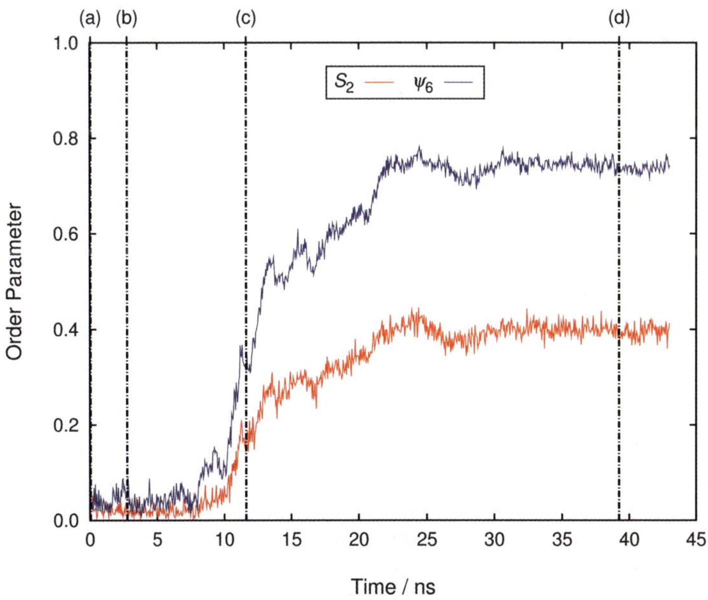

Fig. 10 Time evolution of planar order parameter, S_2, and planar orientational order parameter, ψ_6, during the MD simulation of solvent–bolaamphiphile mixture, with $\upsilon = 0.67$, initiated from a homogeneous-isotropic state. Points (a–d) correspond to the states depicted in Fig. 8.

distribution of solvent density and the low ψ_6 value demonstrate that the system was initiated from a fully homogeneous and isotropic state. The process of phase separation into type 2 bead rich and bolaamphiphile rich homogenous and isotropic phases is relatively rapid, demonstrated by two distinct peaks in the type 2 solvent by time $t = 2.75$ ns (point (b) in Fig. 8–10), whilst the ψ_6 and S_2 order parameters remain close to zero. After phase separation, the measure of global hexagonal columnar order, ψ_6, slowly increases. This results from both increased hexagonal columnar ordering of the bolaamphiphiles locally and the emergence of a dominant ordered domain from the separate ordered domains initially formed. As the director is defined globally it is impossible to decipher their contributions to ψ_6, though it is clear from simulation they both have a role in increasing the order parameter. Inspection of the simulation snapshot in Fig. 8(c) at $t = 11.65$ ns show multiple domains of the hexagonal columnar phase. Eventually during the simulation one domain finally forms and the stable phase equilibria is reach. This is clearly delineated by the metrics, with a constant ψ_6 value and type 2 solvent density distribution. An example of the stable phase formed is given for the simulation time $t = 39.25$ ns in Fig. 8(d). In the diagram removal of one half of the box particles clearly reveals the hexagonal order of the bolaamphiphile phase. Overall, the diagrams clearly show how the system first phase separates and then self-assembles into the final equilibrium phases. The final equilibrium state reached is equivalent to that found using the sequential addition process, suggesting that it is a stable equilibrium morphology.

6 Conclusions

We have presented here a minimalist CG approach, based on a discrete description both of the macromolecular and the solvents. The basic geometric traits are taken from the original LC molecule, such as aspect ratios and volume fractions of the different sections of the molecule. The macromolecule is constructed by assembling spherical segments, although this issue is simply a matter of computational convenience and would not seem to be relevant. The energetic contributions are taken to be simple, either a soft repulsion between incompatible sections, isotropic short range attractions between hydrogen bonding sections or a weak attraction for the lateral chains, mimicking the weak dispersion forces of perfluorinated chains. Similarly, since the solvent is treated explicitly, each spherical solvent molecule representing either a moderate sized molecule or a grouping of a few molecules. Again, nothing in the model restricts the size of the solvent spheres, and they are used here as equal sizes for simplicity.

The lyotropic behaviour of polyphilic liquid crystals has received very little attention in the literature, most likely due to the very complex phase diagrams that may arise (see for example ref. 20). There is, however, a unique report by Qi *et al.* on the effect of adding surface-coated gold nanoparticles to columnar assemblies of T-shaped bolaamphiphiles similar to those modelled here.[21] The nanoparticles had been grafted with hydrophobic (hexane thiolate) monolayers, resembling the solvent used herein. Qi *et al.* reported that the phase stability of the columnar phase critically depends on the concentration of the nanoparticles. The hydrophobic nanoparticles, concentrated on the insides of the columnar phases, had a destabilizing effect on the structure, leading to the disruption of the phase. Furthermore, under certain conditions, there is evidence of the existence of different mesoscopic domains within the sample. The results are in complete agreement with the predictions made herein and confirm the validity of our molecular model and the CG procedure.

We have shown that appropriate coarse-graining has a profound effect on the accessible time scales. Taking order of magnitude estimates from the MARTINI force field,[22] (which has many outstanding common features with this model), the time scale up should be between 3–8 fold, while using estimates from the comparison of diffusivities of similar CG models[23] we calculate a 10–20 fold increase between the real time and the simulated time in the case of the bolaamphiphiles. From this we estimate that an order of magnitude of speedup is gained *i.e.* a ps of simulated time in our model would correspond to around 10 ps of real time. More noticeably, the reduction both in the number of sites and in the complexity of the interactions result in a decrease of several orders of magnitude in the computational effort, of these simulations, allowing us to explore longer time frames. Some of the results shown here cover up to 50 ns (of simulated time), which would loosely correspond to up to 0.5 μs of real time, which explains the relative ease in observing the self-assembly and reassembly of mesophases described herein. It also sheds light on the futility of attempting to observe self-assembly of complex systems with atomistic modelling. As seen from Fig. 10, the self-assembly process has two distinct steps, in the 1st level nanoscale segregation the system separates into two distinct mesophases, however, there is a second, much slower process that corresponds to the relaxation towards a more complex structure. These two processes are separated by a time lag that spans an order of magnitude and may easily lead to confusing the first step segregation with a stable equilibrium state. CG, as described herein, can describe these large time scales without loss of detail and allows the study of the overall dynamic and equilibrium process.

Acknowledgements

Partial financial support for this work has been given by the U.K. Engineering and Physical Sciences Research Council (EPSRC), grant EP/E016340, "Molecular Systems Engineering". The input of Francisco Martínez-Veracoechea is gratefully acknowledged. Some figures were drawn with VMD.[24]

References

1 S. O. Nielsen, C. F. Lopez, G. Srinivas and M. L. Klein, *J. Phys.: Condens. Matter*, 2004, **16**, R481–R512.
2 M. McCullagh, T. Prytkova, S. Tonzani, N. D. Winter and G. C. Schatz, *J. Phys. Chem. B*, 2008, **112**, 10388–10398.
3 E. A. Müller and L. D. Gelb, *Ind. Eng. Chem. Res.*, 2003, **42**, 4123–4131.
4 W. Shinoda, R. Devane and M. L. Klein, *Mol. Simul.*, 2007, **33**, 27–36.
5 S. Izvekov and G. A. Voth, *J. Phys. Chem*, 2005, **109**, 2469; S. Izvekov and G. A. Voth, *J. Chem. Phys.*, 2005, **123**, 134105.
6 M. Bates and M. Walker, *Soft Matter*, 2009, **5**, 346–353.
7 J. P. Hansen and I. R. McDonald, *Theory of Simple Liquids*, Academic Press, London, 3rd edn, 2006.
8 C. G. Gray and K. E. Gubbins, *Theory of Molecular Fluids*, Oxford University Press, Oxford, 1984.
9 W. G. Chapman, K. E. Gubbins, G. Jackson and M. Radosz, *Fluid Phase Equilib.*, 1989, **52**, 31–38; W. G. Chapman, K. E. Gubbins, G. Jackson and M. Radosz, *Ind. Eng. Chem. Res.*, 1990, **29**, 1709–1721. For a review see: E. A. Müller and K. E. Gubbins, *Ind. Eng. Chem. Res.*, 2001, **40**, 2193–2211.
10 C. Tschierske, *Chem. Soc. Rev.*, 2007, **36**, 1930–1970.
11 M. Kölbel, T. Beyersdorff, X. H. Cheng, C. Tschierske, J. Kain and S. Diele, *J. Am. Chem. Soc.*, 2001, **123**, 6809–6818.
12 X. H. Cheng, M. Prehm, M. K. Das, J. Kain, U. Baumeister, S. Diele, D. Leine, A. Blume and C. Tschierske, *J. Am. Chem. Soc.*, 2003, **125**, 10977–10996.
13 X. H. Cheng, M. K. Das, U. Baumeister, S. Diele and C. Tschierske, *J. Am. Chem. Soc.*, 2004, **126**, 12930–12940.
14 A. J. Crane, F. J. Martínez-Veracoechea, F. A. Escobedo and E. A. Müller, *Soft Matter*, 2008, **4**, 1820.
15 G. S. Grest, M. D. Lacasse, K. Kremer and A. M. Gupta, *J. Chem. Phys.*, 1996, **105**, 10583.
16 Z. E. Hughes, L. M. Stimson, H. Slim, J. S. Lintuvuori, J. M. Ilnytskyi and M. R. Wilson, *Comput. Phys. Commun.*, 2008, **178**, 724–731.
17 E. de Miguel, E. M. del Rio and F. J. Blas, *J. Chem. Phys.*, 2004, **121**, 11183–11194.
18 www.ccp5.ac.uk/DL_POLY/. For a recent review see: W. Smith, *Mol. Simul.*, 2006, **32**, 933–1121.

19 M. Parrinello and A. Rahmen, *Phys. Rev. Lett.*, 1980, **45**, 1196–1199.
20 C. Sauer, S. Diele, N. Lindner and C. Tschierske, *Liq. Cryst.*, 1998, **25**, 109–116.
21 H. Qi, A. Lepp, P. A. Heiney and T. Hegmann, *J. Mater. Chem.*, 2007, **17**, 2139–2144.
22 S. J. Marrink, H. J. Risselada, S. Yefimov, D. P. Tieleman and A. H. de Vries, *J. Phys. Chem.*, 2007, **111**, 7812–7824.
23 Based on considering a molecular weight of 721 g/mol and using Fig. 5 of ref. 4. If one considers a typical solvent, the molecular weight would be around 100 g/mol and the speed up closer to 3.
24 W. Humphrey, A. Dalke and K. Schulten, *J. Mol. Graphics*, 1996, **14**, 33–38.

General discussion

Professor Marrink opened the discussion of the paper by Professor Bolhuis: In your SMD simulations, you pull part of the peptide chain away from the β-roll. In the atomistic simulation, which is used to calibrate the CG model, the chain that is pulled away can adopt its secondary structure along the way, quite realistically. For the CG simulation, however, the secondary structure is, for a large part, built in to the model (using optimized dihedral angles). The question is therefore: how meaningful are the SMD simulations to optimize non-bonded interactions, as these will also contain effects arising from the biased bonded interactions in the case of the CG model?

Professor Bolhuis responded: The pulling is done on the glutamate residue perpendicular to the strand. The strand is thus kept under tension during the SMD pulling simulation. Both in the atomic and in the CG representation the strand configuration seems indeed quite robust. Although in both simulations the dihedrals change temporarily during the pulling (due to the direction of the pulling vector), they immediately resume a strand configuration when the four amino acids contributing to the dihedral angle have been pulled off completely. This suggests that we can compare the two simulation outcomes directly. Of course, as the effect of water, as well as of the hydrogen bonds, is condensed into a single non-bonded interaction, we cannot expect to cover all subtleties in the folding. However, our reasoning is that when the PMFs are the same for the atomistic and CG models, in principle, the equilibrium constant between the folded and unfolded should be the same, which is what we aim for in the first place. Here we should note also that the free energy landscape is only probably locally matched, and that we need to test the overall folding equilibrium in the future.

Professor Berendsen queried: In your pulling experiments you compare direct averaging of the work with experimental averaging and 2nd order cumulant expansion as if these quantities are comparable. They are not: the average work including frictional dissipation is unequal to an approximation of the free energy. What I would have liked to see is the dependence of the exponentially averaged work on the pulling rate. Do you have data on that?

Professor Bolhuis responded: We have performed the atomistic and coarse-grained pulling simulation at a single pulling speed of 5 Å per ns. However, it was not our intention to interpret the direct (unweighted) average of the work as a realistic, meaningful quantity. The different averages are computed to make an estimate of the error in the exponential average possible, as is explained in the paper by Park and Schulten.[1]

1 S. Park and K. Schulten, *J. Chem. Phys.*, 2003, **120**, 5946–5961.

Professor Zannoni asked: What is the treatment of charges and partial charges? If partial charges are neglected do you not expect this to be relevant?

Professor Bolhuis answered: All amino residues in the silk block of the protein are neutral at low pH. There are hence no net charges to treat. The partial charges of both protein and solvent are important however, and are treated explicitly in the atomistic force field (here, OPLSAA).

In the coarse-graining procedure all the partial charges, as well as all the collective effects of the water molecules, are combined in the short ranged effective interaction between the beads representing the entire residues. Hydrophobic beads thus attract each other; hydrophilic ones usually are not attractive to each other. The idea of coarse-graining (partial) charges to short ranged effective potentials is very clear in the screened Coulomb potential where the effective interaction comes from a bare Coulombic potential that is influenced dramatically by the dielectric medium (solvent) and the counter ion double layer.

Professor Berendsen opened the discussion of the paper by Professor De Pablo: 1. You start with a self-consistent mean-field approach for the non-bonded free energy; then you solve this by a particle model, where the effective pair interaction is derived from the mean-field approach. Is this not an unnecessary detour? Is the particle model you end up with not simply a superatomic coarse-graining with a mean-field Flory–Huggins type of effective interaction?

2. You use a square cloud function lacking spherical symmetry in 3D. Would it be advantageous to use a spherically-symmetric Gaussian cloud function?

Professor De Pablo answered: The main advantage of the proposed methodology is precisely that one starts from a well defined coarse-grain Hamiltonian, which in some limits can be solved using a self-consistent theoretical approach. This is not a detour but, in our view, a highly desirable feature. The non-bonded interactions of the coarse-grain Hamiltonian are pair-wise interactions between particles. In the grid-based scheme they are computed *via* assigning the local density onto a collocation grid. In the grid-less method they are computed *via* a neighbor list. Only if one treats the model within the mean-field approximation, will the interactions be identical to the Flory–Huggins expression. The method is simple, effective, and efficient, as it scales with the number of particles n. In cases where a constant pressure or constant stress ensemble is of interest (*e.g.* as in the deformation of a polymeric material), we show how the grid-based Hamiltonian can be mapped into a particle model with an effective two-body potential in a continuum, thereby removing any possible limitations introduced by an underlying lattice. In that case one has a grid-less, particle model with "soft" pair-wise interactions not unlike those used in dissipative particle dynamics (DPD) simulations. One can of course switch seamlessly between the grid-based and the grid-less model, resort to volume or shape changes only when strictly necessary, and use a very fast grid-based technique for most of a simulation. To put the computational demands into perspective, in our codes the proposed grid-based approach is roughly 20 to 30 times faster than the grid-less approach. The efficiency of the grid-less approach itself should be comparable to that of DPD, although of course for chain molecules the proposed approach is likely to be highly advantageous for equilibration because it enables implementation of drastic moves (*e.g.* insertion of entire chains into the system) that are not allowed in a strictly dynamic technique. These efficiency figures will of course depend on the degree of discretization, the range of the interactions, density, and the underlying physics that one is trying to address. The advantages or usefulness of the proposed methods are clearly displayed in the calculations presented in our manuscript in the context of pattern interpolation with block copolymers. The magnitude of the calculations considered in that case (over 400 distinct parameters combinations were considered!) are beyond what can be accomplished with traditional particle-based methods. Going one step further, because we start from a well defined coarse-grain Hamiltonian, it becomes possible to apply the ideas outlined in our system to investigate more complex systems, in which the internal structure of the material necessitates that one uses large systems and efficient sampling techniques. A case in point is discussed in the answer to Mark Wilson's question, where we provide an illustration of how our ideas can be used to simulate a cross-linked, liquid crystalline elastomer under deformation. It would be difficult to study such a system with a traditional two-body particle based model.

Dr Milano asked: Regarding equations 4, 5 and 6 in your paper what are the effects of the parameters involved in the gridless implementation on the radial distribution functions also in comparison with the grid implementation?

Professor De Pablo replied: In both the grid-based and the grid-less models, the non-bonded part of the Hamiltonian is equivalent to a sum of pair-wise interactions between the beads and, like in any other model of a fluid, these interactions dictate the local structure of the fluid. The notable difference between the pair-wise interactions of our model and, *e.g.*, a Lennard–Jones fluid, is the softness of the interactions and the density of beads: since one "bead" represents dozens of atomistic repeating units, they strongly overlap and their interaction is very soft. This characteristic behavior is described by small values of K_0. The choice of parameters results in very weak packing effects, *i.e.* the fluid of soft, coarse-grain beads is rather structure-less. For the parameters used in the simulation, this property is rather insensitive to the details of the pair-wise potential, and it allows us to employ a simple, mean-field-like mapping between the parameters of the grid-based model, the grid-less model and the field-theoretic description

Dr Hess said: In Fig. 7 the exponentially weighted curves are higher than the unweighted average, this is impossible. Have you made a simple sign error? If there is only a sign error, the irreversible work is about 10 kJ/mol. In the Gaussian approximation this means the standard deviation of the work values is about 7 kJ/mol. I and others have determined errors for Gaussian distributions.[1] For $\sigma = 3$ kT and 10 work values the standard error estimate is 2.3 kT. (In my question at the meeting I, incorrectly, estimated σ and therefore also the error to be much higher.) An error of 10% might be acceptable for your purposes (if σ is really 3 kT).

1 B. Hess C. Peter, T. Ozal and N. van der Vegt , *Macromolecules*, 2008, **41**, 2283.

Professor Bolhuis responded: It turned out that we made a mistake in the computation of the atomistic PMFs that made the exponential and 2nd cumulant to look very similar. We have amended this mistake and have now calculated new atomistic PMFs which have replaced the original Fig. 7 in the paper. In the new PMFs the unweighted average is now always higher than the exponentially weighted average and 2nd cumulant curves, as you correctly point out. At this moment we cannot estimate the exact error of the calculation.

Professor Tieleman said: In your paper you talk about extending your force field, but the one you are using at the moment is based on a three-letter alphabet, one of the simplest force fields available in the literature. You are already using two interaction levels for glycine and alanine, two chemically very similar side chains. Could you comment on how you would go about extending the amount of detail in your force field?

Professor Bolhuis responded: First a correction: in our final model as presented here we use the same interaction type for alanine and glycine and a second type for glutamic acid. As the force field with its three-letter alphabet is indeed very simple, it would allow for a lot of straightforward extensions by adding new bead types with intermediate repulsion/attraction. Another way of extending the force field would be to look at the excluded volumes of the different amino acids and including them in the non-bonded interactions.

Dr Wilson asked: Could you comment on the relative speed of the model used in comparison to (say) atomistic and more traditional coarse-grained models? Also, could you comment on whether it might be possible to extend the model to contain mesogenic groups (*e.g.* rods) and indicate how this might be done?

Professor De Pablo answered: As alluded to earlier, depending on the details of the system one could anticipate that the relative speed of the grid-based approach *vis a vis* a DPD simulation is between one and 2 orders of magnitude. The grid-less implementation should be comparable to a DPD algorithm. Note, however, that if we take into account the potential benefits of a sampling methodology over a purely dynamic technique, the advantages could be greater.

Professor Kremer addressed Professor De Pablo and Professor M. Müller: The reason to use a chain based MC approach, where the interactions are taken into account through a background field is supposedly the high efficiency compared to a standard MC algorithm. To quantify this, I would like to know the speed up compared to a standard lattice algorithm with occupation numbers used to account for the interactions. To avoid entanglement effects one can easily allow chain crossing in such an algorithm. Also it would be important to get quantitative information about the density fluctuations and how this varies with the phase (*e.g.* ordered, disordered).

Professor M. Müller and Professor De Pablo replied: The advantage of calculating the interactions *via* a collocation lattice in the soft coarse-grained model consists of avoiding a loop over the interaction partners of a segment *via* a neighbor list. The specific speed-up depends on the parameters of the model. Typically, in soft, coarse-grained models, which represent a melt of large invariant number of polymerization, one segment interacts with many 10–100 neighbors and, thus, the speed-up of the computation of the energy is also on the order 10^1–10^2.

If one studies the soft, coarse-grained model by Single Chain in Mean Field (SCMF) simulations rather than Monte Carlo simulations, the use of rapidly updated "background fields" will decouple the molecules during the Monte Carlo part of the SCMF-simulation cycle and will allow for a straightforward and very efficient parallelization of the algorithm.

In dense multi-component polymer systems, composition fluctuations often are orders of magnitude stronger than density fluctuations. This justifies the idealization of incompressibility, which is invoked in many analytical calculations. In our soft, coarse-grained model, the strength of density fluctuations is dictated by the invariant parameter, κN. In the model the compressibility is larger than in an experiment (for a quantitative estimate, see Pike *et al*,[1]) but still sufficiently small to suppress density fluctuations on the length scale of a small fraction of R_e. Thus, the model is able to capture rather non-trivial correlations, which are typical for dense, nearly-incompressible melts, *e.g.*, the correlation hole in the intermolecular pair correlation function and the concomitant, subtle corrections to the Gaussian chain statistics like the power-law decay of the bond–bond correlation function along the chain.[2]

Generally, macrophase separation or microphase ordering will lower the compressibility compared to a disordered system. The strength of the coupling between density and composition fluctuations is controlled by the ratio $\chi N/\kappa N$. The specific behavior depends on the detailed implementation of the repulsive interaction between the components. For instance, if one uses a term of the form, $-(\chi N/4)(\phi_A - \phi_B)^2$, then the ordered system will be more compressible than the disordered one or a non-interacting melt ($\chi N = 0$). If one employs an interaction term of the form $\chi N\ \phi_A\ \phi_B$, instead, the mixed system will have a lower compressibility than the phase-separated one or the non-interacting melt. The compressibility of the non-interacting melt and the phase-separated systems will be similar.

1 D. Q. Pike, F. A. Detcheverry, M. Müller and J. J. De Pablo, *J. Chem. Phys.* 2009, **131**, 084903.
2 K. Ch. Daoulas and M. Müller, *J. Chem. Phys.*, 2006, **125**, 184904.

Dr Milano asked: Is it possible to extend the SCF approach to liquid crystals extending the density scalar field to a density vector field?

Professor De Pablo replied: Liquid crystalline elastomers (LCEs) combine the elastic properties of conventional rubbers with the optical properties of liquid crystals. This coupling gives rise to unique experimental behaviors, including the ability to sustain significant deformations at the polydomain–monodomain transition, a phenomenon which can lead to development of applications that include actuators or artificial muscles. Theoretical models of LCEs have been limited, and are still in their early stages. One such theoretical model has been proposed by Fridrikh and Terentjev.[1] They attribute the existence of polydomains in LCEs to the presence of randomly distributed crosslinks in the elastomeric network. These network junctions have some anisotropic preferences and try to influence the local mesogenic ordering, giving rise to the formation of polydomains in the LCE. Fridrikh and Teretjev proposed a LCE model at the coarse-grain level of nematic domains, with interaction between random crosslinks and directors of local nematic domains. A schematic representation of the model is provided in Fig. 1.

The crosslink-nematic order coupling is included in the model through a free energy F_{rf} of the form:

$$F_{rf} = -\sum_i \frac{\gamma}{2}\left[k_i n(R_i)\right] = -\int \mathrm{d}^3 r \left[\frac{\gamma}{2}\rho(kn)^2\right] \tag{1}$$

where k_i is the unit vector along the axis of the ith cross-link, R_i is the position of this cross-link, and $n(R_i)$ is the local nematic director. Parameter γ characterizes the strength of the crosslink–nematic order coupling and ρ is the density of the cross-links. Combining this free energy with the Frank energy and the mechanical energy of the monodomain nematic elastomer, one can write a complete free energy for the LCE of the form:

$$\begin{aligned} F &= F_{\mathrm{mech}} + F_{\mathrm{cross}} + F_{\mathrm{frank}} \\ &= \int \mathrm{d}^3 r \left[\frac{\mu}{2}\varepsilon_{\alpha\beta}\varepsilon_{\alpha\beta} - \sigma\varepsilon_{\alpha\beta} - UQ_{\alpha\beta}\varepsilon_{\alpha\beta}\right] - \int \mathrm{d}^3 r \left[\frac{\gamma}{2}\rho(kn)^2\right] + \int \mathrm{d}^3 r \left[\frac{K}{2}(\nabla n)^2\right] \end{aligned} \tag{2}$$

where ε is the strain tensor, σ is the stress tensor, U is a coupling constant between the nematic ordering and elastic deformations, $Q_{\alpha\beta} = Q_N(n_\alpha n_\beta - \delta_{\alpha\beta}/3)$ and K is a Frank elasticity constant.

With the above model, Fridrikh and Terentjev provided important insights into the polydomain–monodomain transition in LCEs. However, a complete analysis of the model has not been performed. We have implemented this model using an analog

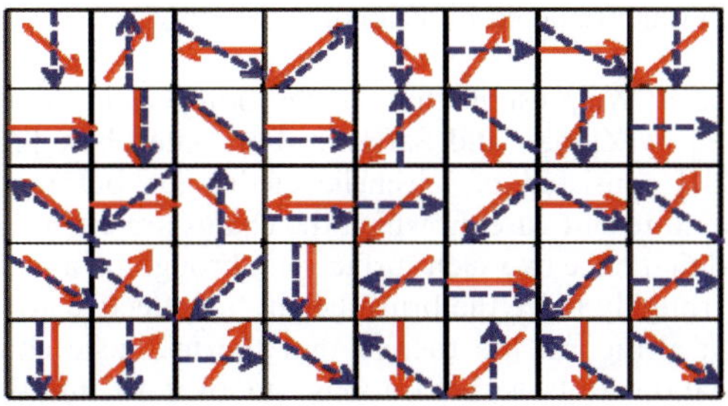

Fig. 1 Schematic of the model proposed by Fridrikh and Terentjev, shown here on our proposed lattice implementation. Each cell represents a nematic domain, with solid arrows showing the local director and dashed arrows showing the anisotropic preference of random crosslinks.

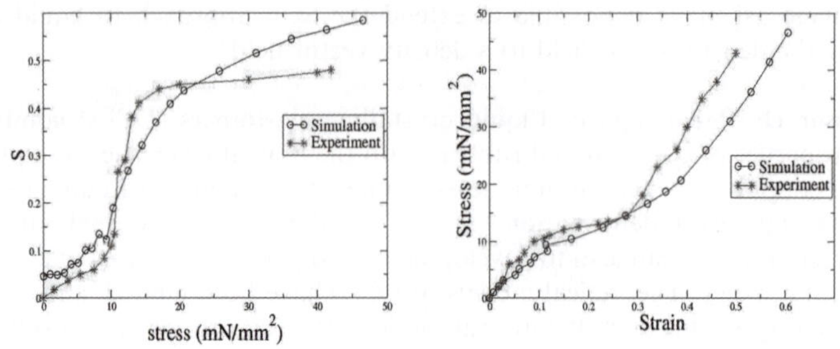

Fig. 2 Comparison of simulation results with experimental results. The grid size is 30 × 30 × 30. Parameter values are $\gamma\rho = 0.45$; U = 2.0; K = 0.85; μ = 2.0. Experimental data points are taken from Fig. 9 (S *vs.* Stress) and Fig. 3 (Stress *vs.* Strain) of Schatzle *et al.*[2]

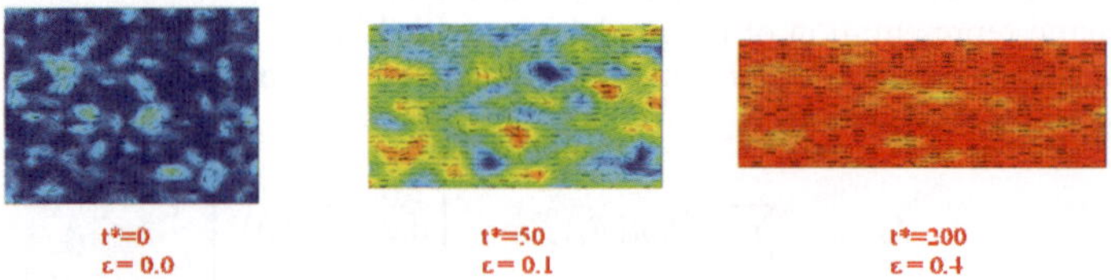

Fig. 3 Polydomain to monodomain transition when the LCE is subjected to stress. t* denotes number of Monte Carlo cycles.

of the grid-based technique described in Fig. 1, and have studied the behavior of the model through Monte Carlo simulations. In our simulations we observe the existence of polydomains in LCEs. When the system is subjected to unidirectional stress, we observed a transition from polydomain to monodomain texture, above some threshold value of the stress. A polydomain texture corresponds to low order parameters (S), and a monodomain texture exhibits a high value of S. The stress–strain curve shows the existence of a plateau corresponding to the polydomain–monodomain transition. In Fig. 2, we compare our simulation results to the experimental results. Fig. 3 shows several representative snapshots from the simulation, that serve to illustrate the transition from polydomain to monodoamin.

1 S. V. Fridrikh and E. M. Terentjev, *Phys. Rev. E*, 1999, **60**, 1847.
2 J. Schatzle, W. Kaufhold, and H. Finkelmann, *Macromol. Chem. Phys.*, 1989, **190**, 3269.

Professor Frenkel opened the discussion of the paper by Dr Lo Verso: Could you make a link between the "energetic" vapour–liquid transitions that you observe and the entropic transitions predicted by Zilman and Safran?

Dr Lo Verso replied: We presume that the question pertains to a comparison with the results reported by Zilman and Safran[1], abbreviated here [ZS01]. The system examined in [ZS01] consisted of telechelic chains (as ours) grafted on planar surfaces—as opposed to our case, in which the chains are connected on a common end. [ZS01] argued that once two such surfaces are brought at a separation D to each other, where D is roughly twice the brush height, an effective attraction shows up. The reason of the attraction lies therein that the telechelic chains, which show intra-brush association at large separations, go over to inter-brush association when the two surfaces are brought sufficiently close. In this way, there is no change in the energy, since association remains, but there is a gain in entropy because intra-brush-associated chains are less confined around their associating ends than inter-brush-associated ones. The situation with telechelic stars is quite similar. As shown in previous publications by some of the authors and also repeated in this work, at low concentrations (analogously to large inter-brush separations), telechelic stars self-associate into a watermelon structure, a configuration that is stabilized by energetic gain but entails an entropic cost. At sufficiently high concentrations (analogously to smaller inter-brush separations), the possibility of inter-star association opens up. Therefore, an entropic gain must also be associated with this phenomenon, although in our case additional energetic factors must play a role, since here we have the possibility that a large number of stars participate in micelle formation (inter-star association), a possibility absent in the case of planar brushes, due to rigid anchoring of the chains on the solid substrate.

1 A. G. Zilman and S. A. Safran, *Eur. Phys. J. E*, 2001, **4**, 467–473.

Professor Pagonabarraga continued the discussion of the paper by Professor De Pablo: The coarse-grained interaction between polymer chains is suggested by a reference free energy model. Since at a mean field level you know the thermodynamics, I wonder how good such an approximation is for the systems you discuss. Correlations can be accounted for both in the thermodynamics and structure in these systems.[1] Can you quantify the role of correlations and hence the departure from a simple mean field approach in your case?

1 S. Merabia and I. Pagonabarraga, *J. Chem. Phys.*, 2007, **127**, 054903.

Professor De Pablo responded: Yes. Figure 7 (and Table 1) of our paper, shows the order–disorder transition temperature (ODT) for a symmetric diblock copolymer. The mean field ODT for diblocks occurs at $\chi N = 10.5$; fluctuations bring the ODT value to 13.5 for the value of the invariant degree of polymerization.

In a previous work,[1] we have studied the coexistence curves of binary polymer blends. Results show that far from the critical point or the ODT, the mean field predictions are in good agreement with the results of our simulations. However, near the critical point and near the ODT, the agreement between mean field solutions and our approach with fluctuations deteriorates considerably. The mean field upper critical solution temperature for the blend occurs at $\chi N = 2$. In contrast, when fluctuations are included, simulations yield larger values for finite degree of polymerization.

Going beyond the predictions of our approach, and closer to the lithographic and patterning scope of our work, it is important for us to predict the behavior of polymeric materials under severe confinement. In such cases fluctuations can have important effects and deviations from mean field behavior can be significant.[2]

1 F. A. Detcheverry, D. Q. Pike, P. F. Nealey, M. Müller and J. J. De Pablo, *Phys. Rev. Lett.*, 2009, **102**, 197801.
2 A. Alexander-Katz, A. G. Moreira and G. H. Fredrickson, *J. Chem. Phys.*, 2003, **118**, 9030.

Professor Theodorou remarked: One should perhaps remark that the fluctuations sampled by the very nice method you presented are fluctuations with respect to a coarse-grained Hamiltonian, and not necessarily representative of the real system. For example, one would not expect the compressibility to be captured very well by the method. Are there guidelines for choosing the parameter κ_0 that governs density fluctuations in your coarse-grained Hamiltonian?

Professor De Pablo replied: For the parameters used in our simulations, the results with other cloud functions (*e.g.* spherically symmetric) are identical to those obtained with a square function. Some of those results are reported elsewhere.[1]

Yes, the fluctuations that arise in the coarse-grain model are, by construction, those pertaining to the coarse-grain Hamiltonian. In the particular case of density fluctuations, these will be governed by the magnitude of κ_0. It is always possible to choose κ_0 so as to match the compressibility of the real system. However, simulations becomes more and more difficult as the level of coarse-graining (*i.e.* the number of real monomers a simulated particle represents) increases. A single "bead" in our model is meant to represent dozens of polymer segments. Therefore, the coarse-grain model cannot capture incompressibility/packing on the scale of an atomistic degree of freedom but it rather describes the reduction of density fluctuations on the scale of soft, coarse-grained "beads". The value of κ_0, which is used in the simulations is, for example, sufficient to create a correlation hole in the intermolecular pair-correlation function and the concomitant corrections to the Gaussian statistics of polymer conformations in a melt.[2] If we enforced an experimental value of the compressibility in our model with a reduced number of degrees of freedom, κ_0 would be very large. This value would result in various artifacts and the simulations would be exceedingly slow. Further discussion on the choice of κ_0 can be found in a different manuscript.[1]

1 D. Q. Pike, F. A. Detcheverry, M. Müller and J. J. De Pablo, *J. Chem. Phys.* 2009, **131**, 084903.
2 K. Ch. Daoulas and M. Müller, *J. Chem. Phys.*, 2006, **125**, 184904.

Professor Theodorou continued the discussion of the paper by Dr Lo Verso: Would the phase behaviour change, if your miktoarm star molecules had the solvophilic part on the outside (corona) and the solvophobic part at the centre (core)? Are there applications, where such molecules may be of interest?

Dr Lo Verso replied: On the first count, *i.e.*, the "inverse" architecture, the phase behavior would be altogether different if the attractive monomers were in the middle and the repulsive in the periphery. Preliminary calculations that are being done by one of us (CNL) show that these inverse miktoarm stars are, to begin with, unstable with respect to coagulation up to an arm number of about $f = 10$ for a 50–50 mixture of attractive/repulsive monomers in every arm. In such cases, two stars would coalesce onto a single one with $2f$ arms and with a rather deformed, prolate solvophobic core, a phenomenon fully absent in the case of telechelic stars, for which molecules maintain their stability but the system undergoes macroscopic phase separation instead. An additional fundamental difference lies therein that for the inverse stars mentioned in the question, a coarse-graining in the form of a single, spherically symmetric pair potential is physically meaningful, since these coalesce and interact as a whole. For telechelics, however, individual arms connect to one another, either in an intra- or in an inter-star fashion. Except for the case of micellar forming systems, a possible coarse-graining and rotational average of the interaction would result in the wrong physical picture because it would obscure the fact that once some arms connect, the conformation of the star is drastically altered and the bonds become quickly saturated, after a few arms have connected. A manifestation of this peculiarity is the stability/formation of lamellae-like phases in our system, which would disappear if an angularly-averaged form of the effective potential had been used. Hence, many-body interactions are manifestly present here and the goal of achieving a coarse-grained description at the macromolecular level remains a challenge. What we have done is a simulation that still operates at the monomer level but employs coarse-grained models for the monomer–monomer interactions and implicitly integrates out the solvent. The possible applications of telechelic molecules are many fold but we would like to emphasize their importance as rheology modifiers. Given the fact that the strength and range of the attractions between the end-monomers can be tuned by, *e.g.*, pH and/or salinity in zwitterionic telechelics, it follows also that the connectivity of the resulting network can be steered and with it the shear properties of the solution. The same properties should be controllable by temperature as well. In a sense, telechelic stars are the polymer analog of the currently popular models of "patchy colloids", with the additional property of flexibility of the repulsive "core", as opposed to the hard nature of the patchy colloids. For the inverse architecture the applications for concentrated solutions are similar: by changing temperature or pH, you can swell the system and switch from flow to arrest. In both cases the applications may also concern drug delivery processes. In the case of attractive ends the individual molecule forms cages (watermelon) which can capture particles in solution. In the opposite case one can capture smaller solvophobic molecules in the internal part of the star and shield them from the environment *via* the solvophilic part.

Dr Vila Verde asked: You observe an unusual behavior of the size of micelles with increasing concentration of polymers: distinct conditions cause the micelles to grow, remain the same size or shrink. Can you relate this to experimental observations?

Dr Lo Verso replied: It is a highly nontrivial task to measure the size of individual molecules in experiments, especially in concentrated solutions. One possibility would be to employ SANS with labeling and zero-average contrast methods to mark individual molecules in the solution and measure the dependence of their size on concentration, but this is a cumbersome and expensive task. Dynamic light scattering would be a more reasonable option and it would deliver information on the hydrodynamic radius of the aggregates, thereby making a direct connection with the dependence of the gyration radius predicted in our simulations. We are not aware of any such measurements, however. At this stage, we are making predictions based on our simulations, hoping that they will stimulate further experimental work on this subject. In fact, a collaboration with the experimental group of Prof. Vlassopoulos (FORTH, Heraklion, Greece) exists, with the goal of exploring these poorly understood systems in more detail.

Dr Wilson continued the discussion of the paper by Professor Bolhuis: You mention in the paper that the Head-Gordon model is one of the few to reproduce protein folding and self-assembly. Could you comment on whether you think this coarse-grained model (with three types of nonbonded interaction only) is really the way to go or should we be looking to develop coarse-grained models with more specific (*e.g.* H-bonding) interactions? I am thinking especially of being able to distinguish between structures such as a β-sheet and a β-roll.

Professor Bolhuis responded: We chose the Head-Gordon model for its simplicity and the fact that it seemed to predict β-sheet formation for small proteins. You are correct in saying that the model might be too simple in general, and one really needs to specifically include hydrogen bonding even when only investigating a simple β-sheet. Coincidentally, T. Bereau and M. Deserno[1] presented a model that does model all 20 amino acids independently and includes hydrogen bonds at this meeting. We are currently investigating this possibility. In general, there is naturally a trade-off between accuracy and efficiency. Models that include hydrogen bonds will probably be less efficient than the model we tried.

1 T. Bereau and M. Deserno, *J. Chem. Phys.*, 2009, **130**, 235106–235115.

Mr Bereau commented: There exists coarse-grained peptide models that try to guess secondary structure rather than constrain it. These typically require a slightly higher resolution in order to reproduce important aspects of local conformations, *e.g.* steric interactions, hydrogen-bonds. A recent model developed in our group is capable of folding simple peptides without *a priori* knowledge of the folded state.[1]

1 T. Bereau and M. Deserno, *J. Chem. Phys.*, 2009, **130**, 235106.

Professor Voth asked: Your coarse-grained model is very nice and also very simple. Experimentalists typically like to study the various effects from simple point amino acid substitutions (mutations). Can your CG model reproduce effects from such experiments?

Professor Bolhuis responded: Point mutation is indeed a much used experimental technique to gain insight in the relevance of specific amino acids on the protein structure or its mode of action. The rather simple Head-Gordon model that we use was certainly not developed to reproduce mutation effects. Since this coarse-grained model is based on grouping the typical 20 different amino acids into a reduced 3-letter alphabet (hydrophilic, hydrophobic, and neutral residues), the type of substitutions that can be done with the model are obviously rather limited. Substitution of a hydrophilic residue by a neutral one can in principle be envisioned, but mutation of a bulky hydrophobic residue by a smaller hydrophobic one would make no difference in the model. Although, mind you that also the backbone dihedrals are parameters that can be "mutated" (changing *e.g.* β-sheet/α-helix/turn propensity). However, for comparison with experimental mutation studies, we would most likely turn to a less coarse-grained model, if we would become interested in that.

Professor Berardi asked: To obtain a folded structure for your model polypeptide you had to invoke a modified form of the Lennard–Jones potential with a shortened range of non-bonding dispersive interactions. Could you comment on that?

Professor Bolhuis responded: The Lennard–Jones (LJ) potential is indeed used for the non-bonded interaction, but does not represent dispersive forces only. It is a combination of (coarse-grained) hydrophobic interactions, hydrogen bonds, van der Waals forces and electrostatics, *etc.* The original Head–Gordon model uses an unscaled LJ potential. The sigma parameter is set by the (fixed) bond length and is sigma = 4.8 Å. This means that the strands attract each other at distances of more than 8 Å. This is quite unrealistic as the hydrogen bond itself is only 2–3 Å. The coarse-grained interaction ranges should therefore be shorter. Indeed, we found that when using the unscaled LJ form non-neighboring strands attracted each other considerably, as illustrated in Fig. 4 which shows the distribution of interstrand distances in the atomistic, shifted CG and unshifted CG simulations. This caused a spurious twist of the beta roll in the CG simulations. When we shortened the range, which as argued above, is more realistic, the twist disappeared and the structure behaved as in atomistic simulations.

Dr Periole said: From the pictures in your paper the β-sheets seem to be right-handed. Is it the case? Note that β-sheets in proteins and amyloid-like fibril are generally left-handed and your model should account for that. It might be difficult on a coarse-grain level though!

Professor Bolhuis replied: We have tested that both left and right-handed images of the β-roll are stable in the atomistic systems. We have chosen to coarse-grain only one of them. We have not checked whether this would be the left or right-handed β-roll, but we will do so in the near future.

In the coarse-grained simulations the sequence renders the model very symmetric and it is hence unlikely that the model can distinguish between these forms.

Professor Marrink asked: As the focus of the work appears to be the study of the peptide self-assembly, why not constrain the secondary structure of the peptide to the atomistic structure using an elastic network model?

This saves a lot of hassle trying to find optimized interaction parameters to ensure the CG structure resembles the atomistic structure.

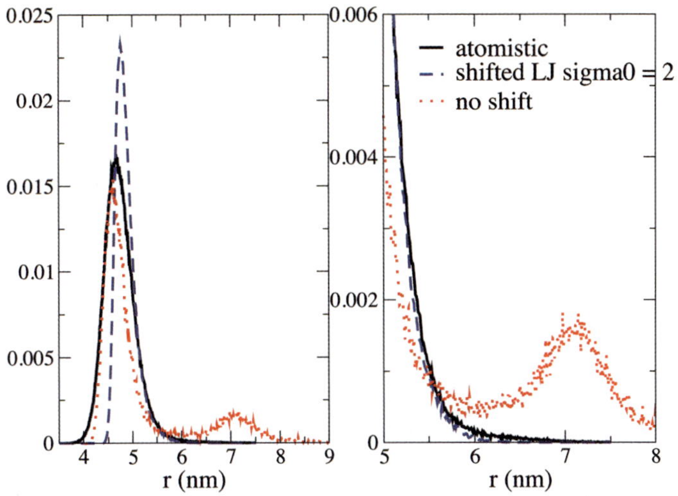

Fig. 4 Effect of shifting the Lennard–Jones potential on the interstrand distance compared to the interstrand distance from an atomistic simulation.

Professor Bolhuis answered: It is true that if one only is interested in the properties of the fiber, one could model the peptide structure with an elastic network model. However, the aim of the work was to provide a possible coarse-graining route to enable the simulation of large polypeptide block copolymer systems, including their self-assembly properties, *i.e.* the formation of the secondary structure. In that case, we cannot impose the secondary structure *a priori*. We therefore think that optimizing the interaction parameters to reproduce atomistic simulations is inevitable. Note, however, that we have not shown (yet) that the optimized model reproduces the desired folding behavior. Note also that our proposed route, *via* the pulling simulation, is not at all the only one, and is probably not the most efficient way to do the optimization.

Professor Allen opened the discussion of the paper by Professor Zannoni: Can you comment on the detailed balance condition for your implementation of the parallel Monte Carlo algorithm? As I understand it, whenever one or more atoms crosses a sub-domain boundary in a set of parallel moves, the set of inverse moves cannot be selected by the algorithm, and so detailed balance is violated. This has been addressed[1] and an algorithm proposed that rigorously obeys detailed balance, by rejecting moves that cross these boundaries. Of course, there is a price to pay in that the boundaries must not be kept fixed.

As a separate issue, it would be interesting to see a molecular dynamics simulation of this kind of system, to see the effects of coupling between orientation and flow, and perhaps compare with mesoscale simulations.[2]

1 A. Uhlherr, S. J. Leak, N. E. Adam, P. E. Nyberg, M. Doxastakis, V. G. Mavrantzas and D. N. Theodorou, *Comput. Phys. Commun.*, 2002, **144**, 1; F. Schmid, D. Düchs, O. Lenz and B. West, *Comput. Phys. Commun.*, 2007, **177**, 168.
2 D. Marenduzzo, E. Orlandini and J. M. Yeomans, *Europhys. Lett.*, 2005, **71**, 604; T. J. Spencer and C. M. Care, *Phys. Rev. E*, 2006, **74**, 061708.

Professor Zannoni responded: We believe the problem to be a minor one, at least for very large domains like our ones and away from a phase transition, so we have privileged fixed domain boundaries and speed and, as we mention in the paper, we have validated the results, *e.g.* checked that the order parameter is within the error bar of previous calculations performed with our standard serial code.

As for the molecular dynamics simulation, this is now possible with effective codes like LAMMPS becoming available and we believe it will be a viable alternative to specific MC codes. Comparing results for different methodologies will of course be interesting.

Professor Müller-Plathe opened the discussion of the paper by Dr E. Müller: Your model has strong repulsion between all unlike beads. On the other hand, hydrocarbons and fluorocarbons are not all that different and they often do mix. What would you change in your results if the interaction between these two moieties were not completely repulsive?

Dr E. Müller answered: Both in the experimental results and in the simulations, the actual energetics of the side chain do not seem to be crucial. Experimental results show that the columnar hexagonal phase (Col_h) can be observed whether the side chains are alkane chains, $C_{14}H_{29}$, (compound 1/14)[1], a branched semi-perfluorinated chain, $(CH_2)_3C_7F_{15}$ or a perfluorinated chain, C_8F_{17} (compounds 2/i7 and 4/8)[2]. Furthermore, we note that these are not the only compounds that produce these Col_h structures. In fact, it is apparent from the experimental data, that the formation of these structures is non-specific to the exact nature of the side chain, and is apparently governed largely by the volume ratio of the side chain to the rigid LC core, *i.e.* steric considerations seem to be dominant over energetic ones. Our simulation results reflect this same non-specificity: the choice of different attraction strengths for the mutual attraction of the flexible side chains affects the transition temperatures between the isotropic and liquid crystal phases, but does not seem to have a significant effect on the global phase diagram.

In the coarse-grained model the overall scale of the interactions is reduced. To illustrate this point we note how the like–like interactions between the type 2 beads (the ones that make up the core of the molecule, corresponding to aromatic-like moieties) are also repulsive in nature.

1 M. Kölbel, T. Beyersdorff, X. H. Cheng, C. Tschierske, J. Kain and S. Diele, *J. Am. Chem. Soc.*, 2001, **123**, 6809–6818.
2 X. H. Cheng, M. Prehm, M. K. Das, J. Kain, U. Baumeister, S. Diele, D. Leine, A. Blume and C. Tschierske, *J. Am. Chem. Soc.*, 2003, **125**, 10977–10996

Dr Yoneya opened the discussion of the paper by Dr Ravnik: What is the crucial requirement to obtain BP temperature range enlargement by the particle insertion?

Quite recently Prof. Kikuchi's group tried the gold nano-particle insertion to the BP, but they could not obtain detectable BP temperature range change. Additionally, the comment of the ref. 26 of the paper on page 2 is incorrect.[1] The system of ref. 26 of the paper is no particle insertion and pure mixture of bimesogenic molecules.[1]

1 H. J. Coles and M. N. Pivnenko, *Nature*, 2005, **436**, 997.

Dr Ravnik responded: The crucial requirement to achieve stabilisation of blue phases by doping with particles is that interaction of particles with the liquid crystal is weak, *i.e.* effective anchoring strength at particle walls should be as small as 10^{-5}–10^{-7} J/m^2. If particles create a perturbation in the ordering of the blue phase through surface interaction which is too large, effectively all beneficial effects of replacing the defect region with the particle are lost. We can comment on experimental results obtained by calorimetric measurements (private communication) which show a strong increase in the temperature range of the blue phases.

Dr Vila Verde continued the discussion of the paper by Dr E. Müller: (1) The liquid-crystal part of your coarse-grained model is linear, but the molecule it aims to represent has two rings. It thus appears that your results are more representative of bolaamphiphiles with a linear liquid-crystal part and less so of the bolaamphiphiles you actually use for comparison. Can you offer a comment on this? In particular, are there experimental reports comparing the behavior of bolaamphiphiles with linear and planar liquid-crystals?

(2) In future work it would be interesting to calculate the free energies of the various structures you observe.

Dr E. Müller answered: The representation of the liquid crystal (LC) core of the bolaamphiphile as a rigid rod-like structure is a result of the aggressive coarse-graining exemplified in this paper. A large part of the fine details are smeared out, leaving out what are understood as the most relevant features. In the case of the T-shaped molecules, the rigid nature of the LC core, the strong hydrogen bonding of the end groups and the size and flexibility (space-filling) nature of the side grafted chains are seen to be the main driving forces for the phase behaviour. In this context, the possible directional attraction of the core rings (stacking) is, amongst others, one of the details which are neglected. We believe this omission is not of significance to the overall global behaviour. As an example, the frequently studied mesogen 5CB (*p*-*n*-pentyl-*p*′-cyanobiphenyl), loosely related in morphology to the LC core studied here, is commonly described by models having axial symmetry, *e.g.* elongated rods or Gay-Berne potentials.[1] With respect to point 2, we are studying the global phase diagrams of the underlying LC bolaamphiphiles by means of thermodynamic integration techniques with a view of extending it to this particular system.

1 M. R. Wilson, *Int. Rev. Phys. Chem.*, 2005, **24**(3–4), 421–455.

Dr Cheung continued the discussion of the paper by Dr Ravnik: (1) As was mentioned by the author in his talk blue phases may give rise to flexoelectric polarization.[1] As colloidal particles in experimental systems are often charged, is the interaction between this spontaneous polarization and charged particles likely to qualitatively change the behaviour of colloids in blue phases?

(2) The interaction energy between colloidal particles in a liquid crystal has been estimated to be of the order of 100–1000 k_{BT},[2] which is comparable to the binding strength of the colloid–disinclination line interaction shown in this paper. Have the authors considered the effect of colloid aggregation as well as trapping at disinclination lines?

1 G. P. Alexander and J. M. Yeomans, *Phys. Rev. Lett.*, 2007, **99**, 067801.
2 H. Stark, J.-I. Fukuda, and H. Yokoyama, *Phys. Rev. Lett.*, 2004, 92, 205502; T. G. Sokolovska, R. O. Sokolovskii, and G. N. Patey, *Phys. Rev. E*, 2006, **73**, 020701.

Dr Ravnik answered: Charging of particles can be in experimental systems fairly well controlled. In our experimental nematic-based systems, the effect of charge is very small and can be ignored.[1]Nevertheless, possible charging of particles will for sure trigger structural transitions within the liquid crystalline ordering, in particular in the defect regions. Detailed interplay between flexoelectric, dielectric and elastic energy will control ordering of the particles which may also result in qualitatively different ordering of particles. We speculate that in particular the relative ratios between particle size, blue phase unit cell size, and Debye-length will determine the qualitative behaviour of the blue phase colloids.

1 I. Musevic, M. Skarabot, U. Tkalec, M. Ravnik, S. Zumer, *Science*, 2006, **313**, 954.

Professor De Pablo asked: What is the range of energy associated with the segregation of the particles to the defects?

Dr Ravnik replied: Two main parameters which determine the segregation of particles into defects are particle size and surface anchoring strength. Depending on exact values, the energies associated with trapping of particles range from kT to several 1000 kT. A more precise analysis is presented elsewhere.[1]

1 M. Skarabot M. Ravnik, S. Zumer, U. Tkalec, I. Poberaj, D. Babic and I. Musevic, *Phys. Rev. E*, 2008, **77**, 061706.

Professor Jackson addressed Dr Ravnik, Professor Carbone and Dr Wilson: Could you comment on the nature of the temperature dependence of the pitch that you obtain with your approach? For example, can you describe the decrease in the pitch with increasing temperature seen in common cholesterics, or do you get a kinetic unwinding of the pitch? What happens to the pitch at the boundaries between the cholesteric and blue phases?

Dr Ravnik responded: Landau-de Gennes model is based on free energy expansion, therefore is intrinsically at constant temperature. The effect of temperature comes as an external parameter (γ), redefining the phenomenological material parameters. Indeed, such an approach effectively does account for the change in the pitch and depends strongly on the choice of phenomenological parameters. For the parameters used in our paper, upon approaching the isotropic phase, the cholesteric pitch increases. Boundary effects between the cholesteric and blue phases far exceed the scope of our paper (and most of the literature). We speculate that interplay will occur between cholesteric pitch and pitch within the blue phases.

Dr Wilson responded: Usually, the decrease in pitch with increases in temperature in a cholesteric (chiral nematic) depends on the proximity of the cholesteric phase to an underlying smectic-A phase. On cooling towards the smectic phase pretransitional effects occur in the cholesteric phase and the helix unwinds. This particular physical effect is not accounted for in the model as formulated in the paper.

Professor Frenkel questioned: In view of the strong interaction of the nano-particles with the disclination lines, one can imagine that in experiments the system could easily get trapped in metastable, disordered phases. Can your simulations shed light on this "kinetic" aspect of blue-phase formation?

Dr Ravnik answered: Numerical modelling presented in our paper is equilibrium-based, meaning that when addressing effective potential that acts on particles upon repositioning, the particles are fixed and the liquid crystal is fully equilibrated. Recent measurements[1] show that equilibrium and dynamically measured potentials in nematics are practically equivalent, at least for micron-sized particles. We therefore believe that by using effective equilibrium dynamics, *e.g.* "quasi-dynamics" that we proposed in nematics ,[2] the kinetic aspect of the structure formation in blue phases could also be addressed.

1 J. Kotar, M. Vilfan, N. Osterman, D. Babic, M. Copic and I. Poberaj, *Phys. Rev. Lett.*, 2006, **96**, 207801.
2 M. Skarabot M. Ravnik, S. Zumer, U. Tkalec, I. Poberaj, D. Babic and I. Musevic, *Phys. Rev. E*, 2008, **77**, 061706.

Professor Berardi commented: Could you describe what would be the effect of shape polydispersity and concentration on the structure and stability of the dispersion of colloidal particles in a blue phase?

Dr Ravnik replied: Effect of shape (which I will more thoroughly address later): By tailoring the shape of the particles to the profiles and structure of defects in blue phases the phase stability of blue phases could be additionally increased, as if compared to spherical particles. Moreover, as a result of complex blue phase structure, complex particle shapes may lead to novel stable crystalline unit cells of colloidal particles. Effect of polydispersity: By varying the particle size, unit cell size of the colloidal crystal changes substantially (up to 10%). Having very polydisperse particles, colloidal unit cells of different sizes will most probably form that will eventually have to pack, either by creating domains or soft variation and deformation of the unit cells.

Effect of concentration: Concentration of particles will, in experiments, be probably the main tuning parameter to achieve good dispersity and avoid formation of metastable structures. We speculate that optimal packing of particles will be achieved at concentrations of ~1 particle/blue phase unit cell.

Professor Dijkstra remarked: Many colloidal particles can be synthesised nowadays with all kinds of shapes, like ellipsoids, plate-like particles, cubes, tetrapods, *etc*. What would be the effect of mixing these kinds of particles within the cholesteric blue phases?

Dr Ravnik responded: The mechanism that controls trapping of particles in liquid crystal blue phases is very universal and is based on replacing energetically expensive regions with particle volume, reducing the total free energy of the system. Tailoring particle shapes to fit more efficiently into defect regions, we speculate it will reduce the total free energy of the system even more, therefore resulting in stronger trapping and even more efficient thermodynamic stabilisation of the blue phases.

Professor Löwen asked: The phenomenological Landau-de Gennes approach with appropriate elastic constants was used to model the free energy of the liquid crystal. Can this be improved by using more microscopic density functional theory? Recent progress has been obtained by constructing a fundamental-measure-theory for arbitrarily-shaped hard bodies.[1] This might be a promising way to go for a more microscopic approach which involves interparticle interaction explicitly.

1 H. Hansen-Goos and K. Mecke, *Phys. Rev. Lett.*, 2009, **102**, 018302 .

Professor De Pablo responded: Yes, depending on the length scales that one is interested in capturing, a more microscopic approach (*e.g.* a density functional theory) might provide additional important information.

Dr Ravnik commented: Despite using a phenomenological model, the modelling results of liquid crystal colloids at micrometer and sub-micrometer scale can be in many liquid crystalline systems almost quantitatively mapped to experiments, in particular if using all three liquid crystal elastic constants. Going to true nanoscale particles (~10 nm), the results of a Landau-de Gennes model become less reliable, although in practice a fairly good agreement with experiments is found. The recent fundamental-measure-theory (FMT) for arbitrarily-shaped hard bodies[1] opens an interesting route also to modelling of liquid crystal colloids, now at a more microscopic level. However, before relying on the new theory, open questions will need to be solved, such as how to incorporate chirality of the liquid crystal, and also how to discriminate between basic liquid crystalline elastic modes (splay, bend, twist), the main elements for more quantitative studies.

Professor Allen said: In your paper you show some plots of free energy against the lattice constant, seeking the minimum, and you also make a correction by hand for the extra volume occupied by colloids. These all seem to involve discrete changes in system volume, although one would hope that the spatial discretization would not be critical in a model of this kind. Would it be possible to avoid this problem, and introduce a formal definition of pressure in this model, and hence work in a constant-pressure ensemble? If the handling of the colloidal interface on a varying mesh proves problematic, I would suggest smoothing out the interface or using something like a pseudopotential.

Dr Ravnik replied: Upon repositioning of particles, discrete changes in the volume of the blue phase unit cell are performed to assure, equivalent position of the particle surfaces with respect to the mesh. Indeed, using the pseudopotential, node distribution method,[1] or the finite element (volume) method, the mesh effects can be avoided to a large extent, all in the course of being implemented in our computer modelling. Having successfully implemented repositioning of particles with respect to the mesh, formally, pressure could also be in principle introduced and further used in the equilibration. We should comment that by introducing the pressure, compressibility of the medium should also be taken into account, unless defining an effective pressure that responds only to liquid crystalline orientational ordering.

1 C. J. Smith and C. Denniston, *J. Appl. Phys.*, 2007, **100**, 014305.

Dr Wilson continued the discussion of the paper by Professor Zannoni: In your simulations there must be a competition between the ease of twist (governed by the strength of the twist elastic constant) and the strength of the surface ordering. Are you sure that the Gay-Berne potential reproduces the balance of these two effects correctly? *i.e.* that the growth of twisted domains out from the surface are not governed by very strong surface interactions in the simulations in comparison to real systems.

Professor De Pablo commented: A generalized Gay-Berne potential should have enough flexibility to provide a reasonable description of the balance between anchoring strength or energy and bulk elastic constants. Note, however, that I am not aware of quantitative studies to have examined that balance and that have attempted to describe experimental data. It would be interesting to conduct such a study.

Professor Zannoni replied: We assume the interaction between surface molecules and mesogen to be the same as the mesogen–mesogen ones. Thus we do not assume very-strong anchoring of the type feared in the question. The extremely large amount of computational resources needed to perform even a few additional studies varying the surface interaction forbids systematically exploring the point.

Professor Allen commented: Elastic constants and surface anchoring coefficients would be of interest in interpreting the results of your simulations, and could be calculated in your system from torque and director profile measurements, and fluctuation measurements.[1] N. H. Phuong, G. Germano, and F. Schmid[2] have also calculated such quantities by inverting the OZ equation in the nematic phase, while H. Steuer and S. Hess[3] have published a paper showing how to compute K_2 from simple averages.

1 D. Andrienko, G. Germano, M. P. Allen, *Phys. Rev. E*, 2000, **62**, 6688; D. Andrienko, M. P. Allen, *Phys. Rev. E*, 2002, **65**, 021704.
2 N. H. Phuong, G. Germano, and F. Schmid, *J. Chem. Phys.*, 2001, **115**, 7227.

3 H. Steuer and S. Hess, *Phys. Rev. Lett.*, 2005, **94**, 027802.

Professor Zannoni replied: We are aware of these methods and in the past we have also done some work of this type.[1] However the calculation of elastic constants and anchoring coefficients was outside the scope of this paper. Indeed, here we aim to show that the molecular and "macroscopic" levels can be directly connected, avoiding the calculation of intermediate quantities as input in the continuum type equations normally used to model displays. The calculation of elastic constants is complicated and the results typically affected by rather large errors and, as far as we are aware, no method has yet found overall consensus and widespread use. Indeed very few results are available in the literature, compared to the variety of model mesogens and systems studied.

1 J. Stelzer, R. Berardi, C. Zannoni, *Chem. Phys. Lett.*, 1999, **9**, 299.

Professor Marrink continued the discussion of the paper by Dr E. Müller: The rescaling of simulated time toward real time has to be considered with care. In MARTINI, as well as in order similar CG models, the conversion factor usually applied, only accounts for the loss of friction from the atomic details embedded within the CG beads. Based on self-diffusion rates of simple liquids and lipid bilayers, a factor of between 3–8 is typically observed. As a first order correction, a simple scaling of the time axis can therefore be applied *a posteriori*.

However, the kinetics of collective processes such as the phase transitions described in the current paper are not dominated by local friction, but rather, by the thermodynamic driving forces and the energy barriers associated with the transformation process. The accuracy of these thermodynamic factors depend on the accuracy of the model, which is at best semi-quantitative for the CG model employed in the current paper (the same is true for MARTINI and other similar models). Considering that a difference of only a couple of kT in the major energy barrier of a process already leads to an order of magnitude difference in kinetics, one should be careful in the interpretation of the time scale.

Dr E. Müller responded: Our simulations describe the aggregation and ordering process and provide only relative timings in which these processes occur. As mentioned in the question, these depend non-trivially on the many choices made in the coarse-graining process (grouping of atoms, choice of energy levels, *etc.*) which will affect the mobility of the molecules, both in terms of the times corresponding to the inherent (equilibrium) self-diffusion and the (non-equilibrium) gradient-driven transport diffusion. We fully agree that the topic of how we should map simulation time with real time is an open and non-trivial question.

1 H. Hansen-Goos and K. Mecke, *Phys. Rev. Lett.*, 2009, **102**, 018302.

Dr Wilson queried: The hydrogen bonding groups at the end of a bolaamphiphile molecule could, in reality, lead to quite specific pairing of molecules. However, in the simulation model no such specific hydrogen bonding interactions exist and it is possible for several terminal "H-bonding" groups to come together. Is this a problem?

Dr E. Müller replied: Each end group is composed of a moiety with several possible bonding sites. So while there is a directionality imposed by the real geometry, one must also consider the flexibility of the end group and the separation between the hydrogen bonding groups. With this in mind, it is not unreasonable that, in the actual molecules, each end group could participate in more than one hydrogen bond with more than one similar molecule. Furthermore, the liquid crystal core seems to impose a natural alignment of the molecules (*e.g.* Fig. 7b of the paper) where these bondings are favored. We see in our simulations that the walls of the bolaamphiphile columns are made of at least two parallel LC cores, meaning that in each node of the columnar structure we have the convergence of at least six type 1 end groups. It is likely that in the real molecules, there would be enough entanglement and flexibility for more than one hydrogen bond to form. The model captures this in a non-specific average way. In some sense, it resembles the more common practice of coarse-graining a cluster of 3 or 4 water molecules into a single isotropic spherical entity.

Professor Jackson continued the discussion of the paper by Professor Zannoni: When the field is switched on in your samples, the main reorientation appeared to occur in the central part of the sample with layers a few tens of molecular dimensions thick left essentially unperturbed close to the surface of the cell walls. Do you think that this is a true effect that would be seen experimentally? Are the voltages that you employ realistic?

Professor Zannoni responded: I would tend to believe the effect to also hold for real systems. For instance, recent experimental results by Charles Rosenblatt[1] show a long decay of the order away from an aligning surface even in the isotropic phase.

The voltage employed is around 75 V/μm. This is sensibly higher than that of a commercial display, but not quite unreasonable. For instance, fields above 100 V/μm have been used in fast electro-optic cells.[2]

1 J.-H. Lee, T. J. Atherton, V. Barna, A. De Luca, E. Bruno, R. G. Petschek and C. Rosenblatt, *Phys. Rev. Lett.*, 2009, **102**, 167801.
2 H. Takanashi, J. E. MacLennan and N. A. Clark, *Jpn. J. Appl. Phys.*, 1998, **37**, 2587.

Dr Wilson questioned: From some of the early work on phase diagrams for Gay-Berne potentials[1] it is evident that there are some undesirable features in relation to modelling thermotropics. For simulations at constant pressure, changes in temperature lead to a big change in the system density when changing phase. Is it possible to make changes to the potential so that the Gay-Berne is more suitable for simulation of thermotropic liquid crystals and (possibly) for other systems such as membranes *etc*?

1 Fig. 1 of J. T. Brown, M. P. Allen, E. Martin del Rio and E. de Miguel, *Phys. Rev. E*, 1998, **57**, 6685.

Professor Zannoni answered: The Gay-Berne potential is a simple generic one, thus it has limitations. However, the simulation of a display depends essentially on the ability of the model potential in reproducing the orientational order parameter within the nematic phase and this is rather well reproduced for thermotropics (see, ref. 19 of the paper).[1] As for the possibility of improving the Gay-Berne model transition properties, introduction of flexibility or a softer repulsive core should help.

1 R. Berardi, A. P. J. Emerson and C. Zannoni, *J. Chem. Soc., Faraday Trans.*, 1993, **89**, 4069–4078.

Dr Baaden opened a general discussion: Preliminary preparation for Faraday Discussion 144 was quite a challenging task, as a broad scope of scientific disciplines is covered, including polymers, liquid crystals and biological macromolecules. Each research field uses slightly different terminology and quoting Kingman Brewster Jr.: "Incomprehensible jargon is the hallmark of a profession". The benefit of such a scientific melting pot is of course to learn something about

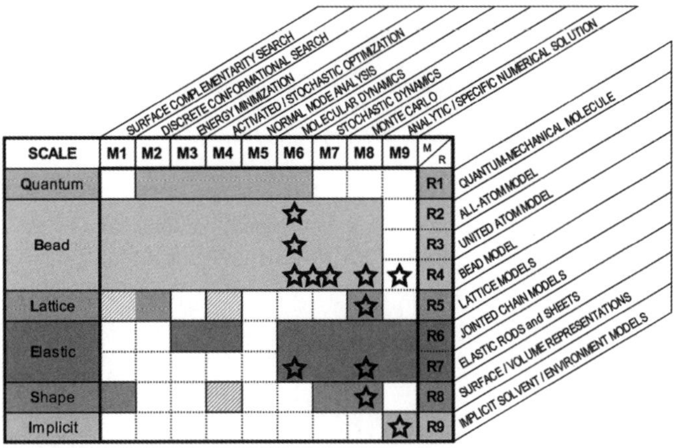

Fig. 5 An attempt to classify multi-scale simulations using a horizontal method axis and a vertical representation axis. Representations are assembled into five scales. Stars indicate simulations discussed at FD 144.

shish kebabs, watermelon structures and blue phases (papers by Dr Graham, Dr Lo Verso and Dr Ravnik, respectively), but still a lot of unnecessary confusion may remain.

As a first example, during Faraday Discussion 144, the process of going from low to high resolution has been described by fine-graining, resolution-exchange, reverse-mapping and variants thereof. After reading the corresponding papers, it appears that these terms refer to a very similar (if not identical) approach, with variations in the protocol that is used. The second example concerns the term multi-scaling. Its meaning is by no means precisely defined. For example one may argue that there is a fundamental difference between concurrent and sequential multi-scaling. Let us define those as using two modeling scales either at the same time or successively for the sake of the argument. Where are the boundaries with respect to multi-resolution simulations, and what is multi-physics?

These examples are meant to illustrate that the rapid progress in the field of multi-scale simulations of soft matter calls for efforts to define a common language and terminology. This language should be shared and agreed upon by different simulation fields such as those treating polymers, liquid crystals and biomolecules.

A related issue concerns the classification of multi-scale simulations themselves. Faraday Discussion 144 has provided a broad overview of current approaches, but how can one describe and compare the resolution, timescale and model description of these simulations? Berendsen's book[1] provides a first guideline, systematically relating different levels of modeling. In the introductory lecture by Professor Kremer, a classification using time and length scale was presented. A multi-scale classification has been suggested for polymer systems.[2] In order to summarize the various representations, methods and target functions presently available to molecular modellers, a finer description might however be needed. An attempt of such a classification for simulations of biological systems, not only taking into account the time- and length scales but also details of the simulations themselves can be found in.[3] Fig. 5 illustrates a two-dimensional approach based on simulation method (M1 to M9) and model representation (R1 to R9). Stars shown in the Figure are an attempt to classify the simulations discussed at Faraday Discussion 144. Colored cells in Fig. 5 correspond to combinations of simulation method and model representation described in the literature.

1 H. J. C. Berendsen, *Simulating the Physical World, Hierarchical Modeling from Quantum Mechanics to Fluid Dynamics*, Cambridge University Press, 2007.
2 F. Müller-Plathe, *Soft Mater.*, 2002, **1**, 1–31
3 M. Baaden and R. Lavery, There's plenty of room in the middle: multi-scale modelling of biological systems, in *Recent Advances in Protein engineering*, ed. A.G. de Brevern, Research Signpost, Trivandrum, Kerala, India, 2007, pp. 173–195.

Coarse-grained simulations of charge, current and flow in heterogeneous media

Benjamin Rotenberg,[a,b] Ignacio Pagonabarraga[c] and Daan Frenkel[b,d]

Received 23rd January 2009, Accepted 19th February 2009
First published as an Advance Article on the web 13th August 2009
DOI: 10.1039/b901553a

We present a coarse-grained simulation method for complex charged systems. This mesoscopic model couples a hydrodynamic description to a free energy functional accounting for the interactions between solvent(s) and charged solutes. It is implemented in a hybrid lattice-based algorithm, whereby the evolution of the overall mass and momentum is taken care of *via* a Lattice Boltzmann scheme, whereas the composition and ionic concentrations are updated using the link-flux method. Several applications illustrate the power of this coarse-grained model for charged heterogeneous media: the transport of charged tracers in charged porous media, the deformation of an oil droplet in water under the effect of an applied electric field, and the distribution of ions at an oil–water interface as a function of their affinity for both solvents.

Introduction

The study of complex fluids and heterogeneous materials offers significant challenges because of the wide range of relevant length and time scales involved. Flow in porous media, the equilibrium and kinetic properties of membranes, oil/water mixtures or the electrokinetics of drops and colloids are examples of situations where the system evolves on scales much larger than molecular ones. Even the equilibrium structures characteristic of these systems are orders of magnitude larger than the molecular scale. Thus a simple atomistic description is not suitable and the use of soft coarse-grained potentials has become a standard alternative.[1] Under the action of external forces, the situation in these heterogeneous materials is even more involved, because of the hydrodynamic coupling between solutes and solvent. Hence, also the dynamic behaviour needs to be accounted for within a consistent coarse-grained approach.

A variety of strategies have been proposed to face these challenges. The combination of soft potentials and local thermostats which conserve momentum locally, as Dissipative Particle Dynamics[2] or the Lowe–Andersen thermostat,[3] has allowed to reach of hydrodynamic scales while keeping track of some microscopic details. Ideas from kinetic theory have also lead to flexible tools. Stochastic Rotation Dynamics couples molecular solutes to a coarse-grained solvent which recovers hydrodynamic behaviour.[4] Lattice Boltzmann (LB),[5] a method evolved from lattice gases to

[a]*CNRS and UPMC-Paris6, Laboratoire PECSA, UMR 7195, 4 place Jussieu, F-75005 Paris, France*
[b]*FOM Institute for Atomic and Molecular Physics, Kruislaan 407, 1098 SJ Amsterdam, The Netherlands*
[c]*Departament de Fisica Fonamental, Universitat de Barcelona, Carrer Martí i Franqués 1, 08028 Barcelona, Spain*
[d]*Department of Chemistry, University of Cambridge, Lensfield Road, Cambridge, CB2 1EW, United Kingdom*

describe the hydrodynamics of fluids at long scales, has been extended to account for complex fluids. In particular, the combination of free energy based models coupled to a hydrodynamic description was introduced by Yeomans *et al.*[6–8] for simulations of non-ideal fluids and binary mixtures and later extended for other systems such as binary mixtures with surfactants,[9–11] liquid crystals,[12–15] ternary mixtures[16] or active fluids.[17] LB methods are particularly well suited for hydrodynamic simulations of fluids, especially in complex media, for it is in principle easy to parallelize the codes and to implement boundary conditions at solid–fluid interfaces.[18] This allowed the simulation of binary fluids in porous media,[19] colloids at an oil–water interface[20–22] and suspensions of charged colloids.[23]

Standard LB schemes for complex fluids often lead to spurious fluxes across solid–fluid boundaries or at liquid–liquid interfaces.[24] This can become catastrophic when one deals with charged solutes, as such fluxes could result in a progressive breakdown of electroneutrality. To overcome this difficulty, Capuani *et al.*[25,26] introduced the link-flux method to reproduce the convective–diffusive dynamics of charged species in an electrolyte. Within this hybrid scheme, the overall mass and momentum of the fluid are evolved using a LB algorithm, whereas the ionic concentrations are updated using the link-flux method (for diffusion and migration) in combination with an advection scheme (for convection). This algorithm satisfies detailed balance at steady-state and makes it possible to rigorously cancel fluxes into the solid, even in the presence of moving boundaries.

We want to put forward a general scheme to deal with complex charged fluids making use of this hybrid strategy, by combining the LB treatment of the solvent with a continuum treatment of binary charged fluids. We show how starting from a general free energy functional it is possible to obtain a dynamic scheme consistent with the prescribed thermodynamics and how it can be coupled to the LB method for the fluid flow. Several applications illustrate the power of this coarse-grained model for charged heterogeneous media: the transport of charged tracers in charged porous media, the deformation of an oil droplet in water under the effect of an applied electric field and the distribution of ions at an oil–water interface as a function of their affinity for both solvents. We compare with analytical predictions in the appropriate limits to illustrate that it is possible to achieve quantitative control of the performance of the model.

Free-energy based model of non-ideal mixtures

In this section, we present the description of a complex mixture of solvents and charged solutes in terms of mesoscopic variables, the associated thermodynamic properties and the hydrodynamic equations governing their dynamics.

Mesoscopic description

Our aim is to describe charged solutes in heterogeneous media involving solid–liquid or liquid–liquid (*e.g.* oil–water) interfaces. For not too concentrated solutions, we can treat the fluid as a continuum, whose state is characterized by its local mass density $\rho(\mathbf{r}, t)$ and concentrations in each solute $\rho_k(\mathbf{r}, t)$. In the following we will always consider 1 : 1 electrolytes and denote the corresponding solutes by + and − for cations and anions, respectively. For oil–water mixtures, we also introduce the local composition

$$\varphi(\mathbf{r}) = \frac{\rho_\mathrm{o}(\mathbf{r}) - \rho_\mathrm{w}(\mathbf{r})}{\rho_\mathrm{o}(\mathbf{r}) + \rho_\mathrm{w}(\mathbf{r})} \in [-1; 1] \tag{1}$$

where the w and o subscripts refer to water and oil, respectively. Immiscible fluids are characterized by regions where ϕ is almost constant ($\phi \sim -1$ in water and $\phi \sim +1$ in oil) separated by a "sharp" interface. The underlying assumption in the above description is to consider that the fluid can be seen as locally homogeneous, although density, composition and solute concentrations can vary on a larger scale.

Thermodynamics

The thermodynamics of the system is determined by its free energy, expressed as the functional:

$$\mathscr{F}[\phi, \rho_+, \rho_-] = \int d\mathbf{r}\ \mathscr{F}_V[\phi(\mathbf{r}), \rho_+(\mathbf{r}), \rho_-(\mathbf{r})] \tag{2}$$

where $\mathscr{F}_V$ is a free energy density. Following Onuki[27–29] we separate the contributions $\mathscr{F}^{\text{mix}}[\phi]$ describing the immiscible solvents and $\mathscr{F}^{\text{ions}}[\phi, \rho_+, \rho_-]$ describing ions in a solvent of composition ϕ. The mixing contribution is chosen of the Landau–Ginzburg form:

$$\mathscr{F}^{\text{mix}} = \int d\mathbf{r}\left[-\frac{1}{2}B\varphi^2 + \frac{1}{4}B\varphi^4 + \frac{1}{2}K(\nabla\varphi)^2\right] \tag{3}$$

The first two terms correspond to the bulk phase behaviour, with minima for $\phi = \pm 1$, while the last reflects the cost of sustaining interfaces. The dimension of B is energy $\times$ length^{-3} and that of K is energy $\times$ length^{-1}. This standard choice gives at equilibrium a planar interface of the form $\phi(x) = \tanh(x/\xi)$ with a width $\xi = \sqrt{2K/B}$ and a surface tension $\sigma = \sqrt{8KB/9}$.

The ionic contribution of the free energy consists of an ideal, a solvation and an electrostatic term:

$$\mathscr{F}^{\text{ions}} = \int d\mathbf{r} \sum_{\alpha=\pm} \rho_\alpha(\mathbf{r})\left[k_BT(\ln\rho_\alpha(\mathbf{r}) - 1) - \mu_\alpha + V_\alpha^{solv}(\mathbf{r}) + \frac{z_\alpha e}{2}\psi(\mathbf{r})\right] \tag{4}$$

where $z_\pm = \pm 1$ is the valency of the ions, μ_α is a reference chemical potential and the electrostatic potential ψ is the solution of the Poisson equation:

$$\nabla\cdot[\varepsilon(\mathbf{r})\nabla\psi(\mathbf{r})] = -[\rho_+(\mathbf{r}) - \rho_-(\mathbf{r})]e \tag{5}$$

The dielectric constant $\varepsilon(\mathbf{r})$ depends on the local composition of the fluid. Although it could be *a priori* an intricate function of ϕ, it is reasonable to assume a linear relation $\varepsilon(\mathbf{r}) = \bar{\varepsilon}[1 - \gamma\phi(\mathbf{r})]$ with $\bar{\varepsilon} = \dfrac{\varepsilon_W + \varepsilon_O}{2}$ the average dielectric constant and $\gamma = \dfrac{\varepsilon_W - \varepsilon_O}{\varepsilon_W + \varepsilon_O} \in [0; 1]$ the dielectric contrast.

The ionic solvation potential $V_\pm^{\text{solv}}$ accounts for the different solvation free energy in the two solvents. It is therefore natural to parameterize it as a function of the composition as $V_\pm^{\text{solv}}(\mathbf{r}) = \Delta\mu_\pm\dfrac{1 + \varphi(\mathbf{r})}{2}$, where we have introduced for each ion the solvation free energy difference between water and oil $\Delta\mu_\pm = \mu_\pm^{\text{o}} - \mu_\pm^{\text{w}}$, also referred to as extraction or Gibbs transfer free energy in the electrochemistry community. As the free energy (3) does not exclude in principle values of the composition parameter outside the $\phi \in [-1, 1]$ range, we have considered that the physical properties (ε, $V_\pm^{\text{solv}}$) for $\phi < -1$ are that of water (ε_W, 0) and for $\phi > 1$ that of oil (ε_O, $\Delta\mu_\pm$).

Let us now briefly analyze the properties of this free energy in terms of the chemical potentials associated to $\rho_\pm$ and ϕ and the corresponding thermodynamic forces acting on the fluid. The ionic chemical potentials are of the usual form:

$$\mu_\pm = \frac{\delta\mathscr{F}}{\delta\rho_\pm} = k_BT\ln\rho_\pm + V_\pm^{\text{solv}} + z_\pm e\psi \tag{6}$$

while the chemical potential corresponding to the solvent mixture reads:

$$\mu_\varphi = \frac{\delta \mathcal{F}}{\delta \varphi} = \mu_\varphi^{\rm mix} + \mu_\varphi^{\rm solv} + \mu_\varphi^{\rm el} \tag{7}$$

The first term is simply:

$$\mu_\phi{}^{\rm mix} = -B\phi(\mathbf{r}) + B\phi(\mathbf{r})^3 - K\nabla^2\phi(\mathbf{r}) \tag{8}$$

When the solvation free energy is taken as $V_\pm^{\rm solv}(\mathbf{r}) = \Delta\mu_\pm \frac{1+\varphi(\mathbf{r})}{2}$ the second term is:

$$\mu_\varphi^{\rm solv} = \frac{[\rho_+(\mathbf{r})\Delta\mu_+ + \rho_-(\mathbf{r})\Delta\mu_-]}{2} \tag{9}$$

Finally, the electrostatic contribution is:

$$\mu_\varphi^{\rm el} = \frac{\mathbf{E}(\mathbf{r})^2}{2}\frac{\delta\varepsilon}{\delta\phi} = -\frac{\gamma\bar{\varepsilon}}{2}\mathbf{E}(\mathbf{r})^2 \tag{10}$$

Excess chemical potential gradients give rise to a thermodynamic force (per unit volume) that can be expressed as a pressure gradient from the Gibbs–Duhem equality:

$$\mathbf{f}_V^{\rm th} = -\nabla P = \phi\nabla\mu_\phi + \rho_+\nabla_{\mu+}^{\rm ex} + \rho_-\nabla_\mu^{\rm ex}- \tag{11}$$

where the ex superscript refers to the excess chemical potentials. Each part of the chemical potentials (mixing, solvation and electrostatic) contribute to this force. For example, gradients of $\mu_\pm^{\rm ex}$ give rise to the force $\rho_{\rm el}\mathbf{E}$, with $\mathbf{E} = -\nabla\psi$ the electric field, and gradients of $\mu_\phi^{\rm el}$ are the source of the dielectrophoretic force $(\varepsilon(\mathbf{r}) - \bar{\varepsilon})\nabla\left(\frac{\mathbf{E}^2}{2}\right)$. The latter drives oil-rich fluid elements ($\phi > 0$, $\varepsilon < \bar{\varepsilon}$) towards the region where $\mathbf{E}^2$ is small and water-rich fluid elements towards the region of higher $\mathbf{E}^2$ (note that $\nabla\mu_\phi^{\rm el}$ also generates a composition flux in addition to this force) and is particularly important in phenomena such as electrowetting. These electrostatic contributions to chemical potential gradients illustrate some new features captured by our free energy model compared to previous ones used in Lattice Boltzmann simulations of binary mixtures.

Hydrodynamics

The thermodynamic description of the system needs to be supplemented by a set of prescriptions for the dynamics. Overall mass conservation of the fluid implies:

$$\partial_t\rho + \nabla\cdot(\rho\mathbf{u}) = 0 \tag{12}$$

with **u** the local barycentric velocity of the fluid. Momentum conservation of the fluid and viscous dissipation then enter in the Navier–Stokes equation, which reads for an incompressible fluid:

$$\partial_t\mathbf{u} + \mathbf{u}\cdot\nabla\mathbf{u} = \nu\nabla^2\mathbf{u} + \frac{\mathbf{f}_V}{\rho} \qquad ((13))$$

where $\nu = \eta/\rho$ is the kinematic viscosity and $\mathbf{f}_V = \mathbf{f}_V^{\rm ext} + \mathbf{f}_V^{\rm th}$ is the sum of the externally applied force and thermodynamic force (eqn (11)). Although we have not considered this in the following, it is in principle possible to introduce a composition-dependent viscosity.[30]

The composition ϕ and ionic concentrations $\rho_\pm$ also satisfy conservation laws:

$$\partial_t \phi + \nabla \cdot (\phi \mathbf{u}) = -\nabla \cdot \mathbf{j}_\phi \tag{14a}$$

$$\partial_t \rho_\pm + \nabla \cdot (\rho_\pm \mathbf{u}) = -\nabla \cdot \mathbf{j}_\pm \tag{14b}$$

where we have introduced fluxes in the barycentric frame $\mathbf{j}_\phi$ and $\mathbf{j}_\pm$ The latter are given by phenomenological equations, namely:

$$\mathbf{j}_\pm = -D_\pm \rho_\pm \nabla(\beta \mu_\pm) \tag{15}$$

with $D_\pm$ the ionic diffusion coefficients, and the Cahn–Hilliard equation for the composition:

$$\mathbf{j}_\phi = -M \nabla \mu_\phi \tag{16}$$

with M a mobility. The units of M differ from those of $D_\pm$ and a diffusivity (in m^2 s^{-1}) of the interface can be defined as $D_\phi = MB(-1 + 3\phi^2)$ with B from eqn (3). Note that near the interface $\phi \sim 0$ and $D_\phi < 0$: this "negative diffusion" maintains the composition jump at the interface.

Discussion

The above description of mixtures of solvent and ions and the particular choice of free energy functional are very similar to the ones adopted by Onuki.[27–29] The functional differs only on two points. First, we follow previous LB studies of binary mixture[18,20,31,32] and use the Landau–Ginzburg (LG) functional (3) instead of the Bragg–Williams one for numerical convenience. The combination of LG and electrostatic free energies (without solvation terms) has also been used to investigate the wetting of a solid substrate by ionic solutions.[33] The second difference with Onuki's approach consists in using $\rho_{el}\psi$ for the electrostatic energy instead of $\varepsilon(\nabla\psi)^2$. This more natural choice, consistent *e.g.* with the DFT work of van Roij *et al.*,[34,35] doesn't require any approximations for the treatment of "image charges".

A major difference with both Onuki's and van Roij's work is that the present model captures not only the equilibrium states of the system, but also its dynamics. It shares many features with Dynamical Density Functional Theory (DDFT),[36–39] since it relies on an expression of fluxes proportional to gradients of chemical potentials. The free energy functional described above is relatively simple, for it neglects *e.g.* the effect of the finite size of the ions and correlations beyond the mean-field level. This description is perfectly valid for dilute solutions of 1 : 1 electrolytes and can be improved if more concentrated solutions or multivalent ions are considered. Moreover, it captures the presence of immiscible solvents and the (possibly asymmetric) affinity of the ions for one of them. The free energy model put forward in this paper can be seen as a simple limiting case of more elaborate free energy functionals. The essential difference with DDFT is that the hydrodynamic behaviour of the fluid is properly described (it satisfies the Navier–Stokes equation resulting from momentum conservation and viscous dissipation), whereas most DDFT studies consider a fluid at rest or mediating hydrodynamic interactions between large solutes *via* effective interactions (Oseen or Rotne–Prager tensors).[40] This latter approach can be efficient for suspensions of solid particles, but is not valid *a priori* for liquid droplets in another liquid.

Lattice simulations

The coarse-grained model introduced in the previous section couples a hydrodynamic description of the fluid to a free energy based representation of its thermodynamic behaviour. The purpose of the present section is to introduce the computational methods used to solve numerically the coupled evolution equations for the composition, ionic concentrations and fluid velocity. The general strategy relies on the use of different lattice models, which are described here successively.

Hybrid lattice scheme

For non-ideal multicomponent fluids, the standard LB approach treats all species on the same footing and populations (see below) are associated to each component. This can become computationally expensive for more than two components. Nevertheless, this approach has recently been applied for a mixture of two solvents and two reactive solutes by Furtado *et al.*[41] to study the convective drop motion driven by non-linear kinetics. An alternative is to use a hybrid LB/finite elements approach, whereby the LB fluid is described by populations evolving as before, but the order parameters (*e.g.* the composition ϕ) are described by scalar fields evolving according to finite elements schemes.[42,43] As mentioned in the introduction, such methods may suffer from spurious fluxes which motivated the development of the link-flux method by Capuani *et al.*[25,26] Here we generalize the link-flux approach to the more complex case of ions in a mixture of solvents. The overall mass and momentum are taken care of *via* a LB scheme, whereas the composition ϕ and ionic concentrations $\rho_\pm$ are updated using the link-flux method. We now develop these two steps.

Lattice Boltzmann

Lattice Boltzmann is a well established method for hydrodynamic simulations based on kinetic theory.[5,44] The Boltzmann equation is a mesoscopic kinetic equation which determines the evolution of the probability density function $f(\mathbf{r}, \mathbf{v}, t)$ of finding a fluid particle with a velocity $\mathbf{v}$ at position $\mathbf{r}$ and time t. The hydrodynamic fluid variables are derived as moments in velocity space of the distribution function starting with the fluid local density $\rho = \int f \mathrm{d}\mathbf{v}$ and mass flux $\rho\mathbf{u} = \int f\mathbf{v}\mathrm{d}\mathbf{v}$. Although the relaxation of the distribution function toward equilibrium is determined by a nonlinear collision operator, the Bhatnagar–Gross–Krook (BGK) model[45] shows that proper hydrodynamics can be already recovered from a linearized collision operator if it is isotropic and conserves mass and momentum.

LB tracks the dynamics of fluid populations, $f_i = f(\mathbf{r}, \mathbf{c}_i, t)$, which evolve on the nodes $\mathbf{r}$ of a lattice of spacing Δx moving to neighbouring nodes at finite time steps through a finite set of allowed velocities $\{\mathbf{c}_i\}_{i\in[1,N\mathrm{max}]}$. The hydrodynamic variables are obtained as appropriate quadratures $\rho = \sum_i w_i f_i$ and $\rho\mathbf{u} = \sum_i w_i f_i \mathbf{c}_i$, where the weights w_i are associated to the chosen set of velocities. The particle distributions f_i also relax at each time step toward a prescribed equilibrium distribution through a linear collision operator which conserves mass and momentum and ensures that the solvent mass density ρ and velocity $\mathbf{u}$ follow the Navier–Stokes equations, eqn (12) and (13), on distances larger than Δx. The natural units in LB simulations are the lattice spacing Δx and the time step Δt. They are fixed by the properties of the system: in the following, the lattice spacing is chosen as a fraction of the Bjerrum length $l_\mathrm{B} = e^2/4\pi\bar{\varepsilon}k_BT$ (approximately 0.7 nm in water at room temperature), while the time step is determined by the value of the solvent viscosity.

The force per unit volume $\mathbf{f}_V$ acting on the fluid also enters in the collision rule. The issue of computing the thermodynamic force (eqn (11)) from the state (ϕ, $\rho_\pm$) of the system will be addressed in the next section. For numerical accuracy and stability reasons, the fluid velocity must remain small compared to the sound velocity c_s. This small Mach number limit implies that the forces are also small, *i.e.* that $\chi_T\mathbf{f}_V\Delta x \ll 1$, with χ_T the fluid compressibility ($\chi_T = 1/\rho c_\mathrm{s}^2$ for the LB fluid). Each term in the excess free energy density contributes to the force (eqn (11)) and we can derive conditions accordingly. This leads for the $\mathcal{F}^\mathrm{mix}$ contribution (eqn (3)) to

$\chi_T B \ll \xi/\Delta x$ and similar requirements are obtained for the solvation and electrostatic ones.

Link-flux method

The composition of the fluid is characterized by the order parameters ϕ and $\rho_\pm$ and evolves according to eqn (14), (15) and (16). The numerical solution of these equations is achieved by the link-flux method. This method was introduced by Capuani *et al.*[25,26] in order to prevent spurious solute fluxes across solid–fluid interfaces. It focuses on solute fluxes between lattice nodes rather than the amount of solute at each node. Integrating the conservation laws (eqn (14)) over a volume corresponding to one lattice node and using Green's formula, we associate the variation of ϕ and $\rho_\pm$ to the fluxes of $\mathbf{j}_\phi$ and $\mathbf{j}_\pm$ across the surface A_0 of the cell (for more details, see ref. 25). The latter can be separated into the contributions of each link between the considered node and all its neighbours:

$$\frac{\rho_\pm(\mathbf{r}, t+\Delta t) - \rho_\pm(\mathbf{r}, t)}{\Delta t}\Delta x^3 = -A_0 \sum_i j_\pm^i(\mathbf{r}) \tag{17}$$

where i refers to the discrete velocities and $j_\pm^i$ to the contribution of link i between $\mathbf{r}$ and $\mathbf{r} + \mathbf{c}_i\Delta t$ to the outward flux of $\mathbf{j}_\pm$. A similar expression can be written for the composition with link-fluxes j_ϕ^i. In order to ensure that the ions follow a Boltzmann distribution at equilibrium, we rewrite eqn (15) as $\mathbf{j}_\pm = -D_\pm e^{-\beta\mu_\pm^{\text{ex}}}\nabla[\rho_\pm e^{\beta\mu_\pm^{\text{ex}}}]$ and express the link-fluxes in the symmetrized form:

$$j_\pm^i(\mathbf{r}) = -d_\pm \frac{e^{-\beta\mu_\pm^{\text{ex}}(\mathbf{r})} + e^{-\beta\mu_\pm^{\text{ex}}(\mathbf{r}+\mathbf{c}_i\Delta t)}}{2} \times \left[\frac{\rho_\pm(\mathbf{r}+\mathbf{c}_i\Delta t)e^{\beta\mu_\pm^{\text{ex}}(\mathbf{r}+\mathbf{c}_i\Delta t)} - \rho_\pm(\mathbf{r})e^{\beta\mu_\pm^{\text{ex}}(\mathbf{r})}}{\Delta_i}\right] \tag{18}$$

with $d_\pm = D_\pm/A_0$ and $\Delta_i = ||\mathbf{c}_i\Delta t||$. For links crossing solid–fluid boundaries (*i.e.* such that $\mathbf{r} + \mathbf{c}_i\Delta t$ is a solid node) we enforce $j_\pm^i = 0$ so that such links do not carry any solute into the solid. For the solvent order parameter ϕ we use the simpler form:

$$j_\varphi^i(\mathbf{r}) = -m_\varphi \frac{\mu_\varphi(\mathbf{r}+\mathbf{c}_i\Delta t) - \mu_\varphi(\mathbf{r})}{\Delta_i} \tag{19}$$

with $m_\phi = M/A_0$ to recover the Cahn–Hilliard expression eqn (16).

The link-flux algorithm just described takes care of the diffusive fluxes $\mathbf{j}_\phi$ and $\mathbf{j}_\pm$. The advective fluxes $\phi\mathbf{u}$ and $\rho_\pm$ $\mathbf{u}$ are treated in a separate step described in detail in ref. 25: it consists of transferring particles according to the overlap between the considered cell (around a node) translated by $\mathbf{u}\Delta t$ and each of its neighbouring cells. Finally, the update of the composition also modifies the local force (eqn (11)) acting on the fluid. The thermodynamic force acting on node $\mathbf{r}$ is the average of the forces on each link $\mathbf{f}_V(\mathbf{r}) = \sum_i w_i F_i(\mathbf{r})\mathbf{c}_i$, with:

$$F_i(\mathbf{r}) = -\varphi(\mathbf{r})\frac{j_\varphi^i(\mathbf{r})}{m_\varphi} - k_B T \sum_{\alpha=\pm}\left[\frac{j_\alpha^i(\mathbf{r})}{d_\alpha} - \frac{\rho_\alpha(\mathbf{r}+\mathbf{c}_i\Delta t) - \rho_\alpha(\mathbf{r})}{\Delta_i}\right] \tag{20}$$

For the ionic part we subtract the ideal term $\rho_\pm\nabla\mu_\pm^{\text{id}} = \nabla\rho_\pm$ from the link-flux, because only the excess (solvation and electrostatic) term $\rho_\pm\nabla\mu_\pm^{\text{ex}}$ contributes to the force.

The computation of the chemical potentials $\mu_\pm$ and μ_ϕ require the evaluation at the lattice nodes of the electrostatic potential ψ and its gradient $\mathbf{E} = -\nabla\psi$, as well as $\nabla^2\phi$ (see eqn (8) and (10)). The electrostatic potential is determined from the local charge and dielectric constant by solving the Poisson eqn (5) with the Successive Over Relaxation (SOR) method,[46] which also requires to compute gradients and

Laplacian of ϕ and ψ. We have implemented a modified SOR algorithm with the additional term arising from the spatial variations of the permittivity. The following stencils are used to compute these differential operators:

$$\nabla\psi(\mathbf{r}) = \frac{1}{\Delta t}\sum_i \frac{w_i}{c_s^2}[\psi(\mathbf{r}+\mathbf{c}_i\Delta t) - \psi(\mathbf{r})]\mathbf{c}_i \tag{21a}$$

$$\nabla^2\psi(\mathbf{r}) = \frac{2}{\Delta t^2}\sum_i \frac{w_i}{c_s^2}[\psi(\mathbf{r}+\mathbf{c}_i\Delta t) - \psi(\mathbf{r})] \tag{21b}$$

This choice was motivated by numerical accuracy considerations, but also by the fact that it is fully consistent with the discretization used for the LB part of the hybrid scheme.

Charged tracers in charged porous media

We now show a first application of the method introduced in the previous section in which no oil is present. To assess the behaviour of a forced, charged fluid, we consider a slit of width L with charged walls (charge density $\sigma^{el} < 0$) and compensating counterions in the fluid as a simple representation of a porous medium. In equilibrium, the counterion concentration is $\rho_+(x) = \frac{\alpha^2}{2\pi l_B}\frac{1}{\cos^2\alpha x}$, where α satisfies $\frac{\alpha L}{2}\tan\frac{\alpha L}{2} = \pi\sigma^{el}Ll_B/e$. If an electric field of strength E_0 is applied parallel to the solid walls, an electroosmotic flow develops in the confined fluid,

$$u_y(x) = u_{ref}\ln\frac{\cos\alpha x}{\cos\alpha L/2} \tag{22}$$

where the amplitude satisfies $u_{ref} \equiv eE_0/2\pi\eta l_B$. In Fig. 1 we compare the theoretical predictions and the computed values for the flow field for increasing charge of the

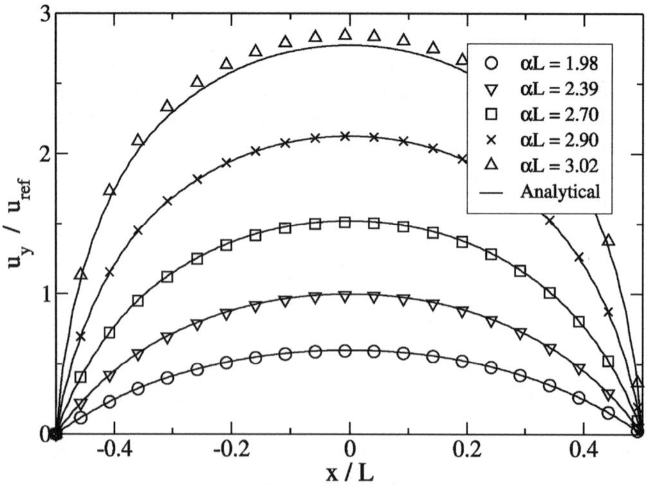

Fig. 1 Steady state flow profile across a slit of width $L = 60\Delta x$ (with $l_B = 0.4\Delta x$) for an applied electric field of magnitude $\beta eE_0L = 3$ as a function of the solid surface charge density. Symbols are simulation results while the continuous curves correspond to the theoretical prediction.

solid walls. One can see that the method describes quantitatively the osmotic flow deep into the non-linear regime of electrostatic coupling ($\alpha L \rightarrow \pi$). Previous work has shown that other dynamic quantities, such as the dispersion of charged tracers by this flow, can be easily recovered.[47]

A very useful quantity to analyze the diffusive dynamics in porous media is the time-dependent diffusion coefficient $D(t)$, that can be measured by NMR.[48] At short times $D(t)$ coincides with the molecular diffusion coefficient D. For neutral tracers, the short-time behaviour of $D(t)$ reflects the geometry of the pores:[49] $\frac{D(t)}{D} \sim 1 - \frac{4}{9\sqrt{\pi}}\frac{S}{V_p}\sqrt{Dt}$ with V_p the pore volume, S the surface of the solid. The long time limit of $D(t)$ is the effective diffusion coefficient:

$$D_e = \lim_{t \to \infty} D(t) \tag{23}$$

which reflects the connectivity between pores throughout the medium. The ratio D_e/D is often referred to as the inverse of the tortuosity. The effective tracer diffusion coefficient through charged porous materials is known to depend on its charge. For example, it has been observed in clays (negatively charged minerals) that the ratio D_e/D for cations is larger than for neutral tracers, whereas that for anions it is smaller.[50–52] This can be at least partly explained by the so-called Donnan effect: the concentration of co-ions (resp. counterions) in the pores of the material is smaller (resp. larger) than the concentration of salt in the reservoirs used to impose a concentration gradient to the sample, and this modifies the effective concentration gradient inside the sample.

Even if we correct for this effect, we expect that ions of different charge will follow different pathways through the pores and this might influence the observed value of D_e. In the framework of the proposed model, we can analyze $D(t)$ as the integral of the tracer's velocity autocorrelation function (VACF) using the moment propagation.[47,53–55] It is worth emphasizing that the present analysis is possible because we are able to evaluate $D(t)$ numerically. Although this has been done for neutral tracers by averaging over trajectories of explicit tracer particles,[56] we found no such analysis for charged tracers.

System

We will analyze the diffusion of charged tracers in a porous medium consisting of a compact FCC lattice of charged spheres of radius R with a surface charge density $\sigma_{el} < 0$ whose pores are saturated with an electrolyte solution of concentration ρ_b. The void fraction (porosity) of $1 - \pi/3\sqrt{2} \sim 26\%$ is divided into large octahedral (O_h) cavities of radius $r_{O_h} \sim 0.41R$ connected by smaller tetrahedral (T_d) pores of radius $r_{T_d} \sim 0.22R$. The size of the bottlenecks between O_h and T_d pores is approximately $0.15R$. The electrostatic potential distribution inside the pore is controlled by the salt concentration, with a typical double-layer thickness $\kappa_b^{-1} = (8\pi l_B \rho_b)^{-1/2}$, with l_B the Bjerrum length. The Debye length κ_b^{-1} corresponds to the exponential decay of the potential near a planar interface, for not too high a surface potential ψ_S (compared to k_BT/e and using the potential of the solution "far" from the surface as a reference). The latter depends on the surface charge density and the salt concentration in the medium.

Simulations were performed on an $a^3 = (100\Delta x)^3$ lattice, with spheres of radius $R = a/2\sqrt{2} \sim 35.4\Delta x$. The lattice spacing is $2.5l_B \sim 1.75$ nm, so that the O_h (resp. T_d) pore size is ~25.8 nm (resp. 13.9 nm). The charge density of the solid is $\sigma_{el} \sim -0.04\ e/\text{nm}^2$, and we considered salt concentrations ρ_b corresponding to $(\kappa_b R)^{-1} \in [0.02;0.57]$. The molecular diffusion coefficients are $D_\pm = 5\times10^{-2}(\Delta x^2/\Delta t)$. The system is initialized with cations and anions distributed homogeneously and evolved until the equilibrium distribution is reached.

Results and discussion

The time-dependent diffusion coefficient $D(t)$ for tracers of valency ±1 and 0 with the same molecular diffusion coefficient D are computed using the moment propagation method. The results show that the charge of the ion influences both the value of D_e and the transient regime to reach this asymptotic value. D_e/D is larger (resp. smaller) for cations (resp. anions) than for neutral tracers. This is in agreement with experimental observations (which also reflect the Donnan effect). $D(t)$ also tends towards D_e faster (resp. more slowly) for cations (resp. anions) than neutral tracers. This can be quantified by a characteristic time:

$$\tau = \int_0^{\infty} \frac{D(t) - D}{D_e - D} \mathrm{d}t \tag{24}$$

Fig. 2 shows the variation of tracer's diffusion and the relaxation time τ with salt concentration ρ_b. The extension of the diffuse layer (approximately κ_b^{-1}) decreases with increasing ρ_b. Both D_e and τ for the charged tracers tend towards the values for neutral tracers at high ρ_b. This effect has been observed in diffusion experiments (also reflecting the smaller Donnan effect in that case). To our knowledge, there are no experimental measurements of time-dependent diffusion coefficient for ions in clays.

The variations of D_e/D and τ with the tracer charge and salt concentrations reflect how tracers go from one cavity to another to explore the whole porosity. Smaller T_d pores act as bottlenecks through which tracers must pass to go from one O_h pore to another. In addition to this purely geometric (entropic) effect, the electrostatic potential distribution in the porosity also affects the motion of charged tracers. The electrostatic potential $\psi(\mathbf{r})$ in the pore is always larger than the surface potential ψ_S and increases with increasing distance from the surfaces. Thus ψ is larger near the center of O_h pores (see the disconnected isopotential surfaces for a large ψ, in red in Fig. 3) than in T_d pores and anions feel a repulsive electrostatic force when approaching the latter. This decreases the probability to go from one O_h cavity to the next and consequently D_e is smaller than for neutral tracers. This also implies that it takes longer for anions to explore the volume accessible to them (although

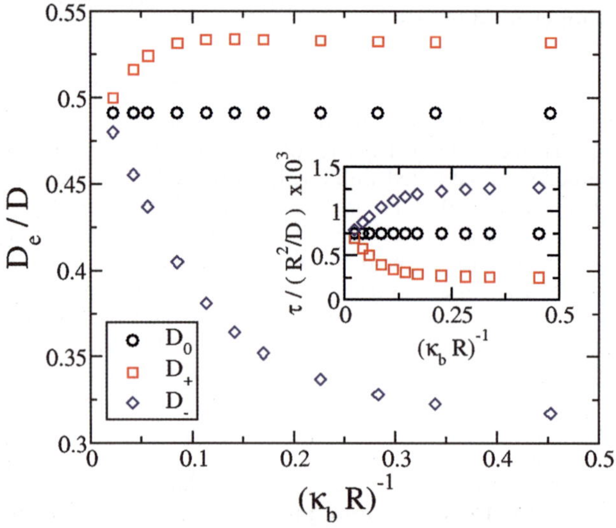

Fig. 2 Effective diffusion coefficient for charged tracers, as a function of the salt concentration. The results for neutral (○), cationic (□) and anionic (◇) tracers, normalized by the molecular diffusion coefficient D, are reported as a function of the equivalent Debye length in a bulk solution $\kappa_b^{-1} = (8\pi l_B \rho_b)^{-1/2}$ divided by the radius R of the spheres. The inset shows the characteristic time to explore the porosity accessible to each tracer, normalized by R^2/D.

this volume is smaller than for neutral species that are not repelled from the surfaces) and the corresponding τ is larger.

As opposed to anions, cations accumulate near the surface. As can be seen in Fig. 3, the diffuse layer forms a continuous volume throughout the porosity and cations can follow preferential pathways along the surfaces. This surface diffusion mechanism is more efficient than for neutral cations, since exploring a smaller volume takes less time than exploring the whole O_h pores. Hence the larger D_e and smaller τ for cations. In both cases, there is a clear interplay between the geometric and electrostatic effects. At higher concentrations, the potential variations are screened and the effect of the above-mentioned mechanisms are less pronounced.

Dielectric droplets under an electric field

In this section, we apply our mesoscopic model to the deformation of an oil droplet in water under an applied electric field, in the absence of ions. Because of surface tension, the equilibrium shape of an oil droplet in water corresponds to the minimal interface area (a disk in 2D, a sphere in 3D) and there exists an excess (Laplace) pressure inside the drop: $P_L = \frac{\sigma}{R_d}$ in 2D and $\frac{2\sigma}{R_d}$ in 3D, with σ the surface tension and R_d the drop radius. When the drop and the suspending liquid have different dielectric constants (*i.e.* when $\gamma = \frac{\varepsilon_w - \varepsilon_o}{\varepsilon_w + \varepsilon_o} \neq 0$), applying an electric field E polarizes the drop and the anisotropic electrostatic stress tensor tends to elongate it in the direction of the field. The final shape is governed by the balance between electrostatic and surface tension forces. For small applied fields the equilibrium shape is an ellipse in 2D (an ellipsoid in 3D) and the deformation is defined as $D = (b - a)/(b + a)$ with b (resp. a) its large (resp. small) axis. In the small E limit, an analytical result for the deformation in the 2D case can be obtained following the lines of ref. 57 for spherical droplets, with the result (see below):

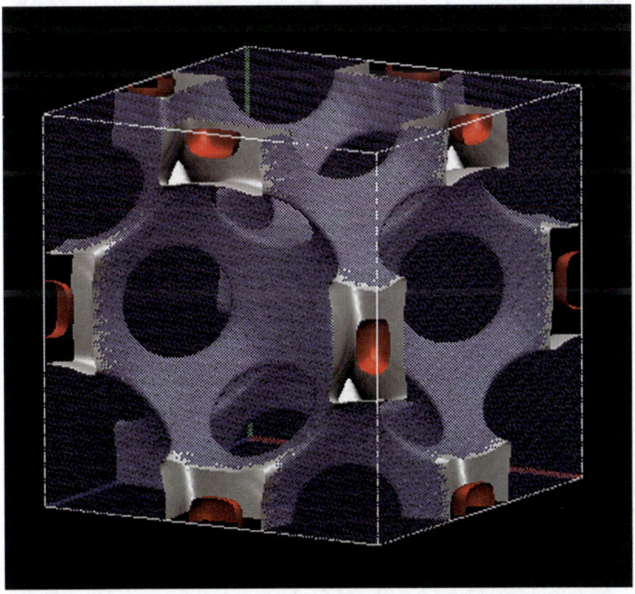

Fig. 3 Isopotential curves for a FCC lattice of charged spheres where the porosity contains solvent, counterions and salt. The grey surface is an electric isopotential curve for a value close to the surface potential while the red surfaces correspond to a larger value of the electrostatic potential. Each lattice node in the solid phase is represented as a blue dot, while fluid nodes are not indicated for clarity.

$$\mathscr{D}_{\text{theor}} = \frac{1}{4}\gamma^2(1+\gamma)\frac{\bar{\varepsilon}E^2 R_d}{\sigma} \tag{25}$$

where $\bar{\varepsilon} = (\varepsilon_w + \varepsilon_o)/2$.

System

We performed simulations of a two-dimensional oil droplet in water. Because of periodic boundary conditions in all directions, this corresponds to an array of infinite cylinders. The box size is $N \times N \times 1$ lattice points, with $N = 50$ or 100. The parameters entering in the free energy (eqn (3)) are $\beta B\Delta x^3 = 10^{-3}$ and $\beta K\Delta x = 3\ 10^{-3}$, giving a theoretical interface width of $\xi = \sqrt{2K/B} \sim 2.45\Delta x$ and a surface tension $\beta\sigma\Delta x^2 = \sqrt{8KB/9} \sim 1.63\ 10^{-3}$. This choice of parameters ensures that the interface is thin while remaining well resolved on the lattice. In particular, we checked that in the case of a planar interface (1D geometry) the simulated systems reproduces accurately the $\phi(x) = \tanh(x/\xi)$ profile with the expected width. These parameters also fulfill the condition $\chi_T B \ll \xi/\Delta x$. The interface mobility M is such that $MB = 10^{-2}\Delta x^2/\Delta t$.

After equilibrating the system, we turn on the electric field E by solving the Poisson equation under the condition of a potential drop $-EL$ between both sides of the simulation box (to be consistent with the periodic boundary conditions). The electrostatic potential inside and outside the droplet is reported in Fig. 4 together with the corresponding electric field lines, for a large dielectric contrast $\gamma = 0.9$ and an applied electric field $\beta eE\Delta x = 10^{-3}$. The average dielectric constant $\bar{\varepsilon} = (\varepsilon_w + \varepsilon_o)/2$ is such that $\bar{l}_B = \beta e^2/4\pi\bar{\varepsilon} = 0.4\Delta x$. Simulation results for two system sizes are shown: the smaller (b)

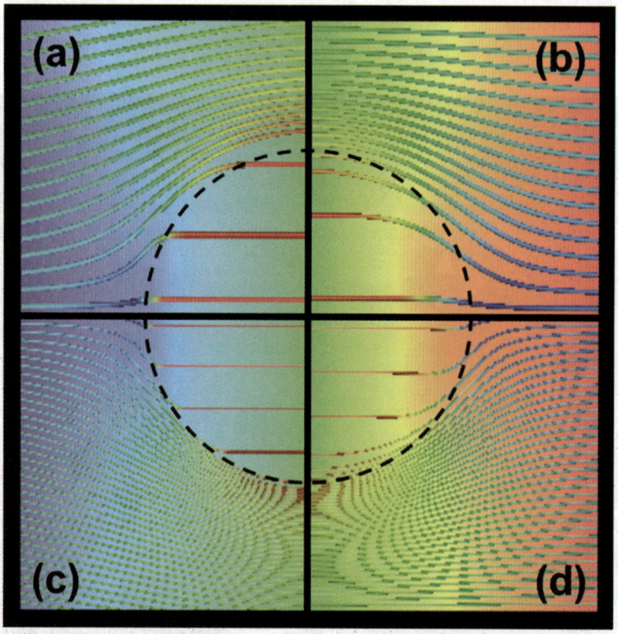

Fig. 4 Electrostatic potential (background) and electric field lines (the color of streamlines indicates the magnitude of the field, with red for large fields) for an oil droplet in water under an electric field. The oil–water interface is indicated as a dashed line. The dielectric contrast is $\gamma = 0.9$. Analytical results for an isolet droplet (a and c) are compared to simulation results with periodic boundary conditions (b and d). The effect of the finite width ($\xi = 2.45\Delta x$ for b and d) is more visible for the smaller system (a,b: box size $L = 50\Delta x$, droplet radius $R_d = 14.1\Delta x$) than with the larger one (c,d: $L = 100\Delta x$, $R_d = 28.2\Delta x$).

corresponds to a box size $L = 50\Delta x$ and a droplet radius $R_d = 14.1\Delta x$, the larger (d) to $L = 100\Delta x$, $R_d = 28.2\Delta x$. These results are compared to the analytical solution for an isolated cylinder:

$$\begin{aligned} \mathbf{E}(\mathbf{r}) &= (1+\gamma)\mathbf{E}_0, & r<R_d \\ \mathbf{E}(\mathbf{r}) &= \mathbf{E}_0 + \gamma R_d^2 \nabla\left(\frac{\mathbf{E}_0 \cdot \mathbf{r}}{r^2}\right), & r>R_d \end{aligned} \quad (26)$$

with $\mathbf{E}_0$ the applied field, evaluated on the nodes of the same lattices (a and c). Each part of the figure represents 1/4 of the simulation box. The figures for the smaller system (a and b) have been magnified to appear of the same size as the larger one (c and d). The field lines are computed numerically from the electrostatic potential and colored according to the magnitude of the local electric field.

The simulation results for a system with periodic boundary conditions are in good agreement with the analytical solution for an isolated droplet. In particular, the field lines tend to bypass the less dielectric droplet and are thus closer to each other in the water phase. The field strength is smaller in the water phase, where the dielectric constant is higher. The analytical solution (eqn (26)) corresponds to an infinitely thin interface ($\xi = 0$) with a discontinuous dielectric constant, whereas in simulations the latter varies smoothly over a distance $\sim\xi$. The effect of the finite width is more pronounced for the smaller system ($\xi/R_d \sim 0.17$) and the agreement with eqn (26) is excellent for the larger ($\xi/R_d \sim 0.09$). Although smaller ξ/R_d can be reached by decreasing ξ, the latter should remain at least of a few Δx for the variations of the composition ϕ to be well resolved on the lattice.

Results and discussion

Under the applied electric field, gradients of μ_ϕ lead to an evolution of the composition ϕ, corresponding to a deformation of the droplet. The precise location of the interface is defined by the curve $\phi(\mathbf{r}) = 0$, which is fitted to an ellipse to obtain the deformation D. Results are reported in Fig. 5 as a function of the dielectric contrast γ for two droplet radii and two electric field strengths. For the simulation parameters used here, the largest deformation is $<10^{-2}$ so that we are always in the small deformation limit assumed in eqn (25). Simulation results are in excellent agreement

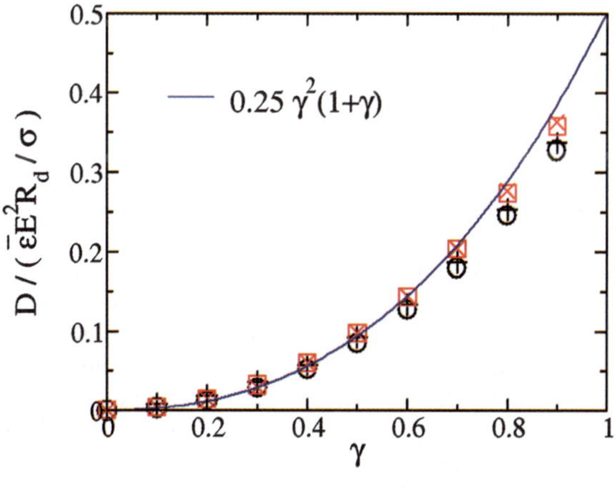

Fig. 5 Deformation of a 2D-oil droplet in water, as a function of the dielectric contrast $\gamma = (\bar{\varepsilon}_w - \bar{\varepsilon}_o)/(\bar{\varepsilon}_w + \bar{\varepsilon}_o)$. The deformation D is normalized by $\bar{\varepsilon}E^2R_d/\sigma$, with $\bar{\varepsilon} = (\varepsilon_w + \varepsilon_o)/2$, E the applied field, R_d the droplet radius and σ the surface tension. Simulation results are for two droplet radii ($R_d = 14.1\Delta x$ for □, × and $28.3\Delta x$ for ○, +) and two electric field strengths ($\beta eE\Delta x = 10^{-3}$ for ○, □ and 3 10^{-3} for ×, +) with the same interface width $\xi = 2.45\Delta x$. The solid line corresponds to the analytical solution eqn (25).

with the analytical ones. This demonstrates that the simulation method faithfully reproduces the prediction of the continuous free energy model.

The deviations observable only for large values of γ depend mainly on the radius R_d and less significantly on the applied electric field $\mathbf{E}$. This can be understood by analyzing the effect of the finite width on the energy balance leading to the final shape of the droplet and the theoretical deformation (eqn (25)). In the case of a thin interface ($\xi = 0$) and assuming that the field at steady-state is still given by eqn (26), we find that the electrostatic energy stored in the droplet when the field is turned on is (per unit length of the cylinder): $\int_{drop} \frac{1}{2}(\varepsilon_o E_{in}^2 - \varepsilon_w E_0^2)\,dS = -\frac{1}{2}\bar{\varepsilon}(1+\gamma)\gamma^2 E_0^2 S$, with $S = \pi R_d^2$ the section area of the droplet. This term tends to increase S by deforming the droplet and is balanced by the increase in surface energy $\sigma 2\pi R_d \mathscr{D}$ (per unit length of the cylinder). The corresponding deformation $\mathscr{D}$ is given by eqn (25).

The main effect of the finite width ξ is that the smooth variation of $\bar{\varepsilon}$ at the interface leads to a $O(\xi R_d \bar{\varepsilon}(1+\gamma)\gamma^2 E^0{}_2)$ correction to the electrostatic term, resulting in a smaller driving force for the deformation. This explains the field-independent $O(\xi/R_d)$ deviation of the simulation results from $\mathscr{D}_{theor}/(\bar{\varepsilon}E^2 R_d/\sigma)$ observed in Fig. 5. Other possible field-dependent corrections do not seem to be observed.

The results shown in this section demonstrate that the lattice simulation scheme is able to reproduce quantitatively the behaviour of a system consisting of two immiscible dielectric fluids under an electric field. We have not considered here the possibility of altering the phase behaviour of the solvent mixture in an inhomogeneous electric field, but this phenomenon is also captured by the free energy model used for our simulations.[58]

Ions at an oil–water interface

In this last section, we further exploit the power of our coarse-grained description by considering ions at a planar oil–water interface. The two solvents may have different dielectric constants and the ions different affinities for the two solvents ($\Delta\mu_\pm = \mu_\pm^o - \mu_\pm^w \neq 0$). The salt is partioned between the two phases and the salt concentration ratio is $\rho^o/\rho^w = \exp(-\beta\Delta\mu_{av})$ with $\Delta\mu_{av} = (\Delta\mu_+ + \Delta\mu_-)/2$. In the case of asymmetric solvation ($\Delta\mu_+ \neq \Delta\mu_-$) there exists at equilibrium an electrostatic potential difference (Donnan potential) across the interface $\psi_D = \psi^o - \psi^w = (\Delta\mu_- - \Delta\mu_+)/2e$. In that case there is a charge separation over distances characterized by two different Debye lengths $\kappa_{o,w}^{-1} = (2\beta e^2\rho^{o,w}/\varepsilon_{o,w})^{-1/2}$. For large values of the dielectric contrast or of $\Delta\mu_{av}$, they can differ by several orders of magnitude.

Experimental investigations of such interfaces at the microscopic scale have revealed that the interplay between solvation and electrostatic forces at the interface is rather complex.[59] This is particularly true if there is a dielectric constrast between the two phases ($\gamma \neq 0$) resulting in image charges interactions. While Onuki accounted for them by an effective "image charge potential" in his free energy description of mixtures of oil, water and ions,[27,28] van Roij *et al.*[35] suggested a Poisson–Boltzmann (PB) treatment for a thin planar interface introducing a shift s between the true solvent interface (where the dielectric constant changes) and the location of the jump $\Delta\mu_\pm$ in solvation potential felt by the ions. When $x < 0$ (resp. $x > 0$) corresponds to the oil (resp. water) phase, assuming $s > 0$ and setting $\psi(\infty) = 0$, we can write their result for the electrostatic potential as:

$$\psi^{PB} = \begin{cases} \psi_D - \dfrac{\psi_D}{A} e^{\kappa_o x}, x<0 \\ \psi_D - \dfrac{\psi_D}{A}[\cosh(\kappa_i x) + n\sinh(\kappa_i x)], x\in[0,s] \\ \dfrac{\psi_D}{A} e^{-\kappa_w(x-s)} p[n\cosh(\kappa_i s) + \sinh(\kappa_i s)], x>s \end{cases} \quad (27)$$

where $A = (1 + np)\cosh(\kappa_i s) + (n + p)\sinh(\kappa_i s)$, $n = \sqrt{\varepsilon_o/\varepsilon_w}$, $p = \sqrt{\rho^o/\rho^w} = \exp(-\beta\Delta\mu_{av}/2)$ and the screening length in the intermediate region is $\kappa_i^{-1} = (2\beta e^2\rho^o/\varepsilon_w)^{-1/2}$. This analytical result is exact only in the linearized regime, *i.e.* potential differences small compared to k_BT/e. Here we report lattice simulations based on our free energy model and compare the resulting ionic profiles to the prediction of the PB treatment for a flat interface.

System

We performed simulations of flat oil–water interfaces in a simulation box consisting of $N \times 1 \times 1$ lattice points with $N = 500$ and containing two interfaces (one O/W and one W/O). Owing to the periodic boundary conditions, the system corresponds to an infinite stack of oil and water slabs of width $2L_o$ and $2L_w$ (we used $L_o = L_w = 125\Delta x$). There are several relevant length scales in the system: the interface width ξ, the Debye screening lengths $\kappa_{o,w}^{-1}$ and the size of both phases $L_{o,w}$. The width of the interface should be small compared to all other length scales: $\kappa_{o,w}\xi \ll 1$ and $\xi \ll L_{o,w}$. In principle, the electric double-layers can overlap, especially in the oil phase where κ_o^{-1} is larger. This is even a crucial point to explain the electrostatic stabilization of surfactant free water droplets in oil.[60,61] In the present paper, we want to assess the validity of our coarse-grained simulation method by comparison with known results for an isolated interface and therefore consider only the limit $\kappa_{o,w}L_{o,w} \to \infty$. In particular, this condition is necessary for the salt concentrations to reach bulk values $\rho^{o,w}$ corresponding to well-defined screening lengths. Otherwise the amount of ions in the double layers could be non-negligible compared to the amount of ions in the bulk oil and water phases and the latter couldn't be considered as reservoirs. All simulations were performed using mixing free energy parameters $\beta B\Delta x^3 = 10^{-2}$ and $\beta K\Delta x = 3\ 10^{-2}$ with a mobility such that $MB = 5\times10^{-2}\Delta x^2/\Delta t$. These values of B and K lead in the absence of ions to an interface width $\xi \sim 2.45\Delta x$. Results are given only for one of the two interfaces in the simulation box, with $x < 0$ (resp. $x > 0$) corresponding to the oil (resp. water) phase.

Results and discussion

We first investigate the influence of the solvation free energy differences $\Delta\mu_\pm$ on the structure of the interface by considering a system without image charges ($\varepsilon_w = \varepsilon_o$, *i.e.* $\gamma = 0$). Fig. 6 compares simulation results to analytical predictions based on eqn (27) for $\beta\Delta\mu_\pm = \pm4$, corresponding to hydrophilic cations and hydrophobic anions. From the PB ionic chemical potential $\mu_\pm^{PB} = \pm e\psi^{PB} + V_\pm^{solv}$ we compute the ionic concentrations either as $\rho_\pm = \rho^w e^{-\beta\mu_\pm^{PB}}$ or as the linear expansion $\rho^w(1 - \beta\mu_\pm^{PB})$. In the following, we refer to the former as the reexponentiated PB result (RPB, see *e.g.* ref. 62) and to the latter as the linearized PB result (LPB). Since $\beta\Delta\mu_{av} = 0$ we expect equal salt concentrations in both phases and equal screening lengths (since we also have $\gamma = 0$) $\kappa_{o,w}^{-1} \sim 10.3\Delta x$, with a relatively good separation of length scales ($\kappa_{o,w}\xi < 1$ and $\kappa_{o,w}L_{o,w} > 1$). We also expect a Donnan potential difference $\psi_D = -4k_BT/e$ across the interface.

This figure shows that the LPB approximation fails to reproduce the ionic profiles, as expected for this large value of $\beta e\psi_D$. It even predicts negative concentrations near the interface. Adding the quadratic term in $\beta e\psi_D$ to approximate $\rho_\pm$ significantly improves the agreement while remaining consistent with the linear approximation for the potential (not shown). The RPB result, although not fully consistent with this approximation, is in quantitative agreement with the simulations, except at the interface where the finite width ξ smoothens the jump in ionic concentrations. The insert of Fig. 6 shows that the composition profile is close to $\phi^{th}(x) = -\tanh(x/\xi)$ and that the electrostatic potential is very well described by the linearized PB solution $\psi^{PB}(x)$ with a shift $s = 0$. Note that eqn (27) for ψ^{PB} is obtained under the assumption $\xi = 0$, and that potential differences are small compared to k_BT/e.

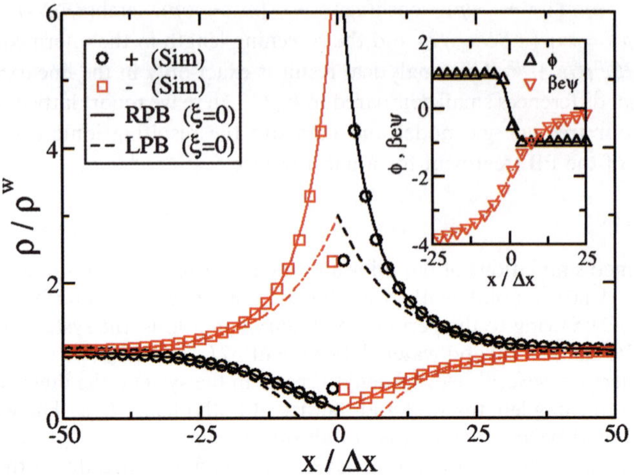

Fig. 6 Ionic profiles at an oil (left)–water (right) interface, for $\gamma = 0$, $\beta\Delta\mu_{\pm} = \pm 4$, $l_B = 0.8\Delta x$ and $\rho^w \Delta x^3 = 4.7\times10^{-4}$. Symbols are simulation results. Lines are analytical results based on eqn (27) with a shift $s = 0$ for the electrostatic potential ψ^{PB}. Dashed lines corresponds to the fully linearized solution, solid lines to the reexponentiated one (see text). The inset compares the simulation results (symbols) for ϕ and ψ to the analytical solutions (lines).

Surprisingly, simulations for a finite ξ with relatively large values of $\beta e\psi_D$ give similar results for the electrostatic potential even in the transition region where the ionic profiles differ. Although it is not easily seen on this inset, the simulated profile is in fact sharper than $\phi^{th}(x)$.

Increasing $\beta e\psi_D$ while keeping $\beta\Delta\mu_{av} = 0$ increases the charge separation and the excess of ions at the interface compared to the bulk concentration. Before analyzing quantitatively the influence of ψ_D on these quantities, let us examine the effect of the finite interface width on the ionic profiles $\rho_{\pm}(x)$ and the influence of the solvation free energy on the composition profile $\phi(x)$. Fig. 7 reports the ionic profiles for $\gamma = 0.0$, $\beta\Delta\mu_{av} = 0$ and Donnan potentials $\beta e\psi_D = -1, -2$ and -3. Simulation results are compared to the RPB solution for a thin interface ($\xi = 0$) and for solvation potentials $V_{\pm}^{solv}$ corresponding to a composition $\phi^{th}(x) = -\tanh(x/\xi)$ expected for an ion-free interface, where $\xi = \sqrt{2K/B}$. The main effect of the finite interface width is to smoothen ionic concentration profiles over a distance $\sim\xi$. Beyond this distance the concentrations are equal to the thin interface result. The profiles are better described by the combination of ψ^{PB} with the smoothed solvation potentials, but the agreement deteriorates as $|\psi_D|$ increases. This is because the composition profile $\phi(x)$ also deviates from $\phi^{th}(x)$ as $|\psi_D|$ increases: the larger the solvation free energy difference (*i.e.* the larger $|\psi_D|$), the sharper the interface.

We now analyze how the finite width ξ of the interface affects the overall excess of ions near the interface, quantified by the adsorption:

$$\Gamma = \int_{-\infty}^{0} [\rho_+(x) + \rho_-(x) - 2\rho^o]\mathrm{d}x + \int_{0}^{\infty} [\rho_+(x) + \rho_-(x) - 2\rho^w]\mathrm{d}x \tag{28}$$

Fig. 8 compares the simulated adsorption to the LPB result:

$$\Gamma^{LPB} = \frac{\kappa}{8\pi l_B}\frac{(\beta e\psi_D)^2}{4} \tag{29}$$

with $\kappa = \kappa_o = \kappa_w$ and l_B the Bjerrum length common to both phases (since $\gamma = 0$). This result is obtained by expanding $\rho_{\pm}(x)$ to second order in $\beta\mu_{\pm}^{PB}$, for the linear expansion yields $\Gamma = 0$. The error bars reported in Fig. 8 correspond to the estimate:

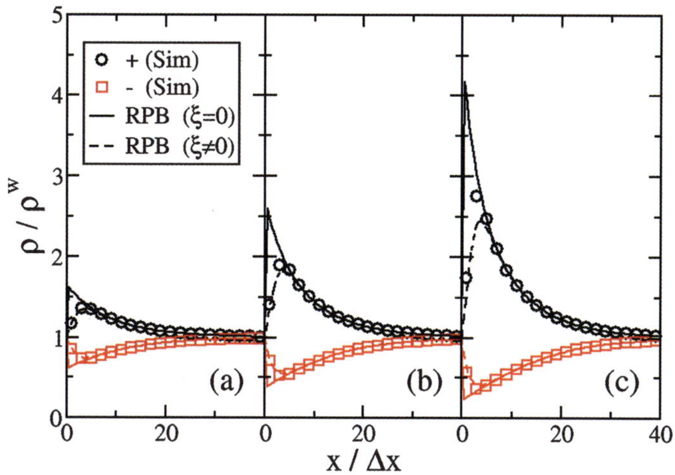

Fig. 7 Ionic profiles in the water phase for $\gamma = 0.0$, $\beta\Delta\mu_{av} = 0$, $l_B = 0.8\Delta x$, $\rho^w\Delta x^3 = 4.9\ 10^{-4}$ and Donnan potentials $\beta e\psi_D = -1$ (a), -2 (b) and -3 (c). Simulation results (symbols) are compared to the RPB result (see text) for a thin interface (solid line) and for $V_\pm^{solv}$ corresponding to $\phi^{th}(x) = -\tanh(x/\xi)$ (dashed line) expected for an ion-free interface.

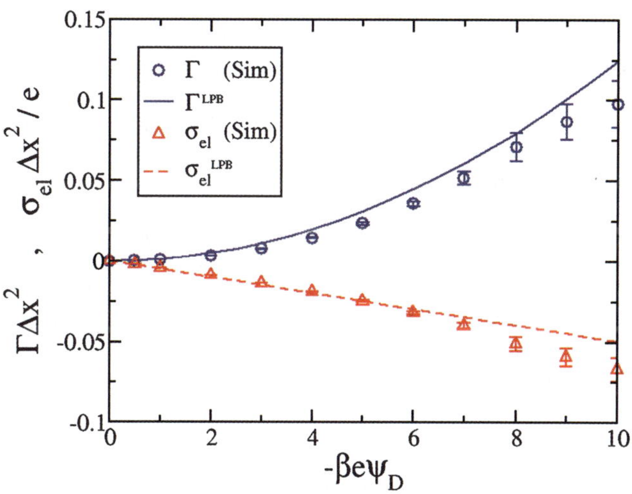

Fig. 8 Ionic adsorption Γ and surface charge density σ_{el} of the oil phase as a function of ψ_D for $\gamma = 0$ and $\beta\Delta\mu_{av} = 0$. Simulation results (symbols) for $l_B = 0.8\Delta x$ are compared to the linearized Poisson–Boltzmann results for $\xi = 0$ (lines) given by eqn (29) and (31).

$$\delta\Gamma = \frac{1}{2}\left(\int_{-\Delta x}^{0}\left[\rho_+(x) + \rho_-(x) - 2\rho^o\right]\mathrm{d}x + \int_{0}^{\Delta x}\left[\rho_+(x) + \rho_-(x) - 2\rho^w\right]\mathrm{d}x\right) \quad (30)$$

with Δx the lattice spacing. The agreement between the simulation and LPB results is seen to be very good, except at very large ψ_D. This performance of the LPB approximation even for $\beta e\psi_D > 1$ where it fails to predict the ionic profiles can be traced back to a compensation of errors. The RPB result, much closer to the simulated ionic profiles, predicts larger adsorptions than the LPB approximation. But the finite interface width smoothens the ionic profiles and therefore diminishes the value of Γ. In addition, for the largest ψ_D, Γ is not small compared to the amount of "bulk" ions $\rho^o L_o + \rho^w L_w \sim 0.12\Delta x^{-2}$ so that the simulated system cannot be considered as an isolated interface in contact with infinite reservoirs.

The good agreement between the simulated ψ and linearized PB result ψ^{PB} suggests that the electric charge density is also close to the PB solution. In particular, the total electric charge of each phase $\sigma_{el} = \sigma_o = -\sigma_w = \int_{-\infty}^{0} (\rho_+ - \rho_-)e\, dx$ can be compared to the PB prediction for $\xi = 0$:

$$\frac{\sigma_{el}^{LPB}}{e} = \frac{\kappa}{4\pi l_B A}(\beta e \psi_D) \tag{31}$$

with here $A = 2$ (see eqn (27)). Fig. 8 also displays σ_{el} as a function of the Donnan potential ψ_D for the simple case $\beta\Delta\mu_{av} = 0$ and $\gamma = 0$. The error bars correspond to $\delta\sigma_{el} = \frac{1}{2}\int_{-\Delta x}^{\Delta x} |\rho_+ - \rho_-| e\, dx$. This figure shows that simulation results are indeed well described by eqn (31) and confirms that the charge separation at the interface is proportional to ψ_D, except at very large ψ_D. The symmetric interface behaves as a capacitor of permittivity $\bar{\varepsilon}$ and width κ^{-1}, with capacitance per unit area $\bar{\varepsilon}\kappa$. When submitted to a potential difference ψ_D, each side builds up a charge per unit area $\sigma_{el} \propto \bar{\varepsilon}\kappa\psi_D$. The good agreement between the simulation and PB results for σ_{el} is consistent with the finding of ref. 35 that global quantities related to the partitioning between the two phases are not influenced by the finite interface width if it is smaller than the interfacial Debye length.

All results presented so far concerned the symmetric solvation case $\beta\Delta\mu_{av} = 0$. Simulations in the asymmetric case also give the expected results. For example, we find that the salt concentration ratio ρ^o/ρ^w decreases from $e^{-1} \sim 0.37$ for $\beta\Delta\mu_+ = +2$ and $\beta\Delta\mu_- = 0$ (hence $\beta\Delta\mu_{av} = +1$) to $e^{-3} \sim 5\times10^{-2}$ for $\beta\Delta\mu_+ = +4$ and $\beta\Delta\mu_- = +2$ ($\beta\Delta\mu_{av} = +3$). While these two conditions correspond to the same $\psi_D = -k_BT/e$, the former yields a larger adsorption Γ. In addition, the ionic profiles are again in quantitative agreement with the RPB solution for a shift $s = 0$. This choice gives the best agreement when there is no dielectric contrast between the two solvents.

When $\gamma \neq 0$ the situation is more complex and even the sign of s depends on γ, $\Delta\mu_+$ and $\Delta\mu_-$. An example of such a situation is illustrated in Fig. 9. The best agreement with eqn (27) was obtained in that case for a shift $s = -1.25\Delta x$. All other parameters being fixed, we find that for increasing $\Delta\mu_+ = -\Delta\mu_-$ the required shift also increases. Simulation results for the electrostatic potential are well described by the PB result for a vanishing width $\xi = 0$ and a finite shift. Therefore the minimal model of van Roij *et al.* to account for image charges seems to be appropriate, at least for the explored range of parameters. Even for $\beta e\psi_D > 1$, where the fully linear PB approximation fails, we find that the reexponentiated PB result gives an accurate description of ionic profiles. Moreover, our results show that the finite width of the interface affects the ionic profiles in such a way that the electrostatic potential is still close to the PB prediction. These profiles are accurately described by the combination of ψ^{PB} with solvation potentials $V_{\pm}^{solv}$ corresponding to an unperturbed composition profile $\phi^{th}(x) = -\tanh(x/\xi)$. In particular, "far" from the interface, the effect of the finite width on solvation forces is negligible and the ionic profiles coincide with the result for $\xi = 0$. As shown previously in the case $\gamma = 0$, we expect for larger ψ_D a perturbation of the composition profile and a corresponding modification of the ionic concentrations.

The results presented in this section demonstrate the ability of our coarse-grained simulation scheme to study the interface between two immiscible solvents in the presence of ions, including the effect of their (possibly asymmetric) affinity for both phases. To assess the validity of the simulation scheme we only showed results in the one-dimensional case, for which analytical predictions are available. Simulation in two or three dimensions is straightforward.

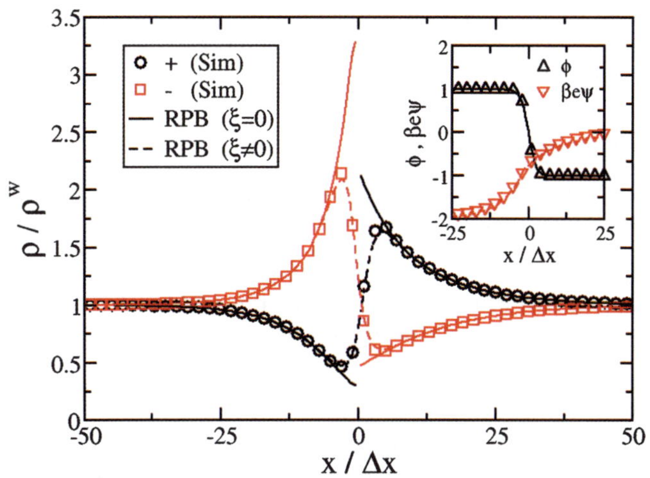

Fig. 9 Ionic profiles at the oil (left)–water (right) interface for $\gamma = 0.5$, $\beta\Delta\mu_{\pm} = \pm 2$, $l_B = 0.8\Delta x$, $\rho^w\Delta x^3 = 4.9\times10^{-4}$ and a shift $s = -1.25\Delta x$. Simulation results (symbols) are compared to analytical results for a thin interface (solid line) and for the combination of ψ^{PB} with solvation potentials $V^{solv}{}_{\pm}$ corresponding to a composition $\phi^{th}(x) = -\tanh(x/\xi)$ (dashed line). The insert compares the simulation (symbols) and analytical (lines) results for the composition ϕ and electrostatic potential ψ.

Conclusion

We have presented a coarse-grained simulation method for complex charged systems. This mesoscopic model couples a hydrodynamic description to a free energy functional accounting for the interactions between solvent(s) and charged solutes. All the parameters entering in the model, such as free energy parameters (*e.g.* related to ion solvation), solvent viscosity or ionic diffusion coefficients are, at least in principle, computable by simulations at the molecular level. We described the implementation of this model in a hybrid lattice-based scheme, whereby the evolution of the overall mass and momentum is taken care of *via* a Lattice Boltzmann scheme, whereas the composition and ionic concentrations are updated using the link-flux method. We presented several applications of the coarse-grained simulation method: the transport of charged tracers in charged porous media, the deformation of an oil droplet in water under the effect of an applied electric field, and the distribution of ions at an oil–water interface as a function of the ions affinity for both solvents. When possible, we compared our simulation results to exact or approximate analytical results to investigate the range of simulation parameters that can be used to recover the continuous results (*e.g.* the effect of a finite interface width).

The proposed method will be very useful to simulate the dynamics of complex mixtures of solvents and ions. In particular, it will be interesting to investigate electrokinetic phenomena at a charged oil–water interface, or of oil–water mixtures in charged porous media. Understanding electrokinetic effects in these systems might be very helpful in designing new electro-acoustic oil recovery techniques or monitoring devices. This coarse-grained simulation method could also be applied to the study of electrowetting and of microfluidic devices, particularly those based on electrokinetic pumping.

Although surfactant free emulsions can be stabilized by the presence of salt, the most usual situation also involves surfactant molecules at the interface and in solution. As free energy based models for mixtures of oil, water and surfactants have already been proposed, it should be rather straightforward to couple them to the one introduced in the present paper. One physical feature not included in the method presented here is the presence of thermal fluctuations. It would be interesting to investigate the possibility of including fluctuations as is done in the fluctuating Lattice Boltzmann method.[63,64]

Acknowledgements

The authors would like to thank Rene van Roij, Jean-Pierre Hansen and Edo Boek for very useful discussions. B.R. acknowledges financial support from the Agence Nationale pour la Gestion des Déchets Radioactifs (ANDRA, France). I.P. acknowledges financial support from CAICYT (FIS2005–01299). The work of the FOM Institute is part of the research program of FOM and is made possible by financial support from the Netherlands organization for Scientific Research (NWO).

References

1 M. Karttunen, I. Vattulainen and A. Lakkarinen, *Novel Methods in Soft Matter Simulations, Lecture Notes in Physics Vol. 640*, Springer Verlag, 2004.
2 P. Español and P. Warren, *Europhys. Lett.*, 1995, **30**, 191.
3 C. Lowe, *Europhys. Lett.*, 1999, **47**, 145.
4 A. Malevanets and R. Kapral, *J. Chem. Phys.*, 2000, **112**, 7260.
5 S. Succi, *The Lattice Boltzmann Equation for Fluid Dynamics and Beyond*, Oxford University Press, 2001.
6 M. Swift, E. Orlandini and J. Yeomans, *Phys. Rev. Lett.*, 1995, **75**.
7 E. Orlandini, M. Swift and J. Yeomans, *Europhys. Lett.*, 1995, **32**, 463.
8 M. Swift, E. Orlandini, W. Osborn and J. Yeomans, *Phys. Rev. E*, 1996, **54**, 5041.
9 O. Theissen, G. Gompper and D. Kroll, *Europhys. Lett.*, 1998, **42**, 419.
10 A. Lamura, G. Gonnella and J. Yeomans, *Europhys. Lett.*, 1999, **45**, 314.
11 R. van der Sman and S. van der Graaf, *Rheol. Acta*, 2006, **46**, 3.
12 C. Denniston, E. Orlandini and J. M. Yeomans, *Phys. Rev. E: Stat., Nonlinear, Soft Matter Phys.*, 2001, **63**, 056702.
13 C. M. Care, I. Halliday, K. Good and S. V. Lishchuk, *Phys. Rev. E: Stat., Nonlinear, Soft Matter Phys.*, 2003, **67**, 061703.
14 A. Dupuis, D. Marenduzzo, E. Orlandini and J. M. Yeomans, *Phys. Rev. Lett.*, 2005, **95**, 097801.
15 T. J. Spencer and C. M. Care, *Phys. Rev. E: Stat., Nonlinear, Soft Matter Phys.*, 2006, **74**, 061708.
16 Q. Li and A. Wagner, *Phys. Rev. E: Stat., Nonlinear, Soft Matter Phys.*, 2007, **76**, 036701.
17 S. Ramachandran, P. Kumar and I. Pagonabarraga, *Eur. Phys. J. E*, 2006, **20**, 151.
18 J.-C. Desplat, I. Pagonabarraga and P. Bladon, *Comput. Phys. Commun.*, 2001, **134**, 273.
19 J. Harting, M. Venturoli and P. Coveney, *Philos. Trans. R. Soc. London, Ser. A*, 2004, **362**, 1703.
20 K. Stratford, R. Adhikari, I. Pagonabarraga and J.-C. Desplat, *J. Stat. Phys.*, 2005, **121**, 163.
21 K. Stratford, R. Adhikari, I. Pagonabarraga, J.-C. Desplat and M. E. Cates, *Science*, 2005, **309**, 2198.
22 K. Stratford and I. Pagonabarraga, *Comput. Math. Appl.*, 2008, **55**, 1585.
23 J. Horbach and D. Frenkel, *Phys. Rev. E: Stat., Nonlinear, Soft Matter Phys.*, 2001, **64**, 061507.
24 C. Pooley and K. Furtado, *Phys. Rev. E: Stat., Nonlinear, Soft Matter Phys.*, 2008, **77**, 046702.
25 F. Capuani, I. Pagonabarraga and D. Frenkel, *J. Chem. Phys.*, 2004, **121**, 973.
26 I. Pagonabarraga, F. Capuani and D. Frenkel, *Comput. Phys. Commun.*, 2005, **169**, 192.
27 A. Onuki, *Phys. Rev. E: Stat., Nonlinear, Soft Matter Phys.*, 2006, **73**, 021506.
28 A. Onuki, *J. Chem. Phys.*, 2008, **128**, 224704.
29 A. Onuki, *Europhys. Lett.*, 2008, **82**, 58002.
30 R. Ledesma-Aguilar, A. Hernández-Machado and I. Pagonabarraga, *Phys. Fluids*, 2007, **19**, 102112.
31 A. Briant and J. Yeomans, *Phys. Rev. E: Stat., Nonlinear, Soft Matter Phys.*, 2004, **69**, 031603.
32 R. van der Sman and S. van der Graaf, *Comput. Phys. Commun.*, 2008, **178**, 492.
33 N. Deneshyuk and J.-P. Hansen, *J. Chem. Phys.*, 2004, **121**, 3613.
34 J. Zwanikken and R. van Roij, *Phys. Rev. Lett.*, 2007, **99**, 178301.

35 M. Bier, J. Zwanikken and R. van Roij, *Phys. Rev. Lett.*, 2008, **101**, 046104.
36 U. M. B. Marconi and P. Tarazona, *J. Chem. Phys.*, 1999, **110**, 8032.
37 U. Marconi and P. Tarazona, *J. Phys.: Condens. Matter*, 2000, **12**, A413.
38 H. Lowen, *J. Phys.: Condens. Matter*, 2003, **15**, V1.
39 A. J. Archer and R. Evans, *J. Chem. Phys.*, 2004, **121**, 4246.
40 M. Rex and H. Löwen, *Phys. Rev. Lett.*, 2008, **101**, 148302.
41 K. Furtado, C. Pooley and J. Yeomans, *Phys. Rev. E: Stat., Nonlinear, Soft Matter Phys.*, 2008, **78**, 046308.
42 D. Marenduzzo, E. Orlandini, M. E. Cates and J. M. Yeomans, *Phys. Rev. E: Stat., Nonlinear, Soft Matter Phys.*, 2007, **76**, 031921.
43 H. Li and H. Ki, *Commun. Comput. Phys.*, 2008, **4**, 337.
44 R. Benzi, S. Succi and M. Vergassola, *Phys. Rep.*, 1992, **222**, 145.
45 P. Bhatnagar, E. Gross and M. Krook, *Phys. Rev.*, 1954, **94**, 511.
46 W. H. Press, S. A. Teukolsky, W. T. Vetterling, and B. P. Flannery, *Numerical Recipes in C: The Art of Scientific Computing*, Cambridge University Press, Cambridge, 2nd edn, 1993.
47 B. Rotenberg, I. Pagonabarraga and D. Frenkel, *Europhys. Lett.*, 2008, **83**, 34004.
48 P. Sen, *Concepts Magn. Reson.*, 2004, **23a**, 1.
49 P. Mitra, P. Sen, S. L. M. and P. Le Doussal, *Phys. Rev. Lett.*, 1992, **68**, 3555.
50 M. Ochs, M. Boonekamp, H. Wanner, H. Sato and M. Yui, *Radiochim. Acta*, 1998, **82**, 437.
51 I. C. Bourg, A. C. M. Bourg and G. Sposito, *J. Contam. Hydrol.*, 2003, **61**, 293.
52 I. Bourg, G. Sposito and A. Bourg, *Clays Clay Miner.*, 2006, **54**, 363.
53 M. Van der Hoef and D. Frenkel, *Phys. Rev. A*, 1990, **41**, 4277.
54 M. Van der Hoef and D. Frenkel, *Physica D*, 1991, **47**, 191.
55 C. Lowe and D. Frenkel, *Physica A*, 1995, **220**, 251.
56 A. Valfouskaya, P. Adler, J.-F. Thovert and M. Fleury, *J. Appl. Phys.*, 2005, **97**, 083510.
57 C. T. O'Konski and H. C. Thacher Jr, *J. Phys. Chem.*, 1953, **57**, 955.
58 G. Marcus, S. Samin and Y. Tsori, *J. Chem. Phys.*, 2008, **129**, 061101.
59 G. Luo, S. Malkova, J. Yoon, D. Schultz, B. Lin, M. Meron, I. Benjamin, P. Vanýsek and M. Schlossman, *Science*, 2006, **311**, 216.
60 M. Leunissen, A. van Blaaderen, A. Hollingsworth, M. Sullivan and P. Chaikin, *Proc. Natl. Acad. Sci. U. S. A.*, 2007, **104**, 2585.
61 J. de Graaf, J. Zwanikken, M. Bier, A. Baarsma, Y. Oloumi, M. Spelt and R. van Roij, *J. Chem. Phys.*, 2008, **129**, 194701.
62 J.-P. Hansen and I. R. McDonald, *Theory of Simple Liquids*, Academic Press, 3rd edn, 2006.
63 R. Adhikari, K. Stratford, M. E. Cates and A. J. Wagner, *Europhys. Lett.*, 2005, **71**, 473.
64 B. Dünweg, U. Schiller and A. Ladd, *Phys. Rev. E: Stat., Nonlinear, Soft Matter Phys.*, 2007, **76**, 036704.

Multi-particle collision dynamics simulations of sedimenting colloidal dispersions in confinement

Adam Wysocki,*[a] C. Patrick Royall,[bd] Roland G. Winkler,[c] Gerhard Gompper,[c] Hajime Tanaka,[d] Alfons van Blaaderen[e] and Hartmut Löwen*[a]

Received 26th January 2009, Accepted 6th March 2009
First published as an Advance Article on the web 2nd September 2009
DOI: 10.1039/b901640f

The sedimentation of an initially inhomogeneous distribution of hard-sphere colloids confined in a slit is simulated using the multi-particle collision dynamics scheme which takes into account hydrodynamic interactions mediated by the solvent. This system is an example for soft matter driven out of equilibrium where various length and time scales are involved. The initial laterally homogeneous density profiles exhibit a hydrodynamic Rayleigh–Taylor-like instability. Solvent backflow effects lead to an intricate non-linear behaviour which is analyzed via the solvent flow field and the colloidal velocity correlation function. Our simulation data are in good agreement with real-space microscopy experiments.

1 Introduction

Mesoscopic colloidal dispersions embedded in a molecular solvent are soft matter systems which need multiscale modelling. For structural equilibrium correlations, this is mainly a problem of different length scales which has been widely addressed and is by now well-understood, *e.g.* by using the concept of effective interactions.[1] For dynamical correlations and non-equilibrium situations, widely different time scales require careful multiscale modelling. The dynamics of a molecular solvent takes place on the picosecond level, while the time a colloid needs to diffuse over its own radius, is in the second time scale for micron-sized colloids. The hydrodynamic interactions[2] between the colloidal particles are mediated by the solvent flow on an intermediate time scale and are long-ranged and of many-body nature. It is only at very low colloidal volume fractions that hydrodynamic interactions can be neglected. Recently, various computational schemes have been developed to tackle hydrodynamic interactions ranging from the lattice Boltzmann technique,[3] fluidized particle methods[4] to multi-particle collision dynamics (MPCD)[5–7] where the solvent flow is modelled as ideal gas particles which exchange momentum locally by stochastic rotation of the relative velocities.

[a]*Institut für Theoretische Physik II, Heinrich-Heine-Universität Düsseldorf, Universitätsstrasse 1, D-40225 Düsseldorf, Germany. E-mail: adam@thphy.uni-duesseldorf.de; hlowen@thphy.uni-duesseldorf.de*
[b]*School of Chemistry, University of Bristol, Bristol, BS8 1TS, UK*
[c]*Institut für Festkörperforschung, Forschungzentrum Jülich, D-52425 Jülich, Germany*
[d]*Institute of Industrial Science, University of Tokyo, 4-6-1 Komaba, Meguro-ku, Tokyo, 153-8505, Japan*
[e]*Soft Condensed Matter Group, Debye Institute for Nanomaterials Science, Utrecht University, PO Box 80000, 3508, TA Utrecht, The Netherlands*

Here we consider a hydrodynamic instability of sterically-stabilized colloids (hard spheres) confined to a slit by inverting gravity which acts perpendicular to the slit. We simulate the hydrodynamic instability and incorporate the crucial hydrodynamic interactions by using the MPCD scheme.

The motion of a colloid is characterised by the Peclet number $Pe = \tau_D/\tau_S$, which is the ratio between the time τ_D it takes a particle to diffuse its own radius and the time τ_S it takes to sediment the same distance. A Peclet number of order unity is the dividing line between colloidal ($Pe \leq 1$) and granular systems ($Pe \gg 1$), *i.e.* Pe measures the importance of Brownian motion. The classical Rayleigh–Taylor instability, which occurs if a heavy, immiscible fluid layer is placed on top of a lighter one has been intensively studied for the case of a simple Newtonian fluid both by theory,[8] simulation[9] and experiment, and is observed in granular matter,[10–12] in surface-tension dominated colloid–polymer mixtures[13] and in a suspension of dielectric particles exposed to an ac electric field gradient.[14] However, except for ref. 15, this instability was never simulated on the particle scale in the context of colloidal sedimentation in confinement including hydrodynamic interactions. Here we present simulation data using the MPCD method. The instability is resolved on the colloidal particle scale and good agreement with real-space experiments is observed. We further show that the instability is accompanied by significant solvent backflow effects. Finally, correlations of the colloidal velocities are calculated which reveal strong lateral correlations and anticorrelations which are time-dependent.

The paper is organized as follows: in chapter II the simulation method is briefly described. Results for the sedimentation problem are presented in chapter III and compared to real-space confocal microscopy data. The instability and the concomitant solvent flow fields and colloidal velocity correlations are discussed. Finally, we present our conclusions in chapter IV.

2 The simulation model

Our model consists of a suspension of N solute particles with mass M and hard sphere diameter σ immersed in a bath of N_s solvent particles with mass m and a number density $N_s = N_s/V$, here V is the volume of the simulation box. The system is confined between two walls with distance L in x-direction and has periodic boundary conditions otherwise. The N colloidal particles with space position $\mathbf{R}_i$ and velocity $\mathbf{V}_i$ propagate according to Newton's equation of motion

$$M\frac{\mathrm{d}\mathbf{V}_i}{\mathrm{d}t} = F_S\mathbf{e}_x - \sum_{j\neq i}\nabla_{\mathbf{R}_j}V(R_{ij}) + F_w(X_i)\mathbf{e}_x$$

The first term on the right hand side is the constant driving force of strength F_S directed perpendicular to the walls in the x-direction and the second one represents the force due to the interaction with other colloids ($R_{ij} = |\mathbf{R}_i - \mathbf{R}_j|$ is the interparticle distance). The third term is a repulsive wall–colloid force. To avoid overlap the colloids interact *via* a screened Coulomb potential which diverges at $R_{ij} = \sigma$, here the reduced inverse screening length is $\kappa\sigma = 40$. A similar potential is used for the colloid–wall interaction. We integrate the equation of motion using a velocity Verlet algorithm with a time step δt. Simultaneously, the solvent particles with space position $\mathbf{r}_i$ and velocity $\mathbf{v}_i$ move ballistically also within the same time step δt, *i.e.*

$$\mathbf{r}_i(t + \delta t) = \mathbf{r}_i(t) + \mathbf{v}_i(t)\delta t \tag{2}$$

To enforce no-slip boundary conditions on the colloid surface a stochastic reflection method[16,17] is applied. If a solvent particle i hits a colloid it gets a new velocity $\mathbf{u}_i = \mathbf{u}_{n,i} + \mathbf{u}_{t,i}$ relative to the velocity of the colloids boundary from a distribution for the normal velocity component $P_n(u_n) = m\beta u_n e^{-m\beta u_n^2/2}$ and the tangential velocity component $P_t(u_t) = \sqrt{m\beta/(2\pi)}e^{-m\beta u_t^2/2}$ with $\beta^{-1} = k_B T$. Then the new velocity of the solvent particle i after collision with the colloid j reads as

$$\mathbf{v}_i(t + \delta t) = \mathbf{V}_j(t) + \mathbf{\Omega}_j(t) \times (\tilde{\mathbf{r}}_i - \mathbf{R}_j(t)) + \mathbf{u}_i \tag{3}$$

where $\mathbf{\Omega}_j$ is the angular velocity of the colloid and $\tilde{\mathbf{r}}_i$ is the point of contact at the colloid surface. After all collisions within δt are completed, the new velocity of the colloid j is updated as

$$\mathbf{V}_j(t+\delta t) = \mathbf{V}_j(t) + \frac{m}{M}\sum_{i\in C}(\mathbf{v}_i(t) - \mathbf{v}_i(t+\delta t)) \tag{4}$$

and the new angular velocity is

$$\mathbf{\Omega}_j(t+\delta t) = \mathbf{\Omega}_j(t) + \frac{m}{I}\sum_{i\in C}(\tilde{\mathbf{r}}_i - \mathbf{R}_j(t)) \times (\mathbf{v}_i(t) - \mathbf{v}_i(t+\delta t)) \tag{5}$$

where C is the set of solvent particles colliding with colloid j in the time interval $[t, t + \delta t]$ and $I = 2/5M(\sigma/2)^2$ is the moment of inertia of the spherical colloids.

After a time $\Delta t = n\delta t$ the solvent particles interact with each other *via* a multi-particle collision.[5] The particles are sorted in cubic cells of size a and the center-of-mass velocity $\mathbf{U}_\xi = N_\xi^{-1}\sum_{j\in\xi}\mathbf{v}_j$ of each cell ξ is calculated (N_ξ is the number of particles in the cell ξ). Then in each cell the relative velocities $\delta\mathbf{v}_i = \mathbf{v}_i - \mathbf{U}_{\xi_i}$ are rotated by an angle α around a random axis, *i.e.*

$$\mathbf{v}_i(t + \Delta t) = \mathbf{U}_{\xi_i}(t) + S_{\xi_i}\hat{\omega}_{\xi_i}(\alpha)\delta\mathbf{v}_i(t) \tag{6}$$

where $\hat{\omega}_{\xi_i}(\alpha)$ is the stochastic rotation matrix and S_{ξ_i} is a thermostat operator (see below). $\hat{\omega}_{\xi_i}(\alpha)$ is equal for all particles within the same cell but uncorrelated between different cells and in time. Due to this operation the particles exchange momentum in the cell while the total kinetic energy and the total momentum in the cell are conserved. It was shown[5] that (2) and (6) lead in equilibrium to a Maxwell–Boltzmann distribution of the velocities and that Navier–Stokes-hydrodynamics is generated.

Since in any kind of a non-equilibrium simulation thermostating is required to avoid viscous heating, we rescale the relative velocities $\delta\mathbf{v}_i$ in each cell by a factor

$$S_\xi = \sqrt{\frac{3(N_\xi - 1)k_B T}{m\sum_{i\in\xi}\delta\mathbf{v}_i^2}} \tag{7}$$

This thermostat acts locally and is unbiased with respect to the flow field and hence does not destroy the hydrodynamic behaviour, see ref. 18.

For a small mean free path $\lambda = \Delta t\sqrt{k_B T/m} \ll a$ the same set of particles interact over several Δt with each other before leaving this cell and hence they remain correlated over several Δt. If one imposes a flow field the degree of correlation and therefore the transport coefficients depends on the magnitude of the flow field. The violated Galilean invariance can be restored if all particle positions are shifted by a random vector before the collision step.[19]

At the wall surfaces we also enforce no-slip boundary conditions using a stochastic reflection method but additionally we fill the wall cells with $n_s a^3 - N_\xi$ "ghost" particles during the collision step ($n_s a^3$ is the average number of solvent particles in the collision cell), because due to the random shift partially occupied boundary collision cells occur.[20] The velocities of the "ghost" particles are drawn from a Maxwell–Boltzmann distribution with zero mean and variance $k_B T$.

We employed the parameters $\Delta t = 0.2\tau$, $\alpha = 3\pi/4$, $n_s a^3 = 5$ with $\tau = \sqrt{ma^2/(k_B T)}$. With these parameters the total solvent kinematic viscosity is $\nu = \eta/\rho \approx 0.5a^2/\tau$

($\rho = mn_s$ is the mass density and η the dynamic viscosity).[21] For the colloids we use $M = 167m$, hard core diameter $\sigma = 4a$ and $\delta t = \Delta t/4$. We calculate the diffusion constant in a bulk simulation from the integral of the velocity autocorrelation function and obtain $D = 0.013a^2/t_0$.[17] With these parameters we achieve the hierarchy of time scales for a colloidal particle $\Delta t = 0.2\tau < \tau_c = 1.5\tau < \tau_B = 2.2\tau < \tau_\nu = 8\tau < \tau_D = 307\tau$, see ref. 22. Here τ_c is the time a sound wave needs to propagate over one colloidal radius (the speed of sound is $c = \sqrt{5k_BT/(3m)}$), τ_B is the time the velocity of a colloid is correlated and τ_ν is the time over which the solvent momentum diffuses over one colloidal radius. We performed simulations up to $Pe_{max} = 4.8$ which correspond to a maximal Reynolds number of $Re_{max} = D/\nu Pe_{max} \approx 0.125 < 1$, such that inertial effects are negligible. To avoid compressibility effects we also ensure that the Mach number is smaller than unity, $Ma_{max} = 2D/(\sigma c)Pe_{max} = 0.024 \ll 1$. With these values of the hydrodynamic numbers and the hierarchy of time scales we are sure that a comparison with a real physical system is reasonable.

The system we consider contains N = 15 048 hard sphere particles and N_s = 14 274 843 solvent particles in a box of $L/\sigma = 18$ and $L_y/\sigma = L_z/\sigma = 54$. We start our simulation with a fully equilibrated system under the influence of a constant force in the x-direction, *i.e.* particles have collected and settled at the top of the box. Then at time $t = 0$ we instantaneously reverse the direction of the force and monitor the evolution of the system until all particles have sedimented to the bottom.

We compare our simulation to single-particle level confocal microscopy experiments with sterically stabilised polymethylmethacrylate colloids. The value of the Peclet number can be adjusted by variation of the density mismatch between colloid and solvent mass density. Prior equilibration was achieved by placing the suspension overnight such that it sedimented across a thin (typically 50 μm) capillary. The capillary was then inverted, and the evolution under sedimentation was followed. More details of the experiment are described in ref. 23,24.

3 Results

A situation where a layer of a heavy fluid is on top of a lighter one is clearly mechanically unstable, *i.e.* the system is inclined to invert. During this process the initially

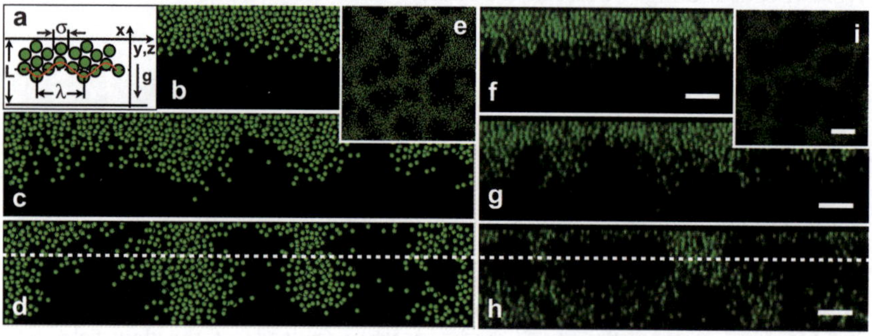

Fig. 1 Simulation and confocal microscopy snapshots. (a) A schematic illustrating the spatial parameters σ, λ and L. (b–e) Simulation snapshots of a system which contains N = 33 858 colloidal particles and N_s = 32 118 397 solvent particles (not displayed) in a simulation box with dimensions $L/\sigma = 18$ and $L_y/\sigma = L_z/\sigma = 81$. The value of the Peclet number is $Pe = 1.6$. (b–d) Time series of the system at time $t/\tau_S = 3.2$ (b), 6.4 (c), 9.6 (d). The snapshots are slices of thickness 2σ done in the xy plane. (e) Slice of thickness 2σ in the yz plane at time $t/\tau_D = 9.6$. The height of the yz plane is $x/L = 2/3$, as indicated by the dashed line in (d). (f–i) Experimental realisation of the Rayleigh–Taylor-like instability in sedimentation of confined colloids. (f–h) Time series of images taken with a confocal microscope in the xy plane for the parameters $\phi = 0.15$, $Pe = 1.1$ and $L/\sigma = 18$ at times $t/\tau_S = 1.43$ (f), 5.5 (g), 11.22 (h). (i) Slice in the yz plane at a height $x/L = 2/3$ (indicated by the dashed line in (h)) at time $t/\tau_S = 11.22$. In (f–h) the scale bars denote 20 μm and in (i) 40 μm.

flat interface starts to develop perturbations with a characteristic wavelength $\lambda = 2\pi/k$, where k is a wave number.

We present snapshots of the time evolution of the system, in Fig. 1b–e from computer simulation, and in Fig. 1f–i from confocal microscopy. The similarity is remarkable, and we note that, at the very least, our simulation qualitatively reproduces the experiment. The time evolution in the development of the instability with a characteristic wavelength is clear. While snapshots in the gravity plane (Fig. 1b, c, d, f, g, and h) illustrate the overall process of sedimentation, snapshots in the horizontal yz plane show the transient pattern or network-like structure that results from the instability (Fig. 1e and i). At later times, the network structure decays and a laterally homogenous density profile develops where the colloids start to form a layer at the bottom of the cell which becomes more compact with time. In the initial regime of the instability, more precisely in a regime where the amplitude of an undulation is smaller than the corresponding wave length, the experimental and the simulation data are in line with the results of a linear stability analysis. A detailed comparison is presented in ref. 23 revealing a fastest growing mode with a wave number k_{max}.†

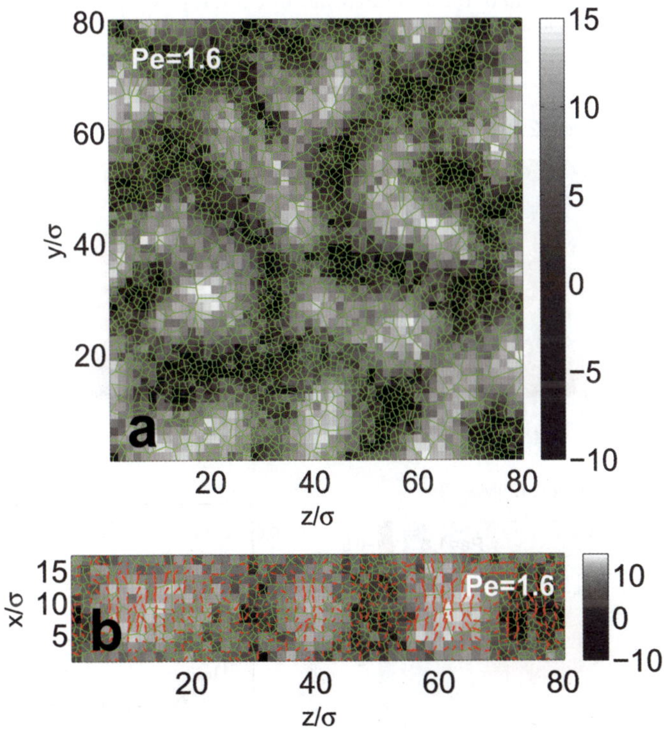

Fig. 2 Solvent flow field. (a) Solvent velocity field in the yz plane. Slice of the simulation box in the yz plane at a height $x/L = 1/2$ at time $t/\tau_S = 11.2$. The parameters are $Pe = 1.6$ and $L/\sigma = 18$. The colour plot represent the magnitude of the solvent velocity $v_x(y, z)/V_s$ in gravity direction, where V_s is the sedimentation velocity at infinite dilution. The Voronoi diagram of the colloid positions in this plane is indicated by green lines. (b) Solvent velocity field in the xz plane. Slice of the simulation box in the xz plane for the same parameters as in (a). The colour plot represent $v_x(x, z)/V_s$, the red arrows the velocity field $(\mathbf{v}_x + \mathbf{v}_z)(x,z)/V_s$ and the green lines the Voronoi diagram of the colloid positions in this plane.

† We remark that both the wave length and the growth rate of the most unstable undulation depends on the slit size L, more details on the slit size dependence are discussed in ref. 23.

Let us now study the solvent flow accompanying the colloidal instability. In Fig. 2 a, the solvent velocity field in the yz plane perpendicular to the driving force is shown at an intermediate height $x = L/2$. Inhomogeneities are revealed on the same length scale λ_{max} as in the colloidal density profile. Actually there is a back-flow effect associated with colloid-rich regions moving downwards and with solvent-rich regions of upward motion. This is clearly revealed in Fig. 2 b where a region rich in colloidal particles is shown as a dense Voronoi tesselation and clearly correlates with a downwards solvent velocity and *vice versa*. Again, the structures exhibit the characteristic length scale $\lambda_{max} = 2\pi/k_{max}$, the fastest growing wavelength in the linear regime.

We finally consider the spatial correlations of colloid-velocity fluctuations in the gravity direction in the plane perpendicular to gravity $C_x(x,r,t) = \langle \delta V_x(x,0,t)\delta V_x(x,r,t)\rangle$, where $\delta V_x(x,r,t) = \langle v_x(x,t)\rangle - v_x(x,r,t)$ are the deviations from the mean velocity in the yz plane at height x from distance r in this plane. We anticipate both positive correlations at short distances (within the same 'branch' of the network) and negative correlations at slightly longer length scales. At longer length scales again, the lack of long-ranged order in the network leads to a loss of correlation and a decay $\lim_{r\to\infty} C_x(x,r,t) = 0$. In Fig. 3 a–c the chronological development of the logarithm of the absolute value of $C_x(x, r, t)$ is shown at the height $x = L/2$ for $Pe = 0.8$, 1.6, 4.8, respectively. The maximum in anticorrelation of $C_x(x, r, t)$ is found at $r \approx \lambda_{max}/2$, see Fig. 3 d, in other words the length scale of the network

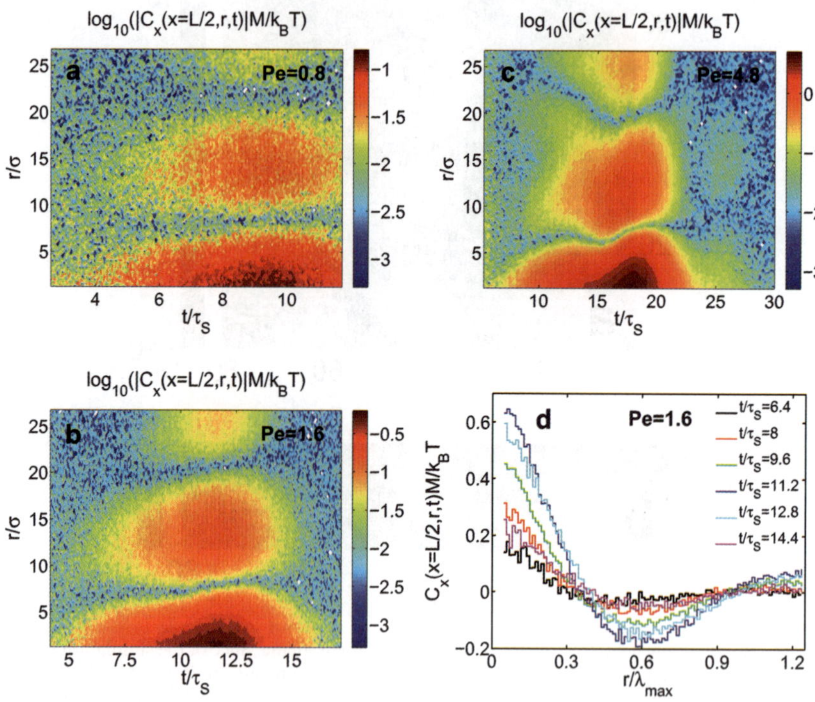

Fig. 3 Spatial correlation functions of the colloid velocity fluctuations. (a–d) The spatial correlation function. The spatial correlation function $C_x(x = L/2, r, t)$ of the colloid velocity fluctuations in the gravity field direction is measured as a function of the distance r perpendicular to gravity. $C_x(x = L/2, r, t)$ was obtained in the yz plane at $x/L = 1/2$ for $Pe = 0.8$ (a), $Pe = 1.6$ (b,d), $Pe = 4.8$ (c) and $L/\sigma = 18$. $C_x(r)$ is scaled by thermal fluctuation strength $k_B T/M$ (M is the colloid mass). In (a–c) r is scaled by the diameter σ of the colloidal particle and in (d) by the wave length λ_{max} of the fastest growing undulation for $Pe = 1.6$. The chronological development of the logarithm of the absolute value of $C_x(x = L/2, r, t)$ is shown in (a–c), whereas in (d) $C_x(x = L/2, r, t)$ is plotted for times $t/\tau_S = 6.4$, 8, 9.6, 11.2, 12.8, 14.4.

structure corresponds very closely to the fastest growing wavelength predicted by the linear stability analysis. The velocity correlations become more pronounced in the non-linear mixing regime which is an indication of fully developed swirls of solvent and particle-rich regions. For $Pe = 0.8$ (Fig. 3 a) and for $Pe = 1.6$ (Fig. 3 b,d) the correlation length slightly increases in time, while for $Pe = 4.8$ (Fig. 3 c) it is markedly *non-monotonic* in time revealing a non-trivial interplay between sedimentation, hydrodynamics and confinement.

4 Conclusions

Using the MPCD simulation technique, we have presented an analysis of a hydrodynamic instability in a colloidal system confined in a slit of micrometre dimensions. Our results show good agreement between experiment and simulation, showing that the latter accurately describes the fundamentally and practically important phenomena caused by hydrodynamic instabilities.

Let us finally discuss some possible extensions of the present work: here we started with a sedimentation density distribution, typical for an experimental situation before turning over the cell. A sharp initial interface as often encountered after flow junctions in microfluidics as well as a linear density gradient, as often found in systems with source–drain stabilized electrolyte concentrations could also be addressed in principle by simulation. We would expect that the instability is most pronounced for high initial density gradients.

Finally, it is tempting to consider more complex interactions between the colloidal particles as realized, for example, for colloid-polymer mixtures where strong attractions can be realized.‡ One would then expect a Rayleigh–Taylor-like instability with surface tension.[27] Also binary systems will establish a playground for sedimentation instabilities where separation and mixing could be tuned by the strength of the drive. External drives which are oscillatory in time could lead to segregation effects in the axial direction[28] and might be interesting for further study. If the initial density profile is crystalline in the lateral direction,[23] surface melting behaviour in non-equilibrium becomes relevant.[29] The recrystallization at the bottom of the cell[30] might be a complex process, in particular for attractive interactions and binary systems.

5 Acknowledgements

We thank T. Palberg for helpful discussions. This work was supported by the SFB TR6 (projects A3, A4 and D3). C. P. R. acknowledges the Royal Society for Funding. H. T. acknowledges a grant-in-aid from MEXT. We acknowledge ZIM for computing time.

References

1 J.-P. Hansen and H. Löwen, *Annu. Rev. Phys. Chem.*, 2000, **51**(1), 209–242.
2 J. K. G. Dhont, *An Introduction to the Dynamics of Colloids*, Elsevier, Amsterdam, 1996.
3 M. E. Cates, K. Stratford, R. Adhikari, P. Stansell, J.-C. Desplat, I. Pagonabarraga and A. J. Wagner, *J. Phys.: Condens. Matter*, 2004, **16**(38), S3903–S3915.
4 H. Tanaka and T. Araki, *Phys. Rev. Lett.*, 2000, **85**(6), 1338–1341.
5 A. Malevanets and R. Kapral, *J. Chem. Phys.*, 1999, **110**(17), 8605–8613.
6 R. Kapral, *Adv. Chem. Phys.*, 2008, **140**, 89–146.
7 G. Gompper, T. Ihle, D. Kroll and R. Winkler, *Adv. Polym. Sci.*, 2009, **221**, 1.

‡ Strong attractions in colloid–polymer mixtures are responsible for a ultralow surface tension and hence for capillary waves.[25] Such long-wavelength interface roughness should enhance the initial growth of the unstable undulations as compared to a fully flat interface. On the other hand it was shown that the non-linear regime is not influenced much by the initial conditions.[26]

8 S. Chandrasekhar, *Hydrodynamic and Hydromagnetic Stability*, Oxford University Press, Oxford, 1961.
9 K. Kadau, T. C. Germann, N. G. Hadjiconstantinou, P. S. Lomdahl, G. Dimonte, B. L. Holian and B. J. Alder, *Proc. Natl. Acad. Sci. U. S. A.*, 2004, **101**(16), 5851–5855.
10 C. Völtz, W. Pesch and I. Rehberg, *Phys. Rev. E*, 2001, **65**(1), 011404.
11 J. L. Vinningland, Ø. Johnsen, E. G. Flekkøy, R. Toussaint and K. J. Måløy, *Phys. Rev. Lett.*, 2007, **99**(4), 048001.
12 E. Kuusela, J. M. Lahtinen and T. Ala-Nissila, *Phys. Rev. E*, 2004, **69**(6), 066310.
13 D. G. A. L. Aarts, R. P. A. Dullens and H. N. W. Lekkerkerker, *New J. Phys.*, 2005, **7**, 40.
14 J. Zhao, D. Vollmer, H.-J. Butt and G. K. Auernhammer, *J. Phys.: Condens. Matter*, 2008, **20**(40), 404212.
15 J. T. Padding and A. A. Louis, *Phys. Rev. E*, 2008, **77**(1), 011402.
16 Y. Inoue, Y. Chen and H. Ohashi, *J. Stat. Phys.*, 2002, **107**(1), 85–100.
17 J. T. Padding, A. Wysocki, H. Löwen and A. A. Louis, *J. Phys.: Condens. Matter*, 2005, **17**(45), S3393–S3399.
18 D. J. Evans and G. P. Morriss, *Phys. Rev. Lett.*, 1986, **56**(20), 2172–2175.
19 T. Ihle and D. M. Kroll, *Phys. Rev. E*, 2001, **63**(2), 020201.
20 A. Lamura, G. Gompper, T. Ihle and D. M. Kroll, *Europhys. Lett.*, 2001, **56**(3), 319–325.
21 N. Kikuchi, C. M. Pooley, J. F. Ryder and J. M. Yeomans, *J. Chem. Phys.*, 2003, **119**(12), 6388–6395.
22 J. T. Padding and A. A. Louis, *Phys. Rev. E*, 2006, **74**(3), 031402.
23 A. Wysocki, C. P. Royall, R. G. Winkler, G. Gompper, H. Tanaka, A. van Blaaderen and H. Löwen, *Soft Matter*, 2009, **5**, 1340–1344.
24 C. P. Royall, J. Dzubiella, M. Schmidt and A. van Blaaderen, *Phys. Rev. Lett.*, 2007, **98**(18), 188304.
25 D. G. A. L. Aarts, M. Schmidt and H. N. W. Lekkerkerker, *Science*, 2004, **304**(5672), 847–850.
26 K. Kadau, C. Rosenblatt, J. L. Barber, T. C. Germann, Z. Huang, P. Carlès and B. J. Alder, *Proc. Natl. Acad. Sci. U. S. A.*, 2007, **104**(19), 7741–7745.
27 A. Wysocki and H. Löwen, *J. Phys.: Condens. Matter*, 2004, **16**(41), 7209–7224.
28 A. Wysocki and H. Löwen, *Phys. Rev. E*, 2009, **79**(4), 041408.
29 H. Löwen and R. Lipowsky, *Phys. Rev. B*, 1991, **43**(4), 3507–3513.
30 T. Biben, R. Ohnesorge and H. Löwen, *Europhys. Lett.*, 1994, **28**(9), 665–670.

Can the isotropic-smectic transition of colloidal hard rods occur *via* nucleation and growth?

Alejandro Cuetos,† Eduardo Sanz‡ and Marjolein Dijkstra*

Received 28th January 2009, Accepted 27th February 2009
First published as an Advance Article on the web 21st August 2009
DOI: 10.1039/b901594a

We investigate the isotropic-to-smectic transformation in a fluid of colloidal hard rods using computer simulations. At high supersaturation, we observe spinodal decomposition: many small clusters are formed at the initial stage of the phase transformation, which form a percolating network that eventually transforms into a stable bulk smectic phase. At low supersaturation, we find that nucleation and growth of the smectic phase is hampered by the pre-smectic ordering in the supersaturated isotropic fluid phase. As the system evolves mainly *via* cooperative motion of these smectic domains, the diffusion and attachment of single particles to the nucleation site is largely hindered.

1 Introduction

The interest in liquid crystalline phases is driven by their great technological potential, but also originates from a fundamental point of view. While the mechanisms that control the stability of the liquid crystalline phases are well-known, the transformation between the different mesophases with positional and orientational ordering is still poorly understood. For instance, the smectic phase is a liquid crystalline state of matter in which elongated particles, such as molecules, colloids, fibers, nanotubes, form parallel stacks of fluid-like layers. The formation of a smectic phase and the nature of the transient structures (clusters, nuclei, droplets) are, however, still unknown. When spherical particles nucleate, the clusters that form, tend to be spherical: *e.g.* the existence of gas-bubbles in a superheated liquid,[1] liquid droplets in a supersaturated gas,[2] or crystallites in an undercooled liquid[3,4] are undisputed transient structures. However, there seems to be no analogue for smectic clusters where both positional and orientational order plays a role and where anisotropic clusters may be formed: simple transient smectic droplets have to the best of our knowledge never been observed.

For instance, a simulation study on the crystal nucleation in fluids of short hard rods shows that the growth of a crystal starts with the formation of a single crystalline membrane with hexagonal order. Subsequently, the growth of the crystal is hampered by rods that are aligned parallel to the bottom and top surface of the crystallite: the surface poisons itself and hence prevents its own growth.[5–7] Consequently, the free energy increases monotonically with cluster size and never crosses a nucleation barrier beyond which the clusters grow spontaneously.

Soft Condensed Matter Group, Debye Institute for NanoMaterials Science, Utrecht University, Princetonplein 5, 3584 CC Utrecht, The Netherlands. E-mail: m.dijkstra1@uu.nl

† Present address: Departamento de Física Aplicada, Universidad de Almería, 04020 Almería, Spain. E-mail: E-mail: acuetos@ual.es

‡ Present address: SUPA, School of Physics and Astronomy, University of Edinburgh, Mayfield Road Edinburgh, UK. E-mail: E-mail: esanz@ph.ed.ac.uk

In addition, very intriguing structures have been observed by Dogic and Fraden in experiments on suspensions of rod-like virus particles and non-adsorbing polymer.[8,9] They observed a fascinating rich phenomenology with novel and interesting metastable structures depending on the polymer concentration, like spindle-shaped nematic droplets, filaments, surface-induced smectic phases on the surface of metastable nematic droplets, individual smectic membranes winding off from tactoids as twisted ribbons, but they never observed simple smectic droplets.[8,9] In addition, the kinetics of the isotropic-to-smectic transformation has also been studied in experiments on attractive β-FeOOH rod suspensions as a function of the aspect ratio, and again no smectic droplets have been observed.[10]

This should be contrasted with the isotropic-to-nematic (I–N) transition where only orientational order plays a role. The (existence of a) crossover between the nucleation and growth regime with a spinodal decomposition regime has been investigated experimentally as a function of supersaturation in systems of colloidal rods,[11] solutions of F-actin[12] or dispersions of rod-like viruses under shear conditions.[13] The location of the isotropic-nematic spinodal have been determined using Brownian dynamics simulations in fluids of hard rods for various aspect ratios.[14] In addition, the structure and the shape of the transient clusters during the isotropic-to-nematic transition have been well-studied, and spindle-shaped elongated nematic droplets, called tactoids, have been observed experimentally.[12,15,16] The shape and the nematic director field of tactoids have been described theoretically and different morphologies of the nematic tactoids have been predicted as a function of the interfacial tension, the anchoring strength, and the bulk elasticity.[17] Very recently, we have studied the isotropic-to-nematic transition in a fluid of hard spherocylinders and in a mixture of rods and non-adsorbing polymer.[18,19] To study the transformation from the isotropic-to-nematic phase, we introduced a new cluster criterion that enabled us to differentiate the nematic clusters from an isotropic fluid phase. Applying this criterion in Monte Carlo simulations, we found at high supersaturation spinodal decomposition, and at low supersaturation, we observed nucleation and growth. We determined the height of the nucleation barrier and we studied the shape and the structure of the cluster as a function of its size. We found that the clusters have an ellipsoidal symmetry with an aspect ratio of about 1.6 and a homogeneous nematic director field. Subsequently, we have explored the validity of classical nucleation theory (CNT) in fluids of hard rods in a mixture of rods and non-adsorbing polymer.[19] Although many questions remain to be answered for the isotropic-to-nematic transformation, *e.g.* on the nontrivial density and order parameter profiles inside the droplets, it is well-accepted by now that the transformation occurs *via* the formation of anisotropic nematic clusters at sufficiently low supersaturations.

In this paper, we investigate the isotropic-to-smectic transformation in a fluid of hard spherocylinders using computer simulations. We investigate whether the transformation proceeds *via* nucleation and growth or *via* spinodal decomposition, and what the shape and structure is of the transient structures along the transition. In addition, we like to investigate whether the growth of the smectic phase is also hampered by self-poisoning as was observed for the isotropic-to-crystal nucleation of short hard rods.[5] We also like to investigate whether the nucleation of the smectic can proceed in multiple steps in which the nucleation starts with the formation of a single layer, and subsequently additional layers are nucleated on top of this layer. The nucleation of additional layers do have their own nucleation barrier, and hence the free energy of formation of a multilayer stacked smectic/crystalline phase consists of a sequence of maxima and minima as was predicted theoretically by Frenkel and Schilling.[7] Finally, we study whether the isotropic-to-smectic transition proceeds *via* nematic tactoids as has been observed experimentally.[8,9]

This paper is arranged as follows. In Section II, we describe the simulation techniques and the model. In section III, we present the results, while in Section IV, we finish with some conclusions and final remarks.

2 Model and simulation method

We study the isotropic-to-smectic (I–Sm) transformation in a suspension of colloidal hard rods. We model the suspension as a fluid of hard spherocylinders (HSC) with diameter σ and length L. In this model, the pair potential of two rods is given by

$$U_{\mathrm{HSC}}\left(\mathbf{r}, \hat{\mathbf{u}}_i, \hat{\mathbf{u}}_j\right) = \begin{cases} \infty & d_{\mathrm{m}} \leq \sigma \\ 0 & d_{\mathrm{m}} > \sigma \end{cases} \tag{1}$$

where d_{m} is the minimum distance between two line segments of length L with orientations $\hat{\mathbf{u}}_i$ and $\hat{\mathbf{u}}_j$, and relative distance vector $\mathbf{r}$. This model has been extensively studied in the past, and its phase diagram is well known as a function of the length-to-diameter ratio of the rods.[20,21] In this paper, we focus on particles with a length-to-diameter ratio of $L^* = L/\sigma = 3.4$. For $L^* = 3.4$, the system exhibits an isotropic (I), smectic A (Sm) and a crystal (X) phase, while the nematic phase is unstable for this elongation. The I–Sm transition occurs at coexistence pressure $P^*_{\mathrm{ISm}} = 2.818$, where the reduced pressure is defined as $P^* = \beta P \sigma^3$ with $\beta = 1/k_{\mathrm{B}}T$, k_{B} Boltzmann's constant, and T the temperature. The packing fractions of the coexisting isotropic and smectic phase are $\eta_{\mathrm{I}} = 0.492$ and $\eta_{\mathrm{Sm}} = 0.552$, respectively.[20] Here, we define the packing fraction $\eta = \rho v_o$ with $\rho = N/V$ the number density and v_o the volume of the rods.

To study the transformation of the isotropic to smectic phase, we perform either standard Monte Carlo simulation in the isobaric-isothermal ensemble, *i.e.*, the number of particles, pressure and temperature are kept fixed (NPT-MC simulations), or we employ the umbrella-sampling method in the isobaric-isothermal ensemble. We perform our simulations in a rectangular box with periodic boundary conditions and we use $N = 8649$ particles. We do not expect appreciable finite-size effects for this system size. The acceptance ratios are kept within 30–40% for the rotational and translational moves of the particles, and within 20–30% for the attempts to change the volume of the box. Box volume changes are attempted by randomly changing the length of each side of the box independently.

We first employ the NPT-MC simulations to equilibrate states in both the isotropic and the smectic phase. We note, however, that our results deviate slightly from those in ref. 20. We find after long equilibration that the packing fraction of the isotropic phase at the coexistence pressure $P^* = 2.818$ reported in ref. 20 is similar to their value, *i.e.*, $\eta = 0.492 \pm 0.001$. However, we find that our value for the packing fraction of the smectic phase, *i.e.*, $\eta = 0.535 \pm 0.001$, at this pressure is slightly lower than in ref. 20. This deviation in the density of the smectic phase can be explained by the fact that our system can relax its stress as the sides of the simulation box have been changed independently in the volume changes.

At high supersaturations, we employ standard MC-NPT simulations to study the I–Sm transformation. We first equilibrate an isotropic fluid phase at the coexistence pressure, and subsequently quench the system to a pressure beyond coexistence and we monitor different order parameters that provide us with information about the intermediate states during the I–Sm transformation. At low supersaturation, *i.e.* at pressures slightly higher than the coexistence pressure, the free energy to form a critical cluster is very high, and hence, a spontaneous fluctuation that brings the system to the top of the free energy barrier, which is needed for spontaneous growth of the cluster, is extremely rare. Standard MC simulations can therefore not be used to study nucleation at low supersaturations. In order to study nucleation at low supersaturations, we employ umbrella sampling that allows us to bias the sampling to configurations that contain clusters with a certain size. We use the number of particles n in the largest nematic cluster as an order parameter. Following the methodology described in ref. 2, we employ the biasing potential W:

$$W(n(\mathbf{r}^N, \mathbf{u}^N)) = 1/2\kappa_n[n(\mathbf{r}^N, \mathbf{u}^N) - n_0]^2 \tag{2}$$

Here the constant κ_n and n_0 determine the width and location of the range of cluster sizes that is sampled. Typical values for κ_n are in the range of 0.13–0.17. In our simulations, we apply the biasing potential only after trajectories of 10 MC cycles, as the computation of the size of the biggest cluster is very time consuming.

If $< N_n >$ denotes the average number of clusters with n particles, one can determine the probability distribution $P(n) = < N_n >/N$, with N the total number of particles. The Gibbs free energy $\Delta G(n)$ for the formation of a cluster of size n is then determined by:

$$\Delta G(n) = -k_{\mathrm{B}}T \ln P(n) \tag{3}$$

When $\Delta G(n)$ is at a maximum, the cluster at the top of the nucleation barrier is defined as the critical cluster. Clusters with a size larger than that of the critical cluster (postcritical clusters) will spontaneously grow and will eventually form the stable smectic phase in order to minimize the free energy, while clusters smaller than the critical cluster will shrink spontaneously. We employ the umbrella sampling technique to calculate the nucleation barrier. In addition, we use this technique to stabilize critical, precritical and postcritical clusters to study the internal structure and shape of these intermediate structures. In order to bias the sampling to configurations that contain clusters with a certain size, we need a cluster criterion that is able to identify the smectic clusters from the isotropic fluid phase. Recently, we developed a cluster criterion that enables us to distinguish the nematic clusters from the isotropic fluid phase. Here, we apply the same cluster criterion to identify the orientationally ordered clusters from the isotropic, orientationally disordered fluid phase. In the cluster criterion, we first make a distinction between particles that have a orientationally ordered and an isotropic environment. Particle i is orientationally ordered if its local environment has an orientational order significantly larger than in the isotropic phase. The local environment of particle i is defined by all particles j with a surface-to-surface distance $\rho_{ij} \leq 1.5\sigma$, *i.e.*, such that it is not only defined by the nearest neighbors, but also by the next-nearest neighbors, thereby taking advantage of the long-ranged orientational order in the nematic/smectic phase. The local orientational order of particle i is defined by

$$S(i) = \frac{1}{n_i}\sum_{j=1}^{n_i}\left(\frac{3}{2}|\mathbf{u}_j \cdot \mathbf{u}_i|^2 - \frac{1}{2}\right) \tag{4}$$

where $\mathbf{u}_j$ is the unit orientation vector of particle j and n_i the number of particles with $\rho_{ij} \leq 1.5\sigma$. We have adopted the cluster criterion that particle i is orientationally ordered if $S(i) > K_1$, where K_1 is a threshold value that has to be optimized for each model. After identifying the orientationally ordered particles in the system, we determine the smectic cluster with the criterion that two particles i and j belong to the same cluster if $\rho_{ij} < 0.5\sigma$ and $|\mathbf{u}_i \cdot \mathbf{u}_j| > K_2$, with K_2 another adjustable threshold value. In contrast with the first step of our cluster criterion, we consider only the first neighbors. In our previous simulation study of the I–N transition in a fluid of hard spherocylinders, we have chosen $K_1 = 0.4$ and $K_2 = 0.85$. In the sequel of this paper, we denote this criterion as C1. The C1 criterion yields valuable information about the structural properties of the intermediate states observed during the I–Sm transition. However, for large cluster sizes and high supersaturations, these threshold values are not optimized to distinguish the smectic clusters from the isotropic fluid phase. We have performed several trial runs to optimize the threshold values and we have chosen $K_1 = 0.5$ and $K_2 = 0.95$, which will be referred to as the C2 criterion. With this choice of parameters, we hardly find any smectic clusters in the coexisting isotropic fluid phase, *i.e.*, only clusters with $n < 20$ particles are found. On the other hand, more than 95% of the particles belong to an unique smectic cluster in the coexisting smectic phase. We wish to make several comments here: (i) the present criterion to distinguish smectic clusters from the isotropic phase is similar to the nematic cluster criterion used previously.[18,19] In fact, our cluster criteria do not make a distinction between nematic- or smectic-like particles, but only differentiate particles with a high local orientational order from particles with an isotropic environment. Therefore, the criterion does not bias the sampling towards cluster with a certain structure, *e.g.* nematic, smectic, or crystalline clusters. (ii) Our cluster criterion deviates from the criterion proposed by Schilling and Frenkel,[5,6] who used their criterion to study the isotropic-to-solid transition in a fluid of hard spherocylinders with elongation $L^* = 2$. In their criterion, two particles belong to the same cluster if the surface-to-surface distance between two rods is smaller than 0.5σ and the absolute value of the dot product of the two unit vectors that define the orientations of the rods is larger than 0.995. However, this criterion cannot be used for the elongation of interest in this work, since only 50% of the particles in the coexisting smectic state belong to the smectic phase.

To analyze the results we employ techniques that are frequently used in liquid crystal studies. For example, we monitor the orientational order in a cluster or in the whole system using the standard nematic order parameter S. The nematic order parameter is defined by the largest eigenvalue of the standard 3×3 nematic order parameter tensor. The corresponding eigenvector defines the preferred orientation of the particles, which defines the nematic director $\mathbf{n}$. To study the structure of the clusters we calculate the density $\rho^*(z, r)$ at distance z from the center-of-mass of the clusters parallel to the nematic director of the clusters and distance r perpendicular to the nematic director of the clusters. In order to study the orientational order inside the cluster, we define a local orientational order parameter by averaging the second Legendre polynomial, $S(z, r) = < 3/2|\mathbf{u}_i(z, r)\cdot\mathbf{n}| - 1/2 >$, for orientationally ordered (smectic) and isotropic particles. Here $\mathbf{n}$ is the nematic director calculated using only the particles that belong to the smectic cluster. In our calculations of the density and nematic order parameter profiles, we consider all the particles in the system.

3 Results

We study the I–Sm transformation using Monte Carlo simulations as a function of supersaturation. To this end, we first equilibrate an isotropic fluid phase of a fluid of hard spherocylinders with $L^* = 3.4$ at bulk coexistence ($P^* = 2.828$, $\eta = 0.495 \pm 0.001$). When the system is well equilibrated, we increment the pressure, and we investigate at which pressure the I–Sm transition occurs spontaneously. We find that the isotropic phase transforms spontaneously into a smectic phase at $P^* \geq 3.1$. When we compress an isotropic fluid phase at $P^* = 3.1$ using NPT-MC simulations, we find that the density of the system increases gradually and that the system is unstable. Phase separation sets in immediately after compressing, as many small clusters appear in the system in the initial stage of the phase separation. These clusters coalesce and form an interconnected cluster (see Fig. 1). The size and the nematic order of this interconnected cluster grow gradually until the whole system is transformed into a single smectic cluster in the final stage of the transformation. These features are characteristic of spinodal decomposition and we conclude that at this supersaturation the system is in the spinodal regime. Fig. 2 presents the evolution of the size of the largest cluster in the system, according to the C1 and C2 cluster criteria (bottom panel). The global order parameter of the whole system as well as the nematic order parameter of the biggest cluster according to both cluster criteria are presented in the top panel of Fig. 2. From this figure, we clearly observe how the system evolves gradually from a state without any orientational order, *i.e.*, the global

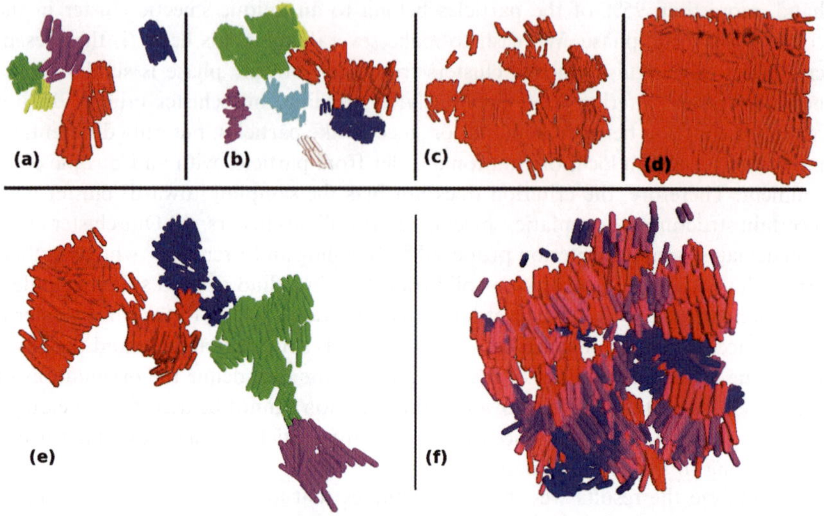

Fig. 1 Typical intermediate configurations of clusters in the I–Sm transformation of a HSC fluid with $L^* = 3.4$ as described in Fig. 2 using the C2 cluster criterion. (a) Clusters of 64 (red), 44 (green), 28 (blue), 20 (yellow) and 18 (magenta) particles are formed simultaneously. The packing fraction of the system is $\eta = 0.500$ here. (b) Many clusters observed simultaneously in a configuration with $\eta = 0.513$. The size of the clusters is 238 particles (red), 209 (green), 118 (blue), 88 (yellow), 77 (cyan), 44 (magenta) and 24 (light pink). (c) In a later stage, when $\eta = 0.518$, the clusters coalesce and form a bigger cluster of 1587 particles. (d) In the final stage, the systems transform into a bulk Smectic A phase with a packing fraction of $\eta = 0.550$. (e) Same configuration as in (a) but now analyzed by using the C1 cluster criterion. We now observe clusters with 375 particles (red), 257 (green), 132 (blue) and 89 (green). (f) Same configuration as in (b), but analyzed using the C1 cluster criterion. An open, interconnected labyrinth-like structure of 2228 particles is found. The red particles have an orientation close to the nematic director n of the cluster, while the blue particles are perpendicular to $\mathbf{n}$.

nematic order parameter S is close to zero, to a state with a high orientational order, *i.e.*, S close to unity. From Fig. 2, we observe that the global nematic order parameter is clearly correlated with the size of the biggest cluster in the system as detected by both cluster criteria. This should be contrasted with the nematic order parameter of the biggest cluster, which display a non-monotonous behavior during the transition. We can detect basically two regimes. In the initial stage of the phase separation, the nematic order parameter of the biggest cluster shows an irregular behavior with large and rapid oscillations around an averaged value, while in a later stage these oscillations almost disappear. In the final stage, the nematic order parameter of the biggest cluster shows a slight increase to a value, that is slightly higher than that of the global nematic order parameter. The trends as described above do not depend strongly on the C1 or the C2 cluster criteria. However, we do find some remarkable differences. First, we observe that the clusters already start to grow at 0.5×10^6 MC cycles according to the C1 cluster criterion, while according to the C2 cluster criterion the clusters start to grow much later, at about 1.5×10^6 MC cycles. At this stage, the C1 cluster criterion detects clusters of about 2000 particles. The C1 cluster criterion has been used to study the I–N transition[18,19] and is less strict than the C2 cluster criterion. Second, we observe from Fig. 2 that the nematic order parameter of the biggest cluster according to the C1 cluster criterion is very noisy in the initial stage of the transformation, decreases monotonically with time, and shows a minimum at about the same stage when the clusters start to grow according to the C2 cluster criterion. Subsequently, the nematic order parameter of the biggest cluster starts to grow till a maximum value is reached. In order to explain the trends observed in Fig. 2, we analyze the clusters observed during the transformation

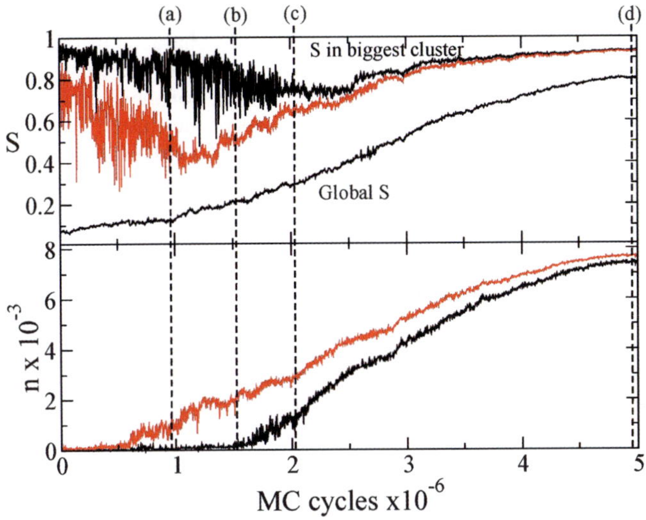

Fig. 2 Evolution of the size of the biggest cluster n in the system (bottom panel), the nematic order parameter S of the biggest cluster and of the total system (top panel), for a HSC fluid with $L^* = 3.4$ at $P^* = 3.1$. The black curves correspond to the results using the C2 cluster criterion, while the gray curves have been obtained using the C1 cluster criterion. The configurations corresponding to the snapshots of panels (a), (b), (c) and (d) of Fig. 1 are indicated with the vertical dashed lines.

with the C1 and the C2 cluster criterion. We find that the I–Sm transition is a two-step process. In the initial stage of the transition, many small clusters are formed throughout the whole system. These clusters are irregular, not very compact, and do have a smectic-like character, although it is hard to distinguish the smectic layers. Clusters with about 1–3 smectic layers are observed. We analyze a typical configuration in the initial stage of the phase separation according to both cluster criteria. The packing fraction of this configuration is $\eta = 0.500$, which is lower than the densest metastable isotropic fluid phase ($\eta = 0.508$ at $P^* = 3.0$). Fig. 1a and e, display the analysis according to the C2 and C1 criterion, respectively. Only a few tiny clusters are detected by the C2 cluster criterion (see Fig. 1a), while a much larger interconnected cluster is found by the less strict C1 criterion (see Fig. 1e). We find that the formation of a labyrinth-like, percolated, interconnected structure is detected in an earlier stage according to the C1 criterion than to the C2 criterion. We like to stress, that the clusters, even the smallest ones, do have a pronounced smectic-like character. This should be contrasted with the observations of Dogic and Fraden in suspensions of rod-like virus particles and non-adsorbing polymer. These authors observed that the transition to the smectic phase starts with the formation of nematic tactoids, and subsequently smectic layers are formed inside or on the surface of the tactoids. In a later stage of our simulations, the clusters grow, coalesce and form larger clusters, but also smaller clusters can diffuse away from the main cluster. As the clusters can have different orientations when they coalesce, the resulting cluster might have a nematic order parameter that is smaller than that of the original clusters. Again, we analyze a typical configuration with $\eta = 0.515$ at this stage of the phase separation using both cluster criteria. Fig. 1b and f, display the analysis according to the C2 and C1 criterion, respectively. In this case, the C2 cluster criterion detects isolated clusters, while the C1 criterion finds an interconnected cluster. As the main mechanism for the growth of the clusters is by coalescence of smaller clusters at this stage, both the nematic order parameter and the size of the biggest cluster are very noisy and display huge sudden fluctuations. At a later stage, the smectic clusters coalesce and form an interconnected cluster. This cluster is an open and branched structure, but with a clear smectic-like character as the smectic layers can clearly be distinguished in Fig. 1c. Finally, the size of the cluster grows by the addition of individual particles and the nematic order of the interconnected cluster grows gradually by reorientation of different parts of the cluster. Fig. 1d shows that eventually the whole system has been transformed into a structure of parallel stacks of rods, which are typical for the smectic phase. In conclusion, the observed phase separation process has all the typical features of spinodal decomposition: phase separation sets in immediately after compressing as many small clusters are formed throughout the system. These clusters coalesce to form an interconnected structure, which eventually transforms into the stable smectic bulk phase.

A similar kinetic pathway has been observed experimentally by Maeda and Maeda in a systematic study of the isotropic-to-smectic transition as a function of the particle elongation in β-FeOOH rod suspensions.[10] For rods with a length-to-width ratio of 3.5, they find at the beginning of the transition, the spontaneous formation of smectic clusters. When the density is increased the clusters grow laterally, coalesce on top of each other to form multilayer stacks until a single stable smectic structure is formed. A similar behavior has also been found for the isotropic-to-nematic transition in experimental[11] and computer simulation[18,19] studies at sufficiently high supersaturations.

At lower supersaturations ($P^* \leq 3.0$) the system does not change spontaneously from the isotropic to the smectic phase, and we need umbrella sampling to study the transition. In this work, we study the I–Sm transition at supersaturations $P^* =$ 2.828, 2.85, 2.90, 2.95 and 3.0, which correspond to metastable isotropic states with packing fractions η = 0.496, 0.498, 0.504 and 0.508 ($\pm$0.001), respectively. Before we discuss our results on the I–Sm transition, we wish to make a few comments on the structure of the metastable isotropic fluid phase. Visual inspection of typical configurations of the isotropic fluid shows that the fluid exhibits pronounced pre-smectic ordering as smectic-like groups of about 5–10 particles are already present in the isotropic bulk phase. A typical configuration at $P^* =$ 2.9 is shown in the inset of Fig. 3. The different colors reflect the different orientations of the rods. In addition, we also show the orientational pair correlation function $g_2(r) = < P_2(\cos \theta (r)) >$[21] for various pressures, where r denotes the

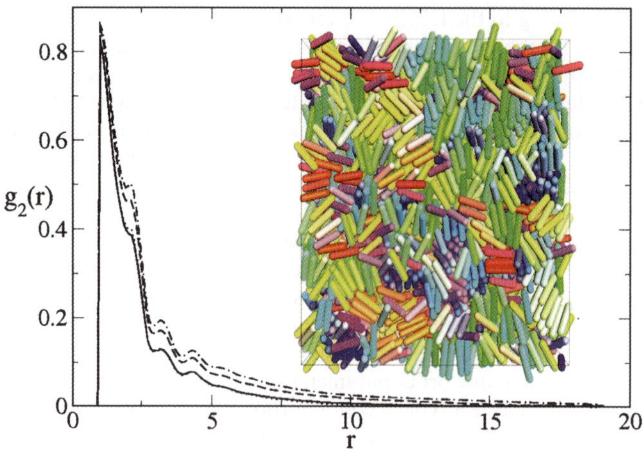

Fig. 3 Orientational pair correlation function $g_2(r)$ at coexistence pressure $P^* = 2.828$ (solid line), $P^* = 2.9$ (dotted line, coincides with the previous one), $P^* = 2.95$ (dashed line) and $P^* = 3.0$ (dotted-dashed line), where r denotes the center-of-mass distance between two particles in units of σ. The inset shows a typical isotropic configuration at $P^* = 2.9$ with pronounced pre-smectic ordering, color is used to distinguish the different orientations of the particles.

center-of-mass distance between the particles in units of σ. The orientational pair correlation function tends to zero at large distances as the orientational order vanishes at long-range as expected for the isotropic phase. On the other hand, we find strong orientational order at short distance, as can be observed from the pronounced peaks for small r, thereby providing evidence of the presence of small smectic-like clusters in the isotropic fluid phase. The peaks in the $g_2(r)$ are at distances $r \approx 1$, 2 and 3, corresponding to the first, second and third neighbor in the same layer. The fourth peak appears at a distance $r \approx 4.4$ corresponding to a particle in a different layer. The short-range orientational order becomes slightly stronger upon increasing the supersaturation. Fig. 3 shows the orientational pair correlation for the coexistence pressure $P^* = 2.828$ and $P^* = 2.9$, which are indistinguishable from each other, and for $P^* = 2.95$ and 3.0. This short-range pre-smectic order may play an important role in the isotropic-to-smectic transformation, which we will study in the sequel of this paper.

We calculate the Gibbs free energy $\beta\Delta G$ as a function of the number of particles n in the cluster using the umbrella sampling technique and the C2 cluster criterion. We did not use the C1 criterion in the umbrella sampling, as this criterion already detects huge clusters at the supersaturation of interest even if we do not introduce the bias in our sampling. However, we do use the C1 criterion to analyze our results. The free energy barriers are shown in Fig. 4 for $P^* = 2.85$, 2.90, 2.95, and 3.00. We clearly observe from Fig. 4 that the free energy of formation grows linearly with the size of the cluster n without reaching a maximum for pressure $P^* = 2.85$, 2.90 and 2.95. For $P^* = 3.0$, we find that the free energy reaches a plateau or decreases slightly for $n > 100$.

Our results are very similar to the simulation studies of the isotropic-to-crystal nucleation of short hard rods of ref. 5,6 where the free energy also increases monotonically with cluster size. They attribute the absence of a nucleation barrier to self-poisoning of the crystallite by rods that lie parallel to the surface of the crystal clusters, thereby hindering the crystal growth. In order to study whether self-poisoning can explain our results, we study in more detail the structure of the smectic clusters and the surrounding particles of these clusters. We calculate the average number of particles $N_{\|}$ as a function of the distance z from the center-of-mass of the cluster in the direction parallel to the nematic director for clusters with $n = 100$, 200,

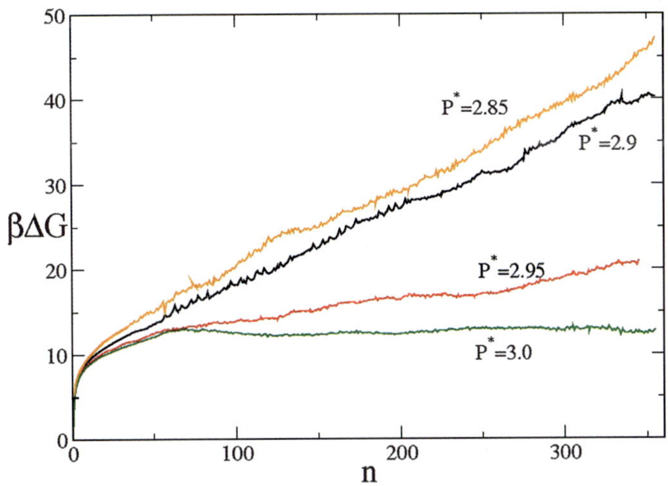

Fig. 4 Gibbs free energy $\beta\Delta G$ of a HSC fluid with $L^* = 3.4$ as a function of the number of particles in the biggest cluster n, as calculated by umbrella sampling MC simulations using the C2 cluster criterion at pressures $P^* = 2.85$, 2.90, 2.95 and 3.00.

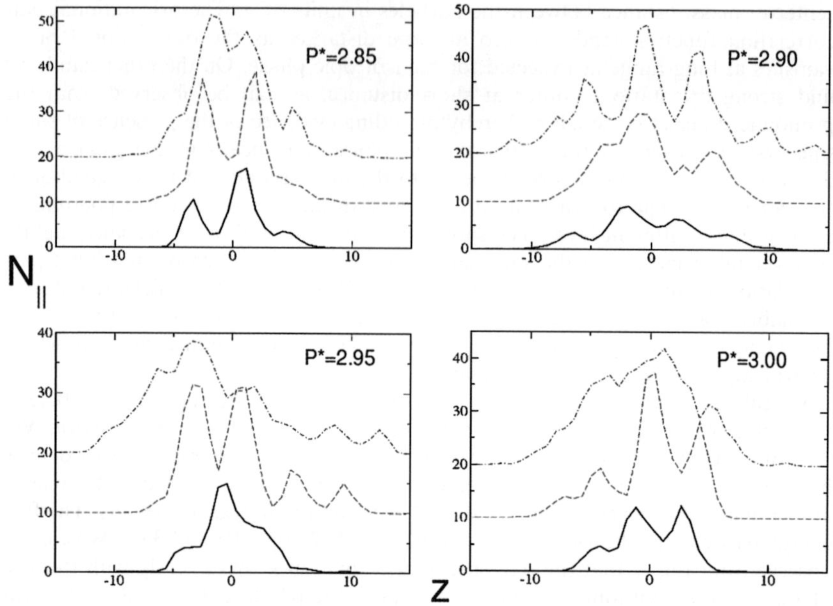

Fig. 5 Averaged number of particles $N_{\|}$ as a function of the distance z from the center-of-mass of the cluster in the direction parallel to the nematic director for clusters with $n = 100$ (black solid lines), 200 (red dashed lines), and 300 (blue dotted-dashed) particles at pressures $P^* =$ 2.85. 2.90. 2.95 and 3.00. The unit of length is taken to be σ. Only the particles detected by the C2 cluster criterion have been considered.

and 300 particles. The unit of length is taken to be σ. We only consider here the particles that are detected by the C2 cluster criterion. We show the results in Fig. 5 for P^* = 2.85, 2.90, 2.95, and 3.00. Fig. 5 shows pronounced peaks in the number of particles, thereby providing clear evidence that the clusters consist of different layers. The distance between the peaks (and between the layers) is about $L^* + 1$, which is equal to the length of the rods and corresponds to the distance that one expects in the bulk smectic phase. Comparing Fig. 4 with Fig. 5 shows that there is no correlation between the number of layers and the Gibbs free energy. For example, we find very similar Gibbs free energy curves for $P^* = 2.85$ and $P^* = 2.90$, but for $P^* =$ 2.85, we find clusters of two layers for $n = 200$ and 300, and three layers for $n =$ 300, while for $P^* = 2.90$ the number of layers ranges from four to six layers. Interestingly, we also find that the number of layers grows with the cluster size for all supersaturations. This should be contrasted with the results of ref. 5 where self-poisoning avoids the growth of additional layers to the cluster. Moreover, ref. 7 predicts a sequence of maxima and minima in the Gibbs free energy as a result of the nucleation of additional layers. Although we do find very weak oscillations in our Gibbs free energy curves, we were not able to find any correlations of the oscillations with the formation of additional layers to the cluster. We attribute the oscillations more to statistical fluctuations.

We also employ the umbrella sampling technique to study the shape and structure of the cluster as a function of its size. In Fig. 6, we show contour plots of the density profiles $\rho(z, r)$ and nematic order parameter profiles $S(z, r)$ for clusters of size $n =$ 150, 250 and 350 at $P^* = 2.85$. The contour plots at $P^* = 2.9$ are shown in Fig. 7. Comparing Fig. 6 and 7, we find pronounced layering of the particles in the contour plots at both pressures. In the density profiles, we find layers with high density and nearly zero density in between the layers. As already observed in Fig. 5, we again find that the number of layers in the clusters is higher at $P^* = 2.9$ than at $P^* = 2.85$.

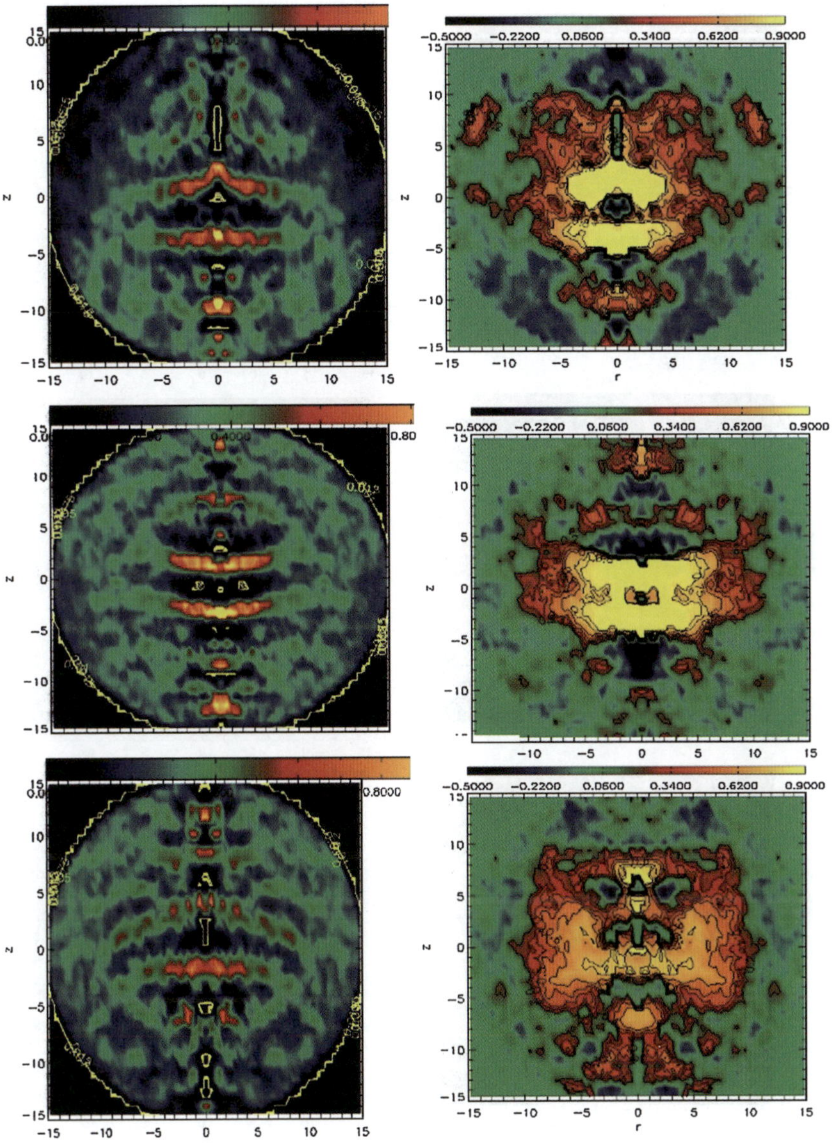

Fig. 6 Contour plots for the density profiles $\rho(z, r)$ (left) and nematic order parameter profiles $S(z, r)$ (right) as a function of the distance from the center-of-mass of the cluster in the direction parallel (z) and perpendicular (r) to the nematic director for clusters of size $n = 150$ (top), $n = 250$ (middle) and $n = 350$ (bottom) at $P^* = 2.85$. The unit of length is σ.

The nematic order parameter profiles show high orientational order in the cluster and hardly any orientational order outside the cluster. From the nematic order parameter contour plots, we also observe that the layers are almost perpendicular to the nematic director (z direction). Moreover, we also find evidence for particles that are aligned parallel to the top and bottom surface of the clusters, as the dark blue regions in the $S(z, r)$ contour plots of Fig. 6 and 7 correspond to $S(z, r) = -0.5$. The observation of particles that are aligned parallel to the top and bottom surface of the clusters has also been observed in ref. 5.

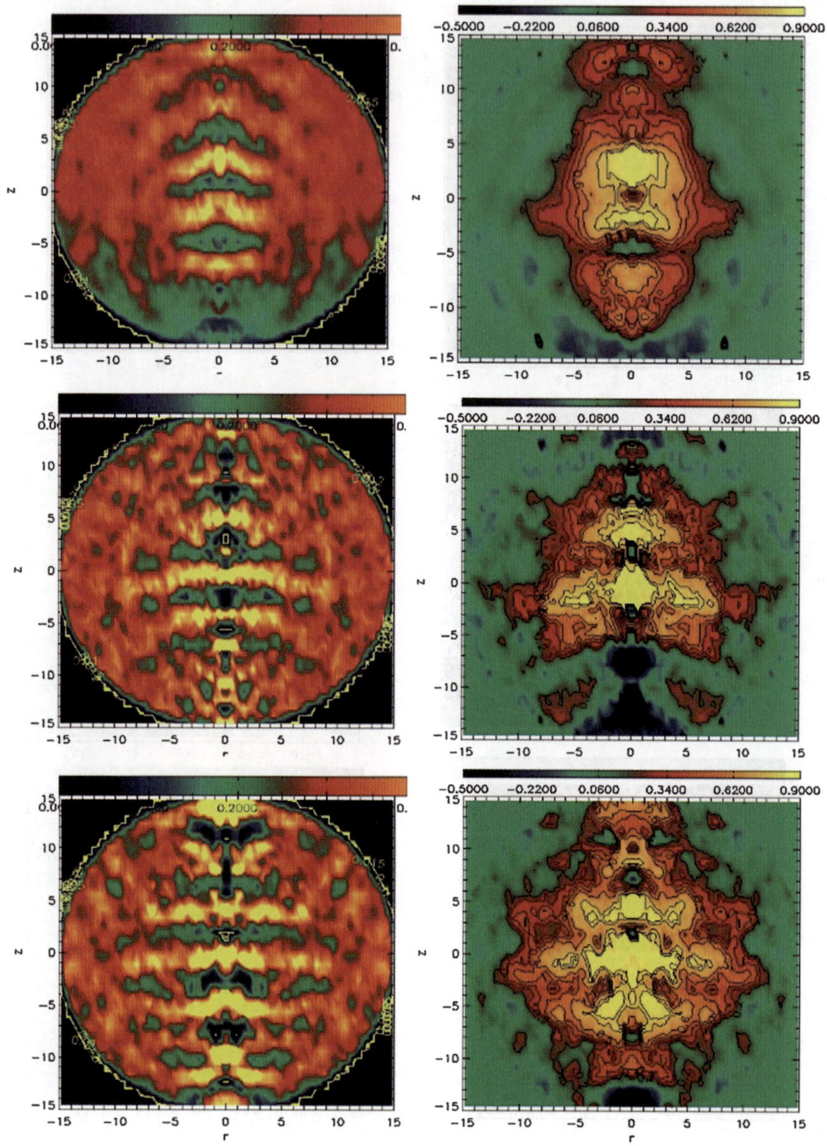

Fig. 7 Contour plots for the density profiles $\rho(z, r)$ (left) and nematic order parameter profiles $S(z, r)$ (right) as a function of the distance from the center-of-mass of the cluster in the direction parallel (z) and perpendicular (r) to the nematic director for clusters of size $n = 150$ (top), $n = 250$ (middle) and $n = 350$ (bottom) at $P^* = 2.9$. The unit of length is σ.

A closer inspection of Fig. 6 shows that we find isolated smectic clusters that are disconnected from the main cluster by a region of particles that are parallel to the surface of both clusters. These isolated smectic clusters can be observed for example in Fig. 6, for a cluster of size $n = 150$ at $z \approx -10$, or at $n = 250$ at $z \approx 7$, 13 and -8. The isolated smectic clusters are also visible in, for instance, Fig. 9d. We already mentioned that the isotropic fluid phase shows pronounced pre-smectic ordering as small smectic-like clusters are already present in the isotropic phase. The isolated smectic clusters observed in Fig. 6 are probably the small smectic-like clusters that are already present in the isotropic fluid phase. Our simulations also show that

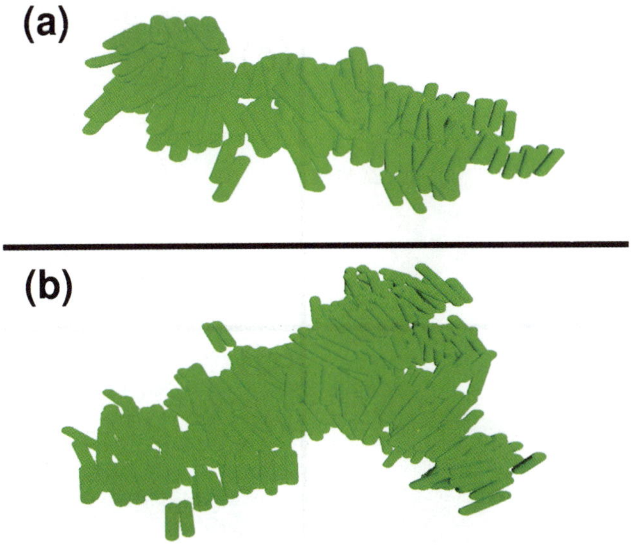

Fig. 8 (a) A typical intermediate configuration in the I–Sm transformation of a HSC fluid with $L^* = 3.4$ at $P^* = 2.85$ analyzed with the C2 cluster criterion resulting in a cluster of 250 particles. (b) Same configuration as in (a), but analyzed by the C1 cluster criterion resulting in a cluster with 631 particles.

the growth of the clusters proceeds more by the reorientation and coalescence of small groups of particles than by individual particles. The growth of the clusters seems to be kinetically inhibited by the pre-smectic ordering, as the dynamics of the particles slows down because the particles can only diffuse by cooperative motion of the whole cluster.

We analyze typical configurations according to the C1 and C2 cluster criteria. Fig. 8 displays the analysis of a typical configuration at $P^* = 2.85$. We find that the C2 cluster criterion detects a cluster of 250 particles, while the less restrictive C1 cluster finds a cluster of 631 particles. In Fig. 9 we show typical configurations at $P^* = 2.90$ analyzed with the C2 and C1 cluster criterion. The C2 cluster criterion detects clusters of 100 and 350 particles, respectively, while the C1 cluster criterion finds clusters of 221 and 846 particles. Again, the less restrictive C1 cluster criterion detects as expected clusters consisting of more particles. We also wish to note that the shape of the clusters can be very different. The clusters in Fig. 8 at $P^* = 2.85$ are oblate, while the clusters in Fig. 9 are more prolate at $P^* = 2.90$. Fig. 5 also shows that the shape of the clusters can be very different and that there is no correlation between the cluster shape (or number of layers) and the supersaturation. According to classical nucleation theory, which is based on the assumption that the clusters are in quasi-equilibrium with the metastable parent phase, the Gibbs free energy ΔG is given by a surface term and a bulk term. The bulk term is given by the volume of the cluster and the chemical potential difference between the phase inside the cluster and the supersaturated bulk phase. While the surface term for elongated particles depends on the interfacial tension, the anchoring strength and the surface and shape of the droplet. The "equilibrium" shape of the droplet can be determined by minimizing ΔG with respect to the cluster shape for a given supersaturation and cluster size.[17,19] Our results show that there is no correlation between the shape of the cluster and the supersaturation and the size of the cluster. This might be explained by either the fact that the clusters are not in quasi-equilibrium or that the surface tensions of the droplets are extremely low. Hence, a theoretical description of the height and shape of the nucleation barrier by classical nucleation theory seems to be doomed to fail.

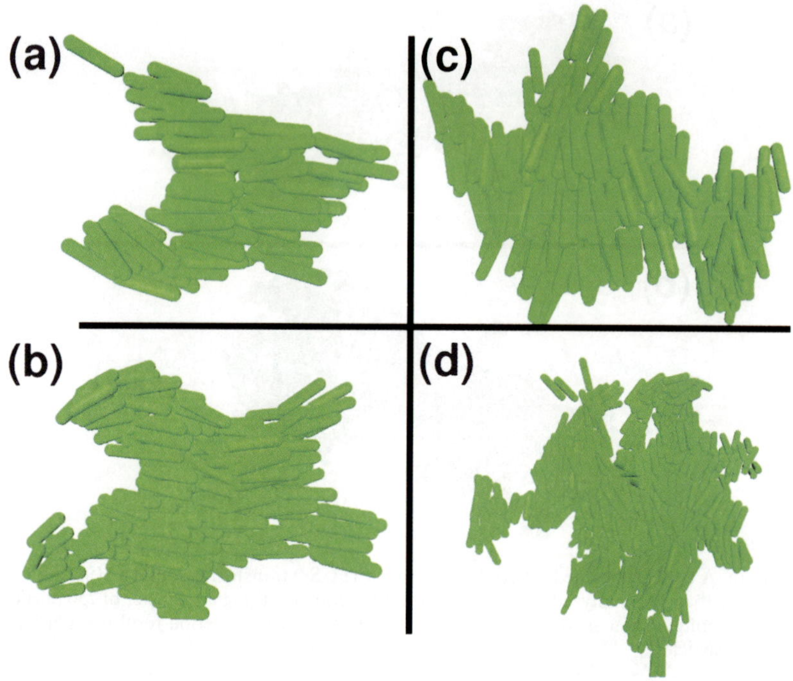

Fig. 9 (a) A typical intermediate configuration in the I–Sm transformation of a HSC fluid with $L^* = 3.4$ at $P^* = 2.9$ analyzed with the C2 cluster criterion resulting in a cluster of 100 particles. (b) Same configuration as in (a), but analyzed with the C1 cluster criterion resulting in a cluster with 221 particles. (c) A typical intermediate configuration analyzed with the C2 cluster criterion resulting in a cluster of 350 particles. (d) Same configuration as in (c), but analyzed with the C1 cluster criterion resulting in a cluster with 846 particles.

Fig. 4 shows that the slope of ΔG changes enormously if we change the supersaturation from $P^* = 2.9$ to $P^* = 2.95$. We therefore study in more detail typical intermediate configurations at $P^* = 2.95$ and 3.00. We first determine the contour plots of the density profiles $\rho(z, r)$ and nematic order parameter profiles $S(z, r)$ for clusters of size 150, 250, and 350. Fig. 10 presents the contour plots at $P^* = 2.95$. We again find similar features in the contour plots as at lower supersaturations. We again observe pronounced layering of the particles in the clusters and we also find particles that are parallel to the top and bottom surface of the clusters. Visual inspection of the cluster shows that the cluster is oblate at this supersaturation (not shown). We now analyze typical configurations using the C2 and C1 cluster criteria. Fig. 11 shows the analysis of a typical configuration at $P^* = 3.00$. The C2 cluster criterion detects a compact smectic cluster of three layers and with $n = 200$ particles, while the C1 cluster criterion finds a very open interconnected labyrinth-like cluster of 1630 particles. The cluster detected by the C2 cluster criterion is located in one of the most ordered branches of the larger (almost percolating) network. The structure of this open network is still smectic-like as the individual smectic layers are clearly visible. However, the orientational order is not long-ranged as the layers can have different orientations in the network. The same observations can be made at $P^* = 2.95$. In conclusion, we find that at low supersaturation ($P^* = 2.85$ and 2.95), a random cluster starts to grow in a metastable isotropic fluid phase. The shape and the number of layers of the cluster is determined by the chance to find a certain smectic cluster shape in a metastable isotropic fluid phase with pronounced pre-smectic ordering. At high supersaturations ($P^* = 2.95$ and 3.00), we find that the metastable isotropic fluid phase shows a more pronounced pre-smectic ordering, where

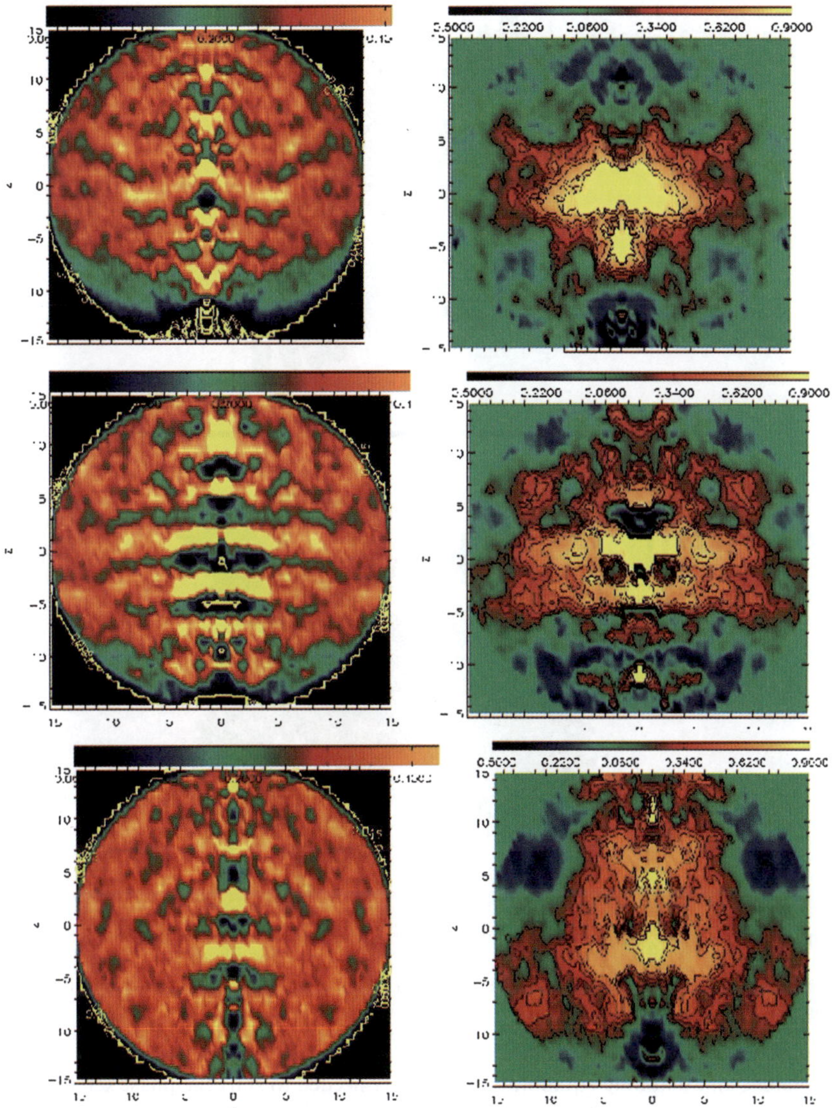

Fig. 10 Contour plots for the density profiles $\rho(z, r)$ (left) and nematic order parameter profiles $S(z, r)$ (right) as a function of the distance from the center-of-mass of the cluster in the direction parallel (z) and perpendicular (r) to the nematic director for clusters of size $n = 150$ (top), $n = 250$ (middle) and $n = 350$ (bottom) at $P^* = 2.95$. The unit of length is σ.

smectic-like clusters merge together to form an interconnected network. The increase in pre-smectic ordering upon increasing the pressure results in a smaller slope of the free energy curves. The slope of the Gibbs free energy curve is nearly zero at $P^* = 3.00$. If we compress a well-equilibrated isotropic fluid phase to $P^* = 2.95$ and 3.00, we find that the isotropic fluid phase is stable for more than 10^6 MC cycles, while phase separation sets in immediately when we quench the system to $P^* = 3.1$. We therefore conclude that state points $P^* = 2.95$ and 3.00 are in the nucleation and growth regime, while $P^* = 3.1$ is in the spinodal region. However, it is also possible that at $P^* = 2.95$ and 3.00, the spinodal decomposition is kinetically inhibited as the system can only evolve by collective re-arrangements of the smectic

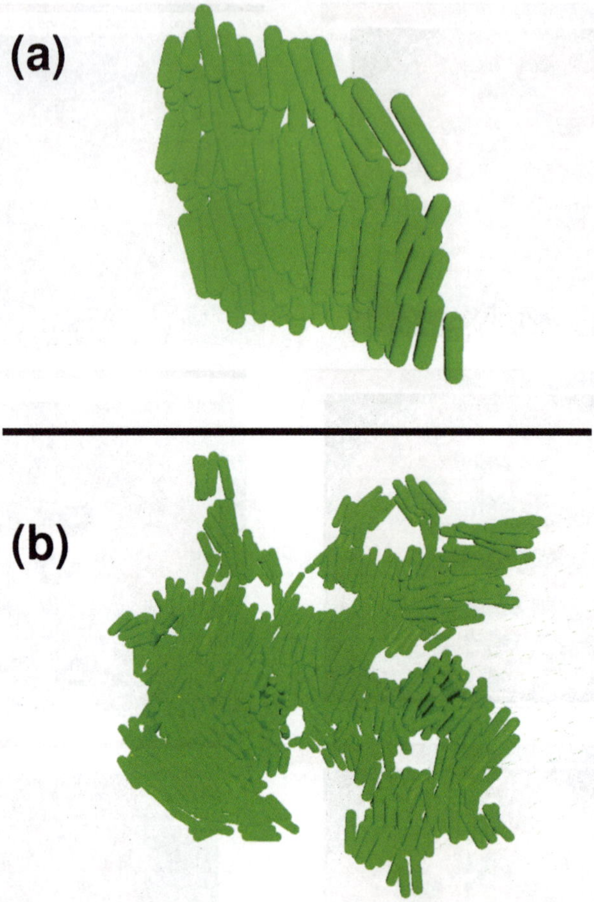

Fig. 11 (a) A typical intermediate configuration in the I–Sm transformation of a HSC fluid with $L^* = 3.4$ at $P^* = 3.0$ analyzed with the C2 cluster criterion resulting in a cluster of 200 particles. (b) Same configuration as in (a), but analyzed with the C1 cluster criterion resulting in a cluster with 1630 particles.

domains, and that long equilibration times are needed to start the phase separation. It seems at odds that the system falls out-of-equilibrium at low supersaturation, where one would expect a faster dynamics of the individual particles, but this can be explained by the fact that collective rearrangements are favored at higher supersaturations.

4 Conclusions and final remarks

In this paper, we have studied the isotropic-to-smectic transformation for a fluid of colloidal hard rods of length $L^* = 3.4$ using computer simulations. We have found spinodal decomposition at high supersaturations, but we were not able to find a nucleation barrier for lower supersaturations as the free energy increases monotonically with cluster size. By analyzing the isotropic fluid phase, we find strong presmectic ordering as many smectic-like clusters are present at $P^* = 2.85$ and 2.9. At $P^* = 2.95$ and 3.0, we observe an open interconnected smectic-like network as the smectic clusters join together. More importantly, we also find that the shape of the intermediate structures can vary from oblate to prolate clusters and from 1 up to 6 layers. As we do not find any correlation between the shape of the cluster and the supersaturation and cluster size, we conclude that the transient clusters are not in quasi-equilibrium with the parent phase as the shape of the clusters is not determined by an "equilibrium" shape of the cluster that can be derived by minimizing the Gibbs free energy as a function of the cluster shape. The cluster shape is more determined by the instantaneous pre-smectic structure of the isotropic fluid phase. We also find that the growth of the clusters proceeds more by the reorientation and coalescence of smectic-like groups than by individual particles as is usual for nucleation. Moreover, due to the pre-smectic ordering of the isotropic phase, the dynamics of individual particles is slow as diffusion can only occur cooperatively. We therefore conclude that the nucleation and growth of the smectic phase is hampered by the pronounced pre-smectic ordering in the isotropic fluid phase. Only when the collective rearrangements are favoured by a higher supersaturation, can we observe the I–SM transition, which occurs in a spinodal-like fashion.

One might think that the inclusion of attractive interactions, *e.g.*, depletion interactions induced by the addition of non-adsorbing polymer,[22] may help the nucleation of tactoids. However, a very recent study of our group has shown that the critical cluster for a system of attractive rods consists of a single layer that only grows laterally, while the nucleation of a second layer on top of the first layer is an extremely rare event.[23] Nucleation of multilayer clusters were never observed during the simulations.

5 Acknowledgements

This work is financially supported by a NWO-VICI grant and a TOP-CW funding of the "Nederlandse Organisatie voor Wetenschappelijk Onderzoek (NWO)" and by the High Potential Programme of Utrecht University. A.C. acknowledges support from *Juan de la Cierva* program of the Ministry of Science and Innovation of Spain.

References

1 Z.-J. Wang, C. Valeriani and D. Frenkel, *J. Phys. Chem. B*, 2009, **113**, 3776.
2 P. R. ten Wolde and D. Frenkel, *J. Chem. Phys.*, 1998, **109**, 9901.
3 D. Moroni, P. R. ten Wolde and P. G. Bolhuis, *Phys. Rev. Lett.*, 2005, **94**, 235703.
4 E. Sanz, C. Valeriani, D. Frenkel and M. Dijkstra, *Phys. Rev. Lett.*, 2007, **99**, 055501.
5 T. Schilling and D. Frenkel, *Phys. Rev. Lett.*, 2004, **92**, 085505.
6 T. Schilling and D. Frenkel, *Comput. Phys. Commun.*, 2005, **169**, 117.
7 D. Frenkel and T. Schilling, *Phys. Rev. E*, 2002, **66**, 041606.
8 Z. Dogic and S. Fraden, *Philos. Trans. R. Soc. London, Ser. A*, 2001, **359**, 997.
9 Z. Dogic, *Phys. Rev. Lett.*, 2003, **91**, 165701.
10 H. Maeda and Y. Maeda, *Phys. Rev. Lett.*, 2003, **90**, 018303.
11 M. P. B. van Bruggen, J. K. G. Dhont and H. N. W. Lekkerkerker, *Macromolecules*, 1999, **32**, 2256.
12 P. Oakes, J. Viamontes and J. X. Tang, *Phys. Rev. E*, 2007, **75**, 061902.
13 M. P. Lettinga, K. Kang, P. Holmqvist, A. Imhof, D. Derks and J. K. G. Dhont, *Phys. Rev. E*, 2006, **73**, 011412.
14 Y.-G. Tao, W. K. Otter, J. K. G. Dhont and W. J. Briels, *J. Chem. Phys.*, 2006, **124**, 134906.
15 Z. X. Zhang and J. S. van Duijneveldt, *J. Chem. Phys.*, 2006, **124**, 154910.
16 J. Viamontes, P. Oakes and J. X. Tang, *Phys. Rev. Lett.*, 2006, **97**, 118103.
17 P. Prinsen and P. van der Schoot, *Phys. Rev. E*, 2003, **68**, 021701.
18 A. Cuetos and M. Dijkstra, *Phys. Rev. Lett.*, 2007, **98**, 095701.
19 A. Cuetos, R. van Roij and M. Dijkstra, *Soft Matter*, 2008, **4**, 757.
20 P. Bolhuis and D. Frenkel, *J. Chem. Phys.*, 1997, **106**, 666.
21 S. C. McGrother, D. C. Williamson and G. Jackson, *J. Chem. Phys.*, 1996, **104**, 6755.
22 S. V. Savenko and M. Dijkstra, *J. Chem. Phys.*, 2006, **124**, 234902.
23 A. Patti and M. Dijkstra, *Phys. Rev. Lett.*, 2009, **102**, 128301.

Multi-scale simulation of asphaltene aggregation and deposition in capillary flow

Edo S. Boek,[*ac] **Thomas F. Headen**[ad] **and Johan T. Padding**[b]

Received 3rd February 2009, Accepted 12th March 2009
First published as an Advance Article on the web 26th August 2009
DOI: 10.1039/b902305b

Asphaltenes are known as the 'cholesterol' of crude oil. They form nano-aggregates, precipitate, adhere to surfaces, block rock pores and may alter the wetting characteristics of mineral surfaces within the reservoir, hindering oil recovery efficiency. Despite a significant research effort, the structure, aggregation and deposition of asphaltenes under flowing conditions remain poorly understood. For this reason, we have investigated asphaltenes, their aggregation and their deposition in capillary flow using multi-scale simulations and experiments. At the colloid scale, we use a hybrid simulation approach: for the solvent, we used the stochastic rotation dynamics (also known as multi particle collision dynamics) simulation method, which provides both hydrodynamics and Brownian motion. This is coupled to a coarse-grained MD approach for the asphaltene colloids. The colloids interact through a screened Coulomb potential with varying well depth ε. We tune the flow rate to obtain $Pe_{\text{flow}} \gg 1$ (hydrodynamic interactions dominate) and $Re \ll 1$ (Stokes flow). Imposing a constant pressure drop over the capillary length, we observe that the transient solvent flow rate decreases with increasing well depth ε. The interactions between the mesoscopic asphaltene colloids can be related to atomistic MD simulations. Molecular structures for the atomistic calculations were obtained using the quantitative molecular representation approach. Using these structures, we calculate the potential of mean force (PMF) between pairs of asphaltene molecules in an explicit solvent. We obtain a reasonable fit using a $-1/r^2$ attraction for the attractive tail of the PMF at intermediate distances. We speculate that this is due to the two-dimensional nature of the asphaltene molecules. Finally, we discuss how we can relate this interaction to the mesoscopic colloid aggregate interaction. We assume that the colloidal aggregates consist of nano-aggregates. Taking into account observed solvent entrainment effects, we deduct the presence of lubrication layers between the nano-aggregates, which leads to a significant screening of the direct asphaltene–asphaltene interactions.

Introduction

The deposition of asphaltenes may cause problems in production facilities, flowlines and oil reservoirs near wellbores. It is the latter issue that we will be dealing with in

[a]*Schlumberger Cambridge Research, High Cross, Madingley Road, Cambridge, CB3 0EL, United Kingdom. E-mail: esb30@cam.ac.uk*
[b]*Computational Biophysics, Department of Science and Technology, University of Twente, P.O. Box 217, 7500 AE Enschede, The Netherlands*
[c]*Department of Chemical Engineering, Imperial College London, SW7 2AZ, United Kingdom*
[d]*Department of Physics and Astronomy, University College London, Gower Street, London, WC1E 6BT, United Kingdom*

this paper. Recently, a number of capillary flow experiments have been carried out to investigate asphaltene deposition. Broseta *et al.*[1] calculated an effective hydrodynamic thickness of a deposited asphaltene layer in flow experiments in a metal capillary, assuming a uniform thickness of the layer deposited. Wang *et al.*[2] studied the deposition of asphaltene on metallic surfaces using the homogeneous deposition hypothesis.

There are several papers in the literature related to the modeling of asphaltene deposition in flowing systems. These include the flow and deposition of asphaltene in production pipe lines[3,4] and formation damage due to deposition in the reservoir. The latter include network models[5] and Darcy scale Deep Bed Filtration continuum models.[6] However, these are all phenomenological models, the parameters of which are difficult to relate to the physics of the underlying deposition process.

In a recent paper[7] we experimentally investigated asphaltene deposition in a glass capillary. We compared our experimental results with stochastic rotation dynamics (SRD) computer simulations. Experiment and simulation were directly compared by calculating the dimensionless conductivity of the capillary. We established the effective value of the interaction potential well depth between the colloidal asphaltene particles by comparison with the capillary flow experiment. The aim of our current work is to relate the depth of the colloidal interaction potential to the molecular interaction of asphaltenes.

In the literature, the mechanism of asphaltene aggregation and precipitation is a matter of lively debate. An excellent overview was given by Porte *et al.*[8] They propose a vesicle model which is consistent with all the experimental data available. In this paper, we will postulate a colloid model consisting of nano-aggregates with a solvent layer which screen the molecular interactions. The model proposed is consistent with capillary flow experiments and computer simulations, on both the molecular and colloid scale.

First, we will describe our simulations of capillary flow at the colloid scale and their comparison with experiments. Then we present simulations at the molecular scale to determine the interactions between the asphaltene molecules. Finally, we will discuss how the colloid scale interactions can be related to the molecular dynamics calculations.

Capillary flow simulations

Recently, we reported experiments of asphaltene deposition in capillary flow.[7] We measured the pressure drop over the capillary as a function of time. For more details, we refer to ref. 7. Here we will discuss the capillary flow simulations in detail.

We have used a particulate simulation technique called stochastic rotation dynamics (SRD). This method has recently been developed to solve the hydrodynamics of complex fluids.[9] Although SRD is a fairly recent method, its statistical mechanical and hydrodynamic fundamentals have been thoroughly studied and validated.[10,11] It is a mesoscopic particulate method where the solvent is represented by N_f point-like (ideal gas) particles. It is similar in spirit to the lattice-Boltzmann (LB) method.[12,13] In contrast to LB however, it includes Brownian motion which emerges naturally from the thermal fluctuations of the SRD particles (N_f). This makes the SRD method particularly useful to study the hydrodynamics of colloids such as oilfield asphaltene and wax particles. The colloidal particles are described by means of a coarse-grained molecular dynamics (MD) method[11] which will be described in the following section. The SRD hydrodynamics emerges from collisions between solvent particles in coarse-grained cells at coarse-grained time intervals. It is a simple and computationally cheap algorithm ($O(N_f)$), which can be easily coupled to solutes such as polymers and colloids. The algorithm proceeds in two steps.[14] In the first of these, a free streaming step, the positions of the solvent particles, $\boldsymbol{r}_i(t)$, are updated simultaneously according to

$$\mathbf{r}_i(t + \delta t) = \mathbf{r}_i(t) + \mathbf{v}_i(t)\delta t \tag{1}$$

where $\mathbf{v}_i(t)$ is the velocity of a particle and δt is the discretised time step of the solvent.

The second part of the algorithm is the collision step. The simulation system is coarse-grained into cells. Stochastic multi-particle collisions are performed within each individual cell, by rotating the velocity of each particle relative to the center-of-mass velocity $\mathbf{v}_{\mathrm{cm}}(t)$ of all the particles within that cell:

$$\mathbf{v}_i\,(t + \delta t) = \mathbf{v}_{\mathrm{cm}}(t) + \mathbf{R}(\mathbf{v}_i(t) - \mathbf{v}_{\mathrm{cm}}(t)) \tag{2}$$

$\mathbf{R}$ is a rotation matrix which rotates velocities by a fixed angle α around a randomly oriented axis. The aim of the collision step is to transfer momentum between the particles while conserving the total momentum and energy of each cell. Because mass, momentum and energy are conserved locally, the thermohydrodynamic equations of motion are captured in the continuum limit.[9] Hence hydrodynamic interactions can be propagated by the solvent. Note, however, that any molecular details of the solvent are excluded—this allows the hydrodynamic interactions to be modelled with minimal computational expense. This procedure conserves mass, momentum and energy and yields the correct hydrodynamic behaviour, in the sense that the Navier–Stokes equations are recovered.[15] The fluid particles only interact with each other through the rotation procedure, which can be viewed as a coarse-graining of particle collisions over space and time. For this reason, the SRD solvent should not be considered as molecules, but rather as a Navier–Stokes solver that naturally includes Brownian noise. The SRD simulation technique has been applied to colloids in solution,[9,11] colloidal sedimentation,[15] clay particles, polymer fluids, amphiphilic systems, flow in porous media, binary fluid mixtures, flow around solid objects and reactive fluids.

Here, we consider the aggregation and deposition of asphaltene colloidal particles in capillary flow in 2 dimensions (2-D) using the SRD method. The solvent is represented by SRD point particles as described above. Asphaltene particles are defined in the fluid as colloidal particles (in 2-D) using the MD algorithm.[11] We fix the length scale of the model by choosing the radius R of the asphaltene colloids.

We will now discuss the colloidal interactions used in the simulations. It is not straightforward to obtain asphaltene colloid–colloid (c-c) and colloid–wall (c-w) interactions. For this reason, we choose a model interaction potential with variable well depth $\varepsilon_{\mathrm{cc}}$, corresponding with a screened Coulomb potential, according to the Yukawa functional form:

$$\phi_{\mathrm{cc}}(r) = \varepsilon_1 \frac{\sigma_1}{r} \exp(\kappa_1(r - \sigma_1)) + \varepsilon_2 \frac{\sigma_2}{r} \exp(\kappa_2(r - \sigma_2)) \tag{3}$$

In this way, the colloid–colloid interactions are defined in a classical van der Waals manner[16] through a combination of attractive and repulsive interactions. Here ε_1 and ε_2 are the parameters for attraction and repulsion, respectively, r is the distance between the colloids, κ^{-1} is the Debye screening length and σ is the colloid diameter. A typical way to parameterize this equation is shown in Fig. 1, where the depth of the potential well at the minimum, $\varepsilon_{\mathrm{cc}}$, is varied between 2, 5, 10, 20 and 50 $k_{\mathrm{B}}T$, by choosing appropriate values of ε_1 and ε_2.

Note that it is the depth of this well that determines the degree of "stickiness" between the colloidal particles; the deeper the potential well, the stronger the shear forces required to break up the aggregate.

Note that we have used a Double Yukawa (DY) functional form for the colloid–colloid interactions. This is a functional form commonly used for the effective interaction between colloidal aggregates.[17] In general, these effective interactions result from a sum over the contributions of the individual molecules, which are normally considered to have a Lennard-Jones 6–12 functional form. Within the colloidal

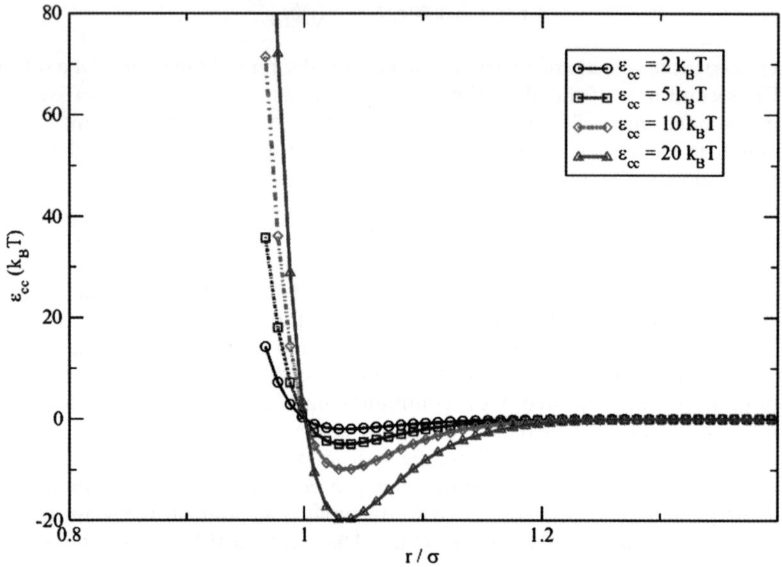

Fig. 1 Colloid–colloid interaction potential. By appropriately choosing the parameters for repulsion and attraction, the depth of the potential well is varied in units of k_BT, between 2 (circles), 5 (squares), 10 (diamonds) and 20 (triangles).

asphaltene particle, the molecules are oriented roughly in a random fashion. Therefore the molecular π–π interactions[3] are spherically averaged and as such included in the DY potential.

In particular, it can be shown that the DY functional form gives a very good fit for the sum over the van der Waals interactions between C60 colloidal particles.[18] We assume here that also the effective interactions between colloidal asphaltene aggregates can be described by a DY potential. In addition to the van der Waals interactions, we have a Coulomb interaction, which is screened by the oil phase. It can be shown that, for an oil phase with a dielectric constant $\varepsilon_r = 3$, the electrostatic interactions will decay to order k_BT over a distance of 19 nm. This shows that the range of the electrostatic interactions is much smaller than the size of our asphaltene aggregate particles with a diameter of 3000 nm.

We note that the DY form of the colloid–colloid interaction may not be exactly the same as the real effective interactions between 3 micrometre sized clusters of asphaltene.[19] However, since the range of attraction is extremely small compared to the size of the colloid, the precise form of the interaction does not matter. It can be shown that only the integral over the attractive well, *i.e.* the second virial coefficient, is of influence on the thermodynamic and aggregation properties.[17] Because a very small range is computationally too expensive (extremely small time steps in MD) we choose a slightly larger range of attraction, but still small compared to the particle diameter.

Obtaining the appropriate value for the effective potential well depth to represent the asphaltenes requires some consideration. In this paper, our main aim is to obtain a correct description of the balance between the deposition of asphaltene aggregates and the stress induced by the fluid flow. In an ideal world, this should be done using a full molecular description of the asphaltenes, with atomistic resolution.[20] Unfortunately, it is very difficult to achieve this goal using MD simulations, due to computational limitations of current computers. In our recent paper,[7] we used the above effective potential and estimated the value of the potential well depth by comparing the results of our capillary experiments to the simulations. In the current paper, we will relate these interactions to the interactions at the molecular level.[21]

For the colloid–wall (c-w) interactions, we consider two different scenarios. First, we consider an infinitely deep potential well ε_{cw}, which means that the particles will stick irreversibly to the capillary wall. This corresponds with adsorption on a metal surface. Second, we assume that the capillary wall is covered by a thin layer of asphaltene particles, so that the c-w interaction is identical to the c-c interaction. This scenario corresponds with adsorption on a mineral (rock) surface or a glass capillary.

Now we will discuss the mechanism of asphaltene aggregation. Yudin *et al.*[19] studied the mechanisms of asphaltene aggregation in toluene–heptane mixtures using dynamic light scattering. They observed that, for asphaltene concentrations above the critical aggregate concentration (CAC), two different mechanisms of aggregation occur, as a function of time. In the initial stages, where the mean radius of the aggregates is smaller than 1 μm, the aggregation can be described by a reaction limited aggretation (RLA) process. This means that not every contact between two particles results in their sticking. In terms of colloidal interaction potentials (see above) this is usually described by means of an activation barrier between the free and aggregated states. In the later stages, where the mean radius of the aggregates is equal to or larger than 3 μm, the aggregation can be described by a diffusion limited aggregation (DLA) process. This means that every contact between the particles results in their association. Based on microscopic and light scattering[19] observations of asphaltene aggregate sizes, we consider here asphaltene particles with a radius of 3 μm, interacting by means of a DLA potential without an activation barrier.

Here we will carry out a dimensional analysis of the problem. In the capillary flow experiment, we consider a volumetric flow rate $Q = 5$ μl/min. Using the capillary surface area $A = (50 \times 500)$ μm^2, this corresponds with a linear flow rate $v_{\text{flow}} = Q/A = 1/300$ m/s.

Assuming a colloid aggregate radius R of 3 μm, we calculate the diffusion coefficient D from the Stokes–Einstein equation

$$D = \frac{k_B T}{6\pi\eta R} = 7.3 \text{ x } 10^{-14} \text{m}^2/\text{s} \tag{4}$$

where η is the solvent viscosity. We calculate the Peclet number for flow Pe^{flow} as

$$Pe^{\text{flow}} = \frac{v^{\text{flow}} R}{D} = \frac{v^{\text{flow}} R^2 6\pi\eta}{k_B T} \tag{5}$$

For the experiment, we find $Pe^{\text{flow}} = 1.4 \times 10^5$. This means that hydrodynamic interactions are much more important than Brownian diffusion. Note that in the simulations, with the current set of parameters we can achieve Peclet numbers up to 10. While not corresponding directly to the experiment, because $Pe^{\text{flow}} \gg 1$ in both simulation and experiment, we expect that hydrodynamic interactions dominate in both cases and it is justified to make qualitative comparisons.

Finally, we calculate the Reynolds number for the particles Re^{p} as

$$Re^{\text{p}} = \frac{v^{\text{flow}} R \rho}{\eta} = 0.01 \tag{6}$$

Because $Re^{\text{p}} \ll 1$, we are in the Stokes flow limit, where inertial effects are unimportant. This regime is typical for flow in porous media.

We will now show that the time scale of the simulation can be matched with the time scale of the experiment, by considering the diffusion of the colloidal particles. First, we consider the time required for an asphaltene colloid to diffuse over its own radius, τ_D:

$$\tau_D = \frac{R^2}{D} \tag{7}$$

where R is the colloid radius and D is the self-diffusion coefficient. Assuming $R = 3$ μm and using the value for D obtained in the previous section, we find $\tau_D = 124$ s.

We determine the self-diffusion coefficient in 2 dimensions consistently from a simulation of a single colloid in a capillary by calculating the velocity autocorrelation function (VACF).[22] From the calculation, we find $\tau_D = 75$ (in time units a_0 $(m/kT)^{1/2}$ used in the program), so that $\tau_{SRD} = 124/75 = 1.65$ s. One MD time step of 0.01 SRD time units therefore corresponds to 0.0165 s. Typically, we carry out a simulation run for 10 million MD time steps, which ensures that a steady state is achieved. This is equivalent to $10^7 \times 0.0165$ s $\cong$ 45 h. This time scale is comparable with the time scale of the flow experiments.

We will now describe the technical details of the simulations. We use a two-dimensional simulation box of dimensions $L_x \times L_y = 75 \times 450$ SRD cells, at a number density ρ of 5 SRD particles per cell.[15] Periodic boundary conditions are used along the direction of the flow, y. This means that we are effectively simulating an infinitely long capillary. We consider a constant number of asphaltene colloidal particles. These are inserted randomly in the SRD fluid at a concentration of 8% by volume. The colloidal particles have a fixed diameter σ_{cc} of 3 SRD cells.

The solvent–colloid radius is always slightly smaller than the colloid–colloid interaction radius (1.3 and 1.5 SRD cells, respectively). This ensures that the simulation is effectively still representative of a three-dimensional system, not in terms of blocking, but hydrodynamically. A similar method has been used successfully in recent LB-DEM simulations.[23] Essentially, this means that solvent flow is still possible through the thin solvent layer around the asphaltene particles. There will always remain a base permeability in this model, even in the case of complete blockage.

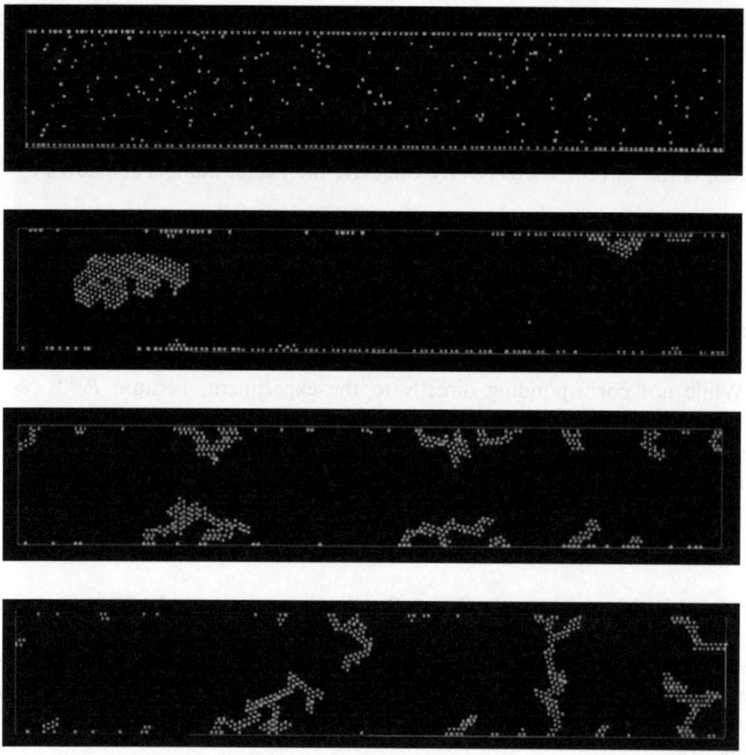

Fig. 2 Simulation snapshots for steady state behaviour, after 10 million time steps, as a function of increasing c-c potential well depths: from top to bottom $\varepsilon_{cc} = 2$, 5, 10 and 20 k_BT. Reprinted with permission from ref. 7. Copyright 2008 American Chemical Society.

We generate Poiseuille flow in the capillary by imposing a body force g_{SRD} on the SRD particles. In the absence of blockage, the required body force to attain a flow velocity v^{flow} is given by:

$$g_{\mathrm{SRD}} = \frac{16\eta v^{\mathrm{flow}}}{\rho L_x^2} \tag{8}$$

This corresponds with a constant pressure drop

$$\frac{\Delta P}{\Delta x} = \rho g_{\mathrm{SRD}} \tag{9}$$

and the corresponding solvent velocity v_{solv} is measured. Note that in the experiment, we impose a constant flow rate and measure the corresponding pressure drop. Direct comparison between the experimental and simulation results is possible by calculating the dimensionless conductivity. This will be shown in the following section.

Now we will present results of the simulations. We consider two different scenarios for the colloid–wall (c-w) interactions. First, we present a set of simulation results with an infinitely deep potential well $\varepsilon_{\mathrm{cw}} = -\infty$, which means that the particles will stick irreversibly to the capillary wall. This corresponds with adsorption on a metal surface. In this set of simulations, we vary the colloid–colloid well depth $\varepsilon_{\mathrm{cc}}$ systematically from 2, 5, 10, 20 to 50 $k_{\mathrm{B}}T$.

We find that the transient solvent flow rate decreases when the asphaltene particles become more "sticky". For a well depth $\varepsilon_{\mathrm{cc}} = 2\ k_{\mathrm{B}}T$, we find the deposition of a monolayer on the capillary wall. With increasing well depth, we find that the capillary becomes totally blocked. The clogging is transient for $\varepsilon_{\mathrm{cc}} = 5\ k_{\mathrm{B}}T$, but appears to be permanent for $\varepsilon_{\mathrm{cc}} = 10$–$20\ k_{\mathrm{B}}T$. Simulation snapshots taken at the end of the runs of 10 million steps each show are presented in Fig. 2.

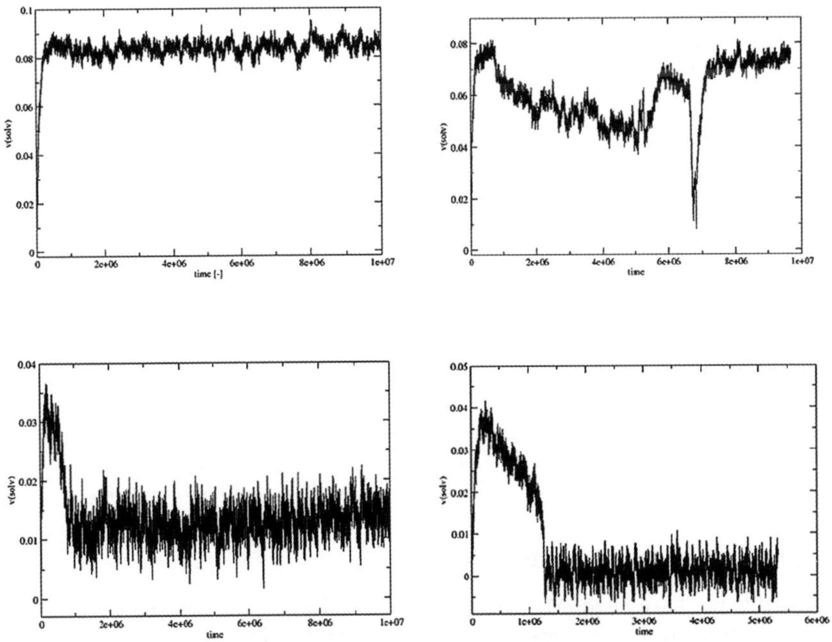

Fig. 3 Solvent velocity as a function of time for $\varepsilon_{\mathrm{cc}} = 2\ k_{\mathrm{B}}T$ (top left), $5\ k_{\mathrm{B}}T$ (top right), $10\ k_{\mathrm{B}}T$ (bottom left) and $20\ k_{\mathrm{B}}T$ (bottom right), respectively. Reprinted with permission from ref. 7. Copyright 2008 American Chemical Society.

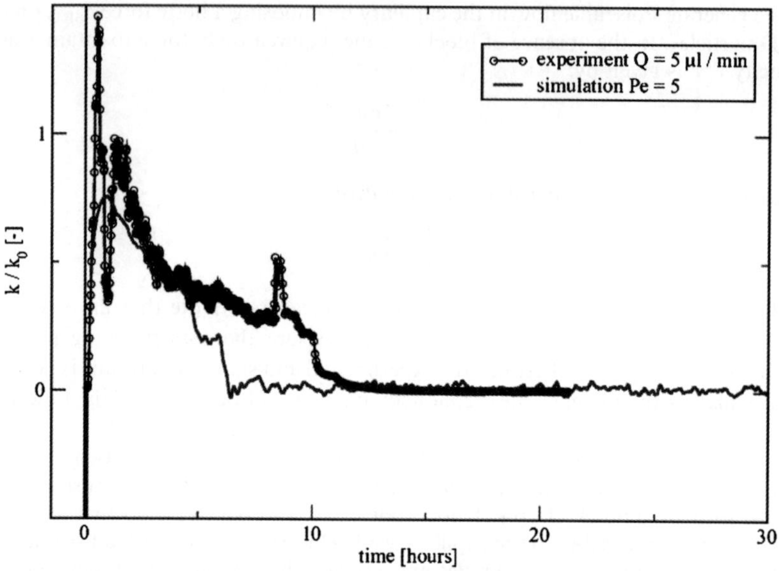

Fig. 4 Dimensionless conductivity k/k_0 of the capillary as a function of time for both experiment ($Q = 5$ µl/min, circular symbols) and simulation ($Pe = 5$, no symbols).

In Fig. 3, we present the numerically measured solvent velocity as a function of time, for $\varepsilon_{cc} = 2$, 5, 10 and 20 k_BT, respectively. It can be clearly observed that the final fluid velocity decreases with increasing potential well depth ε_{cc}.

To make a more quantitative comparison, we will make use of the time scale analysis presented in the previous section. We transform both the experimental time series for the pressure drop dp/dx and the simulation time series for the measured velocity v to dimensionless conductivity k/k_0 using Darcy's law. First, we consider the experimental results for a volumetric flow rate $Q = 5$ µl/min. The experimental values of the dimensionless conductivity are presented in Fig. 4, together with the corresponding simulation results for the case $\varepsilon_{cw} = \varepsilon_{cc} = 10\ k_BT$ and $Pe = 5$.

From this figure, we observe that full blocking of the capillary occurs after roughly 7 h in the simulation and 10 h in the experiment. Furthermore, we observe that the time evolution of the dimensionless conductivity for early times is similar for both experiment and simulation. Finally, we note that, once deposition occurs, it is permanent.

In addition, we consider the experimental results for the double volumetric flow rate $Q = 10$ µl/min. We now find that the best comparison with simulation data is for the case $\varepsilon_{cw} = \varepsilon_{cc} = 10\ k_BT$ and $Pe = 10$, at a flow rate doubled compared with the previous simulation case and the same value of ε_{cc}. More details can be found in ref. 7.

Summarising, we find good agreement with experiments for an interaction potential well depth of the order of 10 k_BT. In the following, we will try to explain this value by upscaling from molecular interactions to the colloid length scale.

Asphaltene molecular interactions

Recently, we have used MD simulations to calculate the interaction between different asphaltene molecules.[24] The asphaltene molecular structures were recently obtained in a systematic fashion from experiments, using a modified version of the quantitative molecular representation (QMR) algorithm. We refer to ref. 21 for more details on the QMR results.

In the MD simulations, we consider three different QMR pairs of asphaltene molecules dissolved in either toluene or heptane. The three different molecules A, B and C are presented in Fig. 5–7, respectively.

Here A represents a typical archipelago structure, B is a typical resin and C is a typical peri-condensed structure. For more details, we refer to ref. 24.

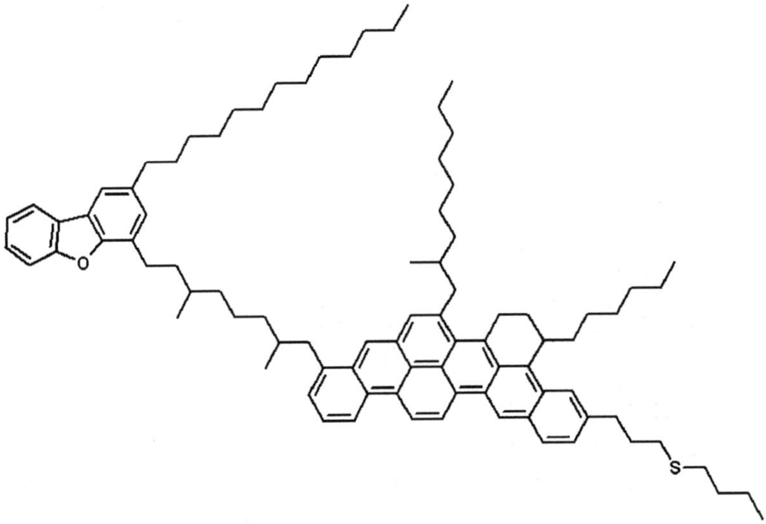

Fig. 5 Chemical structure of Asphaltene A – an "archipelago" type asphaltene with two condensed aromatic cores connected by an aliphatic chain.

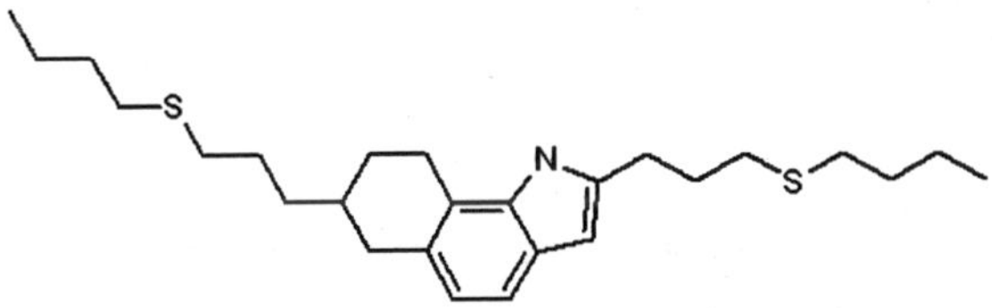

Fig. 6 Chemical structure of Resin B – a structure at the resin end of the asphaltene spectrum.

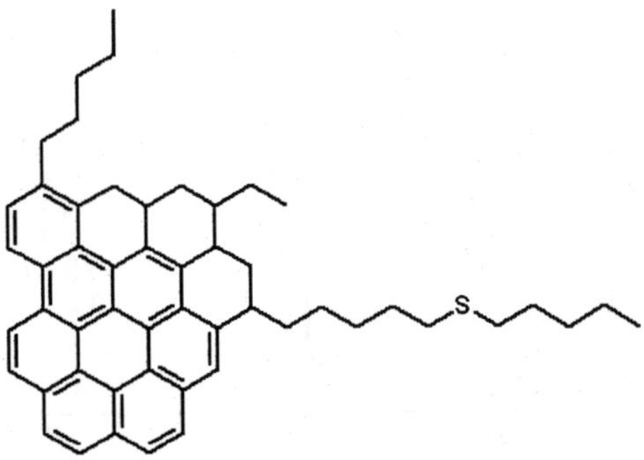

Fig. 7 Chemical structure of Asphaltene C – An "island" type asphaltene structure with one large peri-condensed aromatic core.

From the simulations, we calculate the radial distribution functions $g(r)$ for molecule pairs A, B and C. The potential of mean force (PMF) W can be calculated from the $g(r)$ using

$$W = -kT\ln(g(r)) \quad (10)$$

The PMF is equivalent to the Helmholtz free energy (plus a constant). We can therefore calculate the free energy of dimer formation by taking the difference of the potential of mean force at maximum separation and at equilibrium separation (where the PMF is a minimum). This has been done for the molecule pairs A, B and C.

The results for minus PMF for the three molecular structures are presented in Fig. 8, Fig. 9 and Fig. 10, respectively. The quality of a power law fit for the

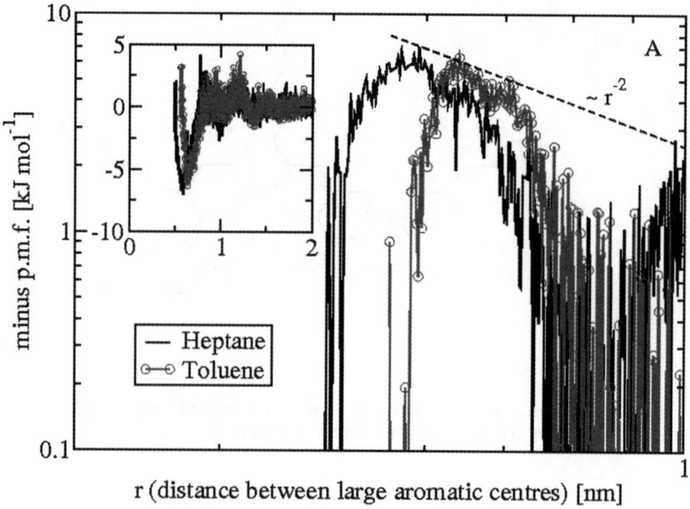

Fig. 8 Minus the potential of mean force as a function of separation distance for Asphaltene pair A on a double logarithmic scale, for a solvent of heptane (no symbols) or toluene (circular symbols). The inset shows the p.m.f. on a linear scale. The $-1/r^2$ scaling behaviour is shown as a dashed line.

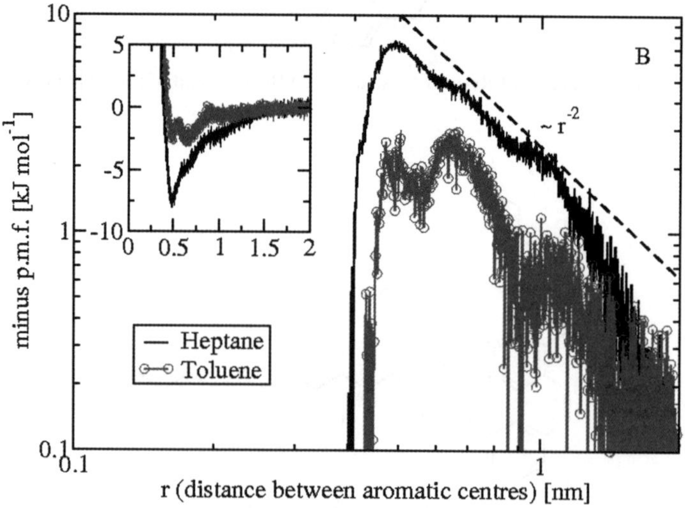

Fig. 9 As Fig. 8, now for a pair of Resin B molecules.

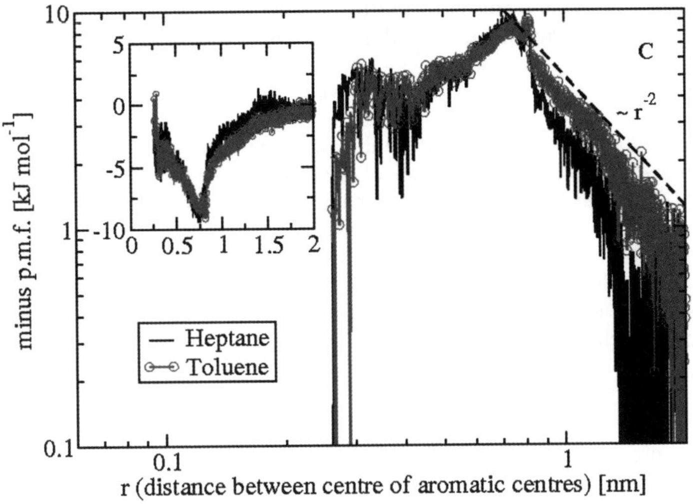

Fig. 10 As Fig. 8, now for a pair of Asphaltene C molecules.

attractive tail depends on the particular molecule. For most asphaltenes, we observe a reasonable fit using a $-1/r^2$ attraction for the attractive tail at intermediate distances. We speculate that this is due to the two-dimensional nature of the asphaltene molecules; Israelachvili[25] has shown that the interaction free energy between two flat parallel surfaces scales as $-1/r^2$. Indeed, at short distances the planes of asphaltene molecules preferably align. At larger distances this alignment is less strong, leading to a faster than $1/r^2$ decay. Note that Asphaltene C, which is generally argued to be more "representative" of the asphaltene structure,[21,26] gives the best agreement with the $-1/r^2$ scaling. This is due to the rigid 2-D planar structure of this molecule, see Fig. 7. In comparison, Asphaltene A gives the least convincing fit to the $-1/r^2$ behaviour, which may be explained by the archipelago, more flexible, nature of the molecule.

For all three cases, we observe a minimum around $a = 0.5$–0.75 nm at a depth of $E = 2.8$–3.2 kT. The tail has an effective range (negative 1 kT) of another 0.5 nm

Discussion: relating the molecular asphaltene interactions to colloidal aggregate interactions

Our goal is to relate the molecular (and rotationally averaged) asphaltene–asphaltene PMF to an effective potential between colloidal, 3 micron-sized spherical aggregates of asphaltenes. For this purpose, we use the PMFs presented in the previous section.

In the SRD simulations, we used a short range potential (see eqn (3)), consisting of a hard core of radius $R = \sigma/2$, and an attractive well of width w of the order of 0.2σ (or $0.1R$) and depth ε. We have fixed the radius R to 3000 nm because this corresponds to the smallest asphaltene cluster size for which the dynamics is diffusion limited instead of reaction limited.[19] The only remaining undetermined parameter is ε.

From the QMR simulations we have found that the asphaltene molecular potential of mean force has a typical 'hard core' diameter a of about 0.5 nm, and an attractive well of width $w' = 1$ nm and depth $\varepsilon' = 3\ kT$.

To choose the value of ε we need to consider what forces are important in our flow experiments and simulations. A large cluster tends to break when the drag force induced by the flow becomes larger than the cohesive force of the cluster. To be more specific, consider a cluster of width and height h, attached to a wall or another cluster. The hydrodynamic drag on this cluster is of the order

$$F^{\text{drag}} = 6\pi\eta hv \tag{11}$$

where η is the solvent viscosity and v the flow velocity. In order to break, the cluster needs to be displaced over a distance of the order of the width w of the interaction potential. This may be accomplished by having fast enough flow. Let us compare the breaking-point in simulation and experiment.

In the simulation, a cluster is just about to break off when the work done by the drag force equals the number of broken bonds times the energy ε. In the simulations the clusters are quite dense, so the number of broken bonds is of the order of h/σ. Therefore the cluster is about to break off when:

$$F^{\text{drag}}w = \frac{h}{\sigma}\varepsilon \tag{12}$$

We can express this balance also in terms of the Peclet number of a cluster of radius R, where

$$Pe = \frac{vR}{D} = \frac{6\pi\eta R^2 v}{kT} = \frac{F^{\text{drag}}R^2}{hkT} \tag{13}$$

The $h \times h$ cluster is just about the break off when

$$2PekT\frac{w}{R} = \varepsilon \tag{14}$$

Experimentally, without detailed information about the structure and organization of asphaltene, resin, and solvent molecules in an asphaltene aggregate, it is difficult to precisely determine the energy required to break off a piece of aggregate. In very rough terms we may say that the cluster is just about to break off when the work done by the drag force equals the number of broken (physical) bonds times the energy ε'. We take into account that the number of broken bonds is very much influenced by the fractal nature of asphaltene clusters. It has been argued that asphaltene molecules self-assemble into quite compact and strongly bound nanoclusters of ~2.5 nanometres in size.[27] On larger length scales these nanoclusters interact with weak Van der Waals forces, and form loosely bound clusters of fractal dimension near 2.[28,29] Such clusters have a large amount (30–50% by volume) of entrained solvent,[30] including surfactants and water.

We will now try to estimate the dimensions of the solvent shell surrounding the nano-aggregate. We assume that the nano-aggregate is spherical with radius R and surrounded, on average, by a spherical shell of solvent with inner radius R and outer radius R'. When a fraction $(1 - f)$ is occupied by solvent, and therefore a fraction f by asphaltene, then

$$\left(\frac{R'}{R}\right)^3 = \frac{1}{f} \tag{15}$$

The average thickness d of the solvent layer is therefore

$$d = 2(R' - R) = 2R(f^{-1/3} - 1) \tag{16}$$

Using the estimate of $f = 0.7$–0.5[30] and $R = 2.5$ nm, we find $d = 0.5$–0.75 nm.

So if we imagine the solvent to form lubrication layers between the nano-aggregates, a 30–50% solvent entrainment corresponds to an average solvent layer thickness of 0.5–0.75 nanometres. This may lead to a significant screening of the direct asphaltene–asphaltene electrostatic interactions.

Lumping the fractal and solvent screening effects together into a factor ϕ, we find that a cluster of size h is about the break off when

$$F^{\text{drag}}w' = \frac{h}{a}\varepsilon'\phi \tag{17}$$

Taking into account the experimental Peclet number Pe' for a cluster of size R, we find that the breaking point is achieved when

$$Pe'\,kT\frac{w'a}{R^2} = \varepsilon'\phi \tag{18}$$

We can now compare the simulation and experimental expressions. The experiments have been performed at a Peclet number $Pe' = 1.4 \times 10^5$, whereas for computational ease the simulations were performed at $Pe = 10$. In order to have the same physics, we should choose the ratio of ε to ε' as:

$$\frac{\varepsilon}{\varepsilon'} = \frac{w}{w'}\frac{2R}{a}\phi\frac{Pe}{Pe'} \tag{19}$$

Using $w = 300$ nm, $w' = 1$ nm, $R = 3000$ nm, $a = 0.5$ nm, $Pe = 10$, and $Pe' = 1.4 \times 10^5$, we find $\varepsilon/\varepsilon' = 260\phi$. As already mentioned, we lack detailed information to estimate ϕ. The good agreement between experimental results with $\varepsilon' = 3\ kT$ and simulations with $\varepsilon = 10\ kT$ suggests that ϕ is about 1.3%. This is a reasonable number, when we consider screening by a solvent layer around nano-aggregates, as described. We are planning to put more detail into this calculation by considering the composition of the solvent layer.

Conclusions

We have investigated asphaltenes, their aggregation and deposition in capillary flow using multi-scale simulations. We used a hybrid simulation approach: for the solvent, we used the stochastic rotation dynamics simulation method, which provided both hydrodynamic and Brownian motion. This is coupled to a coarse-grained MD approach for the asphaltene colloids. The colloids interacted through a screened Coulomb potential with varying well depth ε. We tuned the flow rate to obtain $Pe_{\text{flow}} \gg 1$ (hydrodynamic interactions dominate) and $Re \ll 1$ (Stokes flow). Imposing a constant pressure drop over the capillary length, we observed that the transient solvent flow rate decreases with increasing well depth ε. The interactions between the mesoscopic asphaltene colloids can be related to atomistic MD simulations. Molecular structures for the atomistic calculations were obtained using the quantitative molecular representation approach. Using these structures, we calculated the potential of mean force (PMF) between pairs of asphaltene molecules in an explicit solvent. We obtained a reasonable fit using a $-1/r^2$ attraction for the attractive tail of the PMF at intermediate distances. We speculate that this is due to the two-dimensional nature of the asphaltene molecules; the interaction free energy between two ideal flat parallel surfaces scales as $-1/r^2$. Indeed, at short distances the planes of asphaltene molecules preferably align. Finally we discussed how we can relate the microscopic molecular asphaltene interaction (MD) to the mesoscopic colloid aggregate interaction (SRD) with well depths of 3 kT and 10 kT, respectively. We assumed that the colloidal aggregates consist of nano-aggregates. Taking into account observed solvent entrainment effects, we deduced the presence of lubrication layers between the nano-aggregates, which may lead to a significant screening of the direct asphaltene–asphaltene interactions. Combining the screening effect with the fractal nature of the aggregates, and using the observed value of 10 kT for the colloidal interaction, we found an effective retardation factor of the interaction ϕ of about 1.3%. Such a number would agree with screening by a solvent layer around nano-aggregates.

Acknowledgements

We are grateful to J. Crawshaw and D.S. Yakovlev for helpful discussions. J.T.P. acknowledges The Netherlands Organisation for Scientific Research (NWO) for financial support.

References

1 D. Broseta, M. Robin, T. Savvidis, C. Féjean, M. Durandeau and H. Zhou, *Detection of Asphaltene Deposition by Capillary Flow Measurements*, SPE 59294, 2000.
2 J. Wang, J. S. Buckley and J. L. Creek, Asphaltene Deposition on Metallic Surfaces, *J. Dispersion Sci. Technol.*, 2004, **25**, 287–298.
3 A. Hammami and J. Ratulowski, Precipitation and Deposition of Asphaltenes in Production Systems: A Flow Assurance Overview, in *Asphaltenes, Heavy Oils, and Petroleomics*, ed. O. C. Mullins, E. Y. Sheu, A. Hammami and A. G. Marshall, Springer, New York, 2007.
4 E. Ramirez-Jaramillo, C. Lira-Galeana and O. Manero, *Energy Fuels*, 2006, **20**, 1184–1196.
5 J. E. P. Monteagudo, K. Rajagopal and P. L. C. Lage, *Chem. Eng. Sci.*, 2002, **57**, 323–337.
6 S. Wang and F. Civan, *J. Energy Resour. Technol.*, 2005, **127**, 310.
7 E. S. Boek, H. K. Ladva, J. P. Crawshaw and J. T. Padding, Deposition of colloidal asphaltene in capillary flow: Experiments and mesoscopic simulation, *Energy Fuels*, 2008, **22**(2), 805–13.
8 G. Porte, H. Zhou and V. Lazzeri, *Langmuir*, 2003, **19**, 40–47.
9 A. Malevanets and R. Kapral, *J. Chem. Phys.*, 1999, **110**, 8605; A. Malevanets and R. Kapral, *J. Chem. Phys.*, 2000, **112**, 7260.
10 T. Ihle and D. M. Kroll, *Phys. Rev. E*, 2001, **63**, 020201; T. Ihle and D. M. Kroll, *Phys. Rev. E*, 2003, **67**, 066705–066706.
11 J. T. Padding and A. A. Louis, *Phys. Rev. E*, 2006, **74**, 031402.
12 S. Succi, *The LB Equation for Fluid Dynamics and Beyond*, Oxford University Press, 2001.
13 M. Venturoli and E. S. Boek, *Physica A*, 2006, **362**, 23–29.
14 N. Kikuchi, C. M. Pooley, J. F. Ryder and J. M. Yeomans, *J. Chem. Phys.*, 2003, **119**, 6388.
15 J. T. Padding and A. A. Louis, *Phys. Rev. Lett.*, 2004, **93**, 220601; J. T. Padding and A. A. Louis, *Phys. Rev. E*, 2008, **77**, 011402.
16 J. N. Israelachvili, *Intermolecular and Surface Forces*, Academic Press, San Diego, 2nd edn, 1991.
17 H. N. W. Lekkerkerker, W. C. K. Poon, P. N. Pusey, A. Stroobants and P. B. Warren, *Europhys. Lett.*, 1993, **20**(1992), 559; C. F. Tejero, A. Daanoun, H. N. W. Lekkerkerker and M. Baus, *Phys. Rev. Lett.*, 1994, **73**, 752; C. F. Tejero, A. Daanoun, H. N. W. Lekkerkerker and M. Baus, *Phys. Rev. E*, 1995, **51**, 558.
18 H. Guerin, *J. Phys.: Condens. Matter*, 1998, **10**, L527–L532.
19 I. K. Yudin, G. L. Nikolaenko, E. E. Gorodetskii, V. R. Melikyan, E. L. Markhashov, D. Frot and Y. Briolant, *J. Pet. Sci. Eng.*, 1998, **20**, 297–301.
20 H. Groenzin and O. C. Mullins, *Energy Fuels*, 2000, **14**, 677–684.
21 E. S. Boek, T. Headen and D. Yakovlev, *Energy Fuels*, 2009, **23**, 1209–1219.
22 M. P. Allen and D. J. Tildesley, *Computer Simulation of Liquids*, OUP, 1990.
23 B. K. Cook, D. F. Boutt, O. E. Strack, J. R. Williams and S. M. Johnson, in Numerical Modeling in Micromechanics Via Particle Methods, ed. Y. Shimizu, R. Hart, and P. Cundell, Taylor & Francis Group, London, 2004.
24 Thomas F. Headen, Edo S. Boek and Neal T. Skipper, *Energy Fuels*, 2009, **23**, 1220–1229.
25 J. N. Israelachvili, *Intermolecular and Surface forces*, Academic Press, 1992.
26 *Asphaltenes, Heavy Oils, and Petroleomics*, ed. O. C. Mullins, E. Y. Sheu, A. Hammami and A. G. Marshall, Springer, New York, 2007.
27 F. Mostowfi, K. Indo, O. C. Mullins and R. McFarlane, *Energy Fuels*, 2009, **23**, 1194–1200.
28 L. Barre, S. Simon and T. Palermo, *Langmuir*, 2008, **24**, 3709–3717.
29 T. F. Headen, E. S. Boek, J. Stellbrink and U. M. Scheven, Small Angle Neutron Scattering (SANS and V-SANS) Study of Asphaltene Aggregates in Crude Oil, *Langmuir*, 2009, **25**, 422–428.
30 Keith L. Gawrys, George A. Blankenship and Peter K. Kilpatrick, *Langmuir*, 2006, **22**, 4487–4497.

The crossover from single file to Fickian diffusion

Jimaan Sané,[ab] Johan T. Padding[c] and Ard A. Louis[a]

Received 19th March 2009, Accepted 18th May 2009
First published as an Advance Article on the web 26th August 2009
DOI: 10.1039/b905378f

The crossover from single-file diffusion, where the mean-square displacement scales as $\langle x^2\rangle \sim t^{1/2}$, to normal Fickian diffusion, where $\langle x^2\rangle \sim t$, is studied as a function of channel width for colloidal particles. By comparing Brownian dynamics to a hybrid molecular dynamics and mesoscopic simulation technique, we can study the effect of hydrodynamic interactions on the single file mobility and on the crossover to Fickian diffusion for wider channel widths. For disc-like particles with a steep interparticle repulsion, the single file mobilities for different particle densities are well described by the exactly solvable hard-rod model. This holds both for simulations that include hydrodynamics, as well as for those that do not. When the single file constraint is lifted, then for particles of diameter σ and pipe of width L such that $(L - 2\sigma)/\sigma = \delta_c \ll 1$, the particles can be described as hopping past one-another in an average time t_{hop}. For shorter times $t \ll t_{hop}$ the particles still exhibit sub-diffusive behaviour, but at longer times $t \gg t_{hop}$, normal Fickian diffusion sets in with an effective diffusion constant $D_{hop} \sim 1/\sqrt{t_{hop}}$. For the Brownian particles, $t_{hop} \sim \delta_c^{-2}$ when $\delta_c \ll 1$, but when hydrodynamic interactions are included, we find a stronger dependence than δ_c^{-2}. We attribute this difference to short-range lubrication forces that make it more difficult for particles to hop past each other in very narrow channels.

I. Introduction

When particles are confined to channels so narrow that mutual passage is excluded, the geometric constraints restrict the particles to a single file and a fixed spatial sequence. For short times the mean-square displacement may still take the Fickian form $\langle x^2\rangle = 2D_0t$, with a self-diffusion coefficient D_0, but at longer times the motion is strongly suppressed by collisions with neighbouring particles, leading to an asymptotic scaling of the form:

$$\langle x^2\rangle = 2Ft^{1/2} \qquad (1)$$

first derived by Harris[1] in a more mathematical context, and independently by Levitt[2] for particles diffusing in a narrow pore. Here F is the single file diffusion (SFD) mobility.

There has been an increasing interest in transport through highly confined pores,[3] stimulated in part by biological realisations such as ion channels,[4] and aquaporins.[5]

[a]*Rudolf Peierls Centre for Theoretical Physics, 1 Keble Road, Oxford, OX1 3NP, United Kingdom*
[b]*Department of Chemistry, Cambridge University, Lensfield Road, Cambridge, CB2 1EW, United Kingdom*
[c]*Computational Biophysics, University of Twente, PO Box 217, 7500 AE Enschede, The Netherlands*

Water under extreme nanoscale confinement exhibits behaviour that differs markedly from the bulk.[6] Single file flow of water is also important in artificial materials like carbon nanotubes.[7,8] Transport of simple molecules through porous materials such as zeolites also show single file sub-diffusive behaviour.[9–11]

Although the biological and synthetic nanoscale systems described above show signatures of SFD, their interpretation is complicated by numerous other factors such as the interaction of the particles with the walls of the confining pore. By contrast, well defined model systems can be created with micron sized colloidal particles. One of the major advantages is that the particles can be directly imaged in real time with digital video microscopy. By using lithography[12–14] or optical tweezers[15] to create the one-dimensional confinement for colloidal particles, unambiguous evidence of asymptotic SFD $\langle x^2 \rangle \sim t^{1/2}$ scaling was observed. Lin *et al.*[14] measured the SFD mobility F for different one-dimensional packing fractions $\eta = \rho\sigma$, where the density $\rho = N/L_p$, the number of particles is N, the length of the pipe is L_p, and σ is the colloidal hard-sphere radius. They found good agreement between their measured F and the SFD mobility for a hard-rod fluid (also known as a Tonks gas):[2]

$$F^{HR} = l_c\sqrt{\frac{D_0}{\pi}} = \frac{\sigma(1-\eta)}{\eta}\sqrt{\frac{D_0}{\pi}} = D_0\sqrt{\frac{2t_c}{\pi}} \tag{2}$$

where l_c is defined as the average inter-particle separation and $t_c = l_c^2/2D_0$ is the average time between collisions. On time scales t much less than the collision time t_c, one expects ordinary Fickian diffusion, whereas for time scales $t \gg t_c$ one expects to observe the asymptotic SFD diffusion of eqn (1).

An obvious question raised by the experiments on SFD is what happens as the confinement becomes less severe. At some point the system should cross over to ordinary Fickian diffusion at long time scales. This problem was first studied by coupling two lattice gas models so that particles could jump between chains.[16] Allowing the jumps produced a crossover in the mean-square displacement from sub-diffusive $\sqrt{t}$ scaling to diffusive $\langle x^2 \rangle \sim t$ asymptotic scaling at long times. Difficulties in interpreting measurements of SFD transport in zeolites also inspired simulations that exhibited a similar long time crossover to Fickian diffusion.[17,18]

As shown by Mon and Percus[19] it is convenient to analyze this crossover in terms of a Markov process with a hopping time t_{hop} that measures the average time for two particles to pass one another. If $t_{hop} \gg t_c$, so that the system has reached the SFD regime between hops, then the average mean-square displacement of the particles scales as $l_{hop}^2 \sim F t_{hop}^{1/2}$ between hops. Within this picture a particle makes on average t/t_{hop} hops in a time t so that its mean-square displacement scales as

$$\langle x^2 \rangle \sim l_{hop}^2 \left(\frac{t}{t_{hop}}\right) \sim D_{hop}\, t \tag{3}$$

which defines an effective Fickian diffusion coefficient of the form

$$D_{hop} \sim l_{hop}^2/t_{hop} \approx F t_{hop}^{-1/2} \tag{4}$$

The larger t_{hop}, the smaller D_{hop}, since particles passing events become more rare. Nevertheless, on time scales $t \gg t_{hop}$ the final asymptotic scaling will still be Fickian. On the other hand, if t_{hop} decreases to the point where it is of the order of the collision time t_c, then the SFD picture is expected to break down and D_{hop} will approach the self-diffusion coefficient D_0. So for the hopping picture to be useful, we require $t_{hop} \gg t_c$.

The hopping time t_{hop} has been studied in some depth for the case of two discs diffusing between hard walls a distance L apart. Mon *et al.*[19–22] found a scaling of the form

$$t_{hop} \sim (L - 2\sigma)^{-\nu} = \delta^{-\nu} \tag{5}$$

as the pore width L approaches the limit where two particles can no longer pass. An exponent $\nu = 2$ was found both for Molecular dynamics[20] and Brownian dynamics (BD),[21] but the interpretation of the simulations is subtle.[23] This exponent agrees with direct calculations of the diffusion equation,[22] and a simpler transition state theory (TST),[20,24] but not with the effective one-dimensional Fick–Jacobs equation,[20,24,25] which predicts that $\nu = \frac{3}{2}$.

Given that hydrodynamic interactions (HI) can have subtle effects on the dynamics of colloidal particles,[26,27] one might expect there to be differences between SFD behaviour measured for colloidal suspensions, and the SFD behaviour of particles where hydrodynamics is not important. A careful theoretical study by Kollmann[28] predicted that the HI do not change the asymptotic scaling of eqn (1), and moreover that the mobility F can be connected to the short-time collective diffusion coefficient of the system. Indeed, the experiments with colloidal particles cited above exhibit SFD sub-diffusive behaviour, as expected. Nevertheless, it would be interesting to see if other aspects of SFD behaviour are sensitive to HI. Moreover, the subtle cross-over from SFD to Fickian long-time diffusion for narrow pipes where particles can hop past each other, but where $t_{\rm hop} \gg t_{\rm c}$, could well be affected by HI.

To study these questions, we employ a hybrid MD and stochastic rotations dynamics (SRD) computer simulation method that has been shown to accurately reproduce Brownian and hydrodynamic behaviour for colloidal suspensions.[29,30] SRD, also known as multi-particle collision dynamics, was first described by Malevanets and Kapral.[31] SRD has been applied to a wide number of different systems, including fluid vesicles in shear flow,[32] clay-like colloids,[33] sedimentation of colloids,[34,35] colloidal rods in shear flow,[36] knots in viral DNA[37] and many other examples.[38] We have recently used this method to study the role of confinement on two-dimensional diffusion[39] and the role of finite sized particles on Taylor dispersion.[40]

To capture the correct asymptotic trends for the mean-square displacement, very long simulation runs are needed, which explains why so many of the simulations in the literature have been on two-dimensional models, which are generally faster to simulate than three-dimensional systems. In this paper we study strongly repulsive colloidal discs in confinement, using both the hybrid MD-SRD method as well as simpler BD to compare what happens when HI are ignored. We find, in agreement with experiments on colloidal suspensions,[14] that the simple hard-rod model for the SFD mobility provides a good fit both for BD and for the simulations that include HI.

When the pipe diameter is such that $\delta = L - 2\sigma > 0$, so that the mutual passage constraint is lifted, we observe a crossover from SFD sub-diffusion to Fickian diffusion at longer times $t \gtrsim t_{\rm hop}$. We measure the distribution of hopping times $t_{\rm hop}$, showing that they follow Poissonian statistics for simulations with and without hydrodynamic interactions. For both models we also find a scaling consistent with $D_{\rm hop} \propto 1/\sqrt{t_{\rm hop}}$. For the BD simulations, we find that $t_{\rm hop} \sim \delta^{-2}$ for small δ/σ, but $t_{\rm hop}$ depends more strongly on $1/\delta$ when hydrodynamic interactions are included. We attribute this to repulsive lubrication forces, induced when particles approach each other or the walls of the container.

The paper is organised as follows: in section II, we describe the model, our hybrid MD-SRD method and our implementation of BD. In section III we study the SFD behaviour of colloids confined to one-dimensional flows for different packing fractions η. Section IV considers the case when the mutual passage constraint is lifted, so that at long times Fickian diffusion is recovered. Finally, in section V, we discuss the main conclusions of this study.

II. Model and simulation methods

A. Model

For the colloid–colloid interactions, we use a very strongly repulsive potential of the form:

$$\varphi_{cc}(r) = \begin{cases} 4\varepsilon_{cc}\left(\left(\frac{\sigma}{r}\right)^{48} - \left(\frac{\sigma}{r}\right)^{24} + \frac{1}{4}\right) & (r \leq 2^{1/24}\sigma) \\ 0 & (r \geq 2^{1/24}\sigma) \end{cases}$$

where σ is the colloid–colloid diameter. The colloid–wall interaction is taken to be strongly repulsive as well, and a stick-boundary condition is applied to the colloid–wall, as well as the fluid particle–wall interaction in the SRD simulations. We use periodic boundary conditions in the direction parallel to the pipe walls.

B. Hybrid MD and SRD simulation method

To describe the hydrodynamic behaviour of colloids induced by a background fluid of much smaller constituents, some form of coarse-graining is required. The hydrodynamics can be described by the Navier Stokes equations that coarse-grain the fluid within a continuum description. The downside of going directly through this route is that every time the colloids move, the boundary conditions on the differential equations change, making them computationally expensive to solve.

An alternative to computing a direct solution of the Navier Stokes equations is to use particle based techniques that exploit the fact that only a few conditions, such as (local) energy and momentum conservation, need to be satisfied to allow the correct (thermo) hydrodynamics to emerge in the continuum limit. Simple particle collision rules, easily amenable to efficient computer simulation, can therefore be used. Boundary conditions (such as those imposed by colloids in suspension) are easily implemented as external fields.

In this paper we implement the SRD method first derived by Malevanets and Kapral.[31] An SRD fluid is modelled by N point particles of mass m, with positions $\mathbf{r}_i$ and velocities $\mathbf{v}_i$. The coarse graining procedure consists of two steps, streaming and collision. During the streaming step, the positions of the fluid particles are updated *via*

$$\mathbf{r}_i(t + \delta t_s) = \mathbf{r}_i(t) + \mathbf{v}_i(t)\delta t_s \qquad (6)$$

In the collision step, the particles are split up into cells with sides of length a_0, and their velocities are rotated around an angle α with respect to the cell centre of mass velocity,

$$\mathbf{v}_i(t + \delta t_s) = \mathbf{v}_{c.m,i}(t) + \mathcal{R}_i(\alpha)[\mathbf{v}_i(t) - \mathbf{v}_{c.m,i}(t)] \qquad (7)$$

where $\mathbf{v}_{c.m,i} = \sum_j^{i,t}(m\mathbf{v}_j)/\sum_j m$ is the centre of mass velocity of the particles in each cell, $\mathcal{R}_i(\alpha)$ is the rotational matrix and δt_s is the interval between collisions. The purpose of this collision step is to transfer momentum between the fluid particles while conserving the energy and momentum of each cell.

The fluid particles only interact with one another through the collision procedure. Direct interactions between the solvent particles are not taken into account, so that the algorithm scales as $\mathcal{O}(N)$ with particle number. The carefully constructed rotation procedure can be viewed as a coarse-graining of particle collisions over space *and* time. Mass, energy and momentum are conserved locally, so that on large enough length-scales the correct Navier Stokes hydrodynamics naturally emerges.[31] It is important to remember that for all these particle based methods, the particles should not be viewed as composite supramolecular fluid units, but rather as coarse-grained Navier Stokes solvers (with noise in the case of SRD).[30] Another advantage of SRD is that transport coefficients have been analytically calculated,[41–43] greatly facilitating its use.

The SRD fluid particles can easily be coupled to a solute as first shown by Malevanets and Kapral,[29] and studied in detail in a recent paper.[30] To simulate the behaviour of colloids of mass M, we use the colloid–colloid interaction defined above, while the solvent particles interact with the colloids *via* an interaction of the form:

$$\varphi_{cs}(r) = \begin{cases} 4\varepsilon_{cs}\left(\left(\frac{\sigma_{cs}}{r}\right)^{12} - \left(\frac{\sigma_{cs}}{r}\right)^{6} + \frac{1}{4}\right) & (r \leq 2^{1/6}\sigma_{cs}) \\ 0 & (r \geq 2^{1/6}\sigma_{cs}) \end{cases}$$

where σ_{cs} is the colloid–solvent collision diameter. We propagate the ensuing equations of motion with a Velocity Verlet algorithm[44] using a molecular dynamic time-step Δt

$$R_i(t + \Delta t) = R_i(t) + V_i(t)\Delta t + \frac{F_i(t)}{2M}\Delta t^2 \tag{8}$$

$$V_i(t + \Delta t) = V_i(t) + \frac{F_i(t) + F_i(t + \Delta t)}{2M}\Delta t \tag{9}$$

where R_i and V_i are the position and velocity of the colloid, and F_i is the total force exerted on the colloid. Coupling the colloids in this way leads to slip boundary conditions. Stick boundary conditions can also be implemented,[45] but for qualitative behaviour, we don't expect there to be important differences. In parallel the velocities and positions of the SRD particles are streamed in the external potential given by the colloids and the external walls and updated with the SRD rotation-collision step every time-step δt_s.

The larger the ratio σ/a_0, the more accurately the hydrodynamic flow fields will be reproduced. In ref. 30, it is shown that using $\sigma/a_0 = 4.3$, and $\sigma_{cs} = 2a_0$ reproduces the flow fields with small relative errors for a single sphere in a 3D flow. Because we are interested in processes where the colloids can just barely pass each other, we chose a finer grid ($\sigma/a_0 = 8.6$). This should in particular enhance the resolution of lubrication forces.[30] Other parameter choices taken from ref. 30,39 include $\varepsilon_{cc} = \varepsilon_{cs} = 2.5k_BT$ for the colloids, an SRD particle density of $\gamma = 5$ particles per a_0^2 and a rotation angle of $\alpha = \frac{1}{2}\pi$. The time-steps for the MD and SRD steps are set by different physics,[30] and we chose $\Delta t = 0.025t_0$ and $\delta t_s = 0.1t_0$ for SRD, where $t_0 = a_0\sqrt{\frac{m}{k_BT}}$ is the unit of time in our simulations.

C. Brownian dynamics simulation method

The particle motion for the colloids can also be solved using BD. The positions are updated *via* the equation of motion:

$$r(t + \Delta t_{BD}) = r(t) + \frac{\Delta t_{BD}}{m\xi}F(t) + \delta r^G \tag{10}$$

where $F(t)$ is the force acting on the colloids, which arises from the colloid–colloid interaction, as well as the colloid–wall interactions, and each component δr^G is chosen from a Gaussian distribution with zero mean and variance $\langle(\delta r^G)^2\rangle = 2D_0\Delta t$,[44] where D_0 is the single particle diffusion coefficient. With BD, the particles execute a random walk with a physical time scale

$$t_D = \frac{\sigma^2}{2D_0} \tag{11}$$

called the diffusion time that measures how long it takes the particle to diffuse over its diameter. The BD time-step was chosen to be $\Delta t_{BD} = 2 \times 10^{-5}t_D$, slightly smaller than the value used, for example, in a careful study by Lodge and Heyes.[46] We use this conservative measure of the time-step because there are indications that scaling of the hopping time of colloidal particles simulated with BD is sensitive to the time-step;[23] we checked that our time-step is small enough. In contrast to the SRD method, which also exhibits diffusive motion, the BD method does not include HI.

III. Single file diffusion

We first investigate the SFD behaviour of colloidal discs in pipes narrow enough that they cannot pass each other. At times less than the collision time t_c we expect the particles to behave in a diffusive manner, and for times $t \gg t_c$ we expect SFD sub-diffusion. Lin *et al.*[14] showed that the following ansatz:

$$\frac{1}{\langle x^2 \rangle} = \frac{1}{2D_0 t} + \frac{1}{2Ft^{1/2}} \tag{12}$$

provides a good approximation to the crossover from Fickian to SFD sub-diffusion. Solving for the mean-square displacement gives:

$$\langle x^2 \rangle = \frac{2D_0 t}{1 + (D_0/F)t^{1/2}} = \frac{2D_0 t}{1 + (t/t_x)^{1/2}} \tag{13}$$

where we have defined the crossover time

$$t_x = (F/D_0)^2 \tag{14}$$

a measure of the time needed for the system to transition from Fickian diffusion to the asymptotic SFD regime. For particles that behave like hard-rods with a mobility given by eqn (2), this crossover time scales as:

$$t_x = t_D \frac{2}{\pi} \left(\frac{1-\eta}{\eta} \right)^2 \tag{15}$$

and is connected to the collision time by $t_x = 2t_c/\pi$. That the two times are essentially the same is not surprising, given that the $\langle x^2 \rangle \sim \sqrt{t}$ behaviour is generated by the collisions with neighbours and the long-ranged correlations these induce. The higher the one-dimensional packing fraction η, the lower t_x, and so the more rapidly the system should transition to asymptotic SFD sub-diffusion. For very small packing fractions the crossover time scales with $t_x/t_D \sim 1/\eta^2$, and so very long simulations are necessary to observe asymptotic SFD scaling. For example, for $\eta = 0.5$, $t_x \approx 0.63t_D$, for $\eta = 0.1$, $t_x \approx 51.5t_D$ and for $\eta = 0.01$, $t_x \approx 6240t_D$.

Simulations were carried out with BD for $N = 200$ colloidal particles diffusing between parallel plates a distance $L = 1.40\sigma$ apart. The one-dimensional packing fraction was varied in the range $\eta = \dfrac{N\sigma}{L_p} = 0.1 - 0.7$ and, after equilibration, simulations were run for total times ranging between $100t_D$ up to $300t_D$ to gather data on the mean-square displacement.

In Fig. 1, we plot the mean-square displacement of particles undergoing single file diffusion at various packing fractions. With the exception of the data for $\eta = 0.1$, the crossover time $t_x \ll 10t_D$ so that SFD $\langle x^2 \rangle \sim \sqrt{t}$ scaling is expected for the entire range of data plotted. The measurements are consistent with this scaling. The same data is plotted on a linear plot in Fig. 2, where it is compared to the ansatz of eqn (13). Since D_0 is given by the simulation parameters, there is only one fit parameter, namely the SFD mobility F. The plots demonstrate the $\sqrt{t}$ scaling for the mean-square displacement *vs.* time.

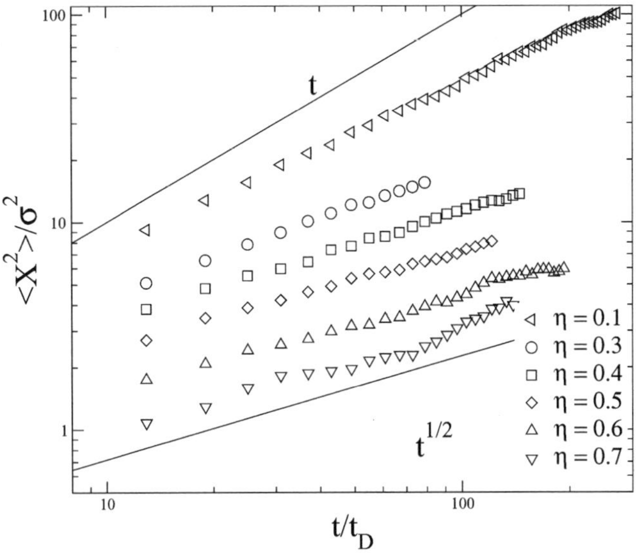

Fig. 1 Log–log plot of the mean-square displacement of colloids calculated with BD for a pipe of width $L = 1.4\sigma$. Results are shown for simulations with increasing line packing fractions $\eta = 0.1, 0.3, 0.4, 0.5, 0.6, 0.7$. The straight lines have slope 1 and 1/2, respectively, and serve as a guide to the eye. At long times the colloids show clear signatures of SFD sub-diffusive scaling.

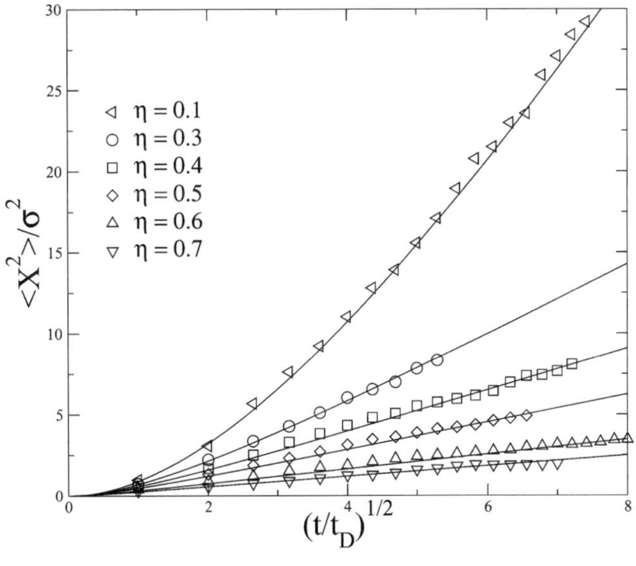

Fig. 2 Linear plot of the mean-square displacement of colloids as a function of $\sqrt{t/t_D}$, calculated with BD for a pipe of width $L = 1.4\sigma$. Results are shown for simulations with increasing line packing fractions $\eta = 0.1, 0.3, 0.4, 0.5, 0.6, 0.7$. The solid lines are a fit from eqn (13). Since D_0 is fixed by the BD simulation parameters the only fit parameter is the SFD mobility F.

Simulations were also carried out with SRD. The explicit inclusion of a background fluid induces HI between the particles, and between the particles and the walls. The combination of strong confinement and HI has an important qualitative effect on the velocity autocorrelation functions[39,47] and on the Fickian diffusion coefficient[48,49] of individual particles, especially in two dimensions,[39,50] where HI are particularly important. But once the system is confined so strongly that particles can no longer pass, the long-ranged HI interactions can be strongly screened.[51] That raises the question: what is the effect of including HI on the SFD mobility F? The SRD method allows us to address these issues.

We performed simulation runs for $N = 200$ colloidal particles in a SRD fluid between parallel plates a distance $L = 1.40\sigma$ and $L = 1.98\sigma$ apart. The longitudinal packing fractions were the same as for the BD simulations. After equilibration, simulation data was gathered for runs of approximately $10–30t_D$ in length. Because the SRD method includes many solvent particles ($\mathcal{O}(10^5)$ here), it is computationally more expensive than the BD simulations, and so the runs are shorter than those performed for BD. The mean-square displacements for the $L = 1.4\sigma$ case are shown in Fig. 3, and $\langle x^2 \rangle \sim \sqrt{t}$ behaviour is still observed. In this case a fit of the data to eqn (13) was used to extract both D_0 and F. The values of D_0 are consistent with those of ref. 39.

The values of the fits to F for both BD and SRD are shown in Fig. 4 as a function of the packing fraction η. These are compared to the hard-rod F^{HR} of eqn (2). At all but the lowest packing fractions, good agreement is found, suggesting that the inclusion of HI does not significantly affect the value of F/D_0. Moreover, in contrast to what is found in two dimensions where HI interactions have an important impact on the value of D_0 (ref. 39,51), here the measured value of the normalised SFD mobility $F/(\sqrt{D_0}\sigma)$ does not seem to depend significantly on the confinement, since differences between what is measured at $L = 1.4\sigma$ and at $L = 1.98\sigma$ appear to be small, even though D_0 changes considerably.[39] Note that F itself does scale with $\sqrt{D_0}$, so the changes in D_0 due to confinement translate into changes in F, but there are no further effects of the HI. Finally, we should point out that some further changes in F are expected when L moves away from the pure hard-rod limit $L = \sigma$, but these are expected to be small.[52]

It is only for the lowest packing fraction $\eta = 0.1$ that we observe significant differences with F^{HR} for the SRD simulations. Interestingly, this deviation is very similar

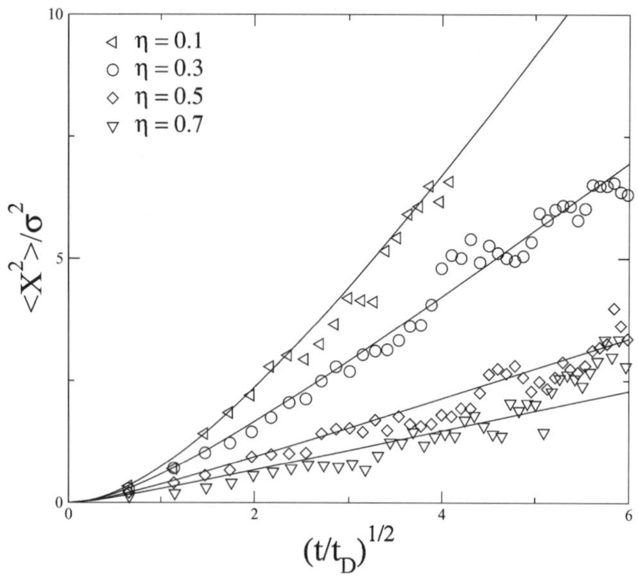

Fig. 3 Linear plot of the mean-square displacement of colloids as a function of $\sqrt{t/t_D}$, calculated with SRD for a pipe of width $L = 1.4\sigma$. In contrast to the BD simulations, these include HI. The solid lines are the fit to eqn (13).

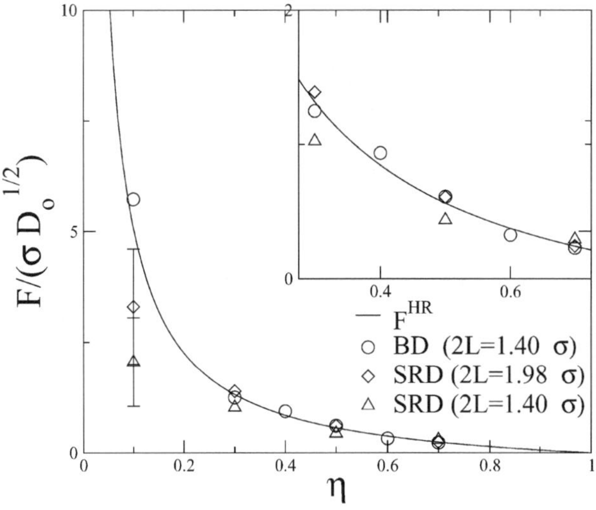

Fig. 4 Comparison of the SFD particle mobility F, extracted from BD and SRD simulations, to the F^{HR} for a hard-rod fluid from eqn (2). Inset: data at higher packing fractions. The results suggest that, within simulation errors, the inclusion of HI interactions does not strongly affect the 1D mobility F.

to that found by Lin *et al.*[14] in their experiments. Unfortunately, at this lower packing fraction very long simulations are needed to reach the asymptotic SFD regime, and so we don't believe that enough data has been gathered here to reliably extract F. The error bars shown in the graph are from a standard non-linear fitting procedure, but it is very likely that the real error is higher. A similar conclusion about the errors was made by Lin *et al.*[14] for their data, which also extended out to $t \approx t_x$ for the lowest packing fraction. To resolve this question, the simulations and the experiments should be performed for at least an order of magnitude longer time than they have been.

IV. Crossover from SFD to Fickian diffusion

Having worked out some properties of SFD when particles cannot pass each other, we now focus on the dynamics that emerge when the single file constraint is lifted. Fig. 5 illustrates how the parameter $\delta_c = \delta/\sigma = L/\sigma - 2$ describes the maximal distance between the particles when they can pass each other. In our case we do not quite have hard sphere particles, but the $1/r^{48}$ repulsion is hard enough that the differences are expected to be very small. So we will still use the parameter δ_c to denote how easy it is for particles to pass.

A. Scaling of the mean-square displacement

As shown in the introduction, it is very useful to consider the hopping time t_{hop} that it takes a particle to switch order with one of its neighbours. From eqn (4) it follows that the effective Fickian diffusion coefficient is expected to scale as $D_{\text{hop}} \sim Ft_{\text{hop}}^{-1/2}$.

To test this concept we performed a number of simulations at different values of δ_c. As an illustrative example, we show the mean-square displacement of $N = 200$ Brownian particles at $\eta = 0.5$ and $\delta_c = 0.093$ in Fig. 6. In this case the hopping time $t_{\text{hop}} \approx 86t_D \gg t_x \approx 0.12t_D$, so that the crossover from the initial Fickian diffusion with a diffusion coefficient of D_0 to SFD at $t \approx t_x$ is at much shorter times than an eventual longer time crossover back to diffusive motion due to the hopping. We simulate for up to $t = 1040t_D$ so that there are on average about 11.4 hopping

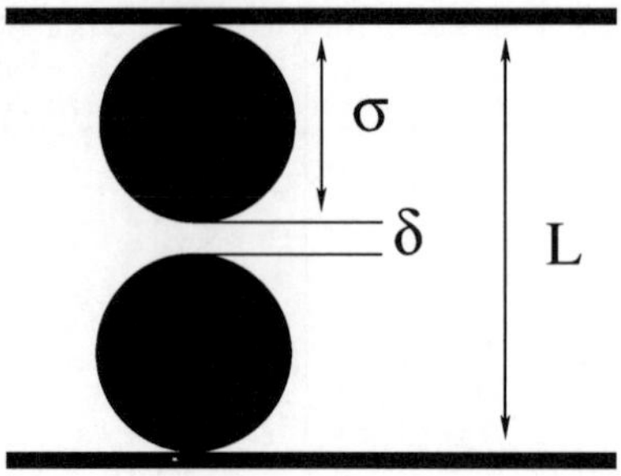

Fig. 5 Schematic depiction of the clearance δ available to particles attempting to switch places with one another. Particles can hop to pass their neighbours if $\delta > 0$.

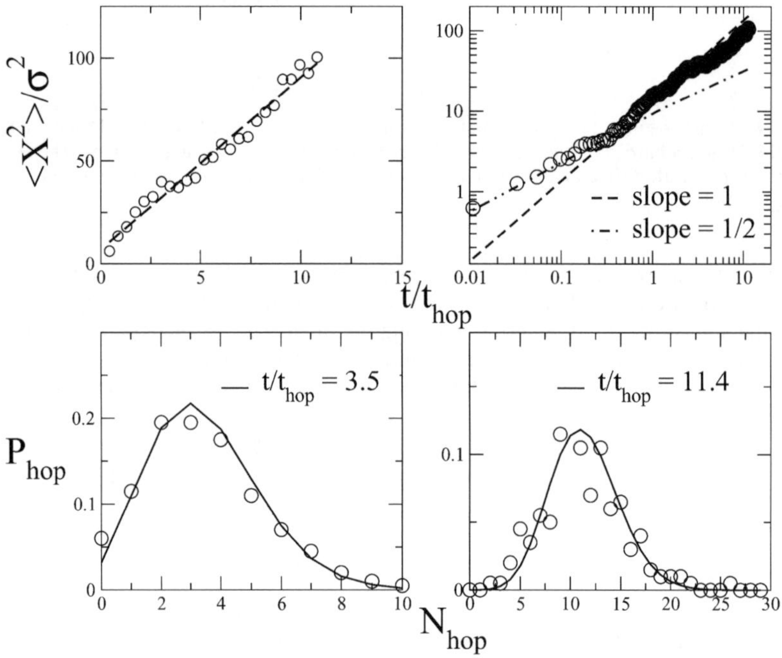

Fig. 6 Crossover from SFD to Fickian diffusion. The top two plots show the mean-square displacement of the particles calculated with a BD simulation at $\eta = 0.5$ for $\delta_c = 0.093$. The mean hopping time was measured to be $t_{hop} \approx 86t_D$. On time scales of the order of the hopping time, we observe a crossover to Fickian diffusion $\langle x^2 \rangle \sim t$ with an effective diffusion coefficient $D_{hop} \approx 0.095D_0$. This crossover is especially clear in the log plot on the top right. The bottom two plots show the probability P_{hop} that a particle has performed N hops in the time intervals, $t \approx 3.45t_{hop}$ and $t \approx 11.43t_{hop}$, respectively.

processes per particle. At times on the order of t_{hop} we observe a crossover from $\langle x^2 \rangle \sim \sqrt{t}$ to the asymptotic linear Fickian diffusive scaling $\langle x^2 \rangle \sim 2D_{hop}t$ with an effective diffusion coefficient $D_{hop} \approx 0.095D_0$. Although in this case a hopping event occurs only about once in every thousand collisions, the net effect is to generate a linear diffusion that at longer times leads to a significantly larger mean-square displacement than if the particles could not pass.

Fig. 7 shows a similar set of simulations at $\eta = 0.5$, but now with the SRD methods that includes HI. Although the runs are shorter (SRD is computationally more expensive), they nevertheless show the same trends as the BD simulations,

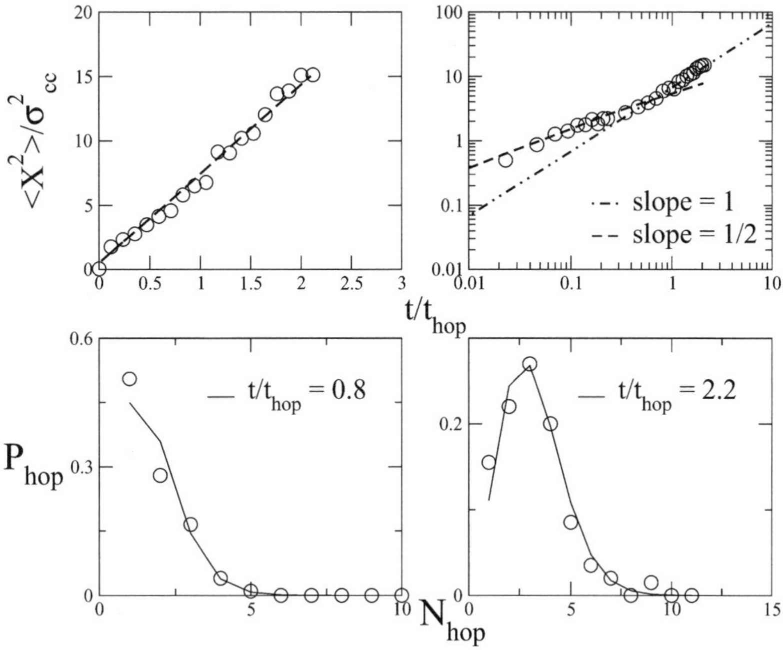

Fig. 7 Crossover from SFD to Fickian diffusion for simulations that include hydrodynamic interactions. The top two plots show the mean-square displacement of the particles calculated with an SRD simulation at $\eta = 0.5$ for $\delta_c = 0.093$. The mean hopping time was measured to be $t_{hop} \approx 79t_D$. The bottom two plots show the probability P_{hop} that a particle has performed N hops in the time intervals, $t \approx 0.8t_{hop}$ $(6.3t_D)$ and $t \approx 2.2t_{hop}(17t_D)$, respectively. The straight line shows the predicted Poissonian distribution (eqn (16)) and the circles are the results from the simulation.

with a crossover to Fickian diffusive motion at $t \gtrsim t_{hop}$ with an effective diffusion coefficient $D_{hop} \approx 0.087D_0$. This scaling behaviour is especially clear in the log plot on the top right. For this set of parameters, the SRD and the BD give similar results for t_{hop} and D_{hop} within the expected errors of the simulations.

B. Poissonian statistics of hopping events

The order of the particles was tracked and the number of times that a particle switched places with its neighbour was recorded. This was done on an interval large enough to avoid double counting short switches during the hopping process. If the hopping events are independent, then the probability that a particle makes N hops is proportional to the time interval over which a measurement is made and should follow Poissonian statistics. The probability that such an event will occur n times in a given time interval can be expressed as

$$P(n;\lambda) = \frac{\lambda^n e^{-\lambda}}{n!} \tag{16}$$

where $\lambda = t/t_{hop}$ in our case. The bottom two plots of Fig. 6 and 7 show the distributions at two different values of λ. The solid lines denote the distributions calculated from the Poisson distribution,[16] and the circles denote those obtained from the simulation. Within the simulation error, these show good agreement for both the BD and the SRD simulations, suggesting that the hopping events are indeed independent and not correlated with each other.

C. Scaling of t_{hop} and D_{hop} with plate separation

Further simulations were performed for 1000 Brownian particles diffusing in pipes of various sizes. In Fig. 8 we plot the hopping times t_{hop} and effective diffusion coefficients D_{hop} for BD simulations at $\eta = 0.7$. The top panel shows how hopping time varies with the distance δ_c and is consistent with $t_{hop} \propto \delta_c^{-2}$. This exponent agrees with direct calculations of the diffusion equation,[22] and a simpler transition state theory,[20,24] but not with the effective one-dimensional Fick–Jacobs equation.[20,24,25] Nevertheless, we do not consider our simulations to be on a large enough range of δ_c to conclusively determine the scaling behaviour. Such simulations become increasingly difficult for smaller δ_c because the hopping events become more rare. However, recent Monte Carlo simulations by Mon[23] for just two discs are consistent with transition state theory over a sufficiently larger range of δ_c to confirm the scaling law for that special geometry. His results suggest that if full simulations of many discs were performed over a wider range of δ_c they would confirm the scaling law we observe over a limited range of δ_c.

In the bottom panel in Fig. 8, we plot the effective diffusion coefficient D_{hop} *vs.* t_{hop}. The solid line has the slope $-1/2$ which is consistent with the scaling of eqn (4), as postulated by Mon and Percus.[19] Again, the value range of δ_c is not large enough to firmly fix the scaling behaviour, but it is at least consistent with $D \propto t^{-1/2}$ scaling, and inconsistent with an earlier prediction of t^{-1} scaling.[17] As the pipes become wider and t_{hop} becomes smaller, we expect D_{hop} to eventually increase to D_0, the self-diffusion coefficient. Thus at larger δ_c this simple scaling law should no longer hold.

Simulations were also performed using the SRD method to test the effect, if any, of hydrodynamics on the hopping process. We simulated a total of 1000 freely diffusing colloids in $2d$ pipes with widths characterized by $\delta_c \approx 0.035$–0.093 The longitudinal colloid density was $\eta = 0.5$ in all cases. A separate set of BD simulations was also performed so that the effect of HI could be clearly contrasted and compared.

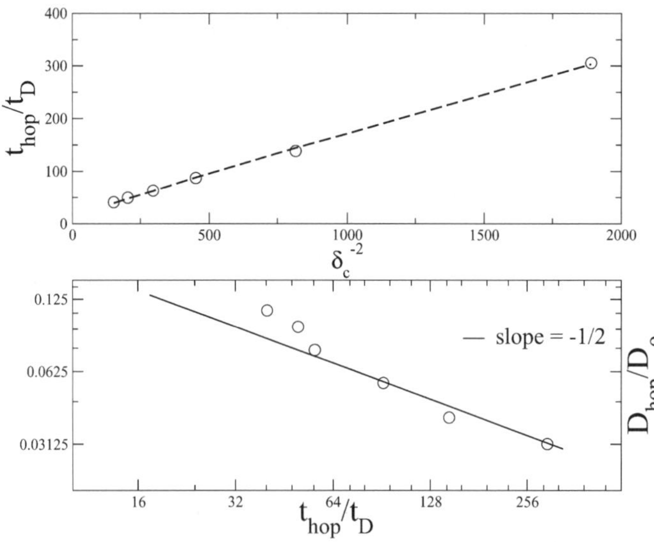

Fig. 8 BD simulations of the hopping time t_{hop} and the effective diffusion coefficient D_{hop} for packing fraction $\eta = 0.7$ and separations $\delta_c = 0.023, 0.035, 0.047, 0.058, 0.070$, and 0.081. Top: average hopping time plotted *versus* $1/\delta_c^2$. The straight line is a fit to the data. Bottom: effective diffusion coefficient D_{hop} *vs.* t_{hop} The straight line has a slope $-1/2$ and serves as a guide to the eye.

The top panel of Fig. 9 shows that the BD exhibits the same δ_c^{-2} scaling that was found in Fig. 8 for a higher packing fraction. For the larger values of δ_c shown, the BD and SRD simulations give almost the same value of t_{hop}/t_D. However, for smaller δ_c, the SRD simulations find a significantly higher value of t_{hop}, which suggests that the HI suppress hopping events. We attribute this to the following processes. When two particles come close together, or close to a wall, the solvent must be displaced for the particle to move past. At very short distances, this gives rise to so-called lubrication forces.[53] These are repulsive for two particles approaching each other, and attractive when two particles are close together and then move away from each other (because now solvent has to flow into the space between them). SRD simulations reproduce the lubrication forces, even on distances considerably less than a_0.[30] We thus attribute the increase of t_{hop} with respect to the Brownian simulations to lubrication forces, since they make it harder for two particles that diffuse towards each other to pass when δ_c is very small.

The bottom panel of Fig. 9 shows the effective diffusion coefficient plotted against the hopping time for the four largest separations. Within simulation errors, we recover the $D_{hop} \sim t_{hop}^{-1/2}$ scaling that was also observed for the BD simulations. Although the scaling law will break down for larger values of δ_c, Fig. 8 and 9 suggest that D_{hop} will reach a value close to D_0 well before $\delta_c = 1$. The exact value of δ_c where D_{hop} fully converges to that of a bulk system will of course depend on packing fractions and other details of the system.

V. Conclusion

We have carried out computer simulations to investigate single file diffusion and the crossover to Fickian diffusion when the particle passing constraint is lifted. By comparing BD and SRD simulations, we can study the effect of HI on these

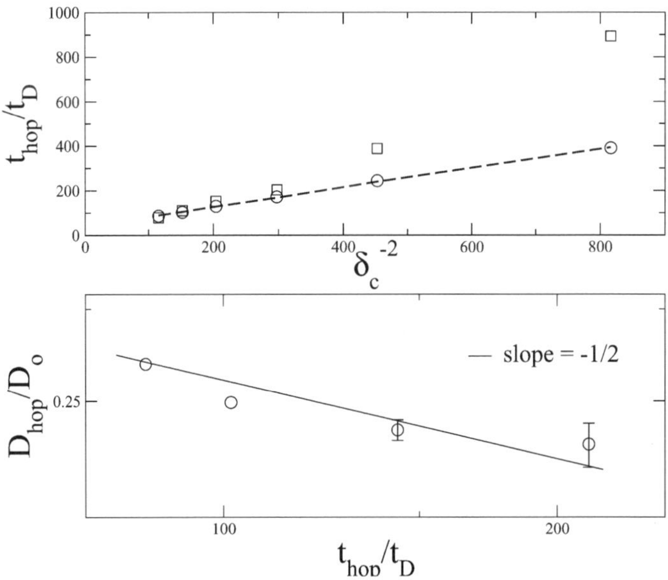

Fig. 9 Top graph: SRD and BD simulations of the hopping time t_{hop} for packing fraction $\eta = 0.5$ and separations $\delta_c = 0.035$, 0.047, 0.058, 0.070, 0.081, and 0.093. For larger δ_c both simulation methods give the same value of t_{hop}, but at smaller separations the simulations with HI have larger t_{hop} values. Bottom: effective diffusion coefficient D_{hop} *vs.* t_{hop} from the SRD simulations. The straight line has a slope $-1/2$ and serves as a guide to the eye; the data is consistent with $D_{hop} \sim t_{hop}^{-1/2}$.

processes. Whereas HI have an important effect on the single particle diffusion coefficient under confinement, especially in 2D,[39] given D_0, we were unable to measure any further effects on the SFD mobility F for particles that cannot pass each other. The value we measure for F at different packing fractions η is consistent with the exactly solvable hard-rod model.[1,2] Experiments on colloidal suspensions,[12–14] also find $\langle x^2 \rangle \approx 2Ft^{1/2}$ scaling and in one case[14] quantitative agreement with F from the hard-rod model, so simulations and experiments both suggest that HI have at best a very small effect on the SFD mobility.

When the no passing constraint is lifted, we can measure the average hopping time t_{hop} that it takes a particle to switch order with a neighbour. While at shorter times $t_x \lesssim t \ll t_{\mathrm{hop}}$ the particles still exhibit SFD, on time scales $t \gtrsim t_{\mathrm{hop}}$, Fickian diffusion emerges with $\langle x^2 \rangle \approx 2D_{\mathrm{hop}}t$. The effective hopping diffusion coefficient scales as $D_{\mathrm{hop}} \sim t_{\mathrm{hop}}^{-1/2}$ for small δ_{c}.

We find that for the Brownian simulations $t_{\mathrm{hop}} \sim \delta_{\mathrm{c}}^{-2}$ for small δ_{c}, as predicted by TST, but when HI are included, a stronger dependence becomes evident at small δ_{c}. We attribute this increase of t_{hop} to the influence of hydrodynamic lubrication forces that make it harder for the particles to pass one another.

The hopping time t_{hop} and diffusion coefficient D_{hop} depend strongly on δ_{c}. This means that measuring how t_{hop} or D_{hop} depend on δ_{c} may be difficult for experiments, because very small relative errors in the channel width can lead to large effects. But this sensitivity to changes in δ_{c} could also be used as an advantage, since it could be exploited, for example, by microfluidic applications[54] as a very sensitive way to measure the size of particles in artificial channels.

Acknowledgements

J.S. thanks Schlumberger Cambridge Research and IMPACT FARADAY for an EPSRC CASE studentship which supported this work. A.A.L. thanks the Royal Society (London), J.T.P. thanks the Netherlands Organisation for Scientific Research (NWO) for financial support. We thank L. Bocquet for helpful conversations.

References

1 T. E. Harris, *J. Appl. Probab.*, 1965, **2**, 323.
2 D. G. Levitt, *J. Stat. Phys.*, 1973, **8**, 3050.
3 P. S. Burada, P. Hanggi, F. Marchesoni, G. Schmid and P. Talkner, *ChemPhysChem*, 2009, **10**, 45.
4 R. Mackinnon, *Angew. Chem., Int. Ed.*, 2004, **43**, 4265.
5 P. Agre, *Angew. Chem., Int. Ed.*, 2004, **43**, 4278.
6 R. Allen, S. Melchionna and J. P. Hansen, *Phys. Rev. Lett.*, 2002, **89**, 175502; O. Beckstein and M. S. P. Sansom, *Proc. Natl. Acad. Sci. U. S. A.*, 2003, **100**, 7063; J. Dzubiella, R. J. Allen and J.-P. Hansen, *J. Chem. Phys.*, 2004, **120**, 5001; J. C. Rasaiah, S. Garde and G. Hummer, *Annu. Rev. Phys. Chem.*, 2008, **59**, 713.
7 G. Hummer, J. C. Rasaiah and J. P. Noworyta, *Nature*, 2001, **414**, 188; M. Whitby and N. Quirke, *Nat. Nanotechnol.*, 2007, **2**, 87.
8 H. Fang, R. Wan, X. Gong, H. Lu and S. Li, *J. Phys. D: Appl. Phys.*, 2008, **41**, 103002.
9 J. Karger, D. M. Ruthven, *Diffusion in Zeolites and Other Microporous Solids*, Wiley-Interscience, New York, 1994.
10 K. Hahn, J. Karger and V. Kukla, *Phys. Rev. Lett.*, 1996, **76**, 2762.
11 B. Smit, *Chem. Rev.*, 2008, **108**, 4125.
12 Q. Wei, C. Bechinger and P. Leiderer, *Science*, 2000, **287**, 625.
13 B. Lin, B. Cui, J.-H. Lee and J. Yu, *Europhys. Lett.*, 2002, **57**, 724.
14 B. Lin, M. Meron, B. Cui and S. A. Rice, *Phys. Rev. Lett.*, 2005, **94**, 216001.
15 C. Lutz, M. Kollman and C. Bechinger, *Phys. Rev. Lett.*, 2004, **93**, 26001; C. Lutz, M. Kollmann, P. Leiderer and C. Bechinger, *J. Phys.: Condens. Matter*, 2004, **16**, 4075.
16 R. Kutner, H. van Beijeren and K. W. Kehr, *Phys. Rev. B*, 1984, **30**, 4382.
17 K. Hahn and J. Kärger, *J. Phys. Chem. B*, 1998, **102**, 5766.

18 H. L. Tepper, J. P. Hoogenboom, N. F. A. van der Vegt and W. J. Briels, *J. Chem. Phys.*, 1999, **110**, 11511.
19 K. K. Mon and J. K. Percus, *J. Chem. Phys.*, 2002, **117**, 2289.
20 R. K. Bowles, K. K. Mon and J. K. Percus, *J. Chem. Phys.*, 2004, **121**, 10668.
21 K. K. Mon, *J. Chem. Phys.*, 2008, **129**, 124711.
22 K. K. Mon and J. K. Percus, *J. Chem. Phys.*, 2006, **125**, 244704.
23 K. K. Mon, *J. Chem. Phys.*, 2008, **128**, 197102.
24 P. Kalinay, *J. Chem. Phys.*, 2007, **126**, 194708.
25 P. Kalinay and J. K. Perkus, *J. Chem. Phys.*, 2008, **129**, 154117.
26 W. B. Russel, D. A. Saville and W. R. Showalter, *Colloidal Dispersions*, Cambridge University Press, Cambridge, 1989.
27 J. K. G. Dhont, *An Introduction to the Dynamics of Colloids*, Elsevier, Amsterdam, 1996.
28 M. Kollmann, *Phys. Rev. Lett.*, 2003, **90**, 180602.
29 A. Malevanets and R. Kapral, *J. Chem. Phys.*, 2000, **112**, 7260.
30 J. T. Padding and A. A. Louis, *Phys. Rev. E.*, 2006, **74**, 031402.
31 A. Malevanets and R. Kapral, *J. Chem. Phys.*, 1999, **110**, 8605.
32 H. Noguchi and G. Gompper, *Phys. Rev. Lett.*, 2004, **93**, 258102; H. Noguchi and G. Gompper, *Phys. Rev. Lett.*, 2007, **98**, 128103.
33 M. Hecht, J. Harting, T. Ihle and H. J. Herrmann, *Phys. Rev. E*, 2005, **72**, 011408.
34 J. T. Padding and A. A. Louis, *Phys. Rev. Lett.*, 2004, **93**, 220601; J. T. Padding and A. A. Louis, *Phys. Rev. E.*, 2008, **77**, 011402.
35 A. Wysocki, C. P. Royall, R. G. Winkler, G. Gompper, H. Tanaka, A. van Blaaderen and H. Löwen, *Soft Matter*, 2009, **5**, 1340.
36 M. Ripoll, P. Holmqvist, R. G. Winkler, G. Gompper, J. K. G. Dhont and M. P. Lettinga, *Phys. Rev. Lett.*, 2008, **101**, 168302.
37 R. Matthews, J. M. Yeomans and A. A. Louis, *Phys. Rev. Lett.*, 2009, **102**, 088101.
38 R. Kapral, *Adv. Chem. Phys.*, 2008, **140**, 89.
39 J. Sané, J. T. Padding and A. A. Louis, *Phys. Rev. E*, 2009, **79**, 051402.
40 J. Sané, J. T. Padding and A. A. Louis, DOI:arXiv:0903.3493.
41 T. Ihle and D. M. Kroll, *Phys. Rev. E*, 2003, **67**, 066705; T. Ihle and D. M. Kroll, *Phys. Rev. E*, 2003, **67**, 066706.
42 N. Kikuchi, C. M. Pooley, J. F. Ryder and J. M. Yeomans, *J. Chem. Phys.*, 2003, **119**, 6388.
43 C. M. Pooley and J. M. Yeomans, *J. Phys. Chem. B*, 2005, **109**, 6505.
44 M. P. Allen and D. J. Tildesley, *Computer simulation of liquids*, Clarendon Press, Oxford, 1987.
45 J. T. Padding, A. Wysocki, H. Löwen and A. A. Louis, *J. Phys.: Condens. Matter*, 2005, **17**, S3393.
46 J. F. M. Lodge and D. M. Heyes, *J. Chem. Soc., Faraday Trans.*, 1997, **93**, 437–448.
47 M. H. J. Hagen, I. Pagonabarraga, C. P. Lowe and D. Frenkel, *Phys. Rev. Lett.*, 1997, **78**, 3785.
48 P. M. Bungay and H. Brenner, *Int. J. Multiphase Flow*, 1973, **1**, 25.
49 L. Bocquet and J.-L. Barrat, *J. Phys.: Condens. Matter*, 1996, **8**, 9297.
50 P. G. Saffman and M. Delbruck, *Proc. Natl. Acad. Sci. U. S. A.*, 1975, **72**, 3111; P. G. Saffmann, *J. Fluid Mech.*, 1976, **73**, 593.
51 B. Cui, H. Diamant and B. Lin, *Phys. Rev. Lett.*, 2002, **89**, 188302.
52 K. K. Mon and J. K. Percus, *J. Chem. Phys.*, 2007, **127**, 094702.
53 J. Happel and H. Brenner, *Low Reynolds Number Hydrodynamics*, Springer, 1973.
54 T. S. Squires and S. R. Quake, *Rev. Mod. Phys.*, 2005, **77**, 977.

Mori–Zwanzig formalism as a practical computational tool

Carmen Hijón,[a] Pep Español,*[ab] Eric Vanden-Eijnden[c] and Rafael Delgado-Buscalioni[d]

Received 5th February 2009, Accepted 24th February 2009
First published as an Advance Article on the web 12th August 2009
DOI: 10.1039/b902479b

An operational procedure is presented to compute explicitly the different terms in the generalized Langevin equation (GLE) for a few relevant variables obtained within Mori–Zwanzig formalism. The procedure amounts to introducing an artificial controlled parameter which can be tuned in such a way that the so-called projected dynamics becomes explicit and the GLE reduces to a Markovian equation. The projected dynamics can be realised in practice by introducing constraints, and it is shown that the Green–Kubo formulae computed with these dynamics do not suffer from the plateau problem. The methodology is illustrated in the example of star polymer molecules in a melt using their center of mass as relevant variables. Through this example, we show that not only the effective potentials, but also the friction forces and the noise play a very important role in the dynamics.

I. Introduction

The theoretical basis of non-equilibrium statistical mechanics were lay down in the middle of the 20th century by Onsager, Kirkwood, Green, Kubo, Mori, Zwanzig, and many others.[1–3] One of the major achievements of the theory, nowadays referred to as Mori–Zwanzig formalism, is the derivation *via* a projection operator technique of a generalized Langevin equation (GLE) which describes the non-equilibrium evolution of any set of functions defined on the phase-space of the microscopic system. These functions have been named as relevant variables, coarse-grained variables, collective variables, or collective modes in different contexts. While exact, the GLE is unfortunately difficult to use as a computational tool, for two main reasons. First, the various terms in the equation involve the so-called "projected dynamics", which is not the real dynamics that one can calculate *e.g. via* molecular dynamics (MD) simulations and it is difficult to write down explicitly in general. This means that the GLE is not explicit in general. Second, even if it were explicit, the GLE is an integro-differential equation with random coefficients, and the numerical integration of such an equation is very difficult.

[a]*Departamento de Física Fundamental, Universidad Nacional de Educación a Distancia, Aptdo. 60141, E-28080 Madrid, Spain*
[b]*Freiburg Institute for Advanced Studies, Albertstrasse 19, Freiburg i. Br., D-79104, Germany*
[c]*Courant Institute of Mathematical Science, New York University, 251 Mercer Street, NY, 10012, USA*
[d]*Departamento de Física Teórica de la Materia Condensada Universidad Autónoma de Madrid, Madrid, 28049, Spain*

To get around these difficulties, it is usually assumed that both the projected and the real dynamics are equivalent, and that the relevant variables are such that they evolve on a time scale much larger that the correlation time of the memory kernel so that the GLE can be rendered Markovian and turned into a standard stochastic differential equation (SDE). However, there is no clear mathematical justification for these approximations, and they often fail. For instance, they lead to the well-known "plateau problem" that requires the introduction of another *ad-hoc* approximation in which one truncates the limit of the time integrals.[4–6] For these reasons, the GLE is not often used in practical computations. This is unfortunate considering the increasing need of formulating simplified coarse molecular models that have sufficient molecular detail and yet remain amenable of simulation.

In the present paper, we propose a procedure to make explicit the projected dynamics and formalize the Markovian approximation, and thereby derive a workable equation from the GLE. The basic idea is to alter the underlying microscopic dynamics by including an artificial parameter ε which controls the ratio between time scales of the relevant variables and the rest. In the limit as $\varepsilon \rightarrow 0$ it can be shown using limiting theorems for singularly perturbed Markov processes that the modified dynamics leads to an *exact* Markovian description for the relevant variables. Thus, by introducing the control parameter ε, we transform the problem of approximating non-rigorously a model into a problem of assessing the adequacy of a well-defined modified model to reality. A similar strategy was used in a different context in ref. 7. The construction of the modified dynamics is one of the main purposes of this paper. In particular, we show how to compute all the objects in the procedure *via* ergodic sampling of appropriately constrained molecular dynamics simulations. We also show that the resulting Green–Kubo expressions obtained in the new formulation do not suffer from the plateau problem, which is inherent to the use of the real dynamics.[4–6]

The procedure is subsequently applied to the coarse-graining of complex molecules. There is at present a great interest in constructing coarse-grained models in which each molecule is represented by a smaller number of degrees of freedom.[8] Towards this aim much attention has been paid to obtaining *coarse-grained (CG) potentials* governing the interaction of the coarse degrees of freedom. Usually, certain functional forms for the CG potentials are assumed and they are parametrized in order to obtain structural properties like the radial distribution function. Several methods have been developed in order to obtain the appropriate parameters: iterative adjustment of potential parameters starting from an approximation based on the potentials of mean force,[9,10] through the solution of the Ornstein–Zernike equation,[11] by using the inverse Monte Carlo technique,[12,13] by matching thermodynamic properties,[14] or by using directly the underlying all-atom interactions through a force matching procedure.[15,16] The CG potentials produce correct results for the study of equilibrium and structural properties, as they have been designed just to recover such behaviour. However, dynamic properties like diffusion or, more generally, time correlations, are not necessarily recovered from a simple molecular dynamics simulation using the CG potentials. The basic reason for this failure is that a coarse-graining procedure eliminates degrees of freedom that should appear in the coarse-grained dynamics in the form of *dissipation* and *thermal noise*, both connected through the fluctuation–dissipation theorem. The construction of CG potentials is only part of the story of coarse-graining. This has been recognized recently by considering the generalized Langevin equation for the coarse-grained variables[17] which includes velocity dependent friction forces and thermal fluctuations. In the present paper, we show how our procedure can be applied to the coarse-graining of star polymer molecules, by computing not only coarse-grained effective conservative forces but also dissipative friction forces. The resulting dynamic equations have the structure of dissipative particle dynamics.[18,19]

The remainder of this paper is organized as follows. In Sec. II we review Mori–Zwanzig theory, and re-derive and interpret the GLE. In Sec. III we present the standard Markovian approximation which is used to turn the GLE into an SDE, and we discuss the caveats with this approximation, in particular the well-known plateau problem. In Sec. IV we introduce our procedure to modify the original dynamics and thereby obtain directly an SDE instead of a GLE in suitable limit. We also show that the coefficients in the SDE do not suffer from the plateau problem and we discuss how to compute them in practice using constrained MD simulations. In Sec. V we show how to apply the procedure to coarse-grain the dynamics of big complex molecules and in Sec. VI we consider the specific example of star polymer in a melt. Finally, in Sec. VII we give a few concluding remarks and indicate how the procedure proposed here could be generalized.

II. Review of Mori–Zwanzig theory

In this section we review Mori–Zwanzig theory by working directly with the equation of motion for the relevant variables rather than the one for the probability density function of these variables.[3,20,21] We choose this approach because it is slightly more expeditive and, we believe, more transparent. We will work with Zwanzig's projector instead of Mori's projector.[3] The former leads to a general non-linear GLE while the latter produces linear GLE. It can be shown that the linear GLE of Mori's is an approximation near equilibrium of the more general Zwanzig's GLE.[3]

We shall focus on systems whose microscopic state is characterized by the instantaneous positions and momenta of the N atoms of the system, $\{\boldsymbol{q}_i(t),\boldsymbol{p}_i(t)\}$ with $i = 1, \ldots, N$. We denote the collection of these variables by $\boldsymbol{Z}(t) = (Z_1(t),\ldots,Z_{6N}(t))$, which is a vector of $6N$ components. In terms of $\boldsymbol{Z}(t)$, the Hamiltonian dynamics of the system can be written as by

$$\frac{\mathrm{d}\boldsymbol{Z}(t)}{\mathrm{d}t} = J\frac{\partial H(\boldsymbol{Z}(t))}{\partial z}, \qquad \boldsymbol{Z}(0) = z \tag{1}$$

where z denotes the initial condition, H is the Hamiltonian and J is the symplectic matrix with a block diagonal structure with the blocks given by

$$J = \begin{pmatrix} \mathbf{0} & \mathbf{1} \\ -\mathbf{1} & \mathbf{0} \end{pmatrix} \tag{2}$$

Suppose that we are interested not in the evolution of $\boldsymbol{Z}(t)$ *per se*, but rather that of $\boldsymbol{A}(\boldsymbol{Z}(t))$ where $\boldsymbol{A}(z) = (A_1(z),\ldots,A_M(z))$ is a specific observable, *i.e.* any set of M functions defined on phase-space. Even more specifically our aim is to calculate the statistical properties of $\boldsymbol{A}(\boldsymbol{Z}(t))$ for $t \geq 0$ for the ensemble of initial conditions $\boldsymbol{Z}(0) = z$ satisfying $\boldsymbol{A}(z) = \boldsymbol{\alpha}$ for some fixed $\boldsymbol{\alpha}$ and with z distributed according to the equilibrium density $\rho^{\mathrm{eq}}(z)$ conditional on $\boldsymbol{A}(z) = \boldsymbol{\alpha}$. Zwanzig's approach is a way to write an integro-differential equation with random coefficients whose solutions in different realizations generate the desired ensemble of $\boldsymbol{A}(\boldsymbol{Z}(t))$.

To see how this equation is derived, let us first make explicit the dependency in the initial condition z in $\boldsymbol{A}(\boldsymbol{Z}(t))$ by denoting $\boldsymbol{A}(\boldsymbol{Z}(t)) \equiv \boldsymbol{a}(t,z)$. This function can be formally expressed as

$$\boldsymbol{a}(t,z) = \exp\{tL\}\boldsymbol{A}(z) \tag{3}$$

in which the exponential operator is defined through its Taylor series expansion and L is the Liouville operator

$$L = -\frac{\partial H}{\partial z} J \frac{\partial}{\partial z} \tag{4}$$

Eqn (3) shows that $\boldsymbol{a}(t,z)$ satisfies the following equation

$$\partial_t\, \boldsymbol{a}(t,z) = L\boldsymbol{a}(t,z), \quad \boldsymbol{a}(0,z) = A(z) \tag{5}$$

Next introduce the conditional expectation operator P_α whose action to an arbitrary phase function $F(z)$ gives the conditional equilibrium expectation of $F(z)$ at $A(z) = \boldsymbol{\alpha}$ fixed, *i.e.* the function of $\boldsymbol{\alpha}$ defined as

$$P_\alpha F = \frac{1}{\Omega(\boldsymbol{\alpha})} \int F(z) \rho^{\text{eq}}(z) \delta(A(z) - \boldsymbol{\alpha}) \mathrm{d}z \tag{6}$$

Here $\rho^{\text{eq}}(z)$ is the equilibrium probability density (*e.g.* the microcanonical density $\rho^{\text{eq}}(z) = \Omega_0^{-1}\delta(H(z) - E)$ where E is the energy and Ω_0 is the normalization factor, assuming that $H(z)$ is the only invariant of motion), and we defined

$$\Omega(\boldsymbol{\alpha}) = \int \rho^{\text{eq}}(z)\, \delta(A(z) - \boldsymbol{\alpha}) \mathrm{d}z \tag{7}$$

$\Omega(\boldsymbol{\alpha})$ is the probability density of $A(z)$ or, loosely speaking, "the number of microstates compatible with the macrostate $A(z) = \boldsymbol{\alpha}$." Let $Q_\alpha = 1 - P_\alpha$, and in eqn (5) use $L\boldsymbol{a}(t,z) = L\exp\{tL\}A(z) = \exp\{tL\}LA(z)$ and insert $1 = P_{A(z)} + Q_{A(z)}$ to transform this equation into

$$\partial_t \boldsymbol{a}(t,z) = \exp\{tL\} P_{A(z)} LA + \exp\{tL\} Q_{A(z)} LA \tag{8}$$

Using the Duhamel–Dyson identity

$$\exp\{tL\} = \exp\{tQ_{A(z)}L\} + \int_0^t \mathrm{d}s \exp\{(t-s)L\} P_{A(z)} L \exp\{sQ_{A(z)}L\} \tag{9}$$

the second term at the right-hand side of eqn (8) can be written as

$$\partial_t \boldsymbol{a}(t,z) = \exp\{tL\} P_{A(z)} LA + \int_0^t \mathrm{d}s \exp\{(t-s)L\} P_{A(z)} L\tilde{\boldsymbol{R}}(s,\bullet) + \tilde{\boldsymbol{R}}(t,z) \tag{10}$$

where we defined

$$\begin{aligned} \tilde{\boldsymbol{R}}(t,z) &= \exp\{tQ_{A(z)}L\} Q_{A(z)} LA \\ &= Q_{A(z)} \exp\{tQ_{A(z)}L\} LA \end{aligned} \tag{11}$$

and we used a dot instead of a z as second argument for $\tilde{\boldsymbol{R}}$in $P_{A(z)}L\tilde{\boldsymbol{R}}(s,\cdot)$ to stress that this term depends on z only through $A(z)$ (the same is true for $P_{A(z)}LA$ but not for $\tilde{\boldsymbol{R}}(t,z)$ which is a general function of z). The second term at the right-hand side of eqn (10) can be simplified by means of the following identity which, for clarity, we write component-wise using the indices μ, $\nu = 1, \ldots, M$ to denote the components of $\boldsymbol{A}$ and $\boldsymbol{\alpha}$ and Einstein sum convention over repeated indices

$$
\begin{aligned}
P_{\alpha} L\tilde{R}_{\mu}(s,\cdot) &= \frac{1}{\Omega(\boldsymbol{\alpha})}\int \mathrm{d}z\rho^{\mathrm{eq}}(z)\delta(\boldsymbol{A}(z)-\boldsymbol{\alpha})L\exp\{sQ_{A(z)}L\}Q_{A(z)}LA_{\mu} \\
&= -\frac{1}{\Omega(\boldsymbol{\alpha})}\int \mathrm{d}z\rho^{\mathrm{eq}}(z)\left[\exp\{sQ_{A(z)}L\}Q_{A(z)}LA_{\mu}\right]L\delta(\boldsymbol{A}(z)-\boldsymbol{\alpha}) \\
&= \frac{1}{\Omega(\boldsymbol{\alpha})}\int \mathrm{d}z\rho^{\mathrm{eq}}(z)\left[\exp\{sQ_{A(z)}L\}Q_{A(z)}LA_{\mu}\right][LA_{\nu}(z)]\frac{\partial}{\partial\alpha_{\nu}}\delta(\boldsymbol{A}(z)-\boldsymbol{\alpha}) \\
&= \frac{1}{\Omega(\boldsymbol{\alpha})}\frac{\partial}{\partial\alpha_{\nu}}\int \mathrm{d}z\rho^{\mathrm{eq}}(z)\delta(\boldsymbol{A}(z)-\boldsymbol{\alpha})[LA_{\nu}(z)] \\
&\quad \left[\exp\{sQ_{A(z)}L\}Q_{A(z)}LA_{\mu}\right][LA_{\nu}(z)] \\
&= \frac{1}{\Omega(\boldsymbol{\alpha})}\frac{\partial}{\partial\alpha_{\nu}}(\Omega(\boldsymbol{\alpha})P_{\alpha}([\exp\{sQ_{A}L\}Q_{A}LA_{\mu}][LA_{\nu}])) \\
&= M_{\mu\nu}(\boldsymbol{\alpha},s)\frac{\partial S(\boldsymbol{\alpha})}{\partial\alpha_{\nu}} + k_{B}\frac{\partial M_{\mu\nu}(\boldsymbol{\alpha},s)}{\partial\alpha_{\nu}}
\end{aligned}
\tag{12}
$$

Here and below the operators inside the brackets $[\cdot]$ only act on the terms at their right in these brackets and we have introduced the entropy

$$
S(\boldsymbol{\alpha}) = k_{B}\ln\Omega(\boldsymbol{\alpha}) \tag{13}
$$

as well as the memory matrix $M(\boldsymbol{\alpha},t) = M^{T}(\boldsymbol{\alpha},-t)$ whose components are given by the following conditional expectation

$$
M_{\mu\nu}(\boldsymbol{\alpha},t) = \frac{1}{k_{B}}P_{\alpha}\Big([LA_{\nu}][\exp\{tQ_{A}L\}Q_{A}\,LA_{\mu}]\Big) = \frac{1}{k_{B}}P_{\alpha}\Big(\tilde{R}_{\mu}(t,\cdot)\tilde{R}_{\nu}(0,\cdot)\Big) \tag{14}
$$

Inserting eqn (12) in eqn (10) and using the property that for any $f(\boldsymbol{A}(z))$, we have $\exp\{Lt\}f(\boldsymbol{A}(z)) = f(\boldsymbol{a}(t,z))$ and so

$$
\begin{aligned}
\exp\{tL\}P_{A(z)}L\boldsymbol{A}(z) &= P_{\boldsymbol{a}(t,z)}L\boldsymbol{A}(z), \exp\big\{(t-s)L\big\}M(\boldsymbol{A}(z),s)\frac{\partial S(\boldsymbol{A}(z))}{\partial\boldsymbol{\alpha}} \\
&= M(\boldsymbol{a}(t-s,z),s)\frac{\partial S(\boldsymbol{a}(t-s,z))}{\partial\boldsymbol{\alpha}}, \exp\big\{(t-s)L\big\}k_{B}\frac{\partial M(\boldsymbol{A}(z),s)}{\partial\boldsymbol{\alpha}} \\
&= k_{B}\frac{\partial M(\boldsymbol{a}(t-s,z),s)}{\partial\boldsymbol{\alpha}}
\end{aligned}
$$

we arrive at the following equation for $\boldsymbol{a}(t,z)$

$$
\begin{aligned}
\partial_{t}\boldsymbol{a}(t,z) &= \boldsymbol{v}(\boldsymbol{a}(t,z)) \\
&\quad + \int_{0}^{t}\mathrm{d}sM(\boldsymbol{a}(t-s,z),s)\frac{\partial S}{\partial\boldsymbol{\alpha}}(\boldsymbol{a}(t-s,z)) + k_{B}\int_{0}^{t}\mathrm{d}s\frac{\partial M}{\partial\boldsymbol{\alpha}}(\boldsymbol{a}(t-s,z),s) + \tilde{\boldsymbol{R}}(t,z)
\end{aligned}
\tag{15}
$$

where $\boldsymbol{v}(\boldsymbol{\alpha})$ is the following conditional expectation

$$\boldsymbol{v}(\boldsymbol{\alpha}) = P_{\alpha} L A \tag{16}$$

Eqn (15) is a formally exact rewriting of eqn (5) and it may not be immediately clear what we have gained with it. Recall however that we are not interested in solving this equation for a specific initial condition $\boldsymbol{a}(0,z) = \boldsymbol{A}(z)$ but rather for an ensemble of initial conditions z satisfying $\boldsymbol{a}(0,z)=\boldsymbol{A}(z) = \boldsymbol{\alpha}(0)$ for some fixed $\boldsymbol{\alpha}(0)$ and with z distributed according to the equilibrium density $\rho^{\mathrm{eq}}(z)$ conditional on $\boldsymbol{A}(z) = \boldsymbol{\alpha}(0)$. In this case, $\tilde{\boldsymbol{R}}(t,z)$, which is the only term in eqn (15) which is not a function of $\boldsymbol{a}(s,z)$ for $0 \leq s \leq t$, can be interpreted as a noise term whose statistics must be consistent with eqn (14). With this in mind, we can introduce the shorthand notation $\boldsymbol{a}(t,z) = \boldsymbol{\alpha}(t)$, and rewrite eqn (15) as an integro-differential equation with a random-source term which is usually referred to as the generalized Langevin equation (GLE):

$$\begin{aligned} \frac{\mathrm{d}\boldsymbol{\alpha}(t)}{\mathrm{d}t} &= \boldsymbol{v}(\boldsymbol{\alpha}(t)) \\ &+ \int_0^t \mathrm{d}s M(\boldsymbol{\alpha}(t-s),s)\frac{\partial S}{\partial \boldsymbol{\alpha}}(\boldsymbol{\alpha}(t-s)) + k_B \int_0^t \mathrm{d}s \frac{\partial M}{\partial \boldsymbol{\alpha}}(\boldsymbol{\alpha}(t-s),s) + \boldsymbol{R}(t) \end{aligned} \tag{17}$$

where $\boldsymbol{R}(t)$ is now viewed as a zero-mean random process whose statistical properties are specified by eqn (11) in which z is random and distributed according to the equilibrium density $\rho^{\mathrm{eq}}(z)$ conditional on $\boldsymbol{A}(z) = \boldsymbol{\alpha}(0)$. By solving eqn (17) with the initial condition $\boldsymbol{\alpha}(0)$ in different realizations of $\boldsymbol{R}(t)$ we can then generate the exact statistics of $\boldsymbol{A}(\boldsymbol{Z}(t))$ for $t \geq 0$ along an ensemble of trajectories consistent with $\boldsymbol{A}(\boldsymbol{Z}(0)) = \boldsymbol{\alpha}(0)$.

While formally exact within the statistical interpretation above, the GLE (eqn (17)) is unfortunately rather useless in practice. Indeed, while $\boldsymbol{v}(\boldsymbol{\alpha})$ and the gradient $\partial S/\partial \boldsymbol{\alpha}$ are conditional expectations which can in principle be computed using constrained molecular dynamics, we cannot calculate $M(\boldsymbol{\alpha},t)$ and $\boldsymbol{R}(t)$ since they involve the projected dynamics associated with $Q_{A(z)}L$ which we do not know how to generate. (Note also that the process $\boldsymbol{R}(t)$ is non-Gaussian in general, *i.e.* it is *not* specified completely by its correlation function in eqn (14).[20]) On top of this, even if we knew how to compute $M(\boldsymbol{\alpha},t)$ and the full statistics of $\boldsymbol{R}(t)$, eqn (17) would remain very challenging to integrate numerically because of its non-Markovian character.

III. The standard Markovian approximation and its caveats

The usual way the GLE (eqn (17)) is made practical is *via* a set of approximations which collectively go under the name of "Markovian approximation." This approximation is introduced as follows. Assuming that the time scale of variation of the relevant variables is much larger than the time scale of decay of the memory matrix $M(\boldsymbol{\alpha},t)$ defined in eqn (14), one approximates this matrix by

$$M(\boldsymbol{\alpha},t) \approx M_T(\boldsymbol{\alpha})\delta(t) \tag{18}$$

where the time-independent friction matrix $M_T(\alpha)$ is defined as

$$M_T(\boldsymbol{\alpha}) = \int_0^T \mathrm{d}t M(\boldsymbol{\alpha},t) = \int_0^T \mathrm{d}t P_{\alpha}\left(\tilde{\boldsymbol{R}}(t,\cdot)\otimes\tilde{\boldsymbol{R}}(0,\cdot)\right) \tag{19}$$

Note that the integral is capped at a finite time T rather than extended to infinity; we will explain shortly why it is necessary to do so. Consistent with the approximation in eqn (18), one then assumes that the random term $\boldsymbol{R}(t)$ in (17) can be modelled as a white-noise, *i.e.* a Gaussian process with mean zero and whose correlation at $\boldsymbol{\alpha}(t) = \boldsymbol{\alpha}$ fixed is given by

$$\langle \boldsymbol{R}_T(t) \otimes \boldsymbol{R}_T(s) \rangle = k_B M_T(\boldsymbol{\alpha}) \delta(t - s) \tag{20}$$

Under these assumptions, the GLE (eqn (17)) becomes the stochastic differential equation (SDE)

$$\frac{\mathrm{d}\boldsymbol{\alpha}(t)}{\mathrm{d}t} = \boldsymbol{v}(\boldsymbol{\alpha}(t)) + M_T(\boldsymbol{\alpha}(t)) \frac{\partial S}{\partial \boldsymbol{\alpha}}(\boldsymbol{\alpha}(t)) + k_B \frac{\partial M_T}{\partial \boldsymbol{\alpha}}(\boldsymbol{\alpha}(t)) + \boldsymbol{R}_T(t). \tag{21}$$

The Fokker–Planck equation which is mathematically equivalent to eqn (21) was derived by Zwanzig.[2] The only thing left open to make eqn (21) fully explicit is how to compute the friction matrix in eqn (19). This is done by assuming that the projected dynamics $\exp\{tQ_{A(z)}L\}$ can be replaced by the real dynamics $\exp\{tL\}$ in eqn (11), *i.e.*

$$\tilde{\boldsymbol{R}}(t,z) \approx \exp\{Lt\} Q_{A(z)} LA = LA(\boldsymbol{Z}(t)) - \boldsymbol{v}(\boldsymbol{\alpha}(t)) \tag{22}$$

where $\boldsymbol{v}(\alpha)$ is defined in eqn (16), $\boldsymbol{Z}(t)$ is the solution of the original Hamilton eqn (1) and we used again the shorthand notation $\boldsymbol{\alpha}(t) = \boldsymbol{a}(t,z)$. The right-hand side of eqn (22) can in principle be computed using a combination of standard MD simulations (to compute $\boldsymbol{Z}(t)$) and constrained molecular simulation (to compute $\boldsymbol{v}(\boldsymbol{\alpha})$). Eqn (22) is valid for short times when the time integrals in eqn (15) can be neglected: indeed, using $\partial_t \boldsymbol{a}(t,z) = LA(\boldsymbol{Z}(t))$ in eqn (15) and solving this equation with the integrals set to zero for $\tilde{\boldsymbol{R}}(t,z)$ gives eqn (22). Unfortunately, eqn (22) is harder to justify at later times. In particular, this approximation is the reason why the integral in eqn (19) must be capped at a finite T: if one extends the limit of integration T in eqn (19) to infinity using the approximation (eqn (22)) for $\tilde{\boldsymbol{R}}(t,z)$, then the integral vanishes. This is the well-known plateau problem[4,5] and the current practice is to select for the upper time of integration a time T which is large compared to the correlation time of the (unspecified) orthogonal dynamics, but small compared to the time scale of evolution of the macroscopic variables.[3,22] This intermediate time scale is assumed to exists, at least in situations where there is a clear separation of time scales, but the specific value for T is not provided by the theory, and it is difficult to predict how the results depend on T.

In summary, both approximations eqn (18) and (22) are uncontrolled and this clearly diminishes the confidence that one can have in eqn (21). In the next section, we introduce another procedure to derive an equation similar to eqn (21) but whose validity is easier to assess and which does not suffer from the plateau problem.

IV. Modified dynamics and Markovian limiting equation

In order to replace eqn (18) and (22) by more controlled approximations, consider the time integrals appearing in the GLE (eqn (17)) and perform the change of variables $s = \varepsilon^2\tau$, where ε is a non-dimensional control parameter. This leads to

$$\begin{aligned}\frac{\mathrm{d}\boldsymbol{\alpha}(t)}{\mathrm{d}t} &= \boldsymbol{v}(\boldsymbol{\alpha}(t)) \\ &+ \varepsilon^2 \int_0^{t/\varepsilon^2} \mathrm{d}\tau M\left(\boldsymbol{\alpha}(t - \varepsilon^2\tau), \varepsilon^2\tau\right) \frac{\partial S}{\partial \boldsymbol{\alpha}}\left(\boldsymbol{\alpha}(t - \varepsilon^2\tau)\right) + \varepsilon^2 k_B \int_0^{t/\varepsilon^2} \mathrm{d}\tau \frac{\partial M}{\partial \boldsymbol{\alpha}}\left(\boldsymbol{\alpha}(t - \varepsilon^2\tau), \varepsilon^2\tau\right) + \boldsymbol{R}(t)\end{aligned} \tag{23}$$

Now observe that if the following limit exists

$$\lim_{\varepsilon \to 0} \varepsilon^2 M\left(\boldsymbol{\alpha}\left(t - \varepsilon^2 \tau\right), \varepsilon^2 \tau\right) \equiv m(\boldsymbol{\alpha}(t), \tau) \tag{24}$$

then, in the limit as $\varepsilon \to 0$, eqn (23) reduces to the SDE

$$\frac{\mathrm{d}\boldsymbol{\alpha}(t)}{\mathrm{d}t} = \boldsymbol{v}(\boldsymbol{\alpha}(t)) + \bar{M}(\boldsymbol{\alpha}(t)) \frac{\partial S}{\partial \boldsymbol{\alpha}}(\boldsymbol{\alpha}(t)) + k_B \frac{\partial \bar{M}}{\partial \boldsymbol{\alpha}}(\boldsymbol{\alpha}(t)) + \bar{\boldsymbol{R}}(t) \tag{25}$$

Here the friction matrix $\bar{M}(\boldsymbol{\alpha})$ has the Green–Kubo form

$$\bar{M}(\boldsymbol{\alpha}) = \int_0^\infty m(\boldsymbol{\alpha}, \tau) \mathrm{d}\tau \tag{26}$$

and (as discussed below) the random term $\bar{\boldsymbol{R}}(t)$ is an Itô white-noise, *i.e.* a Gaussian process with mean zero and whose correlation at $\boldsymbol{\alpha}(t) = \boldsymbol{\alpha}$ is given by

$$\langle \boldsymbol{R}(t) \boldsymbol{R}(s) \rangle = k_B \bar{M}(\boldsymbol{\alpha}) \delta(t - s) \tag{27}$$

In view of the above, it is of great value to find out under what conditions the limit in eqn (24) does indeed exist. By using the definition in eqn (14) we may write

$$\varepsilon^2 M\left(\boldsymbol{a}\left(t - \varepsilon^2 \tau, z\right), \varepsilon^2 \tau\right) \quad = \quad \frac{\varepsilon^2}{k_B} P_{\boldsymbol{a}(t - \varepsilon^2 \tau, z)} \Big(\left[\exp\{\varepsilon^2 \tau Q_A L\} Q_A L A\right] \otimes \left[Q_A L A\right] \Big) \tag{28}$$

From this expression, it can be checked by direct calculation that one way to ensure that the limit eqn (24) exists is by assuming that the Liouville operator has the following form

$$L = L_0 + \frac{1}{\varepsilon} L_1 + \frac{1}{\varepsilon^2} L_2 \tag{29}$$

with the operators L_1, L_2 satisfying

$$\begin{aligned} P_{A(z)} L_2 &= 0 \\ P_{A(z)} L_1 P_{A(z)} &= 0 \end{aligned} \tag{30}$$

Indeed, in this case, eqn (28) simply becomes

$$\varepsilon^2 M\left(\boldsymbol{a}\left(t - \varepsilon^2 \tau, z\right), \varepsilon^2 \tau\right) = \frac{1}{k_B} P_{\boldsymbol{\alpha}(t)} \left(\left[\exp\{\tau L_2\} L_1 A\right] \otimes \left[L_1 A\right] \right) + O(\varepsilon) \tag{31}$$

and the limiting dynamic equation obtained from eqn (17) when $\varepsilon \to 0$ precisely is eqn (25) with the Green–Kubo friction matrix $\bar{M}(\boldsymbol{\alpha})$ now given explicitly by

$$\bar{M}(\boldsymbol{\alpha}) = \frac{1}{k_B} \int_0^\infty \mathrm{d}\tau P_{\boldsymbol{\alpha}} \left(\left[\exp\{\tau L_2\} L_1 A\right] \otimes \left[L_1 A\right] \right) \tag{32}$$

and the drift term is given by

$$\boldsymbol{v}(\boldsymbol{\alpha}) = P_{\boldsymbol{\alpha}} L_0 \boldsymbol{A}. \tag{33}$$

While the derivation above is heuristic, it can be made rigorous using the techniques discussed in ref. 23,24 (see also the Appendix in ref. 25), *i.e.* under a suitable ergodicity assumption of the dynamics associated with the operator L_2, it can be proved that eqn (25) captures exactly (though in a statistical sense) the dynamics of $A(Z(t))$ in the limit as $\varepsilon \to 0$. This derivation also explains why the noise term $\bar{R}(t)$ in eqn (25) is indeed a white-noise to be interpreted in Itô sense. Finally, it shows that the matrix $\bar{M}(\alpha)$ does not, in general, have the plateau problem: indeed the integral in eqn (32) converges to a nontrivial value provided only that the dynamics associated with L_2 is mixing sufficiently fast beside being ergodic.

The next important question is in which sense is eqn (25) useful. Indeed for a general selection of the relevant variables $A(z)$, the Liouville operator will not have the form given in eqn (29) and so it is not *a priori* clear how to derive eqn (25). There is a way to do so, however, based on the observation that it is always possible to decompose the Liouville operator as $L = L_0 + L_1 + L_2$ by defining

$$\begin{aligned} L_0 &= P_{A(z)}(L - \mathscr{R}) \\ L_1 &= Q_{A(z)}(L - \mathscr{R}) \\ L_2 &= \mathscr{R} \end{aligned} \tag{34}$$

and letting $P_{A(z)}$ be the expectation with respect to the equilibrium distribution associated with the operator $\mathscr{R}$. This operator, to be specified more fully later, should be similar to L, except that it leaves both the Hamiltonian *and* the relevant variables invariant, that is

$$\begin{aligned} \mathscr{R}f(H(z)) &= 0 \\ \mathscr{R}g(A(z)) &= 0 \end{aligned} \tag{35}$$

for any functions f and g. By construction, the operators L_0, L_1, L_2 in eqn (34) satisfy the properties in eqn (30). This suggests the introduction of a modified dynamic operator L^{ε} as in eqn (29)

$$L^{\varepsilon} \equiv L_0 + \frac{1}{\varepsilon}L_1 + \frac{1}{\varepsilon^2}L_2 \tag{36}$$

The dynamics associated with L^{ε} coincides with the real dynamics when $\varepsilon = 1$ and produces a dynamics of the relevant variables which is governed by the SDE (eqn (25)) when $\varepsilon \to 0$. By inserting the operators eqn (34) into eqn (32), (33) we obtain

$$\begin{aligned} \boldsymbol{v}(\boldsymbol{\alpha}) &= P_{\alpha}(LA) \\ \bar{M}(\boldsymbol{\alpha}) &= \frac{1}{k_B}\int_0^{\infty} \mathrm{d}\tau P_{\alpha}\left([\exp\{\tau\mathscr{R}\}LA] \otimes [LA]\right) \end{aligned} \tag{37}$$

The advantage of the above procedure, which differs from the usual prescription in that the projected dynamics $\exp\{tQ_{A(z)}L\}$ is approximated by $\exp\{tL_2\} \equiv \exp\{t\mathscr{R}\}$ rather than $\exp\{tL\}$, is that we have now an explicit and practical method to compute the constrained averages once we specify the operator $\mathscr{R}$. A natural candidate for the dynamics associated with $\mathscr{R}$ is one which is obtained from the original Hamilton's eqn (1) by adding the constraint that $A(Z(t)) = \alpha$. How to do so in practice in the special case when $A(z)$ is a linear function of the positions and the momenta is straightforward and an example will be given in Sec. V. The general case when $A(z)$ is a nonlinear function of z is more complicated to handle and will be left for future work.

If we then assume that this constrained dynamics is ergodic and denote compactly by $Z_{\mathscr{R}}(t) = \exp\{t\mathscr{R}\}z$ the constrained trajectory with initial condition $Z_{\mathscr{R}}(0) = z$ with

z such that $A(z) = \alpha$ and $H(z) = E$, then the conditional expectations in eqn (37) can be expressed as time averages

$$P_\alpha LA = \lim_{T\to\infty} \frac{1}{T}\int_0^T \mathrm{d}t LA(Z_{\mathcal{R}}(t)) \tag{38}$$

and

$$\bar{M}(\alpha) = \lim_{T'\to\infty} \frac{1}{k_B}\int_0^{T'} \mathrm{d}t' \lim_{T\to\infty}\frac{1}{T}\int_0^T \mathrm{d}t[LA(Z_{\mathcal{R}}(t+t'))]\otimes[LA(Z_{\mathcal{R}}(t))] \tag{39}$$

In practice the limits in these expressions have to be approximated but it is important to stress that this can be done to arbitrary precision by letting T and T' grow bigger, *i.e.* eqn (38) and (39) do not suffer from the plateau problem.

In summary, if we replace L by L^ε and let $\varepsilon \to 0$, then the dynamics of relevant variables is governed exactly by the limiting eqn (25) with the coefficients in this equation given explicitly by eqn (38) and (39). In doing so, the price we pay is a *modelling error*, that is, we model the original system by one whose dynamics is associated with L^ε in the limit as $\varepsilon \to 0$ rather than with L. In effect this amounts to accelerating the non-relevant degrees of freedom. The validity of this modeling approximation requires that the real dynamics L and the projected dynamics $\mathcal{R}$ be similar, *i.e.* that the effect of accelerating the dynamics orthogonal to that of the variables $A(z)$ only has a small effect in the evolution of these variables. In other words, it requires that the relevant variables $A(z)$ be comparatively slow.

V. Coarsening big complex molecules

In this section, we show how the above methodology can be applied to the coarse-graining of a collection of big molecules by using center of mass (CoM) variables as relevant variables. We will consider later as a concrete example that these molecules are star polymers but the framework is to a large extend independent of what kind of molecule we have, provided that they are made of many atoms and that they are isotropic. For non-isotropic molecules, further orientational information may be required in addition to the CoM variables.

We assume that the fluid system is composed by M molecules and each molecule is made of N_m atoms whose positions and momenta are $\boldsymbol{r}_{i_\mu}$, $\boldsymbol{p}_{i_\mu}$ where the index i_μ runs from 1,..., N_m, while the index μ runs from 1,..., M, *i.e.* Greek indices label molecules. The Hamiltonian of the system is

$$H(z) = \sum_{\mu=1}^{M}\sum_{i_\mu=1}^{N_m}\frac{\boldsymbol{p}_{i_\mu}^2}{2m_{i_\mu}} + \phi \tag{40}$$

where m_{i_μ} is the mass of the atom i_μ and ϕ is the potential energy. The Liouville operator is given by

$$L = \sum_{\mu}\sum_{i_\mu}\frac{\boldsymbol{p}_{i_\mu}}{m_{i_\mu}}\frac{\partial}{\partial \boldsymbol{r}_{i_\mu}} + \sum_{\mu}\sum_{i_\mu}\boldsymbol{F}_{i_\mu}\frac{\partial}{\partial \boldsymbol{p}_{i_\mu}} \tag{41}$$

where $\boldsymbol{F}_{i_\mu} = -\partial\phi/\partial\boldsymbol{r}_{i_\mu}$ is the force on the atom i_μ.

At a coarse-grained level, we will represent the complex molecule by just the position $\boldsymbol{R}_\mu$ and momentum $\boldsymbol{P}_\mu$ of its center of mass. These relevant variables are the following functions of the atomic variables

$$\begin{aligned} \boldsymbol{R}_\mu(z) &= \frac{1}{M_\mu}\sum_{i_\mu=1}^{N_m} m_{i_\mu}\boldsymbol{r}_{i_\mu} \\ \boldsymbol{P}_\mu(z) &= \sum_{i_\mu=1}^{N_m} \boldsymbol{p}_{i_\mu} \end{aligned} \tag{42}$$

where $M_\mu = \sum_{i_\mu=1}^{N_m} m_{i_\mu}$ is the total mass of the molecule μ. Note that we have $L\boldsymbol{R}_\mu = \boldsymbol{P}_\mu/M_\mu$, this is, the Liouville operator applied to a relevant variable is (proportional to) a relevant variable itself. As a consequence, the conditional average of $\boldsymbol{P}_\mu$ conditioned to $\boldsymbol{P}_\mu$ and $\boldsymbol{R}_\mu$ is just $\boldsymbol{P}_\mu$ itself. This means that there are no dissipative terms (nor noise terms) in the evolution of the positions, and eqn (25) reduces to

$$\begin{aligned} \frac{\mathrm{d}\boldsymbol{R}_\mu}{\mathrm{d}t} &= \frac{\boldsymbol{P}_\mu}{M_\mu} \\ \frac{\mathrm{d}\boldsymbol{P}_\mu}{\mathrm{d}t} &= \langle \boldsymbol{F}_\mu \rangle + T\gamma_{\mu\nu}\frac{\partial S}{\partial \boldsymbol{P}_\nu} + k_B T\frac{\partial \gamma_{\mu\nu}}{\partial \boldsymbol{P}_\nu} + \tilde{\boldsymbol{F}}_\mu \end{aligned} \tag{43}$$

where we recall that the sum over repeated indices is implied. Here we use the shorthand notations $\boldsymbol{R} = (\boldsymbol{R}_1,\ldots,\boldsymbol{R}_M)$, $\boldsymbol{P} = (\boldsymbol{P}_1,\ldots,\boldsymbol{P}_M)$ and we denote by $\langle\cdot\rangle$ the conditional expectation $P_{(\boldsymbol{R},\boldsymbol{P})}$ at $(\boldsymbol{R},\boldsymbol{P})$ fixed. The friction tensor is defined by

$$\gamma_{\mu\nu}(\boldsymbol{R},\boldsymbol{P}) = \frac{1}{k_B T}\int_0^\infty \mathrm{d}t\langle \delta\boldsymbol{F}_\mu \exp\{t\mathscr{R}\}\delta\boldsymbol{F}_\nu\rangle \tag{44}$$

where $\delta\boldsymbol{F}_\mu = \boldsymbol{F}_\mu - \langle\boldsymbol{F}_\mu\rangle$ and $\boldsymbol{F}_\mu$ is the total force acting on the molecule μ:

$$\boldsymbol{F}_\mu = \sum_\nu \boldsymbol{F}_{\mu\nu} \equiv \sum_\nu \sum_{i_\mu j_\nu} \boldsymbol{F}_{i_\mu j_\nu} \tag{45}$$

Here $\boldsymbol{F}_{i_\mu j_\nu}$ is the force that atom j_ν exerts on atom i_ν, and $\boldsymbol{F}_{\mu\nu}$ is the total force that molecule ν exerts on molecule μ. The entropy has the form

$$S(\boldsymbol{R},\boldsymbol{P}) = k_B \ln\int \mathrm{d}z\frac{1}{Z}\exp\{-\beta H(z)\}\prod_\mu \delta(\boldsymbol{R}_\mu(z)-\boldsymbol{R}_\mu)\delta(\boldsymbol{P}_\mu(z)-\boldsymbol{P}_\mu) \tag{46}$$

The momentum integrals involved in the entropy function can be performed explicitly with the result

$$S(\boldsymbol{R},\boldsymbol{P}) = S_0 - \frac{1}{T}V(\boldsymbol{R}) - \frac{1}{T}\sum_\mu \frac{|\boldsymbol{P}_\mu|^2}{2M_\mu} \tag{47}$$

where S_0 is a constant and $V(\boldsymbol{R})$ is the so called effective potential defined by

$$V(\boldsymbol{R}) \equiv -k_B T\ln\int \mathrm{d}z\frac{1}{Q}\exp\{\beta\phi(z)\}\prod_\mu \delta(\boldsymbol{R}_\mu(z)-\boldsymbol{R}_\mu) \tag{48}$$

This effective potential satisfies

$$-\frac{\partial V}{\partial \boldsymbol{R}_\mu} = \langle \boldsymbol{F}_\mu \rangle \tag{49}$$

which justify its name. By using eqn (47) in eqn (43), this equation reduces to

$$\begin{aligned}\frac{\mathrm{d}\boldsymbol{R}_\mu}{\mathrm{d}t} &= \frac{\boldsymbol{P}_\mu}{M_\mu}\\ \frac{\mathrm{d}\boldsymbol{P}_\mu}{\mathrm{d}t} &= \langle \boldsymbol{F}_\mu\rangle - \gamma_{\mu\nu}\frac{\boldsymbol{P}_\nu}{M_\nu} + k_BT\frac{\partial\gamma_{\mu\nu}}{\partial\boldsymbol{P}_\nu} + \tilde{\boldsymbol{F}}_\mu\end{aligned} \tag{50}$$

Note that we have $\sum_\nu \boldsymbol{F}_\nu = 0$, because of Newton's Third Law. From eqn (44) this implies that the friction coefficient defined in eqn (44) satisfies $\sum_\mu \gamma_{\mu\nu} = 0$ and, therefore,

$$\gamma_{\mu\mu} = -\sum_{\nu\neq\mu}\gamma_{\mu\nu} \tag{51}$$

While we expect that $\gamma_{\mu\mu}$ will be a positive quantity (because it is the time integral of an autocorrelation function), this equation shows that $\gamma_{\mu\nu}$ may be negative.

Using eqn (45) and (51), eqn (50) can be written as

$$\begin{aligned}\frac{\mathrm{d}\boldsymbol{R}_\mu}{\mathrm{d}t} &= \frac{\boldsymbol{P}_\mu}{M_\mu}\\ \frac{\mathrm{d}\boldsymbol{P}_\mu}{\mathrm{d}t} &= \langle \boldsymbol{F}_\mu\rangle + \sum_\nu\gamma_{\mu\nu}\left(\frac{\boldsymbol{P}_\mu}{M_\mu} - \frac{\boldsymbol{P}_\nu}{M_\nu}\right) + k_BT\sum_\nu\frac{\partial\gamma_{\mu\nu}}{\partial\boldsymbol{P}_\nu} + \tilde{\boldsymbol{F}}_\mu\end{aligned} \tag{52}$$

where we wrote the sums explicitly to avoid confusions (not all repeated indices are summed in eqn (52)). The stochastic force $\tilde{\boldsymbol{F}}_\mu$ can be expressed in terms of a linear combination of Wiener processes as, for example, $\boldsymbol{F}_\mu = \sum_\alpha B_{\mu\nu}\mathrm{d}\boldsymbol{W}_\nu(t)/\mathrm{d}t$ with

$$\sum_\alpha B_{\mu\alpha}B_{\nu\alpha} = 2k_BT\gamma_{\mu\nu} \tag{53}$$

This is the Fluctuation–Dissipation theorem for this problem. Note that eqn (52) has the structure of dissipative particle dynamics (DPD).[18,19] These equations have also been obtained recently in ref. 26 in a re-derivation of Mori–Zwanzig theory. However, an important difference with the usual DPD equations is that the effective force $\langle \boldsymbol{F}_{\mu\nu}\rangle$ and the friction tensor $\gamma_{\mu\nu}(\boldsymbol{R},\boldsymbol{P})$ depend, in principle, on the CoM variables of *all* the molecules in the system and not only on $\boldsymbol{R}_\mu - \boldsymbol{R}_\nu$ as in DPD.

VI. Simulation results for a melt of star polymer

As a concrete application, we now consider a system of star polymers that form a polymer melt. A single typical molecule in the melt is shown in Fig. 1.

The star polymers have $f = 12$ arms and $m = 6$ monomers per arm. Each arm is connected to a central monomer so that the total number of monomers per polymer is $N_m = f \times m + 1$. The melt contained a collection of M polymer molecules, with typically $M = 160$. Excluded volume interactions between monomers were taken into account by the purely repulsive (truncated and shifted) Lennard–Jones potential also known as the WCA potential. Bonded interactions were modelled by linear springs, *i.e.*, $\boldsymbol{F}_{ij}(\boldsymbol{r}_{ij}) = -k\delta r\, \boldsymbol{e}_{ij}$ where $\boldsymbol{r}_{ij} = \boldsymbol{r}_i - \boldsymbol{r}_j$, the unit vector $\boldsymbol{e}_{ij} = \boldsymbol{r}_{ij}/|\boldsymbol{r}_{ij}|$ is the unit vector along $\boldsymbol{r}_{ij}$, and $\delta r \equiv r_{ij} - r_{ij}^{\mathrm{eq}}$, where r_{ij}^{eq} is the equilibrium distance and k is the spring constant. The values used for these parameters are reported in the caption of the polymer model sketch in Fig. 1. The polymer volume fraction is $\Phi = N(\pi/6)\sigma^3/L^3$, where L is the length of the (cubic) simulation box and $N = M \times N_m$ is the total number of monomers in the system. The results presented here correspond to a semidilute

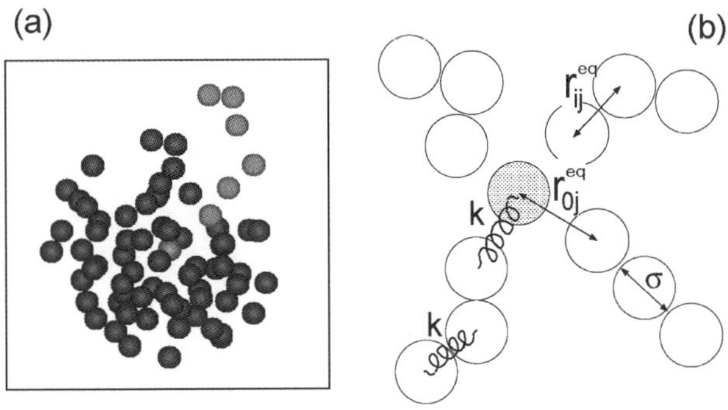

Fig. 1 (a) One of the star polymers in the polymer melt in a typical configuration. Seven monomers are coloured lighter indicating one arm and the central monomer. It has $f = 12$ arms of $m = 6$ monomers each. (b) A sketch of the star polymer model: all monomers interact with the purely repulsive Lennard–Jones potential (truncated at $r_c = 2^{1/6}\sigma$ and shifted to zero potential energy, $\phi(r_c) = 0$) with units such that $\varepsilon = 1$, $m = 1$ and $\sigma = 2.415$. Neighbour monomers are attached by springs of stiffness $k = 20\varepsilon/\sigma^2$ and move around the equilibrium distance r^{eq}_{ij}, with $r^{eq}_{ij} = 1.147\sigma$ if i and j are non-center monomers and $r^{eq}_{0j} = 1.615\sigma$, if i (= 0) is the center monomer.

state $\Phi = 0.1962$ and the simulations were done in periodic cubic boxes of size $L = 76\sigma$ yielding a monomer number density of $n = 0.314\sigma^{-3}$. Throughout the paper, results are given in Lennard–Jones units: energy, mass, length and time units being $\varepsilon = 1$, $m = 1$, $\sigma = 2.415$, $\tau = \sigma(m/\varepsilon)^{1/2}$, respectively. All simulations were carried out in the microcanonical ensemble NVE (care was taken to set the same total energy in every simulation); and the average temperature was $k_BT = 3.965\varepsilon$. An equilibrated configuration was initially taken from a Monte Carlo bond fluctuation model simulation.[27] In these configurations the monomer positions are set in a lattice. These initial configurations were equilibrated using Langevin thermostatted molecular dynamics at the desired temperature. In order to produce a set of independent initial configurations we ran a long enough MD simulation and saved one configuration per each diffusion time, typically about $(R_g^2/D)\tau$, where $R_g = 7.64\sigma$ is the gyration radius and $D \simeq 0.08\sigma^2/\tau$ the diffusion coefficient. In doing so we collected a set of 35 configurations, each one having an independent set of positions and momenta $(\boldsymbol{R},\boldsymbol{P})$ of the CoM of the molecules. Starting from each of these 35 configurations we ran a short simulation (1000τ) to collect the required CoM molecular dynamics raw data. These set of short runs were carried out using the constrained dynamics discussed in Sec. IV and specified below.

First, we estimated whether we have a separation of time scales by measuring the velocity autocorrelation function of the CoM and comparing the result with the force autocorrelation function on the molecules. The former gives the time scale of the CoM velocity, which is a relevant variable. The latter provides an estimate of the typical time scale at which the memory function in the definition of the friction coefficient decays. Both autocorrelation functions are plotted in Fig. 2. There is a large difference between correlation times, which leads us to believe that the CoM are indeed slow variables and that the modification of the dynamics proposed in Sec. IV should have a small impact on their dynamics. The reason is that the molecules are very massive (they contain 73 atoms). While the time scale of the velocity scales with the mass, the time scale of the force does not, because it is given by an atomic collision time scale. Note, however, that the force in Fig. 2, being the time derivative of the momentum (*i.e.* of the mass times the velocity), has also a negative part that decays slowly, in the same time scale as the velocity. This negative part of the force could invalidate the procedure proposed in Sec. IV. However, this part is responsible for the "plateau problem" and disappears with a correct treatment with

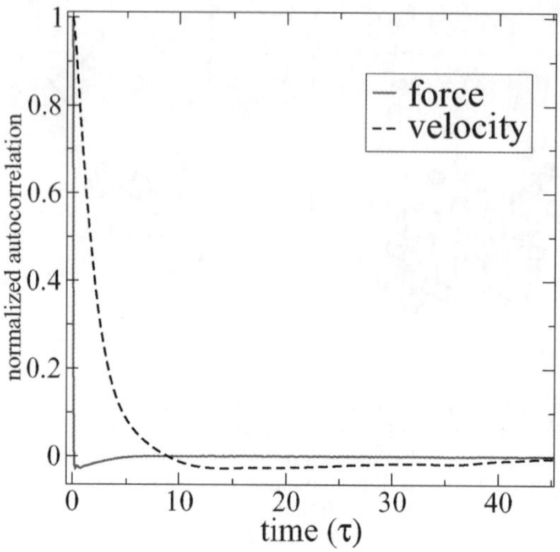

Fig. 2 The velocity autocorrelation function of the CoM and the autocorrelation function of the force on a molecule. These functions have correlation times well-separated and a Markovian behaviour is expected.

the constrained dynamics. We will discuss the plateau problem by the end of this section.

The constrained dynamics $\mathscr{R}$ that we use in order to compute the mean force $\langle \boldsymbol{F}_{\mu\nu} \rangle$ and the friction coefficient $\gamma_{\mu\nu}(\boldsymbol{R},\boldsymbol{P})$ is the following

$$\begin{aligned} \frac{\mathrm{d}\boldsymbol{r}_{i_\mu}}{\mathrm{d}t} &= \frac{\boldsymbol{p}_{i_\mu}}{m_{i_\mu}} - \frac{\boldsymbol{P}_\mu}{M_\mu}, \\ \frac{\mathrm{d}\boldsymbol{p}_{i_\mu}}{\mathrm{d}t} &= \boldsymbol{F}_{i_\mu} - \frac{m_{i_\mu}}{M_\mu}\boldsymbol{F}_\mu \end{aligned} \tag{54}$$

where $\boldsymbol{P}_\mu$ is the CoM momentum of molecule μ, $\boldsymbol{F}_{i_\mu}$ is the total force on monomer i_μ and $\boldsymbol{F}_\mu$ is the total force on molecule μ. Because the constraints to maintain the positions and momenta of the CoM are linear functions, the corresponding Lagrange multipliers can be easily identified and this leads to eqn (54). These equations conserve the total energy, they leave the positions and momentum of the CoM invariant, and they also conserve the volume in phase space. For these reasons, this dynamic samples the constrained ensemble that appears in the definition of $\langle \boldsymbol{F}_{\mu\nu} \rangle$ and $\gamma_{\mu\nu}(\boldsymbol{R},\boldsymbol{P})$, and these averages may be computed as time averages.

Consistent with eqn (38), by running eqn (54), one can compute $\langle \boldsymbol{F}_{\mu\nu} \rangle$ as the time average of the force $\boldsymbol{F}_{\mu\nu}$ that molecule ν exerts on molecule μ. In principle $\langle \boldsymbol{F}_{\mu\nu} \rangle$ depends on all the CoM positions $\boldsymbol{R}$. If it happens, as we expect, that the force that molecule ν exerts on molecule μ depends only on the CoM positions $\boldsymbol{R}_\mu$ and $\boldsymbol{R}_\nu$ of these two molecules and does not depend much on where the rest of molecules are located, then a pair-wise approximation should be valid. By translational and rotational symmetry we expect that the average force will be of the form

$$\langle \boldsymbol{F}_{\mu\nu} \rangle \approx F(R_{\mu\nu})\boldsymbol{e}_{\mu\nu} \tag{55}$$

where $F(R_{\mu\nu}) = \langle \boldsymbol{F}_{\mu\nu} \cdot \boldsymbol{e}_{\mu\nu} \rangle$, $\boldsymbol{e}_{\mu\nu} = (\boldsymbol{R}_\mu - \boldsymbol{R}_\nu)/R_{\mu\nu}$ and $R_{\mu\nu} = |\boldsymbol{R}_\mu - \boldsymbol{R}_\nu|$. Consistent with this assumption, we computed the modulus of the average force $F(R_{\mu\nu})$ by averaging the result of $\langle \boldsymbol{F}_{\mu\nu} \cdot \boldsymbol{e}_{\mu\nu} \rangle$ over all those pairs μ, ν that are at a certain distance $R_{\mu\nu}$. The result is plotted in Fig. 3. Note that it is highly improbable to find a pair at very short

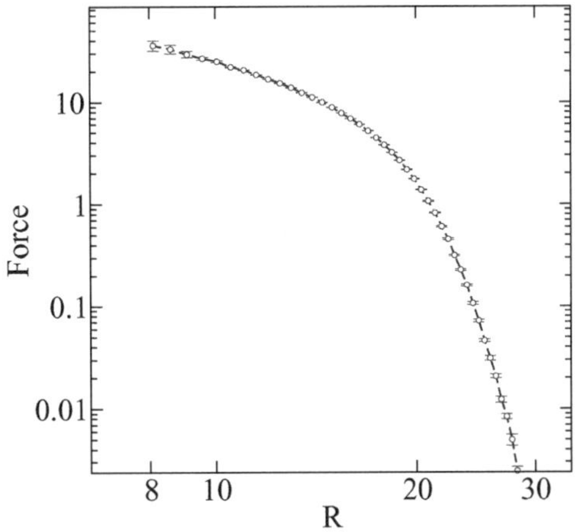

Fig. 3 The conservative force used by the present pair-wise approximation $\langle \boldsymbol{F}_{\mu\nu}\cdot \boldsymbol{e}_{\mu\nu}\rangle$ *versus* the distance $R_{\mu\nu}$ between center of masses of two interacting molecules. Error bars correspond to the standard error of the means of the whole set of pairs.

distances. In fact for $R_{\mu\nu} < 1.1R_g$ (the gyration radius being $R_g = 7.64\sigma$)) there is no data available.

Next we checked that the mean transversal force, $\langle \boldsymbol{F}_{\mu\nu}\cdot \boldsymbol{e}^{\perp}_{\mu\nu}\rangle$ (with $\boldsymbol{e}^{\perp}_{\mu\nu}\cdot \boldsymbol{e}_{\mu\nu} = 0$), averaged over pairs, vanishes within statistical error. We also verified that the average force $\langle \boldsymbol{F}_{\mu\nu}\cdot \boldsymbol{e}_{\mu\nu}\rangle$ for a pair at a distance $R_{\mu\nu}$ is the same, within our statistical accuracy, for any pair of molecules at the same distance. To check this, we first measured the standard error of the mean force $\langle \boldsymbol{F}\cdot \boldsymbol{e}_{\mu\nu}\rangle$ for a given pair μ, ν that is at a distance $R_{\mu\nu}$, and averaged this error over all the pairs at that distance. If all pairs behave similarly in statistical terms, the deviation *between* pairs estimations of the mean force $\langle \boldsymbol{F}_{\mu\nu}\cdot \boldsymbol{e}_{\mu\nu}\rangle$ of pairs at a distance $R_{\mu\nu}$ should be within the single-pair error bar. We observed that this is indeed the case for all values of R, although the standard deviation *between* pairs was found to be slightly larger than the single-pair standard error. A previous computational study has found that the triplet part of the three-star force is only 11% of the pair-wise part even for a close approach of three star polymers.[28] A detailed *significance test* assessing, in probabilistic terms, the validity of the pair-wise hypothesis will be left for future work.

Let us turn now to the friction coefficients $\gamma_{\mu\nu}(\boldsymbol{R},\boldsymbol{P})$. Again, one has to deal with the problem of their many arguments. Our main hypothesis is that the correlation between the forces on molecule μ and ν will depend on the positions of these two molecules but will not depend much on the positions and momenta of the rest of the molecules. We thus introduce the following functional ansatz

$$\gamma_{\mu\nu}(\boldsymbol{R},\boldsymbol{P}) \approx -\gamma_{\perp}(R_{\mu\nu})(\boldsymbol{1} - \boldsymbol{e}_{\mu\nu}\boldsymbol{e}^{T}_{\mu\nu}) - \gamma_{\|}(R_{\mu\nu})\boldsymbol{e}_{\mu\nu}\boldsymbol{e}^{T}_{\mu\nu}. \tag{56}$$

The right-hand side of this equation only depends on $\boldsymbol{R}_{\mu}$ and $\boldsymbol{R}_{\nu}$ and it is a general form for a tensor that is invariant by rotations along the axis joining the particles μ, ν. Compatibility of eqn (56) with eqn (44) then requires that

$$\begin{aligned} \gamma_{\|}(R_{\mu\nu}) &= -\frac{1}{k_B T}\int_0^{\infty} dt \left\langle (\delta \boldsymbol{F}_{\mu}(t)\cdot \boldsymbol{e}_{\mu\nu})(\delta \boldsymbol{F}_{\nu}(0)\cdot \boldsymbol{e}_{\mu\nu}) \right\rangle \\ \gamma_{\perp}(R_{\mu\nu}) &= -\frac{1}{k_B T}\int_0^{\infty} dt \left\langle (\delta \boldsymbol{F}_{\mu}(t)\cdot \boldsymbol{e}^{\perp}_{\mu\nu})(\delta \boldsymbol{F}_{\nu}(0)\cdot \boldsymbol{e}^{\perp}_{\mu\nu}) \right\rangle \end{aligned} \tag{57}$$

Under the assumption that the right-hand side of eqn (57) is the same for all the pairs that are at the same distance $R_{\mu\nu}$ as indicated by the lhs, we may average over all the pairs $\mu\nu$ that happen to be at the distance $R_{\mu\nu}$, this is

$$\gamma_{\parallel}(R_{\mu\nu}) = -\frac{1}{N_{\mu\nu}}\sum_{\mu\nu}{}' \frac{1}{k_B T}\int_0^{\infty} \mathrm{d}t\langle(\delta \boldsymbol{F}_{\mu}(t)\cdot \boldsymbol{e}_{\mu\nu})(\delta \boldsymbol{F}_{\nu}(0)\cdot \boldsymbol{e}_{\mu\nu})\rangle \quad (58)$$

where $\sum'_{\mu\nu}$ is a sum over all those pairs that are at a given distance $R_{\mu\nu}$ and $N_{\mu\nu}$ is the number of pairs at that distance. A similar procedure has been used to extract $\gamma_{\perp}(R_{\mu\nu})$. Both $\gamma_{\parallel}(R_{\mu\nu})$ and $\gamma_{\perp}(R_{\mu\nu})$ are plotted in Fig. 4. Note that the transversal friction $\gamma_{\perp}$ becomes slightly negative at $R_{\mu\nu} \simeq 2R_g = 15.28\sigma$, before tending to a vanishing value around $R_{\mu\nu} \simeq 25\sigma$. Most probably, this is a consequence of a dynamic aspect of depletion forces which tend to join back two neighbouring molecules as soon as they move further away the mean molecular separation due to caging effects. Note that the fact that $\gamma_{\perp}$ may have a small negative part does not compromise that the full friction matrix $\gamma_{\mu\nu}$, which is a $3M \times 3M$ matrix, is positive definite (a positive definite matrix may have indeed some of its elements taking negative values).

Some insight on the validity of this pair-wise assumption is gained from the evaluation of the standard deviation of the force correlation appearing in the integrand of eqn (57), at each lagging time and CoM distance considered. Again, we compared the standard deviation of the mean *between* pairs with the average standard error obtained from one-pair estimations. Our hypothesis requires these two quantities to be similar, otherwise the mean behaviour of different pairs would be different in a significant (measurable) way. Preliminary results indicate that the pair-wise assumption works reasonably well except for short distances and times, $R < 2R_g$ and $t < 0.15\tau$. Note that intermolecular forces decorrelate only after $t > 0.1\tau$. This suggests a possible violation of the pair-wise approximation, mainly within the region of the space–time diagram where the strongest collision events take place. In particular, the star polymers considered here ($f = 12$, $m = 6$) can interpenetrate each other (*i.e.* $R_{\mu\nu} < 2R_g$) and under the present thermodynamic state $\Phi = 0.2$,

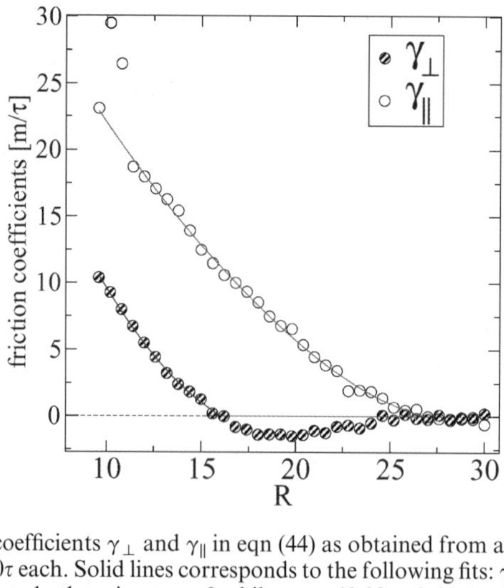

Fig. 4 The friction coefficients $\gamma_{\perp}$ and $\gamma_{\parallel}$ in eqn (44) as obtained from averaging over 35 MD runs over $T_{\mathrm{run}} = 1000\tau$ each. Solid lines corresponds to the following fits: $\gamma_{\perp} = 44.67 - 4.68\,R + 0.12\,R^2$ for $R < 16.05$ and otherwise $\gamma_{\perp} = 0$ while, $\gamma_{\parallel} = 49.38 - 3.61\,R + 0.11\,R^2 - 3 \times 10^{-3}\,R^3 + 5 \times 10^{-5}\,R^4$ for $R < 26.84$ and zero otherwise.

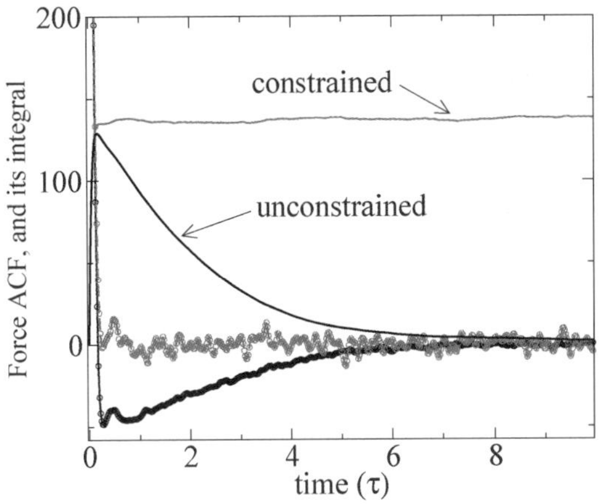

Fig. 5 The force autocorrelation $\langle \delta \boldsymbol{F}_\mu \cdot \delta \boldsymbol{F}_\mu(t) \rangle$ compared with the unconstrained $\langle \delta \boldsymbol{F}_\mu \cdot \delta \boldsymbol{F}_\mu(t) \rangle_{\mathrm{eq}}$. Also shown are their time integrals. The unconstrained dynamics does not have a well-defined plateau, whereas the constrained dynamics allows for a good definition of the friction coefficient.

may reach relatively far into each core (up to $R_{\mu\nu} \sim 1.1 R_g$). This issue shall be studied in detail in a future work, which shall include proper significance tests for the pair-wise hypothesis.

Finally, let us discuss the plateau problem. In Fig. 5 we compare the time integral of the autocorrelation of the force computed with the constrained dynamics and computed with the usual Hamiltonian unconstrained dynamics. We observe that the constrained dynamics produces a correlation function with a time integral that has a well-defined plateau value, while the unconstrained dynamics does not. The constrained and unconstrained dynamics produce identical results for short times. However, the justification of a well-defined value for the friction matrix, with a time integral independent of the upper limit of integration can only be achieved with the constrained simulation.

Comparison of DPD with MD simulations

As a consistency check, we now compare the simulation of the stochastic DPD eqn (52), with the effective forces and frictions computed microscopically, with the simulation of the original MD equations. The first issue to resolve is the fact that the effective force does not have data for short distances. We have smoothly extrapolated this force with a bell shaped function for $R_{\mu\nu} < 8\sigma$ and have checked in the simulations that the results are insensitive to the height of this function at the origin. This is because the DPD particles hardly visit the short distances that are hardly visited by the CoM in the molecular dynamics simulation. A similar situation occurs for the friction coefficients, for which there are no data at short distances. The specific extrapolated values for the friction coefficients do not affect the results.

The stochastic forces that we use in the DPD simulations are detailed in the appendix. Note that in order to obtain the amplitude $B_{\mu\alpha}(\boldsymbol{R})$ of the stochastic forces, we need to extract the square root, in matrix sense, of the $3M \times 3M$ friction matrix, according to the Fluctuation-Dissipation theorem in eqn (53). This is a computationally very expensive task in general. However, for pair-wise forces, it is possible to find a simple solution for the stochastic forces which is a generalization of the usual stochastic forces in DPD. However, this simple solution requires that the coefficients $\gamma_\perp$, $\gamma_\parallel$ be positive. For this reason, we have approximated in the DPD model the coefficient $\gamma_{\perp}(R_{\mu\nu})$ by a strictly positive function obtained by setting to zero the values of $\gamma_{\perp}(R_{\mu\nu})$ that become negative.

The observables selected to compare MD with DPD are the radial distribution function and velocity autocorrelation function for the velocity of the CoM. The former provides information on the static properties of the system while the latter gives insight about its dynamical properties. In Fig. 6 the comparison of the radial distribution function computed from a DPD simulation of eqn (52) with the radial distribution function of the CoM computed from the MD simulation shows a reasonable agreement. For static quantities like the radial distribution function of the CoM in Fig. 6, both a DPD simulation and a coarse-grained molecular dynamics (CGMD) produce identical results. The CGMD is defined as the result of setting to zero the friction (and, consistently, the stochastic force) in eqn (52).

Fig. 7 shows the velocity autocorrelation of the CoM computed with MD, with DPD, and with CGMD. The agreement between the MD and DPD results is very

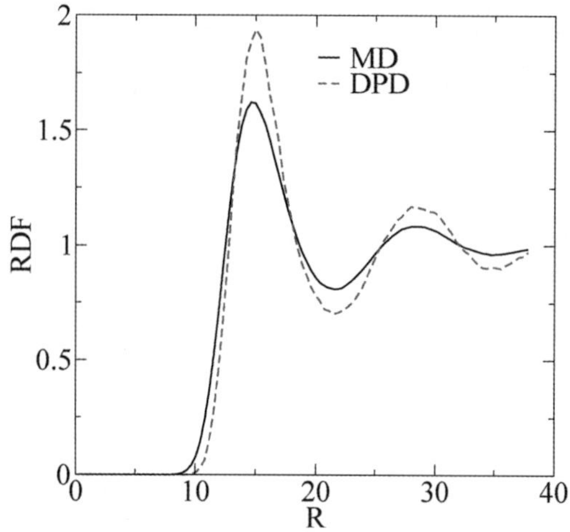

Fig. 6 The comparison of the Markovian prediction (DPD) with the MD simulation results for the radial distribution function of the CoM.

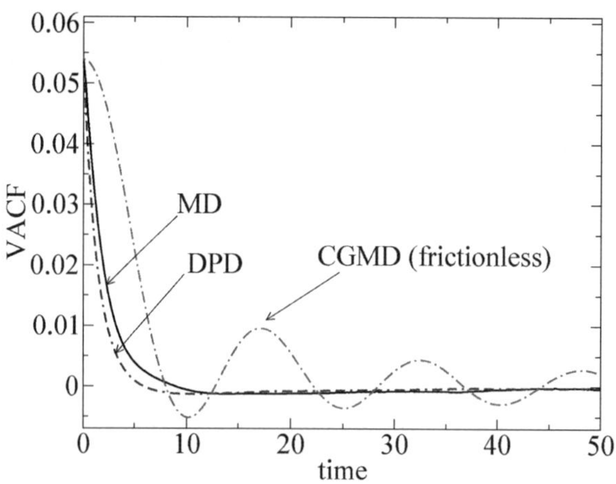

Fig. 7 The velocity autocorrelation function of the CoM computed with MD (solid line), with DPD (dashed line) and with CGMD (dotted-dashed line).

good. As it is apparent from Fig. 7, the CGMD simulation based only on effective potentials provides very nonphysical results. It is necessary to introduce also the dissipation inherent to any coarse-graining procedure in order to obtain realistic results from a coarse-grained simulation. This is one important message to retain from the present paper (see also ref. 29).

The agreement between DPD and MD simulations, although satisfactory, can be certainly improved. The discrepancies between both types of simulations should be attributed exclusively to the pair-wise approximation which is the only approximation that we have taken, apart from the Markovian approximation. The latter, as we discussed, is rather well satisfied. Of course, having the exact expressions eqn (44) and eqn (49) and a method to compute these objects from constrained simulations, opens the door to the investigation of improvements of the approximations eqn (55) and eqn (56) that we have taken in the present work.

VII. Discussion

The main objective of Mori–Zwanzig theory is to obtain the dynamic equations for a set of reduced variables. In principle, the equation for the evolution of these variables is the GLE, which is a complicated integro-differential equation with random coefficients. The GLE is not fully explicit and hence very difficult to use in practice. However, if the relevant variables are slow, their dynamics should be approximately Markovian and, hence, one can expect that it can be captured accurately by an SDE. How to derive this SDE turns out to be non-obvious, however. In particular, the standard Markovian approximation used in the literature suffers from the plateau problem because it used the original dynamics instead of the (unknown) projected dynamics to compute the coefficients in the SDE.

In this paper we gave an operational procedure which permits to alleviate these difficulties. The procedure is based on a modification of the original dynamics which makes precise what one means by the statement that the relevant variables are slow: in fact, the procedure amounts to making all the remaining variables evolve infinitely fast compared to the relevant ones. In this limit, one can define unambiguously what the projected dynamics is, justify the Markovian approximation, and give precise estimates for the coefficients in the SDE.

Being able to derive this SDE is important for two basic reasons. First it provides *understanding* of the dynamics of the system. Second it *facilitates the simulation* of the system. The fact that complex molecules may be represented with equations with the structure of dissipative particle dynamics (DPD) is a very valuable piece of knowledge. At the same time, it is by far less expensive to simulate the system at the level of the CoM through a DPD simulation than to simulate the whole system with molecular dynamics. Both the number of computing units (*i.e.* particles) and the size of the required time steps in DPD makes the coarse simulation the method of choice if only macroscopic time scales are of interest.

One aspect of our approach should be emphasized: the coefficients appearing in the SDE (effective forces and frictions) depend, in general, on the whole set of relevant variables in a nontrivial way. From a theoretical viewpoint this feature is a plus which shows that our approach does not rely on extra assumptions on how the coefficients depend on the relevant variables. From a computational viewpoint, however, this may be an issue. Indeed, to pre-compute these coefficients we need to sample the multidimensional space of the relevant variables, which may not be feasible in practice. For instance, in the particular example of star polymers studied in Sec. VI, for every configuration of the M molecules we should have a different set of effective forces and frictions. The way we got around the multidimensional sampling of the coefficients was through the constitutive assumption of pair-wise additivity which can be made *a posteriori* to simplify the coefficients in the SDE obtained by our procedure. Any such constitutive assumption should be checked through significance tests and before performing any comparison between microscopic and coarse-grained simulations, and in the example in Sec. VI it indeed turned out to be valid and reasonably accurate. But this clearly means that the pre-computing strategy will be limited to situations where it is possible to make sensible assumptions on the functional form of the coefficients of the SDE.

One way to go beyond the pre-computing strategy and not use any extra constitutive assumptions is to compute on-the-fly the coefficients in the SDE for the relevant variables. This fits within the framework of the heterogeneous multiscale method[30–32] and it amounts to do the following. Given the current state $\alpha(t)$ of the relevant variables, a short run of constrained simulations is performed to estimate the value of the coefficients in the SDE at value $\alpha(t)$; these estimates are then used to propagate the relevant variables by one macro-time-step to a new value $\alpha(t + \Delta t)$, and the procedure is repeated. This strategy will be doable even if the dimension of the relevant variable space is large (*i.e.* even if pre-computing the coefficients may not be an option) and still provide a serious computational gain over the vanilla MD simulation. We intend to explore this avenue in future publications.

Appendix A: DPD stochastic forces

In this appendix we show how to formulate the stochastic forces for the DPD model. Instead of performing the square root, in matrix sense, of the $3M \times 3M$ friction matrix $\boldsymbol{\gamma}_{\mu\nu}$, we rather postulate the form of the stochastic forces and demonstrate that their variances have the structure of the friction matrix in eqn (56). The stochastic force is postulated to be of the form $\tilde{\boldsymbol{F}}_\mu = \sum_\nu \tilde{\boldsymbol{F}}_{\mu\nu}$, where the pair-wise stochastic force is

$$\tilde{\boldsymbol{F}}_{\mu\nu}\mathrm{d}t = (2k_BT)^{1/2}\left(\tilde{A}(r_{\mu\nu})\mathrm{d}\overline{W}^S_{\mu\nu} + \frac{1}{3}\tilde{B}(r_{\mu\nu})tr\left[\mathrm{d}W^S_{\mu\nu}\right]\boldsymbol{1}\right)\bullet\boldsymbol{e}_{\mu\nu} \tag{A1}$$

where

$$\mathrm{d}\overline{W}^{S_{\alpha\beta}}_{\mu\nu} = \frac{1}{2}\left[\mathrm{d}W^{\alpha\beta}_{\mu\nu} + \mathrm{d}W^{\beta\alpha}_{\mu\nu}\right] - \frac{1}{3}tr[\mathrm{d}W_{\mu\nu}]\delta_{\alpha\beta} \tag{A2}$$

Here, $\mathrm{d}W^{\alpha\beta}{}_{\mu\nu}$ is a matrix of independent increments of the Wiener process with the symmetries

$$\mathrm{d}W^{\alpha\beta}{}_{\mu\nu} = \mathrm{d}W^{\alpha\beta}{}_{\nu\mu}. \tag{A3}$$

Note that $\tilde{\boldsymbol{F}}_{\mu\nu} = -\tilde{\boldsymbol{F}}_{\nu\mu}$ and momentum will be conserved by the stochastic forces. The matrix $\mathrm{d}W^{\alpha\beta}{}_{\mu\nu}$ satisfies the Ito rule

$$\mathrm{d}W^{\alpha\alpha'}{}_{\mu\mu'}\mathrm{d}W^{\beta\beta'}{}_{\nu\nu'} = (\delta_{\mu\nu}\delta_{\mu'\nu'} + \delta_{\mu\nu'}\delta_{\nu\mu'})\delta_{\alpha\beta}\delta_{\alpha'\beta'}\,\mathrm{d}t. \tag{A4}$$

From this equation one can easily obtain the following identities

$$\begin{aligned}
tr[\mathrm{d}W_{\mu\mu'}]tr[\mathrm{d}W_{\nu\nu'}] &= 3\,(\delta_{\mu\nu}\delta_{\mu'\nu'} + \delta_{\mu\nu'}\delta_{\nu\mu'})\,\mathrm{d}t,\\
\mathrm{d}\overline{W}^{S_{\alpha\alpha'}}_{\mu\mu'}\mathrm{d}\overline{W}^{S_{\beta\beta'}}_{\nu\nu'} &= (\delta_{\mu\nu}\delta_{\mu'\nu'} + \delta_{\mu\nu'}\delta_{\nu\mu'})\\
&\times\left[\frac{1}{2}(\delta_{\alpha\beta}\delta_{\alpha'\beta'} + \delta_{\alpha\beta'}\delta_{\alpha'\beta})\right.\\
&-\left.\frac{1}{3}\delta_{\alpha\alpha'}\delta_{\beta\beta'}\right]\mathrm{d}t,\\
tr[\mathrm{d}W_{\mu\mu'}]\mathrm{d}\overline{W}^S_{\nu\nu'} &= 0
\end{aligned} \tag{A5}$$

These identities allow to show that the variance of the stochastic forces is given by

$$\tilde{\boldsymbol{F}}_\mu \mathrm{d}t \tilde{\boldsymbol{F}}_\nu \mathrm{d}t = 2k_B T(A(r_{\mu\nu})\boldsymbol{1} + B(r_{\mu\nu})\mathbf{e}_{\mu\nu}\mathbf{e}_{\mu\nu})\mathrm{d}t \tag{A6}$$

provided that we choose

$$\begin{aligned} \tilde{A}(r_{\mu\nu}) &= (2A(r_{\mu\nu}))^{1/2} \\ \tilde{B}(r_{\mu\nu}) &= (3B(r_{\mu\nu}) - A(r_{\mu\nu}))^{1/2} \end{aligned} \tag{A7}$$

We see, therefore, that the proposed structure for the stochastic force has a variance with the structure of the friction matrix in the pair-wise approximation, eqn (56).

Acknowledgements

P. E. wants to acknowledge the hospitality and support of the Freiburg Institute for Advanced Studies where part of this work has been done, and BIFI. Financial support from MICINN under projects FIS2007-65869-C03-03 and FIS2007-65869-C03-01 is acknowledged. We appreciate the help of J. J. Freire in providing the initial configurations for our star polymers. Useful discussions with G. Ciccotti are very much appreciated. C. H. wants to acknowledge support from UNED through a scholarship and the Project of Programa Propio de Investigación. R. D.-B benefits from a Ramón y Cajal research contract and also acknowledges funding from MOSNOHO 0505/ESP-0299. E. V.-E. acknowledges funding from NSF grants DMS02-39625 and DMS07-08140, and by ONR grant N00014-04-1-0565.

References

1 M. S. Green, *J. Chem. Phys.*, 1952, **20**, 1281.
2 R. Zwanzig, *Phys. Rev.*, 1961, **124**, 983.
3 H. Grabert, *Projection Operator Techniques in Nonequilibrium Statistical Mechanics*, Springer Verlag, Berlin, 1982.
4 J. G. Kirkwood, *J. Chem. Phys.*, 1946, **14**, 180.
5 P. Español and I. Zúñiga, *J. Chem. Phys.*, 1993, **98**, 574.
6 R. Kubo, M. Toda, N. Hashitsume, and N. Saito, *Statistical Physics II: Nonequilibrium Statistical Mechanics*, Springer, Berlin, 1991.
7 J. Majda, I. Timofeyev and E. Vanden-Eijnden, *Nonlinearity*, 2006, **19**, 769.
8 V. Tozzini, *Curr. Opin. Struct. Biol.*, 2005, **15**, 144.
9 H. Meyer, O. Biermann, R. Faller, D. Reith and F. Müller-Plathe, *J. Chem. Phys.*, 2000, **113**, 6264.
10 J. C. Shelley, M. Y. Shelley, R. C. Reeder, S. Bandyopadhyay and M. L. Klein, *J. Phys. Chem. B*, 2001, **105**, 4464.
11 T. Head-Gordon and F. H. Stillinger, *J. Chem. Phys.*, 1993, **98**, 3313.
12 T. Murtola, E. Falck, M. Patra, M. Karttunen and I. Vattulainen, *J. Chem. Phys.*, 2004, **121**, 9156.
13 H. S. Ashbaugh, H. A. Patel, S. K. Kumar and S. Garde, *J. Chem. Phys.*, 2005, **122**, 104908.
14 S. J. Marrink, A. H. de Vries and A. E. Mark, *J. Phys. Chem. B*, 2004, **108**, 750.
15 F. Ercolessi and J. B. Adams, *Europhys. Lett.*, 1994, **26**, 583.
16 T. D. Hone, S. Izvekov and G. A. Voth, *J. Chem. Phys.*, 2005, **122**, 54105.
17 S. Izvekov and G. A. Voth, *J. Chem. Phys.*, 2006, **125**, 151101.
18 P. J. Hoogerbrugge and J. M. V. A. Koelman, *Europhys. Lett.*, 1992, **19**, 155.
19 P. Español and P. B. Warren, *Europhys. Lett.*, 1995, **30**, 191.
20 P. Español and H. C. Öttinger, *Z. Phys. B: Condens. Matter*, 1993, **90**, 377.
21 J. Chorin, O. H. Hald and R. Kupferman, *Proc. Natl. Acad. Sci. U. S. A.*, 2000, **97**, 2968.
22 H. C. Öttinger, *Beyond Equilibrium Thermodynamics*, J. Wiley & Sons, 2005.
23 G. C. Papanicolaou, *Rocky Mountain J. Math.*, 1976, **6**, 653; *Summer Research Conference on Singular Perturbations: Theory and Applications*, Northern Arizona Univ., Flagstaff, Ariz., 1975.
24 G. C. Papanicolaou, in *Modern modeling of continuum phenomena*, Ninth Summer Sem., *Appl. Math.* Rensselaer Polytech. Inst., Troy, NY, USA, 1975, *Amer. Math. Soc.*, Providence, RI, USA, 1977, pp. 109–147. Lectures in *Appl. Math.*, Vol. 16.
25 L. Maragliano and E. Vanden-Eijnden, *Chem. Phys. Lett.*, 2006, **426**, 168.
26 T. Kinjo and S. Hyodo, *Phys. Rev. E: Stat., Nonlinear, Soft Matter Phys.*, 2007, **75**, 051109.

27 A. Di Cecca and J. Freire, *Macromolecules*, 2002, **35**, 2851.
28 von Ferber, A. Jusufi, C. N. Likos, H. Löwen and M. Watzlawek, *Eur. Phys. J. E*, 2000, **2**, 311.
29 H. C. Öttinger, *MRS Bull.*, 2007, **32**, 936.
30 E. Vanden-Eijnden, *Commun. Math. Sci.*, 2003, **1**, 385.
31 W. E and B. Engquist, *Commun. Math. Sci.*, 2003, **1**, 87.
32 W. E, B. Engquist, X. Li, W. Ren and E. Vanden-Eijnden, *Commun. Comput. Phys.*, 2007, **2**, 367.

General discussion

Dr Padding opened the discussion of the paper by Professor Löwen: We have also studied these sedimentation pattern formations in a slit using Stochastic Rotation Dynamics,[1] but here you have analysed the phenomenon in much more detail. We found that the characteristic wavelength of the instability is always more or less equal to the height of the slit (at least never as small as half or as large as double this length scale). Did you observe this too? If so, can you give a physical argument why the slit height is so dominant in setting the characteristic length scale?

1 J. T. Padding and A. A. Louis, *Phys. Rev. E*, 2008, **77**, 011402.

Professor Löwen answered: In our studies the characteristic wavelength λ_m of the instability is different from the height L of the slit. First of all, it was shown in our work[1] that λ_m depends on the Peclet number for fixed L. Furthermore for other actual parameter combinations ($L/\sigma = 9$ and $Pe = 1.6$), the characteristic wavelength λ_m was considerably smaller than $L/2$. Finally, in the limit L to infinity, at appropriate conditions we think that λ_m stays finite. Therefore $L/2$ is not a good estimate for the characteristic wavelength λ_m in the full parameter space explored. In fact, the theoretical expression of λ_m does not simple scale in L.

1 A. Wysocki, C. P. Royall, R. G. Winkler, G. Gompper, H. Tanaka, A. van Blaaderen and H. Löwen *Soft Matter*, 2009, **5**, 1340.

Professor Bolhuis opened the discussion of the paper by Professor Dijkstra: You found that there is probably no maximum in the free energy of nucleation as a function of cluster size. However, because you only extend the simulation to a certain cluster size, there is always the possibility that the free energy maximum, the critical nucleus, is located in an even larger cluster size. Can you exclude this possibility?

Professor Dijkstra replied: According to classical nucelation theory, the Gibbs free energy to form a cluster of a certain size is given by a bulk term, corresponding to a gain in free energy as the new phase is more stable, and a surface term as it costs free energy to create an interface. For small cluster sizes, the surface term wins and it will cost Gibbs free energy to form a cluster, while for larger cluster sizes, the bulk term wins and the cluster will grow spontaneously. Our results show that the free energy of formation grows monotically with cluster size. Consequently, the free energy cost to grow a certain cluster is independent of the size of the cluster. Apparently the free energy is more related to the probability to reorient a small cluster of rods which can then be attached to the biggest cluster than to a surface and bulk contribution. This finding is corroborated by the fact that we do not find any correlation between the shape of the cluster and the supersaturation and cluster size.

Dr van der Sman opened the discussion of the paper by Professor Frenkel: The authors claim a better prediction of the dynamics compared to the DFT model. Yet, they have taken a constant diffusion coefficient D^+. However, I would expect it to depend on charge density—and the effective friction coefficient as discussed by Professor Holm in his paper.

Professor Frenkel replied: The model discussed in the paper corresponds to the simplest case, where the diffusion constant of the two types of ions is the same constant value. It is possible to introduce a dependence in the solvent and also a dependence on local salt concentration, as we have done for example with the local dielectric constant. The effective friction discussed in Professor Holm's paper refers to the polyelectrolyte. Such an effective friction will also emerge from our model, even at constant D^+, if we introduced a resolved polymer and analyze its motion under the action of an electric field.

Professor van der Vegt addressed Professor Frenkel and Professor Pagonabarraga: Adsorption of ions at interfaces is ion specific with a dependence on ion size. Can you comment on extending your method to include the effect of the finite size of ions?

Professor Frenkel answered: Ionic specificity can be taken into account in a free energy model. The solvation potential reflects the ion-solvent interactions and induces the partitioning of ions between the two phases. Although we considered only a single type of cation and a single type of anion, one could also use different solvation potentials for two types of cations (resp. two types of anions), and thus account for ionic specificity. Another issue is the affinity of ions for the interface, which depends (among other things) on their size. This surface affinity can also be used with several types of ions, thus reflecting the ion-specific adsorption. However, we think that attempts to resolve the ionic profile over molecular distances (ionic size) would be incompatible with the coarse-grained approach that we follow.

Dr van der Sman addressed Professor Frenkel and Professor Pagonabarraga: How easily can your model be extended by ionic surfactants? In your paper you cite my work on the Lattice Boltzmann model of surfactant adsorption on emulsion droplets[1]—which is based on the sharp interface model of Diamant and Andelman. In their paper,[2] they have extended their model to ionic surfactants. Do you foresee the need for special extensions of your model, for example for description of the bulk phase?

1 R. G. M. van der Sman and S. van der Graaf, *Rheol. Acta*, 2006, **46**, 3–11.
2 H. Diamant and D. Andelman, *J. Phys. Chem.*, 1996, **100**, 13732.

Professor Frenkel replied: Accounting for ionic surfactants will require introducing an additional species and generalizing the free energy model to account for the affinity of the new species to the interface. This can be done in principle using the same strategy explained in the paper. This approach does not require, in principle an extension of the bulk phase description for the electrolyte species. One could even think of situations where one of the two electrolytes exhibits surfactant behavior: this situation can be handled generalizing the proposed free energy model with an additional term proportional to the square of the corresponding density gradient. Such a generalization does not affect bulk thermodynamics.

Professor Smit continued the discussion of the paper by Professor Dijkstra: Recent work of Geissler and co-workers[1] show that for the formation of ordered structures one needs to be in a "sweet spot" in parameter space, *i.e.*, for most of the parameters the system gets locked into a kinetically trapped disordered structure and only for a very small set of parameters an ordered structure gets formed. Your models are restricted to hard-core interactions, which can be seen as the limit in which the attractive interactions are zero. Extrapolating Geissler's conclusions to your work would suggest that it is very unlikely that this limit is exactly this sweet spot and you will be in a kinetically trapped part of parameter space. Would it be interesting to add some attractions to your model to see whether these kinetic traps could be avoided?

1 S. Whitelam, E. H. Feng, M. F. Hagan and P. L. Geissler, *Soft Matter*, 2009, **5**, 1251–1262.

Professor Dijkstra replied: This would indeed be very interesting. We have studied many different hard-core systems over the past few years. Hard rods, hard platelets, binary mixtures of hard spheres, *etc.* If there is an interplay between orientational order and positional order, or an interplay of different length scales as in the case of mixtures of large and small hard spheres, nucleation is often hampered as the system gets kinetically trapped. By adding attraction, these kinetic traps can indeed be avoided. We did study a system of hard rods and non-adsorbing polymer, which induces attractive interactions, so-called depletion interactions, between the rods. For this system, we did find a nucleation barrier, but the critical cluster consists of a single layer that grows only laterally, while the nucleation of a second layer on top of the first layer is extremely rare. Nucleation of multilayer clusters were never observed during the simulation.[1]

1 A. Patti and M. Dijkstra, *Phys. Rev. Lett.*, 2009, **102**, 128301.

Professor Allen continued the discussion of the paper by Professor Löwen: You allow the colloidal particles to have angular momentum, and to transfer this to (and from) the solvent *via* a no-slip condition; but there is no such coupling in the direct interaction *between* colloids. Is there a particular reason (physical or technical) for this choice? Does it make a difference to your conclusions?

Professor Löwen responded: In the experiments the particles are almost spherical. Hence an isotropic hard-sphere-like description for their direct interaction is justified as a reasonable approximation. We have not yet studied systems with an anisotropic direct interaction (like sedimenting rods). The angular momentum transfer mediated by the hydrodynamic interactions can be formally turned off in the SRD computer simulations. For sedimentation in a slit, the additional terms involving angular momentum transfer are not expected to change the results much. This may be different for particles in linear shear flow where a single sphere is rotating. In the latter case the angular momentum coupling *via* the hydrodynamic interactions is more crucial.

Dr Padding commented: Real colloids will probably only transfer angular momentum directly when their hard cores touch (at least if the charges on the surface of the colloid are distributed homogeneously). In this work it is assumed that screened charge interactions are also present. Such interactions diverge at the hard core diameter, so the hard cores never actually touch each other. The colloids do exchange angular momentum indirectly *via* the solvent layer in between the hard cores.

Dr Wysocki answered: The physical reason for this choice is that in colloidal suspensions the lubrication force, which diverges at contact, prevents the colloids from touching one another. In contrast to granular rough spheres, here the tangential interaction at contact does not exist. Because in SRD the lubrication force can only be resolved correctly for distances larger than a multiparticle collision cell we have included a short range repulsive interaction to avoid particle overlap. Furthermore, we consider macroscopic smooth colloids, which do not interact sterically with each other. In principle, such an interaction could lead to a direct coupling of the rotational degrees of freedom.

Dr Boek continued the discussion of the paper by Professor Frenkel: Is the dimensional width of the oil–water interface in your simulations consistent with the dimensional lattice-spacing of 1.75 nm? In general, I would expect the interface width to be at least several lattice spacings for reasons of numerical stability. Is this consistent with experimental values of the interface width?

Professor Pagonabarraga answered: The interface is typically a few lattice spacings. A visual width of 3 lattice spacings is enough to avoid numerical instabilities while ensuring that the underlying lattice anisotropy does not perturb the interfacial behaviour. In the paper we select the lattice spacing width according to the Bjerrum length to resolve properly the electric double layer. For such a choice the interfacial width is larger than the one corresponding to a typical oil/water interface away from the critical point. Although in principle realistic values of the interfacial width can be recovered by reducing the length associated to the lattice spacing, such a strategy deceives the coarse-grained approach put forward. In many circumstances it is enough to ensure that the interfacial width is smaller than the typical length scales involved in the dynamics of the process of interest.

Professor Smit asked: What would be the strategy to link your approach to the underlying microscopic system?

Professor Frenkel answered: The first thing one must determine are the physical parameters describing the system, such as the solvent dielectric constant, its viscosity, or the Gibbs transfer free energy and molecular diffusion coefficient of the ions. For the oil/water interface, the B and K parameters of the free energy can be deduced from the surface tension and the interface width. All these quantities can be obtained (at least in principle) using molecular simulation. The second step is to choose the mesoscopic simulation parameters. This is conveniently done by expressing all physical properties in lattice units: the lattice spacing Δx and the time step Δt. The latter are finally set to physical values relevant for the studied system, to ensure that the smallest relevant length scale is correctly resolved on the lattice. For example, charged systems require that the double-layers are spatially resolved, so that the lattice spacing can be chosen of the order of the Bjerrum length. One can then choose a time step to ensure that the fastest mode of interest is resolved, *e.g.* from $\Delta x^2/\nu$ with ν the kinematic viscosity, or $\Delta x^2/D$ with D the ionic diffusion coefficient, depending the ratio ν/D (Schmidt number).

Dr van der Sman addressed Professor Frenkel and Professor Pagonabarraga: The authors state their model can be extended to electrowetting—which was quite well investigated for emulsion droplet.[1] Can the authors give more details on what it takes to model electrowetting—what kind of boundary conditions does it take? Is it advantageous to take the boundary halfway to the lattice nodes, as we have taken in our model on emulsification in microfluidic devices.[2] Here we take an easy implementation of assigning some value of the order parameter to the wall. Can you assign the wall some value of the electrical potential or surface charge?

1 F. Mugele, *J. Phys.: Condens. Matter*, 2005.
2 S. van der Graaf, *Langmuir*, 2006, **22**(9), 4144–4152.

Professor Frenkel answered: In the paper we have considered the case where the two fluids have the same affinity for the solid interface. Including preferential wetting of one of the two fluids can be implemented using standard procedures,[1] and it is a straightforward implementation of the model. To account for electrowetting phenomena we have to take the difference in dielectric constants of the two fluids into account. Again, this is straightforward with the context of the present simulations. Given the way we deal with fluxes, it is technically more convenient to consider the boundary halfway to the lattice nodes. For electro-wetting it would be logical to fix the electrostatic potential of the wall. However, fixing the charge distribution on a non-conducting wall is also possible.

1 J. C. Desplat, I. Pagonabarraga and P. Bladon, *Comput. Phys. Commun.*, 2001, **134**, 273.

Professor Jackson continued the discussion of the paper by Professor Dijkstra: What effect does the aspect ratio (length-to-breadth) have on the nucleation to the smectic phase in such a system of hard-spherocylinders? You have examined

$L/D = 3.4$ systems exclusively. Do you expect to see the same behaviour for longer rods, say $L/D = 4$) where one would expect lower free energy differences between the isotropic and smectic phases? Can you go directly from an isotropic state to a smectic for longer rods by rapid compression?

Professor Dijkstra replied: In this paper, we indeed studied $L/D = 3.4$ exclusively. We did study the isotropic–nematic transition in a system of hard spherocylinders with $L/D = 5$. If we quench the system to high supersaturations, we find spinodal decomposition: nematic clusters appear immediately throughout the whole system. These nematic clusters coalesce and form a spinodal-like structure, which evolves into the nematic phase. Subsequently smectic layers will be formed in this nematic phase and the system transforms into a smectic phase. A direct isotropic–smectic transition can only be found for a very small length-to-diameter ratio range $3.1 < L/D$

Professor Allen said: Bearing in mind the importance of orientational domain formation and domain alignment in the isotropic → smectic nucleation process, how sensitive do you anticipate that this will be to (even microscopically small) fluid flows in the colloidal suspension? In other words, how important would it be to model the dynamics reasonably accurately?

Professor Dijkstra replied: This is an interesting question. Small fluid flows can probably align the nematic/smectic domains, which will facilitate the nucleation process. It would be interesting to investigate what the effect is of hydrodynamic interactions.

Professor Bolhuis asked: One of the conclusions is that nucleation and growth is probably not the process *via* which a smectic phase forms in an isotropic liquid of hard spherocylinders. If that is the case, how does this phase transformation occur instead? Would it be possible to devise any order parameters that are able to describe such a process?

You mentioned that the nucleation process is less likely to occur for phase transitions in hard core particle systems than for phase transitions in systems with other interaction potentials (*e.g.* for attractive potentials such as Lennard–Jones). Can you have some insight in the general rules for when nucleation is the preferred route for crystallisation or liquid crystal formation and when it is not?

Professor Dijkstra replied: Our results show that the transformation proceeds more *via* reorientation and coalescence of small clusters than by the addition of individual particles to a growing cluster as is usual the case for nucleation. The pre-smectic ordering, which is already present in the supersaturated isotropic fluid phase is maybe a precursor of the nematic spinodal instabily. This should be investigated in future work. Moreover, recent work by Geissler and co-workers show that collective motion and attractive interactions can play an important role in the self-assembly process, which should be studied in more detail for these systems in order to gain more insight in the general rules of self-assembly or nucleation.

Dr Wilson commented: Could you comment on the role of polydispersity for colloidal rod systems and how this would be expected to influence nucleation? Its relatively easy to introduce polydispersity into an experimental system (and many real systems are polydisperse in any case) but its unclear to me whether polydispersity will help promote nucleation or not.

Professor Dijkstra replied: In many systems, polydispersity will hamper the nucleation and will prevent crystallisation or liquid crystal formation. Also, for hard-rod systems, the smectic phase will often be destabilized by polydispersity and a columnar phase will be formed instead. For a bidisperse system, one can obtain broad phase coexistence regions, which can promote the nucleation as the supersaturated parent phase is less dense and the system will not be kinetically trapped.

Professor Holm continued the discussion of the paper by Professor Frenkel: I think it would be interesting to apply your method to the problem of electrophoresis of charged colloids and compare it to the Standard Electrokinetic Model developed by R. W. O'Brian and L. R. White[1] and P. H. Wiersema, A. L. Loeb, J. T. Overbeek,[2] which describes the electrophoretic mobility of a single colloid as function of \kappa a (screening parameter of the solution times the size of colloid), and the zeta-potential. Your method should be well suited since analytically the input are also the Poisson–Boltzmann equation and the linearized Navier–Stokes equations.

1 R. W. O'Brian and L. R. White, *J. Chem., Soc. Faraday Trans.*, 1978, **74**, 1607.
2 P. H. Wiersema, A. L. Loeb, J. T. Overbeek, *J. Colloid Interface Sci.*, 1966, **22**, 78.

Professor Frenkel responded: We agree this is an interesting issue that we would like to pursue. In particular, it will be interesting to compare with the results obtained using the methodology explained in Professor Holm's paper[1]and address the relevance of the local ion structure close to the colloid. Elsewhere,[2] we analyzed the sedimentation of charged spheres and were able to compare quantitatively with theoretical predictions by Booth.[3]

1 B. Dünweg, V. Lobaskin, K. Seethalakshmy-Hariharan and C. Holm, *J. Phys., Condens. Matter*, 2008, **20**, 40214.
2 F. Capuani, I. Pagonabarraga, D. Frenkel, *J. Chem. Phys.*, 2004, **121**, 973.
3 F. Booth, *J. Chem. Phys.*, 1954, **22**, 1956.

Dr Boek remarked: If the diffusion coefficient (page 9) is dimensionalised with respect to liquid water/oil, are the transient results *e.g.* for the density profile of the oil–water interface consistent with experimental results?

Professor Frenkel answered: Although Fig. 9 of our paper, refers to time-independent quantities, we did show during the oral presentation a movie of the approach to equilibrium. The results of such dynamical simulations can be compared with experiments. However, to our knowledge, experiments for the specific example that we showed are thus far lacking.

Dr van der Sman continued the discussion of the paper by Professor Löwen: In a review, Jens Harting compared Stochastic Rotation Dynamics (SRD) with Lattice Boltzmann, in which he advises to use Lattice Boltzmann in the regime of higher Reynolds numbers. Up to which Reynolds number would you advise to use SRD?

Dr Wysocki responded: In our paper, the Reynolds number Re is limited to $Re < 1$. D. A. P. Reid[1] and A. Lamura[2] have investigated flow around different objects at intermediate Re (*i.e.*, $Re < 130$) using stochastic rotation dynamics. The computational effort is the factor that limits the magnitude of Re. If we fix the Mach and the Knudsen number then $Re \sim \sigma^2$, where σ is the dimension of the object in flow. If we keep the ratio σ/L constant (L is the dimension of the simulation box) the computational effort is proportional to σ^3 and thus it is proportional to $Re^{3/2}$.

1 D. A. P. Reid, J. T. Padding, H. Hildenbrandt and C. K. Hemelrijk, *Phys. Rev. E*, 2009, **79**, 046313.
2 A. Lamura, G. Gompper, T. Ihle and D. M. Kroll, *Europhys. Lett.*, 2001, **56**, 319.

Professor Löwen responded: For our system of confined colloids, the Reynolds number was small, namely $Re = 0.125$, *i.e.* we are in the regime appropriate for colloidal dynamics.

We have recently also tried to simulate higher Reynolds numbers within SRD up to about $Re = 1$. The technical problem here is to guarantee simultaneously a small Mach number in order to ensure an incompressible fluid.

Professor Jackson queried: Is there a link between the periodicity that you find for the Rayleigh–Taylor instability and the capillary waves present at your interface?

Professor Löwen replied: First of all, what is studied here is a hard sphere system which does not possess a fluid–fluid phase transition in the bulk, hence there is no sharp interface and no surface tension. The latter typically stabilizes short-wave length undulations in the Rayleigh–Taylor instability.

For a colloid-polymer mixture, on the other hand, a fluid–fluid interface is present, separating a polymer-rich from a colloid-rich domain with a non-vanishing surface tension. In this case, capillary waves are on top of the interface. They show up at any length scale with a continuous distribution of wave numbers such that they do not determine the wavelength of the Rayleigh–Taylor instability. However, they represent the initial small perturbative fluctuations which help to grow an undulation after all.

In our hard-sphere there are similar fluctuations in the steep part of the vertical density profiles which initiate the growth of the most unstable wavelength.

Professor Pagonabarraga asked: In your system there is no surface tension, and in the absence of gravity the colloids will disperse forming a single fluid phase. The instability you report then is not analogous to the Rayleigh–Taylor instability. Can you comment which competing physical factors give rise to the instability and how the relevant unstable length scale is initially fixed?

Professor Löwen replied: Indeed, our hard-sphere-like system exhibits Rayleigh–Taylor instability without surface tension. In the traditional Rayleigh–Taylor instability, the growing of short-wave length undulations is prohibited by a cost in surface tension, *i.e.* these undulations die out exponentially with time. For large wavelengths, particle diffusion (*i.e.* limited mass transport) hinders a quick growth of the undulations. Hence, in between there is an optimal wavelength. In our case of vanishing surface tension, however, every undulation is growing. Short-wave-length undulations exhibit a larger friction due to the viscosity contrast and therefore exhibit only a slow growth. Large wavelengths are restricted by particle diffusion. Hence, again, there must be an optimal intermediate wavelength at which the growth rate is maximal. Consequently the relevant unstable length scale is fixed by a competition of the two physical factors viscosity and diffusion.

Dr van der Sman addressed Dr Wysocki and Professor Löwen: Your video of experiments shows a crystalline layer, which is showing a retardation in sedimentation. Is this due to hydrodynamic interaction (lubrication forces) you think? If so, do you think this effect can be captured by SRD?

Professor Löwen answered: First of all, a block of crystals exhibits a finite yield-stress. A hard-sphere bulk crystal possesses a finite shear-modulus at zero frequency. This is an argument for the retarded sedimentation which relies on pure static equilibrium considerations.

On top of that lubrication which is a dynamical effect is expected to enhance the retardation effect for a sedimenting crystal, in particular at high volume fractions. In order to delineate and separate these two effects, more investigations are needed.

Lubrication forces are in principle included in the SRD approach. The practical limitation, however, is the finite multi-particle-collision box size which has to be smaller than the distance between the touching surfaces. This makes the calculation extremely slow for closely touching spheres.

Dr Boek commented: How do you dimensionalise your time scale and how does this compare with experimental results?

Professor Löwen answered: In fact, the time scale is dimensionalized with the $\tau_D = R^2/D$ which is the time a single particle needs to diffuse over its own radius R, D denoting the single particle diffusion constant. If the growth rates are scaled with this time τ_D, there is good agreement between the SRD simulation data and the experimental results. In the experiments, τ_D is 29 seconds. More details can be found in our paper.[1]

1 A. Wysocki, C. P. Royall, R. G. Winkler, G. Gompper, H. Tanaka, A. van Blaaderen and H. Löwen,*Soft Matter*, 2009, **5**, 1340.

Dr van der Sman continued the discussion of the paper by Professor Frenkel: In one of your responses you talked about difficulties of modeling charge fluctuations? Are fluctuations in charge density field not just due to fluctuations in the hydrodynamics, and do the charges not follow that *via* the hydrodynamic drag forces?

Professor Frenkel replied: There will be fluctuations both in the fluid and in the electrolyte density. In an electrolyte solution at rest, correlated ionic motion induces charge density fluctuations over a characteristic distance κ^{-1} (Debye length) with a relaxation time $1/(\kappa^2 D)$ where D is the ionic diffusion coefficient (Debye time). The difficulties we mentioned refer to the need to include electrostatic correlations to avoid the pathological behavior of a Poisson-Boltzmann description of the charge density when including fluctuations due to the attraction of opposed charge species. The possibility to account, at least partially, for such correlations within the free energy approach described in the paper remains a challenge.

Dr Louis opened a general discussion: An important source of progress in computer simulation comes from better coarse-grained models of the underlying materials, that is descriptions that are simpler and more tractable, but nevertheless retain the fundamental underlying physics that one is interested in investigating.[1–3] To most of us it is intuitively obvious that these simplified descriptions throw some information away and that compromises are made. After all, there is no such thing as a free lunch. But how much are we paying, and what can we get away with?

A. Coarse-graining: structure and energy routes To illustrate these questions, consider the example of a one-component reference fluid interacting with a three-body Hamiltonian of the form:

$$H = K + \sum_{i<j} w^{(2)}(r_{ij}) + \sum_{i<j<k} w^{(3)}(\boldsymbol{r}_{ij}, \boldsymbol{r}_{jk}, \boldsymbol{r}_{ki}), \tag{1}$$

where $\boldsymbol{r}_i$ denotes the position of particle i and $\boldsymbol{r}_{ij} = \boldsymbol{r}_i - \boldsymbol{r}_j$ and $r_{ij} = |\boldsymbol{r}_i - \boldsymbol{r}_j|$. K is the kinetic energy operator, $w^{(2)}(r)$ is an isotropic pairwise additive potential, and $w^{(3)}(\boldsymbol{r}_{ij}, \boldsymbol{r}_{jk}, \boldsymbol{r}_{ki})$ is a triplet or three-body potential. Three-body potentials are expensive and cumbersome to simulate, and so one might want to coarse-grain them to a simpler isotropic representation.

There are several ways you could do this. One popular method is to fit a pair-potential such that it reproduces a structural quantity like the radial distribution function $g(r)$ generated by the Hamiltonian of eqn (1). Henderson[4] first showed that "the pair potential $v(r)$ which gives rise to a radial distribution function $g(r)$ is unique up to a constant". An extended proof for orientational correlations can be found in a book by Gray and Gubbins[5] and we recently did this for multi-site potentials.[6] A more rigorous mathematical discussion is provided by Chayes, Chayes and Lieb.[7] Therefore, for a given state-point, the $g(r)$ generated by a Hamiltonian like that of eqn (1) can be reproduced by a unique effective pair potential $v_g(r)$. (The existence of v_g is more subtle, but holds under fairly general conditions[7]) I'll call approaches that derive $v_g(r)$ the structural route to deriving an effective potential. The difference with the bare-pair potential $w^{(2)}(r)$ can be written as:

$$\delta v_g(r) = w^{(2)}(r) - v_g(r). \tag{2}$$

Another popular way of deriving an effective potential is by fitting to a thermodynamic observable like the internal energy. One first calculates the full internal energy $U(N, V, T)$ for the system governed by the Hamiltonian (eqn (1)), and then derives a potential $v_U^{\text{eff}}(r)$ that reproduces the same internal energy by the simpler two-body formula:

$$U(N, V, T) = \frac{1}{2}\rho^2 \int dr_1 dr_2 g(r_{12}) v_U^{\text{eff}}(r_{12}). \tag{3}$$

I'll call this the energy route. The difference with the bare-pair potential $w^{(2)}(r)$ can be written as:

$$\delta v_U(r) = w^{(2)}(r) - v_U(r). \tag{4}$$

It is not hard to show that both $v_g(r)$ and $v_U(r)$ depend on the state point at which they are derived. Thus if they are used at a different state-point, one would expect **transferability problems**, *i.e.* you'd need to re-derive them for your new state-point.

B. Representability problems What is perhaps more worrying is that one can also show that at a given state-point, $v_g(r)$ and $v_U(r)$ cannot be the same. As first demonstrated almost 40 years ago,[8] to lowest order in ρ and $w^{(3)}$, the ratio between the two corrections is:

$$\frac{\delta v_U(r)}{\delta v_g(r)} = \frac{1}{3} + \mathcal{O}\left(\left(w^{(3)}\right)^2; \rho^2\right). \tag{5}$$

In other words the corrections due to the three-body forces that you are coarse-graining out differ by a factor of three! Since $v_g(r)$ is unique, it is therefore impossible to represent all the properties of a system governed by the Hamiltonian of eqn (1) by a single pair potential. There is no free lunch and such **representability problems** are widespread in coarse-grained descriptions of soft-matter systems.[9]

They may also be important in other coarse-grained simulations. For example, we recently used the structural route to derive radially symmetric potentials for water from a more sophisticated underlying model.[6] We explicitly constructed $v_g(r)$ and $v_U(r)$ and they look very different, as anticipated by eqn (5). At some of the state-points we studied, using $v_g(r)$ to calculate the virial pressure resulted in a dimensionless compressibility factor $Z = \beta P/\rho$ that was almost two orders of magnitude larger than that of the original multi-site water model used to parameterise $v_g(r)$. Similar deviations from the underlying water model were also seen elsewhere.[10]

Admittedly, it may not be surprising that an isotropic potential should perform so poorly when the underlying fluid has complex orientational correlations. In fact, treating water as a pair potential is probably not such a good idea.

One way of improving on the thermodynamic performance of the structurally derived potential $v_g(r)$ is to use constraints to simultaneously fit to properties like the virial pressure or the internal energy. Due to the uniqueness of $v_g(r)$, this potential would no longer correctly reproduce the pair correlations, but since the thermodynamic properties are scalar quantities, it might not come at too large a cost for the structure. A nice example of applying this to water can be found in this paper by the Mainz group[10] who showed that the structural properties were affected (and by extension the isothermal compressibility which can be written as an integral over $g(r)$), but not too badly.

C. More general problems with effective potentials In Louis[9] some more general issues with effective potentials are reviewed. For example, most of the time our effective simplified potentials are really state dependent, that is if you were to derive them at different state points they would be different. I argue that you shouldn't naively treat them as if they are real two-body potentials of a Hamiltonian system.

Here is an example of how things can go horribly wrong. Consider a homogeneous fluid in a volume V , with N particles interacting with a spherically symmetric pair potential $v(r;\rho)$, that depends on the state point (that could be density, temperature, *etc.*). For simplicity, here we consider only the density $\rho = N/V$. The standard way to derive the virial equation is directly through the canonical partition function

$$Q(N,V,T) = \frac{\Lambda^{-3N}}{N!}\int \mathrm{d}\mathbf{r}^N \exp\left\{-\beta\sum_{i<j} v(r_{ij};\rho)\right\}, \tag{6}$$

where Λ is the usual thermal de Broglie wavelength. The volume derivative in

$$\beta P = \left(\frac{\partial \log Q(N,V,T)}{\partial V}\right)_{N,T} \tag{7}$$

should also act on $v(r;\rho)$, resulting in a virial equation with an extra $\partial v(r;\rho)/\partial\rho$ term:

$$Z^{\rho}_{vir} = \frac{\beta P}{\rho} = 1 - \frac{2}{3}\beta\pi\rho\int_0^\infty r^2\left\{r\frac{\partial v(r;\rho)}{\partial r} - 3\rho\frac{\partial v(r;\rho)}{\partial\rho}\right\}g(r)\mathrm{d}r, \tag{8}$$

something first pointed out in 1969 by Ascarelli and Harrison[11] in the context of density dependent pair potentials used for modelling liquid metals.

Now on the surface this all looks very kosher—I took my potential, plugged it into my partition function, turned the crank, and out popped eqn (8) with a correction to the pressure due to the state-dependence of the effective potential. However, when we tried this with the $v_g(r)$ derived for the water model,[6] the correction actually made the agreement far worse. In Louis,[9] I give other examples were this correction has the wrong sign (and magnitude). Although this analysis has the veneer of statistical mechanical respectability, it is in fact a disreputable result.

Taking a step back, it is not at all surprising that this derivation cannot be right.[9] The correction term in eqn (8) only takes into account a local dependence on density. If you took the same potential $v(r;\rho)$, and used it in a grand-canonical ensemble, then you would need information from all densities, not just the local derivative. So the two ensembles would not be equivalent, a good sign that there are more problems afoot. Warnings about such difficulties abound, for example, about 40 years ago Barker *et al.* wrote: *We record our opinion that the use of density-dependent eective pair potentials can be misleading unless it is recognised that these are mathematical constructs to be used in specied equations rather than physical quantities*[12] and similar reservations were sounded by other distinguished investigators.[8,13,14]

The point of this exercise is not to say that effective potentials are useless. I've happily used them myself.[15] Sometimes a fine solution to a dubious Hamiltonian is better than a dubious solution to a very fine Hamiltonian. It is just that one must be careful when using coarse-graining procedures not to automatically treat the coarse-grained potentials as if they are real bona-fide potentials of the type one uses in a Hamiltonian. Also, in practice it is also good to remember that these potentials are always compromises and that fitting too strongly to one quantity may generate larger errors in other quantities that one also wants to measure.[6,9]

D. Questions for all Do others have good examples of representability problems generated by their potentials? I'd be interested to know how important they are in other systems.

A related question is: How well can you predict the magnitude of a representability problem? In the case of water it is not surprising that a radially symmetric pair-potential would have problems, but could you have predicted that a potential that generates $g(r)$ so well does so poorly on the internal energy or the virial pressure?

Ignacio Pagonabarraga has an interesting interpretation of density dependent potentials.[16] Do you think this approach could be generalised to the kind of coarse-graining procedures we are applying in this conference?

1 M. Praprotnik, L. Delle Site and K. Kremer, *Annu. Rev. Phys. Chem.*, 2008, **59**, 545.
2 W. G. Noid, J. W. Chu, G. S. Ayton, V. Krishna, S. Izvekov, G. A. Voth, A. Das and H. C. Andersen, *J. Chem. Phys.*, 2008, **128**, 244114.
3 T. Murtola, A. Bunker, I. Vattulainen, M. Deserno and M. Karttunen, *Phys. Chem. Chem. Phys.*, 2009, **11**, 1869.
4 R. L. Henderson, *Phys. Lett. A.*, 1974, **49**, 197.
5 C. G. Gray and K. E. Gubbins, *Theory of molecular fluids*, Oxford University Press, Oxford, 1984.
6 M. E. Johnson, T. Head-Gordon and A. A. Louis, *J. Chem. Phys.*, 2007, **126**, 144509.
7 J. T. Chayes and L. Chayes, *J. Stat. Phys.*, 1984, **36**, 471; J. T. Chayes, L. Chayes and E. H. Lieb, *Commun. Math. Phys.*, 1984, **93**, 57.
8 G. Casanova, R. J. Dulla, D. A. Jonah, J. S. Rowlinson and G. Saville, *Mol. Phys.*, 1970, **18**, 589.
9 A. A. Louis, *J. Phys.: Condens. Matter*, 2002, **14**, 9187.
10 H. Wang, C. Junghans and K. Kremer, *Eur. Phys. J. E*, 2009, **28**, 221.
11 P. Ascarelli and R. Harrison, *Phys. Rev. Lett.*, 1969, **22**, 285.
12 J. A. Barker, D. Henderson and W. R. Smith, *Mol. Phys.*, 1969, **17**, 579.
13 J. S. Rowlinson, *Mol. Phys.*, 1984, **52**, 567.
14 M. A. van der Hoef and P.A. Madden, *J. Chem. Phys.*, 1999, **111**, 1520.
15 A. A. Louis, P. G. Bolhuis, J. P. Hansen and E. J. Meijer, *Phys. Rev. Lett.*, 2000, **85**, 2522.; P.G. Bolhuis, A. A. Louis, and J. P. Hansen, *Phys. Rev. Lett.*, 2002, **89**, 128302
16 S. Merabia and I. Pagonabarraga, *J. Chem. Phys.*, 2007, **127**, 054903.

Dr van der Sman opened a general discussion of all the papers: How well can you predict dynamics after coarse-graining? Do you foresee that we can ever predict p.e. solvent diffusion through polymer matrix *via* coarse-graining of *ab initio* models?

Dr Louis answered: **Coarse-graining dynamics by telescoping down time-scales** The question of how to properly coarse-grain a dynamical simulation is a very interesting one. I think there is no single answer, but want to use an example here from our simulations of colloidal hydrodynamics to make some points that I believe are of more general relevance.

Consider a buoyant colloid of mass M_c and a radius $a = 1$ μm in H_2O. As described in more detail elsewhere,[1] its behaviour is governed by a series of different timescales shown in Table 1. If you are only interested in the behaviour of the colloids, then the two fastest time-scales, the solvent collision time τ_{col} and the solvent relaxation time τ_f, can be ignored as long as they are shorter than any other colloidal time-scales. The first physically relevant time-scale is the Fokker Planck time-scale $\tau_{FP} \approx 10^{-13}$ over which the colloid loses memory of the short-time forces acting on it.[2] For the example colloid, the next time-scale up is the sonic time $t_{cs} \approx 6.7\times10^{-10}$ s. Then comes the Langevin time $\tau_B \approx 2.2 \times 10^{-7}$ s that measures the exponential decay time of the velocity autocorrelation function within the Langevin approximation. Interestingly, for colloids this time-scale is artificial and does not have direct physical meaning (see appendix[1]). Next up is the kinematic time $\tau_\nu \approx 10^{-6}$ s over which vorticity diffuses away from the colloid. If your colloid moves a significant fraction of its radius within the time τ_ν, then the colloid will feel the effects of its own motion from a time τ_ν back, and finite Reynolds number (*Re*) effects start to kick in. For that reason, it needs to be kept small compared to time-scales of colloidal diffusion or advection. The largest time-scale we consider here is the diffusion time $\tau_D \approx 5$ s. However, if the colloid also moves under an external force with a velocity v_s, then there is an additional time-scale $t_s = a/v_s$ that measures how long it takes to advect over its radius, and you can then also define a related Peclet number $Pe = t_s/\tau_D$ that measures the relative importance of convection over diffusion.

From the Fokker Planck time on up to the diffusion time covers 13 orders of magnitude. It is clearly not possible to capture all of these in a simulation. Instead, what is needed is time-scale separation. As long as the time-scales are properly separated, you should still be simulating the correct underlying physics. This process can be visualized in Fig. 1[1] which shows an example of how the hierarchy of time-scales is telescoped down to a more computationally manageable separation in order maximise simulation efficiency, but in such a way that the times are still sufficiently separated to correctly resolve the underlying physical behaviour. A good example of the rationale behind this thinking can be illustrated with the sonic time t_{cs}. Physically it needs to be much smaller than the diffusion time, or else locally you have supersonic behaviour. But if it is too small, the simulation will spend most of its time resolving sound waves that may not be that interesting for the colloidal behaviour you are trying to reproduce, making the simulation very inefficient.

In order to correctly interpret the physical meaning of your simulation you need to telescope the time-scale hierarchy back out to the physical one you want to study. For example, if you are interested in physics that is dominated by diffusion, you would map onto your physical diffusion time. A consequence of this strategy is that a single simulation can map onto many different physical times *e.g.* if your colloid has a radius $a = 1$ μm, then the diffusion time is $\tau_D = 5$ s, and that fixes the time-scales for your simulation. On the other hand, if your colloid has a radius $a = 100$ nm, then $\tau_D = 5\times10^{-3}$ s and the exact same simulation would have a different fundamental time-scale. This means that, for example, the viscosity in your simulations would be different depending on what colloid size you were mapping to, even though the fundamental physics is the same. This fact suggests that physical viscosity is often not such a good parameter to try and fit to in a coarse-grained simulation.

You can also correctly map the same simulation to different time-scales even though the colloidal system is unchanged. Say that you are interested in the

Table 1 Time-scales relevant for colloidal suspension; numerical values are for a buoyant colloid of radius $a = 1$ μm in H_2O. For H_2O, the speed of sound c_s is 1.48×10^9 μm/s. The kinematic viscosity $\nu = 1\times10^6$ μm²/s, and measures the diffusion constant with which vorticity diffuses away

Solvent time-scales	
Solvent collision time over which solvent molecules interact	$\tau_{col} \approx 10^{-15}$ s
Solvent relaxation time over which solvent velocity correlations decay	$\tau_f \approx 10^{-14}$ – 10^{-13} s
Hydrodynamic time-scales	
Sonic time over which sound propagates one colloidal radius	$t_{cs} = a/c_s \approx 6.7 \times 10^{-10}$ s
Kinematic time over which momentum (vorticity) diffuses one colloidal radius	$\tau_\nu = a^2/\nu \approx 10^{-6}$ s
Brownian time-scales	
Fokker-Planck time over which force–force correlations decay	$\tau_{FP} \approx 10^{-13}$ s
Brownian relaxation time over which colloid velocity correlations decay in Langevin eqn	$\tau_B = M_c/\xi_S \approx 2.2 \times 10^{-7}$ s
Colloid diffusion time over which a colloid diffuses over its radius	$\tau_D = a^2/D_{col} \approx 5$ s
Ordering of time-scales for colloidal particles	
$\tau_{col} < \tau_f, \tau_{FP} < t_{cs} < \tau_B < \tau_\nu < \tau_D$	

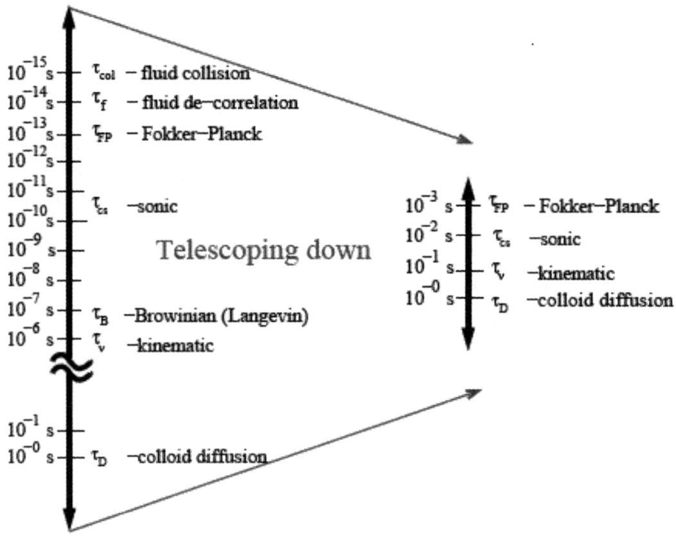

Fig. 1 Telescoping down: The hierarchy of time-scales for a colloid (here the example taken is for a colloid of radius 1 μm in H_2O) is compressed in the coarse-grained simulations to a more manageable separation. As long as the physically important times are clearly separated, the simulation should still generate the correct physical picture. Once the simulations are completed, they can be related in more detail to particular experiments by telescoping back out to the relevant experimental time-scales.

longer-time behaviour of the velocity autocorrelation function that is dominated by the kinematic time τ_ν. In that case if you mapped to a physical system with $a = 1$ μm, the times would be quite different from what you would get when you mapped to the diffusion time of the same system. So here the viscosities *etc.* would have different values depending on what processes you were focusing on. This is a good example of a no free lunch theorem. If you use coarse-grain dynamics, you almost always need to do some kind of telescoping down, and that means that it is hard to simultaneously match multiple time-scales in your system.

The particular example described here concerns a colloid in suspension where it is relatively straightforward to work out what all the time-scales are. Nevertheless, we argue[1] that many other coarse-graining methods for dynamics must make implicit use of the telescoping down process. You can make your method work by carefully analyzing your time-scales, and then making sure you know how to telescope back out to the experimental situation you want to emulate. It helps to do this in terms of dimensionless variables. However, sometimes you cannot compress the hierarchy to a computationally achievable regime without bringing some time-scales too close to each other, or even switching the order of timescales. It is then a matter of subtle judgement if the behaviour that comes out of your simulation is physically correct.

A similar analysis can be used to interpret systems where the dynamics are dominated by energy barriers. It is going to be very hard to know what the real physical time-scales are here because processes depend exponentially on barrier heights. In a coarse-grained dynamical simulation, it may be advantageous to dramatically lower (free) energy barriers in order to speed up the simulation. The hope is that you keep the relative order that characterizes the physical system you want to emulate, so that at least the qualitative dynamic effects are correct. But if, say, you try to interpret your time-scales from a measurement of single particle diffusion, and then use that time to extract a physical time for a different process in your simulation that is dominated by energy barriers, then you will get those numbers completely wrong. Clearly, there are many tricky subtleties that arise when trying to coarse-grain the dynamics of soft-matter systems.

1 J. T. Padding and A. A. Louis, *Phys. Rev. E*, 2006, **74**, 031402.
2 J. K. G. Dhont, *An Introduction to the Dynamics of Colloids*, Elsevier, Amsterdam, 1996.

Professor Voth opened the discussion of the paper by Professor Español: I have always been confused by the mathematical form of the conservative pair potential in DPD. It is very soft at short range and in fact allows the DPD particles to in principle past through each other. I can see how this might somehow make sense for fluid dynamics modeling, but people commonly use DPD at the more "molecular" level to study, *e.g.*, lipids and polymers. These molecules cannot possibly pass through each other. Such behavior in the DPD model is entirely unphysical. How can this feature of the conservative pair potential in the DPD approach be justified or is it simply an *ad hoc* feature to allow the integration time step to be larger?

Professor Español replied: I will give the answer to this question further on.

Professor Löwen responded: Just as a side comment to this question, the level of penetrability is obviously set by the interaction energy at zero separation and the system density. If the former is large enough, the penetrability can be largely avoided.

Dr Louis replied: I think that is an important question. We wrote out an analysis of DPD as a coarse-graining method.[1] Here I summarise the main points we made:

DPD, as introduced by P. J. Hoogerbrugge and J. M. V. A. Koelman,[2] and corrected by P. Español and P. B. Warren[3] should really be viewed as having two independent methodological innovations.

Innovation (1) A Galilean invariant thermostat that generates Navier Stokes hydrodynamics in the continuum limit. Newtonian MD conserves energy, and so naturally generates the microcanonical ensemble. But it is often more convenient to analyze results in the canonical ensemble and so thermostats are introduced. Many of these are not Galilean invariant, and so will screen hydrodynamic interactions (HI) beyond a certain distance. How bad this is depends on how strong the damping is. The DPD thermostat gets around this problem and conserves the HI. However, if you wanted to study these HI, you could also just do microcanonical MD, and you would also have the correct hydrodynamics. Furthermore, you can use a DPD thermostat with any kind of conservative forces, see for example references 4,5 for examples of using the DPD thermostat in this way. The DPD thermostat is independent of the exact form of the DPD potential.

Note also that if you wanted to study colloidal hydrodynamics, as we do in this volume, DPD doesn't help that much, as you would have to use it to simulate the fluid between the colloids, and that is very expensive.

Innovation (2) The use of soft potentials. There are many ways of carefully and systematically coarse-graining over length-scales to reach a potential based picture (see *e.g.* many contributions in this volume). Often, these new effective potentials are softer than those of the underlying soft matter system.[6–9]

However, the interpretation of these potentials is subtle. Even for equilibrium properties you can have serious representability problems.[10] For example, a potential that generates the correct pair structure will not normally reproduce the correct virial pressure. These coarse-grained potentials should really be viewed as a way to simplify the evaluation of the partition function, and as such may very well allow for particle moves that are physically impossible in the original system. That may be alright if you want to derive an approximation to the equilibrium properties. For example, it can be advantageous because you can cross potential barriers more easily.

For non-equilibrium problems the use of effective potentials is much harder to justify theoretically.

For polymer dynamics, for instance, uncrossability constraints must be re-introduced to prevent coarse-grained polymers from passing through one another.[11,12] This is what we concluded in the following paper,[1] *"We are convinced that a correct microscopic derivation of the coarse-grained DPD representation of the dynamics, if this can indeed be done, will show that the interpretation of such soft potentials depends on dynamic as well as static (phase-space) averages. Viewing the DPD particles as static "clumps" of underlying fluid is almost certainly incorrect. It may, in fact, be more fruitful to abandon simple analogies to the potential energy of a Hamiltonian system, and instead view the interactions as a kind of coarse-grained self-energy.[13]"*

The contribution by Pep Español in this volume suggests that using the Zwanzig approach may be a fruitful way forward, but I think a fair amount of further work needs to be done to establish this.

1 J. T. Padding and A. A. Louis, *Phys. Rev. E*, 2006, **74**, 031402.
2 P. J. Hoogerbrugge and J. M. V. A. Koelman, *Europhys. Lett.*, 1992, **19**, 155.
3 P. Español and P. B. Warren, *Europhys. Lett.*, 1995, **30**, 191.
4 B. Dünweg, in *Computational Soft Matter: From Synthetic Polymers to Proteins*, ed. N. Attig, K. Binder, H. Grubmuller, and K. Kremer, NIC Series, 1994, vol. 23, pp. 61.
5 T. Soddemann, B. Dunweg, and K. Kremer, *Phys. Rev. E*, 2003, **68**, 046702.
6 A. A. Louis, P. G. Bolhuis, J. P. Hansen, and E. J. Meijer, *Phys. Rev. Lett.*, 2000, **85**, 2522.
7 A. A. Louis, *Philos. Trans. R. Soc. London, Ser. A*, 2001, **359**, 939.
8 C. N. Likos, *Phys. Rep.*, 2001, **348**, 267.
9 S. H. L. Klapp, D. J. Diestler, and M. Schoen, *J. Phys.: Condens. Matter*, 2004, **16**, 7331.
10 A. A. Louis, *J. Phys.: Condens. Matter*, 2002, **14**, 1987.
11 J. T. Padding and W. J. Briels, *J. Chem. Phys.*, 2001, **115**, 2846.
12 J. T. Padding and W. J. Briels, *J. Chem. Phys.*, 2002, **117**, 925.
13 P. B. Warren, *J. Chem. Phys.*, 2003, **15**, S3467.

Dr Milano asked: Is it necessary in your derivation of DPD from Mori Zwanzig theory to assume the typical functional form of conservative forces (smooth forces allowing chain–chain crossing) or can more stiff potential be considered?

Professor Español replied: The functional form of the conservative forces is not assumed, but rather computed, in our method. For the thermodynamics state considered they turn out to have soft repulsive tails, but note also that we do not have data for short distances, because it is highly improbable to find the CoM of the molecules at short distances. We have checked that the form of the potential at short distances in the DPD simulation does not matter, precisely because it is equally improbable to find the DPD particles at very short distances, at that particular thermodynamic state. It is possible to force the CoM to be closer and compute with the constrained dynamics the resulting effective force. Although in principle it is possible to have the CoM on top of each other, the entanglement and excluded volume of the arms of the star polymers will produce extremely large repulsive forces. This answers the previous question of Professor Voth.

Professor Zannoni addressed Professor Español and Dr Louis: If I understood correctly, the previous comment was that the dynamic, Zwanzig type, evolution can be considered separately from the effective potential, thus avoiding the problem of the potential allowing moving across a polymer (for example). However, the asymptotic limit of the solution to the Zwanzig-equation should be the equilibrium distribution whose log gives the effective potential. Thus it seems that the two aspects cannot be considered independently.

Professor Español replied: I am not sure I understand the question. In Zwanzig theory, the evolution is dictated in part by the effective potential (the other part is friction and noise). So I agree with your last sentence.

Professor Kremer opened the discussion of the paper by Dr Boek: The basis for the present investigation is good force fields for the interactions of the benzene patches.

In the context of organic electronics extensive simulations of the interaction of graphene patches and related molecules, *i.e.* hexabenzene coronenes,[1] have been performed. Did information from that field go into the present study and if not, would this help for a better understanding of the phenomena focused on here?

1 X. Feng, V. Marcon, W. Pisula, M. R. Hansen, J. Kirkpatrick, F. Grozema, D. Andrienko K. Kremer, K. Müllen, *Nat. Mater.*, 2009, **8**(5), 421; and references therein.

Dr Boek responded: In our present study, we used parameters from the OPLS-AA force field (see ref. 21 of our paper and references therein). The OPLS force field has been shown to work well as a general purpose tool for a wide range of aromatic liquids in reproducing experimental data. Because asphaltenes include a wide variety of chemical structures, the OPLS force field was therefore an obvious choice.

For future work, it would be worth considering the approach outlined in the paper on coronenes. However, this approach may turn out to be expensive, as every asphaltene structure needs a separate determination of force field parameters.

Dr van der Sman asked: What is the effect of neglecting the third dimension? In shear flow sticky particles would also have a strong tendency to aggregate in the direction perpendicular to the shear plane?

Dr Boek answered: This is an excellent question. Indeed the third dimension plays an important role. As a matter of fact, we are planning to extend our simulations to three dimensions.

Professor Marrink remarked: According to your asphaltene–asphaltene PMFs (Fig. 8–10 of the paper), the direct contact pair is stabilized considerably (5–10 kJ/mol). No solvent-separated minimum is observed. So why do you assume there would be a lubrication layer formed in between the asphaltene nano-aggregates? Your PMFs do not seem to justify such an assumption.

Dr Boek replied: It is important to distinguish the different length scales in this problem.

At the molecular level, we consider the individual asphaltene molecules. We have observed that there is no lubrication layer between the individual asphaltene molecules. At the second level, we consider nano-aggregates, consisting of several asphaltene molecules. We assume that a lubrication layer is formed between the nano-aggregates, because we do not have direct measurements at the moment. In this context, I refer to the paper by G. Porte *et al.*[1] He showed that a model of nano-clusters, with strong internal interactions (probably without lubrication layers) and weak interactions between different nano-aggregates (partly caused by solvent screening), is consistent with all available experimental data.

Certainly, it would be a good idea to perform direct simulations of nano-aggregates, but these will be computationally much more demanding.

1 G. Porte, H. Zhou and V. Lazzeri, *Langmuir*, 2003, **19**, 40–47.

Professor Pagonabarraga asked: What is the mechanism which prevents clogging? Is it mainly controlled by hindrance in cluster growth due to the stresses induced by the flow? Or is it the change in the cluster structure (for example a change in its fractal dimension)? Which effect hinders more efficiently channel clogging?

Dr Boek replied: We observe that the cluster growth is limited mainly by the flow rate and the associated stresses induced by the flow. We have not yet systematically investigated the cluster structure as a function of increasing flow rate, but this is certainly a very good suggestion. I would expect that the cluster structure will become denser with increasing flow rate, as observed by Kaandorp *et al.*[1]

1 J. A. Kaandorp, C. P. Lowe, D. Frenkel and P. M. A. Sloot, *Phys. Rev. Lett.*, 1996, **77**, 2328.

Professor Allen opened the discussion of the paper by Dr Louis: It is usual to move the lattice of stochastic rotation dynamics cells periodically, so as to maintain translational invariance; Professor Löwen has already mentioned (in the paper by Professor Löwen) the need to introduce "ghost" particles near the confining walls to handle cells that would otherwise be partially occupied. How is this handled in the extremely narrow systems that are of interest here, and is there a danger of artifacts being introduced?

Dr Louis replied: We use a similar method to what Professor Löwen does with the ghost particles.[1]

By using a larger number of SRD cells per colloid diameter, 8 instead of the usual 4, we tried to minimize any discretization artifacts. If you wanted to study details of the flow right next to the wall, you might need to be even more careful, but for our study, focusing on the diffusion of colloids under strong confinement, we don't expect there to be important discretization errors in the physics we report.

1 A. Lamura, G. Gompper, T. Ihle, and D. M. Kroll, *Europhys. Lett.*, 2001, **56**, 319.

Professor Löwen asked: In the limit of very small δ_c when two particles are hopping over each other the details of the potential energy landscape may also become important on top of the hydrodynamic issues (lubrication, boundary conditions on the particles surface). The asymptotic behaviour for small δ will then probe microscopic details of the interaction. Can you comment on the relevance of the details of the potential energy during the hopping process for small δ2?

Dr Louis responded: That is an interesting question that we have only partly investigated and don't discuss in the paper. We used a short range potential in order to separate out the potential effects from hydrodynamic effects. For longer ranged potentials, however, we expect that the hopping time should scale with potential as $\tau \sim \exp(\beta V(\delta))$ for small δ. We think that this effect could be dominant for charge-stabilized colloids with long-ranged repulsive potentials and also for atomistic or molecular systems.

Professor Allen commented: In contrast to the paper presented by Löwen, you chose to apply a slip boundary condition between colloidal particles and fluid, and allow the colloidal particles to carry no angular momentum; hence, because they are circularly symmetric, they interact in a very specific way with the solvent. Do you think that this choice will make a small or large difference to the physics, and which boundary condition best reflects the experimental situation?

As a supplementary (but related) question, does SRD properly model the lubrication force between colloidal particles at a short distance?

Dr Louis answered: We do not expect slip boundary conditions to cause any qualitative change to the basic physics we are studying here. We do include stick boundary conditions for the interaction between the fluid and the wall though.

For colloids in solution, stick boundaries are closer to the real physical situation, and if, for example you wanted to study rotational autocorrelation functions,[1] then you would need to include stick boundary conditions. It is in fact quite interesting to ask yourself what happens to rotational autocorrelation functions near a hard wall.

In this paper we increased the number of SRD cells per colloidal diameter to 8, as opposed to the 4 per colloid diameter that we have previously used in studies of sedimentation *etc*. This was done in order to better resolve the lubrication forces between colloids, and between the colloids and the wall. SRD, like Lattice Boltzmann and other methods, does not completely resolve the lubrication forces at short distances,[2] but for real colloids, the lubrication forces don't diverge because local inhomogeneities, be it a grafted polymer layer or local co and counter ions, change the short ranged hydrodynamic effects.[3] So SRD actually reproduces something more like the expected effects for real colloids.

1 J. T. Padding, H. Löwen, A. Wysocki, and A. A. Louis, *J. Phys.: Condens. Matter*, 2005, **17**, S3393.
2 N.-Q. Nguyen and A. J. C. Ladd, *Phys. Rev. E*, 2002, **66**, 046708.
3 A. A. Potanin and W. B. Russel, *Phys. Rev. E*, 1995, **52**, 730

Dr van der Sman addressed Dr Louis and Dr Padding: The friction of colloids is strongly effected by confinement. This effect is not taken into account in the Brownian Dynamics, which should appear in the fluctuations, due to the Fluctuation-Dissipation Theorem (FDT). Does SRD produce that correctly, and does it also follow the fluctuations of the FDT (with enhanced friction near the walls?)

Dr Padding and Dr Louis responded: In the Brownian Dynamics simulations we assume that both the friction and random forces on a colloid are independent of the distance to the wall or distance to other colloids, thus satisfying the FDT. The constant friction approximation is clearly an oversimplification, but is used surprisingly much in other theoretical and simulation works, which is why we included it for comparison with our SRD method. In SRD the hydrodynamic frictions and fluctuations are not put in by hand, but they emerge from the underlying particle collisions. Hydrodynamic effects such as lubrication and enhanced friction near a wall are faithfully reproduced, with the correct relation between fluctuations and friction.

Professor Pagonabarraga remarked: In one of your replies you have mentioned that you observe a strong hydrodynamic screening when the channel width is smaller than two colloid diameters. In previous work[1] I observed that in this regime the short time diffusion normalized by the single particle diffusion coefficient in the presence of the confining walls exhibits a very weak colloidal concentration dependence, while for wider channels a collapse to a single curve with a stronger sensitivity to colloidal concentration develops. I wonder if you can quantify hydrodynamic screening in such terms, or whether you think that there is a stronger hindrance induced by the walls.

These observations correspond to short time diffusion. Do you have a feeling of the relevance such a change may have at long times? Or do you think that on this longer scale HI will essentially only enter through the lubrication effect you describe in your paper?

1 I. Pagonabarraga, M. H. J. Hagen, C. P. Lowel and D. Frenkel, *Phys. Rev. E*, 1999, **59**, 4458.

Dr Louis answered: We also observed the weak dependence of the diffusion coefficient on colloid concentration for narrow pipes, but didn't do a systematic check as a function of channel width. I don't have a good feeling for what happens on a longer scale. The fact that the only real difference we observe with BD for the crossover from SFD to Fickian diffusion is for very narrow pipes where lubrication forces kick in suggests that HI interactions don't have a strong qualitative effect outside this narrow range of pipe widths. But as always with HI, our intuitions are often wrong, so I think we'd need to measure this to be sure.

Dr Hess continued the discussion of the paper Professor Español: More than a decade ago Professor Berendsen taught me than when constraining a degree of freedom, the average constraint force gives the conservative force (for the PMF) and the autocorrelation time of the constraint force gives the friction coefficient as well as the noise amplitude for this degree of freedom.

Does your work give a more solid foundation for what was already known at that time, or is this another case (which has also happened to me) of rederiving something that was already know to Professor Berendsen long ago?

Professor Español answered: Indeed in the paper we show with certain rigor that using a constrained dynamics is the proper procedure for finding effective potentials and frictions. Unfortunately I did not have the luck to have Professor Berendsen as a teacher.

Professor Berendsen answered: Knowledge about the use of constrained forces for assessing potentials of mean force and friction has been available in the MD community for a long time, but as far as I know, the basis for this knowledge was rather intuitive and there has never been an adequate proof. It is quite nice that such a proof has now been given.

Professor Berendsen asked: The friction forces are pairwise opposite and add up to zero so that linear momentum is conserved. Is that also true for the stochastic forces? Are the stochastic forces also sums of pairwise opposite contributions? This is not directly apparent from your final eqn (52) and (53) in your paper.

Professor Español answered: The Markovian approximation in Zwanzig theory is basically a modelization of the so-called projected current defined in eqn (11) of our paper, with a white noise, as shown in eqn (20). In the case of the coarse description given by the centers of mass of the complex molecules, the projected current is just $\delta\mathbf{F}_\mu = \mathbf{F}_\mu - \langle\mathbf{F}_\mu\rangle$ as can be seen in eqn (44). The forces on the molecules (and their averages) satisfy that all of them add to zero. Obviously, the modelization with a white noise of this projected force should comply with this property. The Dissipative Particle Dynamics modelization of this projected force with a white noise does actually satisfy this property and, therefore, conserves total momentum.

Professor Allen commented: We have heard (*e.g.* in the paper presented by Professor Voth) that including three-body terms in the effective coarse-grained potentials can make a big difference to the accuracy of the description. Can you extract these terms from your test simulations?

Professor Español replied: This is again a question addressing the "multidimensional conundrum of Zwanzig theory", the fact that effective potentials and frictions depend in principle on the positions of all the CoM of the molecules in the system. Our strategy has been to generate a single point in the multidimensional space and recourse to the pair additivity in order to infer from this information the pair-wise forces, by plotting the force felt by a pair against its distance. The resulting cloud of points are fairly concentrated on a line, suggesting that for this system at this particular thermodynamic state, pair-wise is a not so bad assumption. In other systems, where three-body terms are crucial, I would try to run constrained simulations in which I would fix two of the relevant degrees of freedom and vary systematically the third in order to get a sense of the possible functional form of the three-body potential. But in this case, it is probably more efficient to use the force matching method, for example, where you propose sufficiently flexible parametrized expressions that you optimize to get close to microscopically obtained results.

Professor Kremer asked: In the application of your formalism to star polymers, you restrict yourself to two body potentials between the stars. This is known to work well in dilute solutions and otherwise introduces essentially a meanfield approximation on the level of the star polymers, since the influence of pair wise deformations on the interaction with a third molecule is not considered properly. How do the present results compare with previous experimental and theoretical studies on star polymer solutions? What would be the effort to include the three body terms mentioned before?

Professor Español answered: As we emphasize in the paper, the basic difficulty of the coarse-graining approach is the need to deal with effective potentials and frictions that depend on *all* the coarse-variables selected to describe the system. This question is closely related to the previous one by Professor Allen, where I have suggested a possible route to get three body terms. It is possible to run constrained dynamics that allow to explore the functional form of three body interactions, and refine the results with sufficiently flexible parametrized forms for the three body interactions through the force matching method, for example.

Professor Pagonabarraga remarked: The coarse-grained frictional forces you derive for the star polymers are not central (as shown, *e.g.* in eqn (56) of your paper). Therefore, angular momentum is not conserved. What implications does it have on star polymer dynamics? Does it imply that at a coarse-grained level fluid spin will in general have to be taken into account?

The perpendicular component of the friction coefficient is negative, as displayed in Fig. 4. Later, when comparing the coarse-grained and microscopic results you do not allow negative values of the friction coefficient. Is there a physical reason for such a choice? Is it possible that some of the discrepancies observed in Fig. 7 can be attributed to dismissing such negative contributions?

Professor Español replied: Indeed angular momentum is not conserved by these friction forces. We were not concerned with this issue, because our MD simulations with periodic boundary conditions do not conserve angular momentum either. If conservation of angular momentum is absolutely essential in certain applications, then it is necessary to include the angular momentum of the molecules with respect to the CoM as an additional variable. This can be done, at the expense of an increased complexity of the dynamic equations that now couple the dynamics of the CoM with its spin. A discussion of the structure of the resulting equations can be found elsewhere.[1] Regarding the second question, the only reason to approximate the small negative tail with zero is our desire to use the DPD modelization of the random force that requires the amplitude of the noise to be positive. If we do not use this modelization, we would need to construct the noise by, for example, obtaining the square root of the full friction matrix, which is computationally too expensive. This approximation could be at the origin of the discrepancies observed, but this is difficult to test.

1 P. Español, *Phys. Rev. E*, 1998, **57**, 2930.

Professor Berendsen asked: You show that the potential of mean force and the friction can be determined from a constrained dynamics simulation, but only in the limit of $\varepsilon \to 0$, ε being a parameter introduced to separate time scales of the "relevant" and "irrelevant" degrees of freedom. But real systems do not have such a clean separation of time scales! Is the method still useful? Can one use the time correlation function of the force correlation to derive a memory kernel for generalized Langevin dynamics?

Professor Español replied: The Markovian assumption is what makes Zwanzig theory a useful practical theory. This assumption is valid when the evolution of the coarse-grained variables can be decomposed in two parts, a rather systematic part evolving on long time scales plus a small amplitude, rapidly varying part.

The rapidly varying part is modelled then as noise. Of course, as you say, in reality one does not have the infinite separation of time scales suggested above, and the results of the Markovian assumption should be taken as an approximation whose validity must be checked by comparing its predictions with simulations or experiments.

If this separation of time scales does not exist at all, then I would say that Zwanzig theory remains a rather formal and useless one. Beyond the problem of finding a coloured stochastic noise whose correlation is given exactly by the memory kernel, a more fundamental problem is the very computation of this memory kernel. Of course, one can use the correlation of the real forces (as opposed to forces governed by the projected dynamics) or the correlation of the constrained forces (evolving according to a constrained dynamics) to produce models for the kernel, but the mathematical justification for that is unclear to me.

Professor Allen said: In reading this paper I found it useful to make contact with the conventional discussion of Mori-Zwanzig theory, especially as it is applied to hydrodynamic modes.[1] Possibly others will find it helpful. One of the key steps is the introduction of a time scaling parameter ε at the beginning of section IV. In the simplest applications of Mori-Zwanzig, to conserved variables, this role is played by the wavenumber k. In fact, one studies variables $A(k,t)$, microscopically defined in terms of coordinates and momenta, whose time derivatives may be shown rigorously to have a low-k series expansion whose leading term is linear in k: $\mathrm{d}A/\mathrm{d}t = LA = ikX(k,t)$ where X is another microscopic variable. An example is the transverse momentum density $p(k,t) = ik\sigma(k,t)$ where σ is an off-diagonal element of the stress tensor. It then follows that the projected variable QLA also carries this factor ik and, despite its unusual time evolution, the correlation function (memory function) inevitably is proportional to k^2. This automatically leads to a time scale separation between slow and fast variables, in the coarse-grained first-order differential equation that governs the evolution of $A(k,t)$, that can be expressed by defining a scaled time $\tau = k^2t$. Since k may be made arbitrarily small, the plateau problem is avoided, and one gets the expected solution for the correlation function of A, namely $C_{AA}(k,t) \propto \exp(-\nu\tau) = \exp(-\nu k^2 t)$ where ν is a transport coefficient (the kinematic shear viscosity in the case of transverse momentum).

1 *Theory of Simple Liquids*, ed. J. P. Hansen and I. R. McDonald, Academic Press, 3rd edn, 2006, ch. 8, 9.

Professor Español responded: Yes I fully agree.

Dr Hess asked: Is the derivation of a generalized Langevin equation with proper memory functions always problematic, or is it possible in a 1-dimensional case?

A quite rigorous derivation of a 1-D GLE is presented by Lange and Grubmueller.[1] But maybe this derivation is not fully rigorous?

1 O. F. Lange and H. Grubmüller, *J. Chem. Phys.*, 2006, **124**, 214903.

Professor Español replied: The derivation of the generalized Langevin equation is not problematic. What is problematic is the usefulness of this equation. As I have already mentioned previously, I believe that if memory effects are important, then Zwanzig theory is not very useful, because it is difficult to compute explicitly the memory that involves the projected dynamics. However, and at the expense of sacrificing mathematical rigor, nothing refrains you from using your intuition, guess, or whatever and try to offer sensible expressions for the memory that turns out to work in practice. Now, in the particular case of the reference that you mention now, the authors obtained the memory function under the approximation that it does not depend on the collective variable selected, as Zwanzig theory would require in general. This allows them to use the Volterra equation of the first kind (which is linear in the velocity autocorrelation function) and, by computing the velocity autocorrelation function of the selected collective variable, infer the memory. Again, this is an approximation that allows one to proceed and get results. However, the results obtained in this way should always be tested against the full MD simulations.

Professor Voth asked: You have stressed that you view the Mori–Zwanzig theory as the theory of coarse-graining. In principle, your point of view has considerable merit if the Mori–Zwanzig theory could be made to be workable in a general sense for a complex many-dimensional problem, in which a significant number of degrees of freedom are "left behind" as the CG variables. However, it is not clear to me that the Mori–Zwanzig theory is the "only" way to CG a system. There might even be some situation where you would like to formally integrate out the slow variables and be left with the fast ones, though I cannot come up with an example right now. Can you please give your thoughts on this? Do you really think that M–Z theory is the only sensible way to CG a system?

Professor Español answered: I believe that Zwanzig theory is a sufficiently general theory for coarse-graining and it is, essentially, the theory of non-equilibrium statistical mechanics. You can build upon this theory the GENERIC framework[1] and have a rather comprehensive framework for complex fluids. I have tried to be a bit provocative with my position in order to favor a full understanding of this useful theory by our community.

Having said that, fundamentalism is not a good scientific position and I would not refrain myself to explore new possibilities. Indeed there are several issues that deserve much work in Zwanzig theory. A very first basic question is the selection of variables (are CoM the best coarse variables? Are there other "modes" around?). We need a theory of pattern recognition of good coarse variables in the first place. A second problem is the one that you mention, the highly dimensional space of coarse variables, and we need to explore different routes to tackle this problem. Finally, if we are willing to deal with non-Markovian descriptions (for which there would be, in principle, no limitation of which are the coarse variables) we need to devise methods for explicitly computing the projected dynamics.

1 H. C. Ottinger, *Beyond Equilibrium Thermodynamics*, Wiley, 2005.

Dr Wysocki continued the discussion of the paper by Dr Louis: In the paper you have considered diffusion of particles confined to a channel in equilibrium. Do you have an idea as to how the behaviour of the particles would change in presence of an external imposed flow field, *i.e.* in nonequilibrium? (*e.g.* in presence of a shear or a Poiseuille flow).

Dr Padding answered: We have actually already carried out simulations of colloids undergoing flow in a narrow pipe.[1] We find that when the colloid diameter becomes non-negligible compared to the diameter of the pipe, there are important corrections to the original Taylor-dispersion picture. For example, the colloids can flow more rapidly than the underlying fluid, and their Taylor dispersion coefficient is decreased. Also, the long-time tails in the velocity autocorrelation functions are altered by the Poisseuille flow.

1 J. Sané, A. A. Louis, J. Padding, http://arxiv.org/abs/0903.3493.

Dr Louis responded: That is an interesting question that we are thinking about. For the very narrow channels under Poiseuille flow, to first order we expect that the particles will move along with the flow at a faster rate than the average flow because they mainly sample flow lines in the middle of the channel. However, for narrow channels the particles also feel enhanced friction from the walls, which in turn slows them down. You can see an example of how these two effects interact in our paper.[1] For channels with 1d confinement, we would expect that the mean square dispalcement w.r.t. the average flow should still scale as $\sim t^{1/2}$.

1 J. Sané, A. A. Louis, J. Padding, http://arxiv.org/abs/0903.3493.

Dr van der Sman addressed Dr Louis and Professor Löwen: Another interesting problem extending the confinement problem is investigation of two particles in close confinement in gravity field. They show significant enhancement of mixing—a problem in microfluidic devices.

Hierarchical coarse-graining strategy for protein-membrane systems to access mesoscopic scales

Gary S. Ayton, Edward Lyman and Gregory A. Voth*

Received 30th January 2009, Accepted 24th February 2009
First published as an Advance Article on the web 4th August 2009
DOI: 10.1039/b901996k

An overall multiscale simulation strategy for large scale coarse-grain simulations of membrane protein systems is presented. The protein is modeled as a heterogeneous elastic network, while the lipids are modeled using the hybrid analytic-systematic (HAS) methodology, where in both cases atomistic level information obtained from molecular dynamics simulation is used to parameterize the model. A feature of this approach is that from the outset liposome length scales are employed in the simulation (*i.e.*, on the order of ½ a million lipids plus protein). A route to develop highly coarse-grained models from molecular-scale information is proposed and results for N-BAR domain protein remodeling of a liposome are presented.

Introduction

One of the most significant challenges facing the field of biomolecular simulation is the need to access the scales and complexity inherent in real biological systems, which has clearly spurred a rapid growth in the development of coarse-grained (CG) and other multiscale computational methods for biomolecular systems (see, *e.g.*, ref. 1 and contributions therein). Within the context of membrane-protein systems, the ability to accurately simulate systems of proteins and lipids at the length and time-scales where these systems manifest their biological function presents a number of challenges. Real membranes are highly inhomogenous and include multiple lipids, cholesterol, and numerous proteins. The emerging picture of membranes, enhanced from the original fluid mosaic model,[2] is one that emphasizes membrane crowding, varying membrane thickness, lipid spatial organization and even oligomerized proteins.[3,4] It is now appreciated that up to 20% or even higher of the membrane surface area is proteins.[5]

Protein mediated membrane remodeling is an example of a case whereby protein modules (*e.g.*, the BAR (Bin/amphiphysin/Rvs) domain) can remodel entire liposomes having initial diameters of around 200 nm into thin tubulated structures with diameters on the order of 20 to 50 nm[6–8] over time-scales significantly longer than microseconds.[9] The BAR protein domain is a crescent-shaped dimer with a positively charged concave surface.[6–18] The addition of an N-terminal amphipathic helix to this protein module results in an N-BAR domain and these have been observed to strongly tubulate liposomes *in vitro*.[6–8] The N-terminal amphipathic helices "burrow" into the headgroup region of the upper membrane leaflet, resulting in a curvature generating "wedge" mechanism for membrane remodeling.[8,12,13] The wedge mechanism of N-BAR mediated liposome remodeling has been supported by recent theoretical studies,[19] and by large scale molecular dynamics (MD)

Center for Biophysical Modeling and Simulation and Department of Chemistry, University of Utah, 315 S. 1400 E, Room 2020, Salt Lake City, Utah, 84112-0850 simulations.[13] Other remodeling mechanisms, including a "scaffolding"[6,11,16,17] and electrostatic sequestering[4,20] mechanism, have also been proposed. In all of these remodeling mechanisms, there is a strong degree of localized, atomistic level behavior that is collectively and ultimately responsible for the long length and time-scale process of liposome remodeling.

Large-scale atomistic MD simulation has proven invaluable in examining single N-BAR remodeling;[12,13] however, this methodology alone cannot reach the physiological length and time-scales (μm and μs) examined experimentally.[6] CG simulation can potentially examine membrane-protein systems at very large length and time-scales with a degree of quantitative predictability as long as the CG interactions can faithfully represent the averaged underlying atomistic-level interactions. To date, CG-MD has been employed to model both proteins and lipid bilayers (see ref. 1,21–25 for examples and reviews) including so-called "solvent-free" membranes,[26–31] where the effects of the surrounding solvent are somehow incorporated into the CG lipid interactions. A previous CG simulation of N-BAR domain induced membrane remodeling[18] has employed both high-resolution,[32–34] and low-resolution shape-based CG schemes[35] (resolved at about 150 atoms per CG site). These CG simulations were able to examine up to six N-BAR domains on a small slab of bilayer in different spatial arrangements. A generic highly coarse-grained model of protein mediated membrane remodeling has also been used to study vesiculation.[36] Still, despite the various CG schemes employed, the system sizes simulated have still been well below the liposome length scales from experiments.[6]

The 200 nm diameter liposomes employed in N-BAR remodeling experiments[6] contain upwards of half a million lipids with possibly up to 2000 N-BAR domains bound to the surface. However, current CG simulations[28,30,37–41] have modeled at best small vesicles with diameters in the range of 40 to 60 nm[39] having typically less than 5000 lipids.[28,30,37,38] Current CG lipid models typically employ 10 to 15 CG sites per lipid,[32,42–45] resulting in about a 10 to 1 atom to CG site ratio. Thus, if CG methods are going to simulate N-BAR domain remodeling at experimental liposome length scales, some sort of alternative and aggressive coarse-graining scheme, combined with highly parallel computers, are needed. However, even as the degree of coarse-graining is increased, the systematic multiscale nature of the methodology must somehow be retained in order for the molecular level interactions to be bridged to the CG scale.

We have recently developed highly coarse-grained multiscale methods for lipids[46] and proteins.[47] The former is the hybrid analytic-systematic (HAS) approach that employs the multiscale coarse-graining methodology (MS-CG)[45,48–57] to augment and systematically parameterize key aspects of an otherwise generic analytic lipid model. The HAS approach thereby decomposes the total CG interaction into a systematic and an analytic component. The systematic part of the CG interaction is based on the MS-CG algorithm and is employed for those configurations that correspond to well sampled atomistic configurations as generated by the original MD simulation used to parameterize the MS-CG model. The analytic component describes the other configurations based on the global physical behavior.

The partitioning between the analytic and systematic components of the HAS CG interaction is tunable, and for very large systems the CG model can be initially biased towards the analytic component in order to determine the computational feasibility of a particular choice of CG scheme (*e.g.*, is the number and rough location of the CG sites reasonable?). Then, the systematic MS-CG component can be introduced gradually in a sequence of iterations, until the model is biased towards as many interactions as possible obtained from underlying atomistic MD simulations. This approach helps to ensure computational feasibility throughout the process. It can also give important insights into the specific roles of the systematically obtained interactions as they are sequentially introduced. The tunability of the HAS approach will play a key role in the development of the full protein-membrane model to be discussed soon.

The HAS analytic component is based on the Gay-Berne ellipsoid (GB) particle model,[58,59] and was selected as it requires only a single CG site at the center of the ellipsoid (denoted the GB CG site), along with a unit vector, to fully specify the GB ellipsoid's location and orientation. However, the HAS approach is quite general and other analytic models can instead be employed. Small modifications to the analytic component of the interaction can give the model the symmetry of a lipid, *i.e.*, a "head" and a "tail".[46] The result is a low resolution CG lipid model[46] resolved at about a 100 to 1 atom to CG site ratio, but also a CG model that has many of its interactions systematically obtained from actual molecular-scale forces.

The present CG protein model[47] employs a variant of the fluctuation matching method[60] to define a heterogeneous elastic network model (HeteroENM), in this case for an N-BAR protein using our essential dynamics coarse-graining (ED-CG)[61] approach to define the protein CG sites. The level of coarse-graining in this case is also approximately 100 atoms to 1 CG site.

The focus of the remainder of this paper will be to describe our multiscale strategy for large length scale membrane protein systems. The previously introduced HAS CG membrane (which will be referred to simply as the HAS membrane) along with the ED-CG HeteroENM N-BAR domain model (which will be denoted as the ED-BAR model) will be combined to give a unified multiscale CG membrane-protein model capable of operating at liposome length scales. Simulations of the early stages of N-BAR domain induced remodeling of a liposome will be used as a demonstration of the overall methodology.

A "divide and conquer" approach to systematic multiscale simulation is employed. From the outset, a fully realized ED-BAR liposome system is considered. The flexible and tunable HAS approach is applied to all CG interactions in the system, where initially the CG simulation is biased towards the analytic component. The underlying analytic interactions are then sequentially replaced with more fully realized systematic interactions as more atomistically detailed information is made available. The overall process is evolutionary in that aspect and, as will be discussed, there are still components of the interaction that will undergo further refinement in the future.

Simulation

Simulations were composed of 406 092 HAS CG lipids and 52 800 ED-BAR CG sites, the latter corresponding to 2400 ED-BAR domains. The liposome by itself was initially equilibrated, and then the ED-BARs were arranged in a disordered array just beyond the liposome surface. A total of 100 ns of CG simulation was then performed over 512 processors using our in-house CG code TANTALUS, which employs a dual spatial decomposition capable of easily scaling over 1000 processors.

Results and discussion

The ED-BAR and HAS lipid model are graphically depicted in Fig. 1a and 1b, respectively. The algorithmic details and exact functional forms of the interactions for both the protein and lipid model can be found elsewhere.[46,47] Here, aspects pertaining to the hierarchical coarse-graining strategy will be discussed. Two main components of the multiscale CG methodology have been previously developed, these are the ED-BAR protein and the HAS membrane. More details on these models are given below. In order to construct the full ED-BAR HAS membrane system, new cross interactions must be specified, and these will also be presented.

Heterogeneous elastic network model of an N-BAR domain

Elastic network models (ENMs) for proteins typically build an elastic network between pairs of α carbons that are within a fixed cut-off distance[62] with a uniform

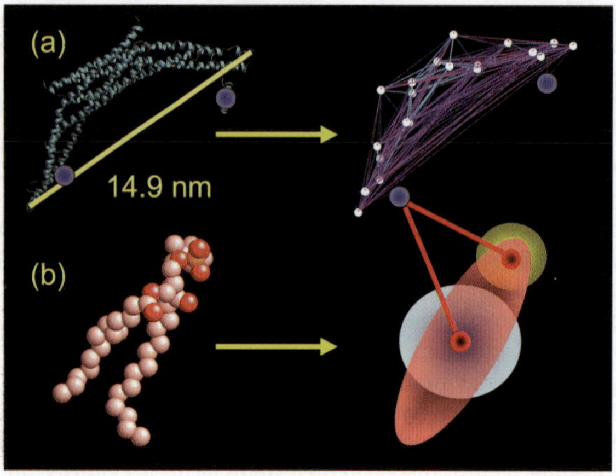

Fig. 1 The CG model used in the ED-BAR-liposome simulation. Part (a) shows the hetero-ENM model of the ED-BAR domain as derived from the fully atomistic representation. The two amphipathic helices (highlighted by blue spheres in the left panel) are modeled by two additional sites, as shown in the right panel image. Part (b) depicts the HAS CG lipid model. The fully atomistic lipid is modeled as a single site GB ellipsoid of revolution, augmented with a radially symmetric MS-CG interaction. See ref. 46 for more details. An additional headgroup site interacts with the ED-BAR domain, but does not alter the total lipid–lipid interaction.

spring constant.[63,64] These ENMs have been employed to examine a number of proteins and macromolecular complexes.[65–68] A more accurate heterogeneous elastic network can instead be constructed by fitting the various spring constants to thermal fluctuations of coarse-grained distances as calculated from atomistic MD.[47] The inclusion of detailed atomistic level information allows for the systematic parameterization of different elastic models of the same protein in different environments, *e.g.*, an N-BAR in solution or bound to a membrane.

The optimal locations of the 20 CG sites of the ED-BAR were determined using the ED-CG method,[61] where the locations of coarse-grained sites were selected such that the essential low frequency dynamics observed in the atomistic simulation are reproduced. An N-BAR bound to a lipid bilayer was employed for the parameterization.[12] Thus, both the location, as well at the interactions, of the ED-BAR CG sites were found from atomistic MD simulation. The N-BAR domain at the atomistic level, as shown in the left panel of Fig. 1a, is thus mapped into an ED-BAR model. For this work, additional CG sites corresponding to the N-terminal amphipathic helices were also included on the ED-BAR; however, at this level of coarse-graining the entire N-terminal helix was reduced to a single CG site. Future work will resolve these particular sites at higher levels of CG resolution, since they account for only about 2% of the total CG sites of the system.

Hybrid analytic-systematic (HAS) lipid bilayer model

The membrane in the present work employs the HAS model as was previously done.[46] The original lipid, as shown in the left panel of Fig. 1b, is mapped into a single ellipsoid of revolution with a different head and tail interaction as shown in the right panel of Fig. 1b. The additional analytic component of the interaction (shown as the white sphere in the center of the ellipsoid) modulates the generic GB interaction to give the desired membrane material properties (*i.e.*, the bending modulus, area compressibility, and diffusion) that are representative of a lipid bilayer in the liquid crystal phase. In fact, without the systematic component of the HAS interaction the membrane generally freezes into a solid.[46] Previous work parameterized the model for a pure dimyristoylphosphatidylcholine (DMPC) membrane, which generally is not what is used in experiments, which typically employ mixtures of dioleoylphosphatidylserine (DOPS) and dioleoylphosphatidylcholine (DOPC).[69,70] We are currently developing a CG mixture formulation for DOPS and DOPC using MS-CG[71] which can be incorporated into the HAS approach. For the present study the membrane can be considered as a single site lipid model with (in part) atomistically determined interactions that result in reasonable material properties for the system under consideration. This HAS lipid model is taken to be a good starting point for the subsequent CG simulation of N-BAR domain induced membrane remodeling.

The HAS approach for protein–lipid cross interactions

With the ED-BAR and HAS membrane in hand, the remaining task is to develop a hierarchical approach whereby systematically obtained cross interactions can be sequentially incorporated into the overall ED-BAR/HAS membrane model. The initial starting point is thus to define a set of reasonable starting analytic interactions between the various CG components (*i.e.*, the ED-BAR/ED-BAR and ED-BAR/HAS lipid interactions).

Each CG site on one ED-BAR (a total of 20 sites plus 2 for the N-termini) interacts with each site on another ED-BAR. Each ED-BAR CG site also interacts with the headgroup and GB CG sites on the HAS lipid (as shown in Fig. 1). Spherical ED-BAR CG sites interact with the GB ellipsoid with a GB sphere-ellipsoid interaction, employing the 3 : 1 aspect ratio of the HAS lipid.[46] This additional analytic component models the excluded volume of the lipid tails and the BAR protein. The interaction strengths for the analytic components were initially varied, and the resulting behavior of the system, in terms of ED-BAR binding and membrane remodeling, was observed. Such preliminary insight into how sensitive are particular initial analytic interactions to the overall process of ED-BAR binding and membrane remodeling can prove quite valuable when the systematic modeling components are introduced later.

The analytic interaction between ED-BARs employs a standard Lennard–Jones (LJ) inverse-power law interaction, expressed as $u(r) = 4\varepsilon\,[(\sigma/r)^{12} - (\sigma/r)^{6}]$ where r is the distance between two CG sites, σ gives the length scale of the interaction, and ε is the interaction strength. Different interaction strengths with $\varepsilon = 0.2$ to 1 kcal mol^{-1} were initially examined in order to determine a physically meaningful interaction strength such that the ED-BARs did not coalesce and condense into an agglomerated phase. The final analytic interaction was chosen to be purely repulsive, obtained with $\varepsilon = 0.24$ kcal mol^{-1} and truncating $u(r)$ at a distance of $2^{1/6}\,\sigma$, with $\sigma = 2$ nm. The behavior of the ED-BAR/HAS membrane system, in terms of ED-BAR binding and remodeling, was found to be relatively insensitive to the ED-BAR/ED-BAR interaction strength. Employing the repulsive interaction at this point greatly reduced the cutoff distance of the interaction, resulting in a measurable computational speed-up. A more rigorous multiscale parameterization would require extensive multi-N-BAR atomistic-level simulations in order to extract a potential of mean force (PMF) between ED-BAR CG sites. This work is currently underway.

The initial analytic cross interactions between the CG sites of an ED-BAR (excluding the N-terminal sites) and the headgroup site of the HAS lipid (see Fig. 1) were also taken to be of a LJ form. The systematic multiscale component was incorporated at this point from an atomistically obtained averaged radial distribution function between the phosphorus atom in the lipid headgroups (as shown in Fig. 1) and the centers of mass of the positively charged residues underneath the arch of the N-BAR. As shown in Fig. 2, a PMF was then constructed, and the resulting well depth of the interaction was employed to parameterize the LJ cross interaction. The well depth was found to be around 0.2 kcal mol^{-1}, and the repulsive wall of the interaction was at about 1.5 nm. The range of the attractive component of the interaction was found to be on the order of 3 to 4 nm. This interaction is not strong

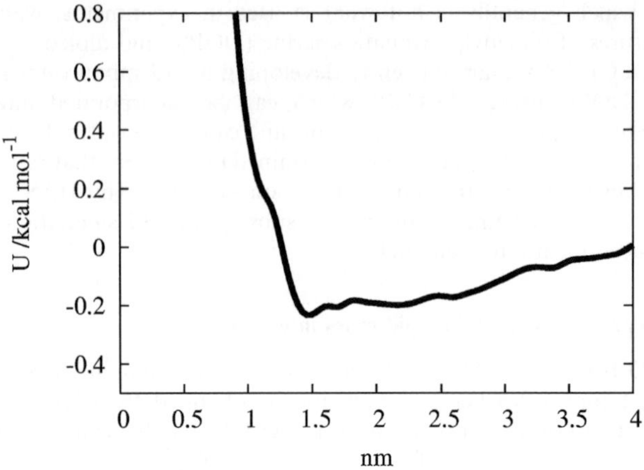

Fig. 2 The potential of mean force (PMF) obtained from the distribution function for the distance between the phosphorus atom in the lipid headgroups to the center of mass of the positively charged residues underneath the BAR domain.

enough alone for the individual ED-BARS to strongly bind onto the membrane, and at least at this level of analysis suggests that electrostatic binding of the BAR domain "arch" to the lipid headgroups may not be the dominant binding interaction for N-BAR domains on membranes. A more advanced form of this interaction would not average over all the residues on the N-BAR arch, and it could instead employ a tabulated potential interaction obtained from the MS-CG method.[45,48–56] This approach requires additional MD simulations in order to obtain statistically reasonable sampling for these interactions.

The N-terminal amphipathic helix CG sites of the ED-BAR were given a LJ interaction such that they can bury themselves and bind into the upper leaflet of the membrane. This is depicted in Fig. 1b where the lines connecting the N-terminal amphipathic CG site on the ED-BAR with the two sites of the HAS lipid designate the interaction between sites. Given that the entire elongated helix is mapped into a single spherical site at this point, it proved difficult to accurately map an atomistically determined PMF onto the single CG interaction site. The approximate "volume" of the helix was thus employed to reproduce the size of the ED-BAR helix, and interaction strengths ranging from $\varepsilon = 5$ to 15 kcal mol^{-1} were tested. The large value of the interaction strength draws from the observation from MD simulation that the N-terminal helices generally remain tightly bound in the upper leaflet of the bilayer.[12] Over this range, the interaction resulted in ED-BARs burying their N-terminal sites into the upper leaflet of the membrane without penetrating through to the other side, but also not consistently floating off. The next level of refinement of this model would be to better resolve the CG N-terminal helix with possibly two or three sites as determined from the ED-CG analysis such that an atomistically obtained PMF can be constructed. This work is currently under way.

Simulation results

The overall behavior of the final ED-BAR HAS membrane model is such that the ED-BARs can bind and unbind to the liposome surface. The N-terminal sites can embed into, and dislodge from, the upper leaflet. The interaction between the ED-BARs and membrane is not so strong that the ED-BARs automatically condense on the membrane surface, yet they are not so weak that they remain unbound.

In Fig. 3, snapshots of the ED-BAR liposome after 100 ns of CG simulation time are shown. After a simulation of this relatively short duration (keeping in mind that

CG time has no relation to "real" time without additional modifications[72] and that we have made no attempt to re-scale time as is often done in CG modeling), only the early stages of liposome remodeling are examined here. The original system shows the cloud of ED-BARs over the liposome surface, with some bound, and some not. It should be noted that the starting configuration set the ED-BARs at a close proximity to the liposome surface, but they were not embedded in the bilayer from the outset. During the course of the simulation, the ED-BAR cloud remained intact, although it can be seen that some ED-BARs float off into the surrounding medium.

It is possible to measure the average N-BAR density, the membrane curvature, and the local membrane density using the density field discretization procedure employed in smooth particle applied mechanics.[73] The color bar at the bottom of Fig. 3 gives an indication of the behavior of various quantities over the liposome surface. The local ED-BAR density gives the density of ED-BAR CG sites in close proximity to the membrane surface. Clearly, the ED-BAR density is far from homogenous and it shows enhancements and depletions over the surface. The regions highlighted in the yellow squares were selected as they visibly show a local ED-BAR density enhancement. When correlated with the mean curvature, H, it suggests that these regions also have an enhanced curvature. Furthermore, the relative membrane

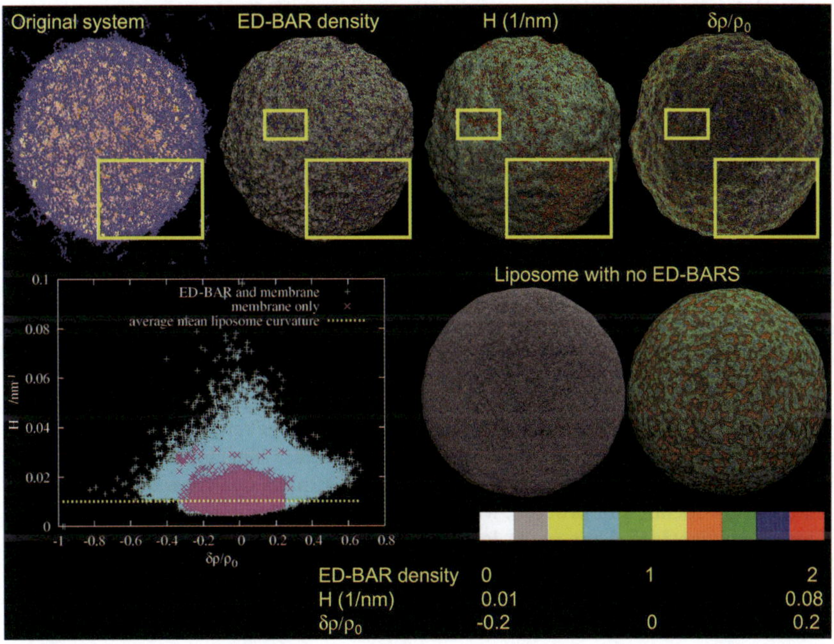

Fig. 3 Analysis of the ED-BAR density, mean curvature, H, and relative membrane density change, $\delta\rho/\rho_0$, in the remodeled and isolated liposomes. The original system is a snapshot of the ED-BAR coated liposome. The local ED-BAR density shows local ED-BAR enhancements and depletions that are correlated with both the mean curvature and membrane density. The yellow squares highlight regions where the correlations are most clearly visible. The lower panel on the right shows corresponding plots for the isolated liposome. The scale bar at the bottom gives the reference for each color in the upper panels. The inset at the lower left plots the mean curvature *versus* liposome density for the two systems. Purple is for the isolated liposome while cyan shows how the local mean curvature is enhanced and distorted because of the ED-BAR membrane remodeling. The upper values of the mean induced curvature correspond to those observed from MD simulations (*i.e.*, about half that observed from MD simulation of a single N-BAR domain,[12] since the mean curvature is the average over two principle curvatures and the single N-BAR domain induces curvature in only one direction).

density change, $\delta\rho/\rho_0$, in this (and similar) regions exhibits depletion. Putting all three together, it suggests that the N-terminal sites have penetrated into the upper leaflet of the membrane, resulting in an induced curvature and subsequently a lower membrane density.

When compared to a liposome with no ED-BARs (*cf.* Fig. 3, lower right), it is clear how significant the initial stages of the membrane remodeling are. Both simulations were performed for the same duration. However, the isolated liposome with no ED-BARs does not exhibit any large curvature deviations and only has small local density fluctuations. It may in fact be possible that the ED-BARs are seeking out regions of the liposome surface with a slightly lower density so the amphipathic helices can be more readily inserted, rather than just seeking out regions with an enhanced curvature.

The correlation between the ED-BAR induced curvature and membrane density is shown in the inset in the lower left of Fig. 3. Each point on the plot corresponds to a HAS CG lipid site at some point in space **r**. The local mean curvature, H, for liposomes with and without ED-BARS is plotted *versus* relative membrane density change, $\delta\rho/\rho_0$, where ρ_0 is the average density $\delta\rho = \rho - \rho_0$, and ρ is the density at **r**. Both the upper and lower monolayers are included in this calculation. The liposome without ED-BARS shows only a small variation of curvature that on average recovers the mean curvature of the liposome. However, the presence of the ED-BARS greatly modifies the liposome structure. The membrane density is spread out and the curvature is increased everywhere. Interestingly, the largest mean curvatures correspond to what has been predicted from MD simulations of a single N-BAR,[12] suggesting that the multiscale nature of the present CG methodology is robust.

Fig. 4 gives some evidence as to how the ED-BARS locally remodel the liposome curvature. The four snapshots of the liposome were found by rotating the liposome, and selecting out regions with distinct structural motifs. The yellow box highlights specific regions for discussion. In (a), a pronounced "double hump" is observed, where the tails of the ED-BARs protrude into the "divot" between the two humps. The N-terminal sites are not right at the end of the ED-BARS and can still embed into the bilayer at the bottom of the divot. The measurement of the mean curvature in these very highly curved divots is suspect due to the fact that the CG membrane structure in these regions is so highly disrupted. This type of ripple structure appears to be evident over the surface of the liposome and seems to correspond to a local mechanism whereby ED-BARS can generate strong curvature, yet still retain the membrane structural integrity. Panel (b) shows a large remodeled section where the embedded helices in the upper leaflet are visible. The lower leaflet remains surprisingly structurally intact. Panel (c) shows just how significantly the ED-BAR induced remodeling can proceed, where the membrane has almost folded in on itself. This particular region, at much longer simulation times, may potentially become a region where tubulation is initiated. Panel (d) shows a relatively flat region of the membrane surface with a correspondingly low ED-BAR density. In this small region, the membrane curvature does not deviate greatly from the average liposome curvature.

It is interesting how the ED-BARs are able to globally increase the local curvatures of the liposome without drastically altering its radius. The radius of the liposome without ED-BARs averaged to 96 $\pm$ 3 nm, while the liposome with ED-BARs was 88 $\pm$ 5 nm, where the enhanced curvature fluctuations arise from the ED-BAR induced membrane sculpting. The overall driving force for remodeling also appears to be related to N-terminal helix insertion. Indeed, some preliminary simulations were performed without the N-terminal helices, and in those cases either the ED-BARs did not bind to the membrane or they elected to, quite often, bind sideways or too deeply in the membrane. In either case, the pronounced and robust membrane sculpting observed in the simulations with ED-BARs containing N-terminal helices was not observed to the same degree.

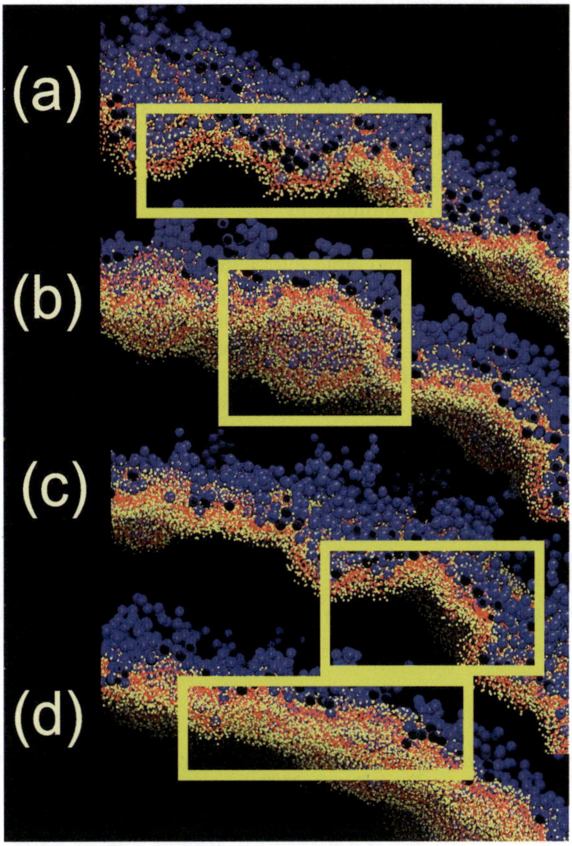

Fig. 4 Selected close-up snapshots of the remodeled liposome, obtained by rotating the liposome and selecting out specific regions. The square boxes highlight specific remodeled regions of interest: (a) shows a double hump, (b) a large curved region where the embedded helix sites are visible, (c) a tightly bent region with a high density of ED-BARs on the surface, and (d) a relatively flat region with a correspondingly low ED-BAR density in that region.

The above observations seem consistent with the "wedge" mechanism of N-BAR remodeling.[8,13] In fact, N-BARs may not be curvature sensors as much as they are density sensors. The N-terminal amphipathic helices seek out regions of the liposome surface with a slightly lower density so that they can more efficiently embed themselves into the outer leaflet of the bilayer. The "arch" of the N-BAR domain, with its screened electrostatic interactions to the lipid headgroups, may exist primarily to "loosen up" the lipid headgroup density and also to scaffold the N-terminal amphipathic helices in place. The result of this is a curvature generation that, when combined with some degree of N-BAR oligomerization, can result in a spontaneous curvature along one direction and hence tubulation. The electrostatic effect gives a soft attractive binding force of the N-BAR to the liposome surface, but it is not enough to drive remodeling without the additional helix insertions into the upper membrane leaflet.

Conclusions

The paper has outlined a general multiscale simulation methodology for protein–membrane systems applicable to very large length scales where highly coarse-grained models are required, but molecular-scale information can still be employed to parameterize key aspects of the CG model. The example of N-BAR induced membrane remodeling at full liposome length scales was employed to demonstrate the features of the methodology. The hybrid analytic-systematic (HAS) CG approach, where the CG interactions contain both analytic and systematic components, provides a means whereby initial analytic components of the overall model can be replaced by systematically determined interactions obtained in a multiscale fashion from underlying MD simulations.

Future work will focus on increasing the overall systematic nature of the model. For example, a more refined lipid–lipid interaction from the MS-CG method will be employed that maps over to the actual lipid mixtures used in experiments. The N-BAR membrane interaction will also be refined based on molecular simulation. Another key component will be to model the N-BAR oligomerization interactions at a higher level of detail. This effort will draw on extensive atomistic-level MD simulations of multi-N-BAR systems, and this work is currently under way.

Acknowledgements

This research was supported by the National Institutes of Health (R01-GM063796). Computational resources were provided by the National Science Foundation through TeraGrid computing resources, specifically the Texas Advanced Computing Center. We acknowledge Professor Vinzenz Unger of Yale University for helpful discussions and insights.

References

1 *Coarse-graining of condensed phase and biomolecular systems*, ed. G. A. Voth, CRC Press/ Taylor and Francis Group, Boca Raton, 2009.
2 S. J. Singer and G. L. Nicolson, *Science*, 1972, **175**, 720–731.
3 D. M. Engelman, *Nature*, 2005, **438**, 578–580.
4 S. McLaughlin and D. Murray, *Nature*, 2005, **438**, 605–611.
5 A. D. Dupuy and D. M. Engelman, *Proc. Natl. Acad. Sci. U. S. A.*, 2008, **105**, 2848–2852.
6 B. J. Peter, H. M. Kent, I. G. Mills, Y. Vallis, P. J. G. Butler, P. R. Evans and H. T. McMahon, *Science*, 2004, **303**, 495–499.
7 K. Takei, V. I. Slepnev, V. Haucke and P. De Camilli, *Nat. Cell Biol.*, 1999, **1**, 33–39.
8 J. L. Gallop, C. C. Jao, H. M. Kent, P. J. G. Butler, P. R. Evans, R. Langen and H. T. McMahon, *EMBO J.*, 2006, **25**, 2898–2910.
9 M. Masuda, S. Takeda, M. Sone, T. Ohki, H. Mori, Y. Kamioka and N. Mochizuki, *EMBO J.*, 2006, **25**, 2889–2897.
10 J. L. Gallop and H. T. McMahon, *Biochem. Soc. Symp.*, 2005, **72**, 223–231.
11 P. K. Mattila, A. Pykalainen, J. Saarikangas, V. O. Paavilainen, H. Vihinen, E. Jokitalo and P. Lappalainen, *J. Cell Biol.*, 2007, **176**, 953–964.
12 P. D. Blood and G. A. Voth, *Proc. Natl. Acad. Sci. U. S. A.*, 2006, **103**, 15068–15072.
13 P. D. Blood, R. D. Swenson and G. A. Voth, *Biophys. J.*, 2008, **95**, 1866–1876.
14 G. S. Ayton, P. D. Blood and G. A. Voth, *Biophys. J.*, 2007, **92**, 3595–3602.
15 K. Futterer and L. M. Machesky, *Cell*, 2007, **129**, 655–657.
16 A. Shimada, H. Niwa, K. Tsujita, S. Suetsugu, K. Nitta, K. Hanawa-Suetsugu, R. Akasaka, Y. Nishino, M. Toyama, L. Chen, Z.-J. Liu, B.-C. Wang, M. Yamamoto, T. Terada, A. Miyazawa, A. Tanaka, S. Sugano, M. Shirouzu, K. Nagayama, T. Takenawa and S. Yokoyama, *Cell*, 2007, **129**, 761–772.
17 A. Frost, R. Perera, A. Roux, K. Spasov, O. Destaing, E. H. Egelman, P. De Camilli and V. M. Unger, *Cell*, 2008, **132**, 807–817.
18 A. Arkhipov, Y. Yin and K. Schulten, *Biophys. J.*, 2008, **95**, 2806–2821.
19 F. Campelo, H. T. McMahon and M. M. Kozlov, *Biophys. J.*, 2008, **95**, 2325–2339.
20 N. Gamper and M. S. Shapiro, *J. Physiol.*, 2007, **582**, 967–975.
21 G. S. Ayton, W. G. Noid and G. A. Voth, *Curr. Opin. Struct. Biol.*, 2007, **17**, 192–198.
22 V. Tozzini, *Curr. Opin. Struct. Biol.*, 2005, **15**, 144–150.
23 E. Lindahl and M. S. P. Sansom, *Curr. Opin. Struct. Biol.*, 2008, **18**, 425–431.
24 P. Sherwood, B. R. Brooks and M. S. P. Sansom, *Curr. Opin. Struct. Biol.*, 2008, **18**, 630–640.

25 A. Liwo, C. Czaplewski, S. Ołdziej and H. A. Scheraga, *Curr. Opin. Struct. Biol.*, 2008, **18**, 134–139.
26 G. Brannigan and F. L. H. Brown, *J. Chem. Phys.*, 2004, **120**, 1059–1071.
27 G. Brannigan, L. C. L. Lin and F. L. H. Brown, *Eur. Biophys. J.*, 2006, **35**, 104–124.
28 I. R. Cooke and M. Deserno, *Biophys. J.*, 2006, **91**, 487–495.
29 I. R. Cooke and M. Deserno, *J. Chem. Phys.*, 2005, **123**, 224710.
30 I. R. Cooke, K. Kremer and M. Deserno, *Phys. Rev. E: Stat., Nonlinear, Soft Matter Phys.*, 2005, **72**, 011506.
31 O. Farago, *J. Chem. Phys.*, 2003, **119**, 596–605.
32 S. J. Marrink, A. H. deVries and A. E. Mark, *J. Phys. Chem. B*, 2004, **108**, 750–760.
33 S. J. Marrink, H. J. Risselada, S. Yefimov, D. P. Tieleman and A. H. deVries, *J. Phys. Chem. B*, 2007, **111**, 7812–7824.
34 A. Y. Shih, A. Arkhipov, P. L. Freddolino and K. Schulten, *J. Phys. Chem. B*, 2006, **110**, 3674–3684.
35 A. Arkhipov, P. L. Freddolino and K. Schulten, *Structure*, 2006, **14**, 1767–1777.
36 B. J. Reynwar, G. Illya, V. A. Harmandaris, M. M. Muller, K. Kremer and M. Deserno, *Nature*, 2007, **447**, 461–464.
37 A. P. Lyubartsev, *Eur. Biophys. J.*, 2005, **35**, 53–61.
38 A. J. Markvoort, R. A. van Santen and P. A. J. Hilbers, *J. Phys. Chem. B*, 2006, **110**, 22780–22785.
39 H. J. Risselada, A. E. Mark and S. J. Marrink, *J. Phys. Chem. B*, 2008, **112**, 7438–7447.
40 A. J. Markvoort, K. Pieterse, M. N. Steijaert, P. Spijker and P. A. J. Hilbers, *J. Phys. Chem. B*, 2005, **109**, 22649–22654.
41 S. Yamamoto, Y. Maruyama and S. Hyodo, *J. Chem. Phys.*, 2002, **116**, 5842–5849.
42 R. Faller and S. J. Marrink, *Langmuir*, 2004, **20**, 7686–7693.
43 M. J. Stevens, *J. Chem. Phys.*, 2004, **121**, 11942–11948.
44 J. C. Shelley, M. Y. Shelley, R. C. Reeder, S. Bandyopadhyay and M. L. Klein, *J. Phys. Chem. B*, 2001, **105**, 4464–4470.
45 S. Izvekov and G. A. Voth, *J. Phys. Chem. B*, 2005, **109**, 2469–2473.
46 G. S. Ayton and G. A. Voth, *J. Phys. Chem. B*, 2009, **113**, 4413–4424.
47 E. Lyman, J. Pfaendtner and G. A. Voth, *Biophys. J.*, 2008, **95**, 4183–4192.
48 W. G. Noid, J. W. Chu, G. S. Ayton, V. Krishna, S. Izvekov, G. A. Voth, A. Das and H. C. Anderson, *J. Chem. Phys.*, 2008, **128**, 244114.
49 W. G. Noid, P. Liu, Y. Wang, J.-W. Chu, G. S. Ayton, S. Izvekov, H. C. Andersen and G. A. Voth, *J. Chem. Phys.*, 2008, **128**, 244115.
50 G. S. Ayton, W. G. Noid and G. A. Voth, *MRS Bull.*, 2007, **32**, 929–934.
51 W. G. Noid, J.-W. Chu, G. S. Ayton and G. A. Voth, *J. Phys. Chem. B*, 2007, **111**, 4116–4127.
52 S. Izvekov and G. A. Voth, *J. Chem. Theory Comput.*, 2006, **2**, 637–648.
53 S. Izvekov and G. A. Voth, *J. Chem. Phys.*, 2005, **123**, 134105.
54 Y. Wang, S. Izvekov, T. Yan and G. A. Voth, *J. Phys. Chem. B*, 2006, **110**, 3564–3575.
55 J. Zhou, I. F. Thorpe, S. Izvekov and G. A. Voth, *Biophys. J.*, 2007, **92**, 4289–4303.
56 I. Thorpe, J. Zhou and G. A. Voth, *J. Phys. Chem. B*, 2008, **112**, 13079–13090.
57 S. Izvekov and G. A. Voth, *J. Phys. Chem. B*, 2009, **113**, 4443–4455.
58 J. G. Gay and B. J. Berne, *J. Chem. Phys.*, 1981, **74**, 3316–3319.
59 J. T. Brown and M. P. Allen, *Phys. Rev. E: Stat., Nonlinear, Soft Matter Phys.*, 1998, **57**, 6685–6699.
60 J.-W. Chu and G. A. Voth, *Biophys. J.*, 2006, **90**, 1572–1582.
61 Z. Zhang, L. Lu, W. G. Noid, V. Krishna, J. Pfaendtner and G. A. Voth, *Biophys. J.*, 2008, **95**, 5073–5083.
62 T. Haliloglu, I. Bahar and B. Erman, *Phys. Rev. Lett.*, 1997, **79**, 3090–3093.
63 D. ben-Avraham, *Phys. Rev. B: Condens. Matter Mater. Phys.*, 1993, **47**, 14559–14560.
64 M. M. Tirion, *Phys. Rev. Lett.*, 1996, **77**, 1905–1908.
65 D. Tobi and I. Bahar, *Proc. Natl. Acad. Sci. U. S. A.*, 2005, **102**, 18908–18913.
66 I. Bahar, A. R. Atilgan, M. C. Demirel and B. Erman, *Phys. Rev. Lett.*, 1998, **80**, 2733–2736.
67 Y. Wang, A. J. Rader, I. Bahar and R. L. Jernigan, *J. Struct. Biol.*, 2004, **147**, 302–314.
68 P. Maragakis and M. Karplus, *J. Mol. Biol.*, 2005, **352**, 807–822.
69 Y. Yoshida, M. Kinuta, T. Abe, S. Liang, K. Araki, O. Cremona, G. Di Paolo, Y. Moriyama, T. Yasuda, P. De Camili and K. Takei, *EMBO J.*, 2004, **23**, 3483–3491.
70 E. Lindahl and O. Edholm, *J. Chem. Phys.*, 2000, **113**, 3882–3893.
71 L. Lu and G. A. Voth, *J. Phys. Chem. B*, 2009, **113**, 1501–1510.
72 S. Izvekov and G. A. Voth, *J. Chem. Phys.*, 2006, **125**, 151101.
73 G. S. Ayton, J. L. McWhirter, P. McMurtry and G. A. Voth, *Biophys. J.*, 2005, **88**, 3855–3869.

Towards an understanding of membrane-mediated protein–protein interactions

Marianna Yiannourakou,[ab] Luca Marsella,[a] Frédérick de Meyer[ac] and Berend Smit*[c]

Received 2nd February 2009, Accepted 20th March 2009
First published as an Advance Article on the web 10th August 2009
DOI: 10.1039/b902190f

We propose a computational framework to study the lipid-mediated clustering of integral membrane proteins. Our method employs a hierarchical approach. The potential of mean force (PMF) of two interacting proteins is computed under a coarse-grained 3-D model that successfully describes the structural properties of reconstituted lipid bilayers of dymiristoylphophatidylcholine (DMPC) molecules. Subsequently, a 2-D model is adopted, where proteins represented as self-avoiding disks interact through the previously computed PMF, which is modified to take into account three body corrections. The aggregation of the proteins is extensively studied under the condition of negative hydrophobic mismatch: the formation of clusters with increasing size agrees with previous computational and experimental findings.

1 Introduction

Biomembranes and membrane proteins are fundamental for the physiology of the cells:[1] lipid-mediated interactions among embedded proteins might form clusters, which are crucial for performing vital biological processes occurring in living cells.[2,3] There are different types of membrane proteins, which all interact in specific ways among each other within the membrane environment.[4,5] Nonetheless, a quite general characterization of their interaction is made possible by a structural property called hydrophobic mismatch,[4,5] which is the difference between the hydrophobic lengths of the protein and the lipid bilayer. Modulations of the bilayer thickness, protein tilting, protein functioning and protein aggregation have been shown experimentally to depend strongly on the protein hydrophobic mismatch.

Several theoretical studies have been published which address the interaction among membrane proteins.[6–10] Molecular simulations of lipid bilayers are valuable for providing insights into the microscopic structure of reconstituted membrane systems.[11–18] The limit in length and time scales afforded by such approaches restricts the number of proteins that might be simulated, and thus hinders the study of membrane protein clustering. Three-dimensional coarse-graining techniques allow simulators to bridge the gap between atomistic and phenomenological descriptions of complex systems and are therefore optimal candidates for an integrated study on

[a]*CECAM - Centre Européen de Calcul Atomique et Moléculaire, 46 Allée d'Italie, 69364 Lyon, France*
[b]*Molecular Thermodynamics and Modelling of Materials Laboratory, Institute of Physical Chemistry, Demokritos, GR-153 10 Agia Paraskevi Attikis, Greece. E-mail: yiannourakou@chem.demokritos.gr*
[c]*Department of Chemical Engineering, University of California, Berkeley, 101B Gilman Hall, Berkeley, CA, 94720-1462, USA. E-mail: Berend-Smit@Berkeley.edu; Fax: +1 (510) 642 8063; Tel: +1 (510) 642 9275/7260*

biological membranes.[19–21] Recently, three-dimensional coarse-grained models and the dissipative particle dynamics (DPD) simulation technique were used to systematically compute the potential of mean force (PMF) between two proteins as a function of the hydrophobic mismatch of the proteins.[22,23] Experimentally, it is difficult to determine the size of the protein clusters in a biological membrane.[4,5] Recent results indicate that the average size of the observed protein clusters could be of the order of 50–100 proteins,[4] and even much higher.[24] Simulating such a large number of proteins is, even for the three-dimensional coarse-grained models, very demanding.

In the present study we explore a hierarchical coarse-grained framework to study membrane-mediated protein clustering. Conventional Monte Carlo simulation in an NVT ensemble[25] is used to study a two-dimensional model in which the proteins interact through effective potentials that are based on the potentials of mean force (PMF) computed using our three-dimensional coarse-grained model.

Our goal is to obtain a better understanding of the lipid-mediated interactions between membrane proteins. In particular, we address the question whether the clustering of membrane proteins can be simulated using the computed PMF as an effective pair potential, which includes a first-order approximation of three-body effects. We illustrate our approach by studying the clustering behavior of a model protein with negative mismatch. The model protein could represent gramicidin, which is well known to cluster as a result of the negative mismatch when embedded in a dymiristoylphophatidylcholine (DMPC) bilayer.

2 Mesoscopic model and simulation details

In this work we introduce two hierarchically connected coarse-grained models. The starting point is a mesoscopic model in which groups of atoms are lumped into a pseudo particle (see Fig. 1). The effect of water is explicitly modeled; three water atoms are regrouped into one water bead. The key aspect is that hydrophilic and hydrophobic interactions are described in terms of differences in repulsive interactions.[26,27] For example, moving a hydrophilic particle from a hydrophobic environment towards a hydrophilic one reduces the net repulsion and hence lowers the total energy of the systems. The parameters of this soft-repulsive interactions model have been obtained from solubility parameters.[28,29] The intramolecular potentials that connect the pseudo atoms of the lipid have been obtained from fitting to all-atom

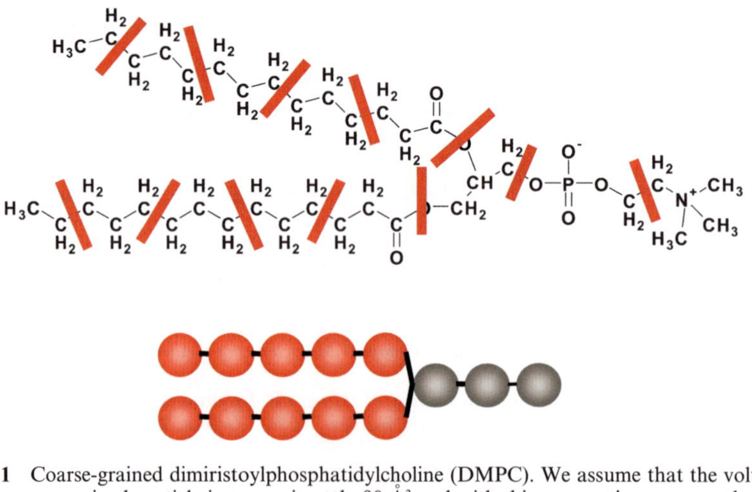

Fig. 1 Coarse-grained dimiristoylphosphatidylcholine (DMPC). We assume that the volume of a coarse-grained particle is approximately 90 Å^3 and with this assumption we map the all-atom representation onto the coarse grained particles.

simulations of a single phospholipid in water.[30] The model lipid that is used in the current study is coarse-grained DMPC. Previous computational studies have shown that this lipid model forms a stable bilayer and displays the typical temperature phase behavior of lipid bilayers.[31]

For the membrane proteins, we focus on the effect of the hydrophobic mismatch. A protein is considered as a rod-like object (see Fig 2). The top and the bottom part of the rod are hydrophilic and the middle hydrophobic. Transmembrane proteins are built by connecting hydrophobic-like beads into a chain and attaching to the ends hydrophilic groups. These chains are then linked together into a bundle of N_P of these amphipathic chains. In each model protein, all the N_P chains are linked to the neighboring ones by springs, to form a relatively rigid body. The diameter of the protein can be changed using different values of N_P. For example, a protein, which mimics the shape[32] of an alpha helix (such as gramicidin A) is constructed by a central chain surrounded by a single layer of six other chains. We denote such a protein by $N_P = 7$. The hydrophobic thickness of the protein can be adjusted by changing the number of hydrophobic beads. The hydrophobic mismatch is defined as the difference between the hydrophobic length of the protein core and the hydrophobic thickness of the pure lipid bilayer. As a consequence, in our model the system can respond to accommodate the hydrophobic mismatch by tilting or changing the thickness of the membrane. In our model the proteins do not have appreciable internal flexibility. We therefore do not allow the proteins to change conformation except for bending. We observe some small bending for positive mismatch and small diameters, but, not for large diameters, because of geometric reasons. However, as we did not optimize these parameters to represent a realistic flexibility of particular proteins, we focus on those systems for which this bending effect is small. Details of the model and the parameters can be found in the literature.[26,30,33]

We used dissipative particle dynamics[29] (DPD) to simulate the properties of our mesoscopic model. The equations of motion were integrated using a modified version of the velocity Verlet algorithm with a reduced timestep of 0.03. The main modification of the standard DPD algorithm is a method we have implemented to ensure that the membrane is simulated in a tensionless state. After on average 15 timesteps a Monte Carlo step was made which involved an attempt to change the area of the lipid in such a way that the total volume remained constant. The acceptance rule for this move involves the imposed interfacial tension,[34] which was set to zero for our simulations. To ensure sufficient hydration, we used a system of 100 000 water molecules for a total of 4,000 lipid molecules.

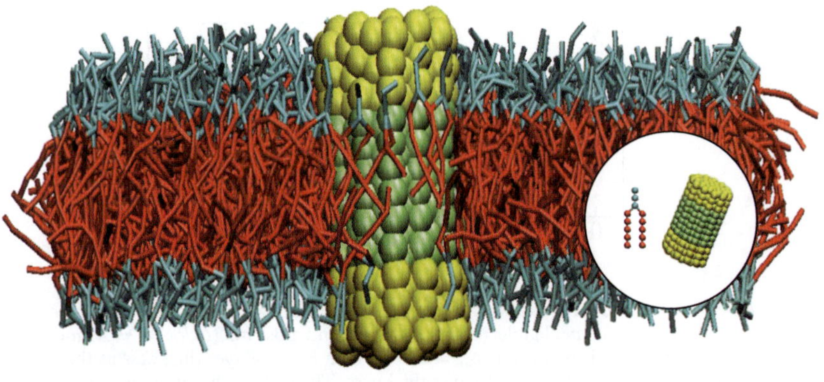

Fig. 2 Schematic representation of model lipids and transmembrane protein.

3 Potential of mean force

The second model that we introduce in this work uses as input the potential of mean force (PMF), which is defined as the reversible work needed to bring two proteins from infinity to a given distance. Before introducing this model we discuss some typical potentials of mean force that we have obtained for this system. Fig. 3 shows some typical potentials of mean force for different values of the hydrophobic mismatch. For negligible mismatch we do not observe any clustering and the PMF is essentially zero. Fig. 3 shows that for large negative and large positive mismatches, however, we see a long-range attraction between the proteins. At this point it is important to emphasize that in our model the intermolecular interactions are short-ranged repulsive. As a consequence, the observed long-range interactions are caused by the perturbation of the membrane as a result of the insertion of the proteins. Indeed, both for a negative and positive mismatch, the membrane around the protein has to change its thickness to accommodate the hydrophobic mismatch. If the proteins cluster, then the total perturbation is less than if the proteins are infinitely far apart. A difference between negative and positive mismatch is found in the range of the interactions. For positive mismatch this range is much shorter.

It is instructive to compare these results with the recent simulations of Schmidt *et al.*,[22] who obtained a potential of mean force that is very different. The results of Schmidt *et al.* suggest that a large energy barrier is keeping two proteins together in a membrane. If we compare our potential of mean force calculations with the theoretical predictions of Dan *et al.*[8] and Kralchevsky *et al.*,[9] our conclusions are the opposite of Schmidt *et al.* We do not observe the high energy barrier observed in the calculations of Dan *et al.* As pointed out by Kralchevsky *et al.*, in the case of zero surface tension, which is imposed in our simulations, the theory of Dan *et al.* should be very similar to the theory of Kralchevsky *et al.* In fact, depending on the choice of parameters, a repulsive barrier can be the result of the model of Kralchevsky *et al.*, if the lipid profile in between the two proteins differs very much from the single protein profile. Our results are in nice agreement with the calculations of Bohinc *et al.*[35] However, in these theories it is assumed that the proteins do not tilt, which is a good approximation for proteins with a large diameter,[33] but may not hold for proteins with a small diameter.

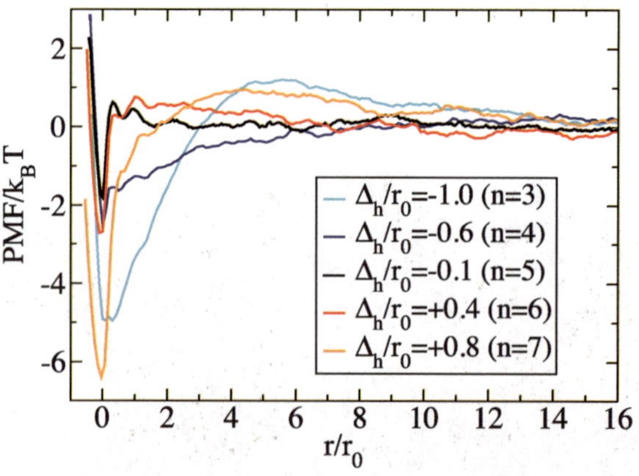

Fig. 3 Potential of mean force as a function of the distance between two proteins with negative ($\Delta_h/r_0 = -1.0$ and -0.6), negligible ($\Delta_h/r_0 = -0.1$), and positive ($\Delta_h/r_0 = +0.4$ and $+0.8$) mismatch. The mismatch is defined as $\Delta_h = h - h_0$, with h the bilayer thickness at the surface of the protein and h_0 the unperturbed bilayer thickness. The mismatch is in units of r_0, which is the interaction cut-off diameter, $r_0 = 6.46$ Å.

Two-dimensional model

Compared to all-atom simulations our mesoscopic model reduces the required amount of CPU time significantly. Extending this model, however, to a very large number of proteins that would allow us to study the clustering behavior would still lead to prohibitively large CPU requirements. Therefore we introduced a two-dimensional model in which the lipids are described as an implicit medium. The interactions between our two dimensional proteins are obtained from the potential of mean force of the mesoscopic model.

Goldman *et al.*[36] have studied the clustering of membrane proteins using a two-dimensional model. In their approach the proteins are modeled as lattice sites having only nearest-neighbor interactions. Implicit in this model is the assumption that protein–protein interactions can be described with a simple pairwise-additive potential. In this work we investigate this assumption in detail and we show that only under negligible hydrophobic mismatch is this assumption reasonable. For any sizeable hydrophobic mismatch, three-body interactions among embedded proteins cannot be neglected, even at low protein concentrations.

To quantify the effect of many body interactions, we compute the PMFs for a protein approaching a cluster of two (Fig 4a) and a cluster of seven proteins (Fig 4b). The hydrophobic mismatch is $(h - h_0)/r_0 = -1$, where r_0 is the cut-off radius of the potential, and h and h_0 the hydrophobic thickness of the protein and the membrane, respectively. For this particular mismatch the attractive forces between the proteins are sufficiently large to compel those proteins that are part of the cluster to remain in the cluster during the entire simulation. All simulations were at the reduced temperature, $T = 0.7$, which corresponds to approximately 60 °C, high enough to ensure that the bilayer is in the fluid phase.[33]

Fig. 4a compares the PMF as obtained from the mesoscopic simulations, with the results from a pairwise two-dimensional model. In this two-dimensional model, in which the proteins are modeled as two-dimensional disks, the intermolecular potential is the PMF as obtained from the mesoscopic simulations for two proteins (*e.g.* see Fig. 3). If we assume pairwise-additive interactions, we can compute the energy of the three proteins cluster. This energy depends on the details of the trimer configuration. If we compute the energy along the dimer axis ($\theta = 0$), we find a lower energy for large distances compared to a perpendicular ($\theta = 90$), while at short distances the triangular

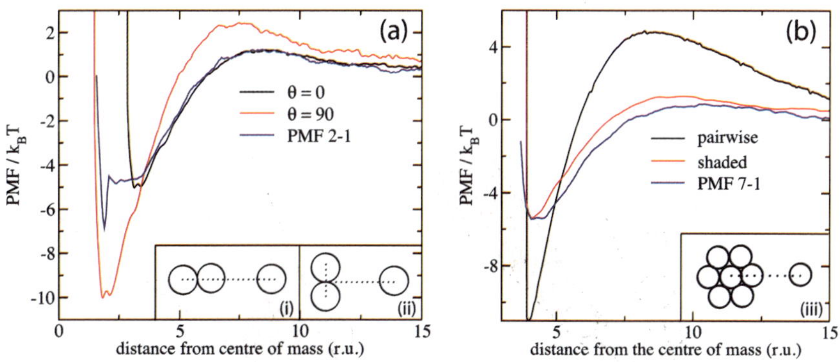

Fig. 4 Potential of mean force (PMF) as a function of the distance between a cluster of proteins and a single protein. Figure (a) shows a cluster of two proteins and figure (b) shows a cluster of seven. (a): PMF 2-1 is computed with the 3-D model the other curves are computed using the pairwise 2-D model in which we use two angles of approach, θ; the black line is for $\theta = 0$, see inset (i), and red for $\theta = 90$, see inset (ii). Right, (b): PMF 7-1 is computed using the 3-D model, the other two lines for the pairwise-additive model and the shaded 2-D model. In both plots, the distance between the proteins is in reduced units (1 r.u. = 6.46 Å). In the two-dimensional model a cut-off radius of around 18 r.u. was used.

configuration has the lowest energy. If we compare these results with the PMF of the mesoscopic model, we see that at large distances there is good agreement with the $\theta = 0$ approach, which is indeed the cluster orientation that has the dominant contribution in the PMF. For short distances the triangular, ($\theta = 90$), orientations are dominant in the PMF. For this case, however, the pairwise-interaction potential in our two-dimensional model overestimates the net attractive interactions. As in our system the mechanism of protein association is the perturbation of the membrane, a pair potential overestimates the total perturbation. Fig. 4b illustrates another situation in which the pairwise-additive potential fails to correctly describe the interactions. Clearly, in a cluster of 7 proteins the middle protein is completely screened from the membrane and therefore does not contribute to the total interactions; as a consequence, the pairwise-additive model overestimates the extent of the repulsive and attractive interactions.

A more realistic description of our two-dimensional model is to introduce three body interactions that take into account that the presence of a third protein screens the interactions with the membrane. For this we introduce a screening parameter for a particular interaction. Let us assume we have three proteins: i, j, and k (see Fig 5). We first compute the polar angles φ_1 and φ_2 of the centers of mass of *i* and *j*, respectively, and the angles θ_1 and θ_2 defined by tangents to *i* and *j* from the center of *k*. A protein is denoted shaded by another one, if the following criterion holds: $(\theta_1 + \theta_2)/2 < |\varphi_1 - \varphi_2|$, where φ_1 and $\varphi_2 \in [-\pi, \pi]$, as shown in Fig. 5. Each protein interacts only with non-shaded ones and does not feel the presence of the shaded proteins. The first neighbor of each protein cannot be shaded, while for all the rest, the above criterion determines whether they are shaded by another protein or not. Fig. 4b shows that if we compute the energy of our two-dimensional system by summing over all non-shaded pairs, we obtain a much better description of the PMF between a cluster of seven and a single protein. This comparison indicates that we have captured in our model the most important contribution of the three-body interactions. A further improvement of this model would be the correction for the double counting of the perturbation in case of the triangular orientation.

We use the screened two-dimensional model to study the clustering. We observe that even at low densities (100 proteins with 0.001 proteins per unit area) all the proteins aggregate and finally form one big cluster, while splitting in smaller aggregates is not observed. Single proteins do continue to escape the cluster and later merge again. It is interesting to contrast these results with our pairwise-additive

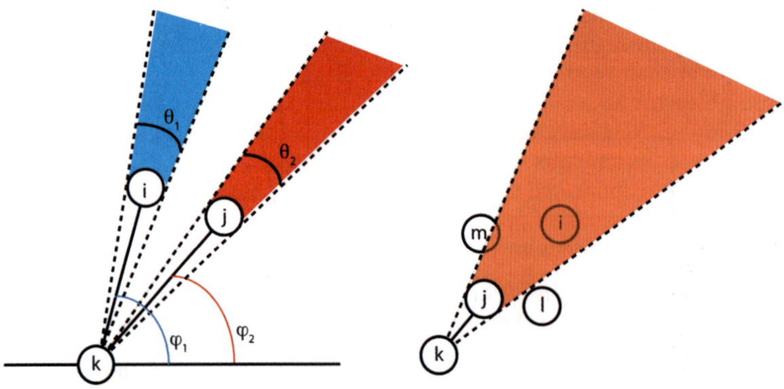

Fig. 5 (Left) Illustration of the screening in the 2-D model: the particles i and j screen all particles in the shaded areas from interacting with particle k. (Right) Because of the screening by protein j, protein k interacts only with j and l.

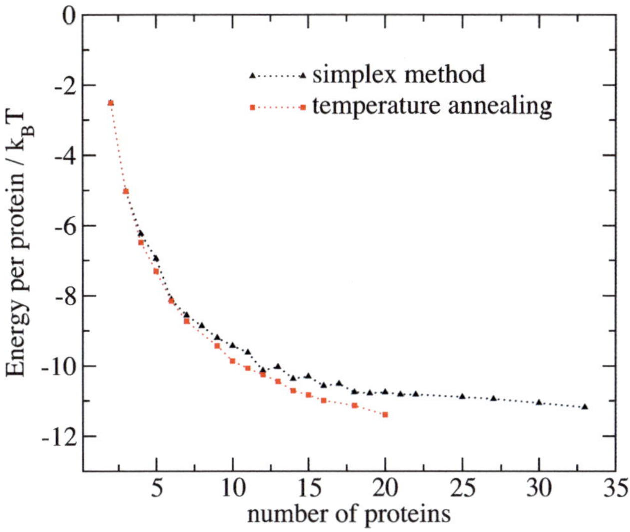

Fig. 6 Ground state energies per protein of the system with negative hydrophobic mismatch are computed using both the simplex method and simulated annealing. The latter is computationally more demanding and it was not easy to obtain reliable minima for systems with more than 20 proteins. Otherwise the two methods show a good agreement.

model. For this system we find a well-defined cluster distribution, which is caused by the repulsive barrier shown in Fig 4b.

To have a better understanding whether phase separation of proteins from the lipids is favored (protein enriched phase), we have studied the ground state configurations of systems of up to 35 proteins. We focused on the question of whether there is a critical cluster size for which the energy is minimal: this would indicate that for any temperature, phase separation would be entropically disfavored and clustering of proteins would take place with an upper limit of the cluster size.[37,38] Searching the ground state of a rugged potential energy surface is a hard task, for which a general solution in a short time scale is not always feasible. Furthermore, there is no way to prove that the ground state is unique.[39] Due to this, we have tried obtaining the ground states of the systems studied here using two different methods: the simplex method[39] and temperature annealing.[40] Fig. 6 shows that with both methods, as the number of proteins in the cluster increases, the energy per protein continues to decrease. This indicates that for systems with comparable densities and number of proteins (*i.e.* up to 100) aggregation, or phase separation, will be favored rather than the formation of a number of clusters.

These observations are in agreement with the known properties of gramicidin within phospholipid bilayers: in fact, previous experimental studies[41,42] have shown the tendency of gramicidin to form big aggregates even in the gel phase. This behavior has been confirmed as expected in the fluid phase, both by computer simulations of elastic models of reconstituted lipid bilayers[6,7,10] and by experiments,[24] notwithstanding the difficulty to detect the tiny mismatch between the lipids in the fluid phase and gramicidin clusters by atomic force microscopy.

Concluding remarks

The methodology developed and used in this study allows us to obtain a better understanding of membrane-mediated interactions. This work illustrates that the protein–protein interactions cannot be described as pairwise additive and that even at low densities three-body interactions are important. We have made the first step in developing such an effective three-body potential, which allows us to extend these calculations to very large systems.

Acknowledgements

The authors wish to acknowledge M. Venturoli for constant support and clarifying discussions; L. M. acknowledges M. M. Sperotto, P. Ruggerone and A. Vargiu for critically reading the draft. This work was partly supported by the EC through the Marie Curie projects: BiMaMoSi (MEXT-CT-2005-023311), EuroSim (MEST-CT-2005-020491) and by the European Science Foundation (ESF) through the activity entitled 'Molecular Simulations in Biosystems and Material Science'

References

1 E. Sackmann, in *Structure and Dynamics of Membranes*, ed. Reinhard Lipowsky and E. Sackmann, Elsevier, Amsterdam, 1995, pp. 1–65.
2 J. A. Killian and B. Dekruijff, *Biochemistry*, 1985, **24**, 7881–7890.
3 E. Wallin and G. von Heijne, *Protein Sci.*, 1998, **7**, 1029–1038.
4 J. J. Sieber, K. I. Willig, C. Kutzner, C. Gerding-Reimers, B. Harke, G. Donnert, B. Rammner, C. Eggeling, S. W. Hell, H. Grubmuller and T. Lang, *Science*, 2007, **317**, 1072–1076.
5 J. J. Sieber, K. I. Willig, R. Heintzmann, S. W. Hell and T. Lang, *Biophys. J.*, 2006, **90**, 2843–2851.
6 K. S. Kim, J. Neu and G. Oster, *Biophys. J.*, 1998, **75**, 2274–2291.
7 H. ArandaEspinoza, A. Berman, N. Dan, P. Pincus and S. Safran, *Biophys. J.*, 1996, **71**, 648–656.
8 N. Dan, P. Pincus and S. A. Safran, *Langmuir*, 1993, **9**, 2768–2771.
9 P. A. Kralchevsky, V. N. Paunov, N. D. Denkov and K. Nagayama, *J. Chem. Soc., Faraday Trans.*, 1995, **91**, 3415–3432.
10 G. Brannigan, L. C. L. Lin and F. L. H. Brown, *Eur. Biophys. J. Biophys. Lett.*, 2006, **35**, 104–124.
11 S. W. Chiu, E. Jakobsson, R. J. Mashl and H. L. Scott, *Biophys. J.*, 2002, **83**, 1842–1853.
12 S. E. Feller, *Curr. Opin. Colloid Interface Sci.*, 2000, **5**, 217–223.
13 S. J. Marrink, M. Berkowitz and H. J. C. Berendsen, *Langmuir*, 1993, **9**, 3122–3131.
14 R. W. Pastor, *Curr. Opin. Struct. Biol.*, 1994, **4**, 486–492.
15 M. C. Pitman, A. Grossfield, F. Suits and S. E. Feller, *J. Am. Chem. Soc.*, 2005, **127**, 4576–4577.
16 A. M. Smondyrev and M. L. Berkowitz, *Biophys. J.*, 1999, **77**, 2075–2089.
17 D. J. Tobias, K. C. Tu and M. L. Klein, *Curr. Opin. Colloid Interface Sci.*, 1997, **2**, 15–26.
18 J. Henin, A. Pohorille and C. Chipot, *J. Am. Chem. Soc.*, 2005, **127**, 8478–8484.
19 S. J. Marrink, A. H. de Vries and A. E. Mark, *J. Phys. Chem. B*, 2004, **108**, 750–760.
20 G. Brannigan and F. L. H. Brown, *Biophys. J.*, 2007, **92**, 864–876.
21 F. Schmid, D. Duchs, O. Lenz and B. West, *Comput. Phys. Commun.*, 2007, **177**, 168–171.
22 U. Schmidt, G. Guigas and M. Weiss, *Phys. Rev. Lett.*, 2008, **101**.
23 F. J. M. de Meyer, M. Venturoli and B. Smit, *Biophys. J.*, 2008, **95**, 1851–1865.
24 V. P. Ivanova, I. M. Makarov, T. E. Schaffer and T. Heimburg, *Biophys. J.*, 2003, **84**, 2427–2439.
25 D. Frenkel and B. Smit, *Understanding Molecular Simulations: from Algorithms to Applications*, Academic Press, San Diego, 2002.
26 M. Kranenburg and B. Smit, *J. Phys. Chem. B*, 2005, **109**, 6553–6563.
27 M. Kranenburg, M. Venturoli and B. Smit, *Phys. Rev. E: Stat., Nonlinear, Soft Matter Phys.*, 2003, **67**, art. no.-060901.
28 R. D. Groot and K. L. Rabone, *Biophys. J.*, 2001, **81**, 725–736.
29 R. D. Groot and P. B. Warren, *J. Chem. Phys.*, 1997, **107**, 4423–4435.
30 M. Kranenburg, J.-P. Nicolas and B. Smit, *Phys. Chem. Chem. Phys.*, 2004, **6**, 4142–4151.
31 M. Venturoli, M. M. Sperotto, M. Kranenburg and B. Smit, *Phys. Rep.*, 2006, **437**, 1–54.
32 J. A. Killian, *Biochim. Biophys. Acta*, 1992, **1113**, 391–425.
33 M. Venturoli, B. Smit and M. M. Sperotto, *Biophys. J.*, 2005, **88**, 1778–1798.
34 M. Venturoli and B. Smit, *PhysChemComm*, 1999, **10**.
35 K. Bohinc, V. Kralj-Iglic and S. May, *J. Chem. Phys.*, 2003, **119**, 7435–7444.
36 J. Goldman, S. Andrews and D. Bray, *Eur. Biophys. J.*, 2004, **33**, 506–512.

37 F. Cardinaux, A. Stradner, P. Schurtenberger, F. Sciortino and E. Zaccarelli, *EPL*, 2007, **77**, 48004.
38 S. Mossa, F. Sciortino, P. Tartaglia and E. Zaccarelli, *Langmuir*, 2004, **20**, 10756–10763.
39 W. H. Press, B. P. Flannery, S. A. Teukolsky and W. T. Vetterling, *Numerical Recipes: The Art of Scientific Computing*, Cambridge University Press, Cambridge, 1986.
40 S. Kirkpatrick, C. D. Gelatt and M. P. Vecchi, *Science*, 1983, **220**, 671–680.
41 J. X. Mou, D. M. Czajkowsky and Z. F. Shao, *Biochemistry*, 1996, **35**, 3222–3226.
42 M. Diociaiuti, F. Bordi, A. Motta, A. Carosi, A. Molinari, G. Arancia and C. Coluzza, *Biophys. J.*, 2002, **82**, 3198–3206.

Measuring excess free energies of self-assembled membrane structures

Yuki Norizoe, Kostas Ch. Daoulas and Marcus Müller

Received 26th January 2009, Accepted 11th February 2009
First published as an Advance Article on the web 19th August 2009
DOI: 10.1039/b901657k

Using computer simulation of a solvent-free, coarse-grained model for amphiphilic membranes, we study the excess free energy of hourglass-shaped connections (*i.e.*, stalks) between two apposed bilayer membranes. In order to calculate the free energy by simulation in the canonical ensemble, we reversibly transfer two apposed bilayers into a configuration with a stalk in three steps. First, we gradually replace the intermolecular interactions by an external, ordering field. The latter is chosen such that the structure of the non-interacting system in this field closely resembles the structure of the original, interacting system in the absence of the external field. The absence of structural changes along this path suggests that it is reversible; a fact which is confirmed by expanded-ensemble simulations. Second, the external, ordering field is changed as to transform the non-interacting system from the apposed bilayer structure to two-bilayers connected by a stalk. The final external field is chosen such that the structure of the non-interacting system resembles the structure of the stalk in the interacting system without a field. On the third branch of the transformation path, we reversibly replace the external, ordering field by non-bonded interactions. Using expanded-ensemble techniques, the free energy change along this reversible path can be obtained with an accuracy of $10^{-3}k_BT$ per molecule in the nVT-ensemble. Calculating the chemical potential, we obtain the free energy of a stalk in the grandcanonical ensemble, and employing semi-grandcanonical techniques, we calculate the change of the excess free energy upon altering the molecular architecture. This computational strategy can be applied to compute the free energy of self-assembled phases in lipid and copolymer systems, and the excess free energy of defects or interfaces.

1 Introduction

The ability to organise on mesoscopic length scales of several nanometres into a diversity of morphologies is a fascinating property of amphiphilic fluids. In some cases, like block-copolymer melts, these morphologies consist of densely packed, ordered structures (*e.g.* lamellae, cylinders, or spheres), which makes them particularly attractive for applications in nanotechnology.[1,2] On the other hand, the self-assembly of amphiphiles in solution can *inter alia* result in structures without long-range order, such as micro-emulsions, micelles or vesicles. The understanding of the kinetics of self-assembly as well as the description of the evolution of interacting supramolecular structures requires the ability to identify preferential

Institut für Theoretische Physik, Georg-August-Universität, 37073 Göttingen, Germany. E-mail: norizoe@theorie.physik.uni-goettingen.de; daoulas@theorie.physik.uni-goettingen.de; mmueller@theorie.physik.uni-goettingen.de

pathways of molecular organisation and to assess the thermal stability of the emerging morphologies.

Typically, the interactions that drive the self-assembly are on the thermal energy scale, k_BT. This "softness" of self-assembling fluids causes the free energy difference between various intermediate structures to be small. Frequently, one finds free energy differences on the order of only $10^{-2}k_BT$ per molecule.

Accurately calculating the excess free energy of self-assembled structures by computer simulation, however, is a challenge because the free energy of a system is not a simple function of the particle coordinates, and special simulation techniques have been devised.[3] In hard-condensed matter systems, *e.g.* crystals, one popular method consists of calculating the free energy by thermodynamic integration along a path that reversibly connects the structure of interest to a reference state of known free energy. For crystalline solids, the Einstein crystal is an appropriate reference state, where non-interacting particles are harmonically tethered to their ideal lattice position. The free energy of the ordered system is derived[4] from thermodynamic integration based on gradually decreasing the strength of the tethers and, in turn, increasing the interactions between particles. In self-assembling fluids, however, there is no analog of the Einstein crystal because even in the defect-free, self-assembled state molecules diffuse and are not constrained to be at some preferential positions; hence the above technique can not be easily generalised to particle-based simulations (*cf.* ref. 5 for a field-theoretic approach).

An alternative technique,[6,7] inspired by similar methods developed for crystalline solids,[8–10] consists of calculating the free energy difference between the structures of interest by transforming them reversibly into each other with the help of an external, ordering field. Like the transition from a liquid to a crystal, self-assembly or transformation between different morphologies in response to a physically relevant control parameter (*e.g.*, temperature, density, or repulsion between amphiphilic entities) occur *via* first-order transitions. Using an external, ordering field, whose spatial structure and strength are adopted to the self-assembled structure and varying the intermolecular interactions, one can avoid the first-order transition and transform one structure into another *via* a reversible path. For the self-assembly from a disordered structure of an ideal gas, such a transformation path is comprised of two branches: along the first branch, one transforms the self-assembled system into an ideal gas that exhibits the same (or very similar) spatial organisation due to the presence of external, ordering fields. Along this branch, the intermolecular interactions are gradually decreased to zero while, simultaneously, the strength of the external, ordering field is increased such that the structural changes along this branch are minimised.[8] Optimally, the morphology remains unaltered during the entire transformation, therefore, this transformation is free of thermodynamic singularities, and the concomitant free energy difference between the self-assembled fluid and the ideal gas in the external fields can be obtained by thermodynamic integration. Along the second branch, we transform the externally structured, ideal gas into a disordered one by progressively reducing the strength of the auxiliary fields. This is also a reversible process because of the absence of collective, ordering effects in the non-interacting system, and the free energy difference along this branch can be obtained by thermodynamic integration (TDI). Along this transformation path one transforms a self-assembled fluid into an ideal gas without passing through a first-order transition.

In this manuscript, we illustrate this computational technique by calculating the free energy of a single, hour-glass shaped connection (stalk) between two, apposed bilayer membranes (see inset of Fig. 1). Dense arrays of these connections have been experimentally observed in diblock copolymer melts[11,12] and aqueous solutions of lipid molecules.[13] The occurrence of stalks in systems with very different microscopic interactions and molecular architectures suggests that their salient properties are universal and can be investigated by minimal, coarse-grained models.[14,15] The structure and free energy of stalks have attracted abiding interest because it is hypothesised that the stalk structure is a key intermediate of membrane fusion.

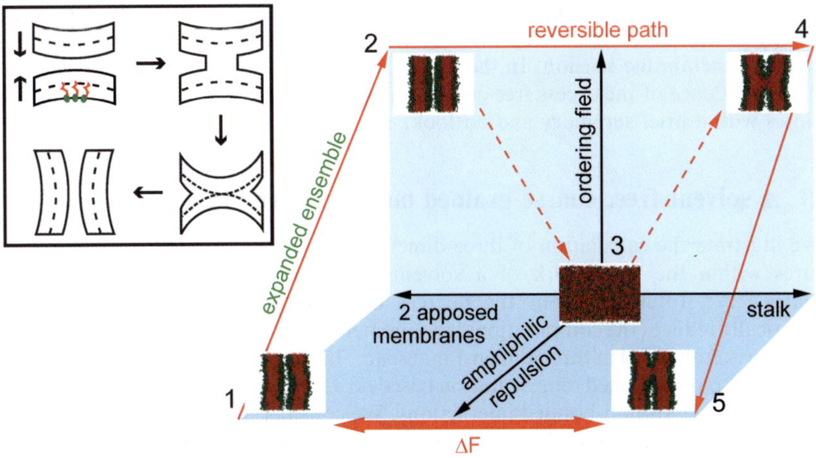

Fig. 1 The inset sketches the classical fusion pathway, encompassing the initial stage of two, apposing bilayer membranes, the stalk, the hemifusion diaphragm, and the final fusion pore. Arrows indicate the expected time order of these fusion intermediates. The main panel depicts the two reversible paths (1 → 2 → 3 → 4 → 5 and 1 → 2 → 4 → 5) used to connect the two, apposed bilayers, state 1, and the final stalk morphology, state 5. The other snapshots show the system of mutually non-interacting amphiphiles (ideal gas) structured by the external, ordering field, states 2 and 4, or the disordered system without ordering field, state 3. In all cases the hydrophobic, *A*, and the hydrophilic, *B*, beads are shown in red and green, respectively.

Membrane fusion is involved in numerous biological processes, such as virial infection, endo- and exocytosis, synaptic release, and cell trafficking.[16–19] Its initial stage involves bringing the membranes into proximity and is regulated by proteins. Once the membranes are in close apposition, however, the proper fusion event, which changes the membrane topology, is thought to be a collective phenomenon, involving a large number of the amphiphilic molecules. Phenomenological theories have assumed a sequence of intermediate structures of the fusion pathway as illustrated in the inset of Fig. 1, among which the stalk morphology plays an essential role in dictating the rate of the fusion process. The subsequent evolution of the stalk into a fusion pore is still a subject of debate.[20]

Early phenomenological calculations estimated the excess free energy of stalk formation to be on the order of $200k_BT$; an unrealistically large value. Subsequent improvements of the theoretical description[21,22] have significantly lowered the estimated excess free energy of the stalk to $30–40k_BT$. Self-consistent field calculations[23,24] have been employed to calculate the free energy of stalks and other intermediate structures along the fusion pathway without assumptions about the detailed geometry and molecular conformations. These mean-field calculations have obtained an even lower value, $\Delta\Omega = 13k_BT$, for the excess free energy of the stalk.[23,24]

Computer simulations are able to observe the fusion process without prior assumptions. The stalk intermediate has been observed in numerous simulations of amphiphilic bilayers using minimal, coarse-grained models,[25–30] systematically coarse-grained descriptions,[31] as well as atomistic models,[32] indicating that the stalk is a universal fusion intermediate. In the following, we will use a minimal, solvent-free, coarse-grained model for amphiphilic bilayer membranes[33] to calculate the excess free energy of a stalk using thermodynamic integration along the reversible path, which is sketched in the main panel of Fig. 1.

The paper is arranged as follows: in Section 2, we describe the coarse-grained model of amphiphilic bilayer membranes with implicit solvent. In the subsequent Section 3, we detail the computational techniques for calculating the excess free energy in the canonical ensemble. Section 4 presents our results for the excess free energy of a stalk in the canonical ensemble and for the free energy difference at constant membrane tension. In the final subsection, we illustrate how to calculate the dependence of the excess free energy on molecular architecture. The manuscript closes with a brief summary and outlook.

2 A solvent-free, coarse-grained model of membranes

We illustrate the calculation of three-dimensional, self-assembled membrane structures within the framework of a solvent-free, coarse-grained model of amphiphiles.[25,34–36] Integrating out the degrees of freedom of the solvent molecules drastically reduces the computational requirements for studying sheet-like membrane structures embedded in three-dimensional space. In this work, we employ an efficient, minimal, coarse-grained representation based on a simple, local density functional for the free energy of non-bonded interactions. We consider an amphiphilic solution in the canonical nVT-ensemble containing n amphiphilic molecules. The presence of solvent will be implicitly taken into account by proper choice of the interactions between the interaction centres (beads) describing the amphiphile. The molecular architecture can be described by a simple, bead-spring Hamiltonian $\mathcal{H}_b$

$$\frac{H_b[\mathbf{r}_i(s)]}{k_B T} = \sum_{s=1}^{N-1} \frac{3(N-1)}{2R_e{}^2}\left[\mathbf{r}_i(s) - \mathbf{r}_i(s+1)\right]^2 \tag{1}$$

where $\mathbf{r}_i(s)$ denotes the coordinate of the sth bead of the ith molecule, and $R_e{}^2$ characterises the mean squared end-to-end distance of the unperturbed molecule. k_B and T are the Boltzmann constant and temperature, respectively. N denotes the number of beads used to describe the molecular contour, of which N_A are hydrophobic and N_B are hydrophilic. Generalisation to more complex architectures, incorporating *e.g.*, chain stiffness or branching, can be envisioned.

The free energy of non-bonded interactions in our solvent-free model is given by a functional, $\mathcal{H}_I[\hat{\rho}_A(\mathbf{r}),\hat{\rho}_B(\mathbf{r})]$ of the molecular densities, $\hat{\rho}_A(\mathbf{r})$ and $\hat{\rho}_B(\mathbf{r})$, of the A and the B beads. In the following, we employ a third-order expansion of the interaction free energy in powers of the molecular densities

$$\frac{H_I}{k_B T} = \int \frac{d\mathbf{r}}{R_e^3}\left\{\frac{1}{2}\sum_{\alpha,\beta=A,B} v_{\alpha\beta}\hat{\rho}_\alpha(\mathbf{r})\hat{\rho}_\beta(\mathbf{r}) + \frac{1}{3}\sum_{\alpha,\beta,\gamma=A,B} w_{\alpha\beta\gamma}\hat{\rho}_\alpha(\mathbf{r})\hat{\rho}_\beta(\mathbf{r})\hat{\rho}_\gamma(\mathbf{r})\right\} \tag{2}$$

The molecular densities, $\hat{\rho}_\alpha(\mathbf{r})$, are defined by

$$\hat{\rho}_\alpha(\mathbf{r}) = \frac{R_e^3}{N}\sum_{i=1}^{n}\sum_{s=1}^{N}\delta(\mathbf{r} - \mathbf{r}_i(s))\gamma_\alpha(s) \tag{3}$$

where $\gamma_\alpha(s) = 1$ if the sth segment is of type α (with $\alpha = A$, or B) and $\gamma_\alpha(s) = 0$ otherwise. In our simulations, the local, molecular density is calculated *via* a collocation lattice. Therefore, the simulation cell is partitioned in a cubic lattice, $\{\mathbf{c}\}$, of grid spacing ΔL. Following related particle-to-mesh methods in electrostatics,[37,38] the densities at each grid point, $\mathbf{c}$, are calculated as:

$$\hat{\rho}_\alpha(\mathbf{c}) = \frac{R_e^3}{N}\sum_{i=1}^{n}\sum_{s=1}^{N}\Pi\left(\mathbf{r}_i(s),\mathbf{c}\right)\gamma_\alpha(s) \tag{4}$$

The function $\Pi(\mathbf{r}, \mathbf{c})$ assigns the particles to grid points and a linear assignment function is used in the following:

$$\Pi(\mathbf{r},\mathbf{c}) = \frac{1}{\Delta L^3}\prod_{\alpha=x,y,z} w(d_\alpha) \quad \text{with} \quad w(d_\alpha) = \begin{cases} 1 - \frac{|d_\alpha|}{\Delta L} & \text{for } |d_\alpha| < \Delta L \\ 0 & \text{otherwise} \end{cases} \tag{5}$$

where $d_\alpha = r_\alpha - c_\alpha$ is the distance between the grid point, $\mathbf{c}$, and the bead position, $\mathbf{r}$, along the Cartesian direction, α. The grid size, ΔL, defines the range of non-bonded interactions and we choose $\Delta L = R_e/6$. Using the grid-based densities, we calculate the non-bonded interactions in eqn (2) from $\hat{\rho}_\alpha(\mathbf{c})$ by replacing the integration $\int \mathbf{dr}$ with summation over the lattice nodes $\sum_\mathbf{c} \Delta L^3$. The calculation of non-bonded interactions *via* the collocation grid is computationally efficient because a bead in our soft, coarse-grained model interacts with many neighbours.

The grid-based version of eqn (2) and the bonded interactions, eqn (1), define a particle-based, soft, coarse-grained model.[39–42] Its statistical mechanics can be studied by a broad spectrum of algorithms traditionally used in Monte-Carlo simulations[43–45] of complex fluids in the framework of conventional atomistic or coarse-grained representations. For example, new configurations can be generated by Monte-Carlo moves proposing random monomer displacements or chain translations, slithering-snake Monte-Carlo moves, identity exchanges of hydrophilic and hydrophobic beads, configuration bias Monte-Carlo techniques,[46–48] and Monte-Carlo algorithms that alter chain connectivity.[49,50] The softness of the interactions, *i.e.*, the absence of harsh, excluded volume, facilitates the efficient implementation of some Monte-Carlo moves (*e.g.*, chain insertions) and reduces relaxation times.

The second- and the third-order coefficients, $v_{\alpha\beta}$ and $w_{\alpha\beta\gamma}$, in eqn (2) are symmetric with respect to permutation of indices, *i.e.*, there are three second-order and seven third-order coefficients. The strategy of their identification has been described in ref. 33 and their values are compiled in Table 1. The model described by eqn (1) and (2) results in stable amphiphilic bilayer with realistic material properties, *i.e.*, the molecular density, $\rho_A = 40$, and compressibility of the hydrophobic interior of the bilayer, which is determined by the coefficients, $v_{AA} < 0$ and $w_{AAA} > 0$. $v_{AA} \ll v_{BB}$ parameterises the solvent preference of the hydrophilic B beads. We use $N = 32$ with $N_A = 28$ and $N_B = 4$. We note that the large difference between the number of hydrophobic and hydrophilic interaction centres per molecule does not give rise to a pronounced wedge-shape of the amphiphiles because the hydrophilic beads, B, of our solvent-free model have a significantly larger effective volume. In fact, the system forms stable bilayers characteristic of amphiphiles with a molecular asymmetry of $f \geq 0.35$.[51]

Selected properties of our solvent-free membrane model are illustrated in Fig. 2. In order to measure the thickness of the membrane in the tensionless state, $\sum = 0$, we consider a configuration, where the membrane spans the periodic box only in one direction (the z-direction in the lower inset of Fig. 2) but not in the other direction, y. Thus, two free edges are formed and, in the canonical ensemble, the extension of the membrane in the y-direction freely adjusts until it neither grows or shrinks. At this stage, the membrane tension, $\sum$, vanishes. Profiles across the membrane in its tensionless state are shown in the main panel of Fig. 2. Integrating

Table 1 List of virial coefficients used in eqn (2)

$v_{AA} = -5.15$
$v_{BB} = -0.01$
$v_{AB} = -1.775$
$w_{AAA} = 0.095625$
$w_{AAB} = 0.095625$
$w_{ABB} = 0.095625$
$w_{BBB} = 0.0$

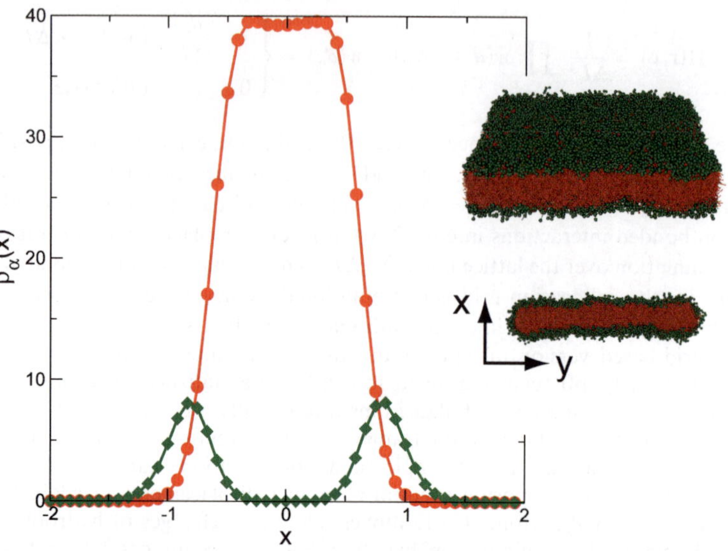

Fig. 2 The top inset shows a snapshot of an isolated bilayer with 5830 amphiphiles, a typical number for the systems studied in this work. The main panel presents the density profiles of the A and B monomers calculated across a membrane patch with a free edge (shown in the lower inset). Length scales are measured in units of R_e. The molecular densities of the hydrophobic and the hydrophilic segments are shown in red and green, respectively.

the profiles, we obtain the area per amphiphile, $A_o = 0.0343R_e^2$ for $N_A = 28$. The lateral self-diffusion coefficient of a single amphiphile in the bilayer is $D \approx 3\times10^{-5}R_e^2$/MCS, where we propose a local random displacement for each bead on the average once in a Monte-Carlo step (MCS).

To calculate the excess chemical potential, μ_0^{ex}, of an amphiphilic molecule in a tensionless membrane, we pre-assemble a bilayer with lateral dimensions, $10R_e \times 10R_e$ and height, $L_x = 5R_e$, comprised of $2L^2/A_o = 5830$ amphiphiles. The bilayer spans the simulation box across the periodic boundary conditions as depicted in the upper inset of Fig. 2. Using about 2×10^4 configurations, which are sampled after a time interval $Dt/R_e^2 \approx 0.3$ in the nVT-ensemble, we accurately calculate the excess chemical potential with respect to a gas of non-interacting molecules described by the bonded interactions, $\mathcal{H}_b$, employing a variation of the Bennet histogram method[52] proposed by Shing and Gubbins.[53] To this end, one generates a conformation of a single molecule according to the bonded interactions, eqn (1), and inserts it at a random position. The insertion of a molecule changes the densities, $\hat{\rho}_\alpha(\mathbf{c})$ (with $\alpha = A, B$), and we monitor the histogram, $f(U_{\text{nb}})$ of the concomitant change, U_{nb}, of non-bonded interactions, $\mathcal{H}_I$. We also sample the distribution, $g(U_{\text{nb}})$, of changes of non-bonded interactions, eqn (2), in response to deleting a random amphiphile. The distributions, g and f, are presented in Fig. 3. The region of their overlap, albeit small, can be used for the calculation of the chemical potential as:[53]

$$\mu^{\text{ex}}(n, V, T) = k_B T \ln\left[\frac{g(U_{\text{nb}})}{f(U_{\text{nb}})}\right] + U_{\text{nb}} \quad (6)$$

The excess chemical potential according to eqn (6) is shown in the inset of Fig. 3 with a thick dashed line. From these data we estimate the chemical potential of an amphiphile in a tensionless bilayer . The result, $\mu_0^{\text{ex}} = -37.745(50)k_BT$, corroborates the

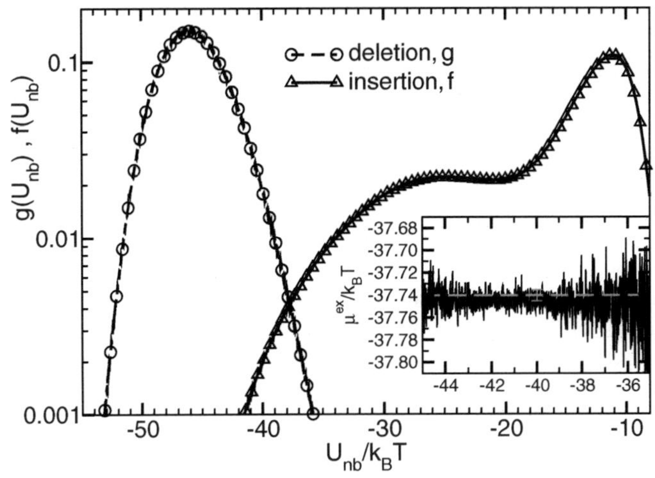

Fig. 3 Probability distributions of the energy change, U_{nb}, upon deleting a randomly chosen amphiphile, $g(U_{\mathrm{nb}})$ (circles), or inserting an amphiphile, $f(U_{\mathrm{nb}})$ (triangles), at a random position in an isolated, tensionless bilayer membrane, respectively. The inset presents the excess chemical potential, calculated according to $\mu^{\mathrm{ex}}/k_BT = \ln[g(U_{\mathrm{nb}})/f(U_{\mathrm{nb}})] + U_{\mathrm{nb}}/k_BT$ in the energy interval where both histograms overlap. The dashed line marks our estimate for the excess chemical potential.

excess chemical potential measured *via* Rosenbluth sampling,[54] $\mu_0^{\mathrm{ex}} = -37.740k_BT$. Adding the translational contribution of the ideal gas to the result of the Rosenbluth sampling, we obtain $\mu_0 = k_BT\ln\left[\frac{n}{V}\right] + \mu_0^{\mathrm{ex}} = -35.284(5)k_BT$.

3 Thermodynamic integration

3.1 Reversible path connecting two apposed bilayers and stalk

Two reversible paths that transform two, apposed bilayers (state 1) to a configuration where the two-bilayers are connected by a stalk (state 5) are sketched in Fig. 1. First, starting from the apposed bilayers, we gradually replace the non-bonded interactions by the external, ordering fields such that the system at the end of this branch (state 2) is an ideal gas of amphiphiles that do not mutually interact, but which are structured by external, ordering fields. Along the branch, 2 → 3, the external fields are gradually turned off and the ideal gas becomes disordered (stage 3). Then, along the branch, 3 → 4, a different external, ordering field is gradually switched on in order to structure the non-interacting amphiphiles into a stalk morphology (stage 4). Along the last branch, 4 → 5, the strength of the external, ordering fields is reduced to zero, while, in turn, the non-bonded interactions are switched on. Alternatively, the branches, 2 → 3 → 4, can be replaced by a gradual change of the external ordering field from a field that creates two, apposed bilayers, to one that orders the non-interacting system into a stalk structure.

The changes of the structure of the ideal gas of non-interacting amphiphiles due to altering the external, ordering fields along the branches, 2 → 3, 3 → 4, or 2 → 4, are completely gradual and free of thermodynamic singularities because of the absence of collective ordering in the non-interacting system. Along the other branches, 1 → 2 and 4 → 5, we choose the external, ordering field such that the changes of the morphology from the interacting system to the ideal gas in the external, ordering field are minimised.[8] The absence of abrupt structural changes indicates that there are no thermodynamic transitions along these branches either.

Since the transformation is reversible, the concomitant free energy change can be obtained by thermodynamic integration. In the following, we formulate the scheme in the canonical ensemble using the strengths, λ_I and λ_E, of the non-bonded interactions and the external, ordering field as additional control variables. Generalisations to other ensembles, *e.g.*, the grandcanonical $\mu VT\lambda_I\lambda_E$-ensemble, can be envisioned.

The Helmholtz free energy, F, of the $nVT\lambda_I\lambda_E$-ensemble has the form:

$$\frac{F}{k_BT} = -\ln\frac{1}{n!}\int\prod_{i=1}^{n}\tilde{\mathrm{D}}[\{\mathbf{r}_i(s)\}]\exp\left[-\frac{\lambda_I \mathrm{H}_I + \lambda_E \mathrm{H}_E^m}{k_BT}\right] \tag{7}$$

where the integration $\tilde{\mathcal{D}}[\{r_i(s)\}]$ sums over all conformations of the i-th amphiphile, taking account of the appropriate weight due to the bonded interactions, *i.e.*,

$$\tilde{\mathrm{D}}[\{\mathbf{r}_i(s)\}] = \prod_{s=1}^{N} \mathrm{d}\mathbf{r}_i(s)\exp\left[-\frac{\mathrm{H_b}[\mathbf{r}_i(s)]}{k_BT}\right] \tag{8}$$

The term $\mathcal{H}_E^m$ in eqn (7) describes the interaction of the amphiphiles with the external, ordering fields, $W_A^m(\mathbf{r})$ and $W_B^m(\mathbf{r})$ (in units of k_BT), and it is defined as:

$$\frac{\mathrm{H}_E^m}{k_BT} = \sum_{\alpha=\mathrm{A,\,B}}\int\frac{\mathrm{d}\mathbf{r}}{R_e^3}W_\alpha^m(\mathbf{r})\hat{\rho}_\alpha(\mathbf{r}) \tag{9}$$

The superscript, m, denotes the morphology, which the external ordering fields create (*i.e.*, m = bilayers or stalk). The parameters, λ_I and λ_E, are conjugated to the corresponding interactions and can be used to control their strength. The different combinations of m, λ_I, and λ_E corresponding to the five states of Fig. 1 are listed in Table 2.

The changes of the free energy with respect to independent variations of the control parameters, λ_I and λ_E, can be calculated as

$$\frac{\partial F^m}{\partial \lambda_I} = \langle \mathrm{H}_I\rangle_{n,V,T,\lambda_I,\lambda_E} \quad \text{and} \quad \frac{\partial F^m}{\partial \lambda_E} = \langle \mathrm{H}_E^m\rangle_{n,V,T,\lambda_I,\lambda_E} \tag{10}$$

Along the branches, 2 → 3 and 3 → 4, only the strength of the external, ordering field, λ_E, varies and the free energy difference is given by

$$\Delta F_{23} = \int_1^0 \mathrm{d}\lambda_E\langle\mathcal{H}_E^{\mathrm{bilayer}}\rangle_{n,V,T,0,\lambda_E} \tag{11}$$

A similar expression holds for ΔF_{34}.

Along the branches, 1 → 2 and 4 → 5, both strengths, λ_I and λ_E, are simultaneously altered according to $\lambda_I = 1 - \lambda_E$, and the free energy change is given by:

Table 2 Combinations of the external, ordering field, m, and the strength of the non-bonded interactions, λ_I, the strength of the external, ordering field, λ_E, for the different states indicated in Fig. 1. If the strength of the external, ordering field vanishes, $\lambda_E = 0$, the type, m, in brackets indicates the type of the external, ordering field along the path that is connected to this state

State	m	λ_I	λ_E
1	(Bilayer)	1	0
2	Bilayer	0	1
3	(Bilayer/stalk)	0	0
4	Stalk	0	1
5	(Stalk)	1	0

$$\Delta F_{12} = \int_0^1 \mathrm{d}\lambda_{\mathrm{E}} \langle \mathcal{H}_{\mathrm{E}}^{\mathrm{bilayer}} - \mathcal{H}_{\mathrm{I}} \rangle_{n,V,T,1-\lambda_{\mathrm{E}},\lambda_{\mathrm{E}}} \tag{12}$$

A similar expression holds for ΔF_{45}. Summing the changes, ΔF_{12}, ΔF_{23}, ΔF_{34}, and ΔF_{45} yields the total, Helmholtz free energy difference between the stalk and the two, apposing bilayers, ΔF.

Alternatively, we can directly calculate the free energy difference between states 2 and 4 by altering the external, ordering field according to the linear superposition

$$\mathcal{H}'_{\mathrm{E}} = \lambda \mathcal{H}_{\mathrm{E}}^{\mathrm{stalk}} + (1 - \lambda)\mathcal{H}_{\mathrm{E}}^{\mathrm{bilayer}} \tag{13}$$

where we introduce an additional, thermodynamic integration parameter, λ, "mutating" the external, ordering field from one that creates two, apposing bilayers, $\lambda = 0$, to one that orders the gas of amphiphiles into a stalk morphology, $\lambda = 1$. The free energy change is calculated according to:

$$\Delta F_{24} = \int_0^1 \mathrm{d}\lambda \langle \mathcal{H}_{\mathrm{E}}^{\mathrm{stalk}} - \mathcal{H}_{\mathrm{E}}^{\mathrm{bilayer}} \rangle_{n,V,T,\lambda} \tag{14}$$

The subscript λ in the average indicates that, for this particular branch, it acts as a thermodynamic integration control parameter. As before, adding the contributions ΔF_{12}, ΔF_{24}, and ΔF_{45}, yields ΔF.

The relation, $\Delta F_{23} + \Delta F_{34} - \Delta F_{24} = 0$, provides an opportunity to gauge the error of the thermodynamic integration scheme.

3.2 External field calculation

An essential prerequisite of our thermodynamic integration technique is the reversibility of the transformation along the path. To this end, the external, ordering fields along the branches, $1 \rightarrow 2$ and $4 \rightarrow 5$, have to be chosen in strength and spatial structure such that the spatial organisation of the system remains unaltered and mimics as closely as possible the density distribution, $\tilde{\rho}_\alpha^{\mathrm{m}}(\mathbf{r})$, at the end points, 1 and 5, respectively. The absence of abrupt changes indicates reversibility and, additionally, it can be shown that minimising structural changes corresponds to the optimal choice that minimises the numerical error of the thermodynamic integration.[8]

In previous applications to dense copolymer systems,[6,7] self-consistent field theory provided an accurate estimate for the external, ordering fields

$$W_\alpha^{\mathrm{m}}(\mathbf{r}) = R_{\mathrm{e}}^3 \left. \frac{\delta \mathrm{H}_{\mathrm{I}}}{\delta \rho_\alpha(\mathbf{r})} \right|_{\tilde{\rho}_\alpha^{\mathrm{m}}(\mathbf{r})} \tag{15}$$

that replace the non-bonded interactions of a molecule with its surrounding in a structure with density distribution, $\tilde{\rho}_\alpha^{\mathrm{m}}(\mathbf{r})$. In the present study of amphiphiles in an implicit solvent, fluctuations turn out to be important, and the density distribution that results from the estimate, eqn (15), significantly differs from the reference distribution, $\tilde{\rho}_\alpha^{\mathrm{m}}(\mathbf{r})$.

In order to calculate the external, ordering fields in the general case, where the mean-field approximation is inaccurate, we propose an iterative strategy. First, we simulate the system at the end points of the transformation path, states 1 and 5, to obtain the density distribution, $\tilde{\rho}_\alpha^{\mathrm{m}}(\mathbf{r})$. Using eqn (15) to obtain an initial estimate for the fields, $W_\alpha^{\mathrm{m},0}(\mathbf{r})$, we calculate the density distributions, $\rho_\alpha^{(1)}(\mathbf{r})$ for $\lambda_{\mathrm{I}} = 0$ and $\lambda_{\mathrm{E}} = 1$. Using these results, we improve the estimates for the external, ordering fields iteratively

$$\Delta W_\alpha^{m,n}(\mathbf{r}) = W_\alpha^{m,(n+1)}(\mathbf{r}) - W_\alpha^{m,(n)}(\mathbf{r}) = \varepsilon[\rho_\alpha^{(n)}(\mathbf{r}) - \tilde{\rho}_\alpha^{\mathrm{m}}(\mathbf{r})] \tag{16}$$

where n denotes the iteration index and ε a small, positive parameter. For our system, we chose $\varepsilon = 0.05$ and simulated the system *via* random, local displacements of beads for 10 000 Monte-Carlo steps to evaluate the average density distribution $\rho_\alpha^{(n)}(\mathbf{r})$ between the iterative adjustments of the external, ordering fields. After about 10 iterations convergence was achieved.

The similarity of the morphology of the self-assembled system (state 5), $\tilde{\rho}_\alpha^{\rm stalk}(\mathbf{r})$, and the ideal gas of the amphiphiles structured by the external fields at the final iteration (state 4) is presented in Fig. 4 for the stalk morphology. The left panel presents a 2D contour plot of the distribution of the hydrophobic A-beads in the self-assembled system, state 5, while the right panel depicts the results for the externally ordered ideal gas, state 4. The distributions have been radially averaged. r denotes the radial distance from the central axis of the stalk, while x is the coordinate along the membrane normal. A similar quality of agreement is achieved for the case of the two, apposed bilayers (not shown).

An alternative technique (or a possibility to optimise ε) consists of re-weighting histograms of the densities. To this end, one stores the density distributions, $\rho_\alpha^{(n)}(\mathbf{r}, t)$, at different stages, $t = 1,\ldots, T$, in the course of the simulation and chooses $\Delta W_\alpha^{{\rm m},n}(\mathbf{r})$ according to

$$\tilde{\rho}_\alpha^{\rm m}(\mathbf{r}) = \frac{\sum_{t=1}^{T} \rho_\alpha^{(n)}(\mathbf{r}, t) \exp\left[- \int \frac{d\mathbf{r}}{R_e^3} \Delta W_\alpha^{{\rm m},n}(\mathbf{r}) \rho_\alpha^{(n)}(\mathbf{r}, t)\right]}{\sum_{t=1}^{T} \exp\left[- \int \frac{d\mathbf{r}}{R_e^3} \Delta W_\alpha^{{\rm m},n}(\mathbf{r}) \rho_\alpha^{(n)}(\mathbf{r}, t)\right]} \tag{17}$$

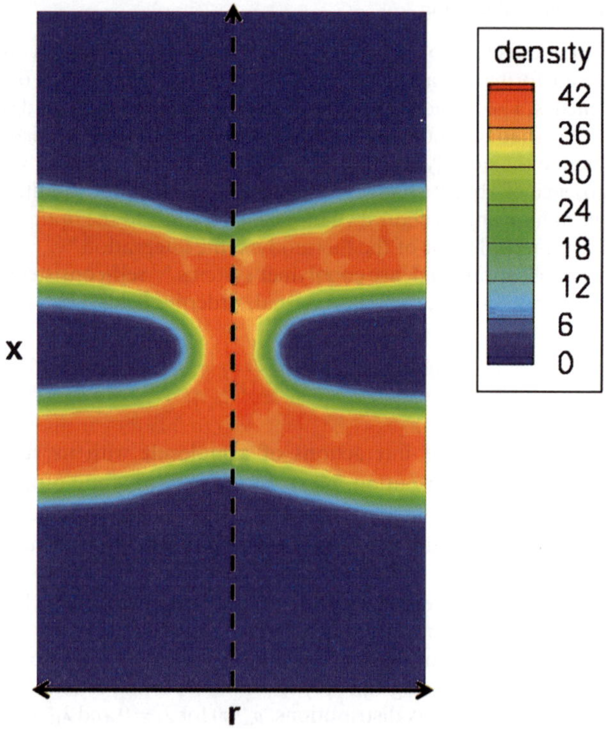

Fig. 4 2D contour plot of the distribution of hydrophobic A-beads in the stalk morphology for the case of the self-assembled system (left) and the ideal gas of amphiphiles structured by external, ordering fields which have been obtained *via* the iteration procedure, eqn (16) (right). The graphs correspond to density distributions radially averaged around the central axis of the stalk, *i.e.*, r is the radial distance from the central axis, while x denotes the coordinate along the membrane normal.

3.3 Absence of thermodynamic singularities and expanded-ensemble simulation

On general grounds, the ideal gas structured by external, ordering fields on branches, $2 \to 3$, $3 \to 4$, and $2 \to 4$, does not exhibit collective, ordering processes that characterise phase transitions. The careful choice of the external, ordering fields presented in Fig. 4 demonstrates the similarity of the self-assembled system and the non-interacting gas in the external, ordering field and suggests that the structure also does not significantly change along the branches, $1 \to 2$ and $4 \to 5$, of the path of integration. It has been argued that the absence of abrupt structural changes indicates the absence of thermodynamic singularities.[8] In order to prove the absence of thermodynamic singularities on the branches, $1 \to 2$ and $4 \to 5$, and accurately calculate the free energy change, we employ expanded-ensemble simulations,[55] where the strength of the ordering field, λ_E, is considered to be a fluctuating, dynamic variable of the expanded system.

In a simple application of the thermodynamic integration scheme, the considered system is simulated at different, fixed values, of the control parameters, λ_I and λ_E, along the integration path. From these simulations the integrands, eqn (10), can be estimated and the integrals, *e.g.* eqn (12), are numerically evaluated.

In an expanded ensemble, the control parameter, λ_E, of the original system, is regarded as a dynamic variable. In addition to the Monte-Carlo moves used to sample the molecular configurations, one uses a Monte-Carlo move that changes the values of the control parameter in the course of the simulation. Thus, during a single simulation run, the system visits different state points, $\{\lambda_E^k\}$, along the integration path. The index, $k = 1,\ldots, K$, enumerates the different, discrete sampling points into which the path of integration is partitioned. In the present study, we discretise the branches $1 \to 2$, $2 \to 4$, and $4 \to 5$ into $K = 58$, 52, and 66 sampling points, respectively. These numbers ensure that the distribution function at neighbouring sampling points overlap. For the thermodynamic integration a slightly different discretization is used.

The configurations at each fixed value, λ_E^k are distributed according to the canonical Boltzmann weight and the integrands of eqn (10) can be obtained at different values of λ_E^k.

The partition function of the expanded ensemble takes the form:

$$Z_{\mathrm{ex}} = \sum_{k=0}^{K} \exp\left[\frac{w(\lambda_E^k)}{k_B T}\right] \frac{1}{n!} \int \prod_{i=1}^{n} \tilde{\mathrm{D}}[\{\mathbf{r}_i(s)\}] \exp\left[-\frac{\lambda_I^k \mathrm{H}_I + \lambda_E^k \mathrm{H}_E^m}{k_B T}\right] \tag{18}$$

where the branches, $1 \to 2$ and $4 \to 5$, are described by $\lambda_I^k = 1 - \lambda_E^k$, while for the other parts of the integration path λ_I^k vanishes.

The set of pre-weighting factors, $\{w(\lambda_E^k)\}$ is chosen to facilitate transitions between neighbouring sampling points, λ_E^k. The probability of finding the system in the sub-ensemble, λ_E^k is given by:

$$P_{\mathrm{ex}}(\lambda_E^k) = \frac{1}{Z_{\mathrm{ex}}} \exp\left[-\frac{F(n, V, T, \lambda_I^k, \lambda_E^k) - w(\lambda_E^k)}{k_B T}\right] \tag{19}$$

Measuring the probability distribution we can estimate the free energy change along the path of integration.

Choosing $w(\lambda_E^k) \simeq F(n,V,T,\lambda_I^k,\lambda_E^k)$ the different sub-ensembles are visited with approximately equal probability. We obtain an estimate for these optimum weights from the free energy that we estimate by sampling the integrands in eqn (10) in the simulations with fixed λ_I and λ_E. We note that the variation of the free energy along the path of integration amounts to $\mathcal{O}(10^4 k_B T)$. Eqn (19) shows that small deviations between the pre-weighting factors and the actual free energy will give arise to large Boltzmann weights and will result in a very non-uniform sampling of the integration interval. In this initial stage, when the pre-weighting factors are inaccurate, the

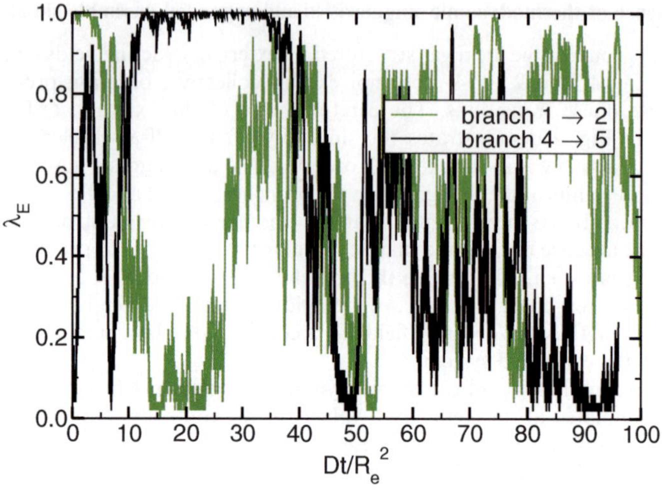

Fig. 5 Evolution of the control parameter, λ_E of the ordering field during expanded-ensembles simulations along the branches 1 → 2 (green line) and 4 → 5 (black line) as a function of time characterizing the lateral motion of a single amphiphile in a tensionless membrane.

system will remain stuck for a large part of the simulation run in a portion of the path of integration.

In order to achieve a uniform sampling of the sub-ensembles, the pre-weighting factors have to be known with an accuracy of the order of k_BT, corresponding to a relative accuracy of the free energy estimate of $\mathcal{O}(10^{-4})$. Several strategies have been devised to optimise the pre-weighting factors.[56–62] In this study, we iteratively use eqn (19) to improve the weights and, simultaneously, accumulate statistics for the integrands, eqn (10), to improve the free energy estimate.

In Fig. 5 we present the evolution of λ_E in the course of the expanded-ensemble simulation of branches, 1 → 2 and 4 → 5 using our final estimates of the pre-weighting factors. From the observation that all sub-ensembles are visited with roughly equal probability, we conclude that the absolute change of the free energy is known with an accuracy of a few k_BT. We also note that the simulation freely diffuses across the different sub-ensembles, $\{\lambda_E^k\}$, and that there are no "kinetic barriers" or a band structure in the "time"-sequence of λ_E^k. This observation demonstrates that there are no hidden free energy barriers along the path of integration, which are not resolved by the specific choice of the reaction coordinate, λ_E. Therefore, the branches, 1 → 2 and 4 → 5, are free of thermodynamic singularities and the application of thermodynamic integration to calculate free energy changes is justified.

4 Excess free energy of a stalk

4.1 Helmholtz free energy, ΔF, of a stalk

The difference in the grandcanonical potential, $\Delta\Omega = \Omega^{\text{stalk}} - \Omega^{\text{bilayer}}$, characterises the (meta)stability of the stalk, *i.e.*, we compare the free energy of the two morphologies at equal chemical potential and therefore at equal membrane tension. As a consequence, the number of amphiphilic molecules in the stalk morphology is larger than their number in the two, apposed bilayers. While the thermodynamic integration technique described in the previous section can be performed in the grandcanonical ensemble, we first calculate the Helmholtz excess free energy, $\Delta F = F^{\text{stalk}} - F^{\text{bilayer}}$, in the nVT ensemble, and, in a second step, we calculate $\Delta\Omega$.

The canonical ensemble turned out to be computationally convenient for two reasons: (i) if we used the grandcanonical ensemble and required that the number of particles approximately remained constant, the strength of the external, ordering fields would have to be extremely fine tuned and the simple linear dependence of the field strengths, λ_I and λ_E, along the branches would not be sufficient. (ii) In our system, stalks are only metastable structures, *i.e.*, $\Delta\Omega > 0$. In the canonical ensemble, they are rather long-lived. A typical lifetime of a stalk in the canonical ensemble is on the order of 10^6 Monte-Carlo steps, which is larger than the relaxation time of λ_E in the expanded-ensemble simulations around the end-point of the branch, $4 \rightarrow 5$. In the grandcanonical ensemble, however, the use of insertion and deletion Monte-Carlo moves reduces the kinetic barrier and the typical lifetime of a metastable stalk is much smaller.

For the calculations, we pre-assemble a stalk between two, apposing bilayers with the help of an external ordering field using a system geometry, $L_x \times L_y \times L_z = 10R_e \times 6R_e \times 6R_e$ with periodic boundary conditions in all directions. Initially, the two membranes are comprised of $4L_yL_z/A_o \approx 4197$ amphiphiles. The dimension along the bilayer normal, $L_x = 10R_e$ is chosen large enough to minimise interactions between the two, apposing membranes across the periodic boundary conditions. Once a stalk has formed in the external field, we remove the auxiliary field and perform a grandcanonical simulation using a chemical potential close to that of a tensionless bilayer. The number of amphiphiles increases to provide the extra material required to form the stalk connection, and their average number is estimated to $n^{\text{stalk}} = n_{\text{TDI}} = 4240$ molecules. We select a configuration with this number of molecules as a starting configuration of the thermodynamic integration scheme in the nVT-ensemble at state 5.

The configuration of the two, apposing bilayers is created using the same number of amphiphiles, $n_{\text{TDI}} = 4240$, assembled by auxiliary fields in a box with the same dimensions as above. The auxiliary fields are then removed and, after an equilibration in the nVT-ensemble, a starting configuration for the thermodynamic integration at state 1 is generated. The system of two, apposing bilayers contains a larger number of molecules than in the tensionless state, *i.e.*, the membranes are characterised by a negative tension, $\sum < 0$. The excess of amphiphiles, however, is too small to create significant bilayer distortions such as buckling.

The results for the integrands, eqn (10), along the different branches of the path of thermodynamic integration, $1 \rightarrow 2 \rightarrow 3 \rightarrow 4 \rightarrow 5$, are presented in Fig. 6 and 7. Fig. 6 depicts the results for the branches, $1 \rightarrow 2$ and $4 \rightarrow 5$, with $\lambda_I = 1 - \lambda_E$. The transformation of the self-assembled, two-bilayer morphology into a non-interacting system structured by external, ordering fields is shown by solid lines, while the transformation of the self-assembled stalk structure to the ideal gas in external, ordering fields is marked by open circles, respectively. The integrand varies more rapidly upon approaching the ends of the branch. For $\lambda_E \rightarrow 0$, the behaviour can be traced back to thermal membrane fluctuations (*e.g.*, undulations), which occur in the self-assembled systems, states 1 and 5, but which we rapidly suppress by turning on the static, external, ordering field. The pronounced dependence of the integrand in the limit, $\lambda_E \rightarrow 1$, can be rationalised by the strong reduction of compressibility and the concomitant growth of fluctuations as the intermolecular interactions are completely turned off. Thus, the system is very susceptible to the external, ordering field in the absence of $\mathcal{H}_I$.

In Fig. 7 we present the results for the branches, $2 \rightarrow 3$ and $3 \rightarrow 4$, where the non-interacting, structured systems with $\lambda_I = 0$ are transformed into a disordered one. The smooth dependence of the integrands on the reaction coordinate, λ_E, corroborates the absence of a first-order transition along the integration path. The dependence of the free energy difference of the stalk and two-bilayer structure on λ_E is shown in the insets of Fig. 6 for branches $1 \rightarrow 2$ and $4 \rightarrow 5$ and Fig. 7 for branches, $2 \rightarrow 3$ and $3 \rightarrow 4$, respectively.

At $\lambda_E = 1$ these data yield, $\Delta F_{12} + \Delta F_{45} = 70.108k_BT$ and $\Delta F_{23} + \Delta F_{34} = -55.038k_BT$, respectively. Thus, the Helmholtz free energy difference between stalk and two-bilayers morphology is $\Delta F = 15.07k_BT$. Alternatively, we can estimate the

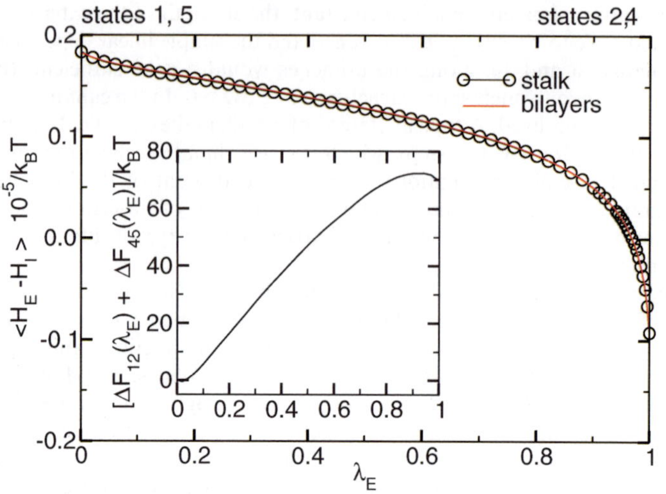

Fig. 6 Variation of the integrands, eqn (12), along the branches with, $\lambda_I = 1 - \lambda_E$, *i.e.* $1 \rightarrow 2$ (bilayer structure, line) and $4 \rightarrow 5$ (stalk, symbols). These branches are discretized in 65 sampling points. The inset shows the sum of the free energy changes for the bilayer and stalk structures, obtained after integrating the data of the main figure (with proper sign to account for the direction of the branch), as a function of the control parameter, λ_E.

free energy difference between states 4 and 2 by altering the external, ordering field from one that creates a two-bilayers structure to one that generates a stalk morphology. The corresponding integrand, eqn (14), is presented in Fig. 8, and the free energy, $\Delta F_{24}(\lambda)$ along the branch, $2 \rightarrow 4$, is plotted in the inset. The accumulated free energy, $\Delta F_{24} = \Delta F_{24}(\lambda = 1) = -53.662k_BT$ compares well with the previous result, $\Delta F_{23} + \Delta F_{34} = -55.038k_BT$. Thus, our result of the thermodynamic

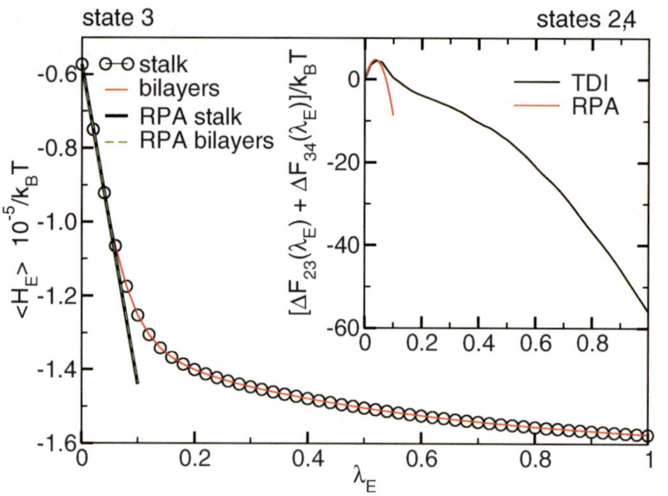

Fig. 7 Variation of the integrands, eqn (11), along the branches with $\lambda_I = 0$, *i.e.* $2 \rightarrow 3$ (bilayer structure, line) and $3 \rightarrow 4$ (stalk, symbols). These branches are discretized in 51 sampling points. The inset shows the sum of the free energy changes for the bilayer and stalk structures, obtained after integrating the data of the main figure as a function of the control parameter, λ_E. The prediction of the Random-Phase-Approximation (RPA), eqn (23), for the behaviour of the integrands near the disordered state ($\lambda_E = 0$) are shown with thick solid and dashed lines for the stalk and the two-bilayers, respectively. The RPA behaviour of the free energy difference for small values of λ_E is marked in the inset with a red line.

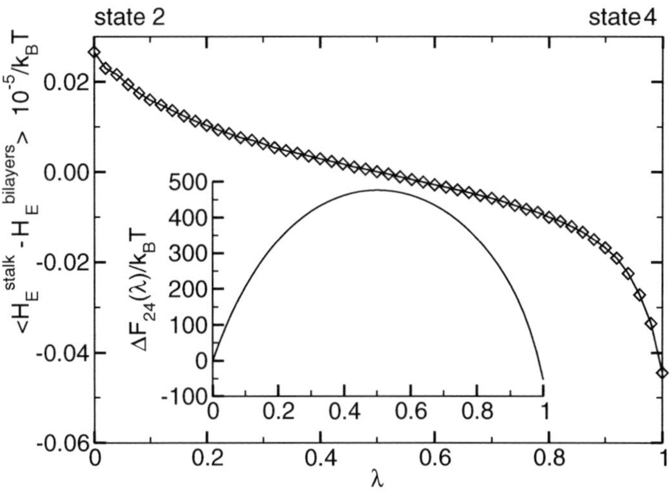

Fig. 8 Variation of the integrand, eqn (14) along the branch, 2 → 4, changing the external, ordering field from one that creates two, apposing bilayers, $\lambda = 0$, to one that orders the gas of amphiphiles into a stalk morphology, $\lambda = 1$. This branch is discretized in 51 sampling points. The inset shows the free energy change as a function of λ obtained after integrating the curve in the main figure.

integration is $\Delta F = 15k_BT$. This corroborates the result, $\Delta F = 15.3k_BT$, accurately determined *via* the expanded-ensemble simulations, eqn (19), along the branches 1 → 2, 2 → 4 and 4 → 5.

An analytical estimate for the integrands around the disordered, ideal gas state, 3, can be obtained by utilising the Random-Phase-Approximation (RPA).[63] Utilising eqn (3), (7), and (9) we can calculate the average density, $\langle\rho_\alpha(\mathbf{r})\rangle$, the external fields generate in the system without the non-bonded interactions, $\lambda_I = 0$

$$\langle\rho_\alpha(\mathbf{r})\rangle_{n,V,T,\lambda_I=0,\lambda_E} = \frac{\int\prod_{i=1}^n \tilde{\mathcal{D}}[\{\mathbf{r}_i(s)\}]\hat{\rho}_\alpha(\mathbf{r})\exp\left[-\frac{\lambda_E \mathcal{H}_E^m}{k_BT}\right]}{\int\prod_{i=1}^n \tilde{\mathcal{D}}[\{\mathbf{r}_i(s)\}]\exp\left[-\frac{\lambda_E \mathcal{H}_E^m}{k_BT}\right]} \tag{20}$$

Expanding the right hand side of eqn (20) in powers of λ_E up to linear order, we obtain for the Fourier components $\langle\rho_A(\mathbf{k})\rangle$

$$\begin{aligned}\langle\rho_A(\mathbf{k})\rangle_{n,V,T,\lambda_I=0,\lambda_E} = {} & \frac{n}{V}\left[\frac{VN_A}{N} + \frac{N_A^2 W_A(\mathbf{0})}{N^2} + \frac{N_A N_B W_B(\mathbf{0})}{N^2}\right]\delta(\mathbf{k}) \\ & - \frac{n\lambda_E W_A(\mathbf{k})S_{AA}(\mathbf{k})}{VN} - \frac{n\lambda_E W_B(\mathbf{k})S_{AB}(\mathbf{k})}{VN}\end{aligned} \tag{21}$$

and a similar expression holds for the density of hydrophilic beads, B. The Fourier transform and its inverse were defined by $W_\alpha(k) = \int d\mathbf{r}\, W_\alpha(\mathbf{r})\exp[-i\mathbf{kr}]$ and $W_\alpha(\mathbf{k}) = \frac{1}{V}\sum_k W_\alpha(\mathbf{k})\exp[i\mathbf{kr}]$, respectively. $S_{\alpha,\beta}(\mathbf{k})$ with α, $\beta = A$, B denote the partial structure factor of a single molecule in state 3, *i.e.* only subjected to the bonded interactions, eqn (1).

$$S_{\alpha,\beta}(\mathbf{k}) = \frac{1}{N}\left\langle\sum_{s,t=1}^{N}\gamma_\alpha(s)\gamma_\beta(t)\exp[-i\mathbf{k}\{\mathbf{r}(s)-\mathbf{r}(t)\}]\right\rangle \tag{22}$$

They have been numerically obtained averaging over a large ensemble of single chain configurations at state 3. Using the Fourier transform of the densities and external, ordering fields, we can rewrite the integrand, eqn (9), along the branches, $2 \rightarrow 3$ and $3 \rightarrow 4$, in the form

$$\frac{H_E^m}{k_B T} = \frac{1}{V} \sum_{\alpha=A,B} \sum_{\mathbf{k}} \langle \rho_\alpha(\mathbf{k}) \rangle_{n,V,T,\lambda_I=0,\lambda_E} W_\alpha(-\mathbf{k}) \tag{23}$$

After Fourier transforming numerically the external fields, the integrands for the stalk and the two-bilayers morphology are obtained according to eqn (21), (23) as a function of λ_E. The result is shown in Fig. 7 for the stalk and the two-bilayers morphology with solid and dashed lines, respectively. The RPA describes well the behaviour for small values of ordering fields, $\lambda_E < 0.05$, but it becomes inaccurate for larger values of λ_E. In the inset, the behaviour of the free energy difference obtained from the RPA is compared with the simulation results and good agreement is found for small λ_E.

4.2 Dependence of the excess free energy on chemical potential

If a stalk connects two membranes of large size, the planar, unperturbed portions of the membranes farther away from the stalk act as a reservoir of amphiphilic molecules. In this situation the free energy excess, $\Delta\Omega$, at constant chemical potential, μ, describes the stability of the stalk. Ω is related to the Helmholtz free energy *via*:

$$\Omega(\mu, V, T) = F(\langle n \rangle, V, T) - \mu\langle n \rangle \tag{24}$$

where $\langle n \rangle$ denotes the average number of amphiphilic molecules at chemical potential, μ.

To obtain $\Delta\Omega$, first, we calculate the chemical potential, $\mu = \ln\left[\frac{n}{V}\right] + \mu^{ex}$, of both morphologies, the stalk structure, state 5, and the two-bilayers morphology, state 1, in the canonical ensemble with $n_{TDI} = 4240$. Employing Rosenbluth sampling, we obtain ${\mu_{TDI}}^{stalk} - \mu_0 = 0.002k_BT$ and ${\mu_{TDI}}^{bilayers} - \mu_0 = 0.039k_BT$. ${\mu_{TDI}}^{stalk} < {\mu_{TDI}}^{bilayers}$ because the bilayers far away from the stalk are thinner than those in the two-bilayers morphology.

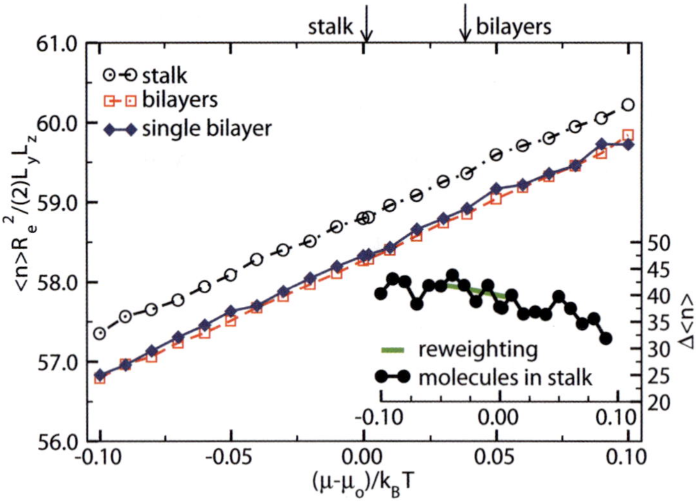

Fig. 9 Number of amphiphiles per area in a single bilayer (diamonds), two-apposed bilayers (squares), and the stalk-morphology (circles) as a function of the chemical potential referred to the chemical potential, μ_0, of the tensionless state. The inset depicts the excess number of molecules of the stalk, $\Delta\langle n \rangle = \langle n \rangle^{stalk} - \langle n \rangle^{bilayers}$ as a function of $\mu - \mu_0$. The green line depicts the result of a weighted histogram analysis.

Second, we calculate $\langle n \rangle^{\rm m}$ as a function of μ for the two morphologies, stalk and two-bilayers, indicated by the superscript, m. The simulation data are presented in Fig. 9 and have been obtained by grandcanonical simulations at different values of μ. The histograms of n at the different values of μ have been combined by the weighted histogram analysis method[64,65] and the results are depicted as lines in the inset of Fig. 9. We observe that the excess number of amphiphiles in the stalk, $\Delta\langle n \rangle = \langle n \rangle^{\rm stalk} - \langle n \rangle^{\rm bilayers}$ is on the order of a few tens and it decreases as we increase the chemical potential, μ, or decrease the tension of the membrane, $\sum$. We also note that the area density of amphiphiles in the single bilayer is slightly larger than in the two, apposed bilayers. We speculate that this effect mirrors the repulsive interactions between the apposed bilayers.

Using the μ-dependence of $\langle n \rangle^{\rm m}$, we obtain

$$\Omega^{\rm m}(\mu,V,T) = F^{\rm m}(n_{\rm TDI},V,T) - n_{\rm TDI}\mu^{\rm m}_{\rm TDI} - \int^{\mu}_{\mu^{\rm m}{\rm TDI}} {\rm d}\mu \langle n \rangle^{\rm m} \tag{25}$$

and the excess free energy, $\Delta\Omega$, is obtained by subtracting the values of the two-bilayers morphology from the value of the stalk.

$$\Delta\Omega(\mu, V, T) = \underbrace{\Delta F(n_{\rm TDI}, V, T) - \int_{\mu^{\rm stalk}_{\rm TDI}}^{\mu^{\rm bilayers}_{\rm TDI}} {\rm d}\mu \left[\langle n \rangle^{\rm stalk} - n_{\rm TDI}\right]}_{=\Delta\Omega\left(\mu^{\rm bilayers}_{\rm TDI}, V, T\right)} - \int_{\mu^{\rm bilayers}_{\rm TDI}}^{\mu} {\rm d}\mu \left[\langle n \rangle^{\rm stalk} - \langle n \rangle^{\rm bilayers}\right] \tag{26}$$

At $\mu_{\rm TDI}{}^{\rm bilayers} - \mu_0 = 0.039k_BT$, we obtain the value, $\Delta\Omega = 14.7k_BT$, *i.e.*, the stalk is metastable. This positive value quantifies the observation that the stalk disappears in very long simulation runs. From the dependence of $\langle n \rangle$ on the chemical potential in the stalk and the two-bilayer structures we compute the μ-dependence of $\Delta\Omega$. The data are presented in Fig. 10.

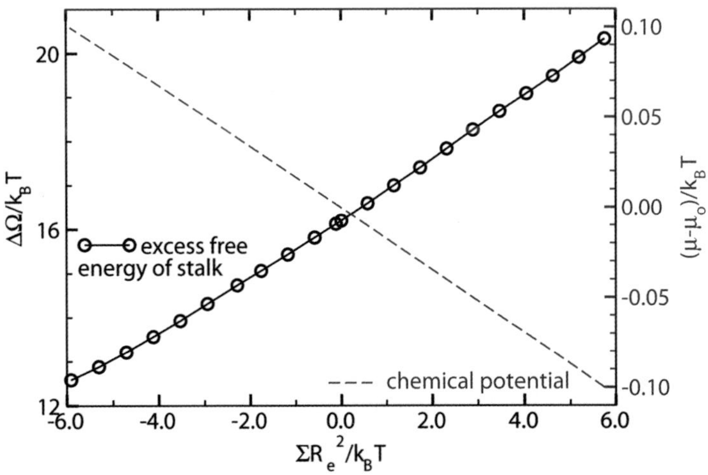

Fig. 10 Excess grandcanonical potential, $\Delta\Omega$ (circles), of the stalk as a function of the membrane tension, $\sum$. On the right hand side, we show the dependence of the chemical potential, μ (dashed line), on the membrane tension, $\sum$.

The excess grandcanonical energy, $\Delta\Omega$, as a function of the chemical potential, μ, also provides information about the dependence of the stability of the stalk on the tension, $\sum$, of the membrane[66] because the tension of a bilayer is related to μ *via* the Gibbs adsorption isotherm

$$\Sigma = \frac{\Omega(\mu) - \Omega(\mu_0)}{L_y L_z} = \int_{\mu}^{\mu_0} \mathrm{d}\mu \, \frac{\langle n \rangle^{\text{singlebilayer}}}{L_y L_Z} \tag{27}$$

where Ω denotes the grandcanonical potential of a single bilayer and $\langle n \rangle$ the μ-dependent number of amphiphilic molecules in a bilayer patch of size, $L_y L_z$. In eqn (27), the tensionless state, $\sum = 0$, at chemical potential, μ_0, has been used as a reference state. Using this information about the dependence of $\sum$ on μ, we plot the simulation results for $\Delta\Omega$ as a function of $\sum$ in Fig. 10. We observe that $\Delta\Omega$ increases with membrane tension.

4.3 Dependence of the excess free energy on molecular architecture

Much of the dependence of the fusion rate on the molecular architecture of the lipids has been rationalised *via* the excess free energy of the stalk. Experiments observe that stronger molecular asymmetries, which give rise to a more pronounced wedge-shape of the molecules, increase the fusion rate. Moreover, stalks or hourglass-shaped passages between membranes may become thermodynamically stable. Under these conditions, many stalks are formed and condense into dense arrays. In diblock copolymer systems this structure is commonly denoted as hexagonally perforated phase. It has been experimentally observed in synthetic polymer melts[11,12] and aqueous solutions of biological lipids[13] and its range of stability has been explored by self-consistent field calculations.[23,67]

Rather than performing the thermodynamic integration for each amphiphilic architecture, we accurately compute the change of the excess free energy of a stalk by using semi-grandcanonical simulations[68] of a mixture of amphiphiles. To illustrate the technique, we consider two lipid species that are represented by the same number of effective interaction centres, $N = 32$. Like in the previous section, lipid species, $L1$, is comprised of $N_A = 28$ hydrophobic beads, while the lipids of species, $L2$, are more symmetric containing $N_A = 27$ hydrophobic beads.

The partition function, Z_{sg}, of the amphiphile mixture in the semi-grandcanonical ensemble takes the form,

$$Z_{sg}(n, \delta\mu, V, T) = \sum_{n_{L1}=0}^{n} \frac{1}{n_{L1}!(n-n_{L1})!} \int \prod_{i=1}^{n} \tilde{\mathrm{D}}[\{\mathbf{r}_i(s)\}] \exp\left[-\frac{\mathrm{H}_{\mathrm{I}} - \frac{\delta\mu\delta n}{2}}{k_B T} \right] \tag{28}$$

where n_{L1} denotes the number of amphiphiles of species, $L1$, while the number of lipids of the second species is given by $n_{L2} = n - n_{L1}$. $\delta\mu = \mu_{L2} - \mu_{L1}$ is the exchange, chemical potential between the lipid species and it controls the number difference, $\delta n = n_{L2} - n_{L1} = n - 2n_{L1}$. For $\delta\mu \to -\infty$ the system is solely comprised of lipids, $L1$, *i.e.*, $\delta n = -n$. In the limit, $\delta\mu \to +\infty$ it only contains amphiphiles, $L2$, *i.e.*, $\delta n = n$. In the Monte-Carlo simulations, we use canonical moves, which update the conformations of the molecules, and semi-grandcanonical Monte-Carlo moves that "mutate" one species into the other and *vice versa*. The structural symmetry between the two lipid species considered in this example greatly facilitates the application of these semi-grandcanonical identity switches, which, in the present mixture, only consist in changing the type of a single bead at the junction between hydrophobic and hydrophilic blocks, keeping the molecular coordinates unaltered.

The semi-grandcanonical free energy, $G(n, \delta\mu)$ (the arguments, V and T being omitted), is related to the Helmholtz free energy, F, *via* the standard thermodynamic relation $G(n, \delta\mu) = F(n_{L1}, n_{L2}) - \frac{\delta\mu\langle\delta n\rangle}{2}$ and its change with respect to variations of the exchange potential, $\delta\mu$, is

$$\frac{\partial G}{\partial \delta\mu} = -\frac{1}{2}\langle\delta n\rangle \tag{29}$$

In Fig. 11 we present the dependence of $\langle\delta n\rangle$, on the exchange chemical potential, $\delta\mu$, for the two, apposed bilayers and the stalk. $\langle\delta n\rangle$ is a smooth function of $\delta\mu$ indicating that the two, similar components are completely miscible. The inset depicts the difference, $\Delta\langle\delta n\rangle = \langle\delta n\rangle^{\text{stalk}} - \langle\delta n\rangle^{\text{bilayers}}$.

Using eqn (29), we can compute the Helmholtz free energy difference, $\delta F_{L1L2} = F_{L2} - F_{L1}$ between a system that is entirely comprised of lipids, $L1$, and one that contains only amphiphiles, $L2$. It is obtained in the limit, $\delta F_{L1L2} = \lim_{\mu\pm \to \pm\infty} \delta F(\delta\mu_+, \delta\mu_-)$ with

$$\begin{aligned} \delta F(\delta\mu_+, \delta\mu_-) &= G(n, \delta\mu_+) + \frac{\delta\mu_+\langle\delta n\rangle_+}{2} - G(n, \delta\mu_-) - \frac{\delta\mu_-\langle\delta n\rangle_-}{2} \\ &= \int_{\delta\mu_-}^{0} \mathrm{d}\delta\mu \; \frac{\langle\delta n\rangle_- - \langle\delta n\rangle}{2} + \int_{0}^{\delta\mu_+} \mathrm{d}\delta\mu \; \frac{\langle\delta n\rangle_+ - \langle\delta n\rangle}{2} \end{aligned} \tag{30}$$

where $\langle\delta n\rangle_+$ and $\langle\delta n\rangle_-$ are the values of the number difference of amphiphiles at $\delta\mu_+$ and $\delta\mu_-$, respectively. Using the asymptotic behaviour, $\langle\delta n\rangle + n = \left[\langle\delta n\rangle_- + n\right]\exp\left[\frac{\delta\mu - \delta\mu_-}{k_B T}\right]$ for $\delta\mu \to -\infty$ and $n - \langle\delta n\rangle = \left[n - \langle\delta n\rangle_+\right]\exp\left[\frac{\delta\mu_+ - \delta\mu}{k_B T}\right]$ for $\delta\mu \to +\infty$, we obtain for the free energy difference of each morphology

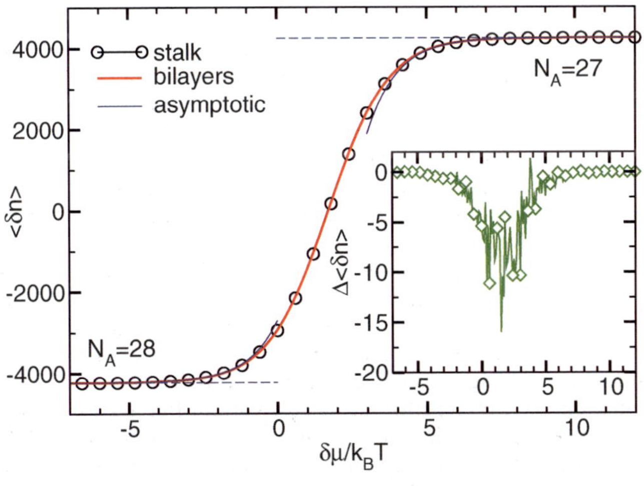

Fig. 11 The main plate shows the dependence of the difference, $\langle\delta n\rangle = \langle n_{L2} - n_{L1}\rangle$, (where n_{L2} and n_{L1} are the numbers of lipids with $N_A = 27$ and $N_A = 28$, respectively) on the exchange chemical potential $\delta\mu$, for the two, apposed bilayers (straight line) and the stalk (symbols). The inset shows the difference $\Delta\langle\delta n\rangle = \langle\delta n\rangle^{\text{stalk}} - \langle\delta n\rangle^{\text{bilayers}}$ as a function of $\delta\mu$.

$$\delta F_{L1L2} = -\int_{\delta\mu_-}^{0} \mathrm{d}\delta\mu \, \frac{n + \langle\delta n\rangle}{2} + \int_{0}^{\delta\mu_+} \mathrm{d}\delta\mu \, \frac{n - \langle\delta n\rangle}{2}$$
$$-k_B T \frac{n + \langle\delta n\rangle_-}{2} + k_B T \frac{n - \langle\delta n\rangle_+}{2} \tag{31}$$

if $\delta\mu_\pm$ are chosen such that the asymptotic behaviour is reached. Subtracting the results of the two-bilayers morphology from those of the stalk, we calculate the change of the excess free energy, $\Delta\delta F_{L1L2}$, in the canonical ensemble

$$\Delta\delta F_{L1L2} = -\int_{\delta\mu_-}^{\delta\mu_+} \mathrm{d}\delta\mu \, \frac{\Delta\langle\delta n\rangle}{2} - k_B T \frac{\Delta\langle\delta n\rangle_- + \Delta\langle\delta n\rangle_+}{2} \tag{32}$$

The negative values of $\Delta\langle\delta n\rangle$ in the inset of Fig. 11 show that the species, $L2$, with the smaller hydrophobic tail destabilises the stalk structure. From eqn (32), we obtain the estimate $\Delta\delta F_{L1L2} = 18.1(6)k_B T$ and $\Delta F = 33.4 k_B T$ for a stalk comprised of $L2$-lipids.

Like in the previous section, we measure the excess chemical potential of a tensionless bilayer, by simulating a membrane with a free edge, measuring its thickness, $A_o = 0.0352R_e^2$, and calculating the excess chemical potential, $\mu^{ex} = -36.044k_BT$, of a bilayer of this thickness in a simulation box of size $5R_e \times 10R_e \times 10R_e$, comprised of $n = 5680$ amphiphiles. The area per amphiphile, A_o, is larger than the previous value, $A_o = 0.0343R_e^2$, for the tensionless bilayer comprised of $L1$-lipids because the hydrophilic segments are effectively larger than hydrophobic segments. Thus, gradually "mutating" a tensionless bilayer of $L1$-lipids into a membrane of $L2$-amphiphiles, we obtain a membrane under negative tension, $\sum < 0$ (*i.e.* compression).

5 Conclusions

Using a thermodynamic integration method,[6,7] we have calculated the excess free energy of a stalk connecting two, apposing bilayer membranes. The technique relies on reversibly transforming one self-assembled structure into another by substituting the non-bonded interactions by external, ordering fields. To ensure reversibility, these external, ordering fields have to be chosen as to generate the structure of the self-assembled system in a system, where the non-bonded interactions have been turned off, *i.e.*, an ideal gas. Previous applications to dense polymer systems, which studied the fluctuation-induced, first-order transition between a disordered and a lamellar phase of a symmetric diblock copolymer[6] and the free energy of grain boundaries in block copolymer materials,[7] exploited the fact that the mean-field theory provides an estimate for the external fields. For the present system, however, the mean-field approximation is not sufficiently accurate and we have devised a numerical strategy for determining the external fields.

The thermodynamic integration along the reversible path has been performed using expanded-ensemble simulations, which allow for an accurate computation of the free energy differences and prove the reversibility of the chosen path. If the reversibility were taken for granted, it would be advantageous to combine this technique with replica-exchange Monte-Carlo simulations.[69–71]

The calculations have been performed using a solvent-free, coarse-grained model[33] which can be efficiently studied by computer simulations because of the lack of solvent particles and the computationally fast evaluation of the soft, non-bonded interactions *via* a collocation grid. Other than speeding up the simulations, none of these beneficial features is essential for applying the thermodynamic integration method. It can be implemented in coarse-grained models with Lennard-Jones interactions, DPD-models with soft, non-bonded potentials and field-theoretic models and employed in conjunction with Monte-Carlo simulations, molecular dynamics, Single-Chain-in-Mean-Field simulations,[40,41] and field-theoretical simulations.[72,73] Generalisations to other soft matter systems, *e.g.* with liquid crystalline order, may be envisioned by using external, ordering fields that couple to the order parameter of the transition.

Since the stalk is formed by a few tens of amphiphilic molecules while the total system is comprised of a hundred times more amphiphiles and since free energy differences between different morphologies in soft matter systems typically are small, generating accurate data for the thermodynamic integration method is computationally challenging, although thermodynamic integration techniques are inherently well-suited for parallel computations.

Once, the excess free energy, ΔF, of a stalk has been computed in the canonical ensemble the dependence of the excess free energy on the chemical potential, μ, or membrane tension, $\sum$, can be easily obtained. Using semi-grandcanonical simulation techniques,[68] the molecular architecture of the amphiphiles can be altered and one lipid species can be gradually "mutated" into another. Provided that the corresponding mixture of amphiphiles is completely miscible, the change of the excess free energy of a stalk can be computed with relative ease. These two examples illustrate that the absolute value of the excess free energy of a stalk is difficult to compute and, with our computational resources, we obtained an accuracy of $4k_BT$. Relative changes of the excess free energy in response to changing the membrane tension or the molecular architecture can be determined rather accurately.

Within our solvent-free, coarse-grained model, we compute the excess free energy of a stalk between two tensionless membranes to be $\Delta\Omega = 16.2k_BT$ with $\mathrm{N_A} = 28$. This value is lower than earlier estimates based on phenomenological models but is in a very good agreement with self-consistent field calculations.[23] This is particularly notable because our solvent-free, coarse-grained model and the model of the self-consistent field calculations significantly differ in their microscopic structure, *e.g.*, we use an implicit solvent while the self-consistent field model represents the solvent by a homopolymer. This finding suggests that the excess free energy is not very sensitive to the specific interactions of the model.

In our simulations, the stalk is comprised of about 40 amphiphiles at $\mathrm{N_A} = 28$. Upon increasing the membrane tension, we observe that the excess number of molecules, of which the stalk is comprised, remains positive and, in turn, the excess free energy, $\Delta\Omega$ increases slightly. This finding differs from the results of self-consistent field calculations,[23] which observe that the free energy of stalks is almost independent from the membrane tension or decreases with $\sum$. As we make the amphiphilic molecules more symmetric, the excess free energy of the stalk increases in agreement with self-consistent field calculations.[23]

Acknowledgements

It is a great pleasure to thank Michael Schick for stimulating and enjoyable discussions and reading of the manuscript. Financial support by the Volkswagen Foundation and the Sonderforschungsbereich 803 "Functionality controlled by organisation in and between membranes" (TP B3) is gratefully acknowledged. Ample computing time has been provided by HLRN II (Hannover/Berlin), JCP (Jülich), and the GWDG (Göttingen).

References

1 I. W. Hamley, *Nanotechnology*, 2003, **14**, R39.
2 M. P. Stoykovich, H. Kang, K. Ch. Daoulas, G. Liu, C. Liu, J. J. de Pablo, M. Müller and P. F. Nealey, *ACS Nano*, 2007, **1**, 168.
3 M. Müller and J. J. de Pablo, *Lect. Notes Phys.*, 2006, **703**, 67.
4 D. Frenkel and A. J. C. Ladd, *J. Chem. Phys.*, 1984, **81**, 3188.

5 E. M. Lennon, K. Katsov and G. H. Fredrickson, *Phys. Rev. Lett.*, 2008, **101**, 138302.
6 M. Müller and K. Ch. Daoulas, *J. Chem. Phys.*, 2008, **128**, 024903.
7 M. Müller, K. Ch. Daoulas and Y. Norizoe, *Phys. Chem. Chem. Phys.*, 2008, in press.
8 S. Y. Sheu, C. Y. Mou and R. Lovett, *Phys. Rev. E*, 1995, **51**, 3795.
9 G. Grochola, *J. Chem. Phys.*, 2004, **120**, 2122.
10 D. M. Eike, J. F. Brennecke and E. J. Maginn, *J. Chem. Phys.*, 2005, **122**, 014115.
11 D. A. Hajduk, H. Takenouchi, M. A. Hillmyer, F. S. Bates, M. E. Vigild and K. Almdal, *Macromolecules*, 1997, **30**, 3788.
12 Y. L. Loo, R. A. Register, D. H. Adamson and A. J. Ryan, *Macromolecules*, 2005, **38**, 4947.
13 L. Yang and H. W. Huang, *Science*, 2002, **297**, 1877.
14 M. Müller, K. Katsov and M. Schick, *J. Polym. Sci., Part B: Polym. Phys.*, 2003, **41**, 1441.
15 M. Müller, K. Katsov and M. Schick, *Phys. Rep.*, 2006, **434**, 113.
16 L. Chernomordik, M. M. Kozlov and J. Zimmerberg, *J. Membr. Biol.*, 1995, **146**, 1.
17 A. Mayer, *Annu. Rev. Cell Dev. Biol.*, 2002, **18**, 289.
18 L. K. Tamm, J. Crane and V. Kiessling, *Curr. Opin. Struct. Biol.*, 2003, **13**, 453.
19 R. Blumenthal, M. J. Clague, S. R. Durell and R. M. Epand, *Chem. Rev.*, 2003, **103**, 53.
20 L. Chernomordik and M. M. Kozlov, *Nature*, 2008, **15**, 675.
21 Y. Kozlovsky and M. M. Kozlov, *Biophys. J.*, 2002, **82**, 882.
22 S. May, *Biophys. J.*, 2002, **83**, 2969.
23 K. Katsov, M. Müller and M. Schick, *Biophys. J.*, 2004, **87**, 3277.
24 K. Katsov, M. Müller and M. Schick, *Biophys. J.*, 2006, **90**, 915.
25 H. Noguchi and M. P. Takasu, *J. Chem. Phys.*, 2001, **115**, 9547.
26 M. Müller, K. Katsov and M. Schick, *J. Chem. Phys.*, 2002, **116**, 2342.
27 M. J. Stevens, J. H. Hoh and T. B. Woolf, *Phys. Rev. Lett.*, 2003, **91**, 188102.
28 M. Müller, K. Katsov and M. Schick, *Biophys. J.*, 2003, **85**, 1611.
29 A. F. Smeijers, A. J. Markvoort, K. Pieterse and P. A. J. Hilbers, *J. Phys. Chem. B*, 2006, **110**, 13212.
30 L. Gao, R. Lipowsky and J. C. Shillcock, *Soft Matter*, 2008, **4**, 1208.
31 S. J. Marrink and A. E. Mark, *J. Am. Chem. Soc.*, 2003, **125**, 11144.
32 V. Knecht and S. J. Marrink, *Biophys. J.*, 2007, **92**, 4254.
33 K. Ch. Daoulas and M. Müller, *Adv. Polym. Sci.*, 2008, in press.
34 J. M. Drouffe, A. C. Maggs and S. Leibler, *Science*, 1991, **254**, 1353.
35 O. Farago, *J. Chem. Phys.*, 2003, **119**, 596.
36 I. R. Cooke and M. Deserno, *J. Chem. Phys.*, 2005, **123**, 224710.
37 J. W. Eastwood, R. W. Hockney and D. N. Lawrence, *Comput. Phys. Commun.*, 1980, **19**, 215.
38 M. Deserno and C. Holm, *J. Chem. Phys.*, 1998, **109**, 7678.
39 M. Laradji, H. Guo and M. J. Zuckermann, *J. Phys. Rev. E: Stat. Phys., Plasmas, Fluids, Relat. Interdiscip. Top.*, 1994, **49**, 3199.
40 M. Müller and G. D. Smith, *J. Polym. Sci., Part B: Polym. Phys.*, 2005, **43**, 934.
41 K. Ch. Daoulas and M. Müller, *J. Chem. Phys.*, 2006, **125**, 184904.
42 F. A. Detcheverry, K. Ch. Daoulas, M. Müller and J. J. de Pablo, *Macromolecules*, 2008, **41**, 4989.
43 K. Binder, *Monte Carlo and Molecular Dynamics Simulations in Polymer Science*, Oxford University Press, New York, 1995.
44 *Simulation Methods for Polymers*, ed. D. N. Theodorou and M. Kotelyanskii, Marcel Dekker, New York, 2004.
45 D. Frenkel and B. Smit, *Understanding Molecular Simulation*, Academic Press, London 1996.
46 D. Frenkel, G. C. A. M. Mooij and B. Smit, *J. Phys.: Condens. Matter*, 1991, **3**, 3053.
47 J. J. de Pablo, M. Laso and U. Suter, *J. Chem. Phys.*, 1992, **96**, 6157.
48 B. Smit, S. Karaborni and J. I. Siepmann, *J. Chem. Phys.*, 1995, **102**, 2126.
49 P. V. K. Pant and D. N. Theodorou, *Macromolecules*, 1995, **28**, 7224.
50 V. G. Mavrantzas, T. D. Boone, E. Zervopoulou and D. N. Theodorou, *Macromolecules*, 1999, **32**, 5072.
51 D. E. Discher and A. Eisenberg, *Science*, 2002, **297**, 967.
52 C. H. Bennet, *J. Comput. Phys.*, 1976, **22**, 245.
53 K. S. Shing and K. E. Gubbins, *Mol. Phys.*, 1982, **46**, 1109.
54 M. N. Rosenbluth and A. W. Rosenbluth, *J. Chem. Phys.*, 1955, **23**, 356.
55 A. P. Lyubartsev, A. A. Martsinovski, S. V. Shevkunov and P. N. Vorontsov-Velyaminov, *J. Chem. Phys.*, 1992, **96**, 1776.
56 G. R. Smith and A. D. Bruce, *J. Phys. A: Math. Gen.*, 1995, **28**, 6623.
57 B. A. Berg, *J. Stat. Phys.*, 1996, **82**, 323.
58 J. S. Wang, T. K. Tay and R. H. Swendsen, *Phys. Rev. Lett.*, 1999, **82**, 476.
59 M. Fitzgerald, R. R. Picard and R. N. Silver, *Europhys. Lett.*, 1999, **46**, 282.

60 F. G. Wang and D. P. Landau, *Phys. Rev. Lett.*, 2001, **86**, 2050.
61 J. S. Wang and R. H. Swendsen, *J. Stat. Phys.*, 2002, **106**, 245.
62 P. Virnau and M. Müller, *J. Chem. Phys.*, 2004, **120**, 10925.
63 L. Leibler, *Macromolecules*, 1982, **15**, 1283.
64 A. M. Ferrenberg and R. H. Swendsen, *Phys. Rev. Lett.*, 1989, **63**, 1195.
65 S. Kumar, D. Bouzida, R. H. Swendsen, P. A. Kollman and J. M. Rosenberg, *J. Comput. Chem.*, 1992, **13**, 1011.
66 M. Müller and M. Schick, *J. Chem. Phys.*, 1996, **105**, 8282.
67 R. R. Netz and M. Schick, *Phys. Rev. E: Stat. Phys., Plasmas, Fluids, Relat. Interdiscip. Top.*, 1996, **53**, 3875.
68 A. Sariban and K. Binder, *J. Chem. Phys.*, 1987, **86**, 5859.
69 K. Hukushima and K. Nemoto, *J. Phys. Soc. Jpn.*, 1996, **65**, 1604.
70 U. H. E. Hansmann, *Chem. Phys. Lett.*, 1997, **281**, 140.
71 Y. Sugita and Y. Okamoto, *Chem. Phys. Lett.*, 1999, **314**, 141.
72 G. H. Fredrickson, V. Ganesan and F. Drolet, *Macromolecules*, 2002, **35**, 16.
73 P. Altevogt, O. A. Evers, J. G. E. M. Fraaije, N. M. Maurits and B. A. C. van Vlimmeren, *J. Mol. Struct. (THEOCHEM)*, 1999, **463**, 139.

Lateral pressure profiles in lipid monolayers

Svetlana Baoukina,[a] Siewert J. Marrink[b] and D. Peter Tieleman*[a]

Received 20th March 2009, Accepted 23rd April 2009
First published as an Advance Article on the web 18th August 2009
DOI: 10.1039/b905647e

We have used molecular dynamics simulations with coarse-grained and atomistic models to study the lateral pressure profiles in lipid monolayers. We first consider simple oil/air and oil/water interfaces, and then proceed to lipid monolayers at air/water and oil/water interfaces. The results are qualitatively similar in both atomistic and coarse-grained models. The lateral pressure profile in a monolayer is characterized by a headgroup/water pressure-interfacial tension-chain pressure pattern. In contrast to lipid bilayers, the pressure decreases towards the chain free ends. An additional chain/air tension peak is present in monolayers at the air/water interface. Lateral pressure profiles are calculated for monolayers of different lipid composition under varying surface tension. Increasing the surface tension suppresses both pressure peaks and widens the interfacial tension in monolayers at the oil/water interface, and mainly suppresses the chain pressure in monolayers at the air/water interface. In monolayers in the liquid-condensed phase, the pressure peaks split due to ordering. Variation of lipid composition leads to noticeable changes in all regions of the pressure profile at a fixed surface tension.

Introduction

A typical lipid molecule consists of a polar head group and apolar long flexible hydrocarbon chains. Amphiphilic molecules self-assemble to reduce unfavorable polar–apolar contacts in a polar or apolar medium or at the interface between media of different polarity.[1] Self-assembly leads to formation of a variety of aggregates, including bilayers, monolayers, micelles, vesicles, tubes, hexagonal and cubic phases. The type of the phase formed depends on a number of factors, including the shape of the lipid molecule (*i.e.* on the effective size of the head group and chains) and the properties of the medium (polarity, geometry, *etc.*). In all phases, the headgroup/headgroup and chain/chain contacts of the neighboring lipids are separated by the hydrophobic/hydrophilic interface. Due to lipid amphiphilicity and anisotropy, the forces along the lipid molecular axis vary in nature and magnitude.[2–4] Dispersion and excluded-volume forces act between all molecular groups. Electrostatic forces originate from charges and dipoles, including hydrogen bonds and salt bridges, in the headgroups and surrounding polar solvent. In the hydrocarbon chain region, entropic interactions arise due to the restriction of accessible conformations for the lipid chains in the presence of their neighbors. The surface tension acts to shrink the hydrophobic/hydrophilic interface. This variation of interactions leads to a strongly inhomogeneous distribution of local pressures on the small length scale of the molecular size of a lipid.

[a]*Department of Biological Sciences, University of Calgary, 2500 University Dr. NW, Calgary, AB, T2N 1N4, Canada. E-mail: tieleman@ucalgary.ca*
[b]*Groningen Biomolecular Sciences and Biotechnology Institute, Zernike Institute for Advanced Materials, University of Groningen, Nijenborgh 4, 9747 AG, The Netherlands*

A lipid membrane (bilayer or monolayer) has a small thickness (~nm) relative to its lateral size (~μm) and can be thought of as a complex interface. Interfacial energy concentrated over the small thickness gives rise to local pressures of high magnitudes.[5] The local pressure distribution (from the lipid head group to the tail) along the membrane normal can be characterized by the lateral pressure profile.

The lateral pressure profile $\Pi(z)$ is defined as the difference between the lateral pressure component P_L and the normal component P_N of the pressure tensor:

$$\Pi(z) = P_L(z) - P_N \tag{1}$$

where the z-axis is directed along the membrane normal. The normal component of the pressure tensor remains constant and is equal to the pressure in the bulk; the lateral component $P_L = (P_{xx} + P_{yy})/2$ can change its amplitude and sign along z depending on the dominant interactions. Based on this definition, positive lateral pressure corresponds to repulsive interactions and negative pressure to attractive interactions. The pressure profile is a local property but is related to several macroscopic membrane properties. The sum of all interactions along the normal, or the integral of the pressure profile over the thickness, h, of the membrane, equals the surface tension with opposite sign, γ:[6]

$$\int_0^h \Pi(z)\mathrm{d}z = -\gamma \tag{2}$$

This integral is zero in a tensionless lipid bilayer in water. The first moment of the pressure profile (in a flat membrane) gives the product of the membrane bending modulus, k_b, and spontaneous curvature, c_0: $\int_0^h \Pi(z)z\mathrm{d}z = k_b c_0$, and the second moment of the pressure profile equals the saddle-splay modulus: $\int_0^h \Pi(z)z^2\mathrm{d}z = k_G$.[7] While the lateral pressure profile is difficult to assess experimentally (due to the strong variation of local pressure on a very small length scale), these macroscopic properties are measurable quantities.

Lateral pressure profiles in lipid bilayers are of specific interest because they can influence the function of proteins in cell membranes.[8] If the protein activity involves non-uniform changes in its cross-sectional area, then variations in the bilayer lateral pressure profile can shift the protein conformational equilibrium.[9] Perturbations of bilayer lateral pressure induced by small amphiphilic solutes have been proposed as a mechanism of general anesthesia.[10–12] Lateral pressure profiles in lipid bilayers have been extensively studied using theoretical and computational approaches.[7,13–25] Theoretical models are usually based on a mean-field approximation and focus on descriptions of the hydrocarbon chain conformations in a lattice or continuum representation. The head groups are not treated explicitly; their contribution enters as a function of the lipid surface density (or area per lipid). Hydrophobic/hydrophilic interactions are localized at the interface by setting an effective interfacial tension. These models allow calculation of the lateral pressure profile in the hydrocarbon chain region for pure or multi-component bilayers.

While theoretical models can take into account such properties as the magnitude of interfacial tension, head group repulsion, and length and unsaturation of the hydrocarbon chains, they still lack molecular details. Computer simulations can provide an atomic level of detail and explicitly include all intermolecular interactions. Molecular dynamics (MD) simulations have been used to investigate lateral pressure profiles in lipid bilayers of varying composition, containing sterols,[26] alcohols,[27,28] and poly-unsaturated hydrocarbon chains.[29] In a recent study,[30] a 3D pressure profile was calculated and a position dependent pressure distribution across the interface was obtained in bilayers with coexisting gel and liquid-crystalline phases, at finite curvature, and with embedded proteins.

In this work, we use MD simulations to study lateral pressure profiles in lipid monolayers. A lipid monolayer is not a mere half of a bilayer: it is formed in anisotropic conditions at a polar/apolar interface, and its properties vary with the surface density, which regulates the total surface tension at the interface. Previous simulation studies investigated monolayers of various surfactants.[31–44] However, the lateral pressure distribution across the monolayer at the interface, to our knowledge, was not calculated before. Here we focus on characterizing the lateral pressure profiles in lipid monolayers at the air/water and oil/water interfaces. To obtain more insight into the nature of pressure distributions at these complex interfaces, we first consider simple oil/air and oil/water interfaces. We then investigate lipid monolayers of different composition under varying surface tension. We compare the pressure profiles for an atomistic model and a coarse-grained model (CG). While the nature and relative contribution of interaction types in the two models differ, in particular in the head-group/water region, both models yield qualitatively similar results. In comparison to lipid bilayers, in monolayers an additional tension peak appears at the chain/air interface. Upon transformation from the liquid-expanded to the liquid-condensed phase, this tension is compensated by positive pressure due to an increased chain density, and the profile is characterized by multiple peaks originating from ordering. Variation of the surface tension induces pressure re-distribution in monolayers at both air/water and oil/water interfaces. Spontaneous curvature of the constituting lipids affects all regions of the pressure profile at a fixed surface tension.

Methods

Simulation details

We simulated 'pure' oil/air and oil/water interfaces, as well as lipid monolayers at these air/water and oil/water interfaces. For comparison with previous work, we also simulated lipid bilayers in water. Simulations were performed with the GROMACS (version 3.3.3) software package[45] using atomistic and coarse-grained (CG) models. All systems were simulated at 310 K (except for a bilayer in the CG model at 270 K to obtain the gel phase) and coupled to a Berendsen heat bath.[46] In the CG model, the coupling constant was 1.0 ps; each monolayer, water (with ions) and oil were included in separate temperature coupling groups. In the atomistic model, the coupling constant was 0.1 ps; molecules of each type from each monolayer were coupled to a separate group. The Berendsen algorithm was used for pressure coupling with a time constant of 4.0 ps in the CG model and 1.0 ps in the atomistic model using the coupling schemes as indicated below. To test the effect of the weak coupling algorithm on the calculated lateral pressure profiles, we also performed simulations with the Nose–Hoover thermostat[47,48] and Parrinello–Rahman barostat,[49] which did not lead to any noticeable changes in the results. This is to be expected as we only use average properties, not the fluctuations in temperature or pressure. The simulations were 1 μs long for the CG model (actual simulation time) and 200 ns long for the atomistic model. The initial 20 ns of each atomistic run were used for equilibration.

Simple interfaces

For simulations of the oil/air and oil/water interfaces, the system setup consisted of an oil slab in vacuum, and water and oil slabs, respectively. Hexadecane molecules were used as oil. To simulate the ordered oil/air interface, position restraints were applied to the first particle in each oil chain; the box lateral size was chosen to approximate the density of the disordered oil. No pressure coupling was used for the oil slab in vacuum. The oil/water interface was coupled to the normal pressure of 1 bar; the box size in lateral direction (parallel to the slabs) was kept constant. The systems included 144 oil molecules, 688 water particles in the CG model and 3255 water molecules in the atomistic model.

Monolayers

The system setup consisted of a water slab in vacuum or oil with two symmetric monolayers at the two polar/apolar interfaces. Surface tensions in the range 0–40 mN/m were applied using the surface-tension coupling scheme. Compressibility in the normal direction was set to zero in the case of the air/water interface; normal pressure was set to 1 bar in the case of the oil/water interface. The following lipid compositions were simulated in the CG model: pure dipalmitoyl-phosphatidylcholine (DPPC), pure diarachidonoyl-phosphatidylcholine (DAPC), a mixture of DPPC and the lyso-lipid palmitoyl-phosphatidylcholine (PPC) in a 1 : 1 ratio, and a mixture of DPPC and palmitoyloleoyl-phosphatidylglycerol (POPG) in a 3 : 1 ratio. In the atomistic model we simulated pure DPPC monolayers. Each monolayer consisted of 36 lipids in all small systems and of 4096 lipids in larger systems in the CG model. Small systems included between 800 and 1220 water particles (depending on the lipid mixture) in the CG model, and 3000 water molecules in the atomistic model. The large system included 150 289 water particles. Na^+ ions were added to compensate for the negative charge of POPG lipids. The system with DPPC monolayers at the oil/water interface in the CG model contained 288 oil molecules.

Bilayers

DPPC bilayers in water were simulated at 310 K and 270 K in the CG model. The system contained 72 lipids and 800 water particles. A semi-isotropic pressure coupling scheme was used with normal and lateral pressures of 1 bar.

Coarse-grained (CG) force field

We used the MARTINI coarse-grained force field for lipids.[50,51] In this model, the molecules are represented by grouping four heavy atoms (two to three in the case of ring structures) into a particle. All considered lipids are standard components of this force field, except for POPG for which the glycerol group in the headgroup was represented by a polar particle (P4). To model less hydrophobic oil, the particles in hexadecane (C1) were substituted by less apolar ones (C5). For non-bonded interactions, the standard cutoffs for the CG force field were used: the Lennard–Jones interactions were shifted to zero between 0.9 and 1.2 nm, the Coulomb potential was shifted to zero between 0 and 1.2 nm. The relative dielectric constant was 15, which is the default for this force-field.[50] A time step of 20 fs was used; the neighbor list was updated every 10 steps.

Atomistic force field

For the DPPC lipids, bonded and non-bonded parameters of lipid tails are adopted from the Berger model[52] with charges from Chiu.[53] Bonded parameters are based on the GROMOS force field;[54] non-bonded parameters are based on the OPLS united atom force field.[55] The headgroup charges were reduced to reproduce the LC phase in monolayers at a surface tension of 0 mN/m. Hexadecane was simulated using Berger parameters. Water was simulated using the SPC model[56] and flexible SPC model where specified. Bonds were constrained with the LINCS algorithm[57] for lipids and the SETTLE algorithm[58] for water. For Lennard–Jones interactions, a cut-off of 1 nm was used. For electrostatic interactions, a cut-off of 1.4 nm was combined with the reaction field method for long-range electrostatics with a dielectric constant of 54, the dielectric constant of SPC water. A time step of 2 fs was used, the neighbor list was updated every 10 steps.

Calculation of the lateral pressure profile

The calculation of lateral pressure was carried out using a procedure analogous to ref. 19. For a system of point-like particles interacting through pair-wise forces the macroscopic pressure tensor is given by:

$$\mathbf{P} = \frac{1}{V}\Big(\sum_{i} m_i\mathbf{v}_i \otimes \mathbf{v}_i - \sum_{i<j} \mathbf{F}_{ij} \otimes \mathbf{r}_{ij}\Big) \tag{3}$$

where m, $\mathbf{v}$, $\mathbf{r}$ and $\mathbf{F}$ are masses, velocities, distances and forces between the particles, respectively. The first term represents the kinetic contribution and the second term is the configurational contribution from the interactions in the volume V. On the scale of molecular interactions, the configurational contribution to the local pressure is defined through the (arbitrary) contour C_{ij} connecting two particles:[59]

$$\boldsymbol{\sigma}(\mathrm{r}) = -\frac{1}{2}\sum_{i<j} \mathbf{F}_{ij} \int_{C_{ij}} \delta(\mathbf{r}' - \mathbf{I})\mathrm{ds} \tag{4}$$

We use the Irving–Kirkwood contour, which connects the pair of interacting particles *via* a straight line. The simulation box is divided into slices of 0.1 nm thickness perpendicular to the z-axis, and the configurational stress tensor is found as a function of z:

$$\boldsymbol{\sigma}(z) = \frac{1}{\Delta \mathrm{V}}\sum_{i<j} \mathbf{F}_{ij} \otimes \mathbf{r}_{ij} \mathrm{f}(z, z_i, z_j) \tag{5}$$

where $\mathrm{f}(z,z_i,z_j)$ is the weighting function depending on the position of the particles with respect to the given slice and ΔV is the volume of the slice.

To calculate the pressure profiles we performed reruns of the trajectories with a modified version of GROMACS which calculates the local pressure tensor in the form of eqn (5). For each slice the pressure tensor was evaluated every 20 ps. The SHAKE algorithm[60] was used for reruns, as LINCS does not directly yield pair-wise forces.[19]

Results

Simple interfaces

To obtain more insight into the nature of the lateral pressure profile in a complex monolayer-covered polar/apolar interface, we first calculated the distribution of lateral pressure for simple interfaces. To this end, we simulated oil/water and oil/air interfaces, the latter for both disordered and ordered hexadecane chains. The lateral pressure profile for a disordered oil/air interface in the CG and atomistic models is shown in Fig. 1a. Negative pressure originates from a positive surface tension (eqn (2)), penalizing exposure of hydrocarbon chains to air. Note that the surface tension of the chain/air interface in the atomistic model (~12 mN/m) is smaller than that in the CG model (~24 mN/m). The lateral pressure distribution for the oil/water interface in the CG and atomistic models is shown in Fig. 1b. Interestingly, the profile is characterized by two negative peaks in the case of the CG model. This is because in this model the surface tension at the oil/water interface (~42 mN/m) is larger than both the surface tension of the oil/air interface and of the air/water interface (~32 mN/m). The first two tensions reproduce well the experimental data; the surface tension at the air/water interface in the CG model is lower than in the atomistic model (~53 mN/m) and than the experimental value (~70 mN/m at 310 K). The least favorable oil/water interface in the CG model thus splits into two with lower tensions. This effect can be removed by substituting the

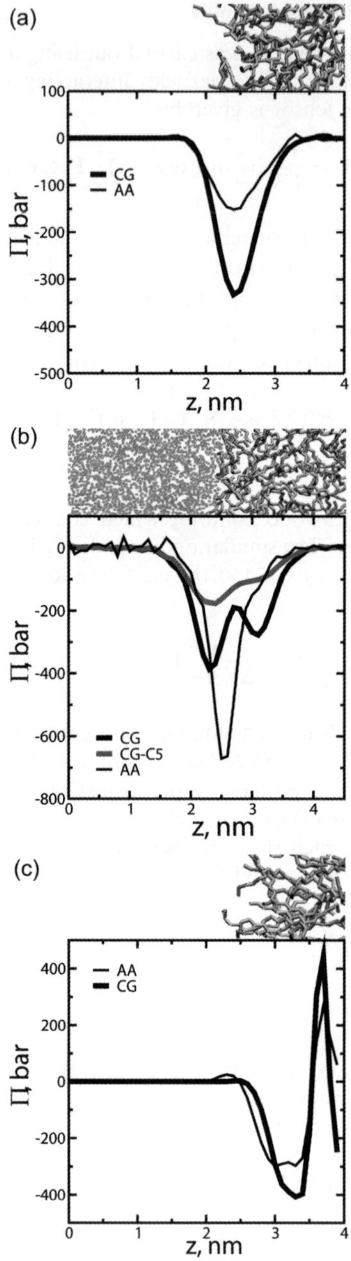

Fig. 1 Lateral pressure profile for simple interfaces in the coarse-grained (CG) and atomistic (AA) models: (a) oil/air, (b) oil/water, (c) ordered oil/water. Oil chains are shown as sticks, water particles as spheres.

oil particles by less hydrophobic ones (see Methods) which lowers the oil/water surface tension (to ~21 mN/m). For the ordered oil/air interface (see Methods), the pressure profile has an additional positive peak originating from an increased density of the hexadecane segments in the imposed orientation normal to the interface, see Fig. 1c.

Monolayer at oil/water interface

The lateral pressure profile for a DPPC monolayer at the oil/water interface (CG model) is shown in Fig. 2. The pressure distribution in the head group region and at the hydrophobic/hydrophilic interface is similar to that of a (DPPC) bilayer. The minimum of negative pressure of the hydrophobic/hydrophilic interface falls into the region where the chain and water density distribution overlap, corresponding to the glycerol/ester region. In the hydrocarbon chain region, in contrast to bilayers, the pressure decreases towards the chain ends. One can distinguish two factors determining the pressure distribution in the chain region: entropic repulsion due to restriction of conformational freedom of the chains, and deviation of chain density from that in the bulk. Entropic repulsion of the flexible chains oriented along the molecular axis increases towards the free ends due to the disorder gradient.[61] In bilayers, this entropic contribution usually dominates as the density of the oriented chains remains high. In the limit of vanishing chain bending stiffness (with no bond angle potentials in the CG model) the chains would lose their preferred orientation, and the pressure in the bilayer center would decrease (results not shown). The requirement of constant density of chain segments in mean-field theoretical models also effectively disorders the chains and leads to a reduction of lateral pressure in the bilayer center. In monolayers, the density of oriented chains decreases noticeably towards the ends. Disordered oil from the bulk partially penetrates the ordered chain region (Fig. 2), and the pressure gradually assumes its bulk value.

Monolayer at air/water interface

The lateral pressure distribution for a DPPC monolayer at the air/water interface is shown in Fig. 3. In contrast to a monolayer at an oil/water interface, the pressure at

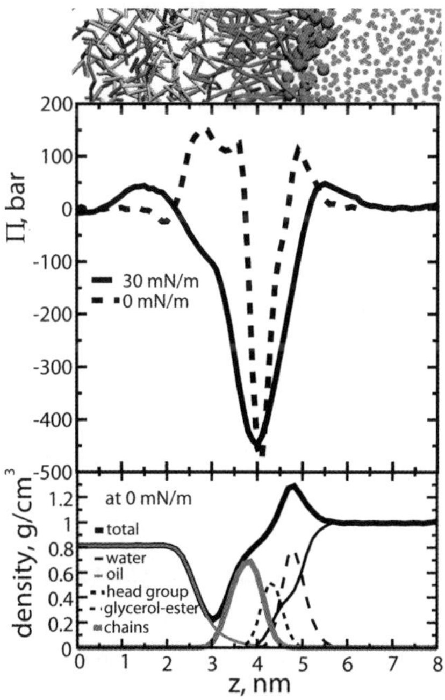

Fig. 2 Distribution of lateral pressure and density for a DPPC monolayer at the oil/water interface in the coarse-grained model. Lateral pressure profiles are shown at surface tensions of 0 and 30 mN/m, the partial density only at 0 mN/m. Lipids are shown as sticks, the head-group particles are marked as spheres, oil and water as in Fig. 1.

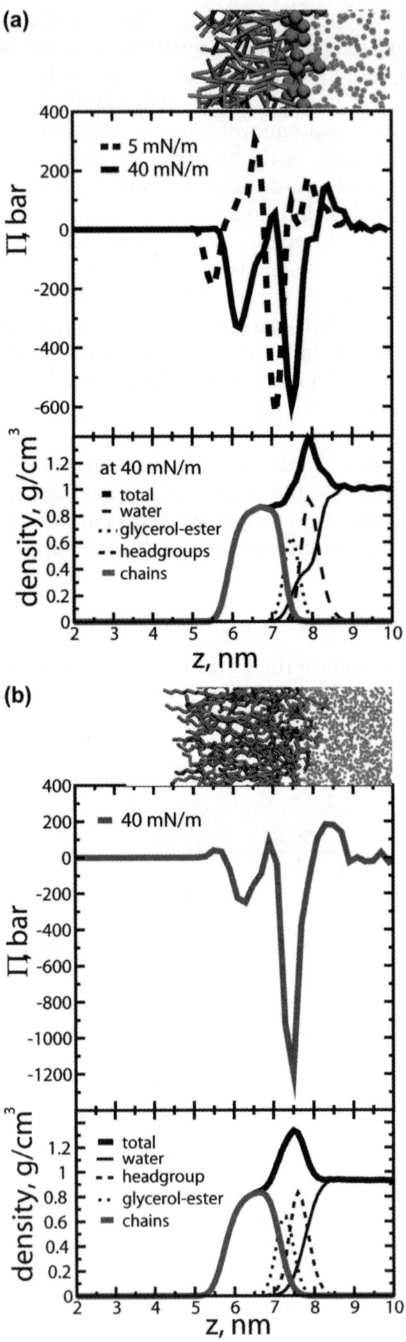

Fig. 3 Distribution of lateral pressure and density for a DPPC monolayer in the LE phase at the air/water interface in the coarse-grained (a) and atomistic (b) models. Lipids are shown as sticks (with the headgroup particles marked as spheres in the coarse-grained model), spheres represent each water atom in the atomistic model and four water molecules in the coarse-grained model.

the chain ends becomes negative. The surface tension at the chain/air interface and a decrease of total density towards the chain ends suppress entropic repulsion. The profiles in the atomistic and CG models are qualitatively similar. Headgroup/solvent interactions are repulsive in total, due to an increased density of the perturbed water in the CG model (Lennard–Jones water/headgroup interactions are stronger than water/water interactions) and ordering of water dipoles in the atomistic model. Headgroup/headgroup interactions are overall attractive in the atomistic model due to partial screening of headgroup dipoles and hydrogen bonding. In the CG model, repulsive interactions between headgroup charges are (partially) compensated by attractive Lennard–Jones interactions with waters solvating the headgroups. As a result of the differences in electrostatic interactions and water representations, the CG and atomistic profiles differ quantitatively in the headgroup/perturbed water and interfacial regions.

Surface tension

A change in surface tension, which changes the monolayer surface density from expanded to compressed, has strong effects on the monolayer pressure profile, both at the oil/water and at the air/water interface. For the monolayer at the oil/water interface (Fig. 2), an increase in surface tension (or decrease in surface density) reduces the magnitude of the pressure in the head group and chain regions. At the same time, the pressure peak corresponding to the hydrophobic/hydrophilic interface significantly widens. Upon increasing the surface tension, the system thus transforms into a single oil/water interface. For the monolayer at the air/water interface (Fig. 3a), a larger surface tension suppresses the pressure in the chain region and between the headgroups, while the pressure peaks at the headgroup/solvent and interfacial regions do not change noticeably. In bilayers, increasing the area per lipid was found to lower the pressure in the chain region and increase interfacial tension.[21] Increasing the bulk pressure, on the other hand, suppressed the pressure magnitudes in all regions in a non-uniform manner.[27] Correspondence of the lateral pressure shifts in bilayers and monolayers likely also depends on the degree of membrane stretching/compression. Besides the above mentioned changes, variation of surface tension and density can induce phase transitions in monolayers.

Phase behavior

We simulated the transition from the liquid-expanded (LE) to the liquid-condensed (LC) phase for the DPPC monolayer at the air/water interface by reducing the surface tension to 0 mN/m. The pressure profile of the LC monolayer (Fig. 4) is characterized by multiple peaks, analogous to a bilayer in the gel phase.[30] In contrast to the LE phase, the pressure at the chain/air interface is positive as a result of the higher chain density. The negative pressure peak from chain ends is shifted towards the mid-chain region. A change in sign of the lateral pressure in the chain region is also similar to a liquid-ordered bilayer with high cholesterol concentration.[30,62] Ordering of the headgroups and adjacent water in the CG model increases oscillations in the density (see Fig. 4a). In the atomistic model (Fig. 4b), the total density of the perturbed water layer does not change noticeably. The headgroup dipoles increase the ordering of the adjacent water dipoles (results not shown), which appears to lead to stronger repulsion in the perturbed water layer. Due to the higher monolayer surface density, solvation of headgroups by water is lower in the LC phase compared to the LE phase. This decreases the electrostatic screening in the atomistic model and attractive headgroup/water interactions in the CG model. This in turn increases the contribution of repulsive electrostatic interactions between the headgroups, which leads to an additional pressure peak in the CG model and a smaller tension in the atomistic model. Combining the CG and atomistic results,

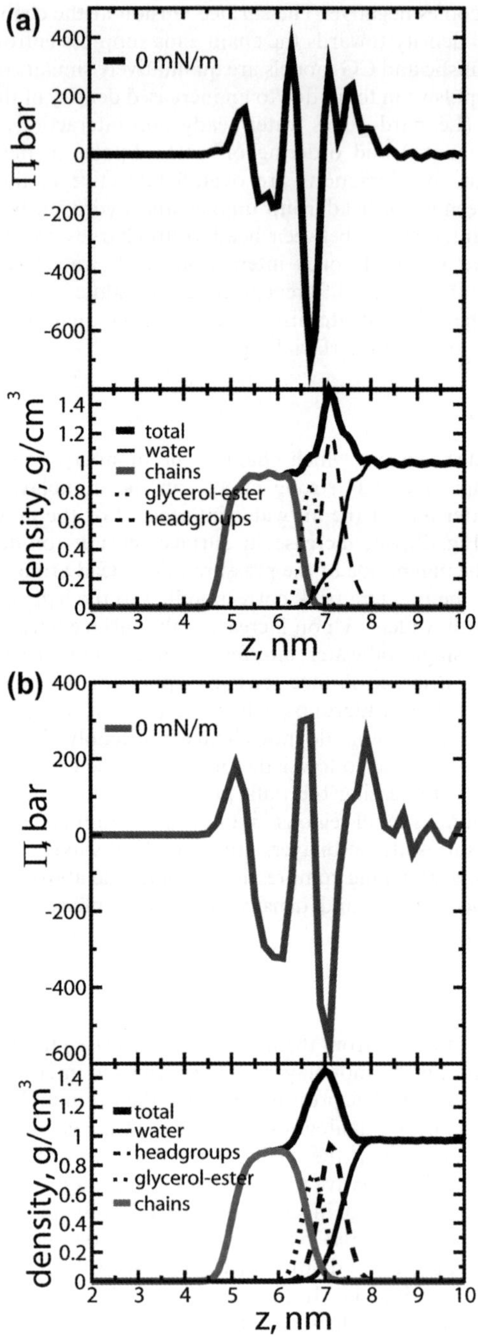

Fig. 4 Distribution of lateral pressure and density for a DPPC monolayer in the LC phase at the air/water interface in the coarse-grained (a) and atomistic (b) models.

the monolayer pressure profile in the LC phase is distinct from that in the LE phase by splitting of the peaks due to ordering and increased density, and repulsive pressure at the chain/air interface.

System size

To test the effect of system size on the calculated pressure distribution in monolayers, we simulated larger systems. This effect can be best demonstrated on a system forming the LE phase (with low bending rigidity) at low surface tension, with the ability to sustain near-zero tension, undergoing strong thermal undulations. As an example of a system with such properties in the CG model, we considered the mixture of DPPC and palmitoyloleoyl-phosphatidylglycerol (POPG) in a 3 : 1 ratio. A comparison between the lateral pressure distribution in a small (36 lipids) and large (4096 lipids) monolayer is shown in Fig. 5. High pressure peaks resolved in the small system are averaged out in the large system due to out-of-plane fluctuations of the monolayer. This favors the use of smaller over larger systems for the calculation of the pressure profiles, if the correct properties in the former can be reproduced (*e.g.* for the laterally homogenous membranes).

Lipid spontaneous curvature

To investigate the influence of spontaneous curvature on the monolayer pressure profile, we simulated cone and inverted-cone shaped lipids, the mixture of DPPC and single chain palmitoyl phosphatidylcholine (PPC) lipids in a 1 : 1 ratio and polyunsaturated diarachidonoyl phosphatidylcholine (DAPC) lipid (in the CG model), respectively. The pressure distributions at a surface tension of 20 mN/m are shown in Fig. 6. The pressure profile in the DPPC : PPC 1 : 1 mixture has smaller peaks in the interfacial and chain regions, as compared to pure DPPC monolayer. The magnitude of the head group peak, however, did not increase, likely due to lower compressibility of the head groups than the chains. In the DAPC monolayer, the pressure becomes dominant in the chain region. Simultaneously, the hydrophobic/hydrophilic interfacial peak decreases noticeably due to the presence of the less hydrophobic unsaturated chain segments. Polyunsaturation of one of the two hydrocarbon chains in (atomistic) bilayers was found to increase the magnitude of the interfacial tension, and to suppress the entropic pressure in the bilayer center.[29] Changes in lipid composition can be expected to have different effects on the lateral pressure distribution in bilayers and monolayers, because of the monolayer asymmetric environment and varying surface density or surface tension. For a monolayer at a fixed surface tension, changes in the lipid hydrocarbon chains which affect the effective size of the chains *versus* the headgroups can introduce noticeable shifts (~100 bar) in all regions of

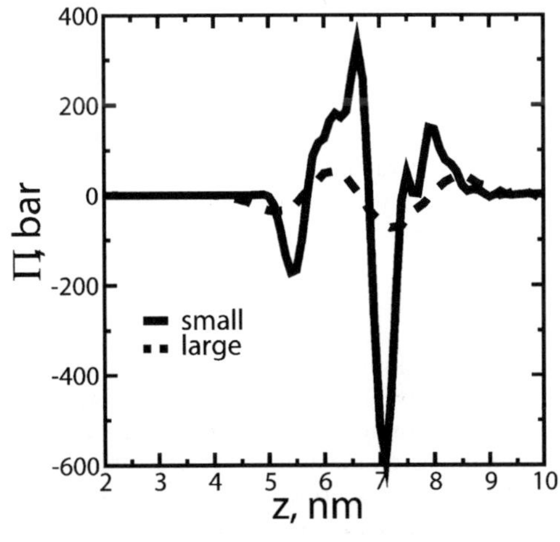

Fig. 5 Lateral pressure profile for a small (36 lipids) and large (4096 lipids) monolayer of DPPC and POPG at the air water interface in a 3 : 1 ratio at a surface tension of 0 mN/m.

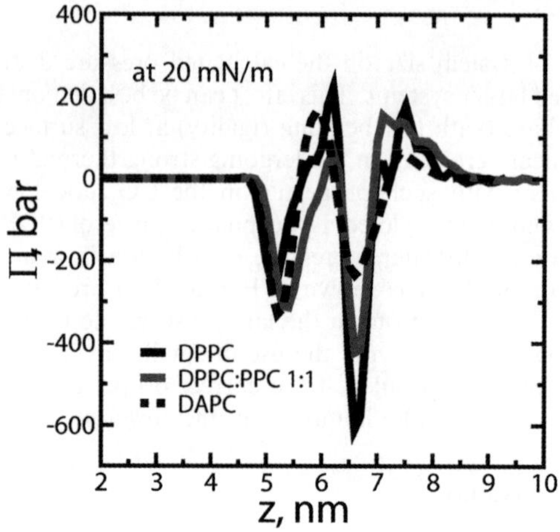

Fig. 6 Lateral pressure profile for DPPC, DPPC : PPC in a 1 : 1 ratio and DAPC monolayers at the air/water interface at a surface tension of 20 mN/m.

the pressure profile. We also calculated the first moment of the pressure profile for these systems at a surface tension of 20 mN/m. As the reference position, we chose the minimum of the tension peak of the hydrophobic/hydrophilic interface. While this choice is arbitrary, it seems logical for assessing relative contributions of the chains and headgroups. The calculated values are 200, 200 and 370 bar·nm^2, for the DPPC, DAPC, and DPPC : PPC mixtures, respectively. Using an estimate of the monolayer bending modulus as $k_b \sim k_A \cdot d^2$, allows characterizing qualitatively the monolayer spontaneous curvature from the fist moment of the pressure profile (see Introduction). Here d is the length of the hydrocarbon chains, and the k_A is the monolayer area compression modulus found from the slope of the monolayer tension-area isotherm in simulations using the formula: $k_A = A_m \partial \gamma_m / \partial A_m$, and A_m is the monolayer area. These estimates give bending moduli of 6, 2 and 5 $\times 10^{-19} J$ and spontaneous curvatures of 0.03, 0.1 and 0.07 nm^{-1}, for the DPPC, DAPC, and DPPC : PPC mixtures, respectively. The monolayer spontaneous curvature clearly depends on the choice of the reference plane and the surface density. Spontaneous curvatures of bilayer leaflets with respect to the bilayer center (in the same CG model) were found[50] to have negative values of comparable magnitudes. Here, all values are positive, even for the DAPC monolayer, which intuitively might be expected to have a negative spontaneous curvature because of the large volume of the unsaturated chains. On the other hand, a positive curvature may be explained by the tendency of polyunsaturated chains to back fold towards the interface observed in liposomes.[63] It is interesting to note that subtracting the chain/air interfacial peak (as obtained for simple chain/air interface) from the monolayer pressure profile results in a negative spontaneous curvature for DPPC and DAPC monolayers (results not shown).

Discussion

We calculated the lateral pressure profiles in a number of interfacial systems, including lipid monolayers, using both CG and atomistic simulations. Previous simulations have extensively studied the pressure profiles in lipid bilayers.

Similar to that of a lipid bilayer, the monolayer pressure profile is characterized by the headgroup/water pressure- interfacial tension–hydrocarbon chain pressure pattern. In contrast to the symmetric conditions of a bilayer, this positive-negative-positive pressure distribution in monolayers is required to balance the bending moments ($\sim \Pi \cdot z$) at the asymmetric interface.

Unlike in lipid bilayers, the pressure decreases towards chain ends in monolayers at the oil/water interface, and an additional chain/air tension peak appears in monolayers at the air/water interface. As with other monolayer properties, the pressure profile in the monolayer depends on the surface density, which determines the total surface tension at the interface. For a monolayer at the oil/water interface, reduction of the surface density widens the interfacial tension region, and suppresses both pressure peaks. For a monolayer at the air/water interface, reduction of the surface density mainly suppresses the chain pressure. A transition from the LE to the LC phase, which can be induced by lowering the total surface tension (or temperature), is characterized by splitting of the pressure peaks in the ordered monolayer.

The shape of the pressure profile can be understood by considering the nature of interactions between different molecular groups. In an earlier simulation study of atomistic bilayers by Lindahl and Edholm,[19] the pressure profile was decomposed into interactions of different origins (including Lennard–Jones, electrostatic, bonded and 1–4 interactions), between headgroups, chains and water, separating entropic and enthalpic contributions. The profile was shown to represent a sum of an order of magnitude larger terms of opposing sign. These terms depend on the simulation force field/details, and their balance is required for the equilibrium structure. Mismatch of these interactions in the simulation setup can lead to a system with incorrect properties (*e.g.* a bilayer with underestimated area per lipid) or to non-lamellar phases. If a monolayer/bilayer with correct properties and phase behavior can be formed, the distribution of lateral pressure is expected to be almost independent of the details, as the self-assembled soft matter is governed by entropic interactions, which are mainly determined by molecular size and shape.

In the atomistic and CG models used in this study, several interactions are different. In comparison to the atomistic model, the CG model does not include explicitly the chain dihedral potentials, carbonyl dipoles, water dipoles, hydrogen bonds, *etc.* The calculated pressure profiles in monolayers in these two models are qualitatively similar, because the CG model appears to capture an essential fraction of the lipid/solvent properties. In the hydrophobic part, the dominating contribution to the profile is the conformational entropy of the chains, which increases towards free ends. This would not be the case for a simpler model with short (*e.g.* two-bead) chain(s) or with zero chain bending stiffness, see the results of Venturoli and Smit[16]). In the headgroups, the treatment of electrostatics in atomistic models may result in overall attractive or repulsive interactions, depending on the distribution of partial charges and the treatment of electrostatic interactions in the simulation, compare previous results.[19,21,26,27] In the absence of charges, the pressure distribution in the headgroup and perturbed water regions would be determined only by the Lennard-Jones/conservative force parameters. Overall, the model parameters can modulate the magnitude and sign of selected pressure regions to an extent comparable to variations in lipid composition. Including more interaction levels, for example, polarizability, can further refine the pressure distribution.

Limitations

We investigated a number of technical limitations of the pressure profile calculations. It is important to keep in mind that a certain range of conditions (*e.g.* surface tensions) applied to systems of small size (~5nm) can correspond to an unphysical state. For example, phase separation cannot be studied and collapse surface tension cannot be assessed using small monolayers. For the long-range electrostatic interactions with the PME scheme, the local pressure integration algorithm with the Irving–Kirkwood contour requires approximations with a cut-off scheme.[64] For this reason we used the RF scheme in all atomistic simulations with long-range electrostatics (see Methods). Another source of errors is the constraint force calculation procedure in the trajectory post-analysis.[19,65] As a test for the validity of the calculated profiles, the integral of the pressure profile is required to converge to the total surface tension

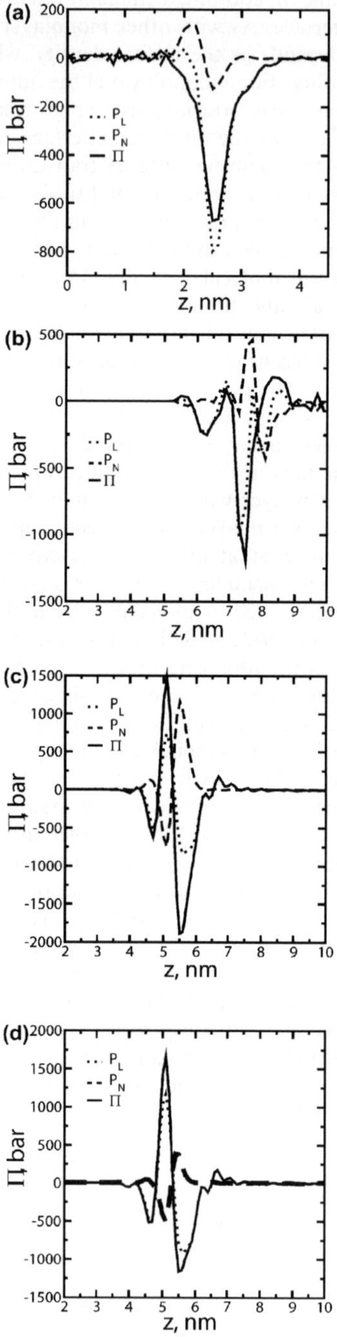

Fig. 7 Distribution of the normal (P_N) and lateral (P_L) pressures and the lateral pressure profile (Π) for (a) the atomistic oil/water interface; (b) an atomistic DPPC monolayer at the air/water interface at a surface tension of 40 mN/m; (c) a coarse-grained monolayer of DPPC : POPG : cholesterol in a 4 : 1 : 4 ratio at a surface tension of 40 mN/m; (d) system as in (c) with constraints in the cholesterol bonds substituted by a harmonic bond potential.

at the interface (with opposite sign, see eqn (2)), and the normal pressure is required to remain constant to satisfy the mechanical equilibrium condition. The normal pressure across the interface deviated from the bulk value in a number of calculated profiles. This was the case for all simulations with constraints for molecular bonds, if orientational ordering of molecular groups with respect to the interface normal was present. The normal pressure across the atomistic oil/water interface (Fig. 7a) is constant at the (disordered) hexadecane boundary but fluctuates at the water boundary as the water dipoles are ordered due to the hydrophobic effect. Using harmonic potential for the bonds in the water model (see Methods) restored the normal pressure to the bulk value.

Deviations of the normal pressure from the bulk are even stronger when a lipid monolayer is present at the interface (Fig. 7b), where molecular segments have a preferred orientation with respect to the monolayer normal. The same effect is observed in the CG monolayer with a high concentration of cholesterol (DPPC : POPG : cholesterol in ratio 4 : 1 : 4), in which several bonds are modeled with constraints, see Fig. 7c. Substituting constraints by the harmonic bonds noticeably reduced the normal pressure (Fig. 7d), but did not provide the bulk value, possibly because of distortions of the cholesterol ring structure. At the same time, the integral of the pressure profile did not converge to the total surface tension in most cases. While in the atomistic systems this could be attributed to the large fluctuations in pressure due to insufficient sampling, this clearly should not be the case in the CG systems. In all systems with non-constant normal pressure across the interface, the contribution of normal pressure is comparable in magnitude to the lateral pressure, and affects significantly the lateral pressure distribution. Due to these factors, the results presented in this work rely mainly on the CG model without constraints.

Implications

In conclusion, we outline the potential significance of the lateral pressure profile in lipid monolayers covering oil/water or air/water interfaces. In such complex interfaces, the "surface of tension" can be determined from the first moment of the pressure profile:[7]

$$\int_0^h \Pi(z)(z - z_0)\mathrm{d}z = 0 \tag{6}$$

Here the integral is proportional to the total bending moment across the interface, and the reference coordinate z_0 is chosen such that the integral becomes zero. We hypothesize that the location of this surface of tension z_0 is related to the stability of the monolayer at a given surface tension, and will pursue a verification of this hypothesis in a further study.

Acknowledgements

SB is supported by postdoctoral fellowships from the Alberta Heritage Foundation for Medical Research (AHFMR) and the Canadian Institutes for Health Research (CIHR). DPT is an AHFMR Senior Scholar and CIHR New Investigator. This work was supported by the Natural Sciences and Engineering Research Council (Canada).

References

1 C. Tanford, *The hydrophobic effect*, Wiley, New York, 1980.
2 J. N. Israelachvili, *Intermolecular and Surface Forces*, Academic Press, London, 1985.
3 D. Marsh, *Biochim. Biophys. Acta*, 1996, **1286**, 183–223.

4 A. Ben-Shaul, Elsevier, Amsterdam, 1995.
5 G. L. Gaines, *Insoluble Monolayers at Liquid-Gas Interfaces*, Wiley (Interscience), New York, 1966.
6 J. S. Rowlinson and B. Widom, *Molecular Theory of Capillarity*, Clarendon, Oxford, 1982.
7 I. Szleifer, D. Kramer, A. Benshaul, W. M. Gelbart and S. A. Safran, *J. Chem. Phys.*, 1990, **92**, 6800–6817.
8 B. de Kruijff, *Nature*, 1997, **386**, 129–130.
9 R. S. Cantor, *J. Phys. Chem. B*, 1997, **101**, 1723–1725.
10 R. S. Cantor, *Biophys. J.*, 1997, **72**, A43.
11 S. M. Gruner and E. Shyamsunder, *Mol. Cell. Mech. Alcohol Anesth.*, 1991, **625**, 685–697.
12 J. M. Seddon and R. H. Templer, Elsevier, Amsterdam, 1995.
13 R. Goetz and R. Lipowsky, *J. Chem. Phys.*, 1998, **108**, 7397–7409.
14 T. X. Xiang and B. D. Anderson, *Biophys. J.*, 1994, **66**, 561–572.
15 D. Harries and A. BenShaul, *J. Chem. Phys.*, 1997, **106**, 1609–1619.
16 M. Venturoli and B. Smit, *PhysChemComm*, 1999, **2**, 45–49.
17 R. S. Cantor, *Toxicol. Lett.*, 1998, **101**, 451–458.
18 R. S. Cantor, *Biophys. J.*, 1999, **76**, 2625–2639.
19 E. Lindahl and O. Edholm, *J. Chem. Phys.*, 2000, **113**, 3882–3893.
20 J. C. Shillcock and R. Lipowsky, *J. Chem. Phys.*, 2002, **117**, 5048–5061.
21 J. Gullingsrud and K. Schulten, *Biophys. J.*, 2004, **86**, 3496–3509.
22 S. I. Mukhin and S. Baoukina, *Phys. Rev. E: Stat., Nonlinear, Soft Matter Phys.*, 2005, **71**.
23 M. Patra, *Eur. Biophys. J. Biophys. Lett.*, 2005, **35**, 79–88.
24 M. Carrillo-Tripp and S. E. Feller, *Biochemistry*, 2005, **44**, 10164–10169.
25 A. L. Frischknecht and L. J. D. Frink, *Biophys. J.*, 2006, **91**, 4081–4090.
26 O. H. S. Ollila, T. Rog, M. Karttunen and I. Vattulainen, *J. Struct. Biol.*, 2007, **159**, 311–323.
27 B. Griepernau and R. A. Bockmann, *Biophys. J.*, 2008, **95**, 5766–5778.
28 E. Terama, O. H. S. Ollila, E. Salonen, A. C. Rowat, C. Trandum, P. Westh, M. Patra, M. Karttunen and I. Vattulainen, *J. Phys. Chem. B*, 2008, **112**, 4131–4139.
29 S. Ollila, M. T. Hyvonen and I. Vattulainen, *J. Phys. Chem. B*, 2007, **111**, 3139–3150.
30 O. H. S. Ollila, H. J. Risselada, E. Lindahl, I. M. Vattulainen and S.J., *Phys. Rev. Lett.*, 2009, in press.
31 P. Ahlstrom and H. J. C. Berendsen, *J. Phys. Chem.*, 1993, **97**, 13691–13702.
32 Y. N. Kaznessis, S. T. Kim and R. G. Larson, *Biophys. J.*, 2002, **82**, 1731–1742.
33 V. Knecht, M. Muller, M. Bonn, S. J. Marrink and A. E. Mark, *J. Chem. Phys.*, 2005, **122**, 024704.
34 S. O. Nielsen, C. F. Lopez, P. B. Moore, J. C. Shelley and M. L. Klein, *J. Phys. Chem. B*, 2003, **107**, 13911–13917.
35 H. Dominguez, A. M. Smondyrev and M. L. Berkowitz, *J. Phys. Chem. B*, 1999, **103**, 9582–9588.
36 J. I. Siepmann, S. Karaborni and M. L. Klein, *J. Phys. Chem.*, 1994, **98**, 6675–6678.
37 S. Baoukina, L. Monticelli, S. J. Marrink and D. P. Tieleman, *Langmuir*, 2007, **23**, 12617–12623.
38 S. L. Duncan and R. G. Larson, *Biophys. J.*, 2008.
39 C. F. Lopez, S. O. Nielsen, P. B. Moore, J. C. Shelley and M. L. Klein, *J. Phys.: Condens. Matter*, 2002, **14**, 9431–9444.
40 D. Rose, J. Rendell, D. Lee, K. Nag and V. Booth, *Biophys. Chem.*, 2008, **138**, 67–77.
41 A. Gupta, A. Chauhan and D. I. Kopelevichc, *J. Chem. Phys.*, 2008, 128.
42 S. Baoukina, L. Monticelli, M. Amrein and D. P. Tieleman, *Biophys. J.*, 2007, **93**, 3775–3782.
43 S. Baoukina, L. Monticelli, H. J. Risselada, S. J. Marrink and D. P. Tieleman, *Proc. Natl. Acad. Sci. U. S. A.*, 2008, **105**, 10803–10808.
44 C. Laing, S. Baoukina and D. P. Tieleman, *Phys. Chem. Chem. Phys.*, 2009, **11**, 1916–1922.
45 E. Lindahl, B. Hess and D. van der Spoel, *J. Mol. Model.*, 2001, **7**, 306–317.
46 H. J. C. Berendsen, J. P. M. Postma, W. F. van Gunsteren, A. DiNola and J. R. Haak, *J. Chem. Phys.*, 1984, **81**, 3684–3690.
47 W. G. Hoover, *Phys. Rev. A*, 1985, **31**, 1695–1697.
48 S. Nose, *Mol. Phys.*, 1984, **52**, 255–268.
49 M. Parrinello and A. Rahman, *J. Appl. Phys.*, 1981, **52**, 7182–7190.
50 S. J. Marrink, H. J. Risselada, S. Yefimov, D. P. Tieleman and A. H. Vries, *J. Phys. Chem. B*, 2007, **111**, 7812–7824.
51 L. Monticelli, S. K. Kandasamy, X. Periole, R. G. Larson, D. P. Tieleman and S. J. Marrink, *J. Chem. Theory Comput.*, 2008, **4**, 819–834.
52 O. Berger, O. Edholm and F. Jahnig, *Biophys. J.*, 1997, **72**, 2002–2013.

53 S. W. Chiu, M. Clark, V. Balaji, S. Subramaniam, H. L. Scott and E. Jakobsson, *Biophys. J.*, 1995, **69**, 1230–1245.
54 W. F. van Gunsteren, *GROMOS. Groningen Molecular Simulation Program Package*, University of Groningen, Groningen, 1987.
55 W. L. Jorgensen and J. Tiradorives, *J. Am. Chem. Soc.*, 1988, **110**, 1657–1666.
56 J. Hermans, H. J. C. Berendsen, W. F. Vangunsteren and J. P. M. Postma, *Biopolymers*, 1984, **23**, 1513–1518.
57 B. Hess, H. Bekker, H. J. C. Berendsen and J. Fraaije, *J. Comput. Chem.*, 1997, **18**, 1463–1472.
58 S. Miyamoto and P. A. Kollman, *J. Comput. Chem.*, 1992, **13**, 952–962.
59 P. Schofield and J. R. Henderson, *Proc. R. Soc. London, Ser. A*, 1982, **379**, 231–246.
60 J. P. Ryckaert, G. Ciccotti and H. J. C. Berendsen, *J. Comput. Phys.*, 1977, **23**, 327–341.
61 K. A. Dill and P. J. Flory, *Proc. Natl. Acad. Sci. U. S. A.*, 1980, **77**, 3115–3119.
62 P. S. Niemela, S. Ollila, M. T. Hyvonen, M. Karttunen and I. Vattulainen, *PLoS Comput. Biol.*, 2007, **3**, e34.
63 H. J. Risselada and S. J. Marrink, *Phys. Chem. Chem. Phys.*, 2009, **11**, 2056–2067.
64 J. Sonne, F. Y. Hansen and G. H. Peters, *J. Chem. Phys.*, 2005, **122**.
65 S. Ollila, *Helsinki University of Technology*, 2006.

Concerted diffusion of lipids in raft-like membranes

Touko Apajalahti,[a] Perttu Niemelä,[b] Praveen Nedumpully Govindan,[c] Markus S. Miettinen,[c] Emppu Salonen,[c] Siewert-Jan Marrink[d] and Ilpo Vattulainen*[efg]

Received 23rd January 2009, Accepted 25th March 2009
First published as an Advance Article on the web 15th August 2009
DOI: 10.1039/b901487j

Currently, there is no comprehensive model for the dynamics of cellular membranes. The understanding of even the basic dynamic processes, such as lateral diffusion of lipids, is still quite limited. Recent studies of one-component membrane systems have shown that instead of single-particle motions, the lateral diffusion is driven by a more complex, concerted mechanism for lipid diffusion (E. Falck *et al.*, *J. Am. Chem. Soc.*, 2008, **130**, 44–45), where a lipid and its neighbors move in unison in terms of loosely defined clusters. In this work, we extend the previous study by considering the concerted lipid diffusion phenomena in many-component raft-like membranes. This nature of diffusion phenomena emerge in all the cases we have considered, including both atom-scale simulations of lateral diffusion within rafts and coarse-grained MARTINI simulations of diffusion in membranes characterized by coexistence of raft and non-raft domains. The data allows us to identify characteristic time scales for the concerted lipid motions, which turn out to range from hundreds of nanoseconds to several microseconds. Further, we characterize typical length scales associated with the correlated lipid diffusion patterns and find them to be about 10 nm, or even larger if weak correlations are taken into account. Finally, the concerted nature of lipid motions is also found in dissipative particle dynamics simulations of lipid membranes, clarifying the role of hydrodynamics (local momentum conservation) in membrane diffusion phenomena.

I. Introduction

The dynamics of membranes have been studied extensively over several decades, see *e.g.* ref. 1–3. Yet, one of the outstanding issues regarding cell membrane properties is the fact that we know far too little about membrane dynamics, and in particular about the underlying molecular mechanisms through which the dynamic processes take place. Experimentally, clarifying this issue is very challenging due to the very short time and length scales associated with many of the dynamic membrane processes, since what one should deal with are processes taking place over time scales of the

[a]*Department of Physics, Tampere University of Technology, Finland*
[b]*VTT Technical Research Center of Finland, Espoo, Finland*
[c]*Department of Applied Physics, Helsinki University of Technology, Finland*
[d]*Groningen Biomolecular Sciences and Biotechnology Institute, Zernike Institute for Advanced Materials, University of Groningen, The Netherlands*
[e]*Department of Physics, Tampere University of Technology, Finland*
[f]*Department of Applied Physics, Helsinki University of Technology, Finland*
[g]*MEMPHYS – Center for Biomembrane Physics, University of Southern Denmark, Denmark. E-mail: Ilpo.Vattulainen@tut.fi.*

order of tens or hundreds of nanoseconds, and length scales of the order of molecular size. For example, while it is well known that lipid translocations (flip-flops) from one membrane leaflet to another are *on average* very slow processes, typical rates being one event per minute or an hour, the actual flip-flop event may happen in tens of nanoseconds through formation of transient membrane defects such as pores.[4,5] The limited understanding of membrane dynamics is in contrast to structural properties of membranes which in turn are reasonably well understood, highlighted by a number of structural models that have been proposed during the last few decades and reviewed very recently.[6–11] Particular interest has been directed to understanding the roles of membrane heterogeneity and lipid rafts, the latter being highly ordered, biologically relevant membrane domains rich in cholesterol and saturated lipids.[7–11]

The limited understanding of membrane dynamics is quite problematic. Membrane dynamics is central to a variety of cellular processes such as the formation of lipid rafts, assembly of membrane–protein complexes as well as their gating mechanisms, and signaling. Understanding of cellular functions is largely incomplete without the proper understanding of membrane dynamics. Rather surprisingly, however, even mechanisms of seemingly simple dynamic processes such as the motion of lipids in the membrane plane (lateral diffusion) are weakly understood.

It would be appealing to think that lipids in membranes diffuse through single-particle motions in terms of nearly instantaneous discrete jumps, where a lipid moves out of its cage, moving a distance comparable to its own size. However, it is not obvious that such a simplified scheme that is typical in solid systems would take place also in soft matter. What is known reliably is the fact that the rate of lipid motion depends on the time scale at which it is observed. At short times of the order of 1 ns, large lateral diffusion coefficients have been reported: in the fluid phase, the reported ones are usually about 10^{-6} cm^2 s^{-1}.[12–14] At longer times, there is a crossover to different dynamic behavior characterized by substantially smaller diffusion coefficients, typically about 10^{-8} to 10^{-7} cm^2 s^{-1} [14–17] in the fluid liquid-disordered phase.

It is commonly thought that short-range techniques such as quasi-elastic neutron scattering (QENS) measure the rapid rattling-in-a-cage motion, characterizing the motion of a lipid confined to a cage formed by its neighbors, while long-range techniques such as fluorescence recovery after photobleaching (FRAP) and fluorescence correlation spectroscopy (FCS) probe the slower motion arising from random-walk-like displacements due to lipid–lipid collisions over much larger scales. The nanoscale rattling-in-a-cage motion has been validated by atomistic simulations,[18], and also the random walk of lipids has been confirmed over macroscopic scales.[19] Yet what happens at intermediate times, and what is the actual mechanism through which lipid diffusion takes place, have remained unsolved.

Thus, is it possible that lipids would diffuse in terms of nearly instantaneous, single-particle jumps? Data from QENS backscattering experiments[12,20] have been interpreted in terms of a "jump model", providing some indirect support for this view. Atomistic simulations of Moore *et al.*[21] have identified "jump" like diffusion events from one cage to another, but the found events were very rare. Summarizing, no direct experimental or simulation evidence for single-particle jumps as a *dominating* mechanism exist. Meanwhile, simulations of Ayton and Voth[22] have proposed that density fluctuations and other collective effects may have a role to play in lateral diffusion. Recent studies by Falck *et al.*[23] showed that this is indeed the case. They considered one-component membranes through atomistic simulations and found the diffusion of lipids to be highly collective. Instead of discrete single-particle jumps, Falck *et al.* observed clusters of lipids moving in unison. The concerted motions were driven by thermal fluctuations and resulted in intriguing flow-like patterns. However, due to the major computational cost associated with atomistic simulations, the understanding of the length and time scales associated with the concerted lipid motions remained largely incomplete.

In this work, we extend the work by Falck *et al.*[23] in quite a number of ways. As in ref. 23, we focus on the mechanisms of lipid lateral diffusion. To overcome the barriers related to nanoscales, we use molecular simulations instead of experiments to elucidate how lipids diffuse in complex membranes. Our main objective is to clarify how the concerted diffusion phenomena take place in many-component rafts over a multitude of scales ranging from nanometers to tens of nanometers, and over times from nanoseconds to microseconds. To cover all these scales, we combine information from atomistic and coarse-grained simulations. We use atom-scale molecular dynamics simulations to characterize lateral diffusion within rafts,[24] and the coarse-grained MARTINI model[25] to elucidate diffusion phenomena in membranes characterized by the coexistence of raft and non-raft domains over larger scales. Additionally, we use dissipative particle dynamics[26] simulations to clarify the importance of hydrodynamic conservation laws for the observed diffusion patterns. The results highlight the emergence of concerted lipid motions in all cases we have studied, and allow us to identify typical length and time scales associated with the correlated lipid motions. The findings emphasize the collective nature of lipid dynamics at mesoscopic scales, stressing the importance of better understanding the role of collective motions in cellular membranes, in particular in the context of protein–lipid dynamics.

II. Models and methods

We performed atomic-scale and coarse-grained (CG) simulations for a number of single- and many-component membranes using three different simulation approaches. First, we conducted dissipative particle dynamics (DPD) simulations for single-component lipid bilayers to consider the influence of hydrodynamic modes (local as well as global momentum conservation) on lipid diffusion. Second, we employed atomistic molecular dynamics (MD) simulations to lateral diffusion within raft-like membranes. Finally, third, we used coarse-grained MD simulations to consider lipid diffusion in the regime where raft and non-raft domains coexist, focusing on diffusion over large scales in time and space.

A. DPD simulations

We used the DPD method[26] to simulate a large single-component lipid bilayer in the fluid phase in the NVT ensemble. The simulation parameters were obtained from a previous study,[27] in which the lipids were constructed from four DPD particles, one representing the hydrophilic head group and three representing the acyl chain. The whole system consisted of 5500 single-chained lipids of type “A”, together with 350 000 water beads in total. The interaction potentials, the parameters used in the Hamiltonian, and further details of the model including its validation are described in ref. 27.

The simulations were carried out in reduced units with a time step of $\Delta t = 0.05$ and a total of 20 000 time steps. The size of the cubic simulation box was $(50\ r_c)^3$, where r_c is the cut-off radius of interactions that follow the standard DPD form for interactions: the weight function for random forces was chosen as $\omega^R(r) \propto (1 - r/r_c)$, if $r < r_c$, and zero otherwise. To convert the reduced units into real units, we determined the lateral diffusion coefficient (see eqn (1) below) through DPD simulations for the model under consideration and compared that with the diffusion coefficient $D = 10^{-7}\ \mathrm{cm^2\ s^{-1}}$ that is typical for lipid bilayers in the fluid phase.[14] This comparison together with a reasonable estimate for the area of 0.25 nm^2 per single-chained lipid results in real units of approximately $\Delta t = 0.04$ ns and $r_c = 0.5244$ nm, thus the total simulated time scale is about 800 ns. We will use these units when displaying the results.

B. Atomistic simulations of rafts

The data for the atomic-scale raft-like bilayer were obtained from our previous study.[24] In particular, here we concentrate on the analysis of the simulation trajectory of a bilayer consisting of 1024 lipids in total, with a 2:1:1 molar mixture of palmitoyl-oleoyl-phosphatidylcholine (POPC), palmitoyl-sphingomyelin (PSM), and cholesterol (Chol), in respective order. This composition and the thermodynamic conditions we have used correspond to a membrane domain in the liquid-ordered phase, and experimental studies of a similar system have concluded the system to be raft-like, displaying coexistence of liquid-ordered and liquid-disordered phase domains.[28] The temperature of the study was $T = 310$ K and the simulation time scale was 100 ns. Details of the simulation protocol, force fields, starting configurations and other relevant simulation conditions are described elsewhere.[24]

Important to the interpretation of the results is to be aware of the analyzed bilayer structures. First, the system size is about 15 nm in linear dimension in the bilayer plane, thus what our bilayer corresponds to is the membrane structure within a raft domain. Second, while there is nanoscale membrane heterogeneity with regard to in-plane distribution of cholesterol (see Fig. 1 and 2 in ref. 24), the lipid distributions are yet rather homogeneous. Though, the most relevant point is that the atomistic simulation data can not provide an insight into the large-scale diffusion phenomena over the entire membrane domains, or across membrane domain boundaries. To this end, we employ CG simulations.

C. Coarse-grained MARTINI simulations of rafts

To access larger scales, we carried out CG simulations on eight different systems. All models are based on the MARTINI force field,[25] see below. The lipids used in the simulations were diarachidonoyl-phosphatidylcholine (DAPC), which is a phospholipid with four double bonds in both chains, the chain length being 20 carbons; dilinoleyl-phosphatidylcholine (referred to as DUPC), which is a phospholipid with two diunsaturated chains with 18 carbons in each chain; dipalmitoyl-phosphatidylcholine (DPPC), a phospholipid with two saturated chains composed of 16 carbons; and cholesterol.

To create initial configurations for the systems, first a pure DAPC system of 1152 lipids (SA1) was created from the end result of a previous (unpublished) simulation of pure DPPC. This was achieved by changing the beads of DPPC to those of DAPC, adding one extra bead needed by DAPC to each of the lipid chains, and moving the leaflets slightly apart to make room for the additional beads. After this, energy minimization was performed on the system. Using this as the base system, three other systems with DAPC were created: one by replacing randomly chosen DAPC molecules from each leaflet with DPPC and Chol molecules so that the system would have molar concentrations of DAPC : DPPC : Chol = 2 : 1 : 1 (S_{A3}), and another similarly but having concentrations of DAPC : DPPC : Chol = 8 : 1 : 1 (S_{A2}). The fourth system with DAPC was achieved by joining four copies of S_{A1} and then replacing randomly chosen DAPC molecules within a chosen radius from the center of the system with DPPC and Chol, thus creating a raft-like circular initial structure, having molar concentrations of DAPC : DPPC : Chol = 2 : 1 : 1 (S_{A4}).

Four more systems were created in the same manner, except using DUPC instead of DAPC. The only difference is that unlike DAPC, DUPC has the same number of beads as DPPC, so there was no need for moving leaflets apart. These systems (S_B) are named analogously to the S_A-systems.

Molar concentrations and initial configurations of all systems are given in Table 1, and snapshots of the initial and final configurations of all ternary systems are depicted in Fig. 1 (for membranes with DAPC) and 2 (for bilayers with DUPC).

All simulations were carried out using the GROMACS software package version 3.3.3,[29,30] except for the systems S_{A4} and S_{B4} which were run using the development version of GROMACS 4[31] because of the substantial system size.

All models are based on the MARTINI force field.[25] Force fields for DAPC and Chol are the same as in ref. 32, and for DPPC and water the same as in ref. 25. The force field for DUPC is described in ref. 33.

Table 1 Molar compositions (in units of mol%) and initial configurations of the simulated systems

Lipid system	DAPC	DUPC	DPPC	Chol	No. of lipids	Initial configuration
S_{A1}	100	—	—	—	1152	Random
S_{A2}	80	—	10	10	1152	Random
S_{A3}	50	—	25	25	1152	Random
S_{A4}	50	—	25	25	4606	Raft-like
S_{B1}	—	100	—	—	1152	Random
S_{B2}	—	80	10	10	1152	Random
S_{B3}	—	50	25	25	1152	Random
S_{B4}	—	50	25	25	4608	Raft-like

For integrating the equations of motion, we used a timestep of 20 fs. The coordinates were written every 0.5 ns. Temperature and pressure coupling were performed using the Berendsen thermostat and barostat.[34] A temperature of 323 K was used in all simulations to match the temperature used in a related study by Falck *et al.*[23] Each simulation was run over 8 μs of real time (using a factor of four to scale simulation time to real units[25]).

D. Analysis and analysed quantities

The definition of the lateral self-diffusion coefficient of lipids is

$$D = \lim_{t\to\infty} \frac{1}{4tN} \sum_{i=1}^{N} \left\langle [\boldsymbol{r}_i(t) - \boldsymbol{r}_i(0)]^2 \right\rangle \tag{1}$$

where the sum runs over all N lipids, and $\{\boldsymbol{r}_i(t)\}$ are the center-of-mass (CM) positions of lipids, projected to the plane of the bilayer. The brackets $\langle\ \rangle$ denote an average over different origins of time. We calculated the diffusion coefficients by following the position of a molecule with respect to the CM of that monolayer, unless mentioned otherwise. In this manner, the influence of the drift of the two monolayers relative to one another is accounted for.[35–37]

To facilitate addressing the issue of diffusion mechanisms, we plotted in-plane (two-dimensional (2D)) displacement vectors of the center of mass (CM) positions of each lipid in a monolayer over different time intervals Δt. Examples are shown in Fig. 3 and 4. The point here is to illustrate the local lateral correlations in lipid motions and to qualitatively address the time dependence of these correlations.

The displacement figures mostly serve for qualitative purposes. To gain more information on the character of these correlated motions, we determined a 2D displacement correlation map as follows. First, we fixed the time interval Δt. Then, for a tagged lipid, the displacement vector determined over Δt is centered in the box, lying along the x-direction, pointing to the right ($+x$), and the displacement vectors of other lipids determined over the same Δt are moved respectively, taking periodic boundary conditions into account. Then, the box surrounding the central lipid is divided into 64×64 bins; additional tests with 128×128 bins yielded consistent results. Next, dot products between the unit central displacement vector and the unit displacement vectors of other lipids within half of the box width are calculated, and the values are allocated to bins according to their relative position to the central lipid; the results are averaged by repeating this procedure for all lipids and over the whole trajectory. For visual purposes, the resulting 2D displacement correlation maps (example shown in Fig. 5 (left)) were smoothed by bilinear interpolation before plotting.

The map depicts dynamical correlations between the lateral motions of the lipid being considered and the other lipids in its vicinity. Clearly, the above approach is

Fig. 1 Snapshots of the (left) initial and (right) final configurations for systems with DAPC based on the coarse-grained MARTINI simulations. In the left column, from top to bottom, there are snapshots from above for the initial configurations of the systems S_{A2}, S_{A3}, and S_{A4}. On the right-hand side, there are respective data for the final configurations at the end of the simulations. DAPC beads are shown in grey, Chol in orange, and DPPC in purple. The small empty spots in the initial configurations are due to system preparation before any dynamics has taken place. Regarding scales of final configurations, the linear system size is about 21.3 nm in S_{A2}, 19.2 nm in S_{A3}, and 42.1 nm in S_{A4}.

just one of a possible number of means to gauge these correlations, but for the present purpose it serves well. When Δt is small, we expect some correlations to emerge. On the other hand, when Δt increases and exceeds some characteristic time, we expect the correlations between lateral motions of diffusing lipids to vanish since in the long-time limit the lipids should act like independent random walkers. The results discussed below show that this is indeed what happens.

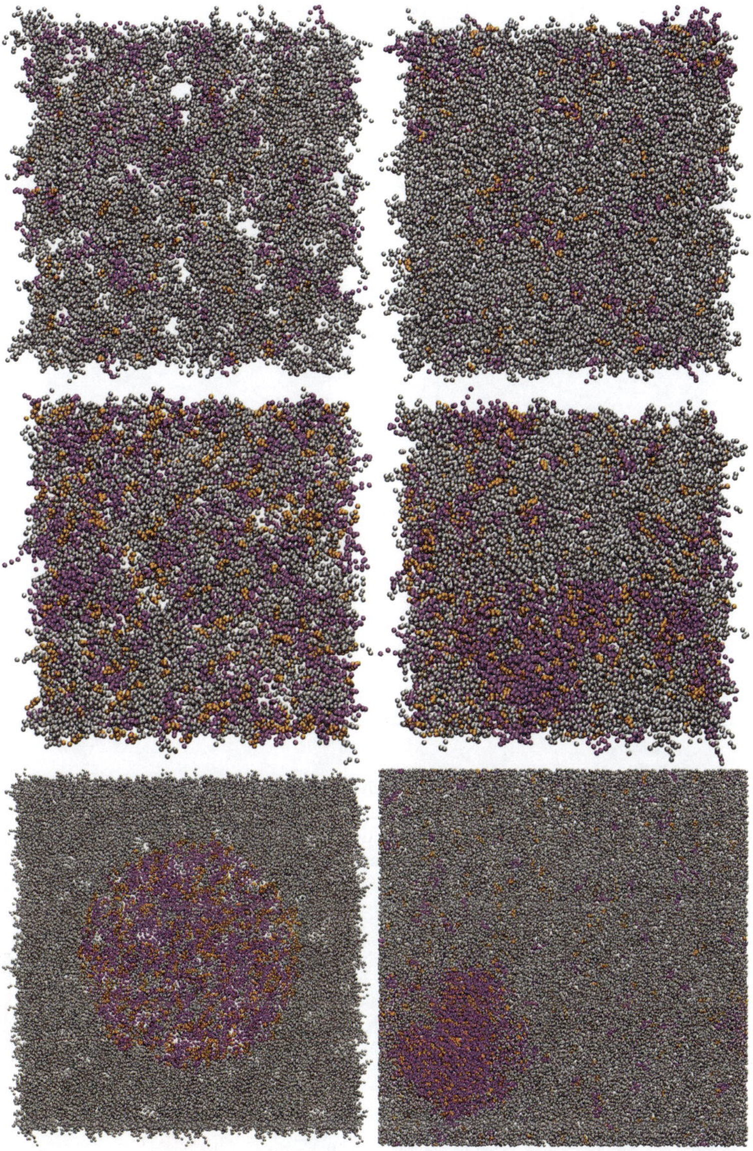

Fig. 2 Snapshots of the (left) initial and (right) final configurations for systems with DUPC based on the coarse-grained MARTINI simulations. In the left column, from top to bottom, there are snapshots from above for the initial configurations of the systems S_{B2}, S_{B3}, and S_{B4}. On the right-hand side, there are respective data for the final configurations at the end of the simulations. DUPC beads are shown in grey, Chol in orange, and DPPC in purple. Regarding scales of final configurations, the linear system size is about 20.1 nm in S_{B2}, 18.3 nm in S_{B3}, and 39.5 nm in S_{B4}.

When studying correlated movement in systems involving rafts, a couple of additional points need to be taken into account. As a raft diffuses in the membrane as an entity, it leads to a situation where movements of raft-forming lipids are highly correlated at long time scales. This obscures the correlation pattern within the raft, leading to a displacement correlation map completely different from the one in Fig. 5 (left). Instead, the result looks more like Fig. 5 (right). This effect can be

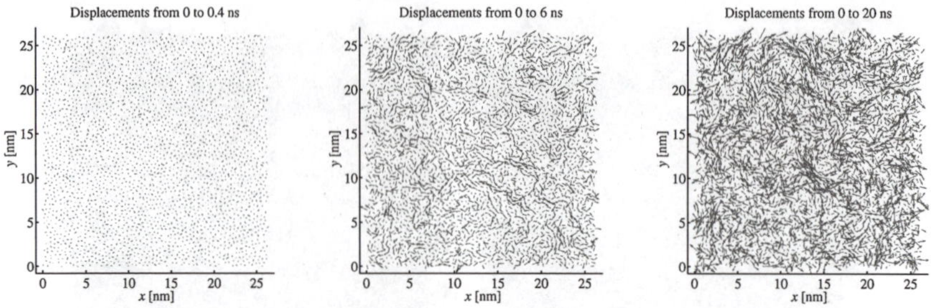

Fig. 3 Lateral displacement vectors of lipid CM positions over different time intervals (DPD simulation). Each arrow/vector describes the in-plane lateral displacement of a single lipid during a given time interval. The plots are from one of the two monolayers in the DPD-simulation.

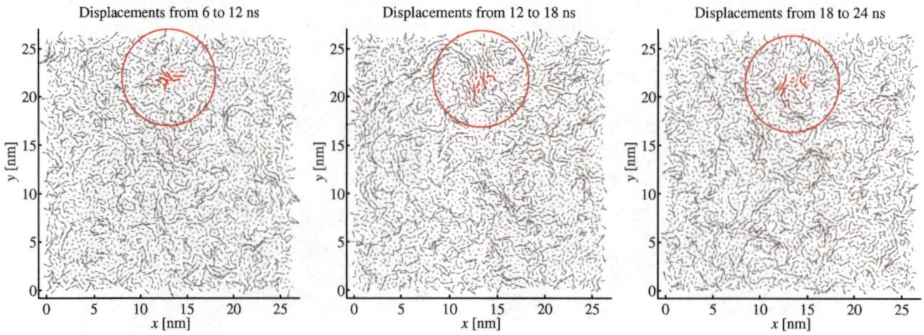

Fig. 4 Three consecutive snapshots of lateral displacement vectors for the same monolayer in the membrane system modeled by DPD, with a time interval of 6 ns. A subset of molecules in the same transient cluster with corresponding displacement vectors is shown in red to better illustrate the time dependence of local spatial correlations in the movements.

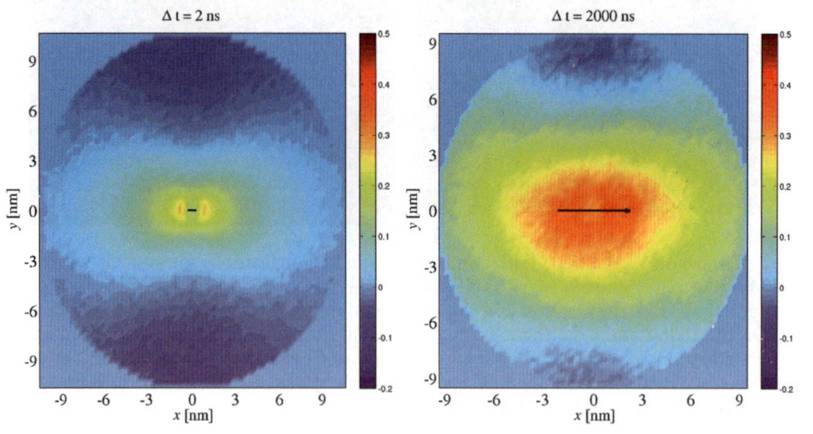

Fig. 5 Left: An example of a correlation plot for DUPC over $\Delta t = 2$ ns simulated through the CG MARTINI model (system S_{B1}). Colour scale represents the strength of correlation. One finds lipids especially in front of and behind the tagged lipid to be positively correlated with the motion of the tagged particle. The lipids beside it and farther away are negatively correlated, moving on average in the opposite direction. Right: An example of a correlation plot for DPPC in a raft over $\Delta t = 2000$ ns (system S_{A3}). Note that in this case the motion of the raft as an entity is not removed from the displacement of particles in a raft. See text for further discussion.

removed by modifying the trajectory so that the center of mass movement of raft-forming lipids is removed, essentially neutralizing the effect due to the motion of the raft as a whole. When this is accounted for, the nature of movement inside the raft can be studied more concretely. However, the non-corrected figure is not useless either. As it depicts the collective movement of the whole raft, it can be used to assess some key quantities of the raft, such as its size and shape. For example, the shape of the raft in Fig. 5 (right) ranges from stripe-like to circular with a radius of about 6 nm.

The correlation map yields information about the shape of the flow-like patterns for a chosen time interval. To assess the lifetimes of these correlations we calculate the average positive and negative values of the dot products as a function of the time interval. It is performed by first calculating the 2D displacement correlation map for a given time interval and then computing the average value of positive and negative bins, successively increasing the interval and repeating the calculation. This yields information on the time evolution of positive and negative correlation in the system. The data are then fitted with the decay function

$$C(t) = C_0 \exp(-t/\tau) + C_n \tag{2}$$

where $C(t)$ is the average (positive or negative) correlation at time t, C_0 is a constant describing the magnitude of correlation in the system, C_n is the random correlation that remains when $t \to \infty$, and τ is the decay constant of the correlation. Using τ, we can compare lifetimes of correlated motions of different lipids in different systems. Note that due to thermal fluctuations, the decay functions $C_+(t)$ and $C_-(t)$ of positive and negative correlations (in respective order) do not decay to zero at long times. However, their sum is expected to vanish for $t \to \infty$.

III. Results for DPD simulations

Fig. 3 illustrates how the lipid CMs move over different time intervals. It is clear that on the short time scale (0.4 ns) the lipids hardly move at all, whereas on the longer time scales there are correlations among lipid motions persisting over a distance of 10 nm. The dynamic correlations in the present system in the liquid-disordered phase are evident for time scales of about 5 ns or more, in agreement with previous findings.[23] Further, the largest correlated areas seem to be queue-like in shape. This observation is also in accordance with the earlier report by Falck *et al.* on atom-scale bilayers.[23]

Fig. 4 illustrates an important aspect of the observed local spatial correlations. The flow patterns are different in each of the consecutive 6-ns time windows, which means that the arrows do not represent conventional fluid flows, but they are rather short-term snapshots of transient lipid motions that vary in time. Therefore, an appropriate way to describe these patterns is to talk about local short-term correlations in lipid diffusion, or concerted transient lipid motions.

The fact that DPD simulations reproduce the recent atom-scale simulation findings[23] is an important checkpoint for the suggested mechanism, since in hydrodynamic flow phenomena the local conservation of momentum drives the formation of flow-like patterns such as vortices.[38] In DPD simulations, the underlying description of pairwise forces especially for the dissipative component guarantees the conservation of local momentum in every particle–particle collision (see, *e.g.*, ref. 39), thus leading to hydrodynamic behavior in the spirit of Navier–Stokes. This suggests that the (local) conservation of momentum could be the principal reason for the emergence of the concerted lipid motions in Fig. 3 and 4. However, in ref. 23 the Nose–Hoover thermostat was employed where local momentum conservation is violated: individual particle–particle collisions do not conserve momentum. In many simulations, the global momentum is made to conserve through the removal of the total system's CM motion, but this does not imply local momentum conservation. More generally, local momentum conservation is violated in basically every thermostat (including Berendsen, Andersen *etc.*) and barostat[39] that are commonly used in atomistic simulations.

The bottom line is that the momentum conservation is not the principal reason for the concerted lipid motions, rather it is due to other factors such as thermal and density fluctuations. The rapidly changing transient lipid motions in Fig. 4 are consistent with this view: if hydrodynamic flows due to conservation of momentum were dominant, the lifetimes of cluster motions would probably be considerably larger.

IV. Results for atomistic raft simulations

The diffusion coefficient for the selected atom-scale raft-like bilayer was around $D = 0.08 \times 10^{-7}$ cm^2 s^{-1} for all three components,[24] which is in rough agreement with pulsed field-gradient experiments in the liquid-ordered phase.[40] Due to the limited time-scale (100 ns) of the study, only small-scale molecular re-arrangements were observed rather than formation of domains.

The results in Fig. 6 reveal that the previously observed local correlations in lipid movements are not limited to one-component bilayers in the liquid-disordered phase. The figure illustrates that the correlations are not dependent on lipid types or the liquid phase of the bilayer, but that the concerted lipid motions emerge also in complex many-component membranes, reflecting the general nature of the phenomenon under study.

V. Results for coarse-grained raft simulations

To consider phenomena over larger scales, we used the MARTINI model. We found cholesterol to prefer saturated lipids, which gave rise to formation of Chol- and DPPC-rich raft domains. The rafts in the two leaflets were clearly coupled to one another (registration), and the conformational order in raft domains was distinctly larger than in the domains enriched in polyunsaturated lipids. This was evident also in cholesterol tilt, which among the S_A systems was smallest in S_{A4}, followed by S_{A3} and S_{A2}. Cholesterol also played a role in membrane thickness, which was larger in rafts compared to polyunsaturated membrane regions. Overall, the results that we found for membrane structural properties, domain formation, and lipid flip-flops are consistent with the very recent results discussed by Marrink *et al.* in ref. 33. Thus, here we will not consider those issues any further but rather concentrate on the lateral dynamics of lipids in the membrane plane.

Snapshots of the final configurations at 8 μs are shown in Fig. 1 and 2. All systems except for S_{B2} are clearly raft-like. In S_{A3} the raft has a stripe-like shape extending

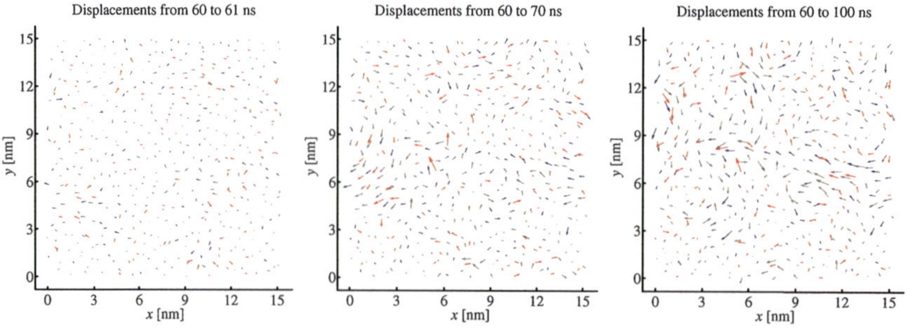

Fig. 6 Lateral displacement vectors of lipid CM positions over different time intervals of the atom-scale raft simulation. The colors indicate different lipid types: POPC (gray), PSM (blue), and Chol (red).

across the system, whereas in S_{A2} and S_{A4} the rafts are circular. In S_{B3} and S_{B4} the shapes are less clear, though S_{B4} seems to retain the round shape of the initial configuration. In S_{B2} one can find temporary clustering, though no permanent phase separation into raft-like behavior takes place.

A. Lateral diffusion

The lateral diffusion coefficients of different lipid types in the systems are given in Table 2. Note that for each of the lipid types in a given system, the diffusion coefficient has been computed over all molecules of that type by following the position of a molecule with respect to the CM of the lipids of the same type in that monolayer. In practice, this means that in single-component systems the positions of lipids in a given monolayer are considered with respect to the CM of that monolayer, as in the studies discussed earlier in this paper. In many-component systems with rafts, the approach used here essentially yields diffusion coefficients that describe diffusion inside a raft; diffusion with respect to the CM motion of the raft. This is most evident for DPPCs that are highly enriched in rafts, being found only rarely outside them. Finally, it is worth pointing out that since cholesterols flip-flop rather often,[33] the lateral diffusion coefficients have been computed only from those Chols that do not flip-flop during the simulation.

The lateral diffusion coefficients found in this work range from 10^{-8} to 10^{-7} cm^2 s^{-1}, which are in line with experimental[40] and simulation studies[24] of three-component PC/SM/Chol raft systems. Further, the lateral diffusion data is consistent with experiments, which generally indicate the lateral diffusion coefficient to be about 10^{-7} cm^2 s^{-1} in the fluid phase for single-component membranes,[15–17,41,42] and to decrease for an increasing concentration of cholesterol.[16,17,41,42]

One finds the lateral diffusion coefficients to follow a number of trends. First, let us bring about the observation that the raft forming tendency follows a pattern (in order of significance) $S_{A4} > S_{A3} > S_{A2} > S_{A1}$, and similarly for S_B. The diffusion results then show that the more raft-like the system is, the slower is diffusion of DPPC and Chol.

The effect of rafts is the clearest when looking at the diffusion of DPPC. In the systems S_{A3} and S_{A4} the rate of DPPC diffusion is ten times slower than the diffusion of other lipids in all systems. What this means is that in these systems DPPC movement is confined into the raft and almost no movement of DPPC outside the raft is observed. DPPC in the raft-system S_{B3} undergoes faster diffusion, suggesting a less strong phase separation in that system.

When looking at diffusion of Chol, the effect of rafts can also be noticed. However, because individual cholesterols that reside outside raft domains diffuse rather fast, increasing the average diffusion rate, the difference is not so pronounced.

Table 2 Diffusion coefficients determined from eqn (1) in units of 10^{-7} cm^2 s^{-1}. Error estimates of the coefficients are of the order of a few percent

System	S_{A1}	S_{A2}	S_{A3}	S_{A4}
D_{DAPC}	4.15	3.83	2.16	4.10
D_{DPPC}	—	0.33	0.16	0.16
D_{Chol}	—	3.39	1.25	1.13
	S_{B1}	S_{B2}	S_{B3}	S_{B4}
D_{DUPC}	4.06	3.58	1.99	1.93
D_{DPPC}	—	2.45	1.02	0.36
D_{Chol}	—	4.61	2.13	0.93

Still, comparison of systems S_{B3} and S_{B2}, of which the latter does not contain rafts, shows a 60% difference in the diffusion coefficients of Chol.

Both S_A and S_B show the same tendency, that is, diffusion slows down upon increasing Chol and DPPC concentration. The change of concentrations from 8:1:1 (S_{A2} and S_{B2}) to 2:1:1 (S_{A3} and S_{A4}, and S_{B3} and S_{B4}) decreases the diffusion rate roughly by 50% for all lipids. Even though the polyunsaturated lipids enter the rafts rarely, the presence of raft domains, where diffusion is slow and order is high, slows down their diffusion considerably, too.

When comparing S_{A3} with S_{B3} and S_{A2} with S_{B2} one observes only a slight difference in the diffusion coefficient of the polyunsaturated lipid. Diffusion of DUPC is a bit slower than that of DAPC in corresponding systems, which is logical given that DUPC is less unsaturated than DAPC, thus being more likely to interact favorably with Chol. Also, the phase separation is less clear or even non-existent in DUPC-systems, which implies that more DUPC are in contact with Chol than in the strongly phase separated systems with DPPC-Chol rafts.

B. Concerted diffusion patterns

Fig. 7 and 8 show the concerted diffusion patterns over chosen time intervals in one of the leaflets of the system S_{A3} used as an example to illustrate the main features.

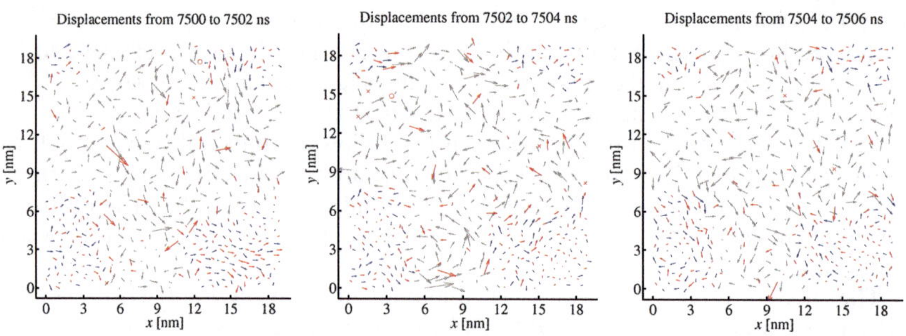

Fig. 7 Three consecutive snapshots of displacement vectors in the system S_{A3} (see also Fig. 1 (2nd row)) studied through the CG MARTINI model. DAPC is shown in grey, Chol in red, and DPPC in blue. To take possible flip-flops into account, the spots where a molecule leaves the leaflet in the beginning of the interval are marked with "o", and the spots where the molecule enters the leaflet in the end of the interval are shown with "x".

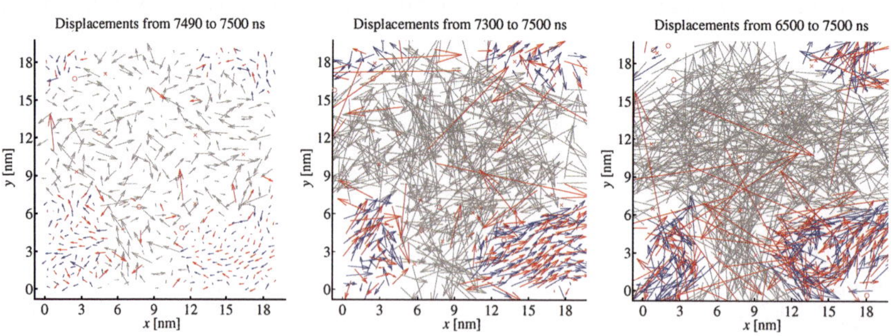

Fig. 8 Snapshots of in-plane displacement vectors in the system S_{A3} (see also Fig. 1 (2nd row)) over time intervals ranging from 10 ns to 1000 ns, studied through the CG MARTINI model. DAPC is shown in grey, Chol in red, and DPPC in blue. To take possible flip-flops into account, the spots where a molecule leaves the leaflet in the beginning of the interval are marked with "o", and the spots where the molecule enters the leaflet in the end of the interval are shown with "x".

The figures show a similar kind of behaviour as that found by Falck *et al.*[23] Movement of lipids seems to happen in streams or flows. As discussed above, the analogy to flows is somewhat problematic, though, since the flow patterns are short lived. For example, when looking at Fig. 7, which shows consecutive snapshots of displacements over a period of 2 ns, one sees that flow patterns change their direction considerably during such a short time scale. Thus, when talking about flows, what we mean are indeed temporary, short-lived collective movements.

Due to the different diffusion rates, the strength of flows depends on the phase. Fig. 7 depicts this clearly for the raft region of the system: lipid displacements in the DAPC-rich liquid-disordered phase are significantly larger than in the liquid-ordered raft region that is in the lower third of the system (see the 2nd row in Fig. 1). This is in line with the diffusion results, as the raft-forming lipids were found to undergo slower diffusion than the polyunsaturated non-raft lipids. Within rafts, the motion of lipids tends to be slower in the middle of the raft than at the edges. The motion is largely circular, which is also reflected in the motion of the non-raft lipids near the boundaries of rafts, as they tend to move more often along the raft boundary than perpendicular to it.

Another concrete finding that can be observed from Fig. 8 is that for times of about 1 μs, the flows are quite large when compared to the size of the system: in quite a few cases there are flows ranging from half of the bilayer size all the way up to the system size. Again, these flows are short-lived; there is no continuous stream revolving in the system.

Over times of several hundred nanoseconds, one finds the diffusion of the raft as a whole. Circular flow patterns transform towards linear movement, and the displacements of in-raft lipids are highly correlated, as at the same time the outside matrix has lost all correlation. In the larger systems S_{A4} and S_{B4} this behavior emerged only at the longest time intervals from 0.5 to 1 μs (data not shown). In smaller systems the lateral diffusion of rafts already started to dominate at 200 ns.

Overall, the flow patterns yield an interesting insight into how membrane dynamics look on varying time scales. However, they provide a mostly qualitative insight. For a more quantitative understanding, we carried out a more concrete analysis on the observed correlated nature of lipid motion.

C. Two-dimensional displacement correlation maps

Fig. 9 depicts the 2D displacement correlation maps for DPPC in S_{A2}, when the center of mass movement of all DPPC molecules is removed. Fig. 10 shows the results obtained for the same system when no CM movement correction is made.

These results for DPPC in the S_{A2} system largely describe the essence of dynamic correlation patterns in raft domains. That is, we focus in the following discussion on DPPC since the ordered domains are highly enriched in this saturated lipid. The S_{A2} system is chosen for discussion since it is one of the appropriate ones where one finds a clear raft-like domain.

Overall, the results show that the nature of correlated movement in rafts is independent of the system in question. The basic pattern is the same in all raft systems: the largest positive correlation is found just in front of and behind the tagged particle, and the largest negative correlation is found in about a quarter of the simulation box width to the side of the molecule under study. Significant positive correlation is observed roughly 3 nm both forward and backward with respect to the tagged lipid, and this distance varies only slightly between different systems and lipid types. Further, the concerted motions mostly take place in the direction of movement, complemented by the negative backflow correlation found along the sides. The negative correlation describing backflows on the sides is much weaker than the positive correlation found in the direction of movement, meaning that backflows do not take place directly opposite to the direction of the central flow. Rather, the large area of negative correlation seems to reflect multiple different backflows

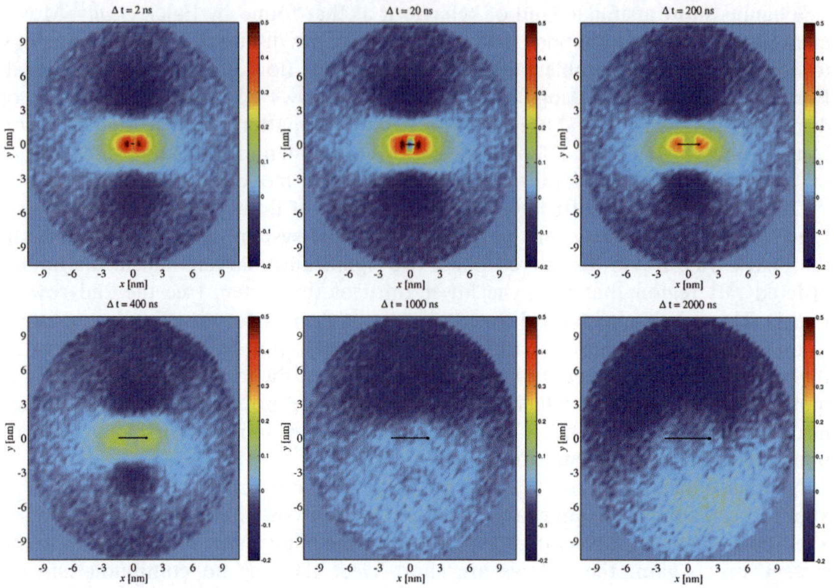

Fig. 9 2D displacement correlation maps of DPPC in the system S_{A2} with center of mass movement of DPPC removed (CG MARTINI model). Colour scale is the same in all figures.

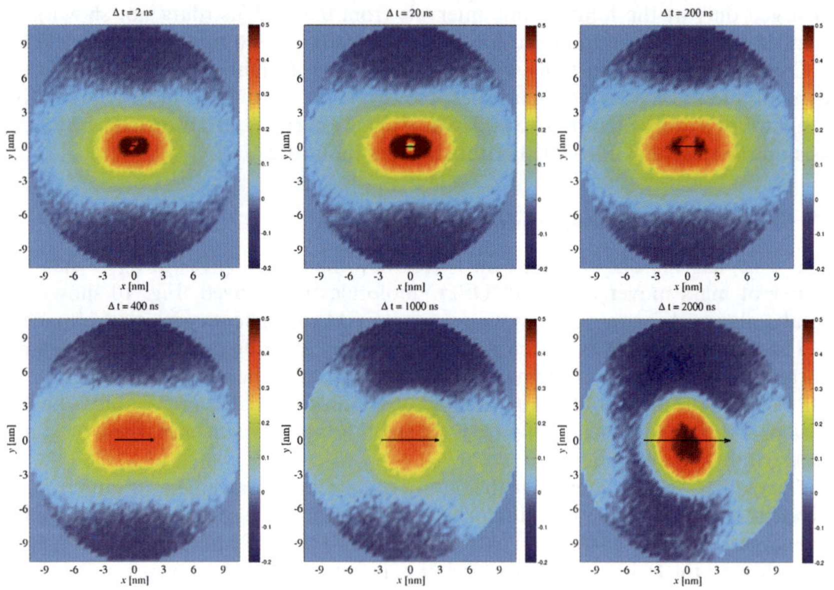

Fig. 10 2D displacement correlation maps of DPPC in the system S_{A2} without center of mass movement correction (CG MARTINI model). Colour scale is the same in all figures.

with varying directions. This is in line with what one can observe from the flow patterns in Fig. 7 and 8. That is, considering any of the instantaneous flows, one cannot find a corresponding backflow which would go exactly to the opposite direction.

The correlation patterns remain relatively constant for times below 100 ns, after which the correlation starts to get notably weaker in time, the correlations fading and the patterns becoming smoother and smoother. The correlation has practically vanished by 1000 ns. The results for Chol (data not shown) in S_{A2} (and in the other raft systems) reflect those of DPPC—not really a surprise, as Chol is the pair of DPPC in raft formation.

Fig. 10 illustrates the case where one essentially considers the motion of a raft domain as a whole. At long times, there is a circular area of very large positive correlation, reflecting the size and shape of the raft in the system, accompanied by a surrounding ring of negative correlation. The data describes the movement of the raft: all lipids in the raft move mainly similarly at long times, causing major correlation, and the surrounding area of negative correlation is due to the outside matrix giving room to the moving raft. It is worth pointing out that Cicuta *et al.* have recently studied the lateral diffusion of entire micrometer-sized membrane domains in giant unilamellar vesicles, considering, *e.g.*, the size dependence of domain diffusion. They found the results to be consistent with membrane-dominated drag in viscous liquid-ordered phases and bulk-dominated drag for less viscous liquid-disordered phases.[43]

When similar 2D displacement correlation maps were constructed for the polyunsaturated lipids, we found the general features to be largely identical to those described above. There are strong forward flow effects in front of and behind the tagged polyunsaturated lipid, and weaker backward flows on the sides.

In essentially all of the CG-systems we have studied, the size of the positively correlated regime is about 10 nm × 6 nm. The total size of the correlated region, including the surrounding negatively correlated area, is roughly 10 nm × 15 nm, being comprised of about 200 lipids. While these correlated regions are relatively large in size, they shrink in time. In every system and for every lipid we considered, the correlations vanished in a few microseconds or less.

Due to the extensive amount of data needed to explain the time dependence of the 2D correlation patterns for all systems and lipid types, we do not deal with them here any further. Instead, we prefer to employ a more quantitative means to characterize the correlation times of the observed collective motions.

D. Correlation times of concerted motions

To gauge the characteristic times of correlated motions, we determined the temporal behavior of positive and negative correlations in the 2D displacement correlation maps and used eqn (2) to fit the data to exponential form. This was performed for each lipid type in each system, resulting in the correlation time τ that describes the lifetime of concerted motions within a loosely bound cluster of lipids.

The resulting plots are shown in Fig. 11. We find that not all systems express exponential decay, see below for discussion. Nonetheless, the results of all systems are qualitatively consistent with regard to typical correlation times, which are found to be of the order of 1 μs.

For more quantitative insight, we focus on those systems and lipid types where exponential decay was most evident. Decay constants τ extracted from exponential fits are shown in Tables 3 and 4. They show quantitatively that the characteristic times of concerted motions are indeed of the order of 1 μs, typical values ranging from about 200 ns to 1.5 μs. These times constitute a major fraction of the simulation time of 8 μs, which readily explains why the data in Fig. 11 fluctuate rather strongly especially in raft systems where sampling is the weakest. Difficulties to establish exponential decay are also related to undulations, which in some of the systems were pronounced, thus the 2D lateral projections used in the analysis inevitably affect the results. Further, for some of the lipids, and especially cholesterols, flip-flops are rather frequent.[33] Since Chol molecules that did flip-flop during an analysis for a given time interval were not included in the computation, the sampling

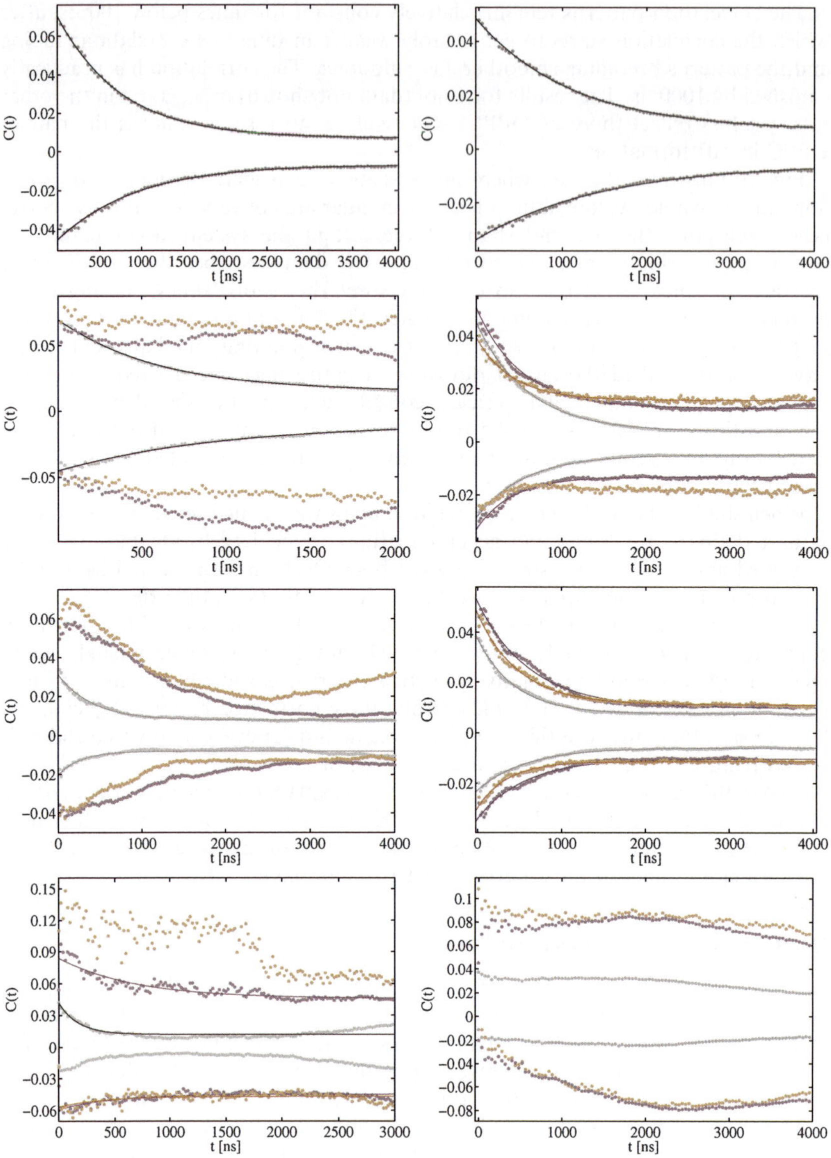

Fig. 11 Time evolutions of positive (+) and negative (−) correlation (CG MARTINI model). In the left-hand columns, from top to bottom, there are data for the systems S_{A1}, S_{A2}, S_{A3}, and S_{A4}. In the right-hand columns, there are corresponding data for the S_B-systems, respectively. The computed data are shown with dots, and the exponential fit is shown with a solid line. DPPC data is presented in purple, Chol in orange, and DAPC/DUPC in grey. In every case, the 2D displacement correlation maps used in the analysis have been computed such that the lipid displacements are determined with respect to the CM of all the lipids of the same type in that monolayer.

gets weaker and weaker for increasing time in Fig. 11. Despite these limitations, exponential decay is quite apparent and in favor of the view that temporal correlations vanish over a microsecond time scale.

For polyunsaturated lipids DAPC and DUPC (see Table 3), the characteristic correlation time is smaller the more ordered the system is. The largest correlation

Table 3 Decay constants of positive (+) and negative (−) correlation in units of nanosecond [ns]. The notion "NA" indicates that no reasonable exponential decay fit could be made. Error estimates are based on 95% confidence limits

Lipid system	S_{A1}	S_{A2}	S_{A3}	S_{A4}
$\tau_{DAPC(+)}$	752 ± 12	600 ± 16	406 ± 15	170 ± 38
$\tau_{DAPC(-)}$	930 ± 40	992 ± 99	261 ± 43	NA
	S_{B1}	S_{B2}	S_{B3}	S_{B4}
$\tau_{DUPC(+)}$	1139 ± 23	603 ± 9	441 ± 10	NA
$\tau_{DUPC(-)}$	1523 ± 84	573 ± 11	572 ± 33	NA

Table 4 Decay constants of positive (+) and negative (−) correlation in units of nanosecond [ns] for the system S_{B3}. Error estimates are based on 95% confidence limits

$\tau_{DUPC(+)}$	$\tau_{DUPC(-)}$	$\tau_{DPPC(+)}$	$\tau_{DPPC(-)}$	$\tau_{Chol(+)}$	$\tau_{Chol(-)}$
441 ± 10	572 ± 33	618 ± 19	560 ± 29	493 ± 13	355 ± 17

time is found in the pure one-component system, and the smallest correlation times are observed in the raft systems (S_{A3} and S_{A4}, and S_{B3}). This trend likely stems from cholesterol, which is partly partitioned in the liquid-disordered phase rich in polyunsaturated lipids, see Fig. 1 and 2. Thus, increasing the amounts of Chol strengthen its local ordering effect in the liquid-disordered phase, suppressing lipid diffusion and emergence of correlated motions. In the same spirit, the most long-lived correlations in the polyunsaturated lipid rich phases are found in the pure single-component systems (S_{A1} and S_{B1}), where disorder is the strongest.

As for correlations in rafts, consider data for DPPC which partitions almost completely into the raft domain. The results (see Table 4) show that correlations in a raft are somewhat more long-lived than in the neighboring matrix composed of polyunsaturated lipids. However, this view should be taken with caution, since the correlation times in Table 4 are almost identical within the error bars. Clearly, the sampling should be substantially more extensive for drawing firmer conclusions. At the moment, what can be said with certainty is that the correlations associated with concerted lipid motions die out in the microsecond time scale.

VI. Concluding remarks

In this work, we have shown that lateral diffusion in many-component lipid membranes (that are characterized by formation of raft and non-raft domains) takes place through a complex concerted diffusion mechanism, a feature that was earlier found in one-component lipid bilayers.[23] Instead of single-particle motions, the lateral diffusion of lipids occurs through the movement of dynamically correlated lipids moving in unison as loosely defined clusters. While the observed mechanism is news considering that it has not been previously observed in many-component lipid membranes, the actual value of the present study lies in the scales of concerted motion that we have observed. The coarse-grained simulations bring about a view that the correlated lipid motions take place over length and time scales that are considerably larger than the nanoscales describing the motion of individual lipids. As for length, the dynamic clusters were found to contain typically a few hundred lipids, the sizes of these transient clusters being roughly 10 nm × 15 nm. As for time scales, the lifetime, which characterizes the decay time of correlations in these clusters, was found to be about a microsecond.

The lifetime of dynamic correlations indicates that the motion of a tagged lipid is influenced by the motion of nearby lipids over this time interval of $\sim$1 μs. Assuming a lateral diffusion coefficient of 10^{-7} cm^2 s^{-1}, an individual lipid moves a lateral distance of about 6 nm during this time window. This length scale and also the time scale of 1 μs are much smaller than the scales probed by most experimental techniques for lateral diffusion, such as FRAP, FCS, and NMR. However, in current atomistic simulations the studied time scales are much smaller than the 1 μs time scale. This implies that the interpretation of lateral diffusion data extracted from atomistic simulations is not straightforward. Further work is clearly warranted to clarify this matter.

The results presented in this work are consistent with experiments concerning values of lateral diffusion coefficients and their trends for varying lipid composition. However, comparison of simulation data with experiments with regard to the diffusion mechanism, the size of the dynamically correlated lipid clusters and their lifetimes is more subtle. This is due to the fact that currently there are no published experimental studies that would have focused on the same topic. Experiments dealing with the concerted diffusion mechanism are likely to be possible through QENS and X-ray scattering, and we are looking forward to forthcoming results, but currently further comparison is not possible.

Are the observed concerted motions specific to lipid membranes, or has similar behavior been observed in other soft matter and liquid systems? At the moment, we can only address this question in part. Similar flow-like features have been observed in simple liquids modeled in terms of 2D Lennard-Jones fluids.[44] Also, collective patterns of the same type have been found in 3D supercooled liquids.[45] However, these studies[44,45] were carried out through simulations in the NVE ensemble, where energy conservation and especially the lack of a thermostat imply conservation of momentum. In the simulations presented and discussed in the present work, however, we found that momentum conservation is not a crucial condition for the emergence of diffusion flows and collective diffusion phenomena that we have found. Studies of Lennard-Jones fluids in the absence of momentum conservation would indicate whether transient collective flow patterns would emerge also in simple liquids under the conditions used in this work. However, as we are not aware of such studies, this issue remains open. It seems yet likely that the concerted diffusion mechanism observed here is an inherent feature found rather generally in soft matter systems.

There is reason to assume that the concerted diffusion phenomena could affect the molecular mechanisms of several processes in membranes. They probably influence membrane fusion[46] and local pore formation in membranes.[47] They are also expected to affect the structure and function of membrane proteins by contributing locally to lateral pressure profiles in the membrane.[48,49] Of particular interest would be to understand the role of these concerted lipid motions on the dynamics of rafts and the joint dynamics of lipid–protein complexes.[7] The latter topic is of profound importance, since while the dynamics of membrane proteins is not well understood, the understanding of the joint dynamics of lipids complexed to a membrane protein is even more limited. Atomistic and coarse-grained simulations such as those presented in this work will be essential in future efforts to clarify the complex dynamic membrane phenomena over a variety of scales in time and space.

Acknowledgements

The authors would like to thank Emma Falck and P. B. Sunil Kumar for discussions and correspondence. This work was supported by the Academy of Finland and the Netherlands Organization for Scientific Research (NWO). The Finnish IT Centre for Science and the HorseShoe (DCSC) supercluster computing facility at the University of Southern Denmark are thanked for computer resources.

References

1 J. T. Groves, R. Parthasarathy and M. B. Forstner, *Annu. Rev. Biomed. Eng.*, 2008, **10**, 311–338.
2 L. Rajendran and K. Simons, *J. Cell Sci.*, 2005, **118**, 1099–1102.
3 *The Structure of Biological Membranes*, ed. P. L. Yeagle, CSC Press, Boca Raton, 2005.
4 A. A. Gurtovenko and I. Vattulainen, *J. Phys. Chem. B*, 2007, **111**, 13554–13559.
5 D. P. Tieleman and S. J. Marrink, *J. Am. Chem. Soc.*, 2006, **128**, 12462–12467.
6 S. J. Singer and G. L. Nicolson, *Science*, 1972, **175**, 720–731.
7 K. Simons and E. Ikonen, *Nature*, 1997, **387**, 569–572.
8 M. Edidin, *Annu. Rev. Biophys. Biomol. Struct.*, 2003, **32**, 257–283.
9 L. J. Pike, *Biochem. J.*, 2004, **378**, 281–292.
10 L. J. Pike, *J. Lipid Res.*, 2006, **47**, 1597–1598.
11 K. Jacobson, O. G. Mouritsen and R. G. W. Anderson, *Nat. Cell Biol.*, 2007, **9**, 7–14.
12 S. König, W. Pfeiffer, T. Bayerl, D. Richter and E. Sackmann, *J. Phys. II*, 1992, **2**, 1589–1615.
13 J. Tabony and B. Perly, *Biochim. Biophys. Acta*, 1991, **1063**, 67.
14 I. Vattulainen; O. G. Mouritsen, in *Diffusion in Condensed Matter: Methods, Materials, Models*, ed. P. Heitjans, and J. Kärger, Springer-Verlag, Berlin, 2nd edn, 2005.
15 J. Korlach, P. Schwille, W. W. Webb and G. W. Feigenson, *Proc. Natl. Acad. Sci. U. S. A.*, 1999, **96**, 8461–8466.
16 A. Filippov, G. Orädd and G. Lindblom, *Biophys. J.*, 2003, **84**, 3079–3086.
17 P. F. F. Almeida, W. L. C. Vaz and T. E. Thompson, *Biochemistry*, 1992, **31**, 6739–6747.
18 J. Wohlert and O. Edholm, *J. Chem. Phys.*, 2006, **125**, 204703.
19 A. Sonnleitner, G. J. Schutz and T. Schmidt, *Biophys. J.*, 1999, **77**, 2638.
20 E. Sackmann, in *Handbook of Biological Physics*, ed. R. Lipowsky and E. Sackmann, Elsevier Science B.V., 1995, vol. 1; pp. 213–304.
21 P. B. Moore, C. F. Lopez and M. L. Klein, *Biophys. J.*, 2001, **81**, 2484–2494.
22 G. S. Ayton and G. A. Voth, *Biophys. J.*, 2004, **87**, 3299–3311.
23 E. Falck, T. Rog, M. Karttunen and I. Vattulainen, *J. Am. Chem. Soc.*, 2008, **130**, 44–45.
24 P. S. Niemela, S. Ollila, M. T. Hyvonen, M. Karttunen and I. Vattulainen, *PLoS Comput. Biol.*, 2007, **3**, 304–312.
25 S. J. Marrink, H. J. Risselada, S. Yefimov, D. P. Tieleman and A. H. de Vries, *J. Phys. Chem. B*, 2007, **111**, 7812–7824.
26 R. D. Groot and P. B. Warren, *J. Chem. Phys.*, 1997, **107**, 4423–4435.
27 M. Laradji and P. B. S. Kumar, *Phys. Rev. Lett.*, 2004, **93**, 198105.
28 R. F. M. de Almeida, A. Fedorov and M. Prieto, *Biophys. J.*, 2003, **85**, 2406–2416.
29 E. Lindahl, B. Hess and D. van der Spoel, *J. Mol.Model.*, 2001, **7**, 306.
30 D. van der Spoel, E. Lindahl, B. Hess, G. Groenhof, A. E. Mark and H. J. C. Berendsen, *J. Comput. Chem.*, 2005, **26**, 1701–1718.
31 B. Hess, C. Kutzner, D. van der Spoel and E. Lindahl, *J. Chem. Theory Comput.*, 2008, **4**, 435–447.
32 S. J. Marrink, A. H. de Vries, T. A. Harroun, J. Katsaras and S. R. Wassall, *J. Am. Chem. Soc.*, 2008, **130**, 10–11.
33 H. J. Risselada and S. J. Marrink, *Proc. Natl. Acad. Sci. U. S. A.*, 2008, **105**, 17367–17372.
34 H. J. C. Berendsen, J. P. M. Postma, W. F. van Gunsteren, A. DiNola and J. R. Haak, *J. Chem. Phys.*, 1984, **81**, 3684–3690.
35 E. Lindahl and O. Edholm, *J. Chem. Phys.*, 2001, **115**, 4938–4950.
36 R. A. Böckmann, A. Hac, T. Heimburg and H. Grubmüller, *Biophys. J.*, 2003, **85**, 1647–1655.
37 M. Patra, M. Karttunen, M. T. Hyvönen, E. Falck, P. Lindqvist and I. Vattulainen, *Biophys. J.*, 2003, **84**, 3636–3645.
38 B. J. Alder and T. E. Wainwright, *Phys. Rev. Lett.*, 1970, **1**, 18–21.
39 S. D. Stoyanov and R. D. Groot, *J. Comput. Chem.*, 2005, **122**, 114112.
40 A. Filippov, G. Orädd and G. Lindblom, *Biophys. J.*, 2006, **90**, 2086–2092.
41 A. Filippov, G. Orädd and G. Lindblom, *Biophys. J.*, 2004, **86**, 891–896.
42 A. Filippov, G. Orädd and G. Lindblom, *Biophys. J.*, 2007, **93**, 3182–3190.
43 P. Cicuta, S. L. Keller and S. L. Veatch, *J. Phys. Chem. B*, 2007, **111**, 3328–3331.
44 C. A. Emeis and P. L. Fehder, *J. Am. Chem. Soc.*, 1970, **92**(8), 2246–2252.
45 C. Donati, J. F. Douglas, W. Kob, S. J. Plimpton, P. H. Poole and S. C. Glotzer, *Phys. Rev. Lett.*, 1998, **80**, 2338–2341.

46 R. Jahn, T. Lang and T. C. Südhof, *Cell*, 2003, **112**, 519–533.
47 H. L. Tepper and G. A. Voth, *Biophys. J.*, 2005, **88**, 3095–3108.
48 R. S. Cantor, *Biochemistry*, 1997, **36**, 2339–2344.
49 O. H. S. Ollila, H. J. Risselada, M. Louhivuori, E. Lindahl, I. Vattulainen and S. J. Marrink, *Phys. Rev. Lett.*, 2009, **102**, 078101.

Membrane poration by antimicrobial peptides combining atomistic and coarse-grained descriptions

Andrzej J. Rzepiela,† Durba Sengupta,† Nicolae Goga and Siewert J. Marrink*

Received 26th January 2009, Accepted 26th March 2009
First published as an Advance Article on the web 18th August 2009
DOI: 10.1039/b901615e

Antimicrobial peptides (AMPs) comprise a large family of peptides that include small cationic peptides, such as magainins, which permeabilize lipid membranes. Previous atomistic level simulations of magainin-H2 peptides show that they act by forming toroidal transmembrane pores. However, due to the atomistic level of description, these simulations were necessarily limited to small system sizes and sub-microsecond time scales. Here, we study the long-time relaxation properties of these pores by evolving the systems using a coarse-grain (CG) description. The disordered nature and the topology of the atomistic pores are maintained at the CG level. The peptides sample different orientations but at any given time, only a few peptides insert into the pore. Key states observed at the CG level are subsequently back-transformed to the atomistic level using a resolution-transformation protocol. The configurations sampled at the CG level are stable in the atomistic simulation. The effect of helicity on pore stability is investigated at the CG level and we find that partial helicity is required to form stable pores. We also show that the current CG scheme can be used to study spontaneous poration by magainin-H2 peptides. Overall, our simulations provide a multi-scale view of a fundamental biophysical membrane process involving a complex interplay between peptides and lipids.

2 Introduction

Antimicrobial peptides (AMPs) exhibit a wide range of antimicrobial and antifungal activity and have attracted significant interest as potential antibiotics.[1–3] Although the details of the many modes of action of AMPs are still unclear, a large number of AMPs function by inducing transmembrane pores that lead to cell death.[4–6] The peptides bind to phospholipid bilayers and above a threshold concentration induce local defects in the bilayer.[7–10] A well studied example of an antimicrobial peptide is magainin, found in the skin of the African clawed frog *Xenopus laevis.*[8,11] The peptide is cationic and unstructured in solution but adopts a predominantly α-helical structure when bound to lipid bilayers.[12] At peptide lipid ratios of about 1/40, magainin peptides have been suggested to permeabilize the lipid matrix, forming water-filled, nanometer-sized toroidal-shaped pores.[7,11] Poration is associated with an increase in lipid flip-flops and the translocation of peptides across the membrane.[13]

Groningen Biomolecular Sciences and Biotechnology Institute & Zernike Institute for Advanced Materials, University of Groningen, Nijenborgh 4, 9747, AG, Groningen, The Netherlands. E-mail: S.J.Marrink@rug.nl

† These authors contributed equally to this paper.

The main characteristic of a toroidal pore is that it is hydrophilic and the peptides are believed to stabilize the pore by interacting strongly with the lipid headgroups that line the pore.[4–6] However, the exact structure of the pore, in particular the arrangement of the peptides and lipid molecules is still debated. The classical model of the toroidal pore postulates a regular structure lined with lipid head-groups and peptides.[5,6] All peptides associated with the pore are thought to remain α-helical and line the pore in a transmembrane orientation. This model of the toroidal pore assumes that peptides are orientated along the membrane surface before poration and perpendicular to the membrane in the porated state. The model does not include pore-formation by peptides with low helicity or β-strand peptides. Alternative models such as the micelle-like aggregate model,[3] disordered-toroidal pore model[14] and chaotic pore model[15,16] have been proposed. These models have helped to interpret recent NMR[17] and fluorescence data[18] and are compatible with kinetic studies.[15] These models all propose a higher degree of disorder in the pore state than had been previously assumed.

Atomistic simulations of magainin had provided the first direct evidence on the disordered nature of the toroidal pore model.[14] The simulation results, though compatible with previous experimental data, pointed to only a few peptides inserting into the toroidal-shaped pore and the other peptides lining the pore edge. Similar pores were also observed in extensive simulations of melittin interaction with DPPC bilayers.[19] The term disordered toroidal pore was coined to describe such pores. In both studies, pores were observed only above a critical peptide to lipid ratio and required local aggregation of peptides. In the two sets of simulations, the peptides showed significant loss of α-helicity and pore formation does not appear to require that the peptides remain helical. However, whether the peptides remain partially unfolded or refold in the pore state can only be addressed by longer simulations. The simulations also shed light on the possible mechanisms and driving forces of pore formation. Removing the positive charges on the AMPs blocked pore formation,[19] pointing to a mechanism similar to electroporation events. The role of electrostatic interactions in AMP action has also been studied in other simulation studies,[20] for a review see ref. 21. However, a framework allowing a comprehensive study of related peptides and lipids to analyze the driving forces of this process is difficult to achieve with atomistic simulations.

Coarse-grain (CG) force-fields allow sampling larger systems at longer time scales and thereby allow faster analyses of different systems (see ref. 22). CG simulations are still in their early stages but have already been successfully applied to analyze lipid–peptide interactions. A simple solvent-free CG simulation technique was used to study interactions between amphipathic peptides and bilayers and to explore the different conditions leading to desorbed, adsorbed and inserted configurations of the peptide.[23] The action of other AMPs have been studied using force-fields based on the MARTINI force-field.[24] A study on the interaction of maculatin on large vesicles concluded that the peptides disrupt the lamellar structures but do not form water-filled channels.[25] Alamethicin, implicated to form regular barrel-stave shaped pores, has been studied in combined atomistic/CG simulations.[26] The authors used CG simulations to equilibrate the distribution of alamethicin within the membrane and then converted the coarse-grained simulation to atomistic to investigate the details of water permeation. The simulations observed quite irregular structures, contrary to the current models in the literature that postulate a highly ordered protein channel. Simulations of a synthetic peptide, LS3 have shown that it assembles in a dehydrated barrel-stave pore.[27] Spontaneous poration and water permeation has been observed in coarse-grain studies for related pore-forming molecules such as dendrimers.[28] These studies show that CG models can be used to study the interplay between peptides and lipids over larger length and time scales; however, at the same time atomistic detail is lost and it remains questionable how realistic the configurations sampled at the CG level are. To fully understand the driving forces of this process, a multi-scale approach is required.

Here, we study the pore-forming propensity of AMPs at multiple scales combining coarse-grain and atomistic simulations. We focus on a member of the magainin family of antimicrobial peptides, magainin-H2 interacting with zwitterionic phosphatidylcholine membranes. The long-time relaxation properties of AMP-pores are studied by evolving pores formed in atomistic descriptions[14,19] with a CG representation using the MARTINI model.[24,29] To test the predictions, key states observed in the CG simulations are subsequently back transformed to an atomistic description using our recently developed resolution–transformation protocol.[30] Similar approaches have been used recently by a number of other groups to study membrane–protein interactions.[26,31–34] Overall, our simulations provide a multi-scale view of a fundamental biophysical membrane process involving a complex interplay between lipids and proteins.

3 Methods

3.1 Simulation protocol and force field

The molecular dynamics simulations were performed using the GROMACS program package[35] under periodic boundary conditions. The temperature was weakly coupled (coupling time 0.1 ps) to $T = 323$ K using a Berendsen thermostat.[36] The pressure was weakly coupled (coupling time 1.0 ps, compressibility 5×10^{-5} bar^{-1}) using a semi-isotropic coupling scheme in which the lateral ($P_{|}$) and perpendicular (P_z) pressures are coupled independently at 1 bar,[36] corresponding to a tension-free state of the membrane.

The atomistic system was described using the GROMOS 43a2 force field[37] for the peptides and the Berger parameters from a previous study[38] for the lipids, identical to our previous work.[14,19] A group-based twin range cut-off scheme (using cut-offs of 1.0/1.4 nm and a pair-list update frequency of once per 10 steps) including a reaction field (RF) correction[39] with a dielectric correction of 78 to account for the truncation of long-range electrostatic interactions was used. We also tested an alternative model where the electrostatic interactions were treated using particle mesh ewald (PME) summation and found that the pores formed using RF were also stable with PME. The water was described using the SPC model.[40] A time step of 2 fs was used. Bond lengths were constrained using the LINCS algorithm.[41]

The MARTINI force-field[24,29] was used to describe the coarse-grain system. The force-field is based on a four-to-one mapping, *i.e.* on average four non-hydrogen atoms are represented by a single interaction center. The force-field has been parametrized based on the reproduction of partitioning free energies between polar and apolar phases and allows an accurate representation of the chemical nature of the underlying atomistic structure. In this force-field, the backbone parameters (backbone bonded terms) are dependent on the secondary structure of the beads but independent of the amino acid. Four different systems with varying helicity—100%, 65%, 40% and 0% were modelled. Secondary structure was imposed by including a dihedral potential between backbone atoms. The force constant used in the system with 100% and 40% helicity was the standard MARTINI parameter of 400 kJ mol^{-1}. The force constant used in the system with 65% helicity was reduced to 70 kJ mol^{-1}. The value was chosen to allow greater conformational flexibility as dictated by our previous results from atomistic simulation.[14,19] In line with the decreased helicity, the polarity of the backbone bead was increased to a P5-particle, similar to the polarity of the fully-coiled system. The back-bone bead (in the helical stretch) of the fully-helical and 40%-helical peptides was of type N0. The LJ (Lennard-Jones) interactions were treated with a switch function from 0.9 to 1.2 nm (pair-list update frequency of once per 10 steps) and PME was used to treat long-range electrostatics. The use of PME is non-standard in the MARTINI force-field since it was parametrized using a shifted potential. However, in simulations of membrane poration events by dendrimers, it has been shown that the use of a PME scheme[28] is required.

Test simulations performed for our systems also indicated that membrane pores are more stable and similar to the pores observed with atomistic models when long range electrostatic interactions are taken into account. Importantly, we found that the use of PME does not significantly affect the equilibrium properties of pure lipid bilayers, including the area per lipid. In all CG simulations, a time step of 30 fs was used. The simulation times reported in the remainder of this manuscript are effective times obtained by the multiplication of the actual simulation time by a factor of four based on the speed-up achieved for diffusion of water and lipids.[24,29]

3.2 Resolution–transformation

Two types of transformation were used, either increasing or decreasing the resolution. The latter case is more trivial; here the CG lipids and peptides were mapped from the atomistic structure using the center of mass to define the CG bead positions. CG water beads were reintroduced to the system afterwards and the system was equilibrated. The case of introducing atomistic detail into a CG system is more demanding. Here we used a resolution–transformation protocol which was recently developed in our group.[30] In this three step method, the goal is to equilibrate an atomistic ensemble within the constraints imposed by the CG configuration, with the additional requirements that (i) the generated atomistic configurations represent an equilibrium ensemble, and (ii) the method is general *i.e.* can be applied to any (bio)molecular system. In short, the method proceeds along the following three steps. First, the initial positions of the atomistic particles were constructed by random insertion in a sphere of radius 0.3 nm centered at their corresponding CG beads. The mapping of the CG bead to the atomistic representation is in accordance with the MARTINI force-field. Four atomistic water molecules are mapped to a single CG bead with additional restraining potentials to keep the four water molecules grouped together. Second, a restrained simulated annealing MD procedure was employed and the temperature was gradually decreased in 40 000 steps from its starting value of 1300 K to the desired target temperature of 323 K. This allowed the system to rapidly cross energy barriers and find a low-energy minimum at the target temperature. To avoid numerical instability caused by the highly strained starting structure, force capping was applied during the simulated annealing, with forces exceeding a maximum value of 15 000 kJ mol^{-1} nm^{-1} reduced to the maximum value. Finally, the CG restraints were gradually removed in 1000 steps to ensure a smooth relaxation of the system in the full atomistic force-field. This method has been tested on various biomolecules such as lipids, cholesterol, amino-acids, short peptides and water. Details of the resolution–transformation protocol together with its verification will be published elsewhere.[30]

3.3 System set-up

Two different set-ups studying the interaction of magainin-H2 (IIKKFLHSIW KFGKAFVGEI MNI) peptides with a phosphatidylcholine lipid bilayer were simulated. The first set focused on the properties of the pore at a multi-scale level in which the resolution of the atomistic and CG structures was exchanged using the above described protocol. The starting atomistic simulations containing 4 magainin-H2 peptides and 128 DPPC lipid molecules were taken from ref. 14. The CG representation of the porated state was then mapped from the atomistic system. We mapped the 50-atom DPPC molecule to a 10-bead CG representation with 3 beads in each tail. Note that the standard MARTINI model maps DPPC to four bead tails. The precise mapping is somewhat arbitrary, however. We found that the thickness of the bilayer with three tail beads is more similar to the atomistic representation of the DPPC bilayer at the relatively high hydration level used in the atomistic studies. The second set of simulations studied poration propensity of the peptides in a CG representation. To study poration after association within the bilayer, 4 magainin-H2 peptides were placed in a transmembrane orientation inside a pre-equilibrated bilayer containing 122 lipids. To study spontaneous pore formation mimicking a biological system, 14 peptides were placed in the aqueous phase close to a bilayer consisting of 304 lipids. Two lipid types were tested—the three tail-bead lipid as described above and a two tail-bead lipid molecule, representing a short-tail lipid. Multiple simulations were performed for each system, starting from different initial random velocity distributions.

4 Results and discussion

4.1 Comparing the atomistic and coarse-grain pores

To explore the possibilities of the CG model to study pores formed by AMPs, CG representations of the starting atomistic pore configuration were obtained and evolved with the MARTINI force-field. We focus on the pores formed by four magainin-H2 peptides in a 128 lipid DPPC bilayer. Snapshots of the starting atomistic structure (taken from the simulations reported in ref. 14), the mapped CG system, and the system evolved for 24 μs are shown in Fig 1 A, B and C. During the CG simulations, a water channel was maintained through the membrane in contrast to previous CG studies on the action of AMPs.[25–27] The topology of the porated state, in particular its disordered nature was also preserved. In the CG representation, lipid molecules and a few peptides continued to line the water channel. The remaining peptides associated with the membrane at the mouth of the pore and did not insert into the pore. The size of the pore, however, was somewhat reduced at the CG level compared to the original atomistic simulation. Toward the end of the CG simulation, after 24 μs, we increased the resolution of the system back to the atomistic level. During the subsequent 50 ns simulation at the atomistic level, the pore remained similar to the transformed atomistic structure (snapshots shown in Fig 1D, E).

In general, comparing the atomistic to CG pore structure, a number of important points become apparent from our multi-scale approach. The first is that the MARTINI model predicts a similar type of pore as seen by the atomistic model, *i.e.* a fully hydrated, toroidally shaped, pore lined by lipids and peptides. Second, as expected, the two models are not fully compatible; quantitative differences exist, evidenced for instance by the size of the pore. The third point which becomes apparent is the limited sampling which can be obtained with the atomistic model. Based on our inverse transformation from the CG level back to the atomistic level we find that the configuration sampled in the CG model, being different from the starting atomistic structure, is also stable at the atomistic level. It suggests that these configurations are either meta-stable states requiring much longer relaxation times or are equilibrium states not sampled in previous atomistic simulations. On the accessible nanosecond time scale the peptide/lipid complex is almost frozen, pointing to the importance of a multi-scale approach.

4.2 Characteristics of the CG pore

At the longer time scales probed in this multi-scale study, the disordered toroidal pore maintained its structure and topology. The main feature of the disordered toroidal pore, a term coined from our atomistic simulations,[14,19] is the presence of some surface-aligned peptides as opposed to only transmembrane orientations. We observe that the peptides did not all insert into the pore and continued to line the mouth of the pore stabilizing the membrane curvature. This pore structure varies considerably from the classical model of a toroidal pore in which the peptides all align along the membrane normal inside the pore and maintain their helicity. Our multi-scale simulations reconfirm that the disordered nature of the pore may be a more realistic model than the classical toroidal pore.

The extended simulation time reached with the CG model allows for a more quantitative analysis of the structure of the pore. Fig. 2 shows the density profile of water

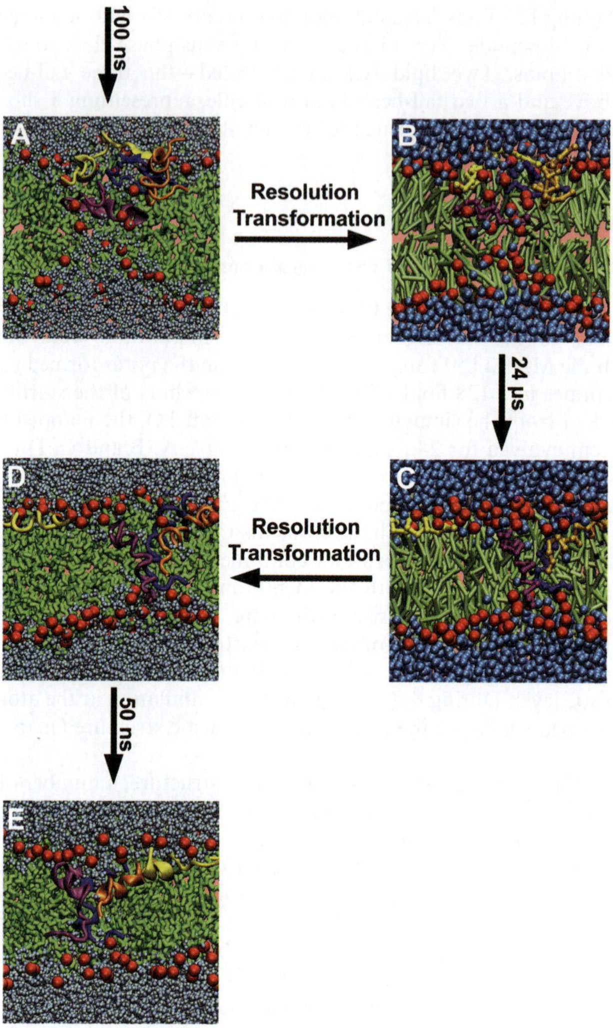

Fig. 1 The structure of the toroidal pore formed by magainin-H2 peptides in DPPC bilayers at multiple scales. A: The toroidal pore in an atomistic representation. B: A snapshot of the CG representation mapped from A. C: The CG pore after evolving the system from B for 24 µs. D: Snapshot of the atomistic representation mapped from C by using the resolution transformation protocol. E: The toroidal pore after evolving with atomistic resolution for 50 ns. All figures are prepared using VMD.[46]

and the phosphate bead of the lipid during the simulation in the porated state, compared to a pure bilayer. From the inset, it is clearly seen that water is present within the membrane in the porated state. The toroidal pores formed in the CG study are therefore hydrated, and a water flux through the pore was observed. Though a water channel is maintained through the bilayer, the flux through the pore is not constant. Large fluctuations were also seen in the position of the lipid head-groups. At a given time, only a few lipid molecules inserted into the pore. The lipid molecules flip flop through the pore from one leaflet to another as observed in other simulations.[42–44] Twelve flip-flop events were counted over the 30 µs trajectory. At the microsecond time scale, the peptides sampled different orientations and transitions between the transmembrane and surface aligned states are observed

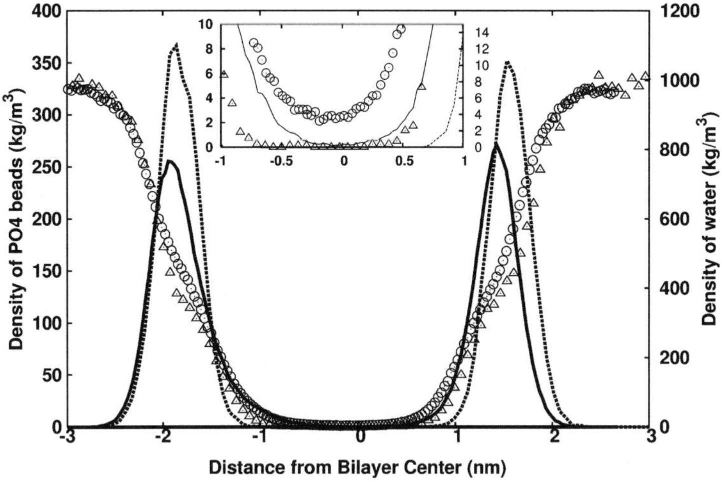

Fig. 2 The density of water and the phosphate bead of the lipids in CG simulations of a pure DPPC bilayer (△ and - - - respectively) and in a porated membrane with 4 magainin-H2 peptides (○ and —, respectively). The inset clearly shows the higher water density in the bilayer in the presence of the peptides.

(Fig. 3). It is seen that one out of the four peptide remains surface-bound on the same leaflet. Another peptide translocates through the pore from an inserted state to the other leaflet. The remaining two peptides insert into the pore from a surface-aligned orientation. Although significant peptide movement is observed, especially compared to motions probed in the atomistic simulations, it appears that a microsecond time scale is still not sufficient to sample equilibrium orientations of the peptides.

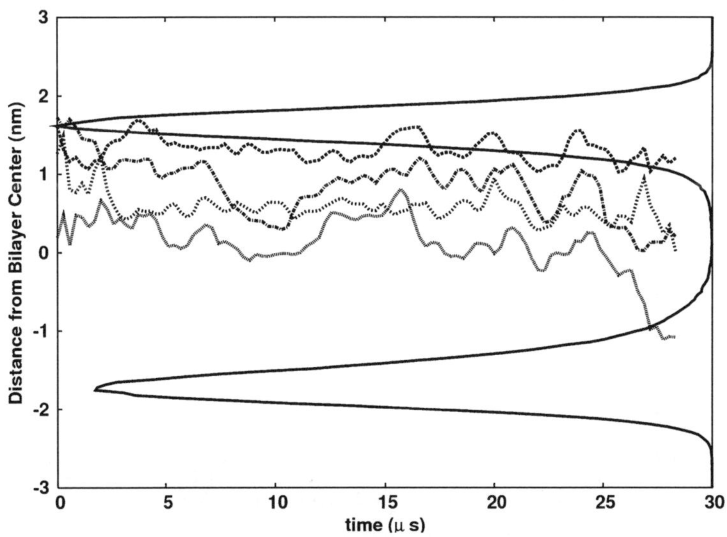

Fig. 3 The center of mass of the four peptides (dashed and dotted lines) in a toroidal pore evolved with a CG representation. The bold line shows the density profile of the phosphate beads of the lipid head-groups during the simulation.

4.3 Effect of helicity on pore stability

In the CG peptide model, secondary structure is imposed on the peptides and the choice of defining the helicity in the peptides was somewhat arbitrary. In the atomistic simulations, widely differing helicity was observed ranging between 30% and 70%. In the first scenario we imposed 65% helicity on all peptides which is slightly higher than the average helicity observed in the atomistic simulations. The pore was stable in this case and the pore state was maintained as described above. Increasing the helicity to full helicity, decreased the stability of the pore. Fewer lipid molecules inserted into the pore and the water flux also decreased substantially. In four independent simulations (starting from different random velocities), the pore closed on a microsecond time scale. Similarly, decreasing helicity completely (fully coil state) or to 40% decreased the stability of the pore, although the pore remained open in these cases. Fig. 4 shows the density of water inside the pores, for different peptide helicity. The water contained within the pore was the highest for the first peptide model (65% helicity) and decreased in the other peptide models. Thus, partial helicity is required for stabilizing pores formed by magainin-H2 peptides in membranes. This is in line with experimental data showing that increasing helicity in synthetic peptides decreased antimicrobial activity.[45] Note that even for the fully unstructured peptides, amphipathicity is maintained and they bind to the membrane. The release of secondary structure allows the peptides to sample a broader range of conformations. This is reflected by the radius of gyration, R_g shown in Fig. 5. The radius of gyration for the fully helical peptide is the largest, with lowest fluctuations, and that of the fully coiled and 40% helical peptides are the lowest with the largest fluctuations. The peptides with 65% helicity have a radius of gyration and fluctuations intermediate to the two. Note that the dihedral potentials applied in this scenario were lower than the standard MARTINI potential applied to the fully helical and 40% helical peptides. It appears that this level of flexibility of the peptide is ideal *i.e.* it can span the membrane in an extended configuration and at the same time favorably interact with the curved membrane, stabilizing a porated state. The current simulations provide evidence that an optimum helicity is required for membrane poration by AMPs.

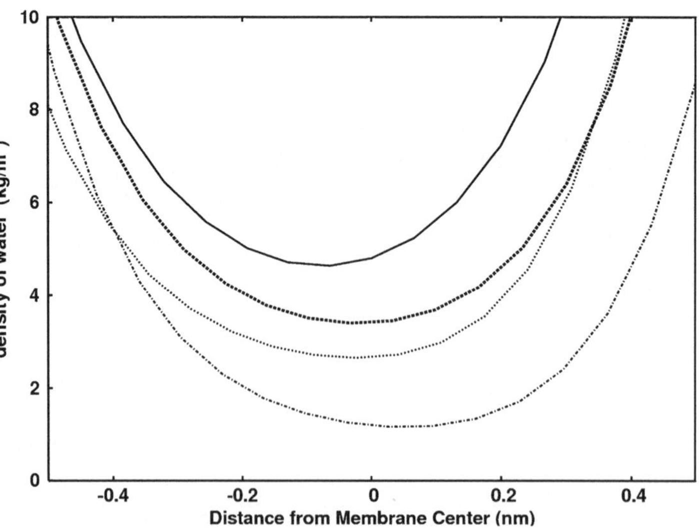

Fig. 4 The density profile of the water beads in the center of the bilayer in the CG simulations at varying peptide helicity: fully helical (dot-dashed), 65% helicity (solid line), 40% helicity (dotted) and fully flexible (dashed). The calculations for the fully helical peptide were performed on the trajectory before the pore closed. The data was fitted with a Bezier curve.

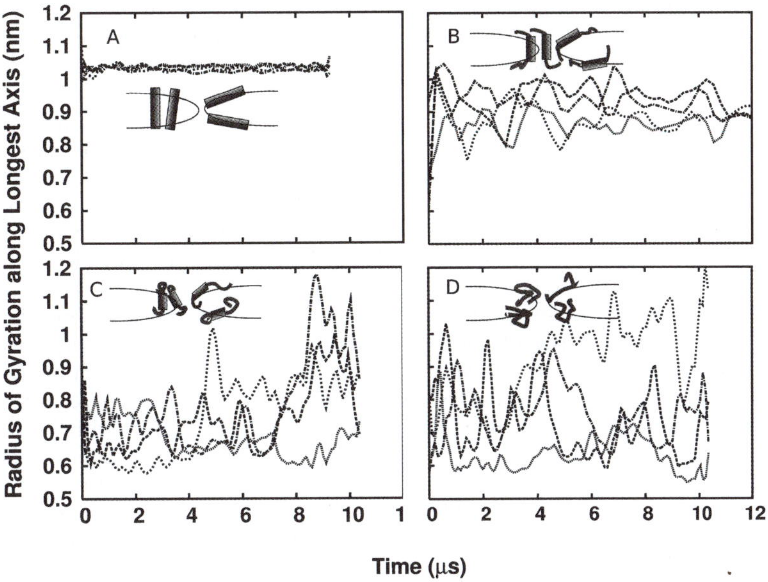

Fig. 5 Radius of gyration, R_g of the four peptides in CG simulations with peptide helicity A: 100% B: 65% C: 40% and D: 0%.

4.4 Spontaneous pore formation in CG models

The CG models in the previous sections were biased towards the starting atomistic structures and we carried out further tests to determine the equilibrium structures predicted by the CG model. Two sets of simulations were performed, either with the peptides pre-inserted into the membrane or with the peptides initially placed randomly in the aqueous phase. In the first set-up, four magainin-H2 peptides, constrained at 65% helicity, were inserted into the membrane in a transmembrane orientation as shown in Fig. 6A. The peptides were observed to diffuse laterally in the membrane to assemble into aggregates of three or more peptides, in four independent simulations. A snapshot of the simulation at the beginning of the aggregation is shown in Fig. 6B. The peptide aggregates perturbed the lamellar state considerably and large fluctuations in the lipid head-group positions were seen. The peptides

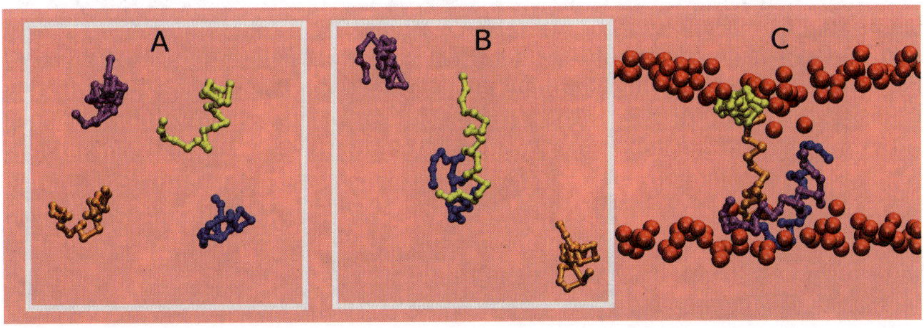

Fig. 6 The time course of events when four magainin-H2 peptides are inserted into a DPPC bilayer modeled at the CG level. A: Top view of the starting structure. Note that the peptides are inserted in a transmembrane orientation. B: Top view of the system after 20 ns. C: Snapshot of the toroidal pore after 5 μs. The phosphate beads of the lipids are shown in red and the four peptides in purple, yellow, orange and blue. The tails of the lipid molecules and the water is not shown for clarity.

along with the fluctuations in the bilayer eventually initiated a water channel and opened a pore (Fig. 6C). In the pore, a few peptides lined the pore and the remaining peptides were surface-aligned. Clear transitions were seen between the transmembrane and surface aligned orientations. The structure of the pore is essentially the same as obtained from the resolution transformation (section 4.1). We also tested cases with peptides constrained at 0%, 40% and 100% helicity. The fully helical structures assembled in the bilayer and opened a water channel with very low water density in the membrane core, reminiscent of the dehydrated barrel-stave pore observed in the simulations of synthetic peptides.[27] No lipid head-groups were inserted into the core of the membrane and lipid flip flop did not occur. The peptides with 40% helicity and fully coiled peptides induced a fluctuating toroidal-shaped water channel through the membrane. However, the flux through the membrane was lower than in the case of the 65% helical peptides. Partial helicity appears to be a criterion to initiate and stabilize pore formation in the coarse-grain model, in agreement with our results for the resolution transformation simulation (section 4.3).

In the second set-up, simulations were performed with the peptides initially placed in water close to one leaflet of the bilayer. This setup is similar to the atomistic simulations and mimics the actual biological process of spontaneous pore formation by the AMPs. The set-up comprised 304 lipids and 14 magainin-H2 peptides (modeled with 65% helicity). The peptides remained membrane bound at the 4 μs simulation period and did not insert into the membrane. Large fluctuations were seen in the head-groups of the lipids but no pore formation occurred. It appears that the barrier associated with opening of a water channel is relatively high in the CG model preventing pore formation even at the high *P*/*L* ratio. Therefore, a two tail-bead lipid model was also tested representative of a short tail lipid. The time course of the

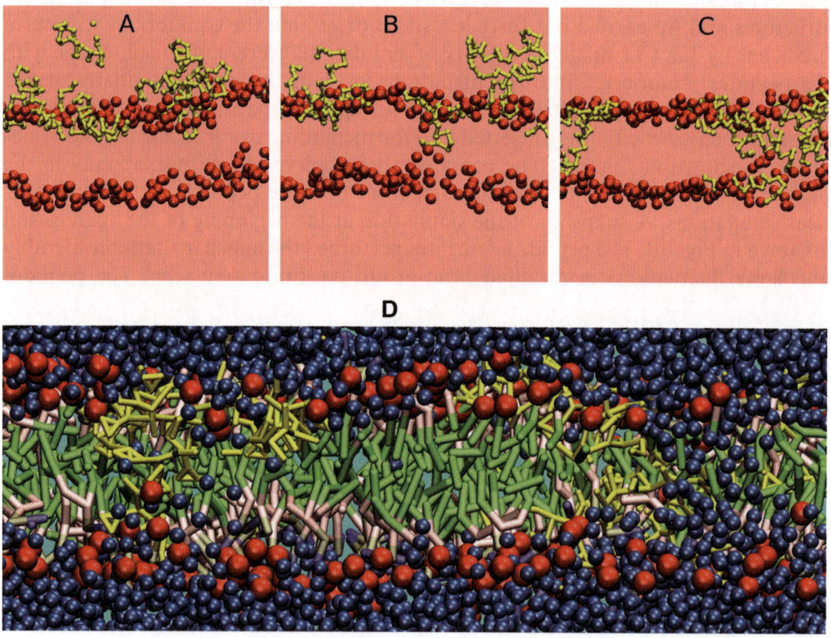

Fig. 7 The time course of events when fourteen magainin-H2 peptides are placed initially in water close to the surface of a short tail PC bilayer. A: Side view of the starting structure, B: The system after 200 ns, C: The toroidal pore after 3 μs. The phosphate beads of the lipids are shown in red and the peptides in yellow. The tails of the lipid molecules and the water are not shown for clarity. D: The toroidal pore in C shown in more detail. The lipid chains are shown in green and the water in blue.

simulation is depicted in Fig 7. The peptides adsorbed at the surface of the membrane within 10 ns. The peptides associated only after adsorption to the membrane surface. After association, they once again induced fluctuations in the head-groups of the lipid molecules and in contrast to the thicker three tail-bead membrane, spontaneous opening of a water pore was observed. In fact two pores were formed, a larger one containing nine peptides and a smaller one with five peptides (see Fig 7D). The structure of the pores was similar to the pores described above. The pores remained stable over the 3 μs simulation. Note that a pure two tail-bead CG bilayer is stable, so the poration is caused by the peptides. The results in this section point out that, even with a CG description, one may easily end up in meta-stable states and sampling remains incomplete on the microsecond time scale.

5 Conclusions

The work presented here is one of the first examples of using a multi-scale simulation framework to study membrane poration by antimicrobial peptides. Using a CG model we are able to approach time-scales compatible with experimental studies of pore formation, whereas a model with atomistic resolution provides a useful check to the accuracy of the CG model. We find that the pore structure obtained with the CG model is similar to the disordered pore seen in atomistic simulations. The pore retains its disordered character and lipid flip flop as well as water permeation through the pore is observed. At a microsecond time scale, peptides translocate *via* the pore to the other bilayer leaflet and both surface aligned and transmembrane orientations are seen in the porated state. Subsequent back-transformation of the newly generated CG structures reveal their stability also at the atomistic level of resolution. From our multi-scale simulations it appears that the disordered nature of the pore is a more realistic model than the classical toroidal pore. However, there remain some problematic sampling issues. Even the microsecond range probed by the CG model is not sufficient to sample the possible peptide orientations inside the pore. Despite these limitations, the CG model opens the way to explore the structure/activity relationship in a systematic way. Here we tested the effect of different peptide helicities on pore stability. We conclude that partial helicity is required to form and stabilize pores. Spontaneous poration events can also be studied at a CG level. Starting from a well separated, transmembrane orientation the peptides associate and induce disordered toroidal pores. A few peptides translocate to either leaflet pointing towards the importance of surface aligned peptides in stabilizing membrane curvature. Mimicking the biological situation, we simulated the attack of magainin-H2 peptides from water. Poration in this case is observed only when the energy barrier is lowered by decreasing the thickness of the bilayer. The high energy barriers to spontaneous pore formation in the CG model is presumably due to the lack of explicit electrostatic screening and polarization effects in the current MARTINI water model. Nevertheless, the sequence of events leading to pore formation, as well as the structure and size of the pore are similar to those observed in atomistic studies. The current work on the action of magainin-H2 peptides on membranes provides a multi-scale view of a fundamental biophysical membrane process involving a complex interplay between peptides and lipids.

6 Acknowledgment

This work was supported by the Netherlands Organization for Scientific Research (NWO) through their TOP and ALW Open programs. Computational access to the supercomputers of the Netherlands National Computing Facilities (NCF) is also acknowledged. The help of Martti Louhivuori in providing access to this facility is greatly appreciated.

References

1 K. L. Brown and R. E. Hancock, *Curr. Opin. Immunol.*, 2006, **18**, 24–30.
2 M. Zasloff, *Nature*, 2002, **415**, 389–395.
3 R. E. W. Hancock and D. S. Chapple, *Antimicrob. Agents Chemother.*, 1999, **43**, 1317–1323.
4 B. Bechinger and K. Lohner, *Biochim. Biophys. Acta*, 2006, **1758**, 1529–1539.
5 Y. Shai, *Biopolymers*, 2002, **66**, 236–248.
6 K. A. Brogden, *Nat. Rev. Microbiol.*, 2005, **3**, 238–250.
7 K. Matsuzaki, S. Yoneyama and K. Miyajima, *Biophys. J.*, 1997, **73**, 831–838.
8 H. Huang, *Biochim. Biophys. Acta*, 2006.
9 A. S. Ladokhin, M. E. Selsted and S. H. White, *Biophys. J.*, 1997, **72**, 1762–1766.
10 C. E. Dempsey, *Biochim. Biophys. Acta*, 1990, **1031**, 143–161.
11 K. Matsuzaki, K. Sugishita, N. Ishibe, M. Ueha, S. Nakata, K. Miyajima and R. M. Epand, *Biochemistry*, 1998, **37**, 11856–11863.
12 B. Bechinger, *J. Membr. Biol.*, 1997, **156**, 197–211.
13 K. Matsuzaki, O. Murase, N. Fujii and K. Miyajima, *Biochemistry*, 1996, **35**, 11361–11368.
14 H. Leontiadou, A. E. Mark and S. J. Marrink, *J. Am. Chem. Soc.*, 2006, **128**, 12156–12161.
15 S. M. Gregory, A. Cavenaugh, V. Journigan, A. Pokorny and P. F. F. Almeida, *Biophys. J.*, 2008, **94**, 1667–1680.
16 P. H. Axelsen, *Biophys. J.*, 2008, **94**, 1549–1550.
17 A. J. Mason, A. Marquette and B. Bechinger, *Biophys. J.*, 2007, **93**, 4289–4299.
18 L. E. Yandek, A. Pokorny, A. Floren, K. Knoelke, U. Langel and P. F. F. Almeida, *Biophys. J.*, 2007, **92**, 2434–2444.
19 D. Sengupta, H. Leontiadou, A. E. Mark and S. J. Marrink, *Biochim. Biophys. Acta, Biomembr.*, 2008, **1778**, 2308–2317.
20 F. Jean-Francois, J. Elezgaray, P. Berson, P. Vacher and E. J. Dufourc, *Biophys. J.*, 2008, **95**(12), 5748–5756.
21 E. Matyus, C. Kandt and D. P. Tieleman, *Curr. Med. Chem.*, 2008, **14**(12), 2789–2798.
22 *Coarse-Graining of Condensed Phase and Biomolecular Systems*, ed. G. A. Voth, CRC-Press, Boca Raton, 2008.
23 G. Illya and M. Deserno, *Biophys. J.*, 2008, **95**, 4163–4173.
24 L. Monticelli, S. K. Kandasamy, X. Periole, R. G. Larson, D. P. Tieleman and S. J. Marrink, *J. Chem. Theory Comput.*, 2008, **4**, 819–834.
25 P. J. Bond, D. L. Parton, J. F. Clark and M. S. P. Sansom, *Biophys. J.*, 2008, **95**(8), 3802–3815.
26 L. Thogersen, B. Schiott, T. Vosegaard, N. C. Nielsen and E. Tajkhorshid, *Biophys. J.*, 2008, **95**, 4337–4347.
27 P. Gkeka and L. Sarkisov, *J. Phys. Chem. B*, 2009, **113**, 6–8.
28 H. Lee and R. G. Larson, *J. Phys. Chem. B*, 2008, **112**, 7778–7784.
29 S. J. Marrink, H. J. Risselada, S. Yefimov, D. P. Tieleman and A. H. de Vries, *J. Phys. Chem. B*, 2007, **111**, 7812–7824.
30 A. J. Rzepiela, L. V. Schäfer, N. Goga, H. J. Risselada, A. H. de Vries and S. J. Marrink, *J. Comp. Chem.*, 2009, in press.
31 T. Carpenter, P. J. Bond, S. Khalid and M. S. P. Sansom, *Biophys. J.*, 2008, **95**, 3790–3801.
32 A. Y. Shih, P. L. Freddolino, S. G. Sligar and K. Schulten, *Nano Lett.*, 2007, **7**, 1692–1696.
33 P. Liu, Q. Shi, E. Lyman and G. A. Voth, *J. Chem. Phys.*, 2008, **129**, 114103–1141038.
34 A. Villa, C. Peter and N. F. A. van der Vegt, *Phys. Chem. Chem. Phys.*, 2009, **11**, 2077–2086.
35 D. van der Spoel, E. Lindahl, B. Hess, G. Groenhof, A. E. Mark and H. J. C. Berendsen, *J. Comput. Chem.*, 2005, **26**, 1701–1718.
36 H. J. C. Berendsen, J. P. M. Postma, W. F. van Gunsteren, A. D. Nola and J. R. Haak, *J. Chem. Phys.*, 1984, **81**, 3684–3690.
37 W. F. van Gunsteren, X. Daura, and A. E. Mark, *Encyclopedia of Computational Chemistry*, vol. 2, John Wiley and Sons, New York, 1998.
38 C. Anézo, A. H. de Vries, H. D. Höltje, D. P. Tieleman and S. J. Marrink, *J. Phys. Chem. B*, 2003, **107**, 9424–9433.
39 G. Tironi, R. Sperb, P. E. Smith and W. F. van Gunsteren, *J. Chem. Phys.*, 1995, **102**, 5451–5459.
40 H. J. C. Berendsen, J. P. M. Postma, W. F. van Gunsteren, and J. Hermans, *Interaction models for Water in Relation to Protein Hydration*, Reidel, Dordrecht, 1981.
41 B. Hess, H. Bekker, H. J. C. Berendsen and J. G. M. Fraaije, *J. Comput. Chem.*, 1997, **18**, 1463–1472.
42 D. P. Tieleman and S. J. Marrink, *J. Am. Chem. Soc.*, 2006, **128**, 12462–12467.
43 A. A. Gurtovenko and I. Vattulainen, *J. Phys. Chem. B*, 2007, **111**, 13554–13559.
44 S. J. Marrink, A. H. De Vries and D. P. Tieleman, *Biochim. Biophys. Acta, Biomembr.*, 2009, **1788**, 149–168.

45 Y. Chen, C. T. Mant, S. W. Farmer, R. E. W. Hancock, M. L. Vasil and R. S. Hodges, *J. Biol. Chem.*, 2005, **280**, 12316–12329.
46 W. Humphrey, A. Dalke and K. Schulten, *J. Mol. Graphics*, 1996, **14**, 33–38.

General discussion

Dr Monticelli opened the discussion of the paper by Professor Smit: (1) how are the 2D and 3D potentials of mean force (PMFs) calculated in your simulations? (unbiased simulations, umbrella sampling or other techniques?)

(2) You wrote that, in your simulations, a tensionless membrane is simulated in the NVT ensemble. How is that achieved?

(3) Zero tension can be the result of pressures that are far from 1 bar in the lateral and normal dimensions. This, in turn, can have an influence on the free energy profiles. How much was the average pressure in your NVT simulations?

Professor Smit replied: Question 1: Depending on the mismatch the proteins can have a very strong attraction, which makes it essential to use biased simulation techniques such as umbrella sampling.[1]

Question 2: We used a combined DPD and Monte Carlo scheme, where the Monte Carlo rules involve a change in the area of the lipid, which is accepted and rejected with an appropriate Monte Carlo rule.[2]

Question 3: In our simulations we use a constant volume so the normal pressure, which corresponds to the bulk pressure of the water, does change as a function of, say, the temperature. Recently, we have carried out some simulations in which we fix the normal pressure with very similar results.

1 F. J. M. de Meyer and B. Smit. *Phys. Rev. Lett.*, 2009, **102**, 219801.
2 M. Venturoli, M. M. Sperotto, M. Kranenburg and B. Smit, *Phys. Rep.*, 2006, **437**, 1–54.

Professor Tieleman asked: I find it quite interesting to see the effects of the helices for different degrees of mismatch on the local thickness of the membrane. We have done similar calculations using the MARTINI model to calculate a potential of mean force between two model helices. In our case, we see clear rings of lipid density changes around the helices. Your paper does not analyze this, but I was wondering if you see something similar? Our PMFs have energies that can be integrated to get binding affinities for dimerization that are in the same ballpark as your experiment, but this is not the case for the energies in your PMFs. Could you comment on your energy scale and its interpretation?

Professor Smit responded: Question 1: We have looked at the density profiles in our model as well. However, we did not observe significant density changes. We found that the hydrophobic shielding gave a more insightful picture. The hydrophobic shielding describes how the presence of the proteins changes the balance between the hydrophobic and hydrophilic segments at a particular position in the membrane.[1]

Question 2: Our calculations have been done at relatively high temperatures (*ca.* 50–70 C). Before we can give a more detailed comment we have to redo our calculations at the conditions of the experiment.

1 F. J. M. de Meyer, M. Venturoli and B. Smit, *Biophys. J.*, 2008, **95**, 1851–1865.

Professor Berendsen said: You use a hydrophobic/hydrophilic balance for the protein–lipid interaction. Real proteins often have more specific interactions at the level of the head groups, *e.g.* tryptophan rings. Is it possible to include such interactions in your model?

Professor Smit answered: In our model we can introduce more specific interactions. At present we focus on a very simple model to see where this simple approach fails and where these specific interactions are needed.

Dr Milano opened the discussion of the paper by Professor M. Müller: Can you comment about the effects of the grid resolution of the applied ordering field on the accuracy of the free energy calculations?

Professor M. Müller replied: The detailed form of the external, ordering field does not affect the thermodynamic integration method provided that the path of integration is reversible, *i.e.*, it does not cross a first-order transition line in the extended space of intermolecular interactions and strength of the external field. Therefore the grid-resolution of the ordering field has a negligible effect on the value of the free energy of the model system. It has been shown that the optimal, external field is the one that results in a minimal change of the structure along the branch where the external, ordering field is replaced by the intermolecular interactions.

A different question is how the grid resolution affects the properties of the computational model. We have discussed that the non-bonded interactions of the grid-based model can be casted in the form of pair-wise interactions.[1] The grid spacing dictates the range of the interactions. Like in any other model with finite-ranged, pair-wise interactions, we expect that the results will be insensitive to the range of interactions if the interaction range is mesoscopic (*e.g.*, one segment interacts with many neighbors and there are no pronounced, fluid-like packing effects) while simultaneously remaining smaller than the smallest physically important length scale of the bilayer (*e.g.*, the interfacial width between the amphiphilic components).

1 K. Ch. Daoulas and M. Müller, *J. Chem. Phys.*, 2006, **125**, 184904.

Dr Sengupta continued the discussion on the paper by Professor Smit: Could you comment further on the 2-D simulation results for the different scenarios: positive, negative and no mismatch.

Professor Smit replied: We have only carried out simulations for a negative mismatch. For these conditions our 2-D model shows a vapor–liquid-like phase separation. For no mismatch the potential of mean force only shows a hard-sphere like potential, so we expect no vapor–liquid-like phase separation. For positive mismatch the potential of mean force shows, depending on the diameter of the protein, a repulsive part. It would be interesting to see whether this would lead to cluster formation instead of phase separation.

Professor Smit opened the discussion of the paper by Professor Voth: In a recent calculation[1] we have used a coarse-grained model to study the phase behavior of lipid cholesterol mixtures. The phase behavior is very complex, which is surprising given that the underlying model is based on purely repulsive interactions. Suppose that we would like to use force matching to use a more realistic description of the coarse-grained interaction. If I understood the force matching procedure correctly, one has to carry out atomic simulations at each state point to obtain the tabulated coarse-grained forces. One could argue that after these atomic simulations a coarse-grained model has very little to add unless there is some form of transferability of these tabulated forces. Could you comment on this? In particular if you have many different coarse-grained groups.

1 F. de Meyer and B. Smit, *Proc. Natl. Acad. Sci. U. S. A.*, 2009, **106**, 3654.

Professor Voth replied: The multi-scale coarse-graining method (the one that you call "force-matching") can in principle define the renormalized, many-dimensional potential of mean force for a given coarse-grained mapping and for a given set of thermodynamic conditions. As far as I know, such a renormalized potential of mean force (and the effective Boltzmann distribution given by it) is the only definition of coarse-graining that is based on rigorous statistical mechanics (there are of course different approximations in the final algorithms to arrive at this renormalization). By definition, this CG many-dimensional PMF function cannot be completely transferable to other thermodynamic conditions, such as to other regions of phase diagram you mention in your question. However, aspects of the PMF will surely be transferable and this is important. Turning your argument around though, I would argue that if the effective potential in a simple CG model such as yours does not substantially agree with the result from a systematic renormalization method like multi-scale coarse-graining, then it has no real fundamental meaning. It is largely a computer model that gives certain desired results based on certain inputs to the model.

Professor Müller-Plathe remarked: Transferability of coarse-grained force fields over a concentration range: in principle, the model (generated by force matching, iterative Boltzmann inversion or other methods) has to be re-coarse-grained at every composition, temperature, *etc.* However, for a polymer in solution, we have found that geometric mixing rules describe the mixture over the whole concentration range.[1]

1 H.-J. Qian, P. Carbone, X. Chen, H. A. Karimi Varzaneh and F. Müller-Plathe, *Macromolecules*, 2008, **41**, 9919–9929.

Professor Voth addressed Professor Müller-Plathe: You mentioned that you have implemented mixing rules for the inverse Boltzmann coarse-graining approach to increase its transferability between systems. Can you explain a little how this is done?

Professor Müller-Plathe answered: Two tabulated potentials for like interactions are combined into a tabulated potential for an unlike interaction as a weighted geometric mean. This is explained in detail elsewhere.[1]

1 H.-J. Qian, P. Carbone, X. Chen, H. A. Karimi-Varzaneh, C. C. Liew, and F. Müller-Plathe, *Macromolecules*, 2008, **41**, 9919.

Dr Vila Verde continued the discussion on the paper by Professor M. Müller: (1) Please make it clear to the reader that nothing in the methods used requires that each lipid contains 32 beads.

(2) Briefly, discuss the implications of using a 1-tail lipid *versus* the 2-tail lipids that normally compose membranes.

(3) Describe more clearly how the initial stalk is constructed and why you expect the particular structure you obtain to be representative.

Professor M. Müller answered: (1) In our coarse-grained model, the discretization of the contour of the amphiphiles, $N = 32$, does not constitute a physical parameter determining the membrane behavior. The specific value, $N = 32$, was chosen according to our previous studies of polymers; a smaller discretization is expected to yield similar results. Essentially, the number of interaction centers is chosen to provide a reasonable characterization of the shape of the amphiphilic conformations. In our model, we use the Gaussian chain model, which is characterized only by a single length scale—the squared end-to-end distance, R_e. All quantities that define the model do not depend on the contour discretization, *i.e.*, are formally invariant under a re-parameterization, $N \rightarrow N'$. For example, the density of hydrophobic or hydrophilic entities, defined by eqn (3) or eqn (4) of the paper, which enter the virial expansion, eqn (2), correspond to the density of A or B blocks per unit volume, R_e^3, and not to the density of A, B interaction centers. A more extensive discussion of this issue can be found elsewhere.[1]

(2) The incorporation of a more complex amphiphilic architecture (*e.g.*, 2-tail lipids or semi-flexible tails), is straightforward within the considered approach. It might require an additional adjustment of the interaction coefficients to ensure

that the asymmetry of the lipids leads to the formation of a planar bilayer membrane. Considering 2-tail lipids is an interesting extension of this work. We expect the 2-tail architecture to facilitate the kinetics of stalk formation *via* conformations, in which the tails are inserted into the different, apposing bilayers. Thus, while a 2-tail lipid might have a lower free energy barrier towards forming a stalk, it is less obvious that the excess free energy of a (metastable) stalk formed by 2-tail lipids, differs significantly from that of a stalk of 1-tail lipids provided that the elastic material constants of the bilayer remain unaltered.

(3) To prepare the stalk morphology, we initially create a pair of parallel, apposed bilayers spanning the simulation box. A typical distance between them is $\sim 1.5\ R_e$. In addition to the segmental interaction potential, we apply an external field, which acts only on the hydrophobic tail beads in a cylindrical region with a diameter $1.4\ R_e$ bridging the two bilayers. After a short Monte-Carlo simulation ($\sim 10^3$ MC steps) the tail segments, driven by the external field, assemble in this cylindrical region. At the same time the hydrophilic segments remain outside the cylinder such that the whole cylindrical structure resembles the stalk geometry. After the formation of this pre-assembled stalk, the external field is switched off and an equilibration run ($\sim 10^4$ MC steps) is performed during which the non-bonded interaction energy adopts its stationary value. This equilibration relaxes the local geometry and curvature of the stalk. At this stage, the structure is considered to be representative of the metastable stalk structure and serves to estimate the external, ordering field of the stalk-morphology in the thermodynamic integration scheme. The fact that this morphology is indeed a metastable morphology is corroborated by its long lifetime during the subsequent TDI simulation runs ($\sim 10^6$ MC steps). Note the time scale separation between the pre-assembling and relaxation of the stalk structure and its metastable lifetime.

1 K. Ch. Daoulas and M. Müller, *Adv. Polym. Sci.*, 2009, http://springerlink.com/content/u8p51lr403224226/fulltext.html.

Professor Holm addressed Professor Voth, Professor Smit and Professor M. Müller: We have now heard talks about 3 different membrane models, and I know there are many more around. Since I am not a specialist on membranes, could the three speakers maybe agree or specify why they think their model is superior to others for the specific problem they are using? Or is there a more generic point of view of when should I use a specific model, and lastly, do they expect that in the future we will even see more membrane models, maybe another 3 per specific problem, or can one expect a convergence of the numbers of models to a finite number?

Professor Smit answered: Coarse-graining implies that one removes degrees of freedom that are unimportant for the properties that one is interested in. This implies that depending on the context on which one would like to use a mesoscopic model, different coarse-graining strategies would imply. For example, our mesoscopic model[1] was developed with the aim to reproduce the phase diagram of phospholipids. It can therefore be expected that our model will do well, if one is interested in those properties that are related to the phase behavior. However, one can argue that if one is interested in the dynamics of a bilayer one would develop a very different strategy. Initially, I expect an increase in the number of models. These different models and approaches will give us useful insights in the merits and limitations of the various strategies to coarse-grain and with these insights it will be possible to develop more systematic approaches and eventually reduce the number of models.

1 M. Kranenburg and B. Smit, *J. Phys. Chem. B*, 2005, **109**, 6553; M. Venturoli, M. M. Sperotto, M. Kranenburg and B. Smit, *Phys. Rep.*, 2006, **437**, 1.

Professor M. Müller responded: The diversity of coarse-grained models available for bilayer membranes underlines the need to describe these systems at different levels of detail, depending on the properties or phenomena that one wants to investigate. Coarse-grained models are well-suited to study the universal behavior of collective membrane phenomena (*e.g.*, pore formation, membrane fusion, phase transformation) involving a large number of amphiphilic molecules. For those phenomena the detailed implementation of the interactions is not expected to be important, but the relevant interactions, which are necessary to bring about the phenomena, have to be captured by the model. The strategy of minimal coarse-grained models consists of implementing only these relevant interactions in the computationally most efficient way. Among the relevant interactions are the repulsion between the hydrophilic and hydrophobic entities and their connectivity along the amphiphilic molecules. However, there may be additional, relevant properties that a model has to incorporate in order to describe specific phenomena. For instance, solvent-free models are extremely efficient from a computational point of view and offer the possibility to investigate large membrane systems, however, they might be inadequate in cases where the dynamics of the solvent becomes important (*e.g.*, vesicle filtration through porous membranes). For some phenomena, the choice of relevant interactions may not be obvious and the comparison of different coarse-grained models, which include different interactions, contributes to identify, what interactions are relevant for a specific phenomenon.

In our work, we investigate the excess free energy of a stalk between two apposed membranes. These hour-glass shaped connections have been observed in systems with very different microscopic interactions (*e.g.*, lipids in aqueous solution or block copolymer melts). Thus, coarse-grained models are well-suited to investigate their universal properties. Since we study the system in (meta-stable) equilibrium, we do not expect the solvent dynamics to be important. Moreover, the free-energy calculation is computationally demanding and, therefore, we choose a solvent-free model. In order to determine the excess free energy as a function of the membrane tension, we accurately calculate the chemical potential of the lipids in the bilayer. Using a model with soft potentials, this calculation can be accomplished by Widom's insertion method. Therefore, we devised a soft, solvent-free model of membranes, which is based on a simple, weighted density-functional for the non-bonded interactions. The parameters of the simple density-functional are directly related to the density and compressibility of the hydrophobic and hydrophilic units. Generalizations of the model can benefit from the substantial knowledge in liquid state theory.

Professor Voth replied: I think the field of coarse-graining is not yet mature enough to have a limited set of CG models for anything. Many people don't even agree on what coarse-graining is. In the end, the CG methods that stand the test of time will be the ones that can predict new phenomena and have those predictions independently confirmed by new experiments.

Professor Smit said: The principles of coarse-graining can be summarized by assuming that the microscopic system has degrees of freedom that are important and degrees of freedom that are unimportant. A coarse-grained description is obtained by integrating out the unimportant degrees of freedom. Important or unimportant depends on the context in which a coarse-grained model is used. For example, if we are interested in the structure, the relevant degrees of freedom are those that describe the free energy minima, while for the diffusion, the free energy barriers may be the dominant degrees of freedom. Hence, the context in which the model is used may result in very different sets of parameters. The bottom-up approaches such as force matching, radial distribution-function matching, or entropy matching each make an implicit choice about the different degrees of freedom that are integrated out. This choice may, or may not, result in a coarse-grained model that can correctly addresses a particular experimental question.

Professor Voth responded: This seems more like a comment than a question. However, we want to pursue coarse-graining as an exercise in rigorous statistical mechanics (to the extent possible). Therefore, we think that various choices for the coarse-grained model development may or may not be based on the same level of rigor. However, I think when one primarily wants to model something on the computer in a phenomenological sense, there may be more flexibility to use whatever method suits your goals and allows you to hit the target that you are after.

Professor Español remarked: I would like to introduce here a concept that may be useful for the discussion of coarse-graining. It is the concept of transitive coarse-graining.

Zwanzig theory leads in a natural way to the concept of level of description, characterized by the coarse-grained variables selected. One single system may be looked at in different levels of detail and each level has its information content and associated time-scales. Typically, the coarser the level the smaller the information and the larger the time scales of the coarse-variables. For example in a colloidal suspension, we can identify the following levels;

(1) Atomic description of the solvent and the colloidal particles.

(2) Hydrodynamics coupled to the colloidal particles. The solvent degrees of freedom are coarse-grained in favor of hydrodynamic fields while the colloids are coarse-grained with a few degrees of freedom (if they move like rigid objects).

(3) Hydrodynamics plus concentration field. The colloidal positional degrees of freedom are coarse-grained in favor of the concentration field (not where the colloidal particles are but how many colloidal particles are in a given region of space)

(4) Thermodynamics.

Another example discussed at length in this discussion is membranes where we can identify at least the following levels;

(1) Full atomistic.

(2) Complex molecules described with blobs, solvent with hydrodynamics.

(3) Elasticity of the membrane coupled with hydrodynamics of the solvent.

(4) Thermodynamics.

Now, we know from Zwanzig theory how to go in principle from level 1 to any other level, while from Gibbs we know how to go from level 1 to level 4 (thermodynamics). Gibbs ensemble theory is a particular case of Zwanzig theory when the selected coarse-grained variables are the dynamic invariants of the system, and the corresponding time scale is infinite. A reasonable requirement is that the result of going from level 1 to level 2, and from level 2 to level 3 should give the same dynamic equations as the ones obtained from a direct coarse-graining from level 1 to level 3. This *transitivity* property should be regarded as an *condition* for having well-defined levels of description. In particular, to get thermodynamic consistency. For the particular case of coarsening complex molecules in terms of centers of mass of portions of them, in order to achieve this transitivity property, I believe that we should introduce *internal energy* variables. Along these lines, I believe we should start thinking in terms of entropies instead of free energies, as already mentioned by Professor Voth, even in those cases that we are only interested in isothermal situations. Only then we may expect transitivity and, therefore thermodynamic consistency.[1]

1 H. C. Ottinger, *Beyond Equilibrium Thermodynamics*, J. Wiley and Sons, 2005, section 6.4.

Professor Faller continued the discussion on the paper by Professor Smit: What would you expect as the effects of mixing model peptides with different hydrophobic core lengths, especially with positive and negative mismatch?

Professor Smit responded: We have done simulations of bilayers containing a mixture of model peptides with both negative and positive mismatch and we observed that those peptides with equal negative or equal positive mismatch selectively cluster. The calculation of the potential of mean force (PMF) between a peptide with negative mismatch and a peptide with positive mismatch shows that the lipid-mediated interaction between both peptides is purely repulsive, while the PMFs between two proteins of equal mismatch is attractive. Thus, the PMFs confirmed the observed selective lipid-mediated interactions between peptides with different mismatch.

Professor Deserno continued the discussion on the paper by Professor Voth: Including more atomic level information into a coarse-grained model makes it more systematic, but not automatically more reliable. It seems desirable to have experimental validations independent of the specific system one is currently interested in, particularly because there are still a variety of free parameters in all these models. For instance, you showed nice results for tubulation, but you also mentioned that your BAR domains porate membranes. I am not familiar with experimental results that show BAR-driven poration, so wouldn't this constitute a problem for your model?

Professor Voth replied: I do not understand how the inclusion of more atomic level information into a CG model would not make it more reliable given the inherent uncertainties in coarse-graining. However, this seems to not be the point of your question. The scientific method obviously requires experimental validation. In fact, N-BAR domain proteins such as endophilin do porate membranes. These results are commonly found by our experimental collaborators in their EM imaging. In the past they have only published images of perfectly remodeled membrane tubules. However, there is very great polymorphism in the N-BAR-driven membrane remodeling outcomes. Such experimental results will be shown, in comparison to our mesoscopic simulations, in our forthcoming paper.[1]

1 G. S. Ayton, R. D. Swenson, C. Mim, V. Unger, and G. A. Voth, *Biophys. J.*, 2009, **97**, 1616.

Professor Deserno commented: During an earlier discussion Dr Baaden lamented the confusing usage of terminology in our field. For instance, what does "multi-scaling" mean? I think most of us would converge on the notion of "studying the same physics on different scales", maybe separately or simultaneously. But you at some point stated that multi-scale coarse-graining is also know as "force-matching". Now for me, "force-matching" is only one example of many for how one could systematically rescue small-scale information on the coarse level. I was wondering whether you want to make a specific point here, for instance, if one is not doing "force-matching", one is not doing systematic multi-scaling?

Professor Voth replied: This is a rather difficult question to address. The phrase "multi-scale" generally means that an overall phenomenon is defined by behavior on multiple, coupled scales. I do not think it means the "same physics" studied on different scales. If it were the "same physics", it would not be manifest on different coupled scales. This would be "different physics". Our term "multi-scale coarse-graining" means that we take forces at the molecular scale, develop CG models at another scale (the reduced resolution scale), and predict phenomena at a third scale (usually a more mesoscopic scale). This approach differs from the so-called "force-matching" because in the end we are actually matching gradients of a thermodynamically defined many-dimensional potential of mean force. The original force matching, developed by Ercolessi and Adams in a nonlinear algorithm and by us with Parrinello in a linear algorithm, does not reduce the resolution of the problem. It instead takes *ab initio* force data on the system nuclei, as defined by the Hellmann–Feynman theorem, and fits this to empirical functions at the same system resolution. There are surely many methods that can be called "multi-scale", including various coarse-graining methods. However, the use of the phrase by us to better characterize our own method is appropriate. We are not trying to claim the phrase "multi-scale" for only our method alone, which would be absurd. On the other hand, I think many coarse-grained models are not "multi-scale", but instead primarily phenomenological computer modeling.

Dr Baoukina continued the discussion on the paper by Professor M. Müller: Formation of stalk during membrane fusion is associated with mechanical deformation of the bilayer. Did you try to calculate elastic constants of the membrane such as bending and stretching moduli? How do they compare to experimental values? This could provide support information regarding the calculated excess energy of the stalk.

Professor M. Müller responded: The bending rigidity, K, and the area elastic modulus, K_A, of our model membranes is $K = 8\ k\mathrm{T}$ and $K_A = 164\ k\mathrm{T}/R_e^2$, respectively. In the latter case R_e^2 is the mean-squared end-to-end distance of the unperturbed amphiphilic polymeric molecule, which can be used to relate the length scale of the model with the one of an experimentally studied membrane system. The value of the bending rigidity, K, is quite close to the value for lipid membranes and somewhat smaller than the value for polymersomes.[1]

In order to compare the area elastic modulus, one has to identify the length scale, R_e. Since the hydrophobic thickness of the bilayer membrane is about 1.3 R_e in our model and 25–30 Å in lipid membranes, we estimate $R_e = 21$ Å and therefore, $K_A = 0.4\ k\mathrm{T}/\mathrm{Å}^2$, which is about a factor 3 larger than typical values for lipid membranes, $K_A = 0.1\ k\mathrm{T}/\mathrm{Å}^2$. If we used the area per amphiphile, $2R_e^2/58$, to estimate the length scale, we would obtain the estimates $R_e = 40$ Å and $K_A = 0.1\ k\mathrm{T}/\mathrm{Å}^2$.

1 H. Bermudez, D. A. Hammer, and D. E. Discher, *Langmuir*, 2004, **20**, 540.

Dr Louhivuori continued the discussion on the paper by Professor Voth: In the paper you mention that the elastic network model of the N-BAR domain reproduces the essential low frequency dynamics. Does the heterogeneity of the elastic network allow the BAR domain to undergo significant conformational changes?

Professor Voth replied: The heterogeneous elastic network is harmonic. It therefore does not allow for large scale conformational changes. The N-BAR dimers are believed to not experience such conformational changes when bound to the membrane (*e.g.*, from experimental EPR data).

Ms Schor said: It surprised me that the N-terminal domain of the protein was left out of the coarse-grained model as it undergoes a conformational change. As this domain is responsible for interaction with the membrane and you mentioned including it in a version 2 of the model I wondered (1) how you described it in combination with the rest of the protein and (2) if including the N-terminus changed the results for the membrane deformation significantly?

Professor Voth replied: The N-terminal amphipathic helix is not left out of the CG model. In the version 1 model (in the paper), it is represented by one CG "bead". In the version 2 model (in the short talk), it has several CG beads and is more realistic.

Yes, a more realistic N-terminal helix describes the membrane "insertion" by the helices more accurately, which in turn drives a greater membrane remodeling through the so-called "wedge" mechanism of the outer leaflet of the bilayer.

Dr Ensing continued the discussion on the paper by Professor Smit: You explain in your lecture that proteins that either both have a negative mismatch or both have a positive mismatch attract each other, while proteins that have opposite mismatches repel each other. Since in your model their is no attraction, the interactions are mediated by the membrane. My question is, whether the interactions that you observe are due to the meniscus formation in the membrane leaflets? I would like to add that we have observed with our coarse-grained model (including attractions) that also "negative mismatch proteins" appear as "positive mismatch proteins" when seen from some distance, because the lipid tail density buildup next to the protein in that case causes a positive meniscus in a second ring around the protein.[1] Do you see a similar fluctuating leaflet meniscus as a function of distance from the protein in your model?

1 S. O. Nielsen, B. Ensing, V. Ortiz, P. B. Moore, and M. L. Klein, *Biophys. J.*, 2005, **88**, 3822–3828.

Professor Smit responded: We found it useful to describe the qualitative behavior in terms of the hydrophobic shielding.[1] However, we have not analyzed the shielding for a case of one protein with a negative and the other with a positive mismatch.

1 F. J. M. de Meyer, M. Venturoli and B. Smit, *Biophys. J.*, 2008, **95**, 1851–1865.

Dr Baaden continued the discussion on the paper by Professor M. Müller: Could the solvent-free, coarse-grained model for amphiphilic membranes presented in the paper by Professor M. Müller be extended in order to include proteins?

This would be particularly interesting—but maybe also particularly complex—with respect to protein-mediated membrane fusion.

Fig. 1 shows an atomistic and a coarse-grained view of the SNARE complex.[1] This four-helical protein bundle is commonly thought to be central to membrane fusion and spans two apposed lipid bilayers.[2] Its complex shape is intimately linked to the biological function of membrane fusion. Very simple protein models such as the cylindrical structures presented in Professor Smit's paper might thus not be sufficient in order to capture the role of SNAREs in an approach as presented in Professor M. Müller's paper.

Important questions that might be tackled within an extended framework of Professor M. Müller's paper concern the number of SNARE complexes required for membrane fusion and the energetics of the intermediate steps of the process. Although more detailed atomistic and coarse-grained models are quite able to capture the detailed interactions between the SNARE proteins themselves and the membrane,[3,4] they cannot currently be extended to the time- and length-scales pertinent to these issues, mainly due to computational limitations. Such limitations don't seem to exist for an extension of the work described in Professor M. Müller's paper. What kind of transformation path would such an approach require?

1 http://www.baaden.ibpc.fr/projects/snaredeci/
2 R. Jahn, T. Lang and T.C. Sudhof, *Cell*, 2003, **112**, 519–533.

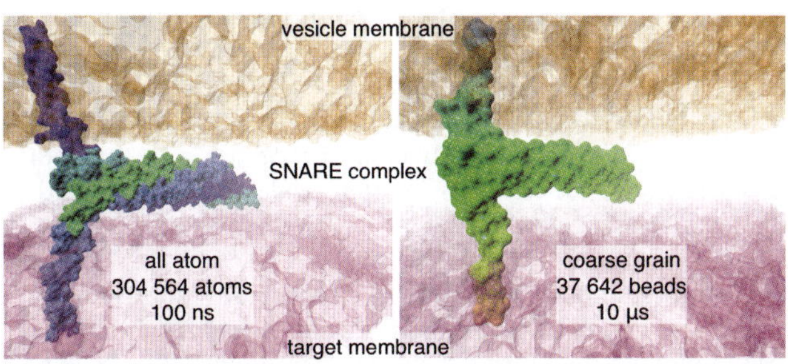

Fig. 1 Illustration of an all-atom (left) and coarse grain (right) molecular model of the SNARE complex inserted into two apposing lipid bilayers.

3 M. P. Durrieu, R. Lavery and M. Baaden, *Biophys. J.*, 2008, **94**, 3436–3446.
4 M. P. Durrieu, P. J. Bond, M. S. P. Sansom, R. Lavery and M. Baaden, *ChemPhysChem*, 2009, **10**, 1548–1552.

Professor M. Müller replied: The inclusion of proteins, in form of generic, geometrical objects with a patchy hydrophilic/hydrophobic surface, can be envisioned within the presented model. A similar problem has been considered before,[1] where the inclusion of nanoparticles into assembling copolymer melts has been considered and details of the implementation in the framework of a grid-based model are described.

We expect the transformation path to obtain the excess free energy of a system with proteins to be qualitatively similar to the one presented here for pure membranes. We speculate that the preferred location of the fusion proteins in the system with and without stalk might differ, but such an effect could be accounted for in the computational scheme by an external field acting on the proteins.

By properly "patterning" the surface of these objects one can create local, preferential interactions with the different components of the amphiphiles. This allows for a mesoscopic description of proteins with hydrophobic and hydrophilic parts. Potentially, such an approach can highlight the local membrane deformation induced by fusion proteins and provide qualitative insights into the role of their conformational rearrangements in bringing together the apposing bilayers. A quantitative description has to include additional information about the interactions and the molecular architecture on the atomistic scale and, thus, requires a closely coupled multi-scale approach.

1 F.A. Detcheverry, H. Kang, K. Ch. Daoulas, M. Müller, P. F. Nealey, and J. J. De Pablo, *Macromolecules*, 2008, **41**, 4989.

Dr Baaden continued the discussion on the paper by Professor Voth: Which algorithm is used in order to propagate the N-BAR/liposome model combining a heterogeneous elastic model with a hybrid analytic–systematic methodology described in the paper by Professor Voth with respect to time? Is it a plain molecular dynamics algorithm? Which simulation conditions such as temperature and pressure control were employed?

Professor Voth responded: It is our own in-house highly scalable (spatial decomposition) MD code for these CG models. They are at constant NVT. The CG model is a solvent-free model, so a constant NPT MD algorithm would not make sense for them.

Professor Español asked: In the force matching method developed in your group, the idea is to take the forces on the coarse-grained degrees of freedom measured from a molecular trajectory and try to fit the functional form of these forces with a set of a conveniently parametrized model forces (coming eventually from a parametrized version of the effective potential), all along the trajectory. My question is: Why not include friction forces in the parametrized model? We know that these friction forces may play a role in the dynamic properties and (maybe) they would improve the correspondence between the time scales being simulated and the real time scales.

Professor Voth responded: We have in fact already shown how to do this.[1] However, if enhanced equilibrium sampling is the desired result in a CG simulation, including friction would be a mistake since it slows down the sampling.

1 S. Izvekov and G. A. Voth, *J. Chem. Phys.*, 2006, **125**, 151101.

Dr Vila Verde continued the discussion on the paper by Professor Smit: (1) Please indicate in what configuration ($\theta = 0$ or $\theta = 90$) was the blue curve in Fig. 4, of the paper, obtained.

(2) The physical motivation for some of the screening rules is unclear. Referring to Fig. 5 of the paper, while it is reasonable to assume that particle i is screened from interacting with particle k, it is less clear why the same should happen to particle m. As a result of these screening rules, it appears that a free particle only interacts with a cluster *via* the closest cluster particle (the red line in Fig. 4.b is almost identical to the blue line in Fig. 3). Comparison of the curves in Figure 4.b makes it clear that the approach works, but it would be useful if the authors could provide an explanation as to why that happens. Since it appears that the PMF is related to the local monolayer curvature due to hydrophobic mismatch between the membrane and the protein, quantifying how the insertion of a protein changes the curvature field between two existing proteins might be helpful.

Professor Smit replied: Question (1): In the potential of mean force calculation the angle was not fixed only the distance.

Question (2): Indeed, the model we propose has many simplifications and the idea of partial screening could lead to an improvement of the model.

Dr Marti opened the discussion of the paper by Professor Tieleman: The authors claim that properties of lipid monolayers must be different to those of lipid bilayers (introduction of Professor Tieleman's paper). My question is about dynamics of lipids, the similarities and differences between mono- and bilayer membranes. In particular I would like to know if in their simulations the authors observed lateral diffusion for the monolayer case.

Professor Tieleman answered: Yes, the coefficients for (long time) lateral diffusion in monolayers depend on the monolayer surface density or surface tension at the interface. The values are comparable to those in bilayers in the liquid–crystalline phase ($\sim 10^{-7}$ cm^2/s) at similar areas per lipid, *i.e.* at a surface tension of 35–40 mN/m, decrease at lower surface tensions, and can become close to those in the bilayer in the gel phase ($\sim 10^{-9}$ cm^2/s) if the monolayer transforms from the liquid-expanded to the liquid-condensed phase.

Professor Deserno opened the discussion of the paper by Professor Vattulainen: We have also recently looked at long ranged correlation effects in lipid diffusion and also found very nice velocity autocorrelation functions. But we got stuck trying to understand what's going on theoretically. What we see looks intriguingly like a 2D-Oseen tensor, but there are subtle differences, and these also depend on whether hydrodynamics is present or not, say by using a DPD *versus* a Langevin thermostat. I was wondering whether you have found a way to get some theoretical handle on your observed patterns?

Professor Vattulainen answered: The simple answer is no. We think that the role hydrodynamics plays in the process is not that of the dominating driving force. The fact that DPD simulations were also found to yield the same diffusion mechanism and the same dynamically correlated regions as the Berendsen thermostat implies that the observed phenomena do not arise from local momentum conservation only. However, even in DPD simulations where local momentum is conserved, the lipids comprising the membrane interact with the solvent molecules. Thus, part of the momentum carried by lipids diffuses to the solvent phase, which in turn implies that even in DPD simulations (and other hydrodynamic approaches that are consistent with Navier–Stokes) the momentum within the membrane itself is not conserved.

Professor Tieleman addressed Professor Vattulainen: You described a form of collective diffusion of the lipids. I was wondering if you or anyone in the audience is aware of similar behavior in other systems, maybe polymer melts or other more complex liquids?

Professor Vattulainen answered: The diffusion we have observed is collective in the sense that there are many lipids moving together. Yet strictly speaking we would not call this collective diffusion, since by definition the collective diffusion coefficient characterizes the decay rate of density fluctuations. In the phenomena that we have considered there is no need for excess local density or its decay but the diffusion processes take place everywhere in the membrane. While raising this point may sound seemingly irrelevant, it may still be better to talk about concerted motions, to avoid misunderstanding. Then, have these concerted motions been observed in other systems? The cases we are aware of deal with supercooled liquids and simple liquids modeled through Lennard–Jones interactions, under the constraint that local momentum is conserved. Further, there are related observations with glasses (see, *e.g.*, the previous questions posed by Professor De Pablo and Professor Allen), suggesting that similar phenomena could also take place in, *e.g.*, polymer melts.

Professor Kremer commented: In polymeric systems such concerted diffusion events are usually not observable. In a typical homopolymer melt the self density of the chains decay like $N^{-1/2}$, N being the chain length, making such motion patterns, which are significantly hindered by the entanglements, almost impossible. The situation can be different in solutions where the chains at most weakly overlap or during domain reorganization and late stage spinodal decomposition. During glass formation such effects have been observed for low molecular weight compounds or very short polymers only.

At the outset of the present study, the authors present the classical view of the hopping diffusion of lipids in a liquid membrane. Taking the flexible nature of the lipids into account, I do not see any argument for such a motion pattern in the liquid phase. What were the physical arguments, which led to this first proposal?

Professor Vattulainen answered: To our knowledge, the idea of jump-like diffusion in lipid bilayers stems from the free-volume theory, which was originally developed for hard spheres.[1] In those early articles which dealt with hard-sphere particles, one used the free volume theory to describe viscosity or other transport properties of simple liquids, and molecular transport was assumed to occur by the movement of molecules into voids in terms of jumps. Later, the free volume theory was extended to two dimensions, and diffusion in lipid bilayers was considered as diffusion of hard disks. Again, in the spirit of hard spheres one assumed that particles undergo jumps from a given site to the neighbouring void.[2] While physical reasons for this assumption were not given, it is plausible that one assumed lipids in a bilayer to be like hard rods with a well-defined cross-sectional area across the membrane. Nowadays we know that this assumption is not correct, but in the early 1980s there were no atomistic simulations available to correct this view. Yet, it is rather surprising that the view of jump-like lipid diffusion processes has remained alive either explicitly, or implicitly in terms of the free volume theories that are partly based on the assumption of lipids undergoing rapid jump-like diffusion events.

1 M. H. Cohen and D. Turnbull, *J. Chem. Phys.*, 1959, **31**, 1164; P. B. Macedo and T. A. Litovitz, *J. Chem. Phys.*, 1965, **42**, 245.

2 H. J. Galla, W. Hartmann, U. Theilen, and E. Sackmann, *J. Membr. Biol.*, 1979, **48**, 215–236; J. E. MacCarthy and J. J. Kozak, *J. Chem. Phys.*, 1982, **77**, 2214; T. J. O'Leary, *Proc. Natl. Acad. Sci. U. S. A.*, 1987, **84**, 429.

Professor De Pablo continued the discussion on the paper by Professor Tieleman: Some of the patterns that you report for the collective motion of phospholipids in a dense bilayer membrane are reminiscent of those observed in glassy materials. This is an analogy that might be worth pursuing. Similarly, a number of useful measures have been developed in the colloids literature to investigate or analyze collective motion.

Professor Vattulainen replied: We agree. There is some interesting work done on understanding the dynamics of glassy systems, as well as supercooled liquids, and there are analogies that are worth following.

Professor Tieleman responded: This is a reply to a question I asked Professor Vattulainen, which was answered by Professor De Pablo and Professor Kremer. The question was whether there are other cases in complex fluids that seem to exhibit the kind of diffusion shown by Professor Vattulainen. Professor Frenkel also added a comment on this, on the danger of filtering to get collective motions.

Professor Frenkel continued the discussion on the paper by Professor Vattulainen: One should be very careful not to over-interpret "flow patterns" in time averaged displacement pictures. Time averaging acts as a "Fourier filter" that suppresses all short wavelength transverse currents, thus revealing the (slowly decaying) long-wavelength patterns. Although I do not wish to imply that this trivial mechanism applies to the case that you show here, one should be aware that similar patterns can even be observed in time-averaged snapshots of 2D Lennard–Jones liquids. The longer the time averaging (t_{av}) the larger λ_c the characteristic length scale of the flow pattern ($\lambda_c^2 \sim t_{av}$).

Professor Vattulainen responded: We understand the concern. The actual diffusion mechanism in terms of concerted motions of many lipids is one of our main results in these complex many-component membranes, and that is not subject to time averaging. The cases where time averaging was used concern our results for displacement correlation plots, whose purpose was to quantify the size and the lifetime of the dynamically correlated regions. The size determined from these plots is in line with visual inspection of lipid displacements in the many snaphots that were shown in the article.

Professor Smit continued the discussion on the paper by Professor Tieleman: The pressure profile is not a well-defined quantity. In the literature one can find different conventions to compute the pressure profile. These methods give a very different pressure profile. Important to mention is that both conventions give the same value for the surface tension. This is exactly what one expects as the surface tension is a well-defined thermodynamic property that can be measured experimentally. These observations are the consequence of the fact that it is not possible to experimentally measure a pressure profile. In your paper you draw many conclusions about a property that is ill defined and cannot be verified experimentally. Is this not a dangerous approach?

A second question is that you speculate on the role of the "surface of tension" as a measure of the stability of the monolayer. Could you elaborate on this?

Professor Tieleman answered: (1) Indeed the local pressure tensor cannot be uniquely defined on the scales below the range of forces, and depends on the choice of a contour connecting the two interacting particles. Previous theoretical studies (*e.g.* ref. 64 of the paper[1]) on lipid bilayers have shown that for the two contours: Irving–Kirkwood and Harasima, the local pressure is independent of the choice of contour and thus the pressure profile is well defined for this type of system. The calculated pressure profiles in monolayers agree with those found previously in bilayers with a difference arising from the anisotropic monolayer environment, and the surface tension peak at the chain/air interface. Several macroscopic properties besides the surface tension, such as the bending and Gaussian moduli and spontaneous curvature, can be calculated based on the pressure profile, and compared with experimental values. Given the above arguments, the calculated profiles carry some degree of uncertainty, but so do many properties derived from theory/experiment if there is no other technique supporting them. (2) At simple interfaces, such as a flat interface formed by two homogeneous phases (oil and water), the densities of the two phases decrease and there is a single surface tension peak. For complex interfaces, such as a surfactant covered oil/water interface, there arise forces of different nature between different molecular groups forming distinct layers along the normal to the interface. The pressure profile across such an interface is characterized by multiple positive pressure and negative pressure (surface tension) peaks. The surface of tension represents the actual interface in the complex interfacial layer. For example, at this actual interface the Laplace equation can be defined for the boundary with finite curvature.

1 J. Sonne, F. Y. Hansen and G. H. Peters, *J. Chem. Phys.*, 2005, **122**, 204901.

Professor Smit continued the discussion on the paper by Professor Vattulainen: What is a raft and do they exist?

Professor Vattulainen answered: In model membranes the current view is that lipid rafts are highly ordered membrane domains rich in cholesterol and saturated (or monounsaturated) sphingo- and phospholipids. The concept of lipid rafts in model membranes is therefore largely based on the liquid-ordered (Lo) phase, the formation of which is characteristic to cholesterol. In real cell membranes the issue is more complicated. Rafts are expected to be biologically relevant, implying that the idea of a membrane domain in the Lo phase is no longer sufficient. This view was accounted for by the Keystone Symposium in 2006, which arrived at the following consensus definition,[1] focusing on microdomains in cells: "Membrane rafts are small (10–200 nm), heterogeneous, highly dynamic, sterol- and sphingolipid-enriched domains that compartmentalize cellular processes. Small rafts can sometimes be stabilized to form larger platforms through protein–protein and protein–lipid interactions." In essence, one refers to rafts as fluctuating and functional nanoassemblies composed of membrane proteins in cholesterol-rich membranes, where the proteins are able to carry out their functions. Discussion on this broad topic can be found elsewhere.[2,3,4]

1 L. J. Pike, *J. Lipid Res.*, 2006, **47**, 1597; L. J. Pike, *J. Lipid Res.*, 2009, **50**, S323.
2 D. Lingwood, J. Ries, P. Schwille and K. Simons, *Proc. Natl. Acad. Sci. U. S. A.*, 2008, **105**, 10005.
3 K. Jacobson, O. G. Mouritsen and R. G. Anderson, *Nat. Cell Biol.*, 2007, **9**, 7.
4 L. J. Pike, *J. Lipid Res.*, 2009, **50**, S323.

Professor Allen replied: Correlated particle motions in two dimensions, such as clusters and chains of moving atoms, have been investigated in computer simulations for many years.[1] There are several more quantitative ways of studying correlated diffusion mechanisms, applied to glassy systems[2] and also in diffusion between smectic layers, in recent work by M. Dijkstra *et al.*[3] These include the measurement of moments of the distribution of atom displacements as functions of time, looking for non-Gaussian contributions, studying the intermediate scattering functions which give an idea of dynamics at different length scales, and observing possible sub-diffusive regimes in which mean squared displacements are not proportional to time. It would be useful to calculate such quantities from your simulations.

1 P. L. Fehder, *J. Chem. Phys.*, 1969, **50**, 2617.
2 W. Kob, C. Donati, S. J. Plimpton, P. H. Poole, S. C. Glotzer, *Phys. Rev. Lett.*, 1997,**79**, 2827.
3 A. Patti, D. El Masri, R. van Roij and M. Dijkstra, arXiv:0906.3093 [cond-mat.soft].

Professor Vattulainen responded: We agree. Some of the mentioned quantities are already under consideration in related projects. Another matter that needs to be emphasized is the importance of coupling simulations to experiments. For example, we are looking forward to data from QENS and other scattering techniques, which would allow one to compare experimental data with simulated scattering functions.

Dr Vila Verde opened the discussion of the paper by Dr Sengupta: (1) The meaning of helicity should be made clearer.

(2) The authors conclude that partial helicity is required to form and stabilize pores. They offer as explanation (section 4.3 of the paper) that partial helicity makes the peptide more flexible and that this both allows it to span the membrane and simultaneously to interact more favorably with the curved membrane forming the pore. While this explanation seems reasonable, the authors have not shown the necessary evidence supporting it, namely differences in the various peptide's ability to span the membrane and to interact favorably with it (the authors also should define what "favorable interaction" is). The paper would be strengthened by either a comment, making clear to the reader that further studies are necessary to decide whether the proposed explanation for the importance of partial helicity is correct, or the actual evidence supporting it.

Dr Sengupta replied: The percentage of helicity is defined by the number of residues that are constrained to be helical by using a dihedral potential on the backbone atoms. The constraints are independent of the amino acid residue. In this particular case, the residues in the central stretch of the peptide were constrained to be helical since unfolding occurs mainly at the termini in atomistic simulations. Comparing different peptides as suggested in the comment is beyond the scope of this paper. Magainin, the focus of this paper is known to have a high antimicrobial activity and the explanation for the importance of partial helicity is naturally specific to this peptide. While experimental and previous simulations have shown that helicity is not required for activity (explained in the introduction of the paper), the strength of this paper is the ability to actually compare between the scenarios. We chose only 4 scenarios and not a systematic study to get the first clues on the action of the peptide. Our results point towards the importance of not having fully helical peptides, as proposed in other models. Though indeed, further studies are necessary to test whether the proposed explanation for the importance of partial helicity is correct.

Professor Deserno remarked: You showed that after back-mapping atomistic detail into the coarse-grained evolved pore, it doesn't close within 50 ns. However, it's common knowledge that "junk goes to defects" and so I am not sure what it means that a peptide-decorated pore defect doesn't vanish. It could take longer to vanish, or it could substantially rearrange on longer time scales. With that in mind, what is the information you gain from the final atomistic 50 ns run?

Dr Sengupta answered: The 50 ns atomistic simulation allows us to probe whether the pores sampled in the CG model are realistic. If the structure sampled in the CG simulation were unrealistic in the atomistic force-field, changes from that structure would be seen such as re-arrangement of lipids and/or peptides, unfolding and possible pore closure.

Dr Wilson asked: Could you comment on the experimental data relating to your simulations, particularly in relation to hydration of pores and flip-flop of lipids. Are their any specific experimental measurements that you are able to compare with?

Dr Sengupta replied: Experimental measures on antimicrobial peptides are on an ensemble of poration events (such as vesicles) and comparison to single pore events studied here is difficult. We do see a large increase in flip-flop events, as also seen in experiments. A quantitative comparison is not possible, since the flip-flop rate per pore can not be measured experimentally. The hydration of the pore is related to the size of the pore which is smaller in our simulations than predicted from dye-leakage experiments. Again, a direct comparison between a macroscopic system and our simulation box is difficult.

Professor Yamamoto continued the discussion on the paper by Professor Tieleman: I would like to know the detailed definition of the weighting function "f" in eqn (5) of the paper. Whether the molecular stress defined for a pair of atoms is distributed locally on positions of atoms or on a line connecting between two atoms. One should use the later definition for investigating local stress with finer resolutions, so I just want to confirm it.

The above definition should contain the thickness of the slice with which the authors have determined the local stress "$\boldsymbol{\sigma}(z)$". Is the thickness larger than (or comparable to) the molecular size? The stress is a mechanical property which can be uniquely defined for a large wave length, but this is not so obvious for molecular scales.

How is the shape like a normal stress profile across the interface? It should be almost completely flat, but quite often it is not, especially if one makes the thickness of the slice comparable to or smaller than the molecular scale. This must be checked.

Professor Tieleman answered: The weighting function used in this work is introduced in ref. 19 of the paper.[1] The stress is distributed uniformly between the two particles. The weighting function depends on the position of the particles with respect to the current slice and if at least one particle is outside of the slice it becomes proportional to the slice thickness. The slice thickness is 0.1 nm and is on the molecular scale. The local stress tensor cannot be unambiguously defined on this scale, and depends on the choice of the integration contour between the two particles. Previous studies have shown that for the two contours: Irving–Kirkwood and Harasima, the pressure profile is well defined. We have also done tests with a larger slice thickness, which did not change the pressure profile. The condition for mechanical equilibrium requires the normal pressure to be constant. In our calculation, the normal pressure was constant if no constraints for molecular bonds were present. In the presence of constraints, the deviation of the normal pressure from the constant is likely related to approximations in the calculation of the constraint forces.

1 E. Lindahl and O. Edholm, *J. Chem. Phys.*, 2000, **113**, 3882–3893.

Professor Deserno remarked: Do you include the contributions of 3- and 4-body terms in the calculations of the local stress tensor?

Professor Tieleman answered: The contributions from angles and dihedrals are included in the configurational contribution of the local stress tensor through the gradients of the potentials. Non-bonded 3- and 4-body interactions are not a part of the force fields used in this study.

Professor Theodorou commented: Are the constraint forces taken into account in your calculation of the local pressure tensor? If they are, and if bending and torsional contributions are also taken into account, it is rather puzzling that your normal pressure profile does not turn out to be flat in the presence of constraints. Perhaps the considerations presented in Theodorou *et al.*[1] may be helpful in resolving this issue.

1 D. N. Theodorou, L. R. Dodd, T. D. Boone and K. F. Mansfield, *Macromol. Chem., Theory Simul.*, 1993, **2**, 191–238.

Professor Tieleman answered: In our calculations of the lateral pressure profiles, the contribution to the forces due to bond constraints is taken into account. We have also done test calculations without including constraints in the trajectory post-analysis, and the resulting pressure profile changed dramatically and became an order of several thousands bars. The deviations of the normal component of the pressure tensor from constant in the presence of constraints could originate from the approximations in the constraint force calculations, see ref. 65 of the paper.[1] Since the meeting there have been some further developments and we hope this issue will be resolved in the near future.

1 S. Ollila, *Helsinki University of Technology*, 2006.

Professor Ollila addressed Dr Baoukina, Professor Tieleman and Professor Marrink: Values for spontaneous curvature of a monolayer in an air/water interface are reported in your paper. The values are all positive in contrast to earlier results from simulations and experiments for lipid water mixtures. Also, the values for a polyunsaturated DAPC monolayer is more positive than others, which is unexpected. Can these differences to earlier reported values be explained by peaks arising from a chain/air interface? You mention that subtraction of this peak gives negative curvatures. Does subtraction also make spontaneous curvature of polyunsaturated bilayers more negative? One could also get rid of air/chain peaks by analyzing the monolayer at an oil/water interface. Did you compare spontaneous curvatures for these systems?

Would you say that spontaneous curvature for a monolayer in an air/water interface is not the same as the spontaneous curvature for a monolayer in a bilayer? This would mean that spontaneous curvature is not an intrinsic property of lipid molecules but depends on the environment.

Professor Tieleman answered: In monolayers, the presence of the tension peak at the chain/air interface leads to noticeable differences of the pressure profile and the properties derived from it, as compared to bilayers. In addition, spontaneous curvature being proportional to the first moment of the pressure profile depends on the choice of the reference plane. In our calculations for monolayers, the reference plane is set to the minimum of the tension peak of the hydrophobic/hydrophilic interface. In earlier calculations of spontaneous curvature for bilayer leaflets, the bilayer center was often used as a reference plane. These factors can explain the differences with respect to previously reported values. From the calculated pressure profiles, the chain/air interface appears to have a higher surface tension for unsaturated lipids at a fixed total surface tension in monolayers. However, the values of chain/air surface tension are similar for disordered saturated and unsaturated chains (in the coarse-grained model used). Differences likely arise due to differences in surface densities of the saturated and unsaturated chains, as well as to differences in the chain entropic repulsive pressure at chain free ends, which is noticeably higher for saturated chains. Subtracting the chain/air tension peak as of disordered oil from the pressure profiles gives a more negative spontaneous curvature of the DAPC monolayer, assuming that the bending modulus remains unchanged. For monolayers at the oil/water interface, the chain/air tension peak disappears. However, the disordered oil penetrates the hydrocarbon chain region. Altogether, it makes sense to define spontaneous curvature of a monolayer/leaflet in a given media, at a fixed surface tension (and temperature).

Professor Ollila said: Authors point out an important technical problem in lateral pressure profile calculations about non-constant normal component in pressure profiles calculated from atomistic simulations. In our own studies we have found similar results: for atomistic simulations normal component is not constant but using harmonic potentials instead of constraints at least reduces deviations from constant value. However, for coarse grained simulations with a MARTINI model we have always found constant normal component. These results indicate that the non-constant normal component would not arise from choice of contour in definition of local pressure but more likely, from example, usage of constraints in atomistic simulations. Also the pressure tensor defined using Irwing-Kirkwood contour have been shown to give functional derivative of the free energy with respect to the local strain tensor.[1] This property of pressure tensor is used to derive connections between pressure tensor and elastic properties of bilayers, which are used in the paper. According to these results, elastic properties calculated from pressure profile should be correct if Irwing-Kirkwood definition is used in definition of pressure profile.

1 L. Mistura, *Int. J. Thermophys.*, 1987, **8**, 397–403.

Dr Baoukina replied: We agree with the comment. In all our calculations the normal pressure deviated from constant if 1) constraints for molecular bonds were used, and at the same time 2) molecules were ordered at the interface. In this case, approximations in the calculations of constraint forces likely sum up and lead to non-zero contribution to all pressure tensor components, which are obvious in the normal component but less evident in the lateral components as they are not constant.

When constraints were substituted by harmonic bonds in the MARTINI model for cholesterol in our calculations, the deviations of the normal pressure were significantly reduced but not zero. This might be related to other factors, such as, for example, distortions of cholesterol ring structure as mentioned in the paper or integration errors due to the use of the large force constant in the harmonic potentials substituting constraints with a large time step.

Professor De Pablo remarked: In the paper, Fig. 5 indicates that the lateral pressure profile (and hence the surface tension determined from the simulations) is strongly dependent on system size. The authors suggest that small systems are better for computation of lateral pressure, but this appears to be a somewhat arbitrary statement. These findings are worrisome, as surface tension is one of the quantities that can be measured experimentally and used to determine the validity of a model. Are there any systematic studies of finite size effects for monolayers? Is there a better way of computing surface tension (*e.g.* by determining the free energy[1])? Similarly, the results of Fig. 7 from the paper, showing that the normal pressure profile is not constant (as required by mechanical stability) are of concern. The authors indicate that the normal profile is not constant when they use bond constraints. But when they use flexible bonds the normal pressure profile continues to be non-uniform, and the pressure is different from the bulk value. Are these behaviors common knowledge? With multiple studies of monolayers and bilayers published in the literature, is it generally acknowledged that constraints introduce artifacts into the pressure calculation, or could there be a problem somewhere in the codes used to generate the numbers shown in Fig. 7?

1 T. S. Jain and J. J. De Pablo, *J. Chem. Phys.*, 2003, **118**, 4226.

Dr Baoukina answered: The paper's Fig. 5 demonstrates that the peaks in the lateral pressure profile resolved in the small monolayer are averaged out in the larger monolayer due to thermal fluctuations, as the monolayer shape deviates from flat. Therefore, the smaller systems in which these out-of-plane fluctuations are suppressed are more suitable for the calculation of the lateral pressure profile. In both cases, however, the surface tension is the same, and the integral of the lateral pressure profile converges to the surface tension (with the opposite sign). The surface tension is calculated directly in simulations as the difference between the normal and lateral pressure in the simulation box multiplied by the box size normal to the interface and divided by the number of interfaces. At the same time, the lateral pressure profile is calculated from the trajectory post-analysis and involves dividing the simulation box in the slabs and calculating the local pressure tensor, see Methods. The surface tension in the simulations does depend on the system size, but this dependence is relatively weak. Fig. 5 illustrates that the dependence of the pressure profile on the system size (at least as calculated with the current algorithm) is dramatic.

In our calculations, the normal pressure deviated from constant if (1) constraints for molecular bonds were used, and at the same time (2) molecules were ordered at the interface. In the absence of constraints, the only case when the normal pressure was not constant (but its deviations from constant significantly reduced), is when constraints were substituted by harmonic bonds in the coarse-grained model for cholesterol. However, we believe it is related to other factors, such as, for example, distortions of the cholesterol ring structure or integration errors due to the use of the large force constant in the harmonic potentials substituting constraints with a large time step. In all other systems, the normal pressure is constant across the interface. This indicates that the deviation of the normal pressure from constant likely originates from the approximations in the calculations of the constraint forces in the trajectory post-analysis.

Dr Ensing continued the discussion on the paper by Dr Sengupta: The water model used in the MARTINI force field is somewhat special, as I understood it, in that it consists of two types of water particles, one that acts as a normal Lennard–Jones liquid parameterized to model liquid water and a second type of larger particles that act as anti-freeze particles to keep the water liquid. My question is whether you also use here the anti-freeze particles and if so whether they can also penetrate into the confined space of the water pore that you observe in your membrane simulations? Also I was wondering if the stability of the water pore that you observe is affected by the concentration (or the omission) of antifreeze particles in the pore region?

Dr Sengupta replied: The simulations were carried out above the transition temperature of DPPC bilayers at 325 K. At this temperature, no freezing is observed and no anti-freeze particles were used.

Professor Faller replied: I would like to point out that both, the old and the new, version of the MARTINI model have problems with water interacting with solid surfaces. While the old one freezes at too high temperatures, especially as a solid acts as a nucleation site the new water model with the anti-freeze particles in some cases actually demixes into two phases with different concentrations of anti-freeze particles: a high order phase with few anti-freeze particles and a lower order liquid phase with more anti-freeze particles. So in order to model, *e.g.*, supported bilayers, the only solution to date is to weaken the water–water interaction which, however, leads to problems with other quantities like the area per molecule in a free bilayer. Details are discussed elsewhere.[1]

1 C. Xing and R. Faller, *J. Phys. Chem. B*, 2008, **112**(23), 7086–7094.

Dr van der Sman continued the discussion on the paper by Professor Tieleman: There has been little discussion about how to coarse-grain the coarse-grained models up to the level of mesoscopic models like Lattice Boltzmann or SRD. There exists Lattice Boltzmann models or phase field models for surfactant adsorption onto oil droplets.[1] This is a kind of (Density Functional) Field description. Coarse-graining is also the art of choosing the right coarse-grained variables. Which coarse-grained parameters of surfactants do you think are relevant at the mesoscopic level? (Having a length scale of droplet sizes of 1 micron).

1 R. G. M. van der Sman and S. van der Graaf, *Rheol. Acta*, 2006, **46**(1), 3–11.

Professor Tieleman responded: We have not given the transition between our coarse-grained model description to a mean-field description, perhaps with partial particle character, much thought. Parameters in a mean-field model would have to reproduce the headgroup pressure–interfacial tension–chain pressure pattern irrespective of the details of the interactions, as well as be able to represent favorable polar–polar (water/headgroup) and apolar–apolar (chain/oil) interactions to give a stable surfactant-covered interface.

Dr Hess continued the discussion on the paper by Professor Vattulainen: The domains that show correlated motion less than an order of magnitude smaller than the system size. I could imagine that (hydro)dynamics in 2D with periodic boundary conditions shows quite strong boundary effects. Have you checked the system size dependence?

Professor Vattulainen responded: We have not studied systematically the system size dependence but we have considered the diffusion phenomena using a large variety of models with varying sizes (see text). In all cases, we have found the same diffusion mechanisms to take place. At present it seems that the quantity that is most sensitive to system size is the lifetime of dynamically correlated membrane regions. In coarse-grained simulations we have found this lifetime to be of the order of one microsecond. When we tried to analyze the lifetime from atomistic models with about 100–200 lipids, we found that the correlations did not decay at all, or they did so very slowly. This is most likely due to the small system size, since if the dynamically correlated region is of the same size as the total membrane system, there is no reason to expect the decay to take place in a realistic manner.

Dr Vacha asked: Have you observed or is known if there is a correlation of diffusion between leaflets of membrane or are the leaflets fully independent in diffusion, particularly in the presence of rafts?

Professor Vattulainen replied: It has been found that a lipid raft domain in one leaflet of a lipid bilayer is matched by another raft domain in the opposing leaflet.[1] As far as diffusion is concerned, we have found that there are possibly minor correlations regarding the directions of concerted motions in the two leaflets. However, this issue has not been studied quantitatively.

1 H. J. Risselada and S. J. Marrink, *Proc. Natl. Acad. Sci. U. S. A.*, 2008, **105**, 17367.

Dr Vacha continued the discussion on the paper by Dr Sengupta: When using PME or RF treatment of electrostatics have you observed any difference in membrane or pore?

Dr Sengupta responded: The structural characteristics of the bilayer or that of the pore do not change on using PME *vs.* RF in either the CG model or in atomistic simulations. However, the poration propensity changes. In CG models, using PME increases poration probably since it allows a direct interaction between the two bilayer leaflets, not implicitly screened as in the shift or RF electrostatic schemes. Interestingly, the opposite is observed in atomistic simulations though we do not understand why the RF electrostatic scheme is important in facilitating poration in atomistic simulations.

Professor Thøgersen remarked: I do not understand why the focus (in the paper's Fig. 4 and 5 and in the discussion) is on the percentage of the helicity. It is stated in the Methods section that the force constant in the 65% helicity case is significantly altered compared to the 40% helicity case (and the bead class correspondingly changed). It therefore seems that the effects seen in Fig. 4 and 5 stems just as much—or even more—from the helicity being less enforced in the 65% case than from the level of helicity itself. It could be both relevant and interesting to see the results for the case of 100% helicity with the force constant and beadtype-setup used in the 65% case, to separate the effects from the level of helicity and the strength of helicity enforcement.

Dr Sengupta replied: Indeed, we change two related parameters—the percentage of helicity and the strength of the helical constraints at the same time. We chose only 4 scenarios and not a systematic study to get the first clues on the action of the peptide. Our results point towards the importance of not having fully helical peptides, as proposed in other models. We do not draw conclusions about the actual level of helicity required and further studies are necessary to test the different models further.

Dr van der Sman remarked: Can you include electric fields into you model, and can you model electroporation? Would electroporation be enhanced with anti-microbial peptides. I am asking this from the perspective of the application of pulsed electric fields, used to electroporate (kill) bacteria in liquid foods like orange juice.

Dr Sengupta responded: Electroporation can not be realistically modeled in the MARTINI model since water is represented without any (partial) charges and is blind to the electric field. As expected and as shown in this paper it leads to unrealistically high barriers to pore formation. We are currently working on a water model with charges on the water. This polarized water model would allow us to study electroporation.

Mr Bereau commented: The term 'resolution exchange' in your paper has been used as a way to transfer a configuration from one resolution to another. However, 'resolution exchange' was previously coined as a more general algorithm, where one simulates in parallel the same system at different resolutions and configurations are periodically exchanged using Monte Carlo moves.[1]

1 E. Lyman, F. M. Ytreberg and D. M. Zuckerman, *Phys. Rev. Lett.*, 2006, **96**, 028105.

Dr Sengupta answered: We now adopt the term "resolution transformation" in our paper.

Dr van der Sman opened the discussion of the paper by Professor Berendsen: At which level in your presented hierarchy do you think it is advantageous to switch from a particle description to a field description? With fields one can much easier capture fluid dynamics, especially if the geometry is complex. In the transition regime, one can of course also use a hybrid description like a field description for fluids and a particle description of solutes/colloids. Today, these models are quite developed *i.e.* in Lattice Boltzmann. How do you regard the potential/future of field descriptions like Self-Consistent Field theory?

Professor Berendsen answered: It seems to me that field descriptions including self-consistent field theory will not survive the competition of coarse-grained particle-based methods when the scale of structural inhomogeneities does not vastly exceed the molecular scale. This will include the whole developing world of nanoscience and almost all of molecular biology. Of course there are really macroscopic scales, say beyond one micrometer, where field methods will be much more efficient. But even for macroscopic scales, I am not sure that a particle description used as a solver of differential equations rather than as a coarse representation of real molecules, will not win out in the long run. Particle-based simulations are easier to program and simpler to understand than finite-element or finite-difference PDE solvers; they can more easily be combined with particle descriptions of solutes.

Dr Goga addressed Professor Berendsen and Professor Kremer: There are limitations of the DPD thermostat because of the complexity of its implementation and some stagnation of the increase of PC power in the last few years and for example it can not be applied to the gradient of temperature. Therefore, simpler thermostats might be preferred. How do you comment on this?

Professor Berendsen answered: The DPD thermostat has the drawback that it needs the velocity at a time when the velocity is not yet known. In principle that requires an iterative solution, but one should think twice before implementing time-consuming precise solutions for this problem. As noise is added anyway, the issue is not precision, but stability and correct temperature regulation.[1] There certainly is room for some clever inventions here! When temperature gradients must be maintained, global thermostats such as Nose–Hoover or weak coupling will not work. However, Langevin-type thermostats, including DPD, will, as they can handle local temperatures.

1 C.P. Lowe, *Europhys. Lett.*, 1999, **47**, 145.

Professor Kremer answered: To implement a DPD thermostat is not really that complex and only marginally increases the computational costs during a simulation. The big advantage is that one has a local, up to the distance of the particles in the considered pair, hydrodynamics preserving thermostat. Thus I think the problem of computational power is not really an issue for this. The situation is somewhat different for nonequilibrium situations, as they might occur for an externally imposed temperature gradient, the problem addressed in the question. There are several ways to deal with that. If hydrodynamic effects do not play a role one can swith *i.e.* to a simple Langevin thermostat or other local thermostats. Otherwise, as is done in many nonequilibrum simulations of sheared systems, one could restrict the thermostat to the components of the relative velocities in the plain perpendicular to the temperature gradient. The situation gets more involved, when one deals with complicated spatial temperature variations, which however would be quite artificial.

Dr Periole asked: It seems that although the meeting is called "Multi-scaling…", we have seen only few applications of methodologies actually mixing different scales. Could you comment on that, and especially on why you don't have this listed as a future "promising" direction?

Professor Berendsen responded: Mixing different scales will be important, especially when interest in details is focused on a small part of the system, *e.g.* the active site of a protein in solution or in a membrane environment. There will be the need to employ a mixed-scale approach in a buffer zone between an atomic and a coarse-grained description, or even between a particle-based and a continuum-based description. Such buffer zones have been used earlier and presently mixed representations are being developed, among others in Kremer's group, at the ETH, and in the Groningen group. I did not list this as a promising direction, as mixing time scales is secondary to the development of new methods.

Concluding remarks

Herman J. C. Berendsen*

Received 18th August 2009, Accepted 25th August 2009
First published as an Advance Article on the web 21st September 2009
DOI: 10.1039/b917077b

1 Introduction

When a retired scientist is asked to present concluding remarks, the risk is high that he will treat you to a nostalgic survey of the past. I shall restrain myself and try to put the developments as witnessed in this conference into a proper future perspective, preceded by and interlaced with some reflections of the past.

For the past ten years I have had the opportunity to watch the development in a field that I have participated in from its conception in the early seventies of the last century: molecular simulation. I have seen the methodological development, but also the enormous expansion in applications to realistic systems. The latter has been largely the result of the incredible increase in available computer power, which has roughly maintained a growth rate of a factor of ten every five to six years. In the mean time, models have gone from descriptions that were, by computational necessity, very much simplified, to more end more detailed molecular descriptions, even including quantum details, to coarse-grained models that allow simulations to reach into the microsecond regime for systems of millions of particles.

This conference is a landmark in the development of multiscale modeling methods and applications. It has shown us where we are now in the field of multiscaling simulation of soft condensed matter, in particular polymers, membranes, proteins and colloids. These are the materials that exhibit self-organization: if properly functionalized they can assemble into organized structures that could form the basis of new classes of functional nanomaterials. Unfortunately, self-organization is—like any other phase change—a slow and highly cooperative process that requires very long simulations of a very large number of particles. Such simulations are still—and will be for a long time to come—far beyond the capabilities of atomistic simulations. They are the ideal playground for the development of coarse-grained methods that aim at describing the system on a coarser spatial and temporal scale. What this conference has not addressed is the important class of nanomaterials that have specific electrical, magnetic or optical properties. The latter invariably require an approach involving quantum mechanical methods in addition to molecular or coarse-grained simulations.

After a nostalgic dip into the first membrane simulations in Section 2, I shall focus in Section 3 on the coarse-graining methodology and classify the various multiscale methods according to their purpose and capabilities, and according to their place in a systematic hierarchy of models. Section 4 elaborates a bit on the effective potentials for reduced systems, while Sections 5 and 6 consider how dynamic and hydrodynamic details can be faithfully incorporated. Section 7 concludes with some trends in the development that emerge from the papers presented in this conference.

Groningen Biomolecular Sciences and Biotechnology Institute, University of Groningen, Nijenborgh 4, 9747 AG Groningen, the Netherlands. E-mail: H.J.C.Berendsen@rug.nl; Web: www.hjcb.nl

2 Early membrane simulations

For someone who has struggled in the early 1980s to simulate simple lipid bilayers with the then available computer power of, say, 10 Mflop (compared to close to 100 Gflop for a modern PC and 1 Petaflop for a large cluster), the achievements today are astonishing. Papers by Voth *et al.*, Vattulainen *et al.* and Marrink *et al.*, are examples.[1–3] With proper coarse-graining, systems with millions of particles can be simulated over real times approaching 100 microseconds. But membrane processes are slow; as is shown in ref. 3, a 45 μs CG simulation did not reach equilibrium distributions for pore-forming peptides in bilayer membranes. Papers by Voth *et al.*[1] and Müller *et al.*[4] had to rely on a coarser description to meet the demands of long timescale events. So did Smit *et al.*,[5] which simplifies membrane proteins to discs.

Fig. 1 shows two snapshots from a bilayer membrane, built from 2 × 16 C_{10} hydrocarbon chains with a head group. The chains were modeled as united atoms, but with proper dihedral interactions, as earlier used for hydrocarbon liquids.[6] The

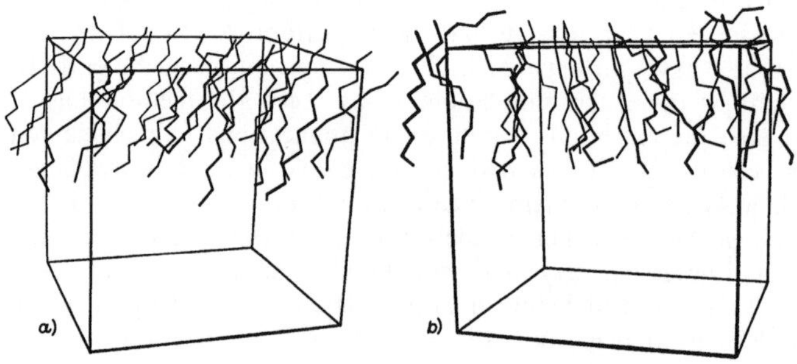

Fig. 1 Two snapshots taken from an ordered (*a*) and less ordered (*b*) state of a decane bilayer with effective head group interactions that restrain the head groups near a plane (reproduced with permission from Berendsen[10]). Copyright 1986, Società Italiana di Fisica.

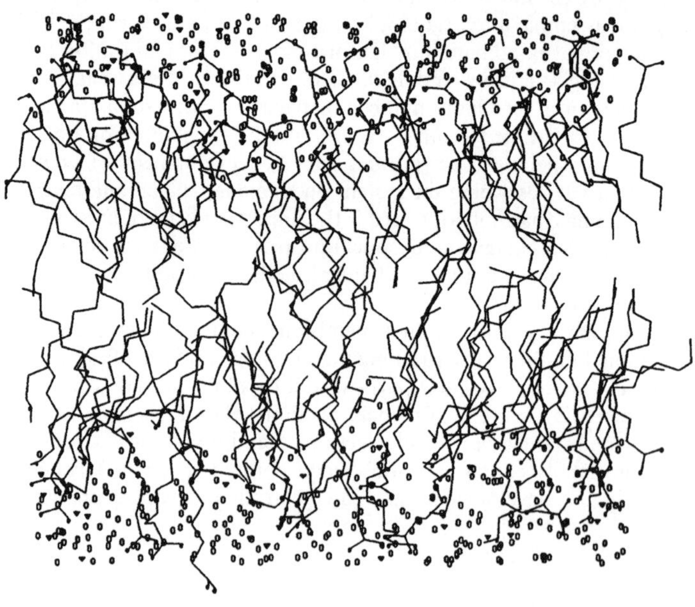

Fig. 2 Snapshot of a sodium decanoate/decanol/water bilayer.[12] Reproduced with permission from Egberts and Berendsen, *J. Chem. Phys.*, 1988, **89**, 3718. Copyright 1988, American Institute of Physics.

head groups were modeled in a coarse-grained fashion *avant la lettre* with a harmonic restraint with respect to the average of the head group positions. This was the first simulation of a realistic bilayer, published in 1980[7] and in 1982.[8] The small system exhibited transient ordering with tilted chains. A 2 × 64 molecule system was studied in more detail,[9] including the molecular tilt, order parameters, lateral pressure and diffusion. The total simulation time was 320 ps, requiring the better part of a PhD research period.

The next challenge was to go to full atomic detail including head groups and water. We studied the simple bilayer system decanol/decanoate/water which was known to form smectic liquid crystals. Fig. 2 shows a snapshot. This was published in 1986[11] and 1988.[12] It was soon followed by a phospholipid bilayer (DPPC) in atomic detail.[13] It appeared to be necessary to adjust details of the force field in order to obtain a gel-to-liquid crystal transition at the correct temperature. Fig. 3 shows a snapshot of the DPPC bilayer in the liquid-crystalline phase.

In the early nineties the emphasis was on enhancing resolution to atomic detail; the growing capabilities of MD were employed to refine the coarse models used before. Much effort was spent on the development of reliable force fields. But it was also realized that many realistic processes could not—and would never—be solvable by atomistic simulations. So coarse-graining in various forms was invented, and methods were developed to parameterize CG models.

I'll conclude the nostalgic part of this talk with an early example of extreme coarse-graining: reduce the system to motion in *one* "reaction coordinate." This was the main topic of the thesis of one of the organizers of this conference, Professor Siewert Jan Marrink. He studied the transport of a single water molecule through a lipid membrane.[15] The average force and the force fluctuation were both determined from simulations of a membrane with single water molecules constrained

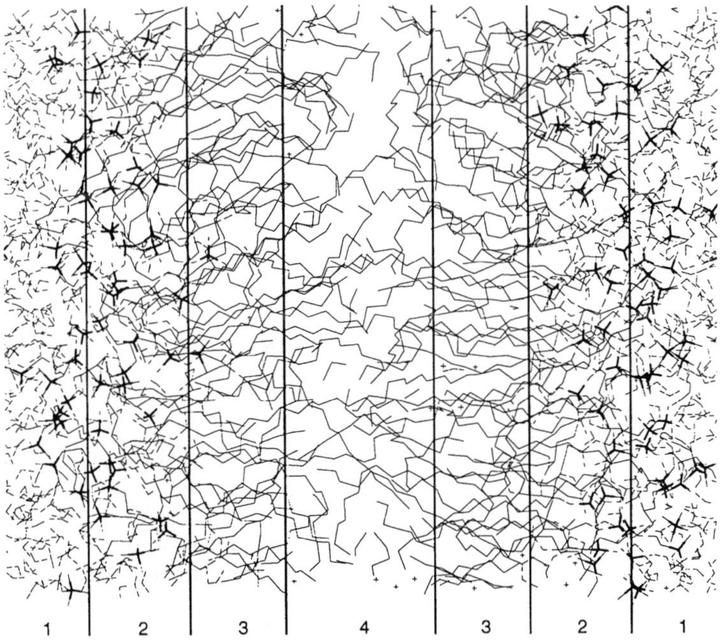

Fig. 3 Snapshot of a DPPC liquid-crystalline bilayer.[13] Reproduced with permission from *Pure Appl. Chem.*[14] (Copyright 1993, IUPAC). Membranes were characterized in four layers: 1. water/headgroups interface (polar, mobile), 2. headgroups/chains interface (dense, weakly polar), 3. ordered chains (dense, nonpolar), 4. disordered tails (open, nonpolar, mobile).

with their centers of mass at a given depth in the membrane. Thus both the potential of mean force and the friction coefficient could be determined along the path of a single water molecule through the bilayer. With these data a stochastic equation can be devised that describes the motion of a water molecule along a path through the membrane. Solving this equation for a steady state of constant difference in thermodynamic potential of water over the membrane yields the water flux and thus the permeability coefficient. We'll return to this example in the following sections.

3 A modeling hierarchy

The heart of any coarse-graining method is the *reduced* description of the system. One distinguishes *relevant* or *important* degrees of freedom from all other degrees of freedom, which are consequentially *irrelevant* or *unimportant*. The next step is to describe the dynamics of the reduced subsystem in such a way that it approaches as faithful as possible the *projection* of the motion of the complete system onto the reduced degrees of freedom. This is only possible when the time scales of the motion of the "relevant" degrees of freedom and of the "irrelevant" degrees of freedom are well-separated (the former being much slower than the latter). When there is overlap in time scales, one must give up on accuracy of the dynamical behavior, but one should at least conserve the probability distribution in configurational space, thus conserving thermodynamic properties of the system. Español *et al.*[16] gives a rederivation of Zwanzig's projection operator technique that gives a formal description of the dynamics in reduced space. The authors show clearly that the reduced description is only valid in the limit of well-separated time scales, in which case the stochastic dynamics in reduced space is Markovian, *i.e.* memoryless, and easy to implement.

The choice of the "relevant" degrees of freedom (d.o.f.) is made on an intuitive basis and depends on the properties one wishes to study. Note that "normal" molecular dynamics is already concerned with a reduced system of atomic coordinates; all electronic coordinates are considered "irrelevant" and a Born–Oppenheimer approximation is assumed (*i.e.*, electrons are infinitely fast with respect to nuclei). In addition, the usual *united atom* treatments which are viewed as accurate atomic reference models, consider the d.o.f. of the nonpolar hydrogen atoms as irrelevant and consider covalent bonds as constraints. Thus, a constrained united-atom model of butane has $3 \times 4 - 3 = 9$ degrees of freedom, which is a considerable reduction with respect to the 42 d.o.f. of the all-atom model and the 144 d.o.f. of the all-electron model. The step to a real coarse-grained model of the *superatom* type, lumping four methylene groups together in one superatom, is relatively moderate as it only reduces the d.o.f. further from 9 to 3. This is the level of coarse-graining that is most extensively used in many applications (*e.g.* 2, 3, 5, 18–21).

A substantial higher level of coarse-graining can be achieved by a continuum representation, such as the Navier–Stokes equations of fluid dynamics, or a density-functional description of the free energy as a functional of the density distribution of components of a composite material. The latter is often applied to block-copolymer melts; the free energy functionals are usually based on a simplified (*e.g.* Gaussian chain) intramolecular model plus a mean-field description for the intermolecular interactions. Dynamics are invoked by the inclusion of linear mobility relations to the gradients of thermodynamic potentials. The continuum equations are usually solved on a regular grid or using irregular finite elements, but they can also be solved by a system of particles obeying specified dynamical rules. The Lattice–Boltzmann method is a special case: the continuum equations are solved on a regular grid by updating attributes of the lattice points as a function of the attribute values at neighboring points. The resolution of these methods depends on the scale at which the simulated system shows structure: thus block-copolymers with structural features on a nanometre scale require nanometre resolution, but fluid flow in macroscopic objects may get away with resolution on a centimetre scale, comprising, say, 10^{22} atoms.

Multiscale modeling methods can be categorized in a modeling *hierarchy* ranging from detailed quantum treatment to macroscopic continuum descriptions. Precisely such a hierarchy is the subject of a book[17] that was published in 2007 entitled "*Simulating the Physical World, Hierarchical modeling from quantum mechanics to fluid dynamics*", from which I shall quote in the following.

Fig. 4 lists a number of approximations, ranging from a complete relativistic quantum-dynamical description to a macroscopic description of fluid dynamics. The former is unworkably complex and the latter has abstracted the system to the very basic level of the conservation laws, together with zero-order thermodynamic and first-order dynamic properties. In between are the practical levels of molecular dynamics and the coarse-grained approaches describing reduced systems of practical interest.

4 Effective potentials for coarse-graining

As mentioned above, it is essential that thermodynamical properties are retained on coarse-graining. This guarantees that equilibrium properties (average structure, free energies, solubilities, partition coefficients, *etc.*) are still validly predicted by the CG methods. Also non-equilibrium properties as driving forces for slow dynamics will be faithfully represented. What does this mean for the effective potential?

Consider a detailed system with (cartesian) coordinates $\{\boldsymbol{r}\} = \boldsymbol{r}_1,\ldots,\boldsymbol{r}_n$. Assume that we have good reasons to distinguish *relevant* coordinates $\boldsymbol{r}'$ and *irrelevant* coordinates $\boldsymbol{r}''$. So the full space consists of a set of coordinates $\{\boldsymbol{r}\} = \{\boldsymbol{r}',\boldsymbol{r}''\}$ and the reduced space consists of the set $\{\boldsymbol{r}'\}$. For simplicity we take the coordinates of the reduced space as a subset of the full space rather than a set of generalized coordinates. The Helmholtz free energy A for a given volume and temperature T is given by

$$A = -k_{\mathrm{B}}T\ln Q \tag{1}$$

$$Q = c\int e^{-\beta V(\boldsymbol{r})}\,\mathrm{d}\boldsymbol{r} \tag{2}$$

Here Q is the partition function and $\beta = (k_{\mathrm{B}}T)^{-1}$; c is a temperature-dependent constant containing the masses of the particles and $V(\boldsymbol{r})$ the conservative potential

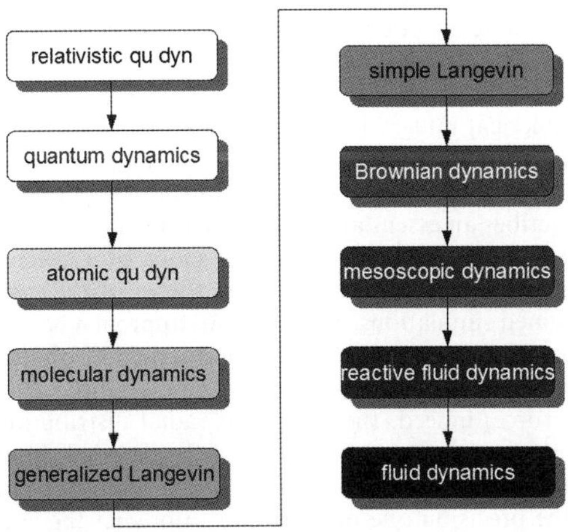

Fig. 4 Hierarchy of models for simulation,[17] ranging from very detailed (white background) to very coarse-grained (black background). Each level has its own description of the reduced system and its own simulation method. Each higher level loses some details of the preceding level.

of the full system. We wish to define an effective potential $V^{\rm mf}(\boldsymbol{r}')$ such that a Boltzmann distribution in $\boldsymbol{r}'$-space is maintained for the reduced system:

$$Q = c' \int e^{-\beta V{\rm mf}(\boldsymbol{r}')}\, {\rm d}\boldsymbol{r}' \tag{3}$$

This is accomplished by defining the effective potential as follows:

$$V^{\rm mf}(\boldsymbol{r}') \stackrel{\rm def}{=} -k_{\rm B}T \ln\left[\int e^{-\beta V(\boldsymbol{r}',\boldsymbol{r}'')}{\rm d}\boldsymbol{r}''\right] + \text{constant} \tag{4}$$

as can be easily verified by inserting eqn (4) into eqn (3). Note that this effective potential is *not* the mean potential, *i.e.* the original potential averaged over the irrelevant degrees of freedom. But its derivative with respect to $\boldsymbol{r}'_i$, which is the *force* acting on $\boldsymbol{r}'_i$ in the reduced system, *is* a mean force, *i.e.* the average of the detailed force over the irrelevant degrees of freedom. This is seen by differentiating eqn (4) with respect to $\boldsymbol{r}'_i$:

$$\boldsymbol{F}_i^{\rm mf} \stackrel{\rm def}{=} -\frac{\partial V^{\rm mf}(\boldsymbol{r}')}{\partial \boldsymbol{r}'_i} = \frac{\int\left(-\dfrac{\partial V(\boldsymbol{r}',\boldsymbol{r}'')}{\partial \boldsymbol{r}'}\right)e^{-\beta V}{\rm d}\boldsymbol{r}''}{\int e^{-\beta V}{\rm d}\boldsymbol{r}''} = \langle \boldsymbol{F}_i\rangle_{\boldsymbol{r}''} \tag{5}$$

Thus the term *potential of mean force* (PMF) is quite appropriate for the effective potential $V^{\rm mf}$. Note that it is not a potential in the Lagrangian or Hamiltonian sense; the potential of mean force is really a free energy with respect to an equilibrium distribution of the irrelevant degrees of freedom. It may (and will) depend on density, constitution and temperature. One has to take such dependencies into account when the normal thermodynamic derivatives are considered to derive thermodynamic quantities. For example, the internal energy U is no longer equal to the ensemble average of $V^{\rm mf}$, but rather

$$U = \partial(\beta A)/\partial\beta = \left\langle V^{\rm mf} + \beta\frac{\partial V^{\rm mf}}{\partial\beta}\right\rangle \tag{6}$$

We see from eqn (5) that the mean force can be generated from an equilibrium simulation in which the relevant degrees of freedom are *constrained* (kept constant). Note that this is true irrespective of the overlap of time scales. When this is done at several values of $\boldsymbol{r}'$, the potential of mean force can be constructed—up to a constant—by numerical integration. In practice this is not as easy as it appears to be because in the multidimensional case many simulations are needed and statistical noise will soon spoil the accuracy. In one dimension (such as a single reaction coordinate that describes an essential event) there is in general no problem. In many dimensions the way to proceed is to devise the shape of a potential and adjust its parameters to minimize the difference between forces of the model and averaged forces from constrained simulations. In ref. 16 this approach has been used to derive a pair-additive potential of mean force for the interaction of star polymers in the melt. By assuming pairwise additivity one may miss essential attributes of the real potential of mean force; indeed, the recovered radial distribution function of the CG simulation is not exactly equal to the rdf of the detailed simulation.

In most cases the *bottom-up* reconstruction of $V^{\rm mf}$ from fine-grained simulations will not produce the precision one may require to predict the adequate thermodynamic properties from CG simulation. A *top-down* approach: adjusting model parameters on the basis of the required thermodynamic properties, will then be necessary. But be aware of a serious pitfall here: if the CG model does not predict any properties beyond the ones you have used for the parametrization, you have achieved nothing! At best you have gained some understanding and insight. Always require your model to predict something yet unknown!

To end this section, we go back to Marrink's work on water permeation through a membrane in 1994.[15] Fig. 5 shows the potential of mean force for a single water molecule as a function of the depth z in a bilayer membrane. The curve has been constructed from three different kinds of simulation:

1. For regions with measurable water density, the PMF was evaluated directly from the local density $\rho(z)$:

$$V^{\mathrm{mf}}(z) = -\,k_{\mathrm{B}}T\ln(\rho(z)/\rho_0) \tag{7}$$

where ρ_0 is the bulk water density, and the PMF is referenced with respect to the bulk water phase (triangles in Fig. 5).

2. The curve was continued by integration of the average constraint force needed to keep a water molecule at a given depth z in the membrane (squares in Fig. 5).

3. In the middle region the density of hydrocarbon chains is much less than in other regions and it is possible to employ Widom's *particle insertion method*.[22] Water molecules are placed at random positions and in random orientations at a depth z in the membrane and their interaction energies with the environment $E_{\mathrm{int}}(z)$ are stored. The water molecules are *ghost particles* that do not influence the system. The potential of mean force $V^{\mathrm{mf}}(z)$ equals $-k_{\mathrm{B}}T\ln\langle\exp[-\beta E_{\mathrm{int}}(z)]\rangle$ plus a known constant[15] (circles in Fig. 5).

5 Dynamics in reduced space

The motion in reduced space (*i.e.*, the primed coordinates of the previous section) can be described in various approximations.[17] The equations of motion are no longer Hamiltonian: the forces depend not only on the present configuration, but also on the past. Forces proportional to velocities give a damping and are non-conservative (kinetic energy is lost). In order to maintain the average kinetic energy and temperature, a noise term must be added, consistent with the fluctuation–dissipation theorem.

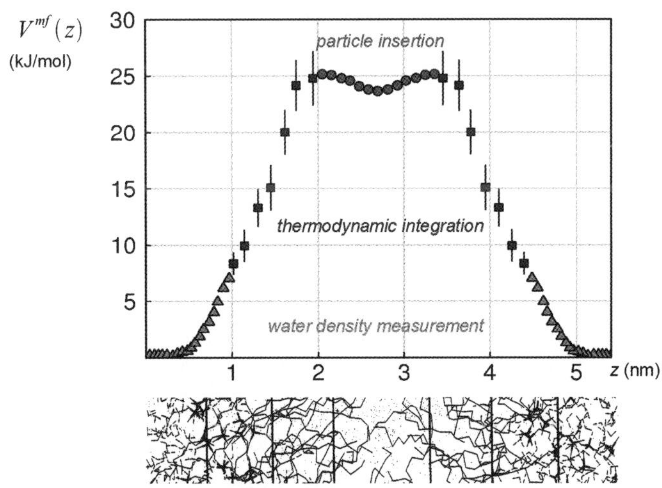

Fig. 5 Potential of mean force for transport of a single water molecule through a bilayer membrane. Triangles: from the water density; squares: from integration of the average constraint force; circles: from Widom's particle insertion method. The layer structure of the membrane (see Fig. 3) is pictured below the graph. Figure redrawn with permission from Marrink and Berendsen.[15] Copyright 1994 American Chemical Society.

The generalized Langevin equation

Taking time-dependent friction into account, the *generalized Langevin equation* is obtained:†

$$m_i \dot{\boldsymbol{v}}_i = -\frac{\partial V^{\mathrm{mf}}}{\partial \boldsymbol{r}_i} - \sum_j \int_0^\infty \zeta_{ij}(\tau)\boldsymbol{v}_j(t-\tau)\,\mathrm{d}\tau + \boldsymbol{\eta}_i(t) \tag{8}$$

where we have dropped the prime in the coordinates and write $\boldsymbol{v}$ for the time derivative of $\boldsymbol{r}\cdot\zeta_{ij}(\tau)$ is the time-dependent friction coefficient between particles i and j and $\boldsymbol{\eta}_i(t)$ is a "coloured" noise force characterized by:

$$\langle \boldsymbol{\eta}_i(t) \rangle = 0; \tag{9}$$

$$\langle \boldsymbol{\eta}_i(t)\boldsymbol{\eta}_j(t+\tau) \rangle = 2k_{\mathrm{B}}T\zeta_{ij}(\tau) \tag{10}$$

The Markovian Langevin equation

The general Langevin equation is not without problems: the equation is not exact when there is overlap between the time range of friction correlation and the characteristic time for the motion due to the systematic force. Also, data on the time dependence of the friction are hard to obtain and algorithms to generate the required colored correlated noise are complicated. Only in the case that the correlation time for the friction is small compared to the characteristic time for the motion are reliable simulations possible. This is the case that the friction (and hence the noise) *has no memory*, usually called the *Markovian limit*. The resulting stochastic equation of motion is the *Markovian Langevin equation*:

$$m_i \dot{\boldsymbol{v}}_i = -\frac{\partial V^{\mathrm{mf}}}{\partial \boldsymbol{r}_i} - \sum_j \zeta_{ij}\boldsymbol{v}_j(t) + \boldsymbol{\eta}_i(t) \tag{11}$$

with

$$\langle \boldsymbol{\eta}_i(t) \rangle = 0; \tag{12}$$

$$\langle \boldsymbol{\eta}_i(t)\boldsymbol{\eta}_j(t+\tau) \rangle = 2k_{\mathrm{B}}T\zeta_{ij}\delta(\tau) \tag{13}$$

A derivation based on Zwanzig's projection operator is given in ref. 16. The authors show that this equation is exact when the time scales of the relevant and irrelevant degrees of freedom are well-separated.

Galilean-invariant frictions

The authors of ref. 16 also point out—and this is a point that has been often overlooked—that the friction and noise forces must conserve the total momentum in order for the dynamics to be correct in the limit of large length and time scales. The validity of the Navier–Stokes equations in that limit requires conservation of

† For simplicity we take i as a selection of particles with cartesian coordinates. When the primed relevant degrees of freedom are generalized coordinates $q_k(\boldsymbol{r})$, the masses m_i should be replaced by the effective mass tensor $\mathbf{M}$, which is defined[17] as the inverse of a matrix $\mathbf{X}$ with elements $X_{kl} = \sum_i (1/m_i)(\partial q_k/\partial \boldsymbol{r}_i)\cdot(\partial q_l/\partial \boldsymbol{r}_i)$. For example, if the relevant coordinates are the centers of mass of a specified group of atoms, the effective mass tensor is diagonal and m_i must be replaced by the total mass of the i-th group.

linear momentum, or, equivalently, invariance under Galilean transformations.‡ This implies that

$$\zeta_{ii} = -\sum_{j \neq i} \zeta_{ij} \tag{14}$$

so that the Markovian Langevin equation can be rewritten as

$$m_i \dot{\boldsymbol{v}}_i = -\frac{\partial V^{\mathrm{mf}}}{\partial \boldsymbol{r}_i} - \sum_{j \neq i} \zeta_{ij} \{ \boldsymbol{v}_j(t) - \boldsymbol{v}_i(t) \} + \boldsymbol{\eta}_i(t) \tag{15}$$

with $\zeta_{ij} = \zeta_{ji}$. The diagonal values of $\boldsymbol{\zeta}$ are positive; the off-diagonal elements are generally negative.

The Langevin dynamics in this form is related to dissipative particle dynamics (DPD)[23,24] which uses Galilean-invariant relative frictions as well. However, the DPD frictional and random forces between particle pairs are restricted to act in a direction *parallel* to the interparticle direction. There is no theoretical need for this restriction: forces acting in the *transverse* direction are allowed as well. There is also no theoretical need for having the same friction coefficient for the parallel and transverse components of the velocity differences, as (eqn (15)) seems to suggest (true if ζ_{ij} would be a scalar, but it isn't: ζ_{ij} is a 3-D tensor or else one can consider i and j to enumerate all components). Junghans *et al.*[25] have shown that transverse friction has a far stronger influence on diffusion and viscosity than parallel friction and can be used to fine-tune the dynamic properties of coarse-grained models. Español *et al.*[16] shows for the example of a star-polymer melt that the parallel friction (between the com's of star polymers) is much larger than the transverse friction.

Simple Langevin

A severe approximation of the Markovian Langevin equation is the assumption that the friction tensor is *diagonal*. The equation of motion then is, for one component of the velocities:

$$m_i \dot{v}_i = -\frac{\partial V^{\mathrm{mf}}}{\partial x_i} - \zeta_i v_i(t) + \eta_i(t) \tag{16}$$

with

$$\langle \eta_i(t) \rangle = 0; \tag{17}$$

$$\langle \eta_i(t) \eta_j(t + \tau) \rangle = 2 k_{\mathrm{B}} T \zeta_i \delta_{ij} \delta(\tau) \tag{18}$$

This is called the *simple Langevin equation*. The friction and noise are simple one-dimensional memoryless additions to the equations of motion, without any coupling between degrees of freedom. This makes implementation in a stochastic dynamics code rather straightforward§ But, of course, this equation is not Galilean-invariant and any velocity deviation tends to die out to zero. The equation makes sense only if the velocity is defined with respect to the center of mass, for example for the motion of a single colloidal particle in a stationary fluid, or for the motion of several particles in a stationary fluid with complete neglect of hydrodynamic interactions.

‡ Recall that a Galilean transformation is a transformation to a reference frame that moves with constant velocity. The laws of classical mechanics are invariant to such a transformation.

§ However: beware that the incorporation of friction and noise in a Verlet algorithm, such that the accuracy is preserved to the same order as the frictionless Verlet scheme,[26] is not trivial!

The simple Langevin equation can be used to act as a thermostat: as noise and friction are designed to maintain a given temperature, deviations from that temperature will be corrected with a first-order kinetics with decay time m_i/ζ_i. In that respect the Langevin thermostat is similar to weak coupling[27] but it introduces a damping that slows down the dynamics of the system.

In the absence of systematic forces, the friction also determines the diffusion constant. Writing $\gamma = \zeta/m$, we obtain the *pure Langevin equation*

$$\dot{v} = -\gamma v + \eta(t) \tag{19}$$

where $\eta(t)$ now is a Markovian random variable with

$$\langle \eta(t)\eta(t+\tau)\rangle = 2\gamma k_B T\delta(t) \tag{20}$$

This stochastic equation¶ is exactly solvable by substituting $v(t)\exp(\gamma t)$ for a new variable. The result is

$$v(t) = v(0)\exp(-\gamma t) + \int_0^t \eta(t-\tau)\exp(-\gamma\tau)\,\mathrm{d}\tau \tag{21}$$

From this the velocity autocorrelation function follows:

$$\langle v(0)\, v(t)\rangle = \langle v(0)^2\rangle \exp(-\gamma t) \tag{22}$$

The displacement $x(t)$ is characterized by a diffusion constant D, given by the integral of the velocity correlation function:

$$D = \int_0^\infty \langle v(0)v(t)\rangle\,\mathrm{d}t = \frac{k_B T}{m\gamma} = \frac{k_B T}{\zeta} \tag{23}$$

Note that this relation between diffusion constant and friction is strictly valid only in the force-less case.

Brownian dynamics

Return to the simple Langevin eqn (16). If the systematic force is constant or slowly changing and the friction is high enough, the *inertial term* $m_i\dot{v}_i$ can be neglected. We then obtain for any degree of freedom x the equation for *Brownian dynamics*:

$$\dot{x} = v = \frac{1}{\zeta}F^{\mathrm{sys}} + \frac{\eta(t)}{\zeta} \tag{24}$$

where

$$F^{\mathrm{sys}} = -\frac{\partial V^{\mathrm{mf}}}{\partial x} \tag{25}$$

Thus the mass and even the velocity drops out of the equation and we can make a time step simply as

$$x(t+\Delta t) = x(t) + \frac{D}{k_B T}F^{\mathrm{sys}}(x) + \sqrt{2D\Delta t}\xi \tag{26}$$

¶ Purists will write the equation as $\mathrm{d}v = \gamma v\,\mathrm{d}t + \sqrt{(2\gamma k_B T)}\mathrm{d}W$, where W is a Wiener process.

where ξ is a random number sampled from a normal distribution with zero mean and variance equal to 1.

The Brownian evolution (eqn (24)) implies[17] an equation for the evolution of the *density* $\rho(x, t)$, which in general is called the *Fokker-Planck equation* and in this special case the *Smoluchowski equation*:

$$\frac{\partial \rho}{\partial t} = \frac{D}{k_B T}\frac{\partial}{\partial x}\left(\rho \frac{dV}{dx}\right) + D\frac{\partial^2 \rho}{\partial x^2} \tag{27}$$

The equilibrium solution ($\partial\rho/\partial t = 0$) is the Boltzmann distribution

$$\rho(x) \propto \exp\left(-\frac{V}{k_B T}\right) \tag{28}$$

and steady-state non-equilibrium solutions are easily derived.

How to determine friction from simulations

The friction coefficient can be obtained from simulations with constrained r'' by monitoring the constraint force F_c acting on the constraint variables. The friction constant ζ in eqn (16) is found from the integral of the correlation function of $\Delta F_c = F_c - \langle F_c \rangle$:

$$\zeta = \frac{1}{k_B T}\int_0^\infty \langle \Delta F_c(t)\Delta F_c(t+\tau)\rangle \, d\tau \tag{29}$$

Instead of computing the friction coefficient, one can derive the diffusion constant from eqn (23):

$$D = \frac{(k_B T)^2}{\int_0^\infty \langle \Delta F_c(t)\Delta F_c(t+\tau)\rangle \, d\tau} \tag{30}$$

This equation is derived in ref. 16, but it was already known and applied[28] almost twenty years ago. We end this section by quoting Marrink,[15] who has applied this equation to compute the friction a water molecule feels in the z-direction when its z-coordinate is constrained in a bilayer membrane. Fig. 6 shows the diffusion constant derived from the force autocorrelation function at various depths in the bilayer. In regions where the water concentration is measurable, the diffusion constant can be measured directly by monitoring the mean-squared displacement of water molecules in the z-direction. There is a smooth connection between the two types of determination, lending credit to the use of constraint forces for friction determination. The friction is high enough for the Brownian limit to be valid; therefore eqn (24) and eqn (27) apply. From the "measured" potential of mean force (Fig. 5) and diffusion constant (Fig. 6) the transport properties for single water molecules, such as the permeability coefficient, can be computed without the need to simulate the stochastic motion of a water molecule through a membrane. Here we give no details,[14,15] but mention that the results agree quite well with experimental values.

6 The importance of dynamics and hydrodynamics

Friction and noise can play various roles in the dynamics of reduced systems, such as providing a thermostat, introducing viscosity, or introducing thermal conductivity. The friction and noise can be tuned to obtain a desired effect. So, before

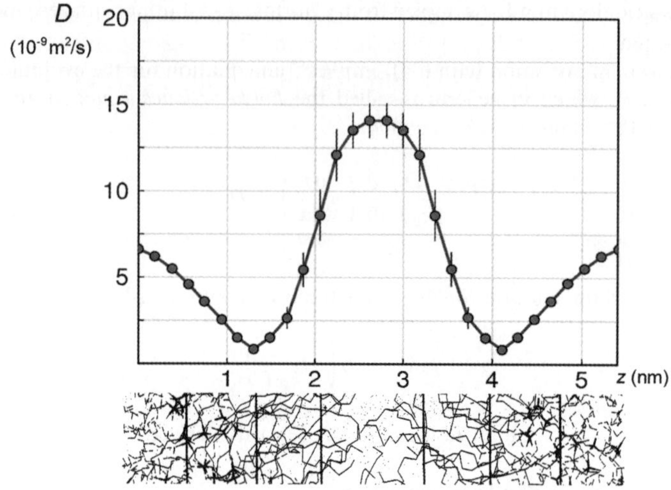

Fig. 6 Diffusion constant for a single water molecule moving through a bilayer membrane. Points without error bars: from mean square displacement; points with error bars: from integration of the autocorrelation function of the constraint force acting in the z-direction in simulations with water molecules at fixed depth z. The layer structure of the membrane (see Fig. 3) is pictured below the graph. Figure redrawn with permission from Marrink and Berendsen.[15] Copyright 1994 American Chemical Society.

introducing friction and noise, ask yourself *what is your purpose?* Let's examine some possibilities.

1. You wish to get the dynamics right

Incorporate friction and noise as accurately as possible. Analyze detailed MD on constrained systems to determine frictional parameters; fine-tune to obtain correct viscosities and/or diffusion constants.

2. You wish to get the hydrodynamics right

Hydrodynamical interactions, *i.e.*, interactions mediated through solvents, have a rather long range. The incorporation of long-range pair interactions, as the Oseen tensor, replacing an explicit solvent, is not very accurate. Better include solvent-like particles with proper DPD-type friction representing accurate viscosity, or else couple your system to a fluid of particles or to lattice points that have the proper limiting Navier–Stokes behavior. Use exclusively Galilean-invariant friction and noise. See below.

3. You are not interested in accurate dynamics, but want to explore configurational space quickly

Omit friction and noise altogether. Configurational probabilities are independent of friction and noise. Use the largest time step that conserves configurational distributions. Friction will generally slow down the dynamics, but noise may be essential to get any dynamics going. Appropriate friction and noise can be inserted to obtain proper thermostat behavior without increasing viscosity. Alternatively, use Monte Carlo sampling.

The *hydrodynamic coupling* to a fluid obeying the Navier–Stokes equations can be accomplished in various ways. The first question you should address is: *which solvent properties are essential for my problem?* Is it possible to mimic the solvent interactions by some implicit model? If not, or if not accurately enough, you better include some coarse-grained model that exerts the essential interactions. Such a model you can augment with Galilean-invariant friction and noise to adjust the viscosity to a desired value. If you don't need other physical characteristics than those that determine hydrodynamic behavior, you can *couple* your solutes to a hydrodynamic fluid. The latter may be realized by *lattice points* or by *particles*. Neither are meant to represent real solvent particles, but they form a framework to solve the hydrodynamic, *i.e.*, the Navier–Stokes equations. The coupling itself will usually represent stick boundary conditions (see *e.g.* Padding *et al.*[29] how to do this).

1. Coupling to a lattice

The lattice points represent the local fluid velocity and possibly other hydrodynamic properties as density and pressure. They are either arranged on a 3D lattice or on a finite-element grid. The properties of the points are updated according to the Navier–Stokes equations. A popular method, used by Grass and Holm,[30] and Fenkel *et al.*,[31] is the *Lattice-Boltzmann* model,[32] which allows a limited range of velocities with a simple update scheme based on nearest neighbors.

2. Coupling to particles

Fluids of particles that interact *via* Galilean-invariant friction and noise will obey the macroscopic equations of fluid mechanics, more or less irrespective of the potential function used (if any) for the conservative inter-particle interactions. Many models are possible, but one particularly simple method,[33,34] invented in 1999, seems quite promising. It is called SRD (Stochastic Rotational Dynamics)‖ and it is used in ref. 35–37. The fluid particles have no conservative interactions (they form an ideal gas) and proceed a time step according to their velocities. After each time step they undergo a stochastic velocity change as follows: partition space into small cubes each containing a few particles. For each cube, subtract the average velocity $\bar{\boldsymbol{v}}$ of the particles in that cube from each velocity $\boldsymbol{v}_i$, yielding $\boldsymbol{v}'_i$. Now define a matrix $\boldsymbol{R}$ representing a rotation over a fixed angle around a randomly chosen axis. Rotate each velocity $\boldsymbol{v}'_i$ by $\boldsymbol{R}$, yielding $\boldsymbol{v}''_i = \boldsymbol{R}\boldsymbol{v}'_i$. Then add the average velocity to each of the $\boldsymbol{v}''_i$. It is easily shown that this procedure conserves both linear momentum and kinetic energy. The dynamics therefore obeys Navier–Stokes, but it does not act as a thermostat. The method is efficient and the viscosity can be adjusted by the choice of parameters.

7 Conclusion

To conclude I will summarize the preferential methods to which simulation methods for complex "soft materials" seem to converge. With complex soft materials I mean the condensed phase with structural inhomogeneities on the nanometre scale, including all biological macromolecular complexes and almost all of the rapidly developing nanomaterials. My summary is a very personally biased view, based on my conviction that simple methods that are easy to understand and to implement—provided they are correct and work—will always prevail over complex methods, even if the latter are more accurate.

1. Superatom models will be the coarse-grained models of choice. They are straightforward and connect naturally to the atomic scale. The alternative density descriptions on a lattice, using free energy density functionals to derive driving forces for the dynamics, are restricted to mean field approximations and include

‖ In ref. 35 SRD is named MPCD (multi-particle collision dynamics).

ad hoc dynamic variables. They are complex and not much more efficient as they require a density of lattice points comparable to the density of superatoms.

2. Parametrization of coarse-grained models will require a combination of *bottom-up* and *top-down* approaches. Bottom-up, *i.e.*, based on detailed atomistic simulations, is the ideal approach. However, the accuracy obtained, especially for multidimensional and non-pair additive interactions, will not be sufficient to determine thermodynamic quantities with the required precision. Therefore, top-down adjustments, based on experimental thermodynamic quantities, will be necessary. But be aware! If the CG simulation does not predict any other experimental quantities than have been used for parametrization, you have achieved nothing! Always test top-down models for predictive power beyond the realm of properties used for parametrization.

3. Thermostats based on DPD-like friction and noise will become more dominant. In general, the incorporation of adjustable Galilean-invariant friction and noise in order to achieve required dynamic properties will become common-place.

4. SRD (stochastic rotational dynamics) or similar variants will become popular for the provision of a hydrodynamic fluid environment. It will make lattice-based methods, including Lattice–Boltzmann schemes, obsolete.

5. Despite the large body of existing codes for fluid dynamics, it is likely that particle-based methods such as SRD will gain importance, even for macroscopic fluid dynamics applications.

References

1 G. S. Ayton, E. Lymon and G. A. Voth, *Faraday Discuss.*, 2010, **144**, DOI: 10.1039/b901996k.
2 T. Apajalahti, P. Niemelä, P. N. Govindan, M. S. Miettinen, E. Salonen, S.-J. Marrink and I. Vattulainen, *Faraday Discuss.*, 2010, **144**, DOI: 10.1039/b901487j.
3 A. J. Rzepiela, D. Sengupta, N. Goga and S. J. Marrink, *Faraday Discuss.*, 2010, **144**, DOI: 10.1039/b901615e.
4 Y. Norizoe, K. Ch. Daoulas and M. Müller, *Faraday Discuss.*, 2010, **144**, DOI: 10.1039/b901657k.
5 M. Yiannourakou, L. Marsella, F. de Meyer and B. Smit, *Faraday Discuss.*, 2010, **144**, DOI: 10.1039/b902190f.
6 J. P. Ryckaert and A. Bellemans, *Faraday Discuss. Chem. Soc.*, 1978, **66**, 95.
7 P. Van der Ploeg and H. J. C. Berendsen, *Biophys. Struct. Mechanisms*, 1980, **6, Suppl.**, 106.
8 P. Van der Ploeg and H. J. C. Berendsen, *J. Chem. Phys.*, 1982, **76**, 3271.
9 P. Van der Ploeg and H. J. C. Berendsen, *Mol. Phys.*, 1983, **49**, 233.
10 H. J. C. Berendsen., in *Molecular-Dynamics Simulation of Statistical-Mechanical Systems*, eds G. Ciccotti and W. Hoover, Soc. Italiana di Fisica, Bologna. North Holland, Amsterdam, 1986, 496–519.
11 H. J. C. Berendsen and E. Egberts, in *Structure, Dynamics and Function of Biomolecules*, eds A. Ehrenberg et al., Springer Verlag 1986, 275–280.
12 E. Egberts and H. J. C. Berendsen, *J. Chem. Phys.*, 1988, **89**, 3718.
13 E. Egberts, S.-J. Marrink and H. J. C. Berendsen, *Eur. Biophys. J.*, 1994, **22**, 423.
14 H. J. C. Berendsen and S.-J. Marrink, *Pure Appl. Chem.*, 1993, **65**, 2513.
15 S.-J. Marrink and H. J. C. Berendsen, *J. Phys. Chem.*, 1994, **98**, 4155.
16 C. Hijón, P. Español, E. Vanden-Eijnden and R. Delgado-Buscalioni, *Faraday Discuss.*, 2010, **144**, DOI: 10.1039/b902479b.
17 H. J. C. Berendsen., *Simulating the Physical World, Hierarchical modeling from quantum mechanics to fluid dynamics*, Cambridge University Press, 2007.
18 A. Lyubartsev, A. Mirzoev, L. Chen and A. Laaksonen, *Faraday Discuss.*, 2010, **144**, DOI: 10.1039/b901511f.
19 M. Schor, B. Ensing and P. G. Bolhuis, *Faraday Discuss.*, 2010, **144**, DOI: 10.1039/b901608b.
20 A. J. Crane and E. A. Müller, *Faraday Discuss.*, 2010, **144**, DOI: 10.1039/b901601e.
21 S. Baoukina, S. J. Marrink and D. P. Tieleman, *Faraday Discuss.*, 2010, **144**, DOI: 10.1039/b905647e.
22 B. Widom, *J. Chem. Phys.*, 1963, **39**, 2808.
23 P. J. Hoogerbrugge and J. M. V. A. Koelman, *Europhys. Lett.*, 1992, **19**, 155.
24 P. Español and P. Warren, *Europhys. Lett.*, 1995, **30**, 191.
25 C. Junghans, M. Prapotnik and K. Kremer, *Soft Matter*, 2008, **4**, 156.

26 W. F. Van Gunsteren and H. J. C. Berendsen, *Mol. Simul.*, 1988, **1**, 173.
27 H. J. C. Berendsen, J. P. M. Postma, W. F. van Gunsteren, A. DiNola and J. R. Haak, *J. Chem. Phys.*, 1984, **81**, 3684.
28 B. Roux and M. Karplus, *J. Phys. Chem.*, 1991, **95**, 4856.
29 J. T. Padding, A. Wysocki, H. Löwen and A. A. Louis, *J. Phys.: Condens. Matter*, 2005, **17**, S3393.
30 K. Grass and C. Holm, *Faraday Discuss.*, 2010, **144**, DOI: 10.1039/b902011j.
31 B. Rotenberg, I. Pagonabarraga and D. Frenkel, *Faraday Discuss.*, 2010, **144**, DOI: 10.1039/b901553a.
32 J. M. Thijssen., *Computational Physics*, 2nd ed., Cambridge University Press, 2007.
33 A. Malevanets and R. Kapral, *J. Chem. Phys.*, 1999, **110**, 8605.
34 R. Kapral, *Adv. Chem. Phys.*, 2008, **140**, 89.
35 A. Wysocki, C. P. Royall, R. G. Winkler, G. Gompper, H. Tanaka, A. van Blaaderen and H. Löwen, *Faraday Discuss.*, 2010, **144**, DOI: 10.1039/b901640f.
36 E. S. Boek, J. T. Padding, T. Headen and J. Crawshaw, *Faraday Discuss.*, 2010, **144**, DOI: 10.1039/b902305b.
37 J. Sané, J. T. Padding and A. A. Louis, *Faraday Discuss.*, 2010, **144**, DOI: 10.1039/b905378f.

Poster titles

Model lipid rafts at atomistic resolution: combining coarse-grained and all-atom MD simulations, **L. V. Schäffer, H. J. Risselada, A. J. Rzepiela, A. H. de Vries, S. J. Marrink**, *University of Groningen, The Netherlands*

Density imbalances and free energy of lipid transfer in supported lipid bilayers, **C. Xing and R. Faller**, *University of California Davis, USA*

Beyond amphiphiles: Coarse grained molecular dynamics simulation of star polyphiles, **J. J. K. Kirkensgaard and S. Hyde**, *Australian National University, Australia*

Coarse-grained simulations of α-helical peptides, **P. Gkeka and L. Sarkisov**, *University of Edinburgh, UK*

Systematic development of a coarse-grained model for 5CB (4-cyano-4′-pentylbiphenyl), **G. Megariotis, A. Vyrkou, A. Leygue and D. N. Theodorou**, *National Technical University of Athens, Greece*

Folding kinetics of an α-helix and a β-hairpin using coarse-grained simulations, **T. Bereau, D. Stone and M. Deserno**, *Carnegie Mellon University, USA*

Predicting porous organic polymers and networks: structures and properties, **A. Trewin and A. I. Cooper**, *University of Liverpool, UK*

Simulating self-assembled porous organic cages, **S. Jiang, A. Trewin and A. I. Cooper**, *University of Liverpool, UK*

FlowVRNano – a virtual laboratory dedicated to interactive simulation of large molecular systems, **N. Férey, O. Delalande and M. Baaden**, *Institut de Biologie Physico-Chimique, France*

Investigating bile salt aggregation using coarse-grained molecular dynamics, **A. Vila Verde and D. Frenkel**, *AMOLF-FOM Institute for Atomic and Molecular Physics, The Netherlands*

Hierachical modelling of polymer permeation, **C. R. Herbers, D. Fritz, K. Kremer and N. F. A. van der Vegt**, *TU Darmstadt, Germany and Max Planck Institute for Polymer Research, Germany*

Precipitation of polycaprolactone nanoparticles *via* solvent-displacement: computational fluid dynamics modeling, **E. Gavi, D. L. Marchisio and A. A. Barresi**, *Politecnico di Torino, Italy*

Influenza HA fusion peptides favor stalk phases in MD simulations, **M. Fuhrmans and S. J. Marrink**, *University of Groningen, The Netherlands*

Encapsulation of local anesthetics into liposomes, **G. Guipponi and M. Pickholz**, *Universitat de Barcelona, Spain*

Coarse-grain modeling of lipid membrane adsorption on nanopatterned surfaces, **M. I. Hoopes, M. L. Longo and R. Faller**, *University of California, Davis, USA*

Effective coarse-grained potentials for DMPC-lipids by Inverse Monte Carlo method: Concentration dependence, **A. Mirzoev and A. Lyubartsev**, *Stockholm University, Sweden*

Determination of protein reduced electrostatic models from smoothed molecular electrostatic potentials, **L. Leherte and D. P. Vercauteren**, *University of Namur, Belgium*

Structure and dynamics of liquid water inside reverse micelles, **J. Martí, E. Guàrdia, J. Rodríguez and D. Iaria**, *Universitat Politècnica de Catalunya, Spain*

Hydrogen bonding and dynamic crossover in Polyamide-66: A molecular dynamics simulation study, **P. Carbone, H. A. Karimi Varzaneh and F. Müller-Plathe**, *University of Manchester, UK*

Back-mapping coarse-grained polymer models under sheared nonequilibrium conditions, **P. Carbone, X. Chen, G. Santangelo, A. Di Matteo, G. Milano and F. Müller-Plathe**, *University of Manchester, UK*

Excess entropy scaling of transport properties of Lennard-Jones chains, **T. Goel, C. Nath Patra, T. Mukherjee, M. Agarwal, R. Sharma, M. Parvez Alam and C. Chakravarty**, *Indian Institute of Technology-Delhi, India*

Agent-based modelling for molecular self-organization, **S. Fortuna and A. Troisi**, *University of Warwick, UK*

Multiscale modeling of polystyrene – Coarse graining methodology, **D. Fritz, V. Harmandaris, K. Kremer and N. F. A. van der Vegt**, *Max Planck Institute for Polymer Research, Germany*

Molecular dynamics study of non-ionic lipids at the air-water interface – structural aspects, **M. Velinova, S. Tzvetanov, A. Ivanova, Ph. Shushkov and A. Tadjer**, *University of Sofia, Bulgaria*

Hybrid particle-field molecular dymanics simulations for soft matter, **A. de Nicola, G. Milano, D. Roccatano and T. Kawakatsu**, *Università di Salerno, Italy*

Simulation studies of model systems for lung surfactant, **S. Baoukina, S. J. Marrink and D. P. Tieleman**, *University of Calgary, Canada*

Liquid crystallinity of the MARTINI coarse grained cholesterol model and its derivatives, **M. Yoneya**, *Nanotechnology Research Institute, AIST, Japan*

Comparison of a microgel simulation to a Poisson-Boltzmann cell model, **G. C. Claudio, C. Holm and K. Kremer**, *Max Planck Institute for Polymer Research, Germany*

Breaking CFTR into pieces: A coarse grained journey into putting together the pieces of the puzzle, **B. Nikolaidi and M. Sansom**, *University of Oxford, UK*

Toward coarse-grained simulations of block copolymers interaction with biological interfaces, **S. Hezaveh, G. Milano and D. Roccatano**, *Jacobs University Bremen, Germany*

Computer simulations of the interaction of fullerene with lipid membranes, **L. Monticelli, E. Salonen, P.-C. Ke and I. Vattulainen**, *INSERM, France and Helsinki University of Technology, Finland*

The counterion effect on the stability of structurally persistent micelles: MD simulations and experimental confirmation, **C. Jäger, H. Lanig, C. Böttcher, A. Hirsch and T. Clark**, *Friedrich-Alexander Universität Erlangen, Germany*

Added hexane controls the size and structure of structurally persistent micelles, **C. Jäger, M. Wildauer, H. Lanig, C. Böttcher, A. Hirsch and T. Clark**, *Friedrich-Alexander Universität Erlangen, Germany*

3D pressure field in lipid membranes and membrane protein complexes, **O. H. S. Ollila, H. J. Risselada, M. Louhivuori, A. Lamberg, A. Cattle, T. Vuorela, E. Lindahl, I. Vattulainen and S. J. Marrink**, *Tampere University of Technology, Finland*

The challenge of extracting atomistic and mesoscopic structural information from neutron diffraction measurement using EPSR, **R. Hargreaves**, *STFC, Rutherford Appleton Laboratory, UK*

Structure and charge transport in polymer semiconductors: Combined quantum, atmomistic, and mesoscale modelling, **D. L. Cheung, D. P. McMahon and A. Troisi**, *University of Warwick, UK*

From achiral to chiral; a coarse-grained simulation study, **J. Lintuvuori and M. R. Wilson**, *Durham University, UK*

Discrete path sampling simulations of peptide folding, **J. M. Carr and D. J. Wales**, *University of Cambridge, UK*

Understanding complex protonation behaviour in mesoscale simulations using a dynamic, variable-dielectric Poisson-Boltzmann solver, **D. Eike, P. Verstraete, B. Murch, J. van Male and J. Fraaije**, *Brussels Innovation Center, The Procter & Gamble Company, Belgium*

Forced reptation revealed by chain pull-out simulations, **M. Bulacu and E. van der Giessen**, *University of Groningen, The Netherlands*

Bridging the scales of simulation and experiment: two scale characterisation of phospholipid bilayers, **A. H. de Vries, A.-P. Kunz, W. F. van Gunsteren and S. J. Marrink**, *University of Groningen, The Netherlands*

Quadrupolar defect structures generated by chiral islands in freely suspended smectic C films, **N. M. Silvestre, P. Patrício, M. M. Telo da Gama, A. Pattanaporkrattana, C. S. Park, J. E. Maclennan and N. A. Clark**, *Universidade de Lisboa, Portugal*

A hybrid molecular dynamics method with preserved Boltzmann distribution using coupling to a scaled thermostat, **N. Goga, A. J. Rzepiela, M. H. Louhivuori, H. J. C. Berendsen, A. H. de Vries and S. J. Marrink**, *University of Groningen, The Netherlands*

A coarse-grained model for a self-assembling dipeptide, **A. Villa, N. F. A. van der Vegt and C. Peter**, *Karolinska Institutet, Sweden*

Direct numerical simulations of colloidal dispersions: electrophoresis and non-linear rheology, **R. Yamamoto and T. Iwashita**, *Kyoto University, Japan*

Translation diffusivity and rotational relaxation in stratified mesosphases, **G. Cinacchi and L. de Gaetani**, *University of Bristol, UK*

Mesophases of concave particles, **G. Cinacchi and J. S. van Duijneveldt**, *University of Bristol, UK*

Protein Domain Model – a preliminary exploration – Modelling the dynamics of a molecular machine, **L. Thøgersen**, *Aarhus University, Denmark*

Computer simulation studies of molecular order in chromonic mesophases, **F. Chami and M. R. Wilson**, *Durham University, UK*

Coarse grained modelling of self-assembling peptide fibres by Monte Carlo simulations, **T. Stedall and S. Hanna**, *University of Bristol, UK*

Conformational sampling of coarse grained peptide models, **O. Bezkorovaynaya, C. Peter and K. Kremer**, *Max Planck Institute for Polymer Research, Germany*

Lattice Boltzmann simulations of suspensions at multiple length scales, **R. G. M. van der Sman, F. Debask, G. Brans, J. Kromkamp, M. Vollebregt and R. M. Boom**, *Wageningen University, The Netherlands*

The Skinner Prize for the best poster was jointly awarded to Svetlana Baoukina of University of Calgary, Canada, for her poster on Simulation studies of model systems for lung surfactant and Juho Lintuvuori of Durham University, UK, for his poster From achiral to chiral; a coarse-grained simulation study.

List of participants

Professor M. Allen, *University of Warwick, United Kingdom*
Professor M. Ashfold, *University of Bristol, United Kingdom*
Dr M.B. Baaden, *CNRS IBPC, France*
Dr S. Baoukina, *University of Calgary, Canada*
Professor R. Berardi, *Università degli Studi, Italy*
Mr T. Bereau, *Carnegie Mellon University, U.S.A.*
Professor H.J.C. Berendsen, *University of Groningen, Netherlands*
Dr O. Bezkorovaynaya, *Max Planck Institute for Polymer Research, Germany*
Dr E.S. Boek, *Schlumberger Cambridge Research, United Kingdom*
Professor P.G. Bolhuis, *University of Amsterdam, Netherlands*
Miss R. Brodie, *Royal Society of Chemistry, United Kingdom*
Dr M.I. Bulacu-Cioceanu, *University of Groningen, Netherlands*
Dr P.C. Carbone, *University of Manchester, United Kingdom*
Dr J. Carr, *STFC Daresbury Laboratory, United Kingdom*
Professor C. Chakravarty, *Indian Institute of Technology Delhi, India*
Dr M. Chapman, *Royal Society of Chemistry, United Kingdom*
Dr D.L. Cheung, *University of Warwick, United Kingdom*
Dr G. Cinacchi, *University of Bristol, United Kingdom*
Professor T.C. Clark, *University of Erlangen-Nuremberg, Germany*
Dr G. Claudio, *Max Planck Institute for Polymer Research, Germany*
Professor M. Deserno, *Carnegie Mellon University, U.S.A.*
Professor M. Dijkstra, *University of Utrecht, Netherlands*
Dr B. Ensing, *University of Amsterdam, Netherlands*
Professor P. Español, *Freiburg University, Germany*
Professor R. Faller, *UC Davis, U.S.A.*
Dr A. Ferrarini, *Padova University, Italy*
Ms S.F. Fortuna, *University of Warwick, United Kingdom*
Professor D. Frenkel, *University of Cambridge, United Kingdom*
Mr D. Fritz, *Max Planck Institute for Polymer Research, Germany*
Mr M. Fuhrmans, *University of Groningen, Netherlands*
Dr L.D.G. de Gaetani, *Università di Pisa, Italy*
Dr G. Giupponi, *Universitat de Barcelona, Spain*
Ms P. Gkeka, *University of Edinburgh, United Kingdom*
Dr N. Goga, *University of Groningen, Netherlands*
Dr R.S. Graham, *University of Nottingham, United Kingdom*
Dr R. Hargreaves, *STFC, United Kingdom*
Ms C. Herbers, *Technical University Darmstadt, Germany*
Dr B. Hess, *Stockholm Center for Biomembrane Research, Sweden*
Ms S. Hezaveh, *Jacobs University Bremen, Germany*
Professor C. Holm, *Institute for Computational Physics, Germany*
Mr M.I.H. Hoopes, *UC Davis, U.S.A.*
Mr D. Hudzinskyy, *Technische Universiteit Eindhoven, Netherlands*
Dr S. Ivanov, *University of Groningen, Netherlands*
Professor G. Jackson, *Imperial College London, United Kingdom*
Dr C.J. Jäger, *University of Erlangen-Nuremberg, Germany*
Ms S. Jiang, *University of Liverpool, United Kingdom*
Mr J.J.K. Kirkensgaard, *University of Copenhagen, Denmark*
Professor K. Kremer, *Max Planck Institute for Polymer Research, Germany*
Dr H.L. Lanig, *University of Erlangen-Nuremberg, Germany*
Dr L. Leherte, *University of Namur, Belgium*

Mr J. Lintuvuori, *Durham University, United Kingdom*
Dr F. Lo Verso, *Johannes Gutenberg-Universitaet Mainz, Germany*
Dr M.J. Louhivuori, *University of Groningen, Netherlands*
Dr A. Louis, *University of Oxford, United Kingdom*
Professor H. Löwen, *Heinrich-Heine-Universität Düsseldorf, Germany*
Professor A. Lyubartsev, *Stockholm University, Sweden*
Dr J. van Male, *Culgi B.V., Netherlands*
Dr D.L.M. Marchisio, *Politecnico di Torino, Italy*
Dr V. Markov, *University of Groningen, Netherlands*
Professor S.J. Marrink, *University of Groningen, Netherlands*
Dr J. Martí, *Technical University of Catalonia, Spain*
Dr G. Milano, *University of Salerno, Italy*
Dr A. Mirzoev, *Stockholm University, Sweden*
Dr L.M. Monticelli, *Helsinki University of Technology, Finland*
Dr E.A. Müller, *Imperial College London, United Kingdom*
Professor M. Müller, *Georg-August Universität, Germany*
Professor F. Müller-Plathe, *Technical Universitat Darmstadt, Germany*
Ms A. Muntean, *Technische Universiteit Eindhoven, Netherlands*
Miss R. Needham, *Royal Society of Chemistry, United Kingdom*
Dr A. de Nicola, *University of Salerno, Italy*
Dr B. Nikolaidi, *University of Oxford, United Kingdom*
Mr S.O.H. Ollila, *Tampere University of Technology, Finland*
Dr S. Osaki, *Industrial Technology Center of Wakayama Prefecture, Japan*
Professor J. De Pablo, *University of Wisconsin-Madison, U.S.A.*
Dr J.T. Padding, *University of Twente, Netherlands*
Professor I. Pagonabarraga, *Universitat de Barcelona, Spain*
Professor X. Periole, *University of Groningen, Netherlands*
Dr M. Ravnik, *University of Ljubljana, Slovenia*
Dr H.J. Risselada, *University of Groningen, Netherlands*
Dr D. Roccatano, *Jacobs University Bremen, Germany*
Mr A.J. Rzepiela, *University of Groningen, Netherlands*
Professor M. Sansom, *University of Oxford, United Kingdom*
Dr L. Schäfer, *University of Groningen, Netherlands*
Ms M. Schor, *University of Amsterdam, Netherlands*
Dr D. Sengupta, *University of Groningen, Netherlands*
Dr N.M. Silvestre, *University of Lisbon, Portugal*
Dr R.G.M. van der Sman, *Wageningen University, Netherlands*
Professor B. Smit, *UC Berkeley, U.S.A.*
Dr T. Stedall, *University of Bristol, United Kingdom*
Professor D. Theodorou, *National Technical University of Athens, Greece*
Professor L.T. Thøgersen, *Aarhus University, Denmark*
Professor P. Tieleman, *University of Calgary, Canada*
Dr A. Trewin, *University of Liverpool, United Kingdom*
Dr R. Vacha, *Academy of Sciences of the Czech Republic, Czech Republic*
Professor I. Vattulainen, *Tampere University of Technology, Finland*
Professor N.F.A. van der Vegt, *Technical University Darmstadt, Germany*
Dr M. Velinova, *University of Sofia, Bulgaria*
Dr P.V. Verstraete, *Procter & Gamble, Belgium*
Dr A. Vila Verde, *AMOLF, Netherlands*
Dr A. Villa, *Karolinska Institutet, Sweden*
Professor G. Voth, *University of Utah, U.S.A.*
Dr A.H. de Vries, *University of Groningen, Netherlands*
Mr M.W. Wildauer, *University of Erlangen-Nuremberg, Germany*

Dr M. Wilson, *Durham University, United Kingdom*
Dr M. Wykes, *University of Mons, Belgium*
Professor A. Wysocki, *Heinrich-Heine-Universität Düsseldorf, Germany*
Professor R. Yamamoto, *Kyoto University, Japan*
Dr M. Yoneya, *National Institute of Advanced Industrial Science and Technology (AIST), Japan*
Professor C.Z. Zannoni, *Università di Bologna, Italy*

Index of contributors*

* The page numbers in **bold** type indicate papers submitted for discussions.

